IV International Scientific and Technical Conference "Energy Systems"

IOP Conference Series: Materials Science and Engineering
Volume 791

Belgorod, Russia
31 October - 1 November 2019

ISBN: 978-1-7138-4720-5
ISSN: 1757-8981

Printed from e-media with permission by:

Curran Associates, Inc.
57 Morehouse Lane
Red Hook, NY 12571

Some format issues inherent in the e-media version may also appear in this print version.

Institute of Physics
Dirac House, Temple Back
Bristol BS1 6BE UK

Phone: 44 1 17 929 7481
Fax: 44 1 17 920 0979

techtracking@iop.org

Additional copies of this publication are available from:

Curran Associates, Inc.
57 Morehouse Lane
Red Hook, NY 12571 USA
Phone: 845-758-0400
Fax: 845-758-2633
Email: curran@proceedings.com
Web: www.proceedings.com

TABLE OF CONTENTS

IOP Conf. Series: Materials Science and Engineering **791** (2020) 011001 doi:10.1088/1757-899X/791/1/011001

Preface

Information about the conference

The IV International Scientific and Technical Conference "Energy systems" (ICES - 2019) was hosted by the federal state budget institution of higher education "Belgorod State Technological University named after V. G. Shukhov" (BGTU named after V.G. Shukhov) on October 31 – November 1, 2019.

The conference suggested the solution to fundamental scientific problems in the field of energy. The main point is a systematic approach. The studied systems are complex and require the management of large systems and integration of knowledge. Also, one of the objectives of the conference was the wide involvement of young scientists – students, undergraduates, and graduate students.

Conference website:

http://energysystem.bstu.ru

The working languages of the conference were English and Russian. 125 reports were selected for the conference. As a result of the conference 29 Russian-language reports were prepared in the main section of the conference and 18 reports in the section of young scientists, 59 reports were in English. 19 reports were issued in both Russian and English versions (the articles are not just a translation of the same text into two languages, but have significant differences due to different volumes of publications, requirements for the structure of the material and number of references in the bibliography). This volume contains reports in English. The reports in Russian can be found on page http://es.bstu.ru/archive/2019.

Fundamental and practical issues in the scientific fields of heat technology, energy, and resource-saving in industrial production have been considered.

The conference demonstrated further approval of innovative approaches which are based on the fundamental research. The identified problems and new challenges that help to strengthen interdisciplinary ties in the energy industry have been shown. It was demonstrated that the domestic industry is in ever-increasing need of modern scientific and technical developments, as well as strengthening of the interaction between universities and industry.

The young scientists have demonstrated a great interest in this conference, which were coauthors of the work of their senior colleagues, and the sole authors of the reports.

Belgorod, Russia, 2019
Prof. Pavel Trubaev
E-mail trubaev@gmail.com

Photos from the conference

IOP Conf. Series: Materials Science and Engineering **791** (2020) 011001 doi:10.1088/1757-899X/791/1/011001

IOP Conf. Series: Materials Science and Engineering **791** (2020) 011001 doi:10.1088/1757-899X/791/1/011001

Conference sections

The main problems of the research were determined by the selected sections based on the analysis of the received applications:

Topic 1. Heat and power engineering (20 reports)

The fundamental and practical issues in the scientific fields of heat and mass transfer processes, heat generation, heat conversion to other types of energy, providing consumers with energy resources were considered.

Speaker
Alexey Biryukov, DSc

Speaker
Michael Averbukh, DSc

Topic 2. Electric power and electrical engineering (18 reports)

Fundamental and practical issues in the scientific fields of production, distribution, transformation, and use of electric energy, development, and optimization of electronic components were considered.

Topic 3. Energy-saving and energy efficiency, power ecology (13 reports)

Fundamental and practical questions in scientific areas of energy resources cost reduction, increases in the energy efficiency of power and heat technical systems, the energy efficiency of buildings, buildings, constructions, impacts of power on the environment, elimination of emissions and pollution, waste recycling for energy were considered.

Speaker
Pavel Trubaev, DSc

Speaker
Evgeniy Gasho, DSc

Topic 4. Alternative and renewable energy sources (10 reports)

Fundamental and practical issues in the scientific fields of production, distribution, transformation, and use of electrical energy, development, and optimization of electronic components were considered.

Topic 5. Energy and resource-saving in technologies (17 reports)

Fundamental and practical issues in the scientific fields of heat technology, energy, and resource-saving in industrial production were considered.

Speaker
Vladimir Sevostyanov, DSc

IOP Conf. Series: Materials Science and Engineering **791** (2020) 011001 doi:10.1088/1757-899X/791/1/011001

Conferees

The conference was attended by 191 people from 52 organizations (including 43 Universities, four Research Institutes and five Enterprises), which are located in 26 Russian regions and four foreign countries:

- Bagdad, University of Technology;
- Barnaul, Polzunov Altai State Technical University;
- Belgorod State National Research University;
- Belgorod State Technological University named after V. G. Shukhov;
- Belgorod, TC “Ekotrans”, LLC;
- Chita, Transbaikal State University;
- Donetsk National Technical University;
- Donetsk National University of Economics and Trade named after Mikhail Tugan-Baranovsky;
- Ekaterinburg, Ural Federal University named after the first President of Russia B.N. Yeltsin;
- Irkutsk National Research Technical University;
- Kazan National Research Technical University named after A. N. Tupolev;
- Kazan National Research Technological University;
- Kazan State Power Engineering University;
- Kazan, M12Systems, LCC;
- Kharkiv , National Technical University “Kharkiv Polytechnic Institute”
- Kharkiv, V. N. Karazin National University;
- Khujand, Polytechnic Institute of Tajik Technical University named after M. S. Osimi;
- Krasnodar, Kuban State University;
- Krasnoyarsk, Siberian Federal University;
- Magnitogorsk, Nosov Magnitogorsk State Technical University;
- Magnitogorsk, PJSC Magnitogorsk Iron and Steel Works
- Makeyevka, Donbas National Academy of Civil Engineering and architecture;
- Moscow National Research University “MPEI”;
- Moscow State University of Civil Engineering;
- Moscow, Financial University under the Government of the Russian Federation;
- Moscow, Research Institute of Agricultural Mechanization;
- Najaf, Kufa University;
- Nizhny Novgorod State Technical University n.a. R.E. Alekseev;
- Novoronezh representative office of JSC ESKM Corporation;
- Novosibirsk State Technical University;
- Novosibirsk State Technical University;
- Novosibirsk, Siberian Transport University;
- Obninsk, I.I. Leypunsky Institute of Physics and Power Engineering;
- Orenburg State University;
- Petrozavodsk State University;
- Rostov-on-Don, Federal Research Centre The Southern Scientific Centre of The Russian Academy of Sciences;
- Samara State Technical University;
- Samara University (Samara national research university of a name of the academician S.P. Korolyov);
- Saratov Scientific Center of the Russian Academy of Sciences;
- Saratov, Yuri Gagarin State Technical University of Saratov

IOP Conf. Series: Materials Science and Engineering **791** (2020) 011001 doi:10.1088/1757-899X/791/1/011001

- Sevastopol State University;
- Simferopol, V.I. Vernadsky Crimean Federal University;
- Smolensk, Branch of the Moscow Power Engineering Institute;
- St. Petersburg, Peter the Great St. Petersburg Polytechnic University;
- Tambov State Technical University;
- Troitsk, South-Ural State Agro University;
- Tver State Technical University;
- Verkhnyaya Pyshma, Technical University UMMC (Ural Mining Metallurgical Company - Holding Corporation);
- Vladimir State University named after Alexander and Nikolay Stoletov;
- Volzhsky, Branch of the Moscow Power Engineering Institute;
- Voronezh State Technical University;
- Voronezh, IDGC of Centre — Voronezhenergo division;
- Yakutsk, M. K. Ammosov North-Eastern Federal University.

The number of authors is as follows:

- Dr. of tech. Sciences (DSc), professors - 22 persons;
- Cand. of tech. Sciences (Phd), associate professors – 92 persons;
- Engineers, assistants, senior teachers – 16 persons;
- Graduate and undergraduate students – 55 persons;
- Persons of industry – 6 persons.

PROGRAM COMMITTEE

IOP Conf. Series: Materials Science and Engineering **791** (2020) 011001 doi:10.1088/1757-899X/791/1/011001

Head of Department of Technology Complexes, Machines and Mechanisms of BSTU named after V. G. Shukhov (Belgorod, Russia)

Zlata Udovichenko, Cand. of tech. Sciences (PhDS)

Associate Professor of the Department of Heat Engineering, Heat and Gas Supply and Ventilation of the Donbass Academy of Architecture and Construction (Makeyevka, Donetsk region)

Marina Kozhukhova, PhD

College of Engineering & Applied Science, University of Wisconsin–Milwaukee (Milwaukee, USA)

ORGANIZING COMMITTEE OF THE CONFERENCE

Sergey Glagolev, Chairman, Dr. of Econ. Sciences

Rector of BSTU named after V. G. Shukhov

Pavel Trubaev, Deputy Chairman, Dr. of Tech. Sciences

Professor of BSTU named after V. G. Shukhov

Alexandr Belousov, Cand. of tech. Sciences (Phd)

Director of the Institute of Energy, Information Technology and Control Systems,
Head of Department of Power and Automation of BSTU named after V. G. Shukhov (Belgorod)

Jury Koshlich, Cand. of tech. Sciences (Phd)

CEO of the STC ASP Ltd, Associate Professor of electricity and automation

Elena Fironova

Head of the Department of Development of Fuel and Energy Complex and Engineering Infrastructure of the Ministry of Housing and Communal Services of the Belgorod region.

Peer review statement

All papers published in this volume of *IOP Conference Series: Materials Science and Engineering* have been peer reviewed through processes administered by the proceedings Editors. Reviews were conducted by expert referees to the professional and scientific standards expected of a proceedings journal published by IOP Publishing.

An experimental investigation on the transient heat transfer characteristics using air/water droplets two-phase flow

A H Abed[1, 2], S E Shcheklein[1] and V M Pakhaluev[1]

[1]Ural Federal University named after the first President of Russia B. N. Yeltsin 19 Mira St., Yekaterinburg, 620002, Russia
[2]Department of Electromechanical Engineering, University of Technology, Al-Sina'a St., Baghdad, 10066, Iraq

E-mail: akraaam82@yahoo.com

Abstract. The present study focused on evaluating the heat transfer behavior and predicting the surface resulting status during air/water droplets two-phase flow. Transient heat transfer based on the lumped capacitance model (LCM) was investigated experimentally under a range of water droplets concentration, surface temperature, and varying Re number. Compared with a single-phase air cooling, the transient surface temperature decreased with the increase in water droplets concentration and Re number. At the same cooling time, the surface temperature decreases about 13.5%, 47%, and 53.2% for (j = 46.79 – 111.68 kg/m^2 hr). It was also noticed that the heat transfer coefficient increased with the increase in water droplets concentration and reach its maximum value at (j = 111.68 kg/m^2 hr). Based on the analysis of the experimental results, the heat transfer mechanism due to the impacting of water droplets on the sphere surface was classified into three important physical regimes. Clear convection heat transfer regime corresponds to the dry region (region I); Convection and evaporation regimes correspond to the dry-out and wet regions (region II and III).

1. Introduction

Air/water droplets cooling technique applied in numerous industrial and technical processes such as the heat exchanger modules in nuclear and thermal power plants, cooling of electronic and electrical equipment, etc., considered to be an excellent technique and distinguished by the higher heat transfer rate compared with those obtainable by conventional single-phase air cooling [1-5]. These high heat transfer rate occurred due to the phase change of water droplets by the direct evaporation process on the heated surface and absorb a large amount of energy in the form of latent heat, increased the specific heat of mixture, and increased the turbulence inside the thermal boundary layer of the heated surface [6-8]. The air/water droplets heat transfer process is a very complicated phenomenon depends on the several factors such as the water droplet diameter size, the water phase concentration, the flow behavior, the water droplet trajectories, the temperature of the heated surface, and the evaporation time of water droplet on the heated surface. The surface temperature and water phase concentration are important factors that have a great influence on the resulting status of the heating surface (dry, dryout, and wet states) and heat transfer behavior. In the present study, an experimental investigation was performed with an aim to analyze and understand the transient heat transfer behavior of a spherical element cooled using air/water droplets two-phase flow. The estimation of heat transfer behavior during the cooling process was employed based on the lumped capacitance model (LCM).

2. The experimental facilities and results treatment

This investigation related to analyzing the heat transfer behavior during the transient -state condition by suspending a fine water droplet on an air main flow. The experimental facilities used in our experiment depicted schematically in figure 1, are the same one was used in references [9, 10]. It consists of the air supply system, water droplets generator system, test specimen, induction heating system, and data acquisition system. The test specimen is represented by a calorimetric copper sphere located inside a cylindrical channel with sphere-to-channel diameter ratio (d/D=0.73). The initial working temperature of the test sphere was 100, 200, and 300 °C. The surface temperature was measured by two calibrated (K-type) thermocouples were implanted inside the sphere. The ambient, inlet and outlet fluid temperatures, and the cooling down time were recorded constantly. All thermocouples were connected to the OWEN MV110-8A data acquisition system with MSD200 data logger. A high-frequency heating induction station was used to heat the sphere as a non-contact heating method with output fluctuating frequency 50 - 100 KHz. The average water droplet diameter was (dp≤10μm) was obtained by an ultrasonic water droplets generator with 1.7 MHz.

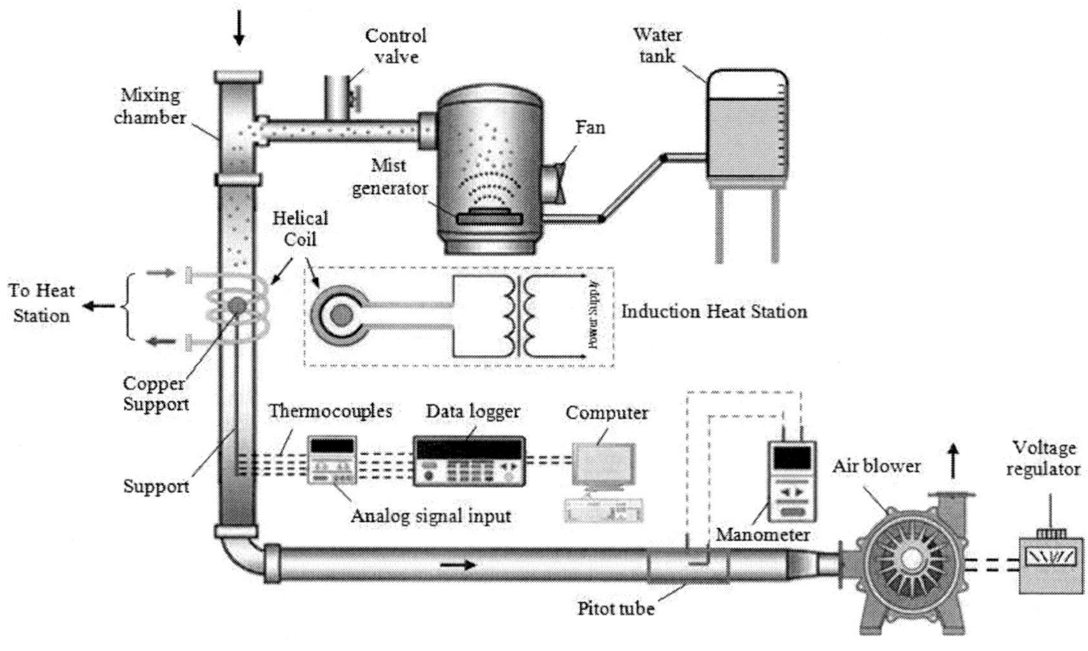

Figure 1. Schematic diagram of the experimental facilities.

The experimental tests were performed under a range of water droplets concentration (j = 46.79 – 111.68 kg/m^2 hr) and Re number varying from ($2.5\cdot10^4$ to $5.5\cdot10^4$). Firstly, the induction heating system was switched on to heat the test sphere and achieved the required temperature. When the test sphere reached the working temperature, the induction heating system was switched off and the test sphere was exposed to air/water droplets two-phase flows. For Biot number ≤ 0.1, the temperature profile inside the test sphere can be assumed uniform and the lumped capacitance model (LCM) can be employed as a heat transfer evaluate method. Based on the change of internal energy inside the test sphere relating to the convection heat transfer rate at the sphere surface, the general heat conduction equation can be defined as: [11]

$$m_s c_s \frac{\partial T}{\partial t} = hA_s\left(T_s - T_{in.}\right) \quad (1)$$

where m_s and c_s are the sphere mass, specific heat, and total surface area respectively. h – is the convection heat transfer coefficient, T_{in} - is the inlet fluid temperature, T_s is the temperature inside the sphere, t - is the cooling time. Eq. (1) was solved analytically and can be written as: [12]

$$h = \frac{C_p \rho d_{sp.}}{6} \sum_{i=1}^{n} \left(\frac{1}{\Delta\tau} \ln \left(\frac{T_{s,i} - T_{in.}}{T_{s,i+1} - T_{in.}} \right) \right) \tag{2}$$

where $\Delta\tau$ – time difference between two readings of the data logger. The Bi number is defined as:[13]

$$Bi = \frac{hL}{k_s} \tag{3}$$

where k_s - the thermal conductivity of the sphere and L - is the characteristic dimension of the sphere, defined as the ratio between the sphere volume V_s and its total surface area A_s. The water droplet diameter can be computed by the Lang equation (4), [14-15].

$$d_p = 0.34 \left(\frac{8\pi\sigma}{\rho F^2} \right)^{1/3} \tag{4}$$

3. Results and discussion

In order to evaluate the effect of air/water droplets cooling process on the test sphere and predicate the resulting status of the heating surface, there were done the experiments under a range of water droplet concentration (j = 46.79 – 111.68 kg/m^2.hr) and Re number varying from ($2.5 \cdot 10^4$ to $5.5 \cdot 10^4$). The temperature profile results of the test sphere cooled by air as well as air/water droplets two-phase flow with the initial temperature equal to 300°C are presented in figure 2.

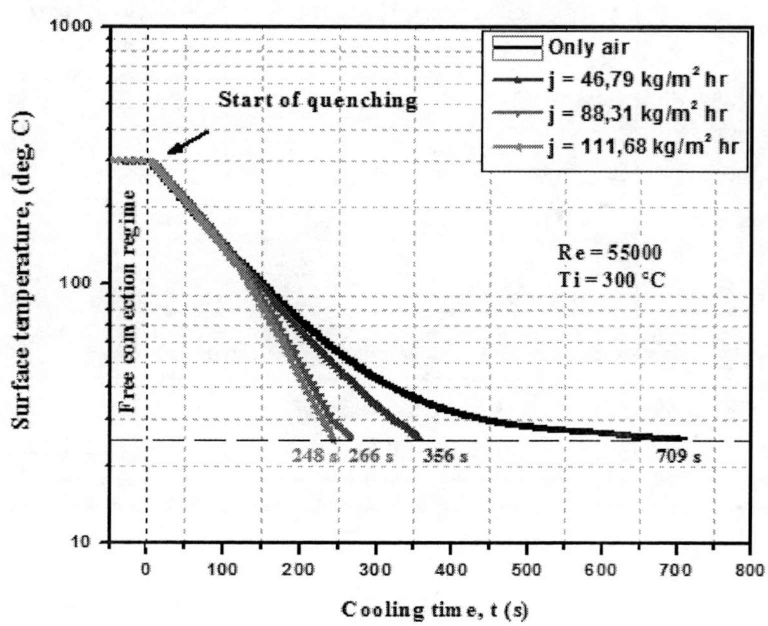

Figure 2. Transient cooling curves for different water mist rate at Re=55000.

Two heat transfer regimes can be distinguished herein. Free convection heat transfer regime occurs when the surface temperature reduced by around 5°C during the water droplet system adjustment to obtain the water droplets concentration required before starting the cooling process. Next, the convective heat transfer was occurred by forced convection. At the starting of quenching, it can be seen that the effect of water droplets on the sphere temperature was insignificant.

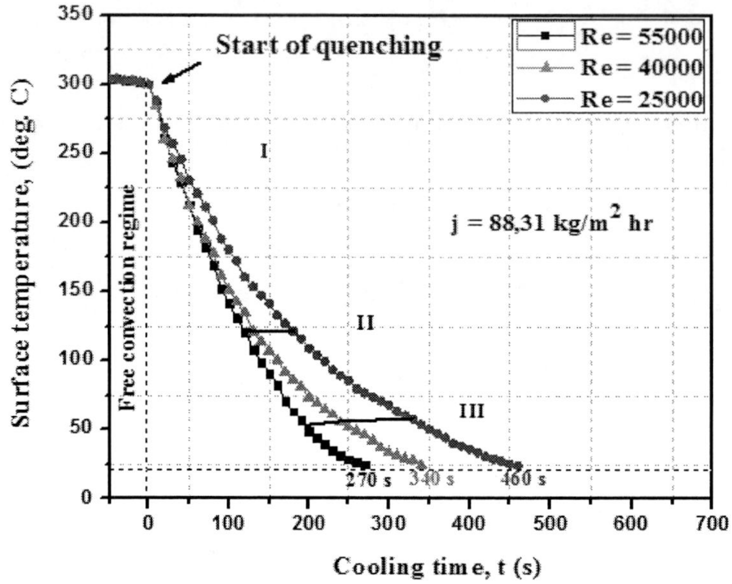

Figure 3. Effect of the Re number on the transient surface temperature for range of water droplet concentration (j=88.31 kg/m^2.hr).

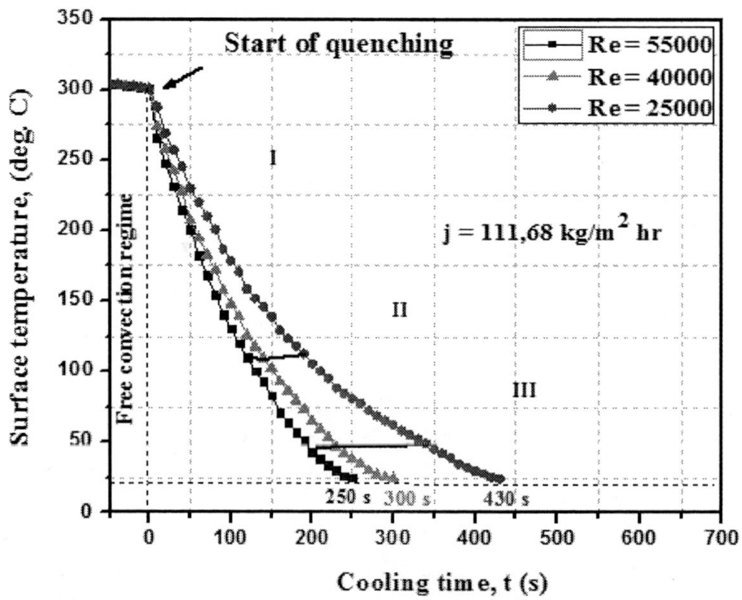

Figure 4. Effect of the Re number on the transient surface temperature for range of water droplet concentration (j=111.68 kg/m^2.hr)

It could be associated with the high surface temperature and force of evaporation that leads to evaporate the water droplets before reaching to the sphere surface. After the temperature of sphere reaches to 100°C the influence of water droplets were appear and it can be noticed that the sphere temperature extremely decreases and the cooling time was reduced from 709s for single-phase air cooling to 248s under (j = 111.68 kg/m^2 hr). The surface temperature decreases about 13.5%, 47%, and 53.2% compared with single-phase air cooling under constant Re number. The cooling time for (j = 46.79 – 111.68 kg/m^2 hr) were found to be 50%, 62%, and 65% less than those obtained by (j=0 kg/m^2 hr).

Figures 3 and 4 represents the effect of the air flow velocity in the terms of Reynolds number (Re) on the cooling process by suspending (j=88.31 kg/m^2 hr) water droplets concentration. In our experiments, high Re number which means high flow acceleration that forcing more and more water droplets reaching to the sphere surface and enhance the cooling process by increasing the surface wettability. At the constant cooling time (t=250 s.) the surface temperature decreases about 67% for Re=55000 compared to the surface tempertature obtained by Re=25000. With increasing in water droplets concentration, the number of water droplets impacting on the surface of the sphere also increases that enhances the extraction of heat by absorbing a large amount of energy in form of latent heat during the evaporation process as shown in figures 4. Based on the temperature gradient and direct observation during the glass channel, the cooling curve can be classified according to the resulting states of sphere surface into three different regions as I- dry region, II- dry-out region, and III- wet region as shown in figures (3-6). The effect of the Re number on the transient heat transfer coefficient for a range of water droplet concentration was depicted in figures 5 and 6. For all cases, it can be noticed that the average heat transfer coefficient was drastically increased with increasing of Re number and water droplets concentration. The heat transfer behavior due to the impacting of water droplets on the sphere surface can be classified into three physical regimes. Clear convection heat transfer regime corresponds to the dry region (I - region), convection and evaporation regimes correspond to the dry-out and wet regions (II and III - regions)

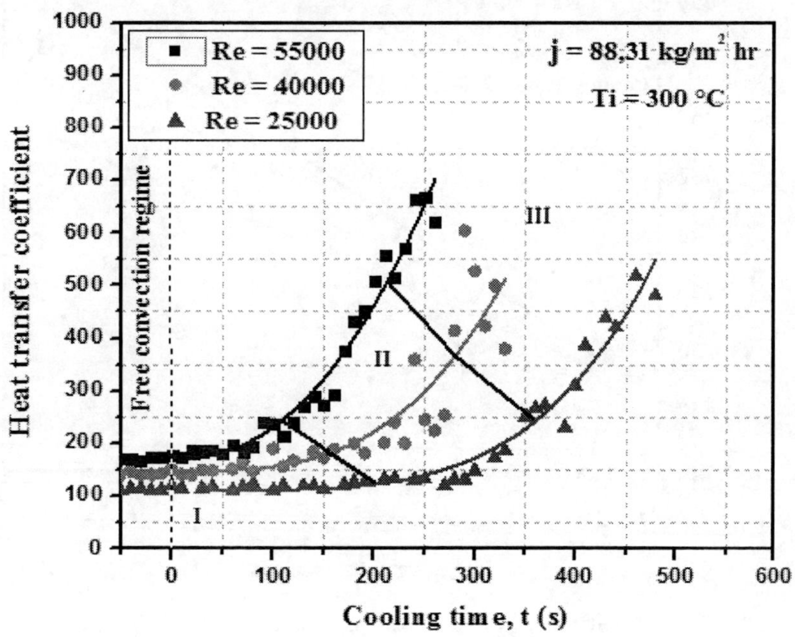

Figure 5. Effect of the Re number on the transient heat transfer coefficient for (j=88.31 kg/m^2 hr)

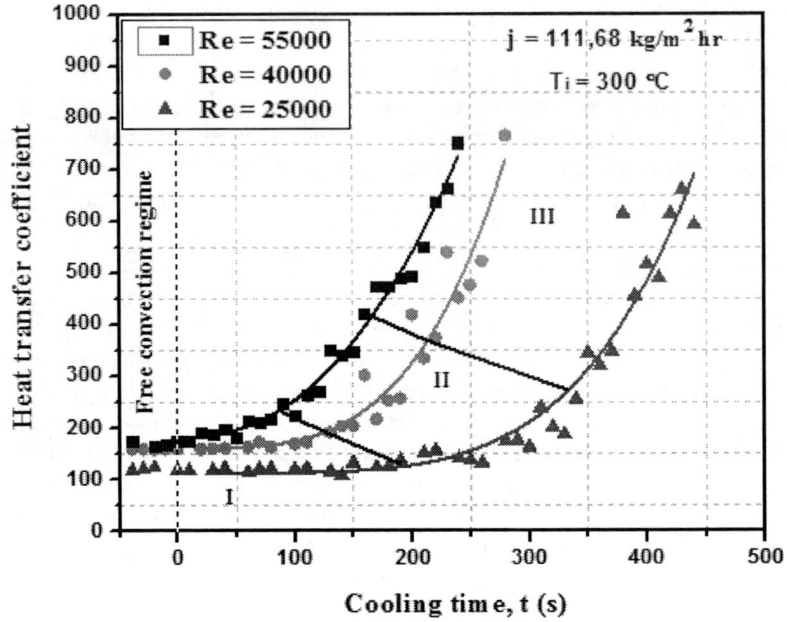

Figure 6. Effect of the Re number on the transient heat transfer coefficient for (j=111.68 kg/m2 hr)

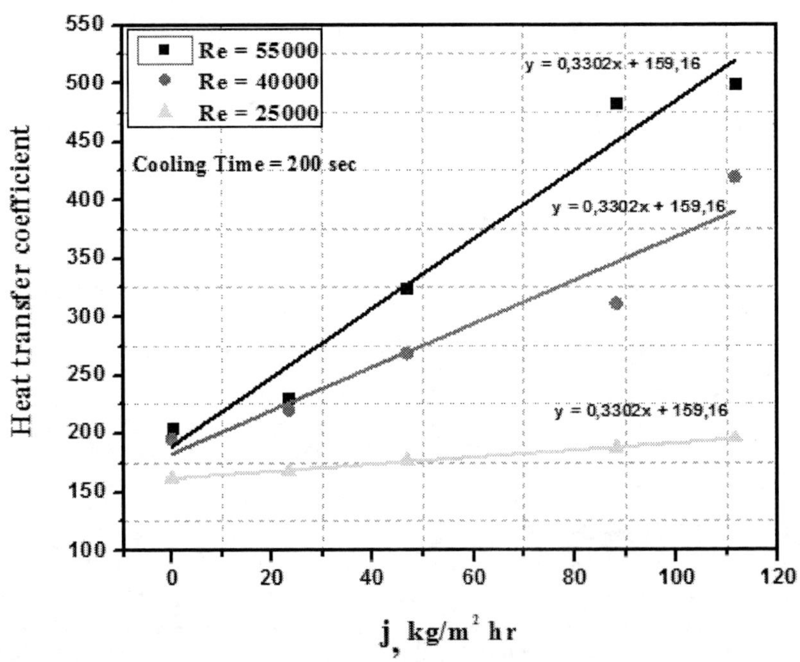

Figure 7. The variations of heat transfer coefficient with water droplets concentration

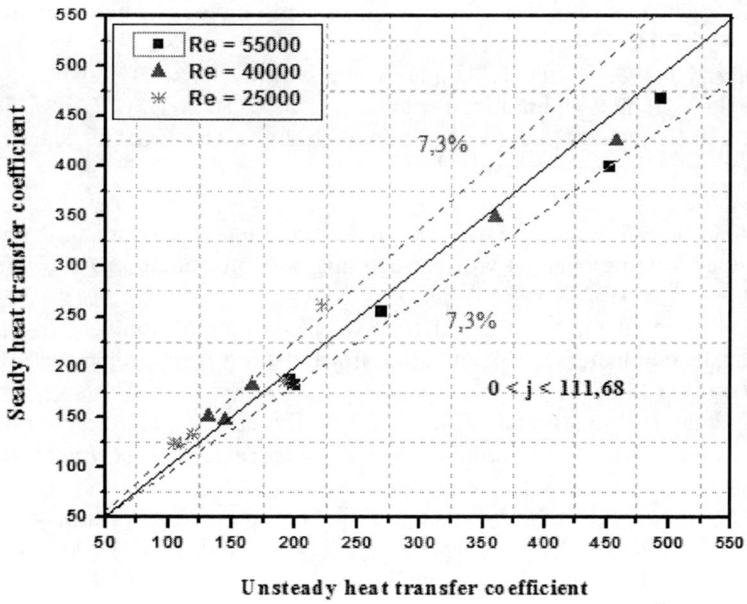

Figure 8. Comparison of steady-state heat transfer coefficient with those obtained by the LCM under corresponding surface temperature conditions.

The variations of heat transfer coefficient with water droplets concentration at constant cooling time (200 sec) is displayed in figure 7. The heat transfer coefficient shows the uptrend with increasing water droplets concentration for all ranges of Re number. In scrutiny of figure 7, the heat transfer coefficient obtained from Re=55000 it's visible to be higher than that obtained by Re=25000 and Re=40000, due to the flow acceleration that forcing more and more water droplets reaching to the sphere surface and enhance the cooling process by increasing the surface wettability as mentioned above. The unsteady-state heat transfer for a varied range of water droplets concentration and different values of Bi numbers were compared with the results obtained by steady-state experiments that were performed in the previous work [9] in terms of corresponding surface temperature of the sphere. The comparison between the steady and unsteady state heat transfer characteristics is shown graphically in figure 8. In this figure, results reasonably agree well within ±7.3%.

4. Conclusions

In this study, an attempt to analyze and understand the heat transfer behavior during air/water droplets cooling process based on the lumped capacitance model (LCM). The major findings can be reported as follows: suspending water droplets on the air main flow providing a significant reduction on the surface temperature. The surface temperature decreases about 13.5%, 47%, and 53.2% for a range of water droplets concentration (j = 46.79 – 111.68 kg/m^2 hr) compared with single-phase air cooling under constant Re number. The heat transfer mechanism due to the impacting of water droplets on the sphere surface was classified into three important physical regimes. Clear convection heat transfer regime corresponds to the dry region (region I). Convection and evaporation regimes correspond to the dry-out and wet regions (region II and III). Also, the heat transfer coefficient obtained by unsteady and steady heat transfer conditions are compared in terms of the corresponding surface temperature of the sphere with an average deviation of 7.3%.

IOP Conf. Series: Materials Science and Engineering **791** (2020) 012001 doi:10.1088/1757-899X/791/1/012001

References

[1] Kumari N, Bahadur V, Hodes M, Salamon T, Kolodner P, Lyons A and Garimella S 2010 Analysis of evaporating mist flow for enhanced convective heat transfer *Int. J. Heat Mass Transf.* **53(15-16)** 3346-56

[2] Agrawal C, Lyons O, Kumar R, Gupta A and Murray D 2013 Rewetting of a hot horizontal surface through mist jet impingement cooling *Int. J. Heat Mass Transf.* **58(1-2)** 188-96

[3] Sharma A and Sahu S 2019 An experimental study on heat transfer and rewetting behavior of hot horizontal downward facing hot surface by mist jet impingement *Applied Thermal Engineering* **151** 459-74

[4] Dhanasekaran T and Wang T 2013 Computational analysis of mist/air cooling in a two-pass rectangular rotating channel with 45-deg angled rib turbulators *Int. J. Heat Mass Transf.* **61** 554-64

[5] Mobtil M, Bougeard D, and Russeil S 2018 Experimental study of inverse identification of unsteady heat transfer coefficient in a fin and tube heat exchanger assembly *Int. J. Heat Mass Transf.* **125** 17-31

[6] Tan X, Zhang J, Liu B and Zhu X 2013 Experimental investigation on heat transfer enhancement of mist/air impingement jet *Science China Technological Sciences* **56(10)** 2456-64

[7] Nirmalan N, Weaver J, and Hylton L 1998 An Experimental Study of Turbine Vane Heat Transfer With Water–Air Cooling *J. of Turbomachinery* **120(1)** 50

[8] Li X, Gaddis J and Wang T 2001 Modeling of Heat Transfer in a Mist/Steam Impinging Jet *J. Heat Transf.* **123(6)** 1086

[9] Abed A, Shcheklein S and Pakhaluev V 2019 Experimental investigation of hydrodynamics and heat transfer of sphere cooling using air/water mist two phase flow *IOP Conf. Series: Materials Science and Engineering* **552** 012001

[10] Abed A, Shcheklein S and Pakhaluev V 2019 Investigation of heat transfer coefficient of spherical element using infrared thermography (IR) and gas – water droplets (mist) as working medium *IOP Conf. Series: Materials Science and Eng.* **481** 012033

[11] Barrow H and Pope C 2007 Droplet evaporation with reference to the effectiveness of water-mist cooling *Applied Energy* **84(4)** 404-12

[12] Yi D and Zhang M 2017 Heat flux investigations during flame thermal spray process using the lumped capacitance method *Applied Thermal Engineering* **123** 554-61

[13] Lienhard J H V, Lienhard J H IV 2013 A Heat Transfer Textbook fourth ed. Dover Publications

[14] Lang R J 1962 Ultrasonic atomization of liquid *Acoust. Soc. Am* **34** 6-8

[15] Kudo T, Sekiguchi K, Sankoda K, Namiki N and Nii S 2017 Effect of ultrasonic frequency on size distributions of nanosized mist generated by ultrasonic atomization *Ultrasonics Sonochemistry* **37** 16-22

Numerical and Experimental Investigation of heat transfer and flow structures around three heated spheres in tandem arrangement

A H Abed[1, 2], S E Shcheklein[1] and V M Pakhaluev[1]

[1]Ural Federal University named after the first President of Russia B. N. Yeltsin 19 Mira St., Yekaterinburg, 620002, Russia
[2]Department of Electromechanical Engineering, University of Technology, Al-Sina'a St., Baghdad, 10066, Iraq

E-mail: akraaam82@yahoo.com

Abstract. The objective of this work is to evaluate the convective heat transfer and flow characteristics around three heated spheres in a tandem arrangement. Numerical simulation and Experimental verification were performed using stationary copper spheres located inside a cylindrical channel with constant channel-to-sphere diameter ratio. Numerical simulation is done for three-dimensional steady-state flow using ANSYS-FLUENT by solving the Reynolds-Averaged Navier Stokes (RANS) equations. Over the test range of Reynolds numbers (2500-55000), the numerical results of the average surface temperature and heat transfer coefficient obtained are a reasonably good agreement with those obtained by experimental measurements. The distributions of the heat transfer coefficient, temperature profiles, velocity field and pressure coefficient around the sphere's surface are calculated and analyzed.

1. Introduction

The heat transfer and flow structure passing around multiple heated bluff bodies are widely investigated problems due to their importance in several engineering such as high temperature gas-cooled nuclear reactor - HTGR [1], bio-film reactor [2] and solar receivers [3], etc. applications such as nuclear power plants, food and chemical processing, and so on. The flow separation complexity and the vortex shedding mechanisms, generated by the sphere arrays have attracted large attention. In the literature, numerical and experimental investigations have been conducted to analyze and understand the wake flow structure around a single-sphere and various arrangements of two spheres [4-7]. In the early studies, Achenbach [8] experimentally investigated the vortical structure of a single sphere wake in the Reynolds number range of $5 \times 104 \leq Re \leq 6 \times 106$. Also, the local static pressure and the skin friction were measured to evaluate the flow separation angle. Sakamoto et al. [9] visualized the wake structure behind a single sphere using dye, and then the vortex shedding was measured by the hot-wire technique. The computational fluid dynamics (CFD) simulation allows an important approach, in which the packing sphere is fully simulated. For example in space between the spheres all local values of relevant variables, such as velocity field, pressure drop, temperature profiles, and so on, is determined by solving the Navies Stokes equations. Such parameters considered to be important details to predict and understanding of the phenomena that occur in such packed spheres. A little investigation has been performed to study the heat transfer and the interaction of the wakes of two spheres in tandem, transverse and staggered arrangements. The Drag force, flow patterns transition and wake structure of two

IOP Conf. Series: Materials Science and Engineering **791** (2020) 012002 doi:10.1088/1757-899X/791/1/012002

interactive spheres were analyzed experimentally and numerically by Chen et al. [10] and Zou et al. [11]. Laser flow visualization has been used to determine the flow separation angle and identify the effect of aligned angle between two spheres. They indicated that the wake structure of a sphere was more affected by the distance and the aligned angle between spheres. Zhang et al. [12] numerically investigated the effect of spacing between two vertically arranged spheres on natural heat transfer for $2 \leq S/D \leq 24$. They noticed that the re-circulation zone is seriously affecting on the convective heat transfer coefficient and as the spacing ratio between two separate spheres increases about 24 times of the sphere diameter, the flow re-circulation zone takes place over the upper sphere for circumferential angle $142.5° \leq \theta \leq 180°$. In the sphere swarms for most engineering and practical applications, the data of the single sphere is not capable to precisely predicate the mass transfer process [8]. Juncu [13, 14] solved numerically the mass transfer between two spheres have the same diameter in the tandem arrangement and incompressible moving fluid. This work aims to extend the analysis of the heat transfer, boundary layer development and flow structure of three interactive spheres under a steady-state condition. The spheres were packed inside the cylindrical channel with a constant channel-to-sphere diameter ratio. The influence of the generated vortices on heat transfer performance and moving flow structure are examined. In order to validate the CFD simulation method, the CFD results were compared with those obtained by experimental measurements.

2. Physical model and CFD Simulation conditions

The computational domain geometry and mesh models for CFD simulation are shown in figure 1. The test spheres are located inside a vertical cylindrical channel in a tandem arrangement with the constant channel-to-sphere D/d ratio.

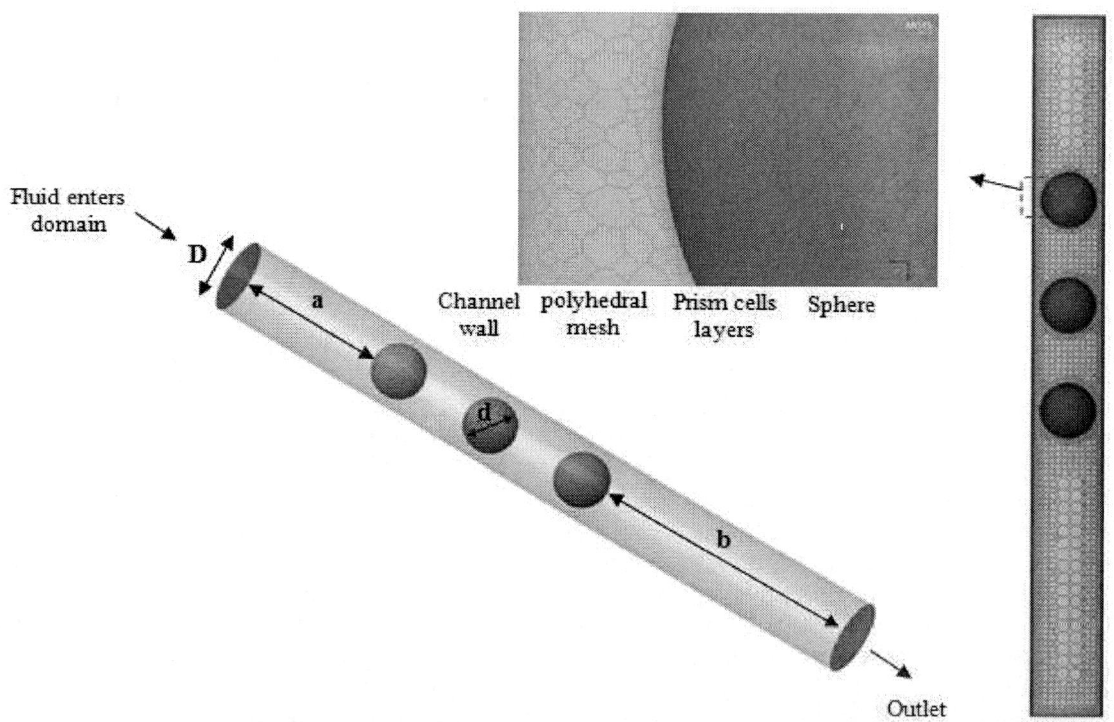

Figure 1. Schematic view of the computational domain geometry and mesh models for CFD simulation.

The computational domain consists of the inlet region, 3 layers of full spheres and an outlet region. The air was used as working fluid and the velocity and temperature at the inlet region were kept at Ti and ui, respectively. In addition, an opening boundary with a constant average static pressure was applied at the outlet region. In the present study, the computational domain is discretized into the fluid region as well as a solid region uses 1.8 million cells and 0.4 million cells for the fluid and solid regions, respectively. Due to the complicated geometry of the flow domain including the gaps region (equal ds) between the spheres, unstructured mesh with tetrahedral cells are adopted for computational domain and then transformed into polyhedral cells in Ansys fluent to enhance the numerical stability of the model. The commercial software ANSYS FLUENT version 17 was used in this work. Governing equations for CFD simulations are as follow [15].

Continuity equation

$$\frac{\partial}{\partial x_i}(\rho u_i)+\frac{\partial}{\partial x_j}(\rho u_j)+\frac{\partial}{\partial x_k}(\rho u_k)=0 \tag{1}$$

Momentum equation

$$\begin{aligned}&\frac{\partial(\rho u_i u_i)}{\partial x_i}(\rho u_i)+\frac{\partial(\rho u_i u_j)}{\partial x_j}(\rho u_j)+\frac{\partial(\rho u_i u_k)}{\partial x_k}(\rho u_k)=\\&=-\frac{\partial p}{\partial x_i}+\frac{\partial(\tau_{ii})}{\partial x_i}+\frac{\partial(\tau_{ij})}{\partial x_j}+\frac{\partial(\tau_{ik})}{\partial x_k}+\rho g_i+F_i\end{aligned} \tag{2}$$

Energy equation

$$\frac{\partial(\rho uT)}{\partial x_i}+\frac{\partial(\rho vT)}{\partial x_j}+\frac{\partial(\rho wT)}{\partial x_k}=\frac{\partial}{\partial x_i}\left(\frac{k}{c_p}\frac{\partial T}{\partial x_i}\right)+\frac{\partial}{\partial x_i}\left(\frac{k}{c_p}\frac{\partial T}{\partial x_j}\right)+\frac{\partial}{\partial x_i}\left(\frac{k}{c_p}\frac{\partial T}{\partial x_k}\right) \tag{3}$$

The SST k-ω turbulence model which combines the advantages of the k-ε and k-ω models were applied in this simulation and the scalable wall-function treatment with an appropriate dimensionless distance of the wall grid elements (y+) were adopted. The SST k-ω model can be described as follows [15]:

Kinematic Viscosity

$$\nu_k=\frac{a_1\kappa}{\max(a_1\omega, SF_2)} \tag{4}$$

Turbulence Kinetic Energy

$$\frac{\partial\kappa}{\partial t}+U_j\frac{\partial\kappa}{\partial x_j}=P_\kappa-\beta^*\kappa\omega+\frac{\partial}{\partial x_j}\left[(\nu+\sigma_k\nu_k)\cdot\frac{\partial\kappa}{\partial x_j}\right] \tag{5}$$

Specific turbulence dissipation rate

$$\frac{\partial\omega}{\partial t}+U_j\frac{\partial\omega}{\partial x_j}=\alpha S^2-\beta\omega^2+\frac{\partial}{\partial x_j}\left[(\nu+\sigma_\omega\nu_k)\frac{\partial\omega}{\partial x_j}\right]+2(1-F_1)\sigma\omega^2\frac{1}{\omega}\frac{\partial\kappa}{\partial xi}\frac{\partial\omega}{\partial xi} \tag{6}$$

3. Experimental Setup and Procedure

The experimental setup details are depicted schematically in figure 2. In the experimental setup, the inlet working fluid was directed to the heated spheres by 1000 W air pump. Using the pitot tube with a digital manometer calibrated by a vane-type anemometer, the airflow rate was measured. The airflow rate was controlled by modifying the pump speed through a variac voltage regulator. The test objects are copper spheres having 34mm diameter were heated independently using 100W - electrical heater with a variac transformer to obtain a uniform heat flux. The test spheres were packed inside an adiabatic cylindrical channel in tandem arrangement with constant pitch ratio (p=ds). For each sphere, two k-type thermocouples were used to directly measure the surface temperature (Tw), were embedded in points referred by the numbers 1, 2 as shown in figure. 2. Further, the inlet and outlet air temperatures were recorded using two thermocouples located immediately in the air entrance region and outlet region of the adiabatic channel. After reaching the steady-state condition, all the temperatures were recorded by using MSD200 data logger.

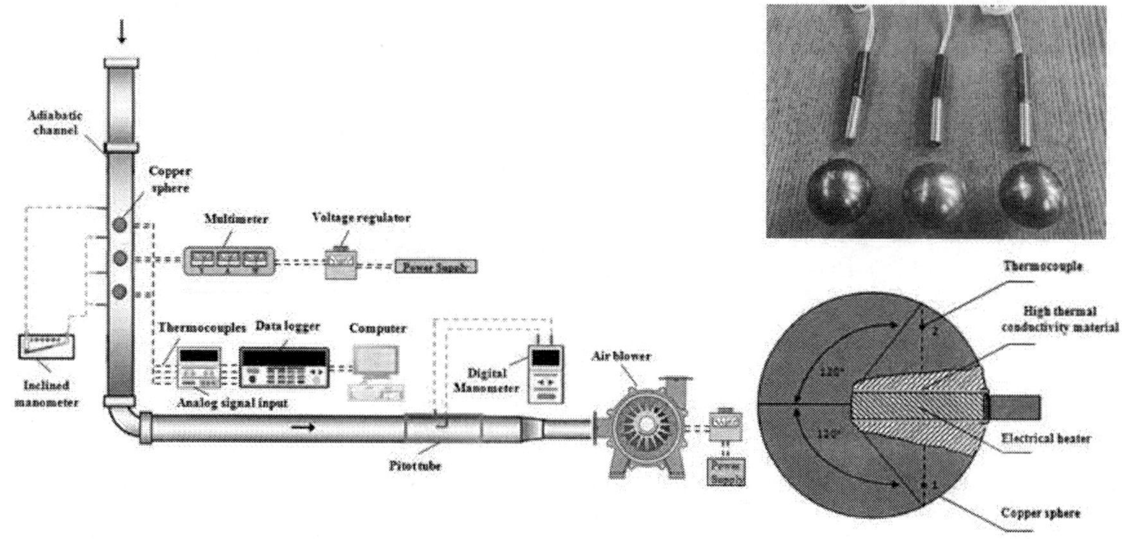

Figure 2. Schematic diagram of the experimental Setup and the construction of heat transfer models.

4. Results and discussion

Herein, the simulation results of the spheres packing inside a vertical cylindrical channel in a tandem arrangement were described and the performed experimental results are presented. First, the numerical results of the average surface temperature and heat transfer coefficient obtained from the present simulation are compared with those obtained by experimental measurements as shown in figure 3. The average deviations of the present results are within 5.8%, 6.1%, and 7.5%, respectively.

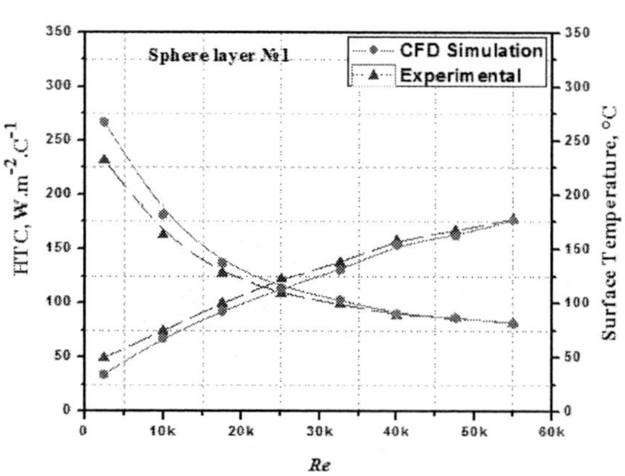

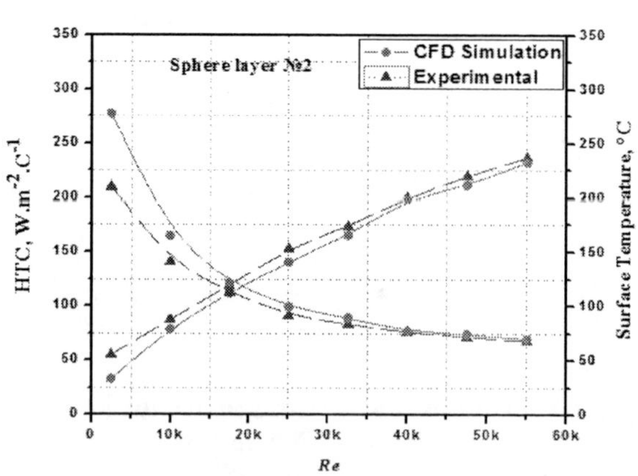

IOP Conf. Series: Materials Science and Engineering **791** (2020) 012002 doi:10.1088/1757-899X/791/1/012002

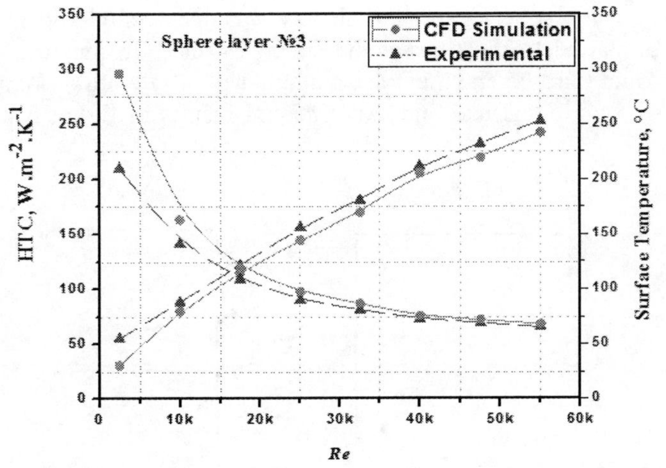

Figure 3. Comparison of average surface temperature and heat transfer coefficient.

It can be observed that the average surface temperature tends to decrease and the heat transfer coefficient tends to increase with increasing of Re for all spheres layers. With the help of the experimental measurements, the experimental results of average Nusselt number are correlated as shown in Eq. (4) below.

$$\mathrm{Nu} = C_i\,\mathrm{Re}^{0.58}\,\mathrm{Pr}^{0.33}$$

where C_i is the constant in the heat transfer rate correlation with $C_1 = 0.47$; $C_2 = 0.65$, and $C_3 = 0.69$.

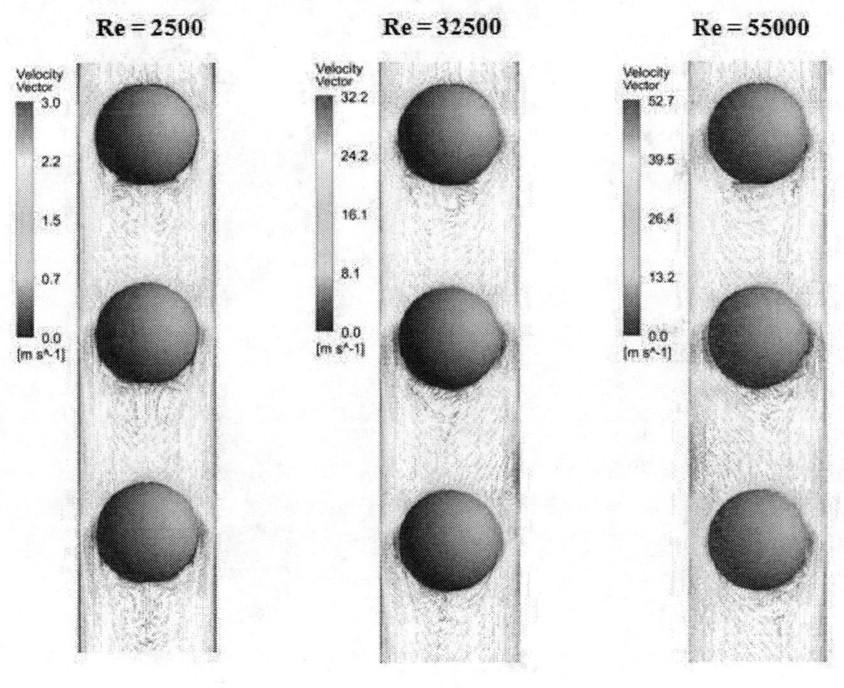

Figure 4. Velocity streamlines around spheres: Re=2500 Re=32500 and Re=55000.

Figure. 4 depicts the flow streamlines characteristics among the spheres layer. The color index represents the magnitude of the velocity at any point. The air flows through the sphere structure and it can be seen that the streamline distribution around the second and third layers are different from those in the first layer. Within the sphere gaps, the airflow was accelerated and leads to increasing of the turbulence intensity imparted to the flow at the sphere surface. In the front region on each sphere surface of the second and third layers, there are a number of accelerating points that may improve the heat transfer rate. From the numerical and experimental results in figure 5, it is observable that the second and third layers yield considerable heat transfer in comparison with the first layer due to the increase of turbulence intensity.

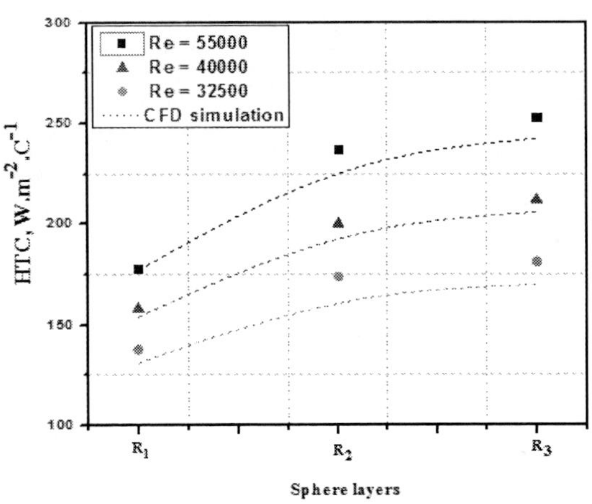

Figure 5. Variations of average heat transfer co-efficient as a function of sphere layers.

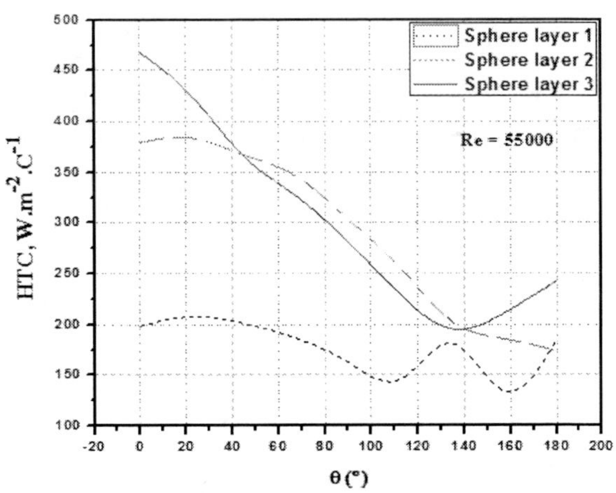

Figure 6. Variations of average heat transfer coefficient as a function of angle measured from stagnation point.

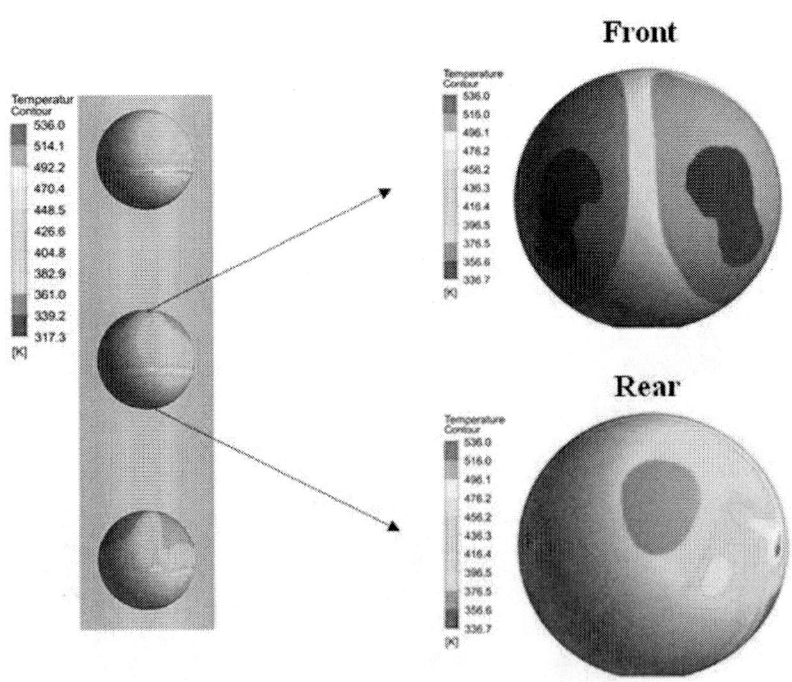

Figure 7. Temperature contours of a selected sphere.

The local heat transfer coefficient as a function of the angle [θ] measured from the front stagnation point is shown in figures 6, for Reynolds number (Re=55000). For the present data, the heat transfer coefficient for the second and third layers around 63–72% higher than the first layer. Figure 7 represents the calculated temperature distribution on both the front and rear regions of the selected sphere and its cross-section. The color index represents the magnitude of the local temperature. From this figure, it is obvious that the heat transfer rate in the front region is higher with comparison to the condition in the rear region due to the high turbulence intensity leading to enhance the convection heat transfer than the rear region as we mentioned above. The variations of pressure coefficient (Cp) for all spheres layer as a function of zenith angle are shown in figure. 8. It can be noticed that the pressure coefficient on the front-upper region (θ = 0° to 60°) of the sphere is positive and that on the lower-front and upper-rear area is negative.

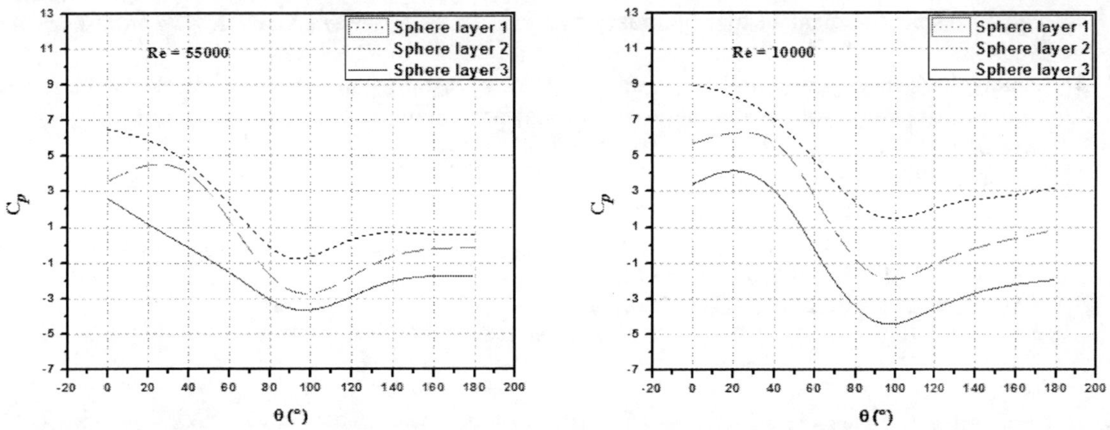

Figure 8. Pressure coefficient distribution as function of angle measured from stagnation point.

5. Conclusions

Numerical and experimental investigation of heat transfer characteristics and flow structures around three heated spheres in a tandem arrangement for Re numbers ranges 2500-55000 has been performed. The highest heat transfer rate is found at the second and third layers due to airflow accelerated that leads to increasing of the turbulence intensity imparted to the flow at the sphere's surface. The heat transfer coefficient for the second and third layers around 63–72% higher than the first layer. Additionally, the new empirical correlation for the Nusselt number based on the experimental data is proposed. The agreement between the numerical results and those obtained from the experimental measurements is reasonable.

References

[1] Filippov G, Bogoyavlenskii R, Ponomarev-Stepnoi N and Gol'tsev A 2004 Vy`sokotemperaturny`j gelievy`j modul`ny`j yaderny`j reaktor s sharovy`mi TVE`Lami dlya proizvod-stva e`lektroe`nergii i vodoroda [Modular High-Temperature Helium-Cooled Nuclear Reactor with Spherical Fuel Elements for Electricity and Hydrogen Production] Atomnaya e`nergiya [*Atomic Energy*] **96(3)** 152-8

[2] Fan C and Wang H 2012 Degradation of Methyl Paraben by the Aerated Pebble-bed Biofilm System *APCBEE Procedia* **1** 299-303

[3] Lu JF, Chen Y, Ding J, Wang WL 2016 High temperature energy storage performances of methane reforming with carbon dioxide in a tubular packed reactor *Appl Energ* **162** 1473-82

[4] Ozgoren M 2013 Flow structures around an equilateral triangle arrangement of three spheres *Int. J. of Multiphase Flow* **53** 54-64

[5] Peng X and Huang G 2017 Effect of temperature difference on the adhesive contact between two spheres *Int. J. of Engineering Science* **116** 25-34
[6] Tsutsui T 2008 Flow around a sphere in a plane turbulent boundary layer *J. of Wind Engineering and Industrial Aerodynamics* **96(6-7)** 779-92
[7] Nagata T, Nonomura T, Takahashi S, Mizuno Y and Fukuda K 2018 Direct numerical simulation of flow around a heated/cooled isolated sphere up to a Reynolds number of 300 under subsonic to supersonic conditions *Int. J. of Heat and Mass Transfer* **120** 284-99
[8] Achenbach E 1972 Experiments on the flow past spheres at very high Reynolds numbers *J. of Fluid Mechanics* **54(3)** 565-75
[9] Sakamoto H and Haniu H 1990 A Study on Vortex Shedding From Spheres in a Uniform Flow *J. of Fluids Engineering* **112(4)** 386
[10] Chen R and Wu J 2000 The flow characteristics between two interactive spheres *Chemical Engineering Science* **55(6)** 1143-58
[11] Zou J, Ren A and Deng J 2005 Study on flow past two spheres in tandem arrangement using a local mesh refinement virtual boundary method *Int. J. for Numerical Methods in Fluids* **49(5)** 465-88
[12] Zhang J, Zhen Q, Liu J and Lu W 2019 Effect of spacing on laminar natural convection flow and heat transfer from two spheres in vertical arrangement *Int. J. of Heat and Mass Transfer* **134** 852-65
[13] Juncu Gh 2007 Unsteady forced convection heat/mass transfer around two spheres in tandem at low Reynolds numbers *Int. J. Thermal Sci.* **46** 1011-22
[14] Juncu Gh 2018 Mass transfer from two spheres in tandem accompanied by chemical reactions *Chemical Engineering Science* **189** 102-13
[15] Fluent Inc. 2005 Fluent 6.2 User *Manual*

IOP Conf. Series: Materials Science and Engineering **791** (2020) 012003 doi:10.1088/1757-899X/791/1/012003

Experimental investigation on the enhancement of heat transfer by using carbon nanotubes CNT taunit m series

A J Ali[1,2], H J Jaber[3], L J Habeeb[4] and E N Tugolukov[1]

[1]Technology and Methods of Nano Products Manufacturing, Tambov State Technical University, 106 Sovetskaya St., Tambov 392000, Russia
[2]Department of Biomedical Engineering, University of Technology, Baghdad 10066, Iraq
[3]Mechanical Engineering Department, Faculty of Engineering, University of Kufa, Al Najaf 54001, Iraq
[4]Training and Workshop Center, University of Technology, Baghdad 10001, Iraq

E-mail : 10678@uotechnology.edu.iq

Abstract. The enhancement of heat transfer is considered as one of the important tasks for controlling the energy dissipation and consumption. In order to secure this task, it is intended to increase the efficiency of heat exchanging systems taking into account the economic and environmental considerations. This work aims to test experimentally the effect of nanoparticle concentration functionalized carbon nanotubes (FCNT Taunit M) in water and Reynolds number on the heat transfer coefficient. Three concentrations of 0.01, 0.05, 0.1 vol.% for FCNT taunit have been used. The heat transfer augmentation of the nanofluid was investigated experimentally through a test rig constructed for this purpose under constant heat flux range (16061-22682) W/m2 at the developing region with Reynold number range 5000, 7000 and 9000 have been used. The particle size 12 nm for FCNT Taunit M. The experimental results revealed an augmentation in the Nusselt number as the volume concentration increased. The economic aspect was taken into account in this study. Pressure drop can be neglected because the increases were very small.

1.Introduction

Nowadays, the location and importance of heat transfer in engineering applications such as heat exchangers, casting processes and cooling systems of electronic equipment is very much considered. Furthermore, traditional fluids such as oil, water, and ethylene glycol EG in general are used for heat transfer. Many of the techniques have been applied to improve heat transfer in cooling systems and heating, but the efficiency of these fluids in the heat transfer applications is low. Dispersion of solid particles in different fluids is a new method increase heat transfer. Using nanofluid is one of the best ways to improve the heat transfer, which was given by Choi [1]. Nanofluids are mixtures of nanometer-sized solid particles such as carbon nanotube (MWCNTs or SWCN, Titanium dioxide (TiO2), Copper (II) oxide (CuO), Aluminium (Al), etc. Suspended in common liquids such as water and EG which recently were used in many fields of thermal engineering [2–5]. Therefore, extensive research has been done to improve heat transfer using active and passive methods [6–11]. In addition, a large number of studies devoted to using thein base fluids. Hussien et al. [12] have reported the enhancement of forced convective heat transfer with different mass fractions of MWCNTs (0.075, 0.125 and 0.25 wt.%). Also, they observed an improvement in heat transfer coefficient with the increase in mass fraction of MWCNTs in the nanofluid reaching to 23.9% improvement at 0.25 wt.% MWCNTs/water at Re from 200 to 500. Kayhani et al. [13] studied experimentally an empirical analysis of pressure drop and forced convection turbulent flow heat transfer of Al2O3-water nanofluid with nanoparticles

IOP Conf. Series: Materials Science and Engineering **791** (2020) 012003 doi:10.1088/1757-899X/791/1/012003

of (40) nm size at (0.1, to 2) % volume concentrations inside a circular horizontal tube at constant heat flux in the tube wall. The experimental result exhibited the enhancement in Nusselt number 22% at Re of 13500 using 2% volume concentration of Al2O3-water nanofluid compared to distilled water. Carbon nanotubes have thermal conductivity reaching (2000- 3500) W/M. K [14]. Carbon nanotube exists as the single type particle such as single (SWNTs), multi-walled (MWCNTs) and doubles (DWCNTs) structures or like hybrid nanoparticles recently appeared [15]. Finally, in this research, we have used functionalized carbon nanotube type (Taunit M) in the base fluid of water to enhance heat transfer, and also it has created an experimental facility, which has high efficiency of heat transfer.

2. Materials and Methods

2.1. Nanofluids

Functionalized carbon nanotubes "Taunit M" was used in base fluid (water) for the enhancement of heat transfer. "Taunit M" carbon nanostructured materials manufactured at Nanotechnology Center (NanoTechCenterLtd) in Tambov, Russia [16]. The functionalized CNTs were prepared by using traditional method of chemical oxidation of CNT with sodium hypochlorite [17, 18]. Three concentrations of 0.01, 0.05, 0.1 vol.% have been used of functionalized carbon nanotubes "Taunit M as shown in figures 1 and 2.

Figure 1. FCNTs Taunit M nanofluids.

Figure 2. Samples of FCNTs.

2.2. The experimental facility

An experimental facility was designed and constructed to the measurement of heat transfer enhancement. The experimental facility, where the heat transfer enhancement was tested, consists of two principle circuits: the nanofluid circuit and the cooling circuit. Each one contains components and measuring devices. The principle components of this experimental facility are the test section, heat exchanger, nanofluid reservoir and centrifugal pump. Figure 3 shows the schematic diagram of the experimental facility.

The test section represents the main part of the test rig where the heat transfer coefficient, the pressure drop and characteristics of the nanofluid will be tested. The high thermal conductivity of the copper made it the first candidate to be the material of the test section pipe with the following details, as listed in table 1. The test section was prepared through the following steps.

a) The nuts for temperature sensors (thermocouples) were fixed on the pipe's outer surface.

b) The pipe was covered by an electrical insulator to prevent the electrical conduction between the heater and the test rig body.

c) The electrical heater type FeCrAl (Iron – Chromium - Aluminum) alloy wire with 0.8 mm diameter, 2.8 Ω/m, 20 American Wire Gauge (AWG) [19] and 8.45 m length was used to generate a constant heat flux. With ceramic beads were wrapped around the pipe and connecting to the electric source with two ceramic joins.

d) To minimize the heat losses and focus the heat on the pipe wall, a thermal-electrical resistance tape was used to wrap over the ceramic beads.

e) To decrease the heat loses, 5 cm thickness of glass wool was used to cover the test section.

f) Finally, an aluminum sleeve with 10 cm diameter was used to cover the test section.

2- The heat exchanger to reject the heat gained from the test section, shell and coil heat exchanger was manufactured and prepared for this task, with the following characteristics:

a) The shell height is 0.7 m with 0.35 m diameter, manufactured from aluminum plate of 0.6 mm thickness.

b) The coil diameter is 0.27 m, manufactured from a copper pipe (0.0127 ×0.0071 ×17 m) diameter, thickness and length, respectively. The number of turns is 19 and the pitch between turns is 0.01 m. Two pipes of size (3/8") were welded on the coil in a diagonal shape to maintain the pitch distance and the coil straightening.

3- The nanofluid reservoir represents the feeding source of the nanofluid to the test rig. The diameter and height of this reservoir are 0.15 m and 0.6 m, respectively. The pump is connected to the lowest point in the reservoir. The desired flow rate of nanofluid that passes to the test section is secured by adjusting the valves and the excess nanofluid return to the reservoir through the bypass line. In addition, the centrifugal pump was selected for the test rig and its specifications to achieve the desired Reynold number.

4- The cooling cycle aims to reduce the temperature of the returned water from the heat exchanger and secure the required inlet temperature to the test section. It consisting of two centrifugal pumps in the cooling cycle, which are identical to the nanofluid's pump characteristics. In addition, the source tank is a cubic shape with 1 m3 capacity. In addition, the air cooler with 2500 m3/hr. capacity receives part of the hot water from the heat exchanger and reduces its temperature by passing through the wood fiber.

5- The air compressor is used to ensure that nanofluids are completely discharged from the working circuit after shutdown it.

6- The voltage stability is important for maintaining on the same work rhythm for the nanofluid pump; this means the same mass flow rate will pass through the test section.

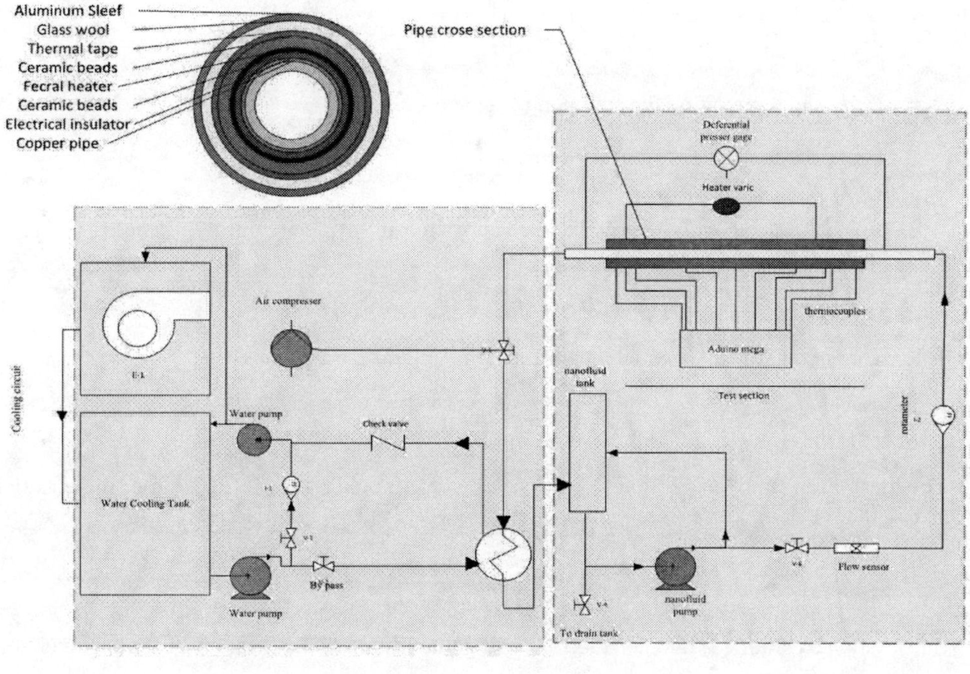

Figure 3. Scheme of the experimental setup.

IOP Conf. Series: Materials Science and Engineering **791** (2020) 012003 doi:10.1088/1757-899X/791/1/012003

2.3. Measuring Devices

1-Thermocouples to measure the variance of pipe surface and nanofluid temperatures along the test section, bolt and pressure spring thermocouples type K were used.
2-Three flow meters were used in the test rig, the first one is a flow sensor model YF-S201 working with Arduino technology at range (0 – 30 l/min) and located after the nanofluid pump discharge.
3-The pressure gauge (GRUNDFOS - Germany) to measure the pressure drop through the test section.
4-The Arduino technology was used with thermocouples, flow sensor and heater power in the test rig.

Table1. Copper pipe details.

Type	*Length*	*Outer diameter*	*Inner diameter*	*Thickness*
ACR-D	*1000* mm	12.7 mm, 1/2"	10. 7 mm	1.0 mm

In this study, the heater is generating a constant heat on the outer surface of the pipe:

$$Q_{heater} = I \times V \tag{1}$$

The generated heat is transferred through the pipe wall by conduction according to Fourier's law.

$$Q_{cond,pipe} = \dot{q}A_{s,out} = 2\pi\Delta xk\frac{[T_{s,out}(x) - T_{s,in}(x)]}{\ln(r_{out}/r_{in})} \tag{2}$$

Therefore, the heat flux on the outer surface of the pipe is given by:

$$\dot{q} = \frac{2k[T_{s,out}(x) - T_{s,in}(x)]}{D_o \ln(r_o/r_{in})} \tag{3}$$

The rate of heat transfer to the nanofluid is given by [20]:

$$Q_{nf} = \dot{q}\,A_{s,in} = \dot{m}_{nf}\,Cp_{nf}\,(T_{nf,out} - T_{nf,in}) \tag{4}$$

where,
$A_{s,in} = \pi D_{in} L$, and rearranging the equation (4) gives:

$$\dot{q} = \frac{\dot{m}_{nf}\,Cp_{nf}}{\pi D_{in} L}(T_{nf,out} - T_{nf,in}) \tag{5}$$

By applying the steady-state energy balance on a slice of the pipe (control volume) [21], the rate of heat that gained from the fluid is:

$$\dot{Q}_{in} - \dot{W}_{in} + \dot{m}h_{in} - \dot{m}h_e = 0 \tag{6}$$

After rearranging, the differential mean fluid temperature can be written as:

$$dT_m = \frac{\dot{q}\,p}{\dot{m}\,Cp}dx \tag{7}$$

The local mean fluid temperature $T_m(x)$ can be determined by the integration of the equation (7) from $x = 0$ to x along the test section and assuming $T_{m,in}$ at x=0 is equal to $T_{nf,in}$

$$T_m(x) = T_{nf,in} + \frac{\dot{q}\,p}{\dot{m}\,Cp}x \tag{8}$$

Substitution equation (5) into equation (8) will get the value of the local mean fluid temperature:

$$T_m(x) = T_{nf,in} + \frac{(T_{nf,out} - T_{nf,in})}{L}x \tag{9}$$

According to Newton's law of cooling:

$$Q_{nf} = h(x)\, A_{s,in}\left[T_{s,in}(x) - T_m(x)\right] \tag{10}$$

The heat flux will be:

$$\dot{q} = h(x)\left[T_{s,in}(x) - T_m(x)\right]. \tag{11}$$

Then, the local heat transfer coefficient is:

$$h(x) = \frac{\dot{q}}{\left[T_{s,in}(x) - T_m(x)\right]}. \tag{12}$$

The Nusselt number will be calculated locally at each section by the equation [22]:

$$Nu(x) = \frac{h(x)\, D_{in}}{k_{nf}}. \tag{13}$$

3.Results and discussion

The experimental results will be presented and discussed in detail the tests of convective heat transfer for the base fluid (water) as well as the nanofluids with particle loading of (0.01, 0.05, 1 and 0.1) vol.%. Experimental conditions are forced convection turbulent developing flow, with Reynolds number range of (5000, 7000, and 9000) and constant heat flux. To confirm the experimental test rig validity in convective heat transfer calculations as well as to provide a reference for comparison, the first test run was conducted on the distilled water only. The predicted Nusselt number calculated by Gnielinski equation (14) [23] for turbulent flow in the developing region and compared to present work versus Reynold number as shown in figure 4. The experimental behavior was conformable to the predicted values with deviation from 11% to 23% due to the experimental actual conditions.

$$u = \frac{\left(\frac{f}{8}\right)(Re - 1000)Pr}{1 + 12.7\left(\frac{f}{8}\right)^{0.5}\left(Pr^{\frac{2}{3}} - 1\right)} \tag{14}$$

The surface temperature varied along the pipe length and increased as the axial distance increased due to the uniform heat flux effect on the surface. Figure 5 illustrates the surface temperature distribution along the test section for different Re numbers for the distilled water. The highest surface temperature appeared at the lowest value of Reynold number (Re = 5000). Also, this figure showed a decrement in the pipe temperature as the Re number increased due to the increase of water velocity.

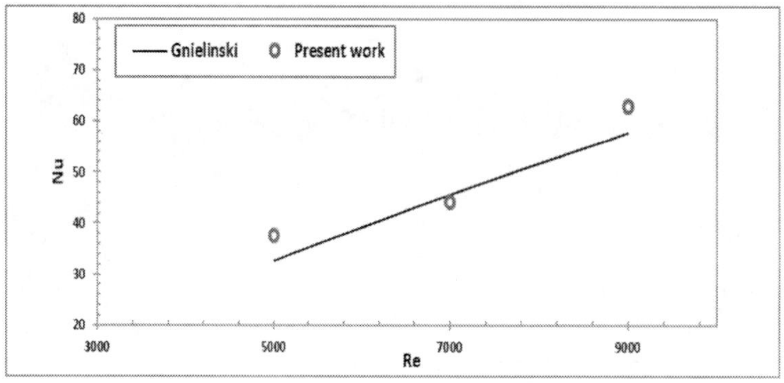

Figure 4. Experimental and theoretical Nusselt number for the distilled water.

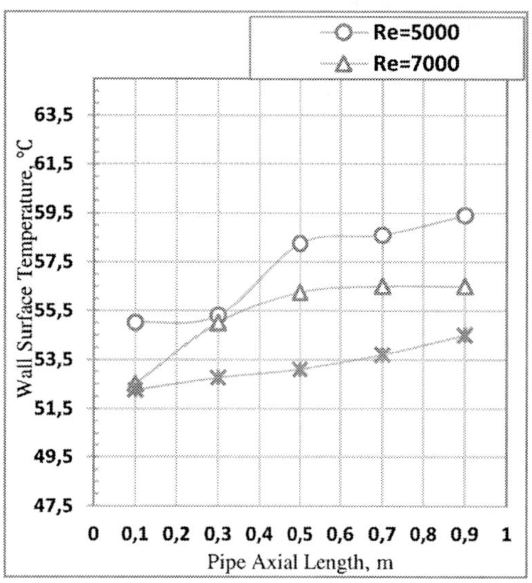

Figure 5. Surface temperature distribution along the pipe axial length for the distilled water at different Reynold numbers.

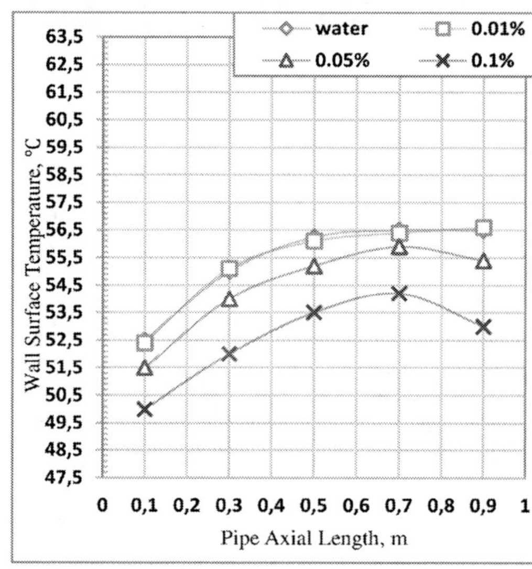

Figure 6. Surface temperature distribution along the pipe axial length for the distilled water and CNT (taunit m) nanofluid at different concentrations at Re =7000.

To make a reference for comparison among the experimental data, Re= 7000 was selected to clarify the nanofluid effect on the heat transfer performance. Figure (6) shows a decreasing in surface temperature with using nanofluid compared with water. The best performance was with 0.1% volume fraction. The average heat transfer coefficient versus Reynold number is explained in figures (7). For figures 8 the enhancement in the average heat transfer coefficient reflected to the Nusselt number, where the maximum augmentation was *21%* at 0.1 vol.% for particle loading, and 9000 Reynold number.

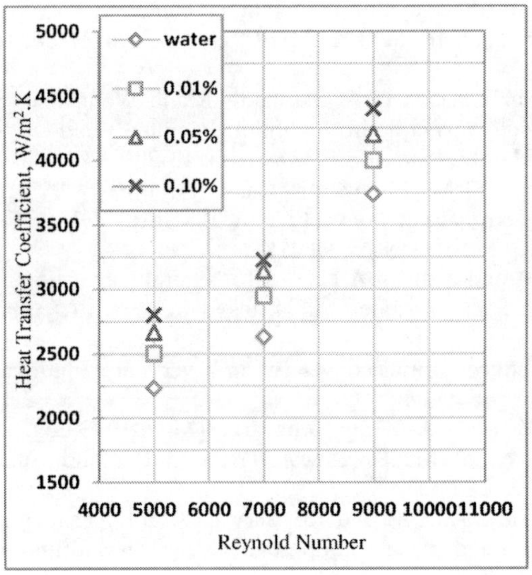

Figure 7. Average heat transfer coefficient for distilled water and CNT taunit nanofluid at different concentrations versus Reynold number.

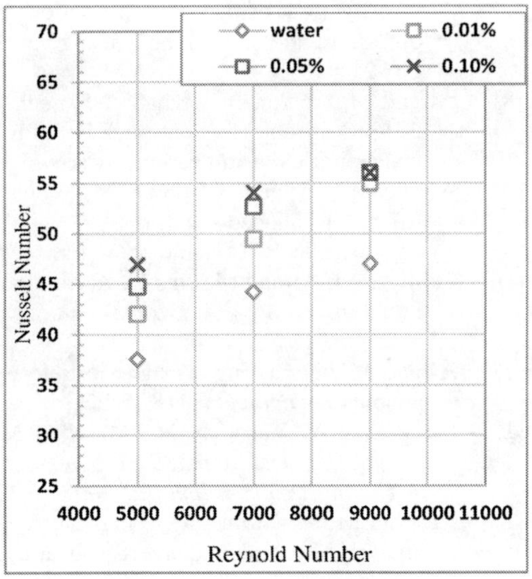

Figure 8. Average Nusselt number for distilled water and (CNT taunit m) nanofluid at different fractions versus Reynold number.

4.Conclusions

In this experimental study, adding the nanoparticles CNT type (taunit m) to the distilled water in small concentrations enhanced the heat transfer characteristic compared to distilled water, where the maximum enhancement in the Nusselt number reached to 21 % at 0.1 vol.% under Reynold 9000. As the volume fraction increased, the convective heat transfers augmented. The following suggestions are recommended for future work: studying the results of the present work theoretically and make a comparison between it and the experimental work, studying the heat transfer performance through the developing region under a turbulent flow regime and investigating the behavior of stable nanofluid under a laminar flow regime.

References

[1] Choi S U and Eastman J A 1995 Enhancing thermal conductivity of fluids with nanoparticles *1995 Int. Mechanical Eng. Congress and Exhibition, CONF-951135-29* (San Francisco)

[2] Ghanbarpour M, Haghigi E B and Khodabandeh R 2014 Thermal properties and rheological behavior of water based Al2O3 nanofluid as a heat transfer fluid *J. Therm. Fluid Sci* **53** 227-35

[3] Kleinstreuer C and Feng Y 2011 Experimental and theoretical studies of nanofluid thermal conductivity enhancement: a review *J. Nanoscale Research Letters* **6(1)** 229

[4] Moghaddam M B, Goharshadi E K, Entezari M H and Nancarrow P 2013 Preparation, characterization, and rheological properties of graphene–glycerol nanofluids *J. Chem. Eng.* **231** 365-72

[5] Hemmat Esfe M, Saedodin S, Mahian O and Wongwises S 2014 Heat transfer characteristics and pressure drop of COOH-functionalized DWCNTs/water nanofluid in turbulent flow at low concentrations *J. Heat Mass Transfer* **73** 186-94

[6] Afrand M, Farahat S, Nezhad A Hossein, Sheikhzadeh G A and Sarhaddi F 2014 Multi-objective optimization of natural convection in a cylindrical annulus mold under magnetic field using particle swarm algorithm *J. Heat Transfer Res.* **45** 749-66

[7] Afrand M, Farahat S, Nezhad A Hossein Sheikhzadeh G A and Sarhaddi F 2014 3-D numerical investigation of natural convection in a tilted cylindrical annulus containing molten potassium and controlling it using various magnetic fields *J. Appl. Electromagn. Mech.* **46** 809-21

[8] Afrand M, Farahat S, Nezhad A Hossein, Sheikhzadeh G A, Sarhaddi F and Wongwises S 2015 Multi-objective optimization of natural convection in a cylindrical annulus mold under magnetic field using particle swarm algorithm *.J. Heat Mass Transfer* **60** 13-20

[9] Afrand M, Rostami S, Akbari M, Wongwises S, M Hemmat Esfe and Karimipour A 2015 Effect of induced electric field on magneto-natural convection in a vertical cylindrical annulus filled with liquid potassium *.J. Heat Mass Transfer* **90** 418-26

[10] Teimouri H, Afrand M, Sina N, Karimipour A and Isfahani A H M 2015 Natural convection of liquid metal in a horizontal cylindrical annulus under radial magnetic field *J. Appl. Electromagn. Mech.* **49** 453-61

[11] Afrand M 2017 Using a magnetic field to reduce natural convection in a vertical cylindrical annulus *J. Therm Sci.* **118** 12-23

[12] Hussien A A, Yusop N M, Abdullah M Z, Moh A N and Khavarian M 2019 Study on convective heat transfer and pressure drop of MWCNTs/water nanofluid in mini-tube *J. Thermal Analysis and Calorimetry.* **135(1)** 123-32

[13] Kayhani M H, Nazari, M, Soltanzadeh H, Heyhat M M and Kowsary F 2012 Experimental analysis of turbulent convective heat transfer and pressure drop of Al2O3/water nanofluid in horizontal tube *J. Micro and Nano Letters* **7(3)** 223-7

[14] Pop E, Mann D, Wang Q, Goodson K and Dai H 2006 Thermal conductance of an individual single-wall carbon nanotube above room temperature *Nano letters* **6(1)** 96-100

[15] Ncibi M C and Sillanpää M 2015 Optimized removal of antibiotic drugs from aqueous solutions using single, double and multi-walled carbon nanotubes *J. of Hazardous Materials* **298** 102-10

[16] Morozova S S , Shubina N A , Voropaeva N L , Tkachev A G , Burakov A E , Romantsova I V and Figovsky O L 2017 Sorption Activity of" Taunit"-Series Carbon Nanomaterials *Int. Letters of Chemistry, Physics and Astronomy* **74** 1-8

[17] Abdolkarimi-Mahabadi M and Manteghian M 2015 Chemical oxidation of multi-walled carbon nanotube by sodium hypochlorite for production of graphene oxide nanosheets *J. Fullerenes, Nanotubes and Carbon Nanostructures.* **23(10)** 860-4

[18] Yang J C, Yen C H, Wang W J, Horng J J and Tsai Y P 2010 Assessment of adequate sodium hypochlorite concentration for pre-oxidization of multi-walled carbon nanotubes *J. of Chemical Technology & Biotechnology* **85(5)** 699-707

[19] Kevin K A T, Field G, Snead M A and Yukinori Yamamoto 2017 Handbook on the Material Properties of FeCrAl Alloys for Nuclear Power Production Applications (*U.S. Department of Energy Nuclear Technology R&D Advanced Fuels Campaign*) URL https://info.ornl.gov/sites/publications/Files/Pub74128.pdf

[20] Cengel Y A 2003 *Heat Transfer a Practical Approach 2nd edition (*New York: McGraw-Hill)

[21] Bergman T L, Incropera F P, Lavine A S and Dewitt D P 2006 *Introduction to heat transfer Fifth* (Hoboken, , United States: John Wiley & Sons)

[22] Kwon Y, Lee K, Park M, Koo K, Lee J, Do Y and Jung Y 2013 Temperature dependence of convective heat transfer with Al2O3 nanofluids in the turbulent flow region *J. of Nanoscience and Nanotechnology* **13(12)** 7902-5

[23] Gnielinski V 1976 New equations for heat and mass transfer in turbulent pipe and channel flow *Chem. Eng.* **16** 359-68

IOP Conf. Series: Materials Science and Engineering **791** (2020) 012004 doi:10.1088/1757-899X/791/1/012004

Estimation of resource capabilities of the NPP turbine unit under the primary frequency control of the current in the power system

R Z Aminov, A N Bairamov and A B Moskalenko

Saratov Scientific Center of the Russian Academy of Sciences, Rabochaya St., 24, Saratov, 410028, Russia

E-mail: oepran@inbox.ru

Abstract. The purpose of the article is to estimate the life cycle on the example of NPP with the K-1000-60/1500 turbine and with the VVER-1000 reactor under cyclic load changes using the primary frequency control. The criterion for estimating the life cycle is the rate of the fatigue crack growth of the most loaded rotor element. The main requirements for the NPP power units under conditions of involvement in the primary frequency control are provided. The cyclic loading of the NPP turbine unit blades under conditions of maintaining the power reserve due to a reduction in the load by 2% by throttling the steam, and in line with the requirements for participation in the primary frequency control by 8% of the nominal capacity is substantiated. At the same time, the cyclic mechanical and thermal stresses are determined for the HPC and the LPC rotor blades. In this regard, it is shown that the first stage blades of the HPC rotor during unloading by 8% of the nominal capacity of the most cyclically loaded. The loading intensity and the rate of the fatigue crack growth of the first-stage blades of the HPC rotor are determined on the basis of the cyclic crack-resistance methodology and the Paris equation. It is shown that the rate of the fatigue cracks growth is negligible which determines their long-term life cycle.

1. Introduction

One of the requirements of power systems is the need to control the frequency of the current and keep it within the acceptable limits. For this purpose, hydropower plants and specially selected thermal power plants are involved. According to the order of JSC SO UPS dated August 19, 2013 No. 314, the main requirements for the primary frequency control by the NPP units were formulated.

The frequency should be within 50 ± 0.2 Hz and at least 95% of the day time and not exceed the maximum permissible 50 ± 0.4 Hz, which is due to the standardization requirements for operational dispatch regulation in the power industry. At the same time, the means of the secondary control together with the primary frequency control should ensure that the current frequency be kept within 50 ± 0.05 Hz (normal level), and within 50 ± 0.2 Hz (allowable level) with the recovery of the normal frequency level within no more than 15 minutes.

Traditionally, nuclear power plants are used in the base part of the load curve. This is facilitated by two important circumstances:

1. Large capital investments, low cost of the generated electricity compared to other thermal power plants due to low proportion of the fuel component.

IOP Conf. Series: Materials Science and Engineering **791** (2020) 012004 doi:10.1088/1757-899X/791/1/012004

2. Technical challenges with unloading and subsequent power ascension in certain periods of the fuel cycle at nuclear power plants, which are caused by xenon poisoning of the reactor core ("iodine well"). In addition, the base-load operation condition stabilizes the NPP reliability indexes at a sufficiently high level and ensures a long life cycle of expensive equipment.

It should be noted that involvement of NPPs in the primary frequency control of the current in the power system is associated with the possible frequent load changes.

The maneuverability of the NPP generating equipment should meet the following requirements for participation in the primary frequency control:

1) a guaranteed participation of the generating equipment in the primary frequency control should be ensured by implementing the required primary power within the control range:

- for the load of up to 2% of the nominal electric capacity of the power unit;
- for the unloading up to 8% of the nominal electric capacity of the power unit;

2) in case of an abrupt frequency deviation, which necessitates realization of the primary capacity within the indicated ranges, it should be ensured:

- implementation of at least half of the required primary capacity within no more than 10 sec.;
- implementation of all the required primary capacity within no more than 2 minutes.

Thus, the dynamics of the changes in the primary capacity of the NPP unit at the maximum required primary capacity per load ΔP_p = 2% P_{nom} should be no worse than 1% P_{nom} in 10 sec. and 2% P_{nom} in 30 sec. in accordance with the requirements; at the maximum required primary capacity for the unloading ΔP_p = -8% P_{nom}, the dynamics of the change in the primary capacity of the power unit should be no worse than -4% P_{nom} in 10 sec. and -8% P_{nom} in 120 sec.

In this regard, under participation of NPPs in the primary frequency control these requirements determine the daily operation of a turbine unit with the cyclic load changes, as well as cause inefficient part-load operation of the power unit to ensure the maintenance of the required capacity margin by 2% of the nominal capacity [1]. In this case, there is a need for additional unloading up to 8% of the nominal capacity during the periods of increasing current frequency.

2. Methods for assessing the life cycle of the turbine unit under conditions of frequency regulation

A cyclical change in the pressure of the working fluid occurs in the flow path of the turbine due to the changes in the steam consumption and its throttling in the steam admission under participation of the NPP units in the primary frequency control. The impact of steam on the blade causes a resultant force that can be decomposed into circumferential and axial components (figure 1) [2].

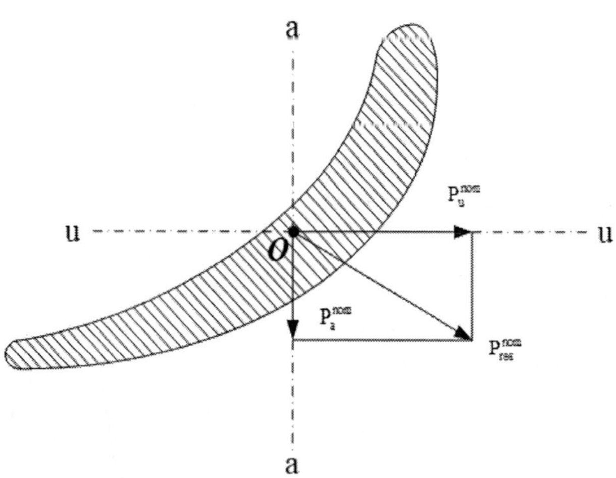

Figure 1. The forces that bend the blade in the nominal mode: P_u^{nom} is the circumferential force in the nominal mode, N; P_a^{nom} is the axial force in the nominal mode, N; P_{res}^{nom} is the resultant force in the nominal mode, N; and O is the gravity center of the profile.

The working blades experience a cyclical change in the resultant force under conditions of the cyclical changes in the pressure of the working fluid. It should be noted that the cyclic loading causes the growth of fatigue cracks [3-12]. In this regard, the rotating blades in the NPP turbine are subjected to the cyclic effects of the resulting effort. A change in the steam pressure leads to a difference in the resultant force affecting the blade, which is defined by the expression, N:

$$\Delta P_{res} = P_{res}^{nom} - P_{main\ res}^{derate}, \tag{1}$$

where P_{res}^{nom}, $P_{main\ res}^{derate}$ is the resultant force from the vapor pressure in the nominal mode, and main resultant force in the part-load operation mode, respectively, N.

The resultant force exerting on the blade in the nominal mode is determined by the expression, N [2]:

$$P_{res}^{nom} = \sqrt{P_u^{nom^2} + P_a^{nom^2}}, \tag{2}$$

where P_u^{nom} is the circumferential force in the nominal mode, N; and P_a^{nom} is the axial force in the nominal mode, N.

The circumferential force exerting on the blade in the nominal mode, N [2]:

$$P_u^{nom} = \frac{G_{nom}}{\varepsilon z_2}\left(c_{1u} - c_{2u}\right), \tag{3}$$

where G_{nom} is the nominal steam consumption through turbine stage, kg/s; ε is the partial admission ratio; z_2 is the number of blades; and c_{1u}, c_{2u} is the circumferential component of the steam release rate from the nozzle vanes channel and from the rotating blades channel, respectively, m/s.

The axial force exerting on the blade in nominal mode, N [2]:

$$P_a^{nom} = \frac{G_{nom}}{\varepsilon z_2}\left(c_{1a} - c_{2a}\right) + \left(P_1^{nom} - P_2^{nom}\right) t_{2opt} l \tag{4}$$

where c_{1a}, c_{2a} is the axial component of the steam release rate from the nozzle vanes channel and from the rotating blades channel, respectively, m/s; P_1^{nom}, P_2^{nom} is the vapor pressure in front of the rotating blade and behind the rotating blade in nominal mode, Pa; t_{2opt} is the optimal pitch of rotating blade, m; and l is the blade height, m.

The expressions (3) and (4) will take the following form under conditions of a reduction in the load by throttling the steam. The circumferential force exerting on the blade in the part-load operation mode, N [2]:

$$P_u^{derate} = \frac{G_{derate}}{\varepsilon z_2}\left(c_{1u} - c_{2u}\right), \tag{5}$$

where G_{derate} is the steam consumption through the turbine stage in the part-load operation mode, kg/s.

The axial force exerting on the blade in the part-load operation mode, N [2]:

$$P_a^{derate} = \frac{G_{derate}}{\varepsilon z_2}\left(c_{1a} - c_{2a}\right) + \left(P_1^{derate} - P_2^{derate}\right) t_{2opt} l \tag{6}$$

where P_1^{derate}, P_2^{derate} is the vapor pressure in front of the rotating blade and behind the rotating blade in the part-load operation mode, Pa.

In accordance with the Flugel formula, a proportional cyclical change of pressure in the working fluid within the flow path of the turbine leads to a cyclic temperature change during the cyclic reduction in the load by throttling the steam with the subsequent loading. This causes the cyclic thermal stresses. Using the Ansys software package, the thermal stresses are represented by three components, including the normal thermal stresses in the direction of three axes beginning from a common point. According to figure 1, the two components of the normal thermal stresses ($\Delta\sigma^{u}_{therm}$ and $\Delta\sigma^{a}_{therm}$ in figure 2) are codirectional with the circumferential and axial forces, respectively. The third component of the normal thermal stresses ($\Delta\sigma^{along}_{therm}$ in figure 2) is directed along the blade in the direction perpendicular to the plane of the circumferential and axial forces. In this regard, when evaluating the life cycle, the cyclical nature of the thermal stresses can be taken into account by adding $\Delta\sigma^{u}_{therm}$ and $\Delta\sigma^{a}_{therm}$.

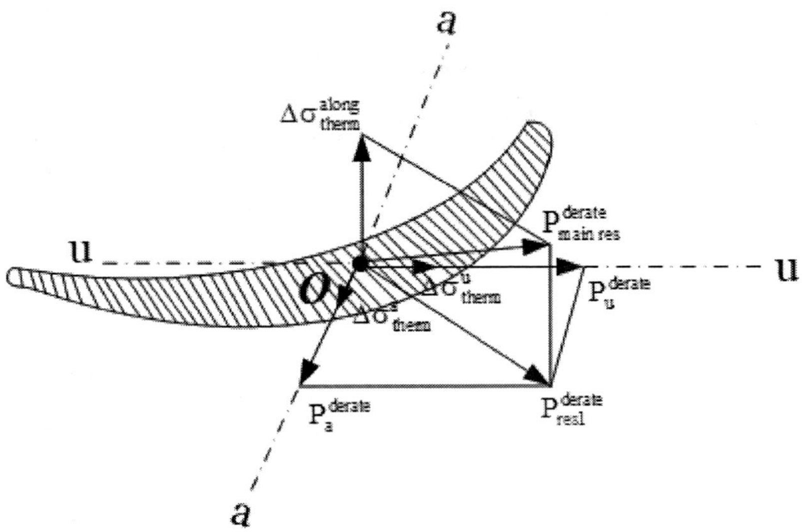

Figure 2. The forces that bend the blade in the part-load operation mode: P^{derate}_{u} is the circumferential force in the part-load operation mode, N; P^{derate}_{a} is the axial force in the part-load operation mode, N; P^{derate}_{res1} is the first resultant force in the part-load operation mode, N; $\Delta\sigma^{along}_{therm}$ is the thermal stresses in the longitudinal direction relative to the blade; and $P^{derate}_{main\ res}$ is the main resultant force in the part-load operation mode, N

The circumferential and axial forces in the cyclic unloading mode taking into account thermal stresses are given as:

$$P^{derate.therm}_{u} = \frac{P^{derate}_{u}}{f} + \Delta\sigma^{u}_{therm} \tag{7}$$

$$P^{derate.therm}_{a} = \frac{P^{derate}_{a}}{f} + \Delta\sigma^{a}_{therm}, \tag{8}$$

where f is the cross-sectional area of blade profile, m^2.

The cross-sectional area of the blade profiles by the turbine stages was determined by the trapezium method.

IOP Conf. Series: Materials Science and Engineering **791** (2020) 012004 doi:10.1088/1757-899X/791/1/012004

Thus, according to figure 2, the first resultant force which affects the blade in the cyclic mode a reduction in the load by throttling the steam, taking into account thermal stresses is calculated as follows, N:

$$P_{res1}^{derate} = \sqrt{P_u^{derate.therm^2} + P_a^{derate.therm^2}} \ . \tag{9}$$

In this case, after determining the first resultant force according to (9), taking into account the third component of normal thermal stresses ($\Delta\sigma_{therm}^{along}$ in figure 2) effecting along the blade, the main resultant force is determined, N:

$$P_{main\ res}^{derate} = \sqrt{P_{res1}^{derate^2} + \Delta\sigma_{therm}^{along^2}} \ , \tag{10}$$

where $\Delta\sigma_{therm}^{along}$ is the thermal stresses in the longitudinal direction relative to the blade, MPa.

With a reduction in the load by throttling the steam, the cyclic temperature changes in the working fluid across the blades of the turbine K-1000-60/1500 for the rotor of high-pressure cylinder, when average, was 2°C, and for a rotor of a low pressure cylinder was 1.6°C. In this case, the components of normal thermal stresses on average amounted to no more than 0.5 MPa for the rotor blades of the HPC and LPC. Thereby, the influence of the thermal stresses on the rate of the fatigue crack growth is negligible. Thus, the greatest pressure change in the flow range occurs when the capacity decreases with the unloading up to 8% of the nominal capacity than when working with the unloading of 2% for the K-1000-60/1500 turbine. According to calculations, the greatest difference in the resultant force, in line with (1) on the blades of the first stage of the HPC rotor was 0.92 MPa when a reduction in the load by 8% by throttling the steam, and 0.4 MPa when working with unloading by 2%. In this regard, based on modeling the steam flow in the turbine stages using the Ansys software package [13] (figure 3), we obtained the field of cyclic thermal stresses arising in the root section of the blade profile for the most loaded blade of the first stage of the HPC when a reduction in the load by 8% of the nominal capacity by throttling the steam. It was determined that the greatest tensile thermal stresses occur in the area of the entrance edge.

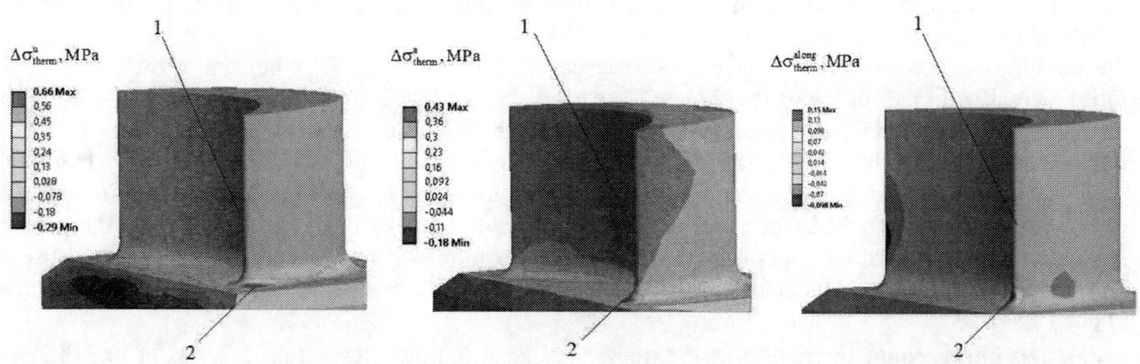

Figure 3. The field of the cyclic thermal stresses in the root section of the blade profile of the first stage of the high-pressure cylinder rotor: 1 – the entrance edge; 2 – the area of the greatest cyclic tensile thermal stresses.

According to (1), the difference in the resultant force is gradually reduced for the remaining stages of the turbine flow path along the steam expansion and reaches 0.1 MPa or less for the blades of the low-pressure cylinder stages. It should be noted that the change in steam temperature in the turbine flow range is practically absent when unloading at 2%; therefore, the cyclic change in the thermal stresses does not occur.

The difference in the resultant force (1) affecting the blades does not lead to appearance of the corresponding changes in the radial and tangential stresses in the turbine disc hub based on the strength analysis performed to estimate the radial and tangential stresses in the turbine rotor disks in nominal mode, and in the part-load operation mode based on the method [2], which indicates maintaining the necessary tension.

The basis for determining the maximum number of cycles to failure, and estimating the resource costs for the steam turbine blades is a flat body model with an applied tensile force along perpendicular to the fatigue crack [14, 15].

The value of the stress intensity coefficient is determined by the expression, MPa$\sqrt{\text{m}}$ [7,14]:

$$K = \sigma\alpha\sqrt{\pi l} \tag{11}$$

where σ is the cyclic stress, MPa; α is the nondimensional coefficient that takes into account the geometric factor and the nature of the stress distribution; and l is the threshold (minimum) crack length, m.

In this case, the magnitude of the stress intensity coefficient of the cycle, MPa$\sqrt{\text{m}}$ [7,14]:

$$\Delta K = K_{max} - K_{\min}, \tag{12}$$

where K_{max} and K_{min} are determined at the corresponding maximum and minimum rate of the cycle stress (σ_{max} and σ_{min}).

According to (2), K_{max} is determined based on the resultant force from the vapor pressure on the blade in the nominal mode (P_{res}^{nom}) in relation to the sectional area of the blade profile. According to (10), K_{min} is determined based on the resultant force from the vapor pressure on the blade, taking into account the thermal stresses in the cyclic unloading mode ($P_{main\ res}^{derate}$) in relation to the sectional area of the blade profile.

It is obvious that intensity of stresses generally occurs where there is a stress difference in the structural elements of the equipment, and this difference is basically caused by a change in either the operating steam pressure or temperature, or both.

According to the fatigue failure diagram [7], an increase in the stress intensity ΔK leads to the fact that the rate of the fatigue crack growth tends from the threshold section to the section with an average amplitude the rate of the fatigue crack growth receives a greater increment per cycle, and the effect of the loading frequency on the crack closure decreases. At the small ΔK, when the growth rate of the fatigue crack is in the region of the threshold section and is very small at the same time, the influence of the loading frequency can have a noticeable effect on its closure. Increasing the frequency can close the fatigue crack and slow down its growth, and declining the frequency will accelerate its growth.

The loading frequency is understood as the number of loading cycles in relation to the time interval over which they occur and is measured in Hz. Under conditions of daily participation in the primary frequency control, the time interval is 24 hours by the example of the received number of adjustments (N) from 1 to 6 per day. Under these conditions, the frequency of cyclic loading is very small and is about 10^{-5} Hz.

Taking into account thermal stresses, the ΔK value was much less, than 1 MPa$\sqrt{\text{m}}$ for the most loaded blades of the first stage of the HPC rotor. The rate of the fatigue cracks growth was determined in accordance with the Paris equation, mm / cycle:

$$\upsilon = \frac{\mathrm{d}l}{\mathrm{d}N} = C\left(\Delta K\right)^n, \tag{13}$$

where C and n are characteristics of cyclic crack resistance of steel.

Fatigue crack length, mm:

$$l_N = l_0 + \upsilon N_{cycles}^{max} \leq 0{,}1...1, \tag{14}$$

where l_0 is the threshold (minimum) value of the crack length, mm (accepted to be equal to 0.002 mm); N_{cycles}^{max} is the maximum number of cycles to failure; and 0,1-1 is the critical value of the fatigue crack length for the steels with a tensile strength of up to 1500 MPa.

Based on (14), the limit number of the loading cycles (N_{cycles}^{max}) is determined by the selection method from the condition for the fatigue crack to reach the lower limit of the critical length (0.1 mm) with a known rate of the fatigue crack growth.

It is accepted that the most loaded blades of the first stage of the HPC rotor are made of the stainless steel 13Cr11H2W2MoV. According to the experimental data, this steel is used for the manufacture of turbine blades. The coefficients C and n in (13) are taken $4.59 \cdot 10^{-11}$ and 4.56, respectively. The rate of the fatigue crack growth was $2.2 \cdot 10^{-23}$ mm / cycle. The maximum number of cycles to failure (N_{cycles}^{max}) is significantly large (over 10^6 cycles) due to the negligibly low rate of the fatigue crack growth (13). Thus, it shows a negligible damage to the most loaded blades of the first stage of the HPC rotor, which determines their long life cycle.

3. Conclusions

Using the NPP units in the primary frequency control is associated with a decrease in the capacity of the steam turbine when operating in the standby mode due to the throttling of steam consumption and subsequent capacity catch up. This causes the cyclical changes in the flow rate, pressure, and steam temperature in the turbine stages, which is associated with the cyclic mechanical and thermal stresses. The cyclic load changes cause the fatigue cracks growth. On the example of the turbine K-1000-60/1500, and using the method of strength calculation of the turbine blades it is shown that the cyclic change in the load is most susceptible to the working blades of the first stage of the HPC rotor. In this regard, the loading intensity of the first stage blades of the HPC rotor is determined using the methodology for estimating the cyclic crack resistance. It is also shown that the rate of the fatigue crack growth within the blade is negligible, which determines their long life cycle.

References

[1] Aminov R Z and Bairamov A N 2018 Participation efficiency of the NPP with the hydrogen production facility in primary frequency regulation of the power system *J. Phys.: Conf. Ser* **1111** 012023

[2] Zhiritsky G S and Strunkin V A 1968 *Konstrukciya i raschet na prochnost` detalej parovy`x i gazovy`x turbin* [The construction and calculations on strength of details steams and gas turbines] (Moscow: Mashinostroenie) [In Russian]

[3] Segura J A, Castro L, Rosales I, Rodriguez J A, Urquiza G and Rodriguez J M 2017 Diagnostic and failure analysis in blades of a 300 MW steam turbine *Engineering Failure Analysis* **82** 631-41

[4] Cano S, Rodríguez J A, Rodríguez J M, García J C, Sierra F Z, Casolco S R and Herrera M 2019 Detection of damage in steam turbine blades caused by low cycle and strain cycling fatigue *Engineering Failure Analysis* **97** 579-88

[5] Katinić M and Kozak D 2018 Steam turbine moving blade failure caused by corrosion fatigue – case history *Procedia Structural Integrity* **13** 2040-7

[6] Bairamov A N 2019 Life cycle assessment of hydrogen energy facility by criterion for maximum load frequency *International Journal of Hydrogen Energy* **44** 5696-703

[7] Panasyuk V V 1990 *Mexanika razrusheniya i prochnost` materialov: spravochnoe posobie* [Mechanics of destruction and strength of materials] (Kiev: Naukova Dumka) [In Russian]

[8] Hyojin K 2011 Crack evaluation of the fourth stage blade in a low-pressure steam turbine *Engineering Failure Analysis* **18** 907-13

[9] Nurbanasari M and Abdurrachim 2014 Crack of a first stage blade in a steam turbine *Case*

Studies in Engineering Failure Analysis **2** 54-60

[10] Rodríguez J A, Castro L, Tejeda A L, García J C, Rodríguez J M, Galindo E and El Hamzaoui Y 2019 Fatigue of steam turbine blades at resonance conditions *Engineering Failure Analysis* **104** 39-46

[11] Kuželka J, Nesládek M, Lutovinov M, Jurenka J, Růžička M, Rund M and Měšťánek P 2019 Numerical simulations of fatigue crack growth in a steam turbine rotor blade groove *Procedia Structural Integrity* **17** 780-7

[12] Akash S and Harsha S P 2015 An experimental and FEM modal analysis of cracked and normal steam turbine blade *Materials Today: Proceedings* **2** 2056-63

[13] Aminov R Z, Moskalenko A B and Kozhevnikov A I 2018 Optimal gas turbine inlet temperature for cyclic operation *J. Phys.: Conf. Ser* **1111** 012046

[14] Kostyuk A G 2007 *Dinamika i prochnost` turbomashin* [Dynamics and strength of turbomachines] ed A G Kostyuk (Moscow: MEI Publising) [In Russian]

[15] Bairamov A N 2017 Evaluation of the operating resource of the most loaded rotor element of the additional steam turbine with steam-hydrogen overheat of the working fluid at a nuclear power station *J. Phys.: Conf. Ser* **891** 012252

IOP Conf. Series: Materials Science and Engineering **791** (2020) 012005 doi:10.1088/1757-899X/791/1/012005

A study of the effect of air temperature on the possibility and rate of formation of nitrogen oxides in boiler flue gases

O A Balandina, S M Puring, D N Vatuzov and E B Filatova

Heat and Gas Supply and Ventilation Department, Samara State Technical University, Academy of Construction and Architecture, Molodogvardeyskaya St., 194, Samara 443001, Russia

E-mail: balandinaolya88@rambler.ru

Abstract. We consider the mechanisms of formation of nitrogen oxides in boiler flue gases. Thermodynamic calculations are used to determine the possibility and rate of formation of nitrogen oxides in the temperature range from 25 to 2000°C. We apply the methods of geometric thermodynamics to demonstrate the changes in the Gibbs free energy of nitrogen oxides formation reactions and determine the conditions of achieving the chemical equilibrium for these reactions. An analysis of concentrations of formed nitrogen oxides and temperatures of the jet flame under different conditions of mixture formation was performed. We produced and analyzed a number of diagrams of distribution of the flame temperatures and nitrogen oxide concentrations in the calculation area for the temperatures of the oxidant of 20, 60, 100, 150 and 200°C. Mathematical simulation of the process of gaseous fuel combustion was performed using FlowVision software package. The analysis of the results showed that the decrease in the temperature of the air supplied as an oxidant results in a considerable drop in the concentration of nitrogen oxides in flue gases without any significant impact on the changes of the flame temperature. The results of the study can be used in addressing the problems of optimization of boiler plants in order to reduce harmful flue gas emissions.

1. Introduction

An increase in energy and heat consumption leads to an increase in harmful emissions polluting the environment [1].

Nitrogen oxides take a prominent place among gaseous toxic components formed during natural gas combustion. These compounds are highly toxic and play an active role in the formation of dangerous atmospheric processes [2]. Nitrogen oxides are considered to be one of the causes of smog and acid rain formation. Increased concentrations of these substances have a harmful impact on human health, in particular, by irritating the mucous membranes of the respiratory tract and the eyes.

The following oxides are formed in the process of combustion as a result of the combination of nitrogen and oxygen, depending on the degree of oxidation: nitric anhydride N_2O_5, dinitrogen tetroxide N_2O_4, nitrous anhydride N_2O_3, nitrous oxide N_2O, nitrogen dioxide NO_2 nitrogen monoxide NO. NO and NO_2 have a practical significance from the environmental point of view.

In order to reduce the impact of combustion products on the atmosphere, it is necessary to optimize the combustion process in such a way as to reduce the concentration of nitrogen oxides in the emissions without reducing the efficiency of the boiler unit. A number of methods aimed at reducing the concentration of oxidant and temperature in the flame zone is applied to reduce the content of nitrogen oxides, as well as methods of recovery of nitrogen oxides in the furnace [3].

IOP Conf. Series: Materials Science and Engineering **791** (2020) 012005 doi:10.1088/1757-899X/791/1/012005

The temperature range in the gas path of a boiler unit is in a wide enough range. Maximum local temperatures in furnace chambers of boilers are as high as about 2000 K [4]. At such temperatures, the reaction of nitrogen oxides synthesis from nitrogen and oxygen proceeds at a very high rate. Even after leaving the flue stack, at temperatures under 373 K, the process of formation or transformation of nitrogen oxides continues.

Thus, the study of nitrogen oxide formation reactions at different temperatures, as well as the assessment of the effect of process parameters on the combustion contribute to the identification of modes and methods that result in a reduction in the environmental impact of boiler units and increase their efficiency.

2. Materials and Methods

2.1. Thermodynamic calculation of the Gibbs free energy of nitrogen oxides formation reactions in flue gases of boiler plants

In order to assess the direction of chemical reactions of thermal nitrogen oxides formation in the course of natural gas combustion, we calculate the change of the Gibbs free energy in the temperature range typical for gas paths of boiler units.

The mechanism developed by Yakov Zeldovich determines the yield of nitrogen oxides formed in a reaction between the oxygen atom and the nitrogen molecule [5]. The oxidation of atmospheric nitrogen by free oxygen in the process of combustion is described by the following reaction:

$$N_2+O_2 \leftrightarrow 2NO \quad (1)$$

The rate of formation of thermal nitrogen oxides is significantly impacted by the process temperature, as breakdown of O_2 molecules requires overcoming a high energy threshold [6].

Equilibrium concentrations of nitrogen oxides are formed over a period of time equal to $10^{-2}\div10^{-3}$ seconds. Such rapid formation of nitrogen oxides is explained by binding of nitrogen molecules to radicals CH and C_2 in reactions with low energy threshold, occurring according to the mechanism [7]:

$$CH+N_2 \leftrightarrow HCN+N \quad (2)$$

$$HCN+(H,OH) \leftrightarrow (CN+H_2,H_2O) \quad (3)$$

$$2C+N_2 \leftrightarrow 2CN \quad (4)$$

$$CN+O_2 \leftrightarrow CO+NO \quad (5)$$

$$CN+OH \leftrightarrow CO+NH \quad (6)$$

$$NH+OH \leftrightarrow NO+H_2 \quad (7)$$

$$NH+NO \leftrightarrow N_2+OH \quad (8)$$

After leaving the flue stack and entering the atmosphere, most of NO is transformed into NO_2. The oxidation mechanism can be represented by the following reactions [5]:

$$NO+O_3 \leftrightarrow NO_2+O_2 \quad (9)$$

$$2NO+O_2 \leftrightarrow 2NO_2 \quad (10)$$

$$NO+RO_2 \leftrightarrow NO_2+RO \quad (11)$$

2.2. Mathematic simulation of nitrogen oxides formation intensity depending on temperature of oxidant fed to the flame

To optimize performance of existing burners, it is necessary to study possible gas combustion modes with varying process parameters which may impact the rate of nitrogen oxides formation. In experimental studies, it is quite difficult to vary parameters in a wide range of adjustment. For this reason, mathematic simulation is utilized to study the combustion modes [8]. The scope of currently known theoretical papers on filtration gas combustion are limited either to analytical review of experiment data, description of physical phenomena and the problem statement or to individual particular calculations based on quite complex models and algorithms or conditions attainable in a laboratory-based experiment [9].

Jet flow simulation is quite relevant especially in using jet gas burners.

A model of gas combustion in a cylindrical burner is proposed and studied below.

FlowVision software package was utilized to study temperature impact of the air supplied to the process as an oxidant on nitrogen oxides formation rate. The mathematical model of the combustion process is based on Navier-Stokes and continuity equations [10]:

$$\left.\begin{array}{l}
\frac{dv_x}{dt}+v_x\frac{dv_x}{dx}+v_y\frac{dv_x}{dy}+v_z\frac{dv_x}{dz}=X-\frac{1}{\rho}\frac{dp}{dx}+v\left(\frac{d^2v_x}{dx^2}+\frac{d^2v_x}{dy^2}+\frac{d^2v_x}{dz^2}\right),\\
\frac{dv_y}{dt}+v_x\frac{dv_y}{dx}+v_y\frac{dv_y}{dy}+v_z\frac{dv_y}{dz}=Y-\frac{1}{\rho}\frac{dp}{dy}+v\left(\frac{d^2v_y}{dx^2}+\frac{d^2v_y}{dy^2}+\frac{d^2v_y}{dz^2}\right),\\
\frac{dv_z}{dt}+v_x\frac{dv_z}{dx}+v_y\frac{dv_z}{dy}+v_z\frac{dv_z}{dz}=Z-\frac{1}{\rho}\frac{dp}{dz}+v\left(\frac{d^2v_z}{dx^2}+\frac{d^2v_z}{dy^2}+\frac{d^2v_z}{dz^2}\right)\\
\frac{dv_x}{dx}+\frac{dv_y}{dy}+\frac{dv_z}{dz}=0,
\end{array}\right\} \quad (12)$$

where t is time in seconds; x,y,z are the particle coordinates; v_x, v_y, v_z are the particle velocity projections; X,Y,Z are volume force projections; p is pressure in Pa; v is kinematic viscosity, m^2/s; ρ is density, kg/m^3.

Combustion process is represented as a gross reaction:

$$1\,kg_{fuel}+i\,kg_{oxidant}\xrightarrow{W}(1+i)\,kg_{combustion\ products} \quad (13)$$

where W is the reaction rate.

For fuel and oxidant flows, which are not preliminary mixed, W gross reaction, rate is determined by turbulent mixing rate (Magnussen model) [11]:

$$W_{max}=23{,}6\left(\frac{\mu\varepsilon}{\rho k^2}\right)^{1/4}\rho\frac{\varepsilon}{k}\min\left(f,\frac{o}{j}\right) \quad (14)$$

where f is fuel concentration; o is oxidant concentration; j is the weight-based stoichiometric factor; μ is turbulent viscosity kg/m·s; ρ is density, kg/m^3; k is pulsation; ε is turbulence scale; W_{max} is the rate of combustion of a flow of unmixed components.

The calculation area was set as 40×200×8 mm with gas jet entry point in the center of the leftmost edge. The gas jet was supplied at the rate of v=0.003374 kg/ (m^2s) through the section of 8×8 mm. The air jet was supplied with velocity of v=0.07445 kg/ (m^2s) through two symmetrically located sections 16×8 mm [11].

Diagrams of concentrations NO_x^{therm} and temperatures in the calculation area were made based on the calculations.

IOP Conf. Series: Materials Science and Engineering **791** (2020) 012005 doi:10.1088/1757-899X/791/1/012005

3. Results

3.1. Results of calculations of ΔG changes in reactions of formation of nitrogen oxides in flue gases of boilers

Thermodynamic calculation of reactions with nitrogen oxides formation during gas fuel combustions in boiler furnaces shows the possibility and direction of processes in the temperature range from 25 to 2000 °C.

Direction of nitrogen thermal oxides formation for the reaction (1) was assessed with the calculation of Gibbs free energy (ΔG) changes.

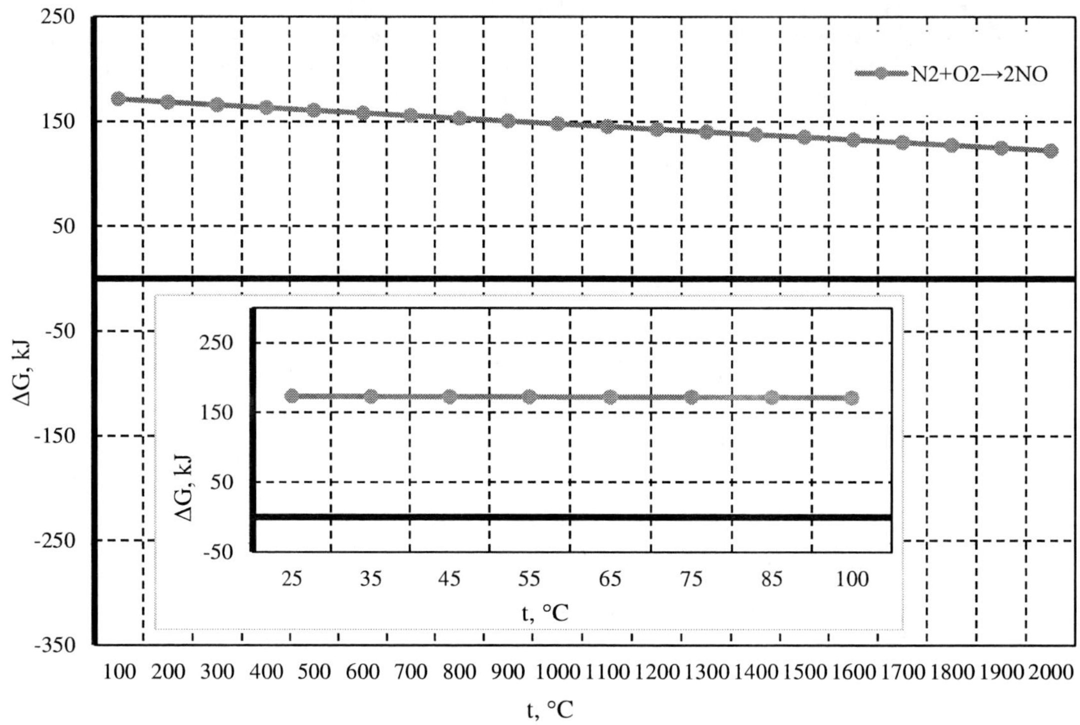

Figure 1. Temperature dependence of ΔG changes in the reaction of thermal nitrogen oxides formation for temperature range of 25÷2000 °C.

For comparative assessment of nitrogen compounds binding with radicals CH and C_2 in low energy barrier reactions, ΔG calculation were done (2÷8) for the following temperature ranges (Figures 2 and 3):

- specific for processes taking place in the furnace chamber and gas duct of the boiler unit (100÷2000 °C);
- specific for processes taking place after the emission of flue gasses into the atmosphere (25÷100 °C).

Results of oxidation reaction direction NO to NO_2 based on the Gibbs energy changes (9÷11) for the temperature range from 25 to 100°C, taking into account temperature decrease in gases exhausted in the atmosphere are shown at Figure 4.

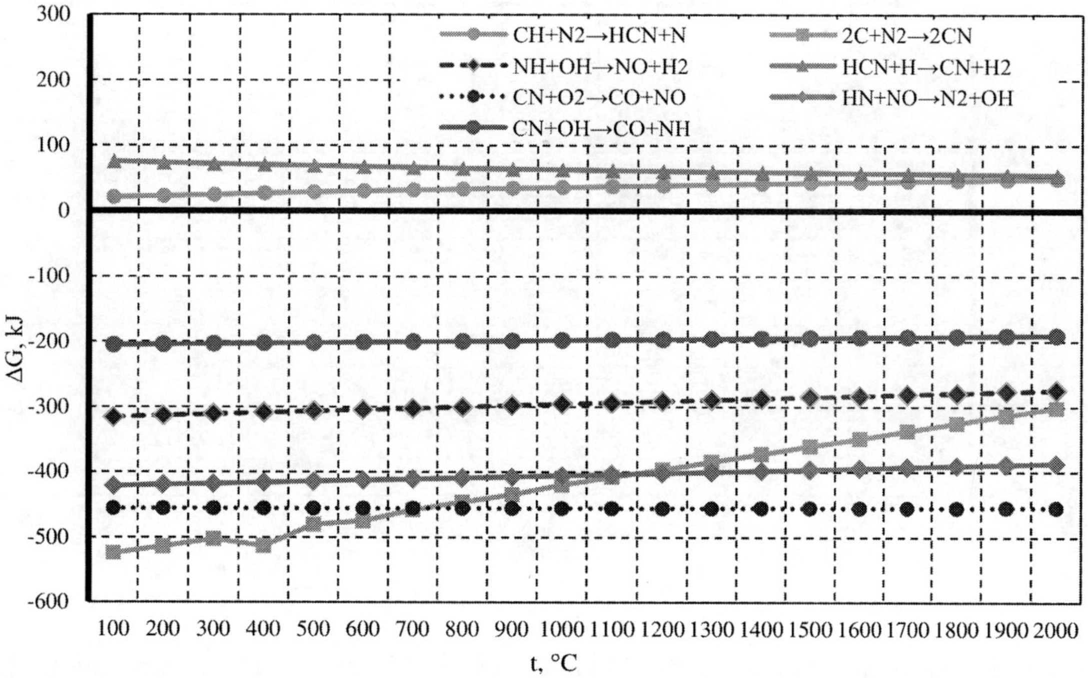

Figure 2. Temperature dependence of ΔG changes in the reactions (2÷8) for the temperature range of 100÷2000 °C.

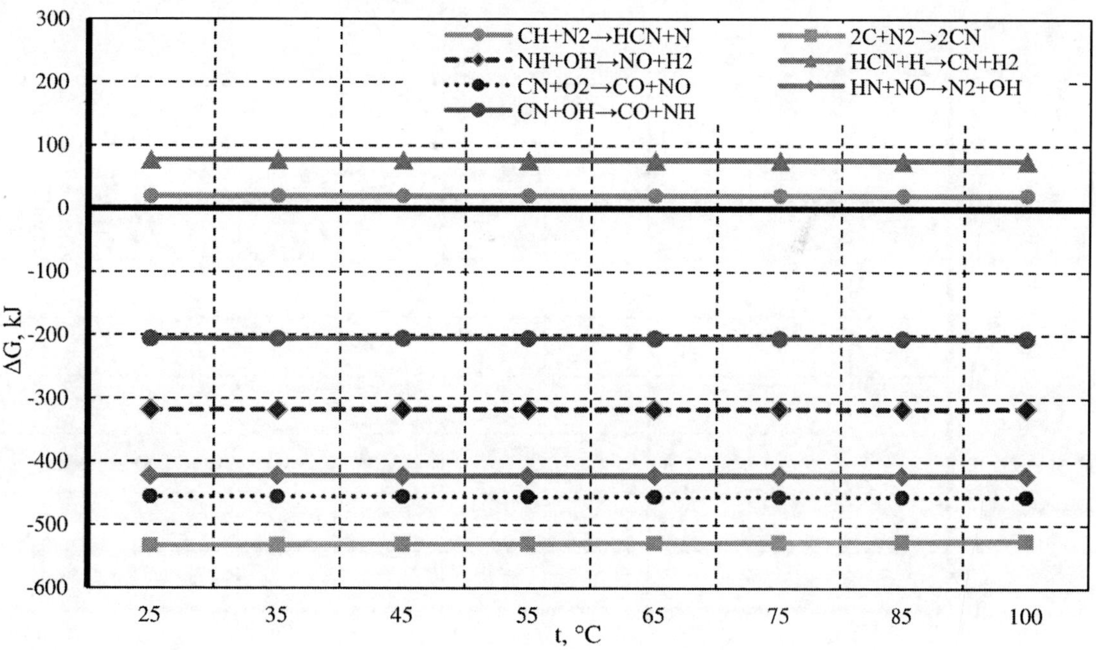

Figure 3. Temperature dependence of the Gibbs free energy changes in Reactions (2÷8) for the temperature range of 25÷100 °C.

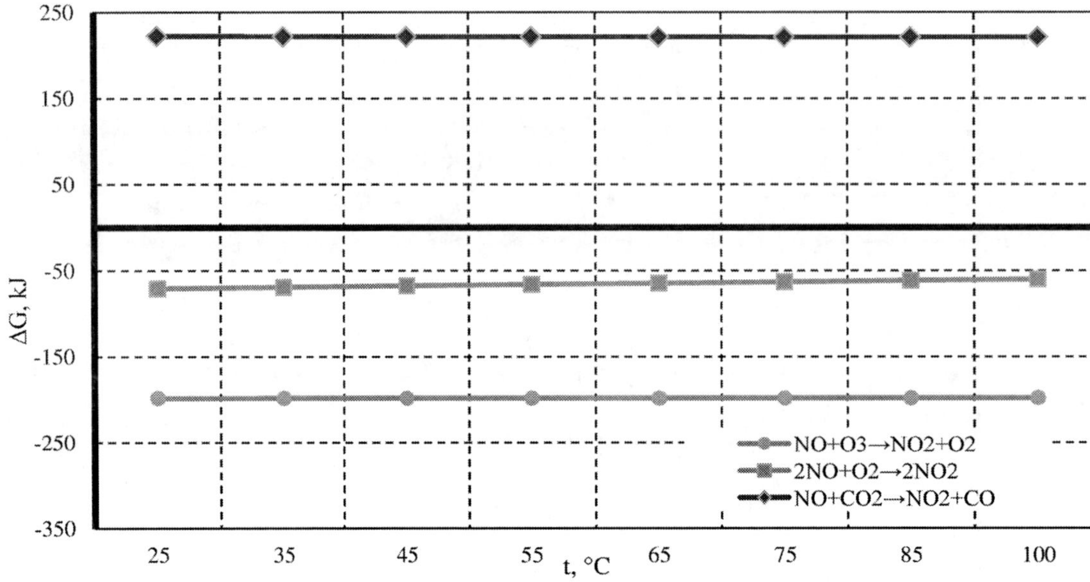

Figure 4. Temperature dependence of the ΔG change in the reactions (9÷11).

3.2. Results of calculation of nitrogen oxides formation rate, depending on the temperature of oxidant supplied to the combustion flame

The simulation included calculations for air temperatures of 20, 60, 100, 150 and 200°C. We obtained distributions of temperatures and concentrations of nitrogen oxide for the five considered options (Figures 5 and 6), depending on the distance from the burner nozzle.

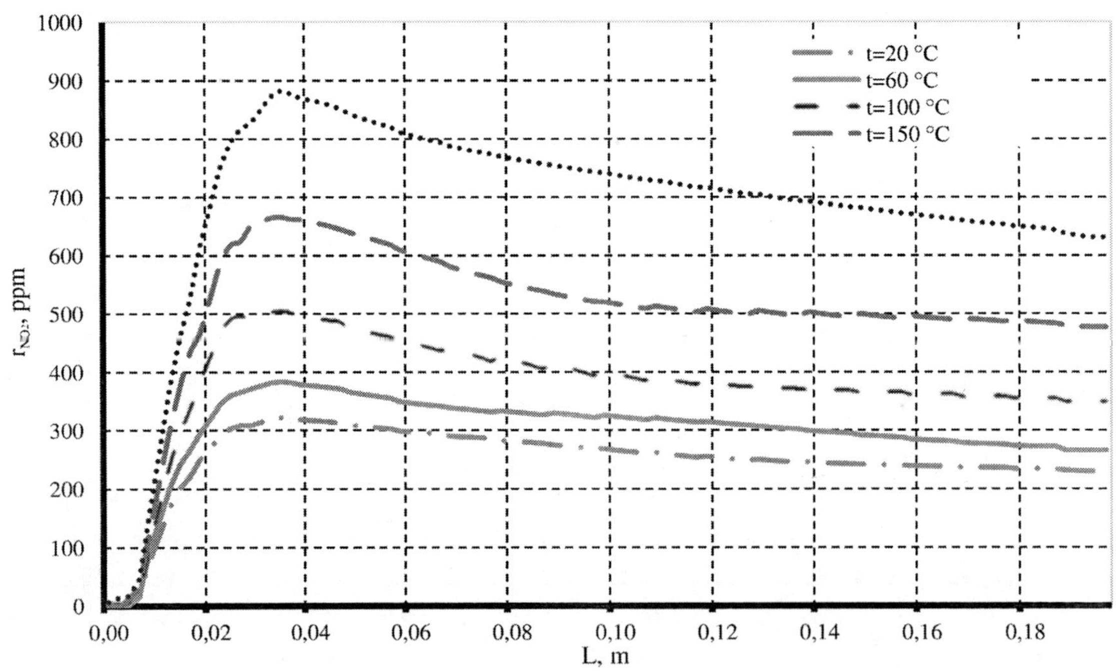

Figure 5. Distribution diagram of nitrogen oxides concentrations along the length at different oxidant temperatures.

IOP Conf. Series: Materials Science and Engineering **791** (2020) 012005 doi:10.1088/1757-899X/791/1/012005

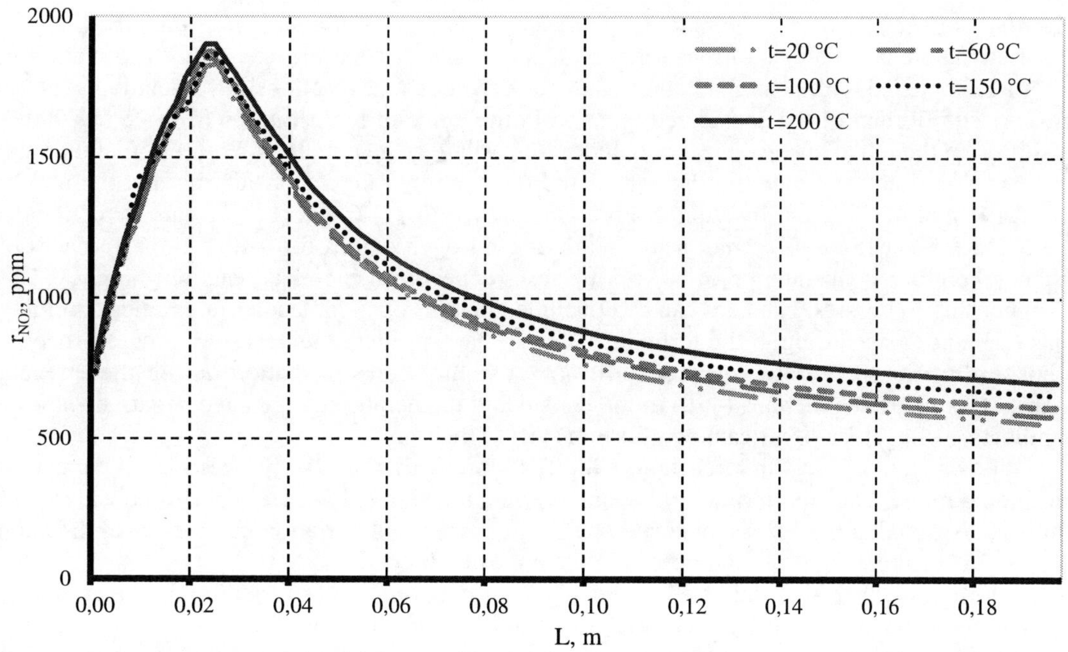

Figure 6. Distribution diagram of flame temperatures along the length in the calculation area at different oxidant temperatures.

The diagram of the dependence of nitrogen oxides concentrations by furnace sections at different oxidant temperatures is shown in Figure 7.

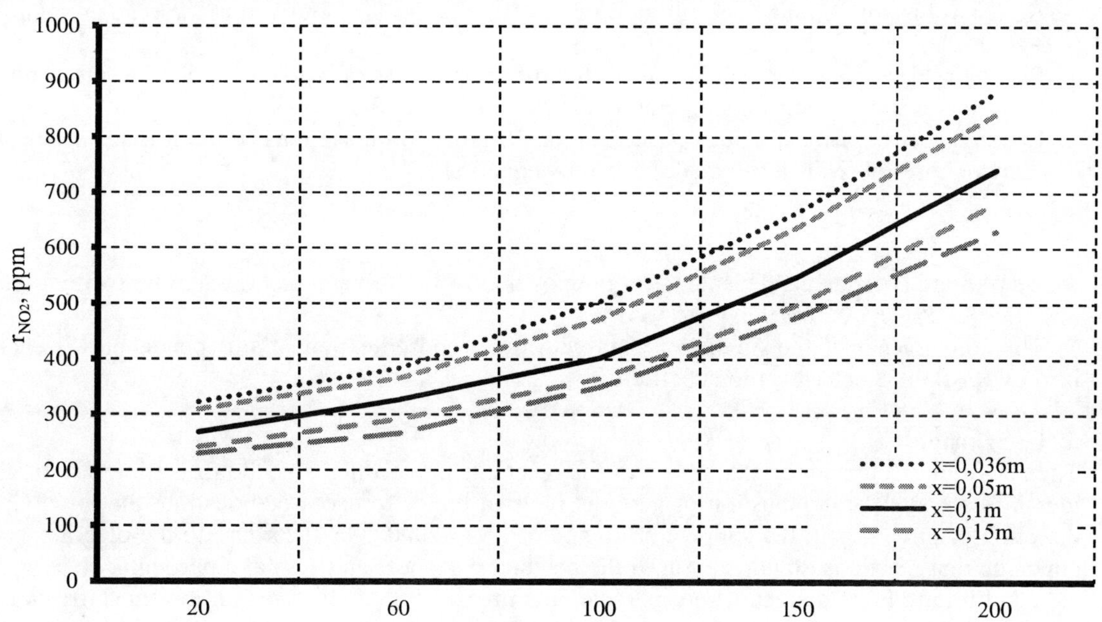

Figure 7. Dependence of nitrogen oxides concentrations by furnace sections at different oxidant temperatures.

4. Discussion

4.1. Study of changes in ΔG of reactions of nitrogen oxides formation in boiler flue gases

Thermodynamic calculations of the reaction (1) show that thermal nitrogen oxides formation process is not accompanied with heat generation, changes of Gibbs free reactions have positive values, therefore, the equilibrium of the considered reaction in the range of temperatures from 25 to 2000 °C will be shifted to the left. Thus, it can be concluded according to the data from Figure 1 that thermal nitrogen oxides will not be formed in the considered temperature range. Considering that thermal NO oxides form mainly in a narrow temperature range[T_{max}÷(T_{max})-60K] [12], it can be concluded that steady-state concentrations are not achieved during nitrogen oxides formation in the furnaces of boilers and, therefore, concentration of NO in flue gases depends on chemical reactions kinetics [13-15].

According to Figures 2 and 3 it can be concluded that nitrogen molecules and radicals binding reactions CH and C_2 according to the formulas (2÷3) in the considered temperature ranges proceed in the opposite direction, therefore, the probability of cyan molecules formation during these reactions is minimal. Exothermic reactions (4÷8) in the considered temperature range have negative values, which is indicative of a rather significant equilibrium shift to the right.

It follows from figure 4 that reactions (9, 10) are prevailing, as they proceed at all temperatures of the studied range. The equilibrium of the endothermic reaction (11) in the range of considered temperatures is displaced to the left, as evidenced by positive values of variation parameters of the Gibbs free energy. Therefore, this reaction proceeds in the opposite direction.

In addition to "thermic" and "fast" mechanisms of nitrogen oxides formation in the combustion area mentioned above, there is another "fuel" mechanism depending on nitrogen content in fuel and the presence of excess air. However, no nitrogen oxides formation according to the later mechanism does occur during flame combustion of natural gas.

4.2. Calculation of nitrogen oxides formation rate depending on the temperature of oxidant supplied to the flame

As it follows from Figure 5, formation of thermal nitrogen oxides in combustion of natural gas significantly depends on the temperature of air supplied to the burner to produce gas and air mixture. The maximum values of volume concentration of nitrogen oxides are observed at the oxidant temperature of t= 200 °C.

Based on Figure 6, it can be concluded that the temperature of oxidant supplied to the combustion area has only a minor effect on temperature distribution in the flame area.

Dependence of nitrogen oxides concentrations (Figure 8) on sections of the furnace space at various temperatures of oxidant can be described by equation

$$r_{NO_2} = ae^{bt_o} \qquad (15)$$

where *a*, *b* are coefficients depending on the position of the furnace section and the oxidant temperature (*a*=239.25÷166.9; *b*=0.2604÷0.2564).

The convergence of the simulation data with the proposed exponential dependencies is characterized by R^2=0.9933 determination coefficient.

5. Conclusion

The above thermodynamic calculations of the reactions proceeding with the formation of nitrogen oxides in the course of combustion of gaseous fuels in boiler furnaces demonstrate the possibility and direction of processes in the temperature range of 25 to 2000 °C. According to the diagrams, we can conclude that reactions do not start until the reactant disappear and stop at a particular state of chemical equilibrium. From a practical perspective, it is important to know the equilibrium shift and at what conditions it happens, making it possible for us to find out the maximum achievable yield of end products. To accomplish this, it is necessary to find the parameters responsible for the chemical equilibrium shift in the required direction. Therefore, by applying the methods of geometrical thermodynamics, we demonstrate that nitrogen oxides formation by the combustion of gas fuel results in the equilibrium being almost completely shifted to the right.

Data obtained from the mathematic simulation of nitrogen oxides formation rate depending on the air temperature indicate that the temperature of oxidant flow greatly impacts the distribution of nitrogen oxide concentration in flue gases: a decrease in temperature of air fed as an oxidant results in nitrogen oxides concentration in flue gases decreasing; the maximum concentration decrease is observed at air temperature of 20°C when temperature in the flame zone decreases insignificantly.

This study considers only one of many factors impacting the level of nitrogen oxides formation in flue gases of boiler units, but the method of numerical calculation of natural gas combustion process characteristics makes it possible to study a wide range of changing parameters.

The proposed model can be used to obtain tentative values of temperatures in the combustion area and flue gas nitrogen oxides concentrations. For boiler unit operation optimization, including the development of effective solutions to reduce nitrogen oxides levels, investigation of a variety of parameters is necessary, which implies further studies.

References

[1] Liang F Y, Ryvak M, Sayeed S and Zhao N 2012 The role of natural gas as a primary fuel in the near future, including comparisons of acquisition, transmission and waste handling costs of as with competitive alternatives *Chemistry Central J.* **6(1)** S4

[2] Smirnov B Yu 2015 Termodinamicheskiy analiz vosstanovleniya oksidov azota v otkhodyashchikh dymovykh gazakh [Thermodynamic analysis of the reduction of nitrogen oxides in exhaust flue gases] *Akademicheskiy zhurnal Zapadnoy Sibiri* **11 1(56)** 122-3 [In Russian]

[3] Kamalova G A, Tulegenov K K, Rakhmetova K B and Ramazanova K M 2018 Three-dimensional modelling of gas-air mixture combustion process *J. of King Saud University* **10** 1-13

[4] Habib M A and Elshafei M 2006 Computer simulation of NO_x formation in boilers *MSV 2006* (Las Vegas, Nevada, USA) 94-100

[5] Michos K, Bikas G and Vlaskos I 2016 A new global algebraic model for NO_x emissions formation in post-flame gases-application to lean premixed combustion systems *SAE 2016 World Congress and Exhibition* [Technical Paper 2016-01-0803]

[6] Srivastava R K, Hall R E, Khan S, Culligan K and Lani B W 2005 Nitrogen oxides emission control options for coal-fired electric utility boilers *J. of the Air & Waste Management Association* **55** 1367-88

[7] Chuang, Kai-Hsiang, Kuo-Chuang Chiu and Kuan-Yi Li Composition and apparatus for purifying nitrogen-oxide-containing gases (U.S. Patent Application 10/252 215)

[8] Satyavada H and Baldi S 2018 Monitoring energy efficiency of condensing boilers via hybrid first-principle modelling and estimation *Energy* **142** 121-9

[9] Alekseenko S V, Kuznetsov V A, Mal'tsev L I, Dekterev A A and Chernetskii M Y 2019 Analysis of combustion of coal-water fuel in low-power hot-water boiler via numerical modeling and experiments *J. of Engineering Thermophysics* **28 (2)** 177-89

[10] Arpaia F 2019 *Laminar flame speed prediction for Natural Gas/Hydrogen blends and application to the combustion modeling in IC Engines (Tesi di Laurea Magistrale)* URL https://webthesis.biblio.polito.it/10725/1/tesi.pdf

[11] FlowVision Version 2.5.4. Primery resheniya tipovykh zadach [Examples of solving typical problems] URL http://flowvision.ru/index.php/public-downloads/category/8-dokumentatsiya-flowvision

[12] Castoldi L, Matarrese R, Morandi S, Righini L and Lietti L 2018 New insights on the adsorption, thermal decomposition and reduction of NO_x over Pt-and Ba-based catalysts *Applied Catalysis B: Environmental* **224** 249-63

[13] Sigal I Y, Smikhula A V, Sigal O I, Marasin O V and Kernazhytska E S 2018 Features of the

influence of recirculation of combustion products on the formation of nitrogen oxides in boilers during burning natural gas *Energy Technologies & Resource Saving* **4** 62-8
[14] Morris W J, Baldrey K E, Senior C and Bisque R 2018 U.S. Patent Application 15/941 522
[15] Guan C 2019 Kinetic study of dry reforming of Methane with Ni over different supports and Pt/Al_2O_3 URL https://doi.org/10.14279/depositonce-8655

IOP Conf. Series: Materials Science and Engineering **791** (2020) 012006 doi:10.1088/1757-899X/791/1/012006

Possibility of using associated petroleum gas as a fuel for a production boiler house

D S Balzamov[1], E Yu Balzamova[1], V V Bronskaya[2] and N V Valitov[1]

[1]Department of Power Supply of Enterprises and Energy Resource Saving Technologies, Kazan State Power Engineering University named after D.S. Balzamov, E. Yu. Balzamova, N.V. Valotov, Krasnoselskaya St., 51, Kazan 420066, Russia.
[2]Department of Chemical Technology of Petroleum and Gas Processing, Kazan National Research Technological University named after V.V.Bronskaya, Karl Marx St., 68, Kazan 420015, Russia

E-mail: dbalzamov@mail.ru

Abstract. Current legislation of the Russian Federation sets requirements for oil companies for utilization of produced associated petroleum gas (APG) at the level of 95%. Thus limiting its burning in torches in the fields. The most convenient way of APG utilization for oil companies is to pump it back into the reservoir, which sometimes allows to increase oil recovery. APG is also used for commercial generation of electricity and heat in fields by burning in gas turbine and gas-piston installations. Given the cheapness of associated gas in comparison with natural gas (NG), it is promising to replace NG with APG for large production boilers located in the immediate vicinity of the associated petroleum gas pipeline. The article provides a comparative analysis of the parameters of APG and NG, a scheme for preliminary preparation of APG for its subsequent combustion in the boiler room, an analysis of the performance of the boiler house to generate thermal energy, the feasibility of using APG as the main fuel for the production boiler room.

1. Introduction

The possibility of substitution of natural gas (NG), which is used as the main fuel in a production boiler with associated petroleum gas (APG), is considered. In this connectionwith it, a comparison of the characteristics of fuels was performed. Comparison of associated petroleum gas and natural gas is made in accordance with the fuel quality certificates. The comparison results are presented in table 1.

As can be seen from the table, NG and APG have significant differences. NG mainly consists of methane; APG also contains a significant amount of C_2-C_6, which may require the replacement of burner devices. If C_1-C_2 is burned relatively fully, C_4-C_6 when are burned together with C_1-C_2 are not completely burnedaccording to the materials of literary sources [1,2].

The hydrogen sulfide content is significantly higher in APG than in NG, which ultimately will lead to increasing the content of sulfur dioxide in flue gases. This component is strictly regulated and controlled by environmental services. The risk of corrosion processes in economizers of boiler units increases [3].

The calorific value of APG is higher by 20-25%, which may lead to the need for modernization of burners.

Table 1. Quality indicatorNGandAPG.

Composition (volumeconcentration,%)	NG	APG
Methane	92.00 – 98.00	42.08
Ethane	-	20.00
Propane	-	13.45
Isobutane	-	1.30
Normalbutane	-	2.79
Isopentane	-	0.48
Normalpentane	-	0.42
Hexanes	-	0.17
Nitrogen	0.699	18.11
Carbondioxide	0.069	0.98
Helium	-	0.03
Hydrogensulphide	-	0.19
Volumefractionofoxygen	-	-
Mass concentration of mercaptan sulfur, g/m^3	0.021	<0.002
Mass concentration of hydrogen sulfide, g/m^3	<0.0001	2.64
Net calorific value (kcal/m^3)	8070	10492.00
Wobbenumber, kcal/m^3	11856	10728.00
Density, kg/m^3	0.6864	1.15
Dewpoint, °C	-	-6.00
Absolute humidity of gas, g/m^3	-	3.00

In the cold season, the use of wet APG can lead to the formation of crystalline hydrates on the inner surface of the pipelines and emergency shutdowns. There is a high probability of ice formation in burner devices and on butterfly valves when gas pressure is released [4, 5].

Thus, the use of APG can be considered a promising type of fuel under the following conditions [6-8]:

1. Purification of APG from hydrogen sulfide to the level of mercaptan sulfur in NG.
2. Drying APG, or treatment with anti-icing agents (alcohols, glycols).
3. Application of universal burner devices.

2. Materials and Methods

A scheme for preparing APG for combustion in the boiler plant (Figure 1) is proposed. It allows to solve the above conditions.

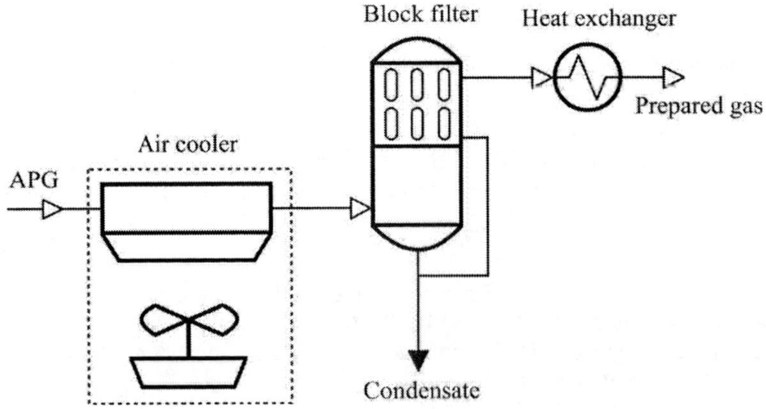

Figure 1. APGP reparation Schemefor Combustion.

The APG with a pressure of 6 kgf/cm^2 enters the air cooler for maximum allowable cooling, then the gas enters the block filter, which includes a separator and a coalescer filter to completely remove droplet moisture, and also mechanical impurities. At the exit from the block filter unit, dried gas and water condensate flows are formed. Gas is supplied to the heater, where it is heated to a temperature of 40°C due to the coolant.

Dried fuel gas enters the burner type MDGG-250B, the results of which are presented in table 2.

Table 2. Gas burner setup resultsMDGG-250B

№	Parameter	Dimension	The method of obtaining indica-tors(values)	Burnerload, stepnumber							
				1	2	3	4	5	6	7	8
1.	Burnerloading	%	Calculation	14.4	25.6	37.6	44.4	56.4	66.0	76.8	85.6
2.	Heatingcapacity. Q_{hc}	MW	Calculation	0.36	0.64	0.94	1.16	1.41	1.65	1.92	2.14
	Fuel is associated petroleum gas, Q_{ncv} = 10343 kcal/nm^3										
3.	Inletgaspressure P_i	kPa	measurement	15.0	15.0	14.5	14.0	13.8	13.0	12.5	12.0
4.	The gas pressure before the burner $P_{b.}$	kPa	measurement	13.5	13.0	12.2	11.5	10.5	9.2	8.0	6.2
5.	Gasconsumption	nm^3/h	measurement	30	54	79	98	119	139	161	180
6.	Gas pressure in burner manifold	kPa	measurement	+7	+14	+32	+47	+64	+90	+117	+145
	Air										
7.	Combustionair-temperature, t_{air}	°C	thermometer	+18							
8.	Air combustion	m^3/h	measurement, Calculation	450	800	1200	1500	1800	2100	2400	2700
9.	Vacuum in the burner air manifold	kPa	measurement	+0.07	+0.1	+0.15	+0.2	+0.25	+0.3	+0.35	+0.4

The APG contains hydrogen sulfide in the amount of 0.19%, which at high humidity can cause corrosion of the supply pipes. Replacing burners and inlet pipelines will prevent such problems. Environmental standards for the content of hydrogen sulfide, sulfur oxide, mercaptans will not be violated when working under the proposed scheme. The proposed burners are designed for tasks with a high content of hydrogen sulfide and are successfully used in the combustion of associated gas with a hydrogen sulfide content of up to 4%.A specially developed configuration of the combustion chamber, as well as a significant excess of oxygen, allows sulfur-containing compounds to be burned to residual H_2S concentrations of less than 1 ppm [9]. The main combustion product of sulfur-containing compounds in excess oxygen is sulfur dioxide [10-12]. The estimated flue gas composition for APG combustion is shown in Table 3.

It is seen from table 3 that the content of sulfur dioxide does not exceed 100 ppm. The indicated sulfur dioxide content corresponds to GOST R 50831-95 "Boiler installations. Thermomechanical equipment.General technical requirements." According to the specified GOST, specific emissions of sulfur dioxide are normalized depending on the thermal capacity of the boiler [13, 14]. The most stringent requirements apply to boilers of maximum thermal power (more than 300 MW), the specific

IOP Conf. Series: Materials Science and Engineering **791** (2020) 012006 doi:10.1088/1757-899X/791/1/012006

emission should not exceed 300 mg/MJ. The specific dioxide emission for boilers based on twenty megawatt burners when using associated petroleum gas will not exceed 150 mg/MJ, the restriction according to the given GOST for these types of boilers is 500 mg/MJ.

Table3. Exhaustgascomposition

Components	Volumeconcentration, %
Methane	< 1 ppm
Ethane	< 1 ppm
Propane	< 1 ppm
Butane	< 1 ppm
Pentane	< 1 ppm
Hexanes	< 1 ppm
Nitrogen	72.6
Helium	0.013
Hydrogensulphide	< 1 ppm
Carbondioxide	9.21
Sulphurdioxide	0.0094
Oxygen	2.55
Watervapor	15.6

To prevent condensation, it is proposed to use a complex preparation method, including a coalescer filter, to separate droplet moisture and mechanical impurities with a filter fineness of 3 microns and further heating the gas stream to a temperature of 40°C. Such a scheme eliminates the possibility of condensation along the entire pipeline line [14]. With the indicated composition, pressure and temperature of associated petroleum gas, the temperature of the hydrocarbon dew point will be minus 28°C, the hydrate formation temperature is lower than the freezing temperature of water, the assessment was carried out using the modern «Aspen Technologies» software package.

Since cooling during the planned operation of the installation is not expected to be below +10°C, the formation of hydrocarbon condensate is excluded. Table 4 shows the approximate composition of the water condensate obtained from the coalescer filter during gas cooling in an air-cooling apparatus to a minimum temperature (+10°C).

Table4. Component composition of condensate.

Components	Massconcentration, %
Methane	less 0.0001
Nitrogen	0.0022
Helium	<0.0001
Hydrogensulphide	<0.0028
Carbondioxide	0.0050
Water	99.99

The estimated cost of this solution taking into account the project for the replacement of burners, equipment delivery will be 150 million rubles.

The maximum possible supply of APG from the supplier to the production boiler house is at the level of 50 million m^3/year, which is confirmed by an official letter from the supplier.

IOP Conf. Series: Materials Science and Engineering **791** (2020) 012006 doi:10.1088/1757-899X/791/1/012006

3. Results

In order to assess the feasibility of switching from NG to APG, an analysis of the operation of the boiler room was carried out.

The dynamics of the fuel consumption of the production boiler room during the year is presented in Figure 2.

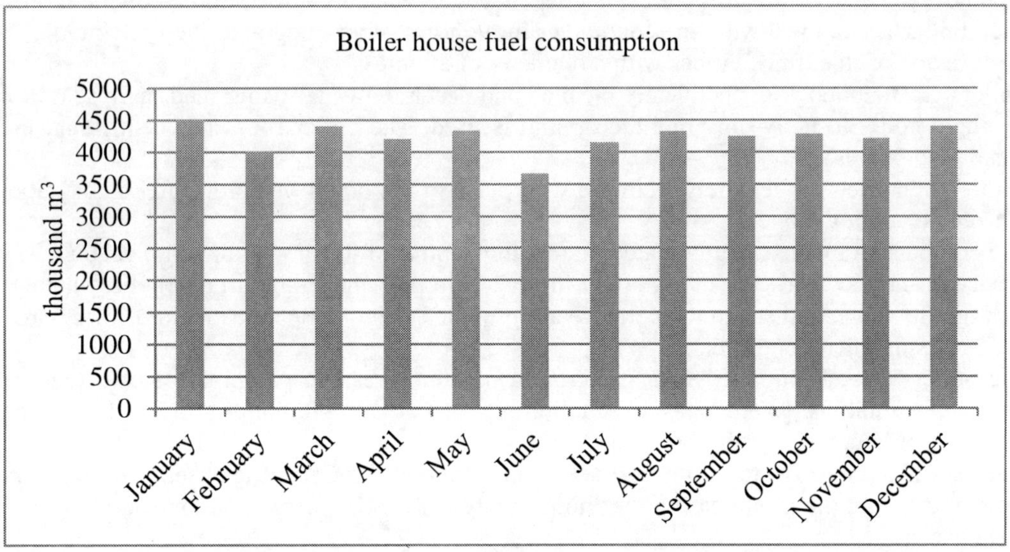

Figure 2. Dynamics of fuel consumption boiler house

From the analysis of the fuel consumption schedule it follows that the total fuel consumption of the boiler house is 50736.062 thousand m^3.

In the summer period, the productivity of the steam boiler house slightly decreases, while the specific fuel consumption for steam generation increases (Figure 3).

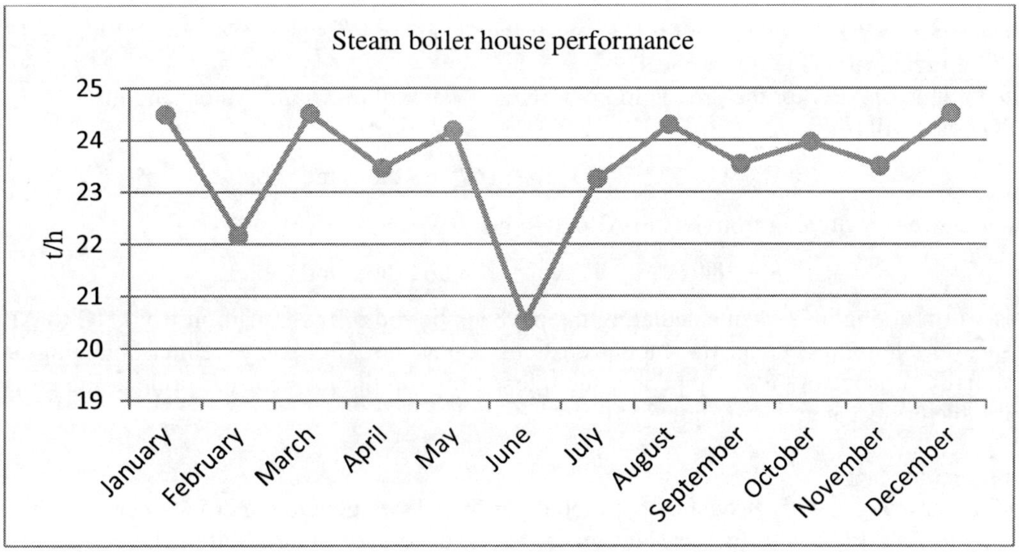

Figure 3. Specific fuel consumption depending on the performance of the boiler house

IOP Conf. Series: Materials Science and Engineering **791** (2020) 012006 doi:10.1088/1757-899X/791/1/012006

From the presented figure it follows that the fuel efficiency at a steam capacity of 25 t / h is 15% higher than at 20.51 t/h. At the same time, indicators for months with incomplete hourly loading of boilers (March, June, November) fit into the general statistical regularity. At the same time, as follows from the regime maps, the efficiency gross boilers in this load range varies slightly.

The boiler house has three BEM-25/16-310G steam boilers with a rated steam capacity of 25 t/h. Combustion products are removed through a metal chimney 60 m high and 1,44 m in diameter. The heat load of the boiler room has no seasonal dependence. There are always three boilers in operation.

Each boiler is equipped with an economizer in the form of an integrated the tube bundle. The tube bundle is made of steel finned tubes with a diameter of 25 mm.

Boilers are supplied with chemically cleaned and deaerated water using feed pumps. When operating a single boiler, usually only one feed pump is used. The boiler room has centrifugal multistage sectional feed pumps.

Ignition is carried out remotely from the workplace of the boiler operator, after stabilization of the gas-air duct operation mode.

The combustion modes are automated: automatic control of the gas-to-air ratio, vacuum in the boiler furnace, average water level in the boiler drum, steam pressure in the drum (fuel regulator), continuous purge flow rate, and steam temperature are provided. Automatic adjustment of these processes is carried out using microcontrollers.

The boiler house is supplied with gas from a medium-pressure gas pipeline of category II $Pg \leq 6{,}0 kgf/cm^2$. Natural gas enters a gas distribution plant located in the boiler house, where it is reduced to a working pressure of $Pg = 0.45–0.5\ kgf/cm^2$.

The main feature of the technological scheme of the boiler room is the absence of connected heating networks. From the production boiler house, only the heating networks connected to the adjacent boiler house are fed.

Release of thermal energy – 356722 Gcal per year.

The cost of natural gas is 5600 rubles/thousand m^3.

Annual fuel costs are:

$$E_{ng} = 50736.063 \cdot 5600 = 284\ 122 \text{ thousandrubles.}$$

The specific fuel consumption in the boiler room amounted to 164,21kg of equivalent fuel/Gcal or 142.3 m^3/Gcal.

The cost ofAPG– 3650rubles/thousand m^3.

The specific APG consumption for heat generation at enterprises using this type of fuel is 170 kg of equivalent fuel/Gcal or 113.33 m^3/Gcal.

The volume of APG for the production of 356722 Gcal will be 40427.3 thousand m^3.

APG costs will be:

$$E_{apg} = 40\ 427\ 300 \cdot 3650 = 147\ 560 \text{ thousand rubles.}$$

Savings when switching from NG to APG will be:

$$S = 284\ 122 - 147\ 560 = 136\ 562 \text{ thousand rubles.}$$

It is worth noting that when calculating the payback period of the transition from NG to APG, it is necessary to take into account the capital costs of laying the APG gas pipeline, obtaining technical conditions for connecting to the pipeline, the interest rate on the loan for the project, the costs of developing the project and other factors [15].

4. Conclusion

Replacing natural gas with associated petroleum gas in order to generate thermal energy in the form of water vapor is a fairly cost-effective solution, given the difference in tariffs at the level of 35-40%.

IOP Conf. Series: Materials Science and Engineering **791** (2020) 012006 doi:10.1088/1757-899X/791/1/012006

References

[1] Vazim A, Romanyuk V, Ahmadeev K and Matveenko I 2015 Associated petroleum gas utilization in Tomsk Oblast: energy efficiency and tax advantages *IOP Conf. Series: Earth and Env. Sci.* **27** 012078

[2] Knizhnikov A Yu and Il'in A M 2017 *Problemy` i perspektivy` ispol`zovaniya poputnogo neftyanogo gaza v Rossii* [Problems and prospects of associated petroleum gas use in Russia] (Moscow: WWF Rossii) [In Russian]

[3] Balzamov D S, Sabitov L S, Timershin B F and Balzamova E Yu 2018 Increase of efficiency of heat sources work due to application of condensation economizers on an example of a boiler PTVM-180 *IOP Conf. Series.: Mater. Sci. Eng.***412** 012007

[4] Balzamov D S, Balzamova E Yu, Ibatullin S R and Sabitov L S 2019 Efficiency increase of gas turbine work in the summer period *IOP Conf. Series.: Mater. Sci. Eng.* **570** 012008

[5] Filimonova I V, Eder L V, Mishenin M V and Mamakhatov T M 2017 Current state and problems of integrated development of mineral resources base in Russia *IOP Conf. Series: Earth and Env. Sci.* **84** 012011

[6] Balzamov DS, Timershin B F 2017 Obzor peredovykh tekhnologiy dlya generiruyushchikh kompaniy [Overview of advanced technologies for generating companies] *Modern Science* **2** 26-9 [In Russian]

[7] Galikeev R M, Leontiev S A and Umrenkov M V 2011 Primenenie poputnogo neftyanogo gaza dlya sobstvenny`x nuzhd na Xoxryakovskom mestorozhdenii [Application of passinggas for domestic use on Khokhryakovskoye field] *Nauka i TE`K* **3** p. 37-43

[8] Barkan M S and Kornev A V 2017 Prospects for the Use of Associated Gas of Oil Development as Energy Product *Int. J. of Energy Economics and Policy* **7(2)** 374-83

[9] Taymarov M A, Akhmetova R V and Akhmetov E A 2019 Efficiency of application of various layout arrangements of oil-gas burners in thermal power plant boilers *IOP Conf. Ser.: Mater. Sci. Eng.* **552** 012008

[10] Vigon G, Rubcov A and Egov C 2012 *Utilizaciy poputnogo neftjnogo gaza: problema* [Associated petroleum gas utilization: a problem] (Moscow: Energeticheskii centr SKOLKOVO) [In Russian]

[11] Tokarev A N 2009 Resheniye problemy utilizacii poputnogo neftjnogo gaza: interesi I rol neftedobivaushih kompani [Solving the problem of associated petroleum gas utilization: interests and role of oil-producing regions] *Region: Economika i Sociologiy* **3** 212-31 [In Russian]

[12] Zubareva V D and Ozdoeva A Kh 2019 Osobennosti e`konomicheskoj ocenki primeneniya poputnogo neftyanogo gazadlya texnologii GTL [features of economic assessment of application of associated oil gas for GTL] *Ximiya i texnologiya topliv i masel* **4(614)** 3-7 [In Russian]

[13] Kozodoev L V, Kuzin N A, Amosov Y I, Kirillov V, Sobyanin V A and Kireenkov V 2011 Novaya texnologiya pererabotki poputny`x neftyany`x gazov v mestax ix doby`chi [New technology of associated petroleum gas processing in oil production fields] *Promy`shlennost` i E`kologiya Severa* **11(19)** 40-3 [In Russian]

[14] Tarasova M Yu, Klevtsov E A and Fakhretdinov I Z 2018 O povy`shenii e`ffektivnosti ispol`zovaniya neftyanogo gaza koncevy`x stupenej separacii [On the raising efficiency of associated petroleum gas utilization at the final separation stages] *Transport i podgotovka nefti* **3** 74-6 [In Russian]

[15] Feigin V I 2011 *Issledovanie sostoyaniya i perspektiv napravlenij pererabotki nefti i gaza, nefte- i gazoximii v RF* [Investigation of trends and prospects of oil and gas, petrochemicals and gas in Russia] (Moscow: Econ-inform) [In Russian]

IOP Conf. Series: Materials Science and Engineering **791** (2020) 012007 doi:10.1088/1757-899X/791/1/012007

Contribution of heat loss by infiltration to energy saving and microclimate in multi-family residential buildings

Y V Elistratova, V M Kireev and A S Seminenko

Department of Heat and Gas Supply and Ventilation, Belgorod State Technological University n. a. V. G. Shukhov, Kostyukov St., 46, Belgorod, 308012, Russia

E-mail: vit31rus@mail.ru

Abstract. The problem of energy performance takes a leading position in world politics today. In the territory of RF a number of regulatory documents have been developed and put in force in construction industry, regulating the activity of design and construction organizations in the field of energy performance, including energy loss issues. The article analyses current methodology of calculating infiltration heat loss in a multi-storey residential building with account of the number of storeys. Calculation results in graphic form have been shown for the determination of infiltration heat loss for a 14-storey typical residential building, and imperfection of the methodology has been shown on the basis of these results. A conclusion has been drawn that infiltration loss calculation using current methodology gives overestimated values leading to unreasonable excess consumption of heat energy, and consequently, to economic losses. The article will be useful for engineers of designers of heating systems, employees of management companies.
Keywords: infiltration, heat loss, energy saving in heating, resource-saving, thermal resource saving, economy heating mode, control of building heating parameters, optimization of building heating parameters.

1. Introduction

Energy saving is one of top priority tasks of national energy politics in Russia today [1]. It is explained by inefficient use of energy and fuel resources and by their ever growing cost. For instance, relative consumption of heat energy for house heat supply systems (heating, ventilation, air conditioning, hot water supply) per 1 m^2 of heated area during a heating season in Russia is 2.9-4.3 times higher that that in countries with similar climatic conditions, e.g., in Sweden and Finland. Obviously, efficiency of energy resources used for house climatisation needs is low, and energy saving is needed. Solution of this problem is possible through development and implementation of energy saving measures, technologies and calculation methods [2, 3].

Residential houses' heating represent the most important sector of power consumption [4, 5]. To determine the required calorific power of the heating system, heat loss of each of the house premises is determined, including heat loss due to outer (infiltrating) air heating [6 - 8]. This heat loss may reach up to 60% and more of the total heat loss of the house, depending on initial conditions. Current methodology concerning calculation of the amount of heat required for infiltrating air heating do not consider a number of factors, resulting eventually in overheated air in premises, excessive thermal energy consumption, and insufficient thermal acceptability.

IOP Conf. Series: Materials Science and Engineering **791** (2020) 012007 doi:10.1088/1757-899X/791/1/012007

We believe that emphasis in achieving energy saving in heat supply of multi-storey buildings should be made on the methodology of calculating the amount of heat required for infiltration air heating, which has determined timeliness of this study.

2. Materials and Methods

For convenience, first let introduce some notations:

- Q_{calc}- calculated heat loss, W;
- Q_i- calculated heat loss by infiltration, W;
- index $_r$ – corresponds to calculation as per regulatory requirements;
- index $_v$ – corresponds to natural exhaust ventilation performance records;
- index $_{tw}$ – corresponds to calculation taking into account wind and thermal pressure.

The amount of heat consumed for infiltration air, taking into account wind pressure $Q_{i\,tw}$, is established with account of consumed heated air G_h which equals the total amount of infiltrating air coming through external enclosure leaks ΣG_{ee} [9 - 11]. Let us note that $\Sigma\ G_{ee}$ depends on the type and character of external enclosure leaks and is determined for each storey and premise separately [8, 12]. Besides, the value of differential air pressure Δp, Pa at outer and inner enclosure surfaces corresponding to a particular premise [8, 12] is taken into account.

Based on the formula for the calculation of infiltrating air consumption [8, 12]:

$$\Sigma G_{ee} = 0.216\sum A_1 \Delta p^{0.67} / R_n + \sum A_2 G_{ee} (\Delta p / \Delta p_1)^{0.67} + \\ + 3456\sum A_3 \Delta p_1^{0.5} + 0.5\sum l \cdot \Delta p / \Delta p_1 \tag{1}$$

where Δp_1, Δp_2, respectively, is the difference in air pressure at inner and outer surface of windows, balcony doors and outer doors in a standard storey and at the first storey level, Pa;
l - length, m, of wall panel joints;
A_1, A_2- area of windows, balcony doors and other enclosures, m^2;
A_3- area of crackage, leaks through window apertures, outer doors, m^2;
R_n- resistance of windows, balcony doors to air infiltration, (m^{2}h Pa)/kg;
and the formula for the calculation of difference in air pressure at inner and outer enclosure surfaces

$$\Delta p = (H - h)(\gamma_{out} - \gamma_{in}) + 0.5 \cdot V^2 \rho_d (c_d - c_u) k_v - p_{\text{int}} \tag{2}$$

H–height of the building, from ground level to cornice, centre of skylight exhaust openings or ventilation shaft, m
h– design height, from ground level to top of the windows, balcony doors or to the axis of wall panel joints, m
γ_{out}, γ_{in} = specific gravity, at outdoor t_{out} and indoor t_{in} air temperature, H/m^3, respectively, defined as

$$\gamma = \frac{3463}{273 + t} \tag{3}$$

G_e – density of external air, kg/m^3;
V – wind velocity, m/s;
C_d, c_u – aerodynamic factors for downwind and upwind enclosure surfaces, respectively, c_d= 0.8, c_n= -0.6.

3. Results

As thermal design of heating system calorific power is made for the lowest temperatures of outdoor air, a possibility to control thermal conditions inside the premises must be provided; on that basis, several possible variants will be analyzed:

1-st design: thermal design taking into account regulated heat flow for the riser passing through the living rooms of single-type flats in a 14-storey residential building with the upper heating agent supply as per current regulatory documents;

2-nd design: thermal design where heating radiator length must cover 50-70% of window aperture length;

3-rd design: we shall calculate expected air temperature taking into account increased consumption of heating agent, without increasing heating unit surface area;

4-th design means increasing both heating unit surface area and heating agent consumption;

5-th design assumes taking into account the impact of the number of radiator sections on temperature head in heat loss calculation.

Let us calculate total consumption of infiltration air ΣG_{ee} and $Q_{i\ tb}$ in order to establish calculated heat loss in a premise Q_{calc} for each of the rooms from the 14th to the 1st storey at t_{out}= -23; -20; -10; -5; and 0 °C. Results are presented as a diagram (Figure 1).

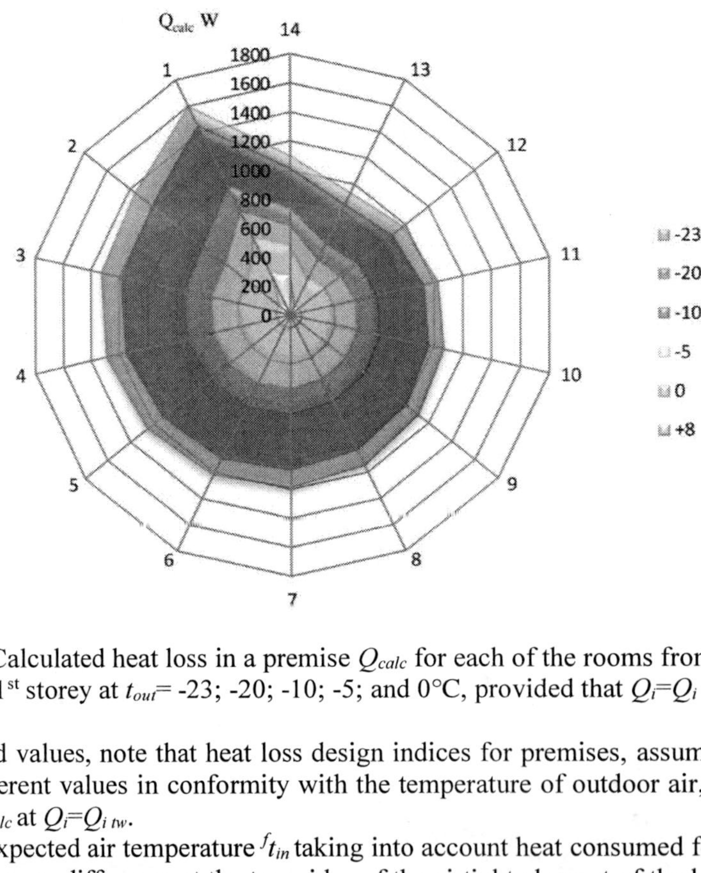

Figure 1. Calculated heat loss in a premise Q_{calc} for each of the rooms from the 14th to the 1st storey at t_{out}= -23; -20; -10; -5; and 0°C, provided that Q_i=$Q_{i\ tw}$.

Comparing obtained values, note that heat loss design indices for premises, assuming that Q_i=$Q_{i\ w}$ and Q_i=$Q_{i\ tw}$, have different values in conformity with the temperature of outdoor air, namely, Q_{calc} at Q_i=$Q_{i\ tw}$ is less than Q_{calc} at Q_i=$Q_{i\ tw}$.

Let us recalculate expected air temperature ${}^{f}t_{in}$ taking into account heat consumed for infiltration air $Q_{i\ tw}$, depending on pressure difference at the two sides of the airtight element of the building at standard storeys using formula:

$$ {}^{f}t_{in} = \frac{n_{sec} f_{sec} t_{pr} K_{ac} + t_{in}(F_{wa}K_{wa} + F_{wall}K_{wall} + 0.28 \cdot G_{inf} \cdot \rho_{in} \cdot c_{in})}{n_{sec} f_{sec} K_{ac} + (F_{wa}K_{wa} + F_{wall}K_{wall} + 0{,}28 \cdot G_{inf} \cdot \rho_{in} \cdot c_{in})} \quad (4) $$

P_{in}, c_{in} – density, kg/m^3, and heating capacity, kJ/kg·°C, of indoor air, respectively.

Number of heating unit sections is taken in compliance with conditions of the1-st thermal design using the formula:

$$n = \frac{Q_{calc}^{st}}{a_{\sec} \cdot q_{ac}} \cdot \frac{\beta_4}{\beta_3} \quad (5)$$

where Q_{calc}^{st} – calculated heat loss of the storey under consideration, which equals to required radiator heat transfer, W;
$a_{\sec}$ – heating surface area of one section, for MS-140 radiators, $a_{\sec}$ =0.244 m^2;
β_4– correction factor, taking into account method of radiator installation [13], assumed β_4=1 [7, page 47];
β_3– factor depending on the number of sections in the radiator, as per recommendations [14];
q_{ac}– actual density of heating unit heat flux for conditions differing from standard ones [10], W/m^2.

Comparing the data under condition of the 1-st design (Figure 2) and condition of 2-nd design (Figure 3), we may make a conclusion that expected air temperature value in corresponding rooms in view of radiator thermal designs at Q_i=$Q_{i\ r}$ is different from those in calculation with Q_i=$Q_{i\ tw}$. So, it should be noted that maximum value of $^f t_{in}$=38.9°C is observed in the 14th storey at t_{out}= -23°C, and minimum value $^f t_{in}$=15.9°C corresponds to t_{in} = +8°C in a room at the 12th storey. Overestimation of $^f t_{in}$ against recommended range of indoor air values $^f t_{in}$ = +18…20 °C is the evidence of excessive heat demand to cover heat loss in the case when real heat loss by infiltration Q_i, W equals to the heat loss with the account of wind pressure impact $Q_{i\ tw}$, not to the largest of $Q_{i\ r}$ and $Q_{i\ tw}$, W (in our case, $Q_{i\ r}$>$Q_{i\ tw}$, W) values.

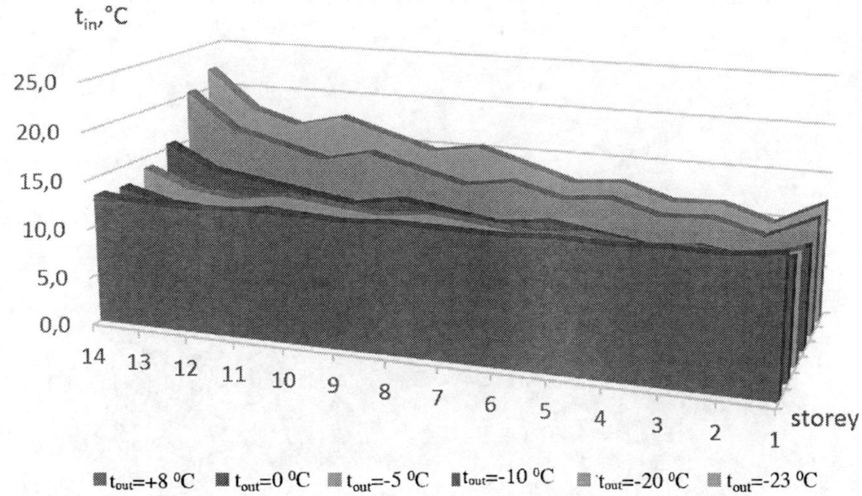

Figure 2. Diagram of expected indoor temperature $^f t_{in}$, °C design data in premises from the 1st to the 14th storey inclusive.

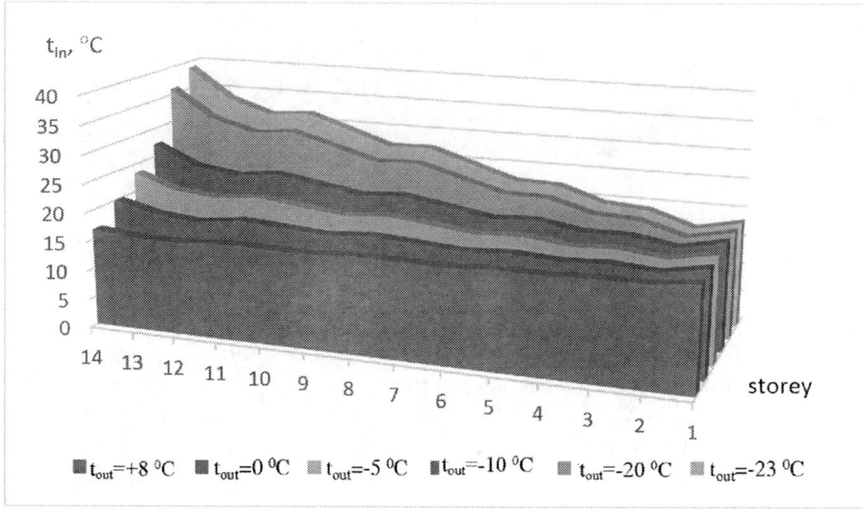

Figure 3. Diagram of expected air temperature value at $^{f}t_{in}$, 0 °C when Q_i=$Q_{i\ tw}$, W and t_{out}= -23; -20; -10; -5; and 0 °C.

Let us calculate the resultant value of expected indoor air temperature $^{f}t_{in}$,°C for premises in storeys 1-14 at t_{out} = -23; -20; -10; -5; 0 and +8 °C, similar to conditions of 2nd and 5th *designs* but taking into account that infiltration loss Q_i equals to heat consumption for infiltrating air heating $Q_{i\ tw}$, where the amount of heated air G_h is considered equivalent to the amount of infiltrating air penetrating through the leaks of external enclosure. Final calculation values are shown in the form of diagrams.

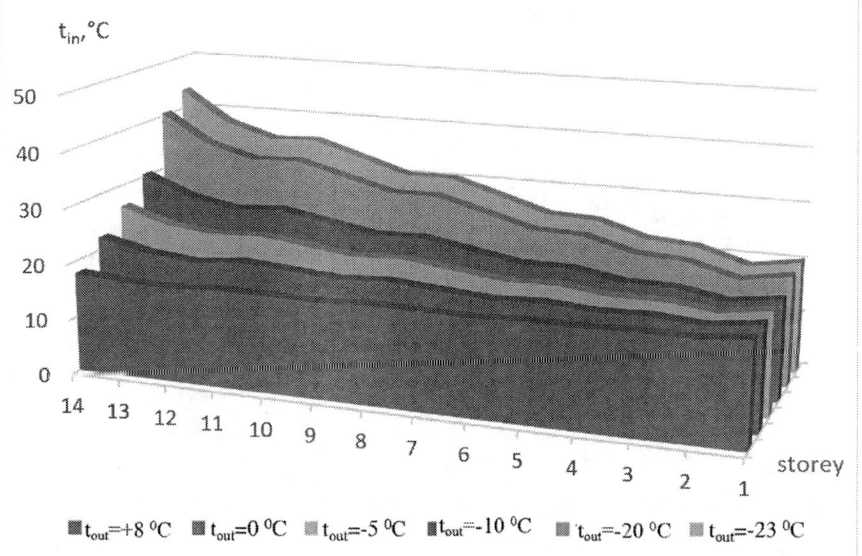

Figure 4. Diagram of expected indoor temperature $^{f}t_{in}$, °C design data in premises from the 1st to the 14th storey inclusive, taking into account changed number of radiator sections under condition of the 2nd *design* and assuming that Q_i=$Q_{i\ tw}$, W

Relying on obtained values $^{f}t_{in}$ (Figure 4), it must be emphasized that condition of increasing heating unit surface area at 50-70% of window aperture has resulted in reaching minimum value of $^{f}t_{in}$=17 °C, while maximum indoor temperature $^{f}t_{in}$ is 44 °C.

According to the diagram of ${}^{f}t_{in}$, °C values (Figure 5), with the increase of heating agent flow, indoor temperature in rooms of storeys 2nd to 14th at t_{out} = +8 °C does not meet the minimum recommended value of t_{in} = +18 °C, while at t_{out} = -20…-23 °C, in some rooms, excessive indoor temperature ${}^{f}t_{in}$ as compared with the comfort range is observed [15, 16].

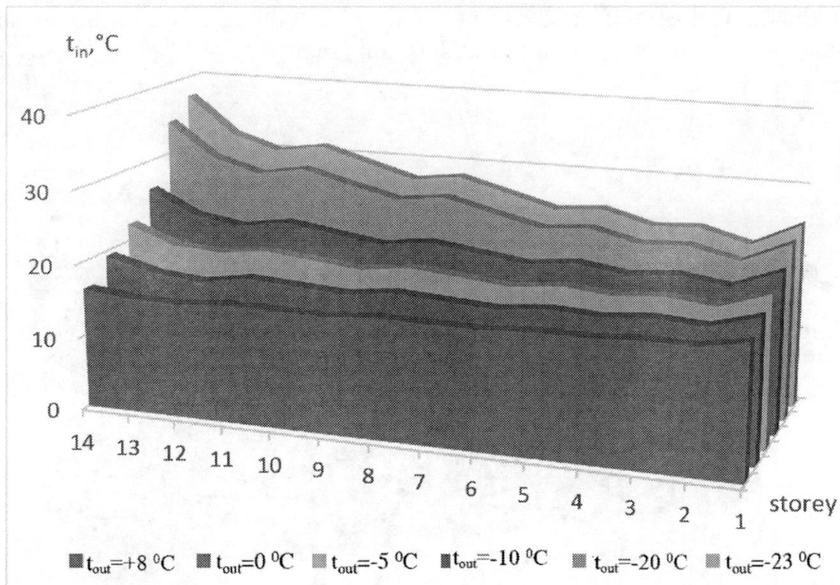

Figure 5. Diagram of expected indoor temperature ${}^{f}t_{in}$, °C design data in premises from the 1st to the 14th storey inclusive, taking into account increased heating agent consumption as per the 3rd design, and assuming Q_i=$Q_{i\ tw}$.

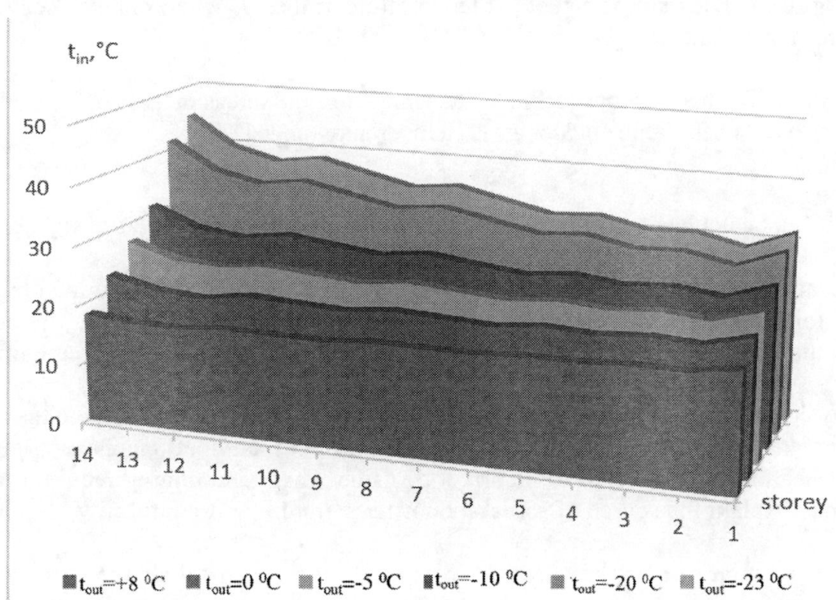

Figure 6. Diagram of expected indoor temperature ${}^{f}t_{in}$, °C design data in premises from the 1st to the 14th storey inclusive, taking into account increased heating agent consumption but without heating unit surface area increase, and assuming Q_i=$Q_{i\ tv}$.

Analyzing the diagram with obtained data of ${}^{f}t_{in}$ changes in rooms at t_{out} = -23; -20; -10; -5; 0; and +8 °C (Figure 6), let us put emphasis on too high value of ${}^{f}t_{in}$ when the value range of t_{out}= -10…-23 °C, which shows the need to apply individual regulation, that is, to shut down a required number of sections n_{sec} until recommended condition of t_{in}= +18…-20 °C is reached. For this purpose we decrease the number of sections n_{sec} until the requirement of ${}^{f}t_{in}$=+18…-20°C is met; it is assumed that Q_i=$Q_{i\ tw}$.

As in the 5th *design*, let us recalculate ${}^{f}t_{in}$ and Q_{ac} due to dependence of average temperature head of the unit Δt_{av} on the number of unit sections, proceeding from the relevant formulae and assuming that Q_i=$Q_{i\ tw}$. Calculation results are shown in the diagram (Figure 7).

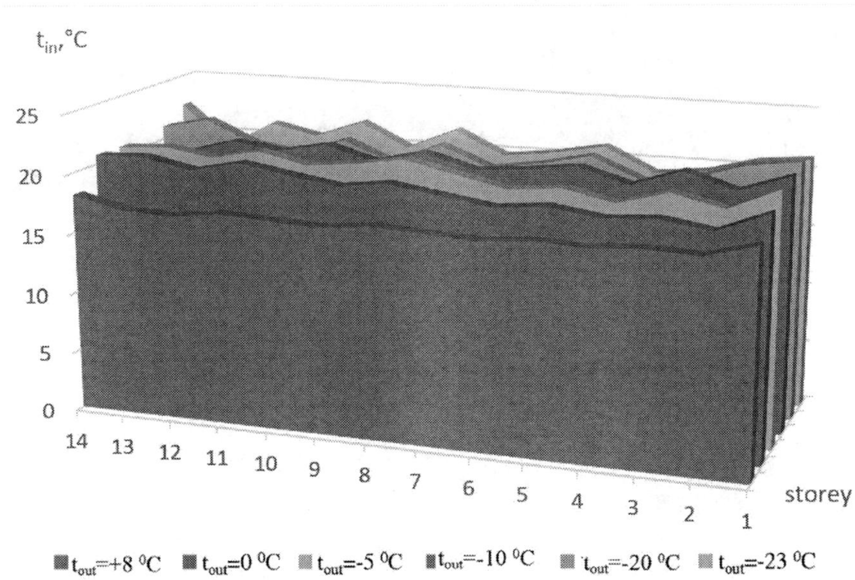

Figure 7. Diagram of actual air temperature values ${}^{f}t_{in}$ when radiator sections are shut off and Q_i=$Q_{i\ tw}$.

Shutting down individual sections facilitates reaching comfort values of ${}^{f}t_{in}$ [16] in residential rooms of all storeys at various values within outdoor air t_{out} temperature range [17].

4. Conclusions

Taking into account calculated data of actual indoor temperature ${}^{f}t_{in}$ under various conditions of design variants *1* and *2,* we may observe dependence of this parameter not only on t_{out}, but on conditions and accuracy of heating units' thermal design. So, a change in ${}^{f}t_{in}$ is directly determined by the amount of heat required for infiltration air heating Q_i, which highlights the need of taking into account the real value of Q_i in the process of heating system design in order to determine a required number of radiator sections.

Thermophysical properties of the heating agent and heating units also determine the indicator of the heating surface area; for this reason, their deviation from nominal coefficients used in standard thermal design of heating systems leads to incorrect decision when determining required radiator surface area, which may explain the source of actual indoor temperature non-conformity with accepted standards.

Individual regulation of units heat transfer by means of mechanical shutdown of unit sections allows reaching comfort values of ${}^{f}t_{in}$ in various outdoor air temperatures with minimum impact on average radiator temperature head Δt_{av}, which does not cause significant changes in thermophysical properties of the heating agent and the heating unit itself.

IOP Conf. Series: Materials Science and Engineering **791** (2020) 012007 doi:10.1088/1757-899X/791/1/012007

Thus, imperfection of current methodology of calculating infiltration heat loss has been analyzed in operation, using a typical residential building as an example. As this methodology is the basis for regulatory documents, this fact should be taken in consideration first and foremost when designing energy efficient facilities.

Acknowledgments
The article was prepared within a development program of the Base University on the basis of BSTU named after V.G. Shoukhov.

References
[1] Zadvinskaya T O and Gorshkov A S 2014 Metodika povy`sheniya e`nergoe`ffektivnosti tipovogo mnogokvartirnogo doma putem vnedreniya sistem ucheta, avtomatizacii i regulirovaniya teplovoj e`nergii [Method of energy efficiency of residential house by implementing of automatic controlled heat metering system] *Stroitel`stvo unikal`ny`x zdanij i sooruzhenij* [*Construction of Unique Buildings and Structures*] **8(23)** 79-92 [In Russian]
[2] Ghiaus C 2013 Causality issue in the heat balance method for calculating the design heating and cooling load *Energy* **50** 292-301
[3] Sasin V I 2002 K voprosu o snizhenii raschetny`x parametrov teplonositelya v sistemax otopleniya [On the problem of reducing the design parameters of the coolant in heating systems] *Akvaterm* **1** 24-6 [In Russian]
[4] Vasilyeva E 2017 Formation of Conceptual Provisions for the Development of the Energy Efficient Housing Construction in Russia *IOP Conf. Ser.: Earth Environ. Sci* **90(1)** 012216
[5] Lauenburg P and Wollerstrand J 2014 Adaptive control of radiator systems for a lowest possible district heating return temperature *Energy and Buildings* **72** 132-40
[6] Lobayev B.N 1996 *Raschet sistem otopleniya* [Calculation of heating systems] (Kiev: Budivel'nik) [In Russian]
[7] Shchekin R V, Berezovskiy V A and Potapov V A *Raschet sistem tsentral'nogo otopleniya* [Calculation of central heating systems] (Kiev: Vishcha shkola) [In Russian]
[8] Yeremkin A I 2008 *Teplovoy rezhim zdaniy* [The thermal regime of buildings] (Rostov on Don: Feniks) [In Russian]
[9] Kubenin A S 2015 Metodika opredeleniya maksimal'nykh raskhodov teploty na nagrev infil'truyushchegosya vozdukha cherez okna zdaniya za ves' otopitel'nyy period [Methods determination of the maximum heat consumption for heating infiltrating air through the windows of the building full heating season] *Naukovedenie* **5(5)** [In Russian]
[10] Bogoslovskiy V N 1991 *Otopleniye* [Heating] (Moscow: Stroyizdat) [In Russian]
[11] Ahn B-C and Song J-Y 2010 Control characteristics and heating performance analysis of automatic thermostatic valves for radiant slab heating system in residential apartments *Energy* **35** 1615-24
[12] Seminenko A S and Elistratova Y V 2017 Analysis of thermal design of heating units with meteorological climate peculiarities *IOP Conf. Ser.: Mater. Sci. Eng.* **327** 042097
[13] Staroverov I G 1990 *Vnutrenniye sanitarno-tekhnicheskiye ustroystva v 3 ch. CH. I. Otopleniye* [Internal sanitary facilities at 3 p.m. I. Heating] ed I G Staroverova and Yu I Shillera (Moscow: Stroyizdat) [In Russian]
[14] Široký J, Oldewurtel F, Cigler J and Privara S 2011 Experimental analysis of model predictive control for an energy efficient building heating system *Applied Energy* **88** 3079-87
[15] Pérez-Lombard I, Ortiz J and J.F. and Coronel J F 2011 A review of HVAC systems requirements in building energy regulations *Energy Build.* **43(2–3)** 255-68
[16] Chutia D, Bhattacharyya D K and Sarma 2017 An effective ensemble classification framework using random forests and a correlation based feature selection technique *Trans. GIS* **21(6)** 1165-78
[17] Bychikhin S A, Svirin M V and Trubaev P A 2019 Evaluation of energy-saving effect during the installing process of automated heating stations in educational institutions *IOP Conf. Ser.: Mater. Sci. Eng.* **552** 012019

IOP Conf. Series: Materials Science and Engineering **791** (2020) 012008 doi:10.1088/1757-899X/791/1/012008

Study on heat transfer between the rod and the environment under conditions of forced convection

A V Eremin

Department of Industrial Heat Power Engineering, Samara State Technical University, Molodogvardeyskaya St., 244, Samara, 443100, Russia

E-mail: a.v.eremin@list.ru

Abstract. Obtaining analytical solutions to unsteady heat conduction problems is of great scientific and practical interest. Such solutions make it possible to do an in-depth analysis of thermal processes, such as isotherm fields analysis, study of the thermally stressed states of structures, parametric identification, etc. This article deals with a simple method for obtaining approximate analytical solutions of one-dimensional heat conduction problems. In particular, an algorithm for solving the problem for a rod (plate) with a given boundary condition of the third kind on one of the surfaces is presented. It is shown that solving the equation at isolated points of the spatial variable allows obtaining the high-precision solutions to this problem with a minimum amount of computational work. The relations for determining the temperature have a simple form and do not contain special functions and parameters. It should be noted that the exact solution to a similar problem based on the Fourier separation method is an infinite series containing eigenvalues (roots of the transcendental equation).The practical application of such solutions is very limited. The paper also contains the convergence analysis of the method, the residuals of the initial differential equation for various approximations. The method developed can be used to solve more complex problems that allow separation of variables in the initial l differential equation.

1. Introduction

Various mathematical methods are used in obtaining solutions to classical problems of heat conduction, in particular, to problems for an infinite rod, cylinder, ball, etc. Methods for obtaining solutions in the form of a formula are usually called analytical. Such solutions are more convenient than numerical ones, which are represented by a data array.

The most commonly used classical methods in the theory of thermal conductivity are the separation of variables (the Fourier method), the Laplace transform, and the Green's functions [1 – 8].The range of tasks that can be solved using these methods is very limited. In this context, the development of approximate analytical methods, such as the Kantorovich theorem [9], the integral heat balance method [10, 11], the method of additional boundary characteristics [12 – 16], the Galerkin method [17 – 20], the Ritz method [21, 22] etc. is important today.

So, this paper focuses on the development of an effective approximate analytical method for solving heat conduction problems. Thus, a solution to the non-stationary heat conduction problem for the rod has been obtained on the basis of the joint use of the variable separation method and the collocation method. Simple solutions that can be used in engineering practice are obtained by directly satisfying the differential equation of the Sturm-Liouville boundary value problem at a certain number of points of the spatial variable.

2. Formulation of the problem

Consider an example of solving the boundary-value heat conduction problem for an infinite plate under symmetric boundary conditions of the first kind as follows

$$\frac{\partial t(x,\tau)}{\partial \tau} = a\frac{\partial^2 t(x,\tau)}{\partial x^2} \qquad (\tau > 0;\quad 0 \le x \le \delta); \tag{1}$$

$$t(x,0) = t_0; \tag{2}$$

$$\partial t(0,\tau)/\partial x = 0; \tag{3}$$

$$t(\delta,\tau) = t_{\text{ст}}, \tag{4}$$

where t – temperature; x – coordinate; τ – time; t_0 – initial temperature; t_w – wall temperature; a –temperature conductivity coefficient; δ – half the thickness of the plate.

Problem (1) - (4) can be written in a dimensionless form

$$\frac{\partial \Theta(\xi,\text{Fo})}{\partial \text{Fo}} = \frac{\partial^2 \Theta(\xi,\text{Fo})}{\partial \xi^2} \qquad (\text{Fo} > 0;\quad 0 < \xi < 1); \tag{5}$$

$$\Theta(\xi,0) = 1; \tag{6}$$

$$\frac{\partial \Theta(0,\text{Fo})}{\partial \xi} = 0; \tag{7}$$

$$\partial\Theta(1,\text{Fo}) = 0, \tag{8}$$

where $\Theta = (t - t_{\text{ст}})/(t_0 - t_{\text{ст}})$ – dimensionless temperature; $\xi = x/\delta$ – dimensionless coordinate; $\text{Fo} = (a\tau)/\delta^2$ – Fourier criterion (dimensionless time).

3. Method of solution

The solution to problem (5) - (8) using the variable separation method can be found in the form

$$\Theta(\xi,\text{Fo}) = \varphi(\text{Fo})\,\psi(\xi). \tag{9}$$

Substituting (9) into (5), we can find

$$\frac{d\varphi(\text{Fo})}{d\text{Fo}} + \nu\varphi(\text{Fo}) = 0; \tag{10}$$

$$\frac{d^2\psi(\xi)}{d\xi^2} + \nu\psi(\xi) = 0, \tag{11}$$

where ν – some constant (eigenvalue).

The solution to equation (10) is known and is as follows

$$\varphi(\text{Fo}) = A\exp(-\nu\text{Fo}), \tag{12}$$

where A – unknown coefficient.

Substituting (9) into (7), (8) we obtain

$$\frac{d\psi(0)}{d\xi} = 0; \tag{13}$$

$$\psi(1) = 0. \tag{14}$$

IOP Conf. Series: Materials Science and Engineering **791** (2020) 012008 doi:10.1088/1757-899X/791/1/012008

The solution to the Sturm-Liouville boundary value problem (11), (13), (14) can be written as follows

$$\psi(\xi) = B_0 + \sum_{i=1}^{r} B_i \xi^{i+1}, \tag{15}$$

where $B_i \, (i = \overline{1, r})$ – unknown coefficients. Note that relation (15) satisfies the boundary condition (13).

Relation (13) allows one more boundary condition

$$\psi(0) = const = 1. \tag{16}$$

Substituting (15) into (16), we find $B_0 = 1$.

We require that the relation (15) satisfies the boundary condition (14) and equation (11) at the points $\xi = 0; 0.25; 0.5; 0.75$. Let's consider five terms of the series ($r = 5$). Substituting (15) into relation (14) and equation (11), for points $\xi = 0; 0.25; 0.5; 0.75$ with respect to unknown coefficients $B_i \quad (i = \overline{1, 5})$, we obtain a system of five linear algebraic equations. Then from the solution of this system we can find the coefficients $B_i \quad (i = \overline{1, 5})$.

$$B_1 = -\nu/2;$$

$$B_2 = (25\nu^4 - 3594\nu^3 + 162080\nu^2 - 2370560\nu + 4915200)/c_1;$$

$$B_3 = -(105\nu^4 - 18134\nu^3 + 847280\nu^2 - 12945920\nu + 27033600)/c_2;$$

$$B_4 = 2(40\nu^4 - 7632\nu^3 + 371712\nu^2 - 5652480\nu + 11796480)/c_1;$$

$$B_5 = -(48\nu^4 - 9760\nu^3 + 494848\nu^2 - 7536640\nu + 15728640)/c_2,$$

where $c_1 = 6\nu^3 - 512\nu^2 + 20480\nu$; $c_2 = 9\nu^3 - 768\nu^2 + 30720\nu$.

Each of the found coefficients B_i depends on some constant - coefficient ν. To find ν, we calculate the integral of the weighted residual of equation (11)

$$\int_0^1 \left[\frac{\partial^2}{\partial \xi^2} \left(1 + \sum_{i=1}^{r} B_i \xi^{i+1} \right) + \nu \left(1 + \sum_{i=1}^{r} B_i \xi^{i+1} \right) \right] d\xi = 0. \tag{17}$$

Then we calculate integral (17) taking into account the found values of the coefficients $B_i \quad (i = \overline{1, 5})$ with respect to the eigenvalues ν_k and obtain an algebraic equation of the fifth degree

$$17901\nu^5 - 1.0173 \cdot 10^{-7} \nu^4 + 1.8109 \cdot 10^{-9} \nu^3 - 1.0253 \cdot 10^{-11} \nu^2 +$$

$$+1.6365 \cdot 10^{-12} \nu - 3.4401 \cdot 10^{-12} = 0. \tag{18}$$

From the solution of equation (18) we get five eigenvalues, two of which are complex self-adjoint. As we accept only real eigenvalues, we get $\nu_1 = 2.467312$; $\nu_2 = 21.896613$; $\nu_3 = 60.857583$. The resulting solution will be called the third approximation (according to the number of real eigenvalues).

Exact values ν [5]: $\nu_1 = 2.467401$; $\nu_2 = 22.206609$; $\nu_3 = 61.685028$.

Substituting (12), (15) into (9), for each eigenvalue we have particular solutions:

IOP Conf. Series: Materials Science and Engineering **791** (2020) 012008 doi:10.1088/1757-899X/791/1/012008

$$\Theta_k(\xi, \mathrm{Fo}) = A_k \exp(-\nu_k)\left(1+\sum_{i=1}^{r} B_i(\nu_k)\xi^{i+1}\right). \tag{19}$$

Each particular solution from (19) satisfies the boundary conditions (7), (8) exactly, and satisfies equation (5) approximately (in the third approximation). However, none of them, including their sum

$$\Theta(\xi, \mathrm{Fo}) = \sum_{k=1}^{n}\left[A_k \exp(-\nu_k)\left(1+\sum_{i=1}^{r} B_i(\nu_k)\xi^{i+1}\right)\right], \tag{20}$$

do not satisfy the initial condition (6). To fulfill the initial condition, we find its residual and require the residual orthogonality to each eigenfunction:

$$\int_0^1 \left\{\sum_{k=1}^{5}\left[A_k\left(1+\sum_{i=1}^{r} B_i(\nu_k)\xi^{i+1}\right)\right]-1\right\}\psi_j(\nu_j,\xi)d\xi = 0. \quad (j=1,2,3;\ r=5) \tag{21}$$

To find A_k $(k=\overline{1,3})$, we calculating integral (21) and get a system of three linear algebraic equations. The solution to the system is as follows

$$A_1 = 1.27798;\ A_2 = -0.4337;\ A_3 = 0.265008.$$

The exact values of the first three coefficients A_k [5] are shown in Table 1.

Table 1. Calculated values of the eigenvalues compared to exact values

№	ν_k (approximate value)	ν_k (exact value)	A_k (approximate value)	A_k (exact value)
1	2.4674011003	2.4674011003	1.2722175645	1.2732395447
2	22.2066099025	22.2066099025	–0.4213383287	–0.4244131816
3	61.6850275068	61.6850275068	0.2495131649	0.2546479089
4	120.902653913	120.902653913	–0.1746716643	–0.1818913635
5	199.859489122	199.859489122	0.1321424687	0.1414710605
6	298.555533133	298.555533133	–0.1042688938	–0.1157490495
7	416.990785946	416.990785946	0.0842665383	0.0979415034
8	555.165247561	555.165247561	–0.0689496574	–0.0848826363
9	713.078917978	713.078917978	0.0566386114	0.0748964438
10	890.731797198	890.731797198	–0.0463412842	–0.0670126076
11	1088.12388521	1088.12388522	0.0374479197	0.0606304545
12	1305.25518217	1305.25518204	–0.0295426615	–0.0553582411
13	1542.12568846	1542.12568767	0.0223389702	0.0509295818
14	1798.73507417	1798.73540209	–0.0156241657	–0.0471570202
15	2075.08408591	2075.08432532	0.0091983751	0.0439048119
16	2371.41373545	2371.17245736	–0.0712573770	–0.0410722434
17	2686.81399536	2686.99979819	0.0682304199	0.0385830165

IOP Conf. Series: Materials Science and Engineering **791** (2020) 012008 doi:10.1088/1757-899X/791/1/012008

Calculation results by formula (20) in the third approximation are shown in Figure 1. Their analysis allows us to conclude that in the range of numbers $0.03 \le \text{Fo} < \infty$ the difference between the solution obtained and the exact one [5] does not exceed 1%. The accuracy decreases at shorter times.

In order to increase accuracy, it is necessary to satisfy equation (11) at a larger number of points. The algorithm remains the same. Table 1 shows the eigenvalues and coefficients A_k $(k=\overline{1,17})$ in the seventeenth approximation. Number of real values ν_k is considered as an approximation number.

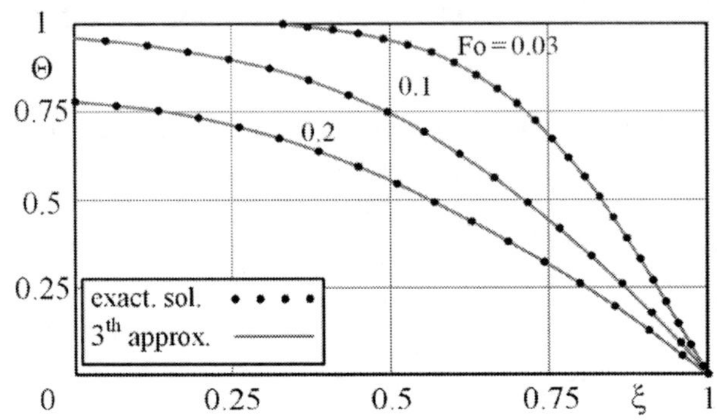

Figure 1. Temperature changes in the plate

Let's also consider the algorithm for using this method to solve the heat conduction problem for a rod with a boundary condition of the third kind on one of its surfaces [23]. In this case the problem includes differential equation (1), boundary conditions (2), (3), as well as the condition

$$-\lambda \frac{\partial t(\delta,\tau)}{\partial x} = \alpha(t(\delta,\tau) - t_{cp}). \tag{22}$$

where t_m – environment temperature

In a dimensionless form, this problem includes equation (5), boundary conditions (6), (7), as well as the condition

$$\frac{\partial \Theta(1,\text{Fo})}{\partial \xi} + \text{Bi}\Theta(1,\text{Fo}) = 0, \tag{23}$$

where $\Theta = (t - t_{cp})/(t_0 - t_{cp})$ – dimensionless temperature; $\text{Bi} = (\alpha\delta)/\lambda$ – dimensionless criterion (Biot number).

Substituting (9) into (23), we obtain

$$\frac{\partial \psi(1)}{\partial \xi} + \text{Bi} \cdot \psi(1) = 0. \tag{24}$$

Now we require that relation (15) should satisfy conditions (14), (16), (24) and equation (11) at points $\xi = 0; 1/3; 2/3; 1$. Substituting (15) (limited to the five terms of the series) in relation (14) and equation (11) for points $\xi = 0; 1/3; 2/3; 1$ with respect to unknown coefficients B_i $(i=\overline{1,5})$, we obtain a system of five linear algebraic equations. Then from the solution of this system we find the coefficients B_i $(i=\overline{1,5})$.

The first two coefficients (at $\text{Bi} = 1$) are as follows

$$B_1 = -\nu/2;$$

$$B_2 = \frac{566.054\nu^4 - 2.356\nu^5 + 39736725\nu^3 - 109618125\nu^2 + 8769375\nu + 59066250}{0.415\nu^4 - 61.713\nu^3 + 3623.437\nu^2 + 16171.875\nu + 2531250}.$$

Calculating the integral of the weighted residual (17) taking into account the found values of the coefficients B_i $(i=\overline{1,5})$ with respect to the eigenvalues, we obtain the algebraic equation of the fifth degree

$$\nu^5 - 747\nu^4 + 105360\nu^3 - 4044240\nu^2 + 36028800\nu - 24494400 = 0. \qquad (25)$$

From the solution of equation (25) we get five eigenvalues: $\nu_1 = 576.117009$; $\nu_2 = 115.852710$; $\nu_3 = 42.670602$; $\nu_4 = 11.619503$; $\nu_5 = 0.740173$.

To fulfill the initial condition, we find its residual and require the residual orthogonality to each eigenfunction, i.e.

$$\int_0^1 \left\{ \sum_{k=1}^{5} \left[A_k \sum_{i=1}^{r} B_i(\nu_k)\xi^{k+1} \right] - 1 \right\} \psi_j(\nu_j, \xi) d\xi = 0. \quad (j = 1, 2, 3; \ r = 5) \qquad (26)$$

Calculating the integrals in (26) to find A_k $(k=\overline{1,5})$, we obtain a system of five algebraic linear equations. The solution is as follows

$$A_1 = 0.010142; \ A_2 = -0.042089; \ A_3 = 0.054192;$$

$$A_4 = -0.137528; \ A_5 = 1.120143.$$

Calculation results by formula (20) in the fifth approximation are shown in Figures 2 and 3. Their analysis allows us to conclude that in the range of numbers $0.01 \le \text{Fo} < \infty$ the difference between the obtained solution and the exact one [5] does not exceed 1%.

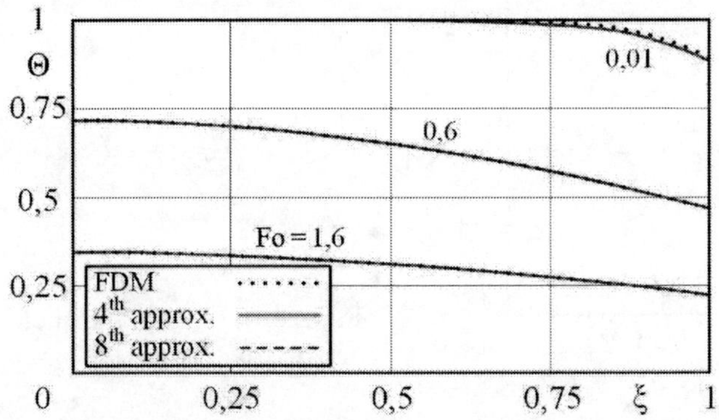

Figure 2. Temperature distribution in the plate

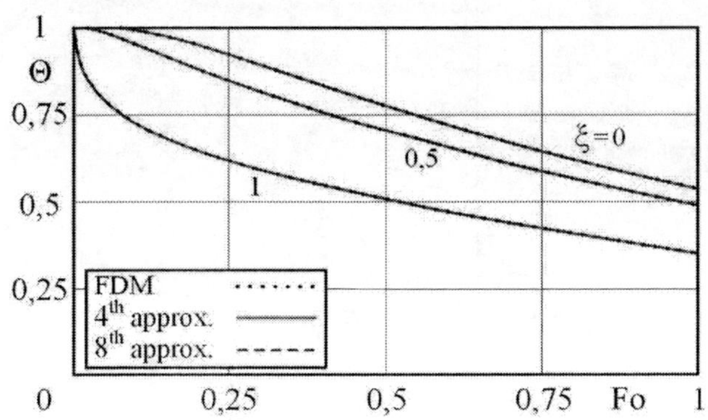

Figure 3. Temperature changes over time in the plate.

IOP Conf. Series: Materials Science and Engineering **791** (2020) 012008 doi:10.1088/1757-899X/791/1/012008

To increase the accuracy of the solution, it is necessary to increase the number of terms of the series (15). To obtain additional equations in order to determine unknown coefficients B_i, we should increase the number of points along the coordinate ξ, where the equation (11) should be performed. And, in particular, if we take eight points (with a step $\Delta\xi = 1/7$, starting from the point $\xi = 0$), with respect to the unknown coefficients B_i, we obtain 9 equations (one more equation is added as a result of the fulfillment of the boundary condition (14)). After determining the unknowns B_i $(i = \overline{1,9})$ from the solution of this system of equations, the algorithm of the solution is repeated.

Determination of the heat flow density on the surface of the body is of particular importance in solving practical problems.

$$|Q| = \left.\frac{\partial\Theta(\xi;\text{Fo})}{\partial\xi}\right|_{\xi=1} = \text{Bi}\,\Theta(1;\text{Fo}), \tag{27}$$

where $Q = -\dfrac{\delta}{\lambda(T_0 - T_{cp})}q$; q – heat flow density on the plate surface, W/m^2.

Using a known value Q, one can determine the power of the heating device, the heat loss through building envelopes, etc. Figure 4 demonstrates the distribution of the dimensionless heat flow over time depending on the Biot criterion value. The analysis shows that the heat flow has a maximum value at the initial moment, then, it intensively decreases, tending to zero. It should also be noted that a period of intense decrease in heat flow is followed by a period of linear decrease in value Q. These results are completely consistent with the known ones [5].

Figure 5 presents the residual of the differential equation (5), from which we can conclude that the solution is being clarified.

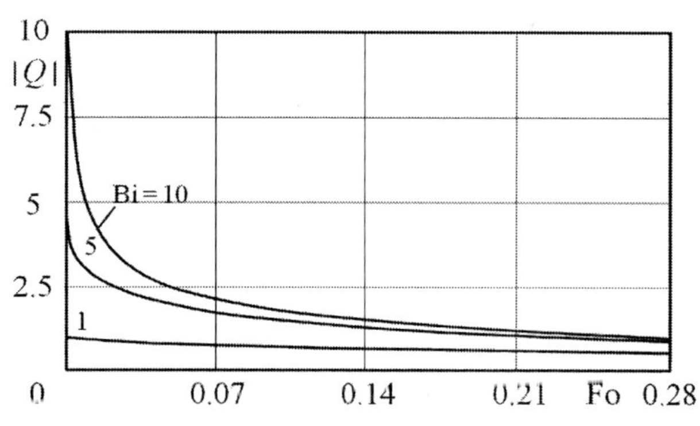

Figure 4. Changes in heat flow density over time

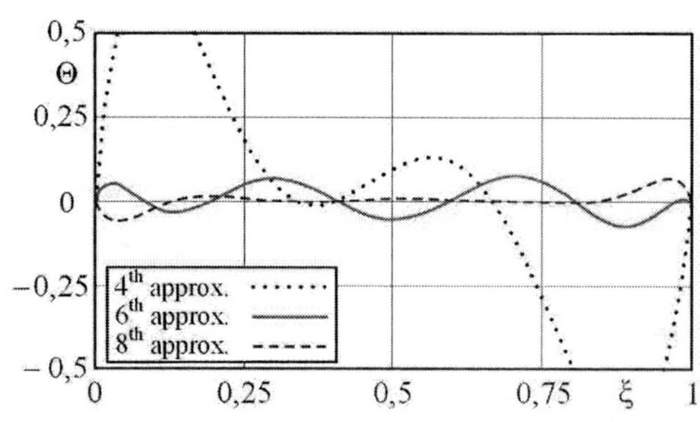

Figure 5. The residual of the differential equation.

4. Conclusion

The development results of an approximate analytical method for solving the one-dimensional heat conduction problems are presented. By solving the boundary value problem for the rod under the boundary conditions of the first and third kind, it is shown that the method developed has high accuracy. So, if the ordinary differential equation is satisfied at only a few points of the spatial variable, the error of the method is no more than 5%.

Acknowledgments

This study was supported by the financial support of the Russian National Science Foundation within the framework of the research project No. 18–79–00171.

References

[1] Sneddon I N 1995 *Fourier transforms* (New York: Dover Publications)

[2] Tsoi P V 2005 *Sistemnye metody rascheta kraevykh zadach teplomassoperenosa* [*System methods for calculating boundary problems of heat and mass transfer*] (Moscow: Izdatel'stvo MEI) [In Russian]

[3] Cherati D Y and Ghasemi-Fare O 2019 Analyzing transient heat and moisture transport surrounding a heat source in unsaturated porous media using the green's function *Geothermics* **81** 224-34

[4] Tranter C J 1966 *Integral transforms in matematikal physics* (London: Methuen)

[5] Lykov A V 1967 *Teoriya teploprovodnosti* [*Heat conduction theory*] (Moscow: Vysshaya shkola) [In Russian]

[6] Kartashov E M 2001 *Analiticheskie metody v teorii teploprovodnosti tverdykh tel* [*Analytical methods in the heat-conduction theory of solids*] (Moscow: Vysshaya shkola) [in Russian]

[7] Tsoi P V 1971 *Metody rascheta otdel'nykh zadach teplomassoperenosa* [*Methods of Calculating Individual Problems on Heat and Mass Transfer*] (Moscow: Energiya) [in Russian]

[8] Belyaev N M, Ryadno A A 1978 *Metody` nestacionarnoj teploprovodnosti* [*Methods of Nonstationary Heat Conduction*] (Moscow: Vy`sshaya shkola) [in Russian]

[9] Kantorovich L V 1934 Ob odnom metode priblizhennogo resheniya differencial`ny`x uravnenij v chastny`x proizvodny`x [On a method for the approximate solution of partial differential equations] Doklady` Akademii Nauk SSSR [*Reports of the USSR Academy of Sciences*] **2(9)** pp 532-6 [in Russian]

[10] Layeni O, Johnson J 2012 Hybrids of the heat balance integral method *Applied Mathematics and Computation* **218 (14)** pp 7431-44

[11] Mitchell S L and Myers T 2010 Improving the accuracy of heat balance integral methods applied to thermal problems with time dependent boundary conditions *Int. J. of Heat and Mass Transfer* **53(17-18)** pp 3540-51

[12] Kudinov V A, Stefanyuk E V 2009 Analytical solution method for heat conduction problems based on the introduction of the temperature perturbation front and additional boundary conditions *J. of Eng. Physics and Thermophysics* **82(3)** 537-55

[13] Kudinov I V, Kudinov V A and Kotova E V 2017 On one method of solving nonstationary boundary-value problems *J. of Eng. Physics and Thermophysics* **90(6)** 1317-27

[14] Fedorov F, Egorov I 2000 *Granichny`j metod resheniya prikladny`x zadach matematicheskoj fiziki* [*Boundary method for solving applied problems of mathematical physics*] (Novosibirsk: Nauka) [in Russian]

[15] Eremin A V 2019 Study of Thermal Exchange with Liquid Flowing in a Cylindrical Channel *Int. Science and Technology Conf. "EastConf"* (Vladivostok, Russia) pp 1-5

[16] Eremin A V, Kudinov V A and Stefanyuk E V 2018 Heat Exchange in a Cylindrical Channel with Stabilized Laminar Fluid Flow *Fluid Dynamics* **53(1)** 29-39

[17] Rao T, Chakraverty S 2017 Modeling radon diffusion equation in soil pore matrix by using uncertainty based orthogonal polynomials in galerkin's method *Coupled Systems Mechanics*

IOP Conf. Series: Materials Science and Engineering **791** (2020) 012008 doi:10.1088/1757-899X/791/1/012008

6(4) 487-99

[18] Nourgaliev R, Luo H and Weston B 2016 Fully-implicit orthogonal reconstructed discontinuous galerkin method for fluid dynamics with phase change *J. of Computational Physics* 305 964-96

[19] Belytschko T, Lu Y Y and Gu L 1994 Element-free galerkin methods *Int. J. for Numerical Methods in Engineering* **37(2)** 229-56

[20] Arnold D N, Brezzi F and Cockburn B 2002 Unified analysis of discontinuous galerkin methods for elliptic problems *SIAM J. on Numerical Analysis* **39(5)** 1749-79

[21] Dutta S, Sil A N and Saha J K 2018 Ritz variational method for the high-lying nonautoionizing doubly excited 1, 3fe states of twoelectron atoms *Int. J. of Quantum Chemistry* **118(14)** e25577

[22] Lotfi A, Yousefi S A 2017 A generalization of ritz-variational method for solving a class of fractional optimization problems *J. of Optimization Theory and Applications* **174(1)** 238-55

[23] Kudinov V A, Eremin A V and Stefanyuk E V 2015 Analytical solutions of heat-conduction problems with time-varying heat-transfer coefficients *J. of Engineering Physics and Thermophysics* **88(3)** 688-98

IOP Conf. Series: Materials Science and Engineering **791** (2020) 012009 doi:10.1088/1757-899X/791/1/012009

The study of the heat transfer process in bodies with internal heat sources of variable power

A V Eremin[1], K V Gubareva[1], A A Ilyasov[1], K V Trubitsyn[2] and P V Iglin[1]

[1]Department of Industrial Heat Power Engineering, Samara State Technical University, Molodogvardeyskaya St., 244, Samara, 443100, Russia
[2]Heat Power Engineering Faculty, Samara State Technical University, Molodogvardeyskaya St., 244, Samara, 443100, Russia

E-mail: a.v.eremin@list.ru

Abstract. An analytical solution to an unsteady heat conductivity problem based on the method of using a new function for a plate under first-order boundary conditions with time-varying internal heat sources was obtained. Obtaining an exact analytical solution for such problems has encountered serious mathematical difficulties. However, analytical solutions have significant advantages compared to numerical ones. For example, solutions obtained in an analytical form make it possible to perform the parametric analysis of the system under study, parametric identification, programming of measuring devices; controlling the production process, etc. Therefore, approximate analytical methods have been widely used. So, this paper is devoted to the development of such a method. The solutions obtained have a simple form of algebraic polynomials without special functions. This allows doing the research in the fields of isotherms and determining their velocities.

1. Introduction

In engineering practice there are cases when heat occurs inside the body due to the internal sources, for example, the electric current flow, exothermic chemical reactions, nuclear transformations, etc. Obtaining accurate analytical solutions to unsteady heat conduction problems with heat sources using classical methods (Fourier, Green's functions, integral transforms [1 – 6]) includes mathematical difficulties, as the solutions obtained are expressed by complex functional dependencies. Thereby, approximate analytical methods for mathematical modeling such as various modications of the heat balance integral method [7 – 11], Ritz method [12, 13], Kantorovich method [14], Galerkin method [15 – 18] etc. have rapid development.

The use of such solutions in engineering practice is extremely difficult. If the power of internal sources is time-varying, the classical methods often cannot be applied.

Consider an unsteady heat conduction problem for an infinite plate with a time-varying internal heat source under symmetric boundary conditions of the first kind:

$$\frac{\partial T(x,\tau)}{\partial \tau} = a\frac{\partial^2 T(x,\tau)}{\partial x^2} + \frac{\omega(\tau)}{c\gamma} \qquad (\tau > 0;\quad 0 < x < \delta); \tag{1}$$

$$T(x,0) = T_0; \tag{2}$$

IOP Conf. Series: Materials Science and Engineering **791** (2020) 012009 doi:10.1088/1757-899X/791/1/012009

$$T(0,\tau)=T_{wall};\tag{3}$$

$$\frac{\partial T(\delta,\tau)}{\partial x}=0,\tag{4}$$

where T – temperature; x – coordinate; τ – time; $a=\lambda/(c\rho)$ – temperature conductivity coefficient; λ – heat conduction coefficient; ρ – density; c – specific isochoric heat capacity; $\omega(\tau)=\omega_0(1+\beta\tau)$ – power of an internal heat source; ω_0 – initial power of an internal heat source(at τ =0); T_0 – initial temperature; T_{wall} – wall temperature; δ– half thickness of the plate.

The problem (1) – (4) can be presented in a dimensionless form [1] (Figure 1)

$$\frac{\partial\Theta(\xi,\text{Fo})}{\partial\text{Fo}}=\frac{\partial^2\Theta(\xi,\text{Fo})}{\partial\xi^2}+\text{PoFo}+\text{Po}_1\qquad(\text{Fo}>0;\quad 0<\xi<1);\tag{5}$$

$$\Theta(\xi,0)=0;\tag{6}$$

$$\Theta(0,\text{Fo})=1;\tag{7}$$

$$\frac{\partial\Theta(1,\text{Fo})}{\partial\xi}=0,\tag{8}$$

where $\Theta=(T-T_{wall})/(T_0-T_{wall})$ – dimensionless temperature; $\xi=x/\delta$ – dimensionless coordinate; $\text{Fo}=(a\tau)/\delta^2$ – Fourier criterion (dimensionless time); $\text{Po}=\text{Po}_1\delta^2/\beta a$ – Pomerantsev criterion; $\text{Po}_1=(\omega\delta^2)/[c\rho a(T_0-T_{wall})]$ – initial value of the Pomerantsev criterion (at Fo = 0)

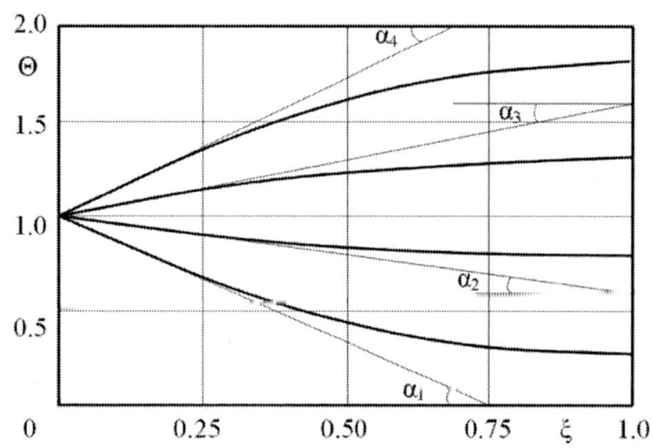

Figure 1. Schematic of heat exchange.

2. Methodology

According to the research methodology, we introduce the new time-dependent function

$$\varphi(\text{Fo})=\frac{\partial\Theta(0,\text{Fo})}{\partial\xi}=tg\,\alpha,\tag{9}$$

where α –angle between the tangent to the graph of the function $\Theta(\xi,\text{Fo})$ at point $\xi=0$ and the coordinate axis.

In dimensional form relation (9) can be written as

$$\phi(\tau)=\frac{\delta}{T_{wall}-T_0}\frac{\partial T(0,\tau)}{\partial x}. \tag{10}$$

According to the Fourier law, the heat flux density on the surface of the plate is determined by the relation

$$q(\tau)=-\lambda\frac{\partial T(0,\tau)}{\partial x}, \tag{11}$$

then

$$\phi(\tau)=\frac{\delta}{\lambda(T_0-T_{wall})}q(\tau)=k\,q(\tau), \tag{12}$$

where $k=const$ – scale factor of the system. Therefore, the new function is the product of heat flux density at the boundary point by a constant.

A feature of the proposed method (in comparison with [8]) is the use of the law of heat flux density change at the point of the third kind boundary condition in the product with a constant as a new function.

The solution to problem (5) – (8) is sought in the form of an algebraic polynomial

$$\Theta(\xi,\mathrm{Fo})=\sum_{i=1}^{n}b_i(\mathrm{Fo})\,\xi^{i-1}, \tag{13}$$

where $n\in N$ – natural number corresponding to the number of terms of the series (13); $b_i(\mathrm{Fo})$ – unknown coefficients that depend on the dimensionless time.

To obtain a solution to problem (5) - (8), substitute relation (13) into boundary conditions (7), (8) and into the additional condition (9) in the first approximation. As a result of substitution, we obtain a system of three algebraic equations

$$\begin{cases} b_1=1; \\ b_2+2b_3=0; \\ b_2-\varphi(\mathrm{Fo})=0, \end{cases} \tag{14}$$

Solving the system, we can obtain the unknown coefficients

$$b_1(\mathrm{Fo})=1;\quad b_2(\mathrm{Fo})=\varphi(\mathrm{Fo});\quad b_3(\mathrm{Fo})=-\frac{\varphi(\mathrm{Fo})}{2}.$$

Relation (13) taking into account the coefficients found can be written as

$$\Theta(\xi,\mathrm{Fo})=f_1(\xi)\varphi(\mathrm{Fo})+1, \tag{15}$$

where $f_1(\xi)=\xi\,(1-0.5\xi)$ – coordinate function.

We now require the solution (15) to satisfy not the initial differential equation (5), but an averaged one - the heat balance integral [7 – 11]

$$\int_0^1\frac{\partial\Theta(\xi,\mathrm{Fo})}{\partial\mathrm{Fo}}d\xi=\int_0^1\left(\frac{\partial^2\Theta(\xi,\mathrm{Fo})}{\partial\xi^2}-\mathrm{PoFo}-\mathrm{Po}_1\right)d\xi. \tag{16}$$

Calculating the integral, an ordinary differential equation can be obtained

$$\frac{d\varphi(\mathrm{Fo})}{d\mathrm{Fo}}+3\varphi(\mathrm{Fo})-3(\mathrm{Po}_1+\mathrm{PoFo})=0, \tag{17}$$

Solving the equation, we can find

$$\varphi(\text{Fo}) = C_1 e^{-3\text{Fo}} + \text{Po}_1 + \text{Po}\left(\text{Fo} - \frac{1}{3}\right), \tag{18}$$

where C_1 – integration constant.

Substituting (18) into (15), we obtain

$$\Theta(\xi, \text{Fo}) = f_1(\xi)\left(C_1 e^{-3\text{Fo}} + \text{Po}_1 + \text{Po}\left(\text{Fo} - \frac{1}{3}\right)\right) + 1. \tag{19}$$

Relation (19) exactly satisfies the boundary conditions (7), (8), additional condition (9), and also the heat balance integral (16). To fulfill the initial condition (5), we find its residual and require the orthogonality of the residual to the coordinate function

$$\int_0^1 [\Theta(\xi, 0)] f_1(\xi)\, d\xi = 6C_1 - 2\text{Po} + 6\text{Po}_1 + 15 = 0. \tag{20}$$

Solving equation (20), we determine the integration constant $C_1 = \frac{1}{3}\text{Po} - \text{Po}_1 - \frac{5}{2}$. Relation (19) taking into account the value found is a solution to problem (5) - (8) in the first approximation and can be written as

$$\Theta(\xi, \text{Fo}) = \left(\left(\frac{1}{3}\text{Po} - \text{Po}_1 - \frac{5}{2}\right) e^{-3\text{Fo}} + \text{Po}_1 + \text{Po}\left(\text{Fo} - \frac{1}{3}\right)\right) \xi(1 - 0{,}5\xi) + 1. \tag{21}$$

To increase the accuracy of the solution obtained, it is necessary to increase the degree of the approximating polynomial (13). When determining unknown coefficients $b_i(\text{Fo})$ in relation (13), in addition to conditions (7), (8), (9), we use additional boundary conditions [8, 9].

The physical meaning of additional boundary conditions is the solving the initial differential equation (5) and the relations obtained after its differentiation at points $\xi = 0$ and $\xi = 1$. Note that solving the equation only at the boundary points leads to its solving within the area as well. [15]

To obtain a solution to problem (5) - (8) in the second approximation, we use six terms of the series (13) ($n = 6$). In this case, to determine unknown coefficients $b_i(\text{Fo})$, in addition to conditions (7), (8), (9), three additional conditions are used (two of them are for the point $\xi = 0$, and one for the point $\xi = 1$).

The first additional boundary condition is obtained by writing equation (5) at the point $\xi = 0$

$$\frac{\partial^2 \Theta(\xi, \text{Fo})}{\partial \xi^2} + \text{PoFo} + \text{Po}_1 = 0. \tag{22}$$

Differentiate the initial differential equation with respect to the spatial variable ξ to obtain the second additional condition

$$\frac{\partial^2 \Theta(\xi, \text{Fo})}{\partial \xi \partial \text{Fo}} = \frac{\partial^3 \Theta(\xi, \text{Fo})}{\partial \xi^3}. \tag{23}$$

Using relation (23) at a point $\xi = 0$ taking into account (9), the second additional condition is obtained

IOP Conf. Series: Materials Science and Engineering **791** (2020) 012009 doi:10.1088/1757-899X/791/1/012009

$$\frac{\partial q(\mathrm{Fo})}{\partial \mathrm{Fo}} = \frac{\partial^3 \Theta(0,\mathrm{Fo})}{\partial \xi^3}. \tag{24}$$

The third boundary condition can be obtained by a single differentiation of the initial equation (5) with respect to the spatial variable as applied to the point $\xi = 1$.Taking into account the boundary condition (8), it can be written as

$$\frac{\partial^3 \Theta(1,\mathrm{Fo})}{\partial \xi^3} = 0. \tag{25}$$

Substituting (13) into (7), (8), (9) and additional conditions (22), (24), (25), we obtain a system of six algebraic equations. Solving the system, we determine the unknown coefficients $b_i(\mathrm{Fo})$ ($n = 6$):

$$b_1(\mathrm{Fo}) = 1; \quad b_2(\mathrm{Fo}) = \varphi(\mathrm{Fo}); \quad b_3(\mathrm{Fo}) = -\frac{1}{2}(\mathrm{Po}_1 + \mathrm{FoPo});$$

$$b_4(\mathrm{Fo}) = \frac{1}{6}\frac{d\varphi(\mathrm{Fo})}{d\mathrm{Fo}}; \quad b_5(\mathrm{Fo}) = -\frac{5}{24}\frac{d\varphi(\mathrm{Fo})}{d\mathrm{Fo}} - \frac{1}{2}\varphi(\mathrm{Fo}) + \frac{1}{2}(\mathrm{Po}_1 + \mathrm{FoPo});$$

$$b_6(\mathrm{Fo}) = \frac{1}{15}\frac{d\varphi(\mathrm{Fo})}{d\mathrm{Fo}} + \frac{1}{5}(\varphi(\mathrm{Fo}) - \mathrm{Po}_1 - \mathrm{PoFo}).$$

After substituting relation (13) into the heat balance integral (16), taking into account the coefficients found, we obtain a second order homogeneous differential equation

$$\frac{d^2\varphi(\mathrm{Fo})}{d\mathrm{Fo}^2} + 39\frac{d\varphi(\mathrm{Fo})}{d\mathrm{Fo}} + 90\varphi(\mathrm{Fo}) - 9(10\mathrm{Po}_1 + 10\mathrm{PoFo} + \mathrm{Po}) = 0. \tag{26}$$

The solution has the following form

$$\varphi(\mathrm{Fo}) = C_1 \exp(K_1\mathrm{Fo}) + C_2 \exp(K_2\mathrm{Fo}) + \mathrm{Po}_1 + \mathrm{Po}\left(\mathrm{Fo} - \frac{1}{3}\right), \tag{27}$$

where $K_1 = \frac{3}{2}(-13 + \sqrt{129}) = 2.4633$; $K_2 = -\frac{3}{2}(13 + \sqrt{129}) = 36.5370$.

Substituting (27) into relation (15) we have

$$\Theta(\xi, \mathrm{Fo}) = f_1(\xi)\, C_1 \exp(K_1\mathrm{Fo}) + f_2(\xi)\, C_2 \exp(K_2\mathrm{Fo}) + 1, \tag{28}$$

where

$$f_1(\xi) = \left(\frac{K_1}{15} + \frac{1}{5}\right)\xi^5 - \left(\frac{5K_1}{24} + \frac{1}{2}\right)\xi^4 + \frac{K_1}{6}\xi^3 + \xi;$$

$$f_2(\xi) = \left(\frac{K_2}{15} + \frac{1}{5}\right)\xi^5 - \left(\frac{5K_2}{24} + \frac{1}{2}\right)\xi^4 + \frac{K_2}{6}\xi^3 + \xi.$$

Then we calculate the residual of the initial condition and require the orthogonality of the residual to each coordinate function $f_1(\xi)$ and $f_2(\xi)$, and obtain a system of two algebraic equations

$$\begin{cases} \int_0^1 \Theta(\xi,0) f_1(\xi)\, d\xi = 0, \\ \int_0^1 \Theta(\xi,0) f_2(\xi)\, d\xi = 0. \end{cases}$$

Solving the system, we can find $C_1 = 0.005\mathrm{Po} - 0{,}1\mathrm{Po}_1 - 2.0$; $C_2 = 0.3\mathrm{Po} - 0.8\mathrm{Po}_1 - 2.0$.

Relation (28) taking into account the integration constants found is the solution to problem (5) - (8) in the second approximation. Results of temperature calculations using formula (28) are shown in Figures 2, 3.

To further increase the accuracy, it is necessary to increase the number of terms of the series (13). So, in the third approximation we use nine terms of the series, in the fourth approximation - twelve, and so on. To find the unknown coefficients, we use additional boundary conditions. The general formulas are as follows

$$\frac{\partial^{k+1}\Theta(0,\mathrm{Fo})}{\partial\xi^{k+1}} + j\mathrm{Po} = 0 \quad (j = 0,\, k > 3;\; j = 1,\, k \le 3).$$

$$\frac{\partial^{k-1}\varphi(\mathrm{Fo})}{\partial \mathrm{Fo}^{k-1}} = \frac{\partial^{2k-1}\Theta(0,\mathrm{Fo})}{\partial\xi^{2k11}};$$

$$\frac{\partial^{2k-1}\Theta(1,\mathrm{Fo})}{\partial\xi^{2k-1}} = 0,$$

where $k = 1, 2, 3 \ldots$ – approximation number.

The results of temperature calculations by formula (21) compared to the numerical solution are presented in Fig. 2, 3. The analysis shows that in the range $0.1 \le \mathrm{Fo} < \infty$ the discrepancy of the results obtained does not exceed 10%.

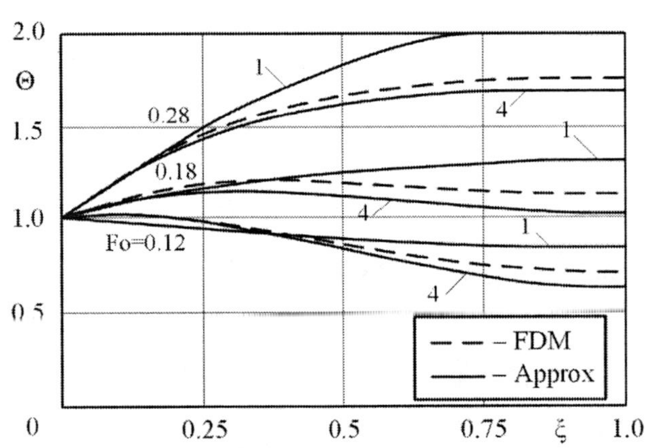

Figure 2. Distribution of dimensionless temperature in the plate ($\mathrm{Po}_1 = 5$; $\mathrm{Po} = 5$). 1 and 4 – approximation numbers.

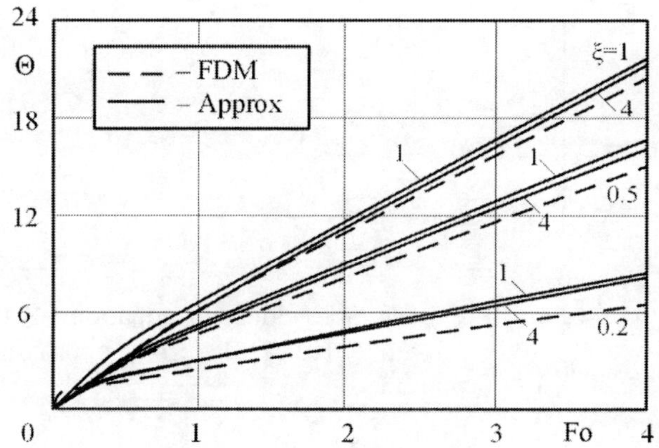

Figure 3. The distribution of dimensionless temperature in the plate over time ($Po_1 = 5$; $Po = 10$). 1 and 4 – approximation numbers.

Figure 4 presents graphs of the function $\varphi(Fo)$. Note that the accuracy of determining the unknown function $\varphi(Fo)$ is much higher than the accuracy of determining the temperature. The discrepancy between the calculation results by formulas (11), (21) in the range of dimensionless time $0{,}1 \le Fo < \infty$ is no more than 5%.

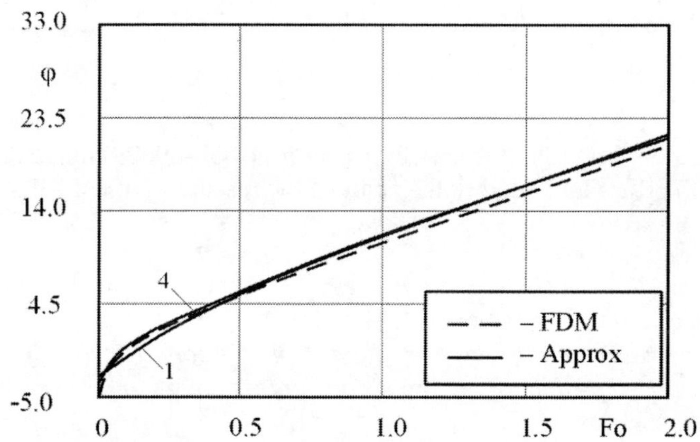

Figure 4. Graphs of measuring the function $\varphi(Fo)$ ($Po_1 = 5$; $Po = 10$). 1 and 4 – approximation numbers.

Figures 2, 3, 4 present the temperature distribution curves $\Theta(\xi, Fo)$ and $\varphi(Fo)$ obtained by formula (13) in the third and fourth approximations. The discrepancy in the interval of dimensionless time $0.1 \le Fo < \infty$ does not exceed 3%.

The analysis (Figure 4) also shows that the value of the function $\varphi(Fo)$ at time point Fo=0 increases to 10 in the first approximation, to 100 in the fourth approximation which is fully consistent with the hypothesis of the infinite rate of heat distribution underlying the derivation of the classical parabolic heat equation (5).

The convergence of the method in its accuracy can also be checked on the analysis of the residual $\varepsilon(\xi, Fo)$ of equation (6)

$$\varepsilon(\xi, Fo) = \frac{\partial\Theta(\xi, Fo)}{\partial Fo} - \frac{\partial^2\Theta(\xi, Fo)}{\partial\xi^2} - PoFo - Po_1 \tag{29}$$

The analysis of the curves presented in Fig. 5 shows that with an increase in the number of approximations, the residual of the equation decreases, which indicates the convergence of the method.

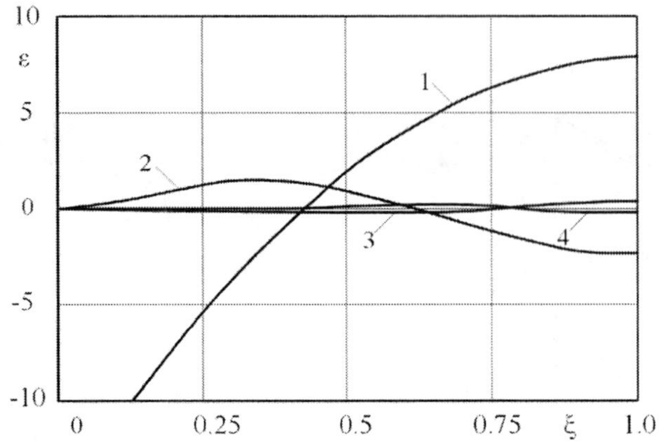

Figure 5. Residual of equation (6) ($Po_1 = 15$; $Po = 10$). 1, 2, 3, 4 – approximation numbers.

3. Conclusion

1. The development results of an approximate analytical method for solving the differential heat equation based on the introduction of a new function and additional boundary characteristics are considered. It is shown that the accuracy of the proposed method depends on the degree of the approximating polynomial.

2. The introduction of a new function (flux density $\varphi(Fo)$) made it possible to reduce the solution of the partial differential equation to the integration of an ordinary differential equation.

3. The solution has a simple form of the product of functions which greatly simplifies the practical use of the solution obtained.

Acknowledgments
The reported study was funded by RFBR according to the research project № 18-38-00029 mol_a and the Council on grants of the President of the Russian Federation as part of the research project MK-2614.2019.8.

References

[1] Sneddon I N 1995 *Fourier transforms* (New York: Dover Publications)
[2] Tsoi P V 2005 *Sistemnye metody rascheta kraevykh zadach teplomassoperenosa* [*System methods for calculating boundary problems of heat and mass transfer*] (Moscow: Izdatel'stvo MEI) [In Russian]
[3] Lykov A V 1967 *Teoriya teploprovodnosti* [*Heat conduction theory*] (Moscow: Vysshaya shkola) [In Russian]
[4] Kartashov E M 2001 *Analiticheskie metody v teorii teploprovodnosti tverdykh tel* [*Analytical methods in the heat-conduction theory of solids*] (Moscow: Vysshaya shkola) [in Russian]
[5] Tsoi P V 1971 *Metody rascheta otdel'nykh zadach teplomassoperenosa* [*Methods of Calculating Individual Problems on Heat and Mass Transfer*] (Moscow: Energiya) [in Russian]
[6] Belyaev N M, Ryadno A A 1978 *Metody` nestacionarnoj teploprovodnosti* [*Methods of Nonstationary Heat Conduction*] (Moscow: Vy`sshaya shkola) [in Russian]
[7] Layeni O, Johnson J 2012 Hybrids of the heat balance integral method *Applied Mathematics and Computation* **218(14)** 7431-44
[8] Kudinov I.V., Kudinov V.A., Kotova E.V., Eremin A.V. On One Method of Solving Nonstationary Boundary-Value Problems *J. of Engineering Physics and Thermophysics* **90(6)** 1317-27
[9] Eremin A V 2019 Study of Thermal Exchange with Liquid Flowing in a Cylindrical Channel *Int. Science and Technology Conf. "EastConf"* (Vladivostok, Russia) pp 1-5
[10] Eremin A V, Kudinov V A and Stefanyuk E V 2018 Heat Exchange in a Cylindrical Channel with Stabilized Laminar Fluid Flow *Fluid Dynamics* **53(1)** 29-39

[11] Kudinov V A, Eremin A V and E V Stefanyuk 2016 Critical conditions for thermal explosion in a plate with a nonlinear heat source *J. of Machinery Manufacture and Reliability* **45(1)** 38-43
[12] Dutta S, Sil A N and Saha J K 2018 Ritz variational method for the high-lying nonautoionizing doubly excited 1, 3fe states of twoelectron atoms *Int. J. of Quantum Chemistry* **118(14)** e25577
[13] Lotfi A and Yousefi S A 2017 A generalization of ritz-variational method for solving a class of fractional optimization problems *J. of Optimization Theory and Applications* **174(1)** 238-55
[14] Kantorovich L V 1934 Ob odnom metode priblizhennogo resheniya differencial`ny`x uravnenij v chastny`x proizvodny`x [On a method for the approximate solution of partial differential equations] *Doklady` Akademii Nauk SSSR* [*Reports of the USSR Academy of Sciences*] **2(9)** pp 532-6 [in Russian]
[15] Rao T, Chakraverty S 2017 Modeling radon diffusion equation in soil pore matrix by using uncertainty based orthogonal polynomials in galerkin's method *Coupled Systems Mechanics* **6(4)** 487-99
[16] Nourgaliev R, Luo H and Weston B 2016 Fully-implicit orthogonal reconstructed discontinuous galerkin method for fluid dynamics with phase change *J. of Computational Physics* **305** 964-96
[17] Belytschko T, Lu Y Y and Gu L 1994 Element-free galerkin methods *Int. J. for Numerical Methods in Eng.* **37(2)** 229-56
[18] Arnold D N, Brezzi F and Cockburn B 2002 Unified analysis of discontinuous galerkin methods for elliptic problems *SIAM J. on Numerical Analysis* **39(5)** 1749-79

IOP Conf. Series: Materials Science and Engineering **791** (2020) 012010 doi:10.1088/1757-899X/791/1/012010

Development of a new scheme for the use of heat in the process of dry quenching coke in order to save energy

I S Gordeeva and E G Necheporenko

[1]Department of Electric Power Engineering and Automation, Nosov Magnitogorsk State Technical University, Lenin St., 38, Magnitogorsk, 455000, Russia

E-mail: 5otl@mail.ru, neshporenkoeg@mail.ru

Abstract. The process of efficient use of the heat of red-hot coke in order to obtain the annealed concentrate of siderite ore in the coke dry quenching plant, with the allocation of the zone of preliminary technological heat treatment (ZPTO) of siderite ore, is proposed and investigated. Thermodynamic analysis was carried out, in which the shares of heat of incandescent coke consumed for the implementation of the processes of heating and decomposition of $FeCO_3MgCO_3$ with the processes of the bell-Boudoir reaction at 490°C and 650°C, respectively, were determined. The thermodynamic analysis resulted in the construction of a temperature-thermal graph of carbon monoxide cooling in the preheating zone, the heat of which is used to preheat the material to the decomposition temperature of 490°C and a graph of coke cooling in the main zone of material processing (ZMTO) - the zone of the forkamera of the coke dry quenching plant. A comparative table of the mass flow rates of coke and sideriteplezite ore is presented.

1. Introduction

The heat of hot coke is a valuable high-potential secondary energy resource.

The use and utilization of hot coke heat is an actual task of coke chemical enterprises from the point of view of both energy saving and environmental protection [1-7].

Efficient use of secondary energy resources allows to replace purchased fuel energy resources, which significantly reduces the energy intensity and cost of production.

For example, the current method of dry quenching in modern coke dry quenching plants, in which coke is cooled by gases, allows to utilize 80 % of the heat carried away by coke from the furnaces, or 40 % of the total heat spent on coking, while, from one ton of extinguished coke, it is possible to obtain 400-500 kg of steam of high energy parameters [8, 9]. Coke, extinguished in this way, is characterized by high strength and less cracking, compared with the wet method of extinguishing it.

Studies have previously determined that it is expedient to allocate a high-temperature area in the dry coke quenching plant, in which it is possible to carry out thermal treatment of various solid process materials [10, 11].

The aim of the work is to find and apply the most effective process of using the heat of hot coke, with the use of sideritoplesite ore in the high-temperature area of the coke dry quenching plant (CDQP), with the possibility of obtaining a larger amount of fired concentrate of siderite ore, which is valuable in the conditions of modern industrial production.

IOP Conf. Series: Materials Science and Engineering **791** (2020) 012010 doi:10.1088/1757-899X/791/1/012010

2. Materials and Methods

For this purpose, in the thermal scheme it was expedient to allocate a new preliminary area, with the delimitation of the zones of the main and preliminary heat treatment of the process material.

As a source of heat in the preliminary zone, carbon monoxide released from the main high-temperature region was considered at its initial temperature of 1200°C.

The analysis was performed for thermodynamically ideal conditions and processes. Calculations have established that for the full heat treatment of 1 kg of sideritoplesite ore on the heat of hot coke only in the high-temperature region, when it is cooled from 1200°C to 740°C, about 2325 kJ/kg of ore is consumed, while the composition of sideritoplesite ore with a fraction of $FeCO_3$ and $MgCO_3$ of 73.36% and 22.8%, respectively, was taken into account [5].

Heat consumption of sideritoplesite ore in high temperature area of coke dry quenching plant

$$m_k \cdot \int_{740}^{1200} c_k(t)\,\mathrm{dt} = \sum_{i=1}^{n} Q_i \tag{1}$$

where i=1,2...n -the number of elementary areas of heat consumption during the heat treatment of siderite ore in the decomposition processes of $FeCO_3$ and $MgCO_3$, ore heating from 490°C to 650°C, bell-Boudoir reactions at 490°C [13-15] and 650°C [13-15] and ore overheating from 650°C to 740°C.

In the study of thermal treatment of siderite ore, it was revealed that in the decomposition of iron carbonate and magnesium carbonate at 490°C and 650°C, about 0.153 m^3 of carbon dioxide is released, followed by their reduction to CO, when interacting with coke carbon by the bell-Boudoir reaction equation, in the volume of 0.307 m^3/kg of ore at a temperature of 1200°C, the temperature potential of which is used for preheating siderite ore.

For a more complete visualization of heat consumption in the process of heat treatment of the material in the zones of preliminary and primary technological operations, a temperature-thermal diagram of the cooling of hot coke and heating of 1 kg of ore was constructed, presented in (Figure 1). The figure shows the individual areas in which the heat of coke is spent on the following reactions: 1-the heat of hot coke, spent on the implementation of the reaction of heating carbon monoxide in the counterflow from 490°C to 1200°C when cooling coke from 1200°C to 740°C; 2-the heat spent on the reaction of Bella - Boudoir, at a temperature of 490C; 3 - the heat spent on the reaction of decomposition of iron carbonate ($FeCO_3$) at 490°C; 4 - the heat, spent on reheating the ore from 490°C to 650°C; 5-heat spent on reheating carbon monoxide in the counterflow from 650°C to 1200°C; 6-the heat expended on the Bella-Boudoir reaction at 650°C; 7-the heat expended on the decomposition reaction of magnesium carbonate ($MgCO_3$) at 650°C; 8-the heat expended on reheating the ore from 650°C to 740°C.

The preliminary heating region of siderite ore - zone I, carried out on the physical heat of carbon monoxide when it is cooled from 1200 °C to 20 °C, where 0 is the cooling of carbon monoxide, was also investigated. The study was carried out for 1 kg of siderite ore.

From the analysis it also follows that when cooling coke from 1200°C to 740°C and heat treatment of 1 kg of siderite ore, the bulk of the heat of hot coke in the high-temperature region (forkamera zone) is spent on the Bell-Boudoir reaction processes at 490°C and 650°C, consuming 30% and 20% respectively of the total heat of hot coke, which is determined by the ore composition; about 12% and 14% of the heat of hot coke is spent on the decomposition reaction of magnesium carbonate, iron carbonate in the process of heat treatment and the minimum amount of heat is spent on the site of overheating of the ore from 650°C to 740°C-3% of the total heat of coke and about 6% of heat-on the process of overheating the ore from 490°C to 650°C. No less costly are the processes of heating carbon monoxide from 490°C and 650°C, which is 10% and 5% of the total heat consumption, when cooling coke. The data on the percentages of the consumed heat of hot coke during the heat treatment of 1 kg of sideritoplesite ore are presented in detail by the diagram on (Figure 2).

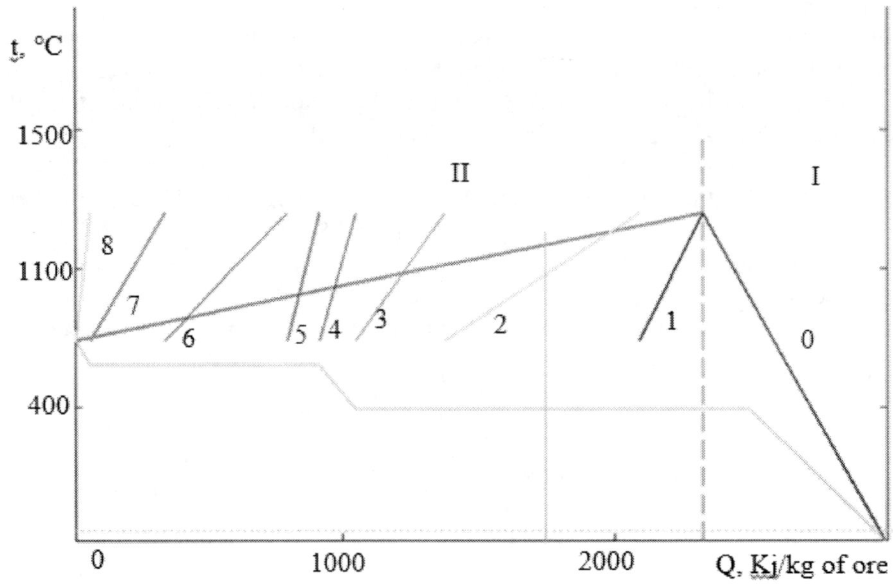

Figure 1. Temperature-thermal graph of coke cooling and heating of 1 kg of siderite ore in (ZPTO) and (ZOTO) process material, where I-preliminary region of siderite ore heating on the physical heat of carbon monoxide, II-high-temperature region of heating and decomposition of process material in the zone of coke dry quenching unit forkamera.

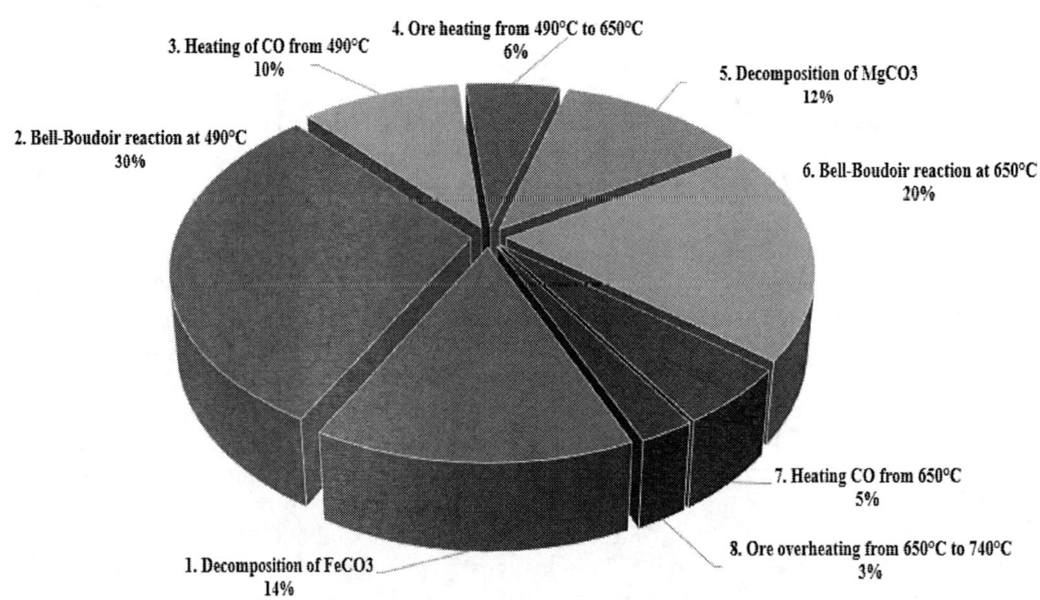

Figure 2. The structure of the consumed heat of hot coke in the processes of heat treatment of siderite ore in the main technological zone (ZMTO).

The calculations also determined that when cooling 1 kg of hot coke from 1200°C to 740°C in the high-temperature region and cooling carbon monoxide from 1200°C to 20°C in the preliminary heating region, it is possible to carry out a total heat treatment of the process material in the amount of 0.296 kg, taking into account the specific heat consumption per 1 kg of the incoming flow of hot coke. At the same time, the total heat consumption of sideritoplesite ore will be 884.2 kJ/kg of coke. The display of these processes, with the cost of heat for the main processing zones are presented in figure 3.

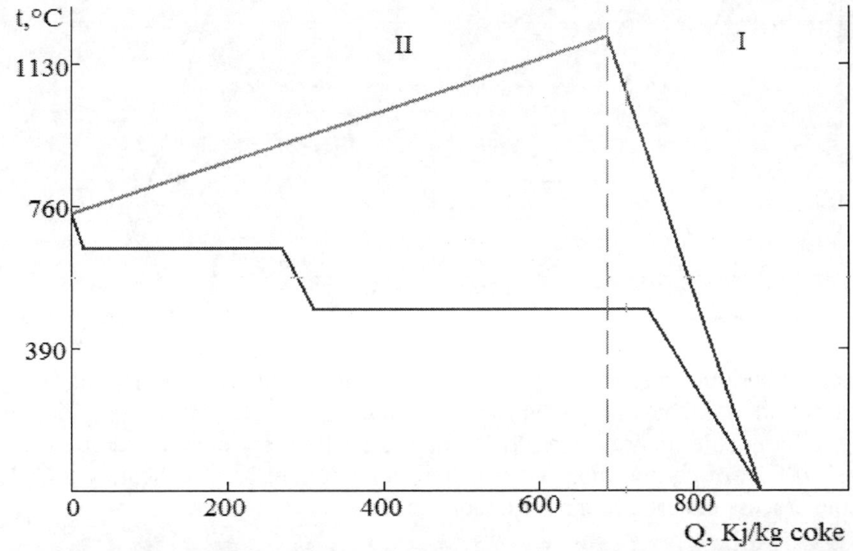

Figure 3. Temperature-thermal graph of cooling of 1 kg of coke and heating of 0.296 kg of siderite ore in (ZPTO) and (ZOTO), where I - preliminary area of heating of siderite ore on physical heat of carbon monoxide, II-high-temperature area of heating and decomposition of process material in the zone of forkamera of coke dry quenching plant.

The heat consumption of the process material, taking into account its specific consumption per 1 kg of annealed concentrate of sideritoplesite ore, will be 852.7 kJ / kg of concentrate, shown in figure 4.

The construction of the proposed thermal scheme of heat use, given to 1 kg of siderite ore and presented in figure 5, was a reflection of the above processes carried out during the heat treatment of ore with the cost of material resources.

This scheme involves the loading of cold siderite ore at a temperature of 20°C into the zone of preliminary technological operation in the amount of 1 kg. Preheating of the process material is carried out on the heat of carbon monoxide in the amount of 0.384 kg/kg of ore, at its initial temperature of 1200°C. Partially fired and heated to 490°C ore enters from the preliminary area into the zone of the main technological operation, where the heat of coke, when it is cooled from 1200°C to 740°C, complete thermal treatment of the ore from 490°C to 740°C.

At the exit from the scheme, according to the results of the process of dry quenching of coke in the zones of the main and preliminary technological operations, a mixture of coke weighing 3,301 kg/ kg of ore and a mixture of annealed concentrate of siderite ore in the amount of 0.602 kg/kg of ore, the mass of the total mixture in this case is 3,903 kg.

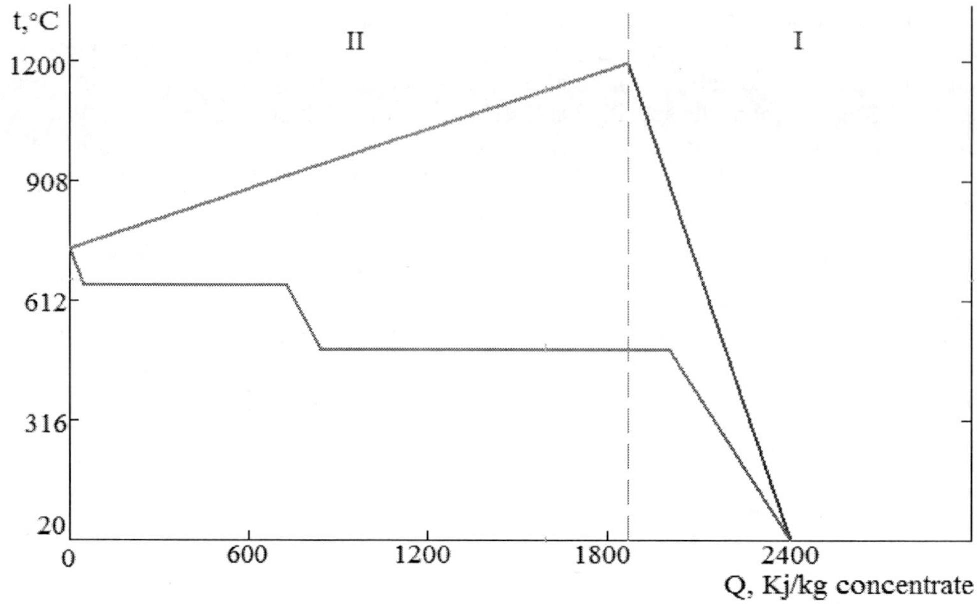

Figure 4 Temperature-thermal graph of coke cooling and siderite ore heating in the zones of preliminary and main technological processing, taking into account the specific ore flow rate per 1 kg of burnt siderite ore concentrate, where I-preliminary area of siderite ore heating on the physical heat of carbon monoxide, II-high-temperature area of heating and decomposition of the process material in the forkamera zone.

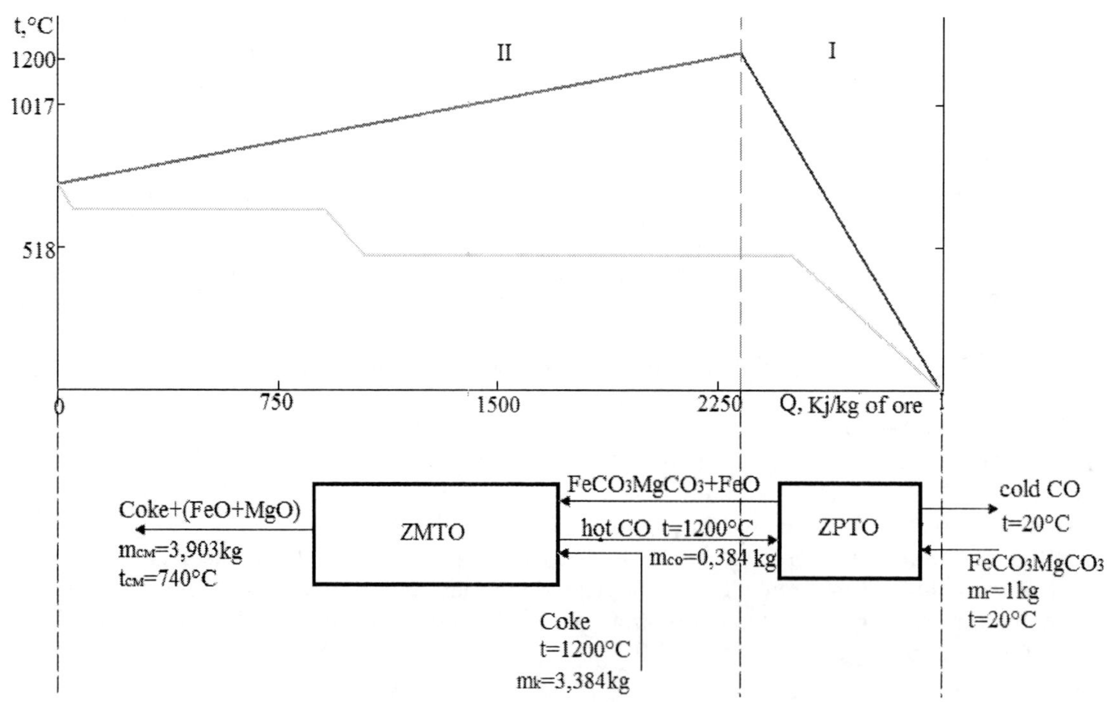

Figure 5. Thermal scheme of coke dry quenching process.

IOP Conf. Series: Materials Science and Engineering **791** (2020) 012010 doi:10.1088/1757-899X/791/1/012010

The results of thermodynamic analysis during the heat treatment of sideritoplesite ore in the zones of preliminary and main technological operations were summarized and presented in table 1.

Table 1. Heat treatment of 1 kg of siderite ore in (ZPTO) and (ZMTO).

Sections of the thermal processing of ore	Required heat, Kj /kg of ore	Coke consumption / kg of ore	The mass of ore, kg	The decrease in the mass of ore, kg	The mass of CO, kg
Area of preliminary technological operation: area I					
0. Heating of ore with CO gas	613.819	-	1	-	0.384
Result area I	613.819	-	1	-	0.384
Zone of the main technological operation: area II					
1. Decomposition at 490° C	331.286	0.524	1	-	-
2. Bell-Boudoir reaction at 490 °C	707.677	0.828	0.817	0.182	-
3. Heating CO from 490 °C	236.204	0.373	0.817	-	0.232
4. Ore heating from 490 °C to 650 °C	137.327	0.217	0.817	-	-
5. Decomposition at 650 °C	273.691	0.433	0.817	-	-
6.Bell-Boudoir reaction at 650 °C	459.746	0.727	0.698	0.119	-
7. CO heating from 650 °C	121.917	0.193	0.698	-	0.152
8. Ore overheating from 650 °C to 740 °C	56.919	0.09	0.698	-	-
Result area I	2325	3.384	0.698	0.301	0.384
Result	2938.586	3.384	1	0.301	0.384

3. Results

According to the results of the calculations. it follows that for the full heat treatment of 1 kg of siderite ore in the high-temperature region (ZMTO). it is possible to cool 3.384 kg of hot coke from 1200°C to 740°C. while for the heat treatment of the process material. the heat of hot coke 2325 kJ/kg of ore is required.

In the pre-processing zone. when the ore is heated from 20°C to 490°C and partially fired at 490°C. 614 kJ/kg of ore will be spent. while the total heat consumption of sideritoplesite ore. taking into account the preliminary and main technological zones. will be about 2938.6 kJ/kg of ore.

4. Conclusion

Thus. it follows from the analysis that the proposed method of cooling coke with the allocation of the preliminary area of heat treatment of the process material. which involves the implementation of preheating sideritoplesite ore on the physical heat of carbon monoxide with subsequent heat treatment of the ore in the high-temperature zone on the heat of hot coke will produce 0.178 kg of the finished fired concentrate of siderite ore. spending 0.296 kg of siderite ore per kilogram of hot coke. In the process of heat treatment of the process material will be achieved coke burnout in 2.43%.

The proposed process of using the heat of incandescent coke allows to obtain a valuable material resource (annealed concentrate of siderite ore) and cooled carbon monoxide. with its further possible use in the scheme with a preliminary area of heat treatment of the process material. The resulting resource-cooled carbon monoxide opens up new possibilities for use in schemes of its complete after-burning. which thus allows to solve the problem of complete utilization of carbon monoxide.

References

[1] Errera M and Milanez L 2000 Thermodynamic analysis of a coke dry quenching unit *Energy Conversion & Management* **41** 109-27.

[2] Liu H, Zhang X, Xu L and Wang M 2002 Mathematical model for fluid flow and heat transfer in the cooling shaft of coke dry quenching unit *J. of Thermal Science* **11** 65

[3] Yang Wei-hua, Xu Tao, Chen Guang, Guo Yue-jia, Jia Li-yue and Li Wei 2009 Coke dry quenching and utilization of its waste heat for electricity in iron and steel industry *Int. Conf. on Power Electronics and Intelligent Transportation System* (Shenzhen, China) pp 227-30

[4] Sun L, Wu J, Xu J and Li Q 2011 Thermodynamic Analysis of a New Coke Dry Quenching Technology *Advanced Materials Research* **225-226** 84-7

[5] Bisio G and Rubatto G 2000 Energy saving and some environment improvements in coke-oven plants *Energy* **25** 247-65

[6] Zaitsev S 2001 Vy`bor racional`noj sxemy` suxogo kontaktnogo tusheniya koksa [Choice of rational scheme of dry contact coke extinguishing] *Chernaya Metallurgiya* 7 23-6 [In Russian]

[7] Strakhov V 2006 O nekotory`x problemax lokalizacii i obezvrezhivanii vy`brosov v koksovom proizvodstve [On some problems of localization and neutralization of emissions in coke production] *Koks i Ximiya* **3** 47-9 [In Russian]

[8] Falkov M 2009 E`nergosberezhenie i e`nergoe`ffektivnost` v proektax GIPROKOKSa na predpriyatiyax chernoj metallurgii Ukrainy` [Energy Saving and energy efficiency in Giprokoks projects at the enterprises of ferrous metallurgy of Ukraine] *Koks i Ximiya* **7** 69-72 [In Russian]

[9] Danilov O and Munts V 2008 *Ispol`zovanie vtorichny`x e`nergeticheskix resursov* [Use of secondary energy resources] (Yekaterinburg: USTU-UPI) [In Russian]

[10] Kartavtsev C, Gordeeva I, Neshporenko E and Demin Yu 2017 Issledovanie e`nergetiki teplotexnologii suxogo tusheniya koksa [Study of energy heat technology of dry coke quenching] *Promy`shlennaya E`nergetika* **5** 38-43 [In Russian]

[11] Neshporenko E, Kartavtsev S and Gordeeva I 2018 Povy`shenie e`nergeticheskoj e`ffektivnosti teplovoj texnologii processa suxogo tusheniya koks [Improving the energy efficiency of thermal technology of dry coke extinguishing process] *Energetik* **2** 44-8 [In Russian]

[12] Krupenin M 2017 Temperaturny`e ogranicheniya metasomatoza Bakal`skix sideritovy`x mestorozhdenij po geoximicheskim danny`m [Temperature limitations metasomatism Bakal siderite deposits in geochemical data] *Vestnik Permskogo Universiteta* **16(2)** 167-77 [In Russian]

[13] Glushko V 1962 *Termodinamicheskie svojstva individual`ny`x veshhestv* [Thermodynamic properties of individual substances] (Moscow: Nauka) [In Russian]

[14] Nikolsky B P 1971 *Spravochnik ximika* [Handbook of the chemist] vol 2 (Moscow: Ximiya)

[15] Melamud C, Shatsillo B, Yuriev B and Zagainov C 2013 Texnologiya vosstanovitel`nogo obzhiga sy`roj i obozhzhennoj sideritovoj rudy` dlya polucheniya vyustitnogo produkta [Technology of reducing roasting of raw and burnt siderite ore to obtain a vustite product] *Stal`* **2** 8-11 (Moscow) [In Russian]

IOP Conf. Series: Materials Science and Engineering **791** (2020) 012011 doi:10.1088/1757-899X/791/1/012011

Determination of the thermotechnical measures of the condensing water heating boiler`s high-temperature section

A V Gubarev

Department of Energy Engineering of Heat Technologie, Belgorod State Technological University named after V.G. Shukhov, Kostyukov St., 46, Belgorod, 308012, Russia

E-mail: artwo0248@mail.ru

Abstract. This paper presents the experimental data processing results, obtained during tests of the fuel-saving condensing water heating boiler, that has been designed by the employees of the Belgorod State Technological University named after V.G. Shukhov. The mentioned boiler can be used in heat supply for residential, industrial and office buildings. The paper presents the justification of the stand-alone systems actuality use for heat supply of consumers with different needs, and also the use of such systems as a heat-generating unit in condensing boiler type as a heat source. It demonstrates the functional diagram and description of the test bench for the condensing water heating boiler specimen with heat production of 250kW, and also gives an operation description of this test bench. The paper presents the heat transfer agent values of flow rates and temperatures, obtained during the boiler`s tests under different loads. It also describes the test result processing methods, presents formulas for calculations. It presents dependences of the energy carrier parameters, and also heat transfer coefficients in the boiler`s high-temperature section elements on its load. The results analysis of the experimental data processing has been conducted. The conclusions about prospective use of the suggested condensing water heating boiler as the heat-generating unit in heat supply stand-alone systems for consumers with different needs have been formulated.

1. Introduction

Rational control of use and conservation of different energy resources, that residents of cities and other communities consume, is the most vital factor, which is aimed at achieving stable and effective fuel and energy resources consumption [1-2]. One of the most energy-intensive system in the city is a heat supply system for consumers with different needs.

At present, in most European countries for the heat supply of the consumers with different needs (residential, social and industrial buildings and facilities) a stand-alone systems of heat supply are being used [3-4]. In recent decades in the Russian Federation use of such systems has increased [5]. Among advantages of the heat supply stand-alone systems can be pointed out the following factors [6-8]:

- no need for long-distance heat supply system network, and as a result losses elimination of heat energy and heat transfer agent inside the network;
- simpler control of heat delivery by the heat source, and as a result heat loss decrease in the heat supply system;
- less expensive maintenance and equipment repair in comparison with the central heating systems.

A tendency of increased use of condensing water heating boilers in the new stand-alone heat supply systems as heat source is being observed in recent years. In this type of boiler products of combustion get cooled down to the temperature when water vapor condensates, which is a part of the organic fuel combustion products. Some countries are trying to replace boilers in stand-alone heat supply systems completely by the boilers and water heaters of condensing type [9-11]. This type of boilers conserves up to 15% of fuel.

The employees of the Belgorod State Technological University have developed the design of the condensing water heating boiler for its use in the stand-alone heat supply systems for consumers [12]. This boiler is a double-circuit boiler. In the first circuit – high-temperature section – fuel burns and it results in a heat transfer agent production, which is then being delivered for the consumer`s heating needs. In the second circuit – contact-recuperative section – the gaseous fuel combustion products experience deep cooling and most of the water vapors, that are part of it, condense, also heat transfer agent is being prepared for consumer`s hot water supply. The high-temperature and contact-recuperative sections are connected by the adiabatic section, where isoenthalpic cooling takes place and combustion products get moisturized at the contact with pulverized into the gas flow water. The functional diagram of the condensing water heating boiler is presented in Figure 1.

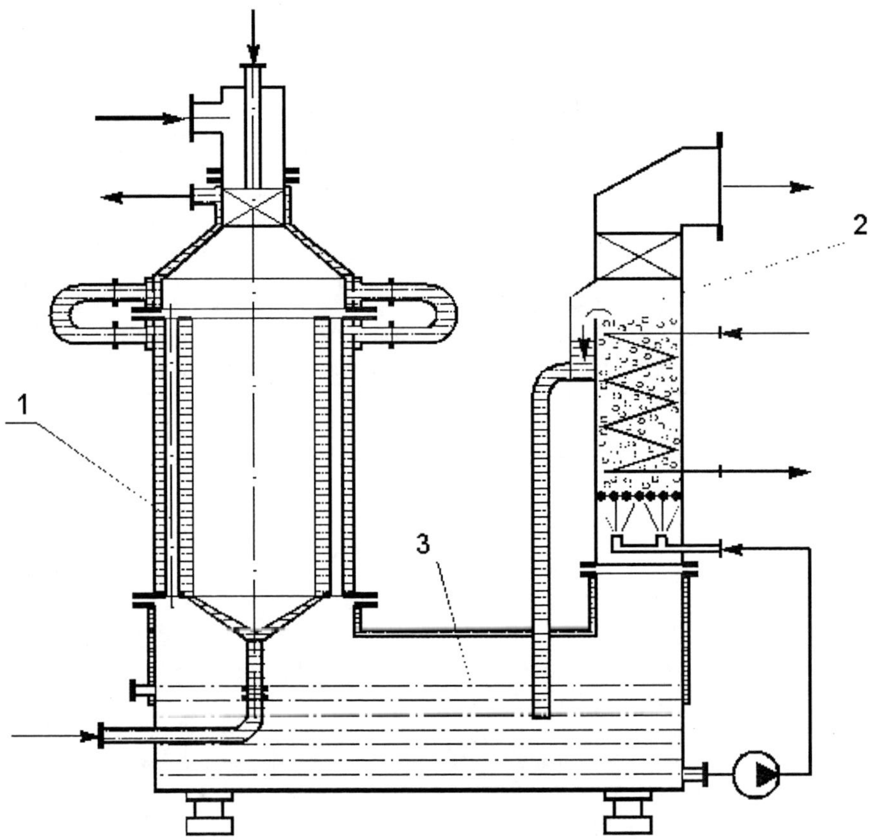

Figure 1. Functional diagram of the condensing water heating boiler:1 – high-temperature section; 2 – contact-recuperative section; 3 – adiabatic section.

To determine thermotechnical and hydraulic characteristics of the condensing water heating boiler tests were conducted, using a mentioned-above boiler with heat production of 250kW. The objective of this paper is the presentation of the results of the considered condensing water heating boiler`s high-temperature section experimental research.

IOP Conf. Series: Materials Science and Engineering **791** (2020) 012011 doi:10.1088/1757-899X/791/1/012011

2. Materials and Methods

To conduct experimental research of the condensing water heating boiler operation and determine thermotechnical measures during its operation a test bench was constructed [13], a description of which can be seen in Figure 2.

Two shell-and-tube heat exchangers, besides the boiler, were introduced into the thermal equipment of the test bench. One of them imitates a consumer of the heating system. Into the shell side of this heat exchanger enters the water, which was heated up in the boiler`s high-temperature section. It gets cooled here by the flowing inside tubes cold tap water to the temperature of the return water in the heating system. Water that was cooled in the shell side of the heat exchanger returns to the boiler.

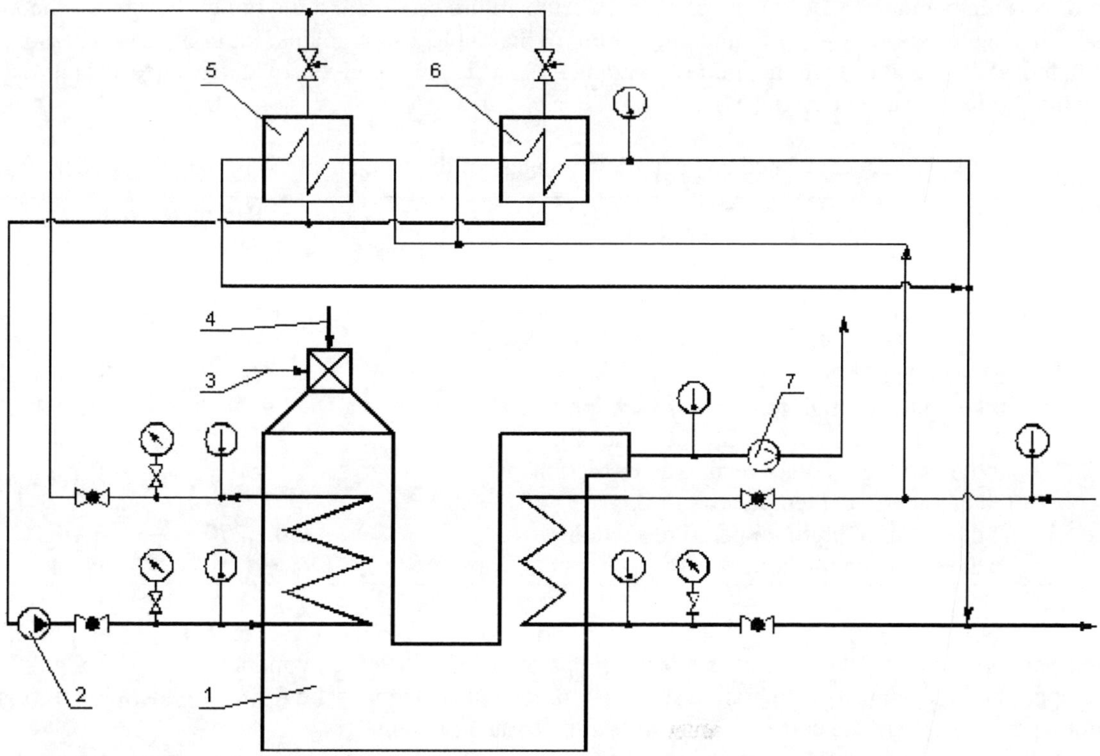

Figure 2. Test bench diagram of the condensing water heating boiler: 1 – condensing water heating boiler; 2 – pump of the heating system; 3 – air; 4 – natural gas; 5 – heat exchanger, which imitates a consumer of the heating system; 6 – water heater for the hot water supply needs; 7 – flue gas fan.

The purpose of the second heat exchanger is water heating for drinking and household needs, which flows in the shell side, to the temperature of water for hot water supply needs. Water is heated with the aid of cooling by water flowing in the shell side, which was prepared in the boiler`s high-temperature section. Afterward, heated in the mentioned heat exchanger water mixes with water that was heated in the contact-recuperative section of the condensing water heating boiler.

Therefore, the second heat exchanger controls the relation of the heat, which is directed by the system to cover the heating load and hot water supply load. This kind of control became an important part in systems with comparable values of heating and hot water supply loads, because at nominal boiler`s load heat production of the boiler`s high-temperature section is near 75%, and heat production of the contact-recuperative section is near 25% from the boiler`s total capacity.

The test bench works in the following order. The high-temperature section of condensing water heating boiler, which basically is a fire-tube and gas-tube water heater, and also shell side of the heat

IOP Conf. Series: Materials Science and Engineering **791** (2020) 012011 doi:10.1088/1757-899X/791/1/012011

exchangers get filled with salt-free water, and into the tubes of the contact-recuperative section`s tube bundle gets cold water of drinking and household quality, which flows in it with constant flow rate. To cool down water in heat exchangers, which was heated in the boiler`s high-temperature section, into its tube side cold tap water is being directed. During the test process water heated in the heat exchanger, which imitates consumers of the heating system, was dumped into a sewage system. The cooling water after exiting the heat exchanger, which is used to control the relation of the heating system and hot water supply loads, gets mixed with water, which was heated in the contact-recuperative section, and then dumped into the sewage also.

During the thermotechnical characteristics determination of the condensing water heating boiler, the following parameter measurements were conducted: natural gas flow rate, excess air ratio, water temperature at the inlet and outlet of the boiler`s contact-recuperative section, and also combustion products temperature at the exit from the high-temperature section and the boiler. Ranges of the measured during tests values are presented in Table 1. This Table also contains, for comparison reason, calculated values of the mentioned-above parameters, which were given or determined during experimental boiler specimen design [14].

Table 1. Main calculated and actual parameters of the condensing water heating boiler.

Parameter name	Parameter value	
	calculated	actual
Natural gas flow rate, m^3/hr.	23.5	11.0...23.28
Excess air ratio	1.10	1.09...1.14
Water temperature, °C:		
at the inlet of high-temperature section	60	50...62
at the outlet of high-temperature section	95	70...92
at the inlet of contact-recuperative section	≈ 10	≈ 10
at the outlet of contact-recuperative section	55	40...64
Combustion product temperature, °C:		
at the outlet of high-temperature section	250...370	177...336
at the exit from the boiler	35	20...42

The results, obtained during tests, are very close to the calculated values. It has to be noted, that temperature values of heat transfer agents, preparation of which was conducted in the boiler`s high-temperature and contact-recuperative sections, meet requirements of the Russian Federation normative documents, that regulate corresponding heat transfer agent parameters.

The boiler`s high-temperature section thermotechnical characteristics determination, using obtained during the experiment values, required calculations of fuel combustion, parameters of gas in the furnace, at the exit from the high-temperature section and at the exit from boiler, boiler`s heat balance preparation, calculation of the heat transfer in a furnace chamber (fire-tube), calculation of the heat transfer in flue-tubes.

During the process of calculation of the fuel combustion using standard methods were determined volumes of dry and wet combustion products and components included in it, that can be found in 1 m^3 of burnt natural gas. This temperature was calculated by the following formula, °C

$$t_a = \frac{Q_i^d + i_f + I_a}{V_g^n c_{p.c}} \tag{1}$$

where Q_i^d – lower heating value of natural gas combustion on a dry basis; i_f – heat content of fuel; I_a – heat content of air delivered for combustion; V_g^n – volume of combustion products per unit volume of burnt gas in normal conditions; $c_{p.c}$ – mean volumetric heat capacity of combustion products.

The heat capacity of the combustion products depends on its temperature values. Therefore, the calculation of the adiabatic combustion temperature is done by the method of successive approximations. The calculation results can be found in Table 2.

Table 2. Dependence of the adiabatic combustion temperature on the excess air ratio.

Parameter name						
Excess air ratio	1.09	1.10	1.11	1.12	1.13	1.14
Adiabatic combustion temperature, °C	1907	1895	1882	1869	1857	1845

Further calculations required determination of the combustion product parameters in normal conditions and operating conditions at boiler`s measuring points: in the furnace at an adiabatic temperature (parameters are given the index G1), at the exit from boiler`s high-temperature section (G3) and at the exit from the boiler (G5). The density of the combustion products in normal conditions with known volumes of dry and wet gases and components, included in it, was determined as the density of the gases mixture using the additivity principle. The combustion products density in the boiler`s measuring points in operation conditions, assuming that gas pressure is changing marginally and close to the atmospheric, were calculated by the formula, kg/m^3

$$\rho_G = \overset{\circ}{\rho}_G \frac{273}{273 + t_G} \tag{2}$$

where $\overset{\circ}{\rho}_G$ – combustion products density in the boiler`s measuring point in normal conditions; $t_G$ – combustion products temperature in the boiler`s measuring point.

The volumetric flow rates in normal and operation conditions, and also the mass flow rate of combustion products in the boiler`s measuring points were determined. Having in mind that in the process of combustion products cooling in the adiabatic and contact-recuperative sections of the boiler their moisture content is changing, so the mass flow rate of gases at the exit from the boiler can be determined by the formula, kg/s

$$G_{G5} = G_{d.g}\left(1 + x_{G5}\right) \tag{3}$$

where$G_{d.g}$ – mass flow rate of dry gases; x_{G5} – gases moisture content at the exit from the boiler.

Enthalpy of the combustion products in the boiler`s measuring points were determined by the formula, kJ/kg of dry gases

$$I_G = \left(c_{d.g} + c_{w.v} x_G\right) t_G + r_0 x_G \tag{4}$$

where$c_{d.g}$ – mean specific isobaric heat capacity of dry gases; $c_{w.v}$ – mean specific isobaric heat capacity of water vapor; x_G – combustion products moisture content; r_0 – heat of water vapor condensation in normal conditions.

Heat flux with combustion products in the boiler`s measuring points were determined by the formula, kW

$$Q_G = I_G G_{d.g} \tag{5}$$

As a result of the boiler`s heat balance preparation were determined efficiency and flow rates of the heat transfer agent, which is directed for the consumer's heating needs. The efficiency of the condensing water heating boiler was determined using indirect heat balance equation

$$\eta = 1 - \left(q_2 + q_3 + q_4 + q_5\right) \tag{6}$$

IOP Conf. Series: Materials Science and Engineering **791** (2020) 012011 doi:10.1088/1757-899X/791/1/012011

where q_2 – portion of heat loss due to the exiting from the boiler gases; q_3 – portion of heat loss due to incomplete combustion; q_4 – portion of heat loss due to unburned carbon; q_5 – portion of heat loss due to external cooling.

The portions of heat loss q_3, q_4, q_5 were taken in accordance with the normative method of the boiler`s thermal design recommendations. The portion of heat loss due to the exiting from the boiler gases was calculated as a relation of the heat flux with combustion products at the exit from the boiler to the heat flux with combustion products in the furnace at the adiabatic combustion temperature

$$q_2 = Q_{G5}/Q_{G1} \tag{7}$$

The heat transfer agent flow rate, which is directed for the consumers heating needs, was calculated by the formula, kg/s

$$G_{h.s} = \frac{Q_{h.s}}{i'_{dnw} - i'_{rnw}} \tag{8}$$

where $Q_{h.s}$ – amount of heat, which is usefully utilized in the boiler`s high-temperature section; $i'_{dnw}$ $i'_{rnw}$ – enthalpy of water at the entrance and exit of the boiler`s high-temperature section.

Amount of heat, usefully utilized, which was transferred from the combustion products to the being heated water in the boiler`s high-temperature section was determined as a difference between the heat fluxes with combustion products in the furnace at an adiabatic combustion temperature and at the exit from the boiler`s high-temperature section with consideration of the heat losses into the atmosphere through the boiler`s walls, kW

$$Q_{h.s} = (Q_{G1} - Q_{G3})\cdot(1 - q_5) \tag{9}$$

The radiation heat transfer in the furnace calculation using standard method, which can be found in the normative method of boiler`s thermal design, resulted in determination of the temperature and heat content of combustion products at the exit from the boiler`s furnace upon condition of radiation heat transfer only, and also amount of heat, transferred in the furnace chamber by radiation, was calculated.

The combustion chamber in the considered boiler is a reverse-flow type and made as a fire-tube, where combustion products speed is very high and the portion of heat, which is being transferred in the furnace by convection from the fire-tube`s inner surface, is considerable. Therefore, it was worthwhile conducting the amount of heat determination, which is being transferred in the fire-tube from gas to the being heated heat transfer agent by convection and also calculate temperature and heat content of the gasses, that are exiting the combustion chamber considering convective heat exchange in it.

The heat exchange in the combustion chamber calculation resulted in the determination of the total amount of heat, which combustion products release in the fire-tube by radiation and convection, and also total heat transfer coefficient in the furnace, which was determined by the basic equation of heat transfer, W/(m^2·K)

$$K_f = Q_f/(F_f \cdot \Delta t_f) \tag{10}$$

where Q_f – amount of heat released by the gasses in the fire-tube by radiation and convection; F_f – fire-tube`s inner surface area; Δt_f – mean temperature difference in the fire-tube.

To determine the heat transfer coefficient in the flue tubes amount of heat was calculated, which is being transferred by gases moving in the flue tubes; water flowing in the shell side of the boiler`s high-temperature section. This amount of heat was determined as a difference between the amount of heat usefully utilized in the high-temperature section and the amount of heat released by gases in the fire-tube.

$$Q_{s.t} = Q_{h.s} - Q_f \tag{11}$$

The heat transfer coefficient in the flue-tubes was determined by the basic equation of heat transfer, W/(m²·K)

$$K_{s.t} = Q_{s.t}/(F_{s.t} \cdot \Delta t_{s.t}) \quad (12)$$

where$F_{s.t}$ – surface area of heat transfer in flue tubes; $\Delta t_{s.t}$ – mean logarithmic temperature difference in the fire-tube.

3. Results

The temperature measurement results of heated water at the high-temperature section inlet and outlet, combustion products at the outlet of the high-temperature section and the boiler, and also temperature calculation results of the combustion products at the furnace outlet with different natural gas flow rates can be found in Table 3.

Table 3. Heat transfer agent values at different natural gas flow rates.

Parameter name	Parameter value						
Natural gas flow rate, m³/s	11.00	13.00	15.00	15.90	19.63	22.02	23.28
Water temperature, °C:							
at high-temperature section inlet	50	53	58	58	57	61	62
at high-temperature section outlet	70	74	88	87	77	93	92
Combustion products temperature, °C, at the exit from:							
boiler	20	22	40	40	35	42	40
high-temperature section	177	227	252	279	302	332	336
furnace	758	804	841	859	913	942	959

Also, Figure 3 presents the results of the combustion product temperature determination.

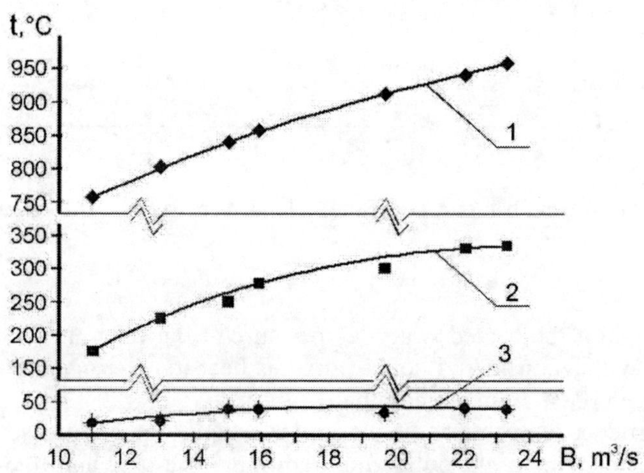

Figure 3. Temperature values of the combustion products at the boiler`s measuring points:1 – at the furnace outlet; 2 – at the high-temperature section outlet; 3 – at the boiler outlet.

Figure 4 presents the calculation results which allow to evaluate values change of the amount of heat, that gets transferred per unit time by combustion products to the heated water in the boiler`s high-temperature section elements, and also in the high-temperature section taken as a whole, at different natural gas flow rates.

IOP Conf. Series: Materials Science and Engineering **791** (2020) 012011 doi:10.1088/1757-899X/791/1/012011

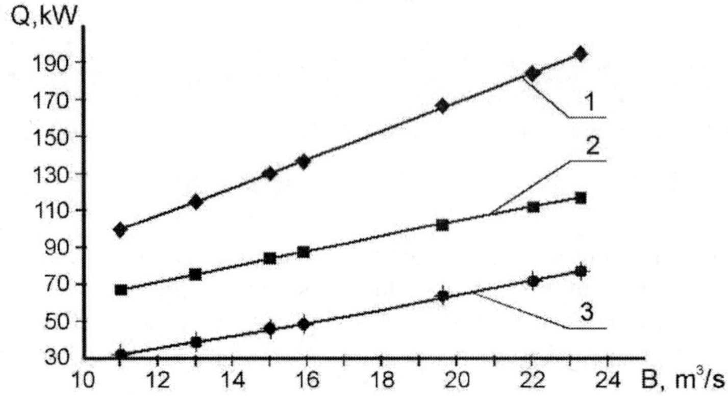

Figure 4. Amount of heat transferred from the combustion products to the heated water: 1 – total in boiler`s high-temperature section; 2 – in fire-tubes; 3 – in flue-tubes.

Figure 5 presents the heat transfer coefficient values in fire-tube and flue-tube at different flow rate values of natural gas delivered into the boiler`s furnace, which were determined by calculations using the mentioned-above method.

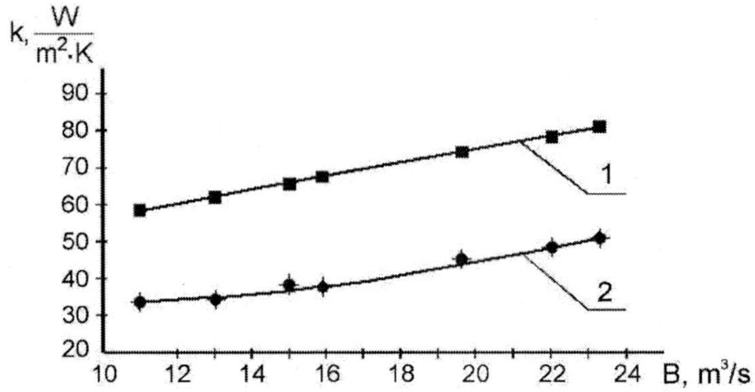

Figure 5. Heat transfer coefficient:1 – in fire-tube; 2 – in flue-tube.

4. Discussion

The measurement results of the heated water temperature at the inlet and outlet of condensing high-temperature boiler show that considered boiler provides heat transfer agent parameters that consumer needs when its load is 50-100% from the nominal.

The combustion product temperature at the boiler`s high-temperature section outlet far exceeds combustion product temperature at the outlet of a traditional fire-tube and flue-tube type water heating boilers, which don`t have a condensation of water vapors, that are part of the combustion products [15, 16]. This specifies an increased value of the mean temperature difference of the heat transfer agents on the heating surfaces. Therefore, on the condition of the same heat production, the heat transfer surface of the considered condensing boiler type is substantially smaller than heating surfaces in the traditional water heating boilers. This allows substantial decrease, in comparison with the analogs, of metal consumption in the considered boiler.

At the same time, combustion products temperature at the outlet of condensing water heating boiler, due to deep cooling of the gases in its contact-recuperative section, and, therefore, heat losses with flue gas are not high. However, there is a drawback, which is defined by the presented here factors, and that is a necessity to use expensive anti-corrosive materials and coating for the boiler`s contact-recuperative section and also flue gas ducts and smokestack production.

The result analysis of the amount of heat calculations, which is being transferred by the combustion products to the heated water in the boiler`s high-temperature section, allow to make a conclusion that the increase of boiler`s load from 50% of nominal to nominal makes the total balance of the boiler`s high-temperature section to decrease by a portion of heat, which gases release in the fire-tube, from 67,3% to 60,3%, and corresponding increase by a portion of heat which combustion products release in the flue-tubes.

This can be explained by several factors. Boiler`s load increase creates an increase of the heat stress in furnace volume and combustion products temperature at the boiler outlet and at the flue-tube inlet, which has been confirmed by the data presented in Table 3. Due to higher temperature at the flue-tube inlets increases mean temperature difference of the heat transfer agents on this heating surface and that intensifies convective heat transfer. Also, the increase in the boiler`s heat production creates an increase in the flow rate and combustion products flow speed also increase, which intensifies convective heat transfer.

The heat transfer coefficient values in the elements of the high-temperature section of condensing water heating boiler match values of this coefficient in the corresponding heating surfaces of the fire-tube and flue-tube type water heating boilers. At the same time, an increase in the boiler`s load creates a more intensive increase in the heat transfer coefficient in the flue-tubes in comparison with the heat transfer coefficient in the fire-tube. This can be explained by the fact that in the flue-tubes portion of the convective heat transfer in comparison with the portion of the radiation heat transfer is much higher, while in the furnace chamber radiation heat transfer prevails.

5. Conclusion

The experimental research results processing has shown that the thermotechnical parameters values, that characterize the heat transfer intensity in the elements of condensing water heating boiler`s high-temperature section, which was designed by the employees of the Belgorod State Technological University named after V.G. Shukhov, match the equivalent parameters' values that the traditional fire-tube and flue-tube type water heating boilers have. However, the ability to keep higher temperature values of the combustion products at the suggested boiler`s high-temperature section outlet in comparison with the exhaust gas temperatures in the traditional boilers allows substantially reduces the size of the high-temperature section as opposed to the analogs. Because the heat transfer in the contact-recuperative section of the condensing water heating boiler proceeds much more intensive than heat transfer in the back-end surfaces of the traditional water heating boilers, its size and metal consumption are much smaller than in the analogs.

The suggested boiler, as opposed to the traditional analogs, has a weakness and that is a necessity to use for manufacture more expensive anti-corrosive materials and coating for boiler`s contact-recuperative section elements and also flue gas ducts behind the boiler. However, this kind of materials are being used in any other type of heat generator or heat recovery unit of the condensing type.

The experiment results have shown that the heat transfer agent parameters, which were produced in the condensing water heating boiler, are in compliance with the standard requirements applied to the corresponding heat transfer agents. Taking into account all the above-mentioned allows to conclude that the use of the suggested boiler as a heat generator in the stand-alone systems for the heat supply of consumers with different needs is very promising.

Acknowledgments
The article was prepared within development program of the Flagship Regional University on the basis of Belgorod State Technological University named after V.G. Shukhov.

IOP Conf. Series: Materials Science and Engineering **791** (2020) 012011 doi:10.1088/1757-899X/791/1/012011

References

[1] Trubaev P A, Bulanin A V, Shirrime K J and Koshlich Yu A 2018 Particular qalities of execution of the program component for energy resources management of the Belgorod region *IOP Conf. Ser.: J. Phys.: Conf. Ser.* **1066** 012020

[2] Gasho E G, Guzhov S V and Krolin A A2019 On the forecasting of trends of specific mixed loads of microdistricts and cities change in the conditions of inadequate data *IOP Conf. Ser.: Mater. Sci. Eng.* **552** 012013

[3] Merkel E, Fehrenbach D, McKenna R and Fichtner W 2014 Modelling decentralized heat supply: An application and methodological extension in TIMES *Energy* **73** 592-605

[4] Buoro D, Casisi M, Pinamonti P and Reini M 2012 Optimal synthesis and operation of advanced energy supply systems for standart and domotic home *Energy Conversion and Management* **60** 96-105

[5] Marchenko O and Solomin S 2014 Economic efficiency of renewable sources in autonomous energy systems in Russia *Inter. J. of Renewable Energy Research* **4(3)** 548-54

[6] Lichter J M and KonstantinovV A 1995 Avtonomnoe teplosnabzhenie zdanij [Independent heating of buildings] *Energetik* **4** 9-10 [In Russian]

[7] Kozhevnikov V P, Kuznetsov V A, Mochalin A A, Titarenko R Y and Sokolova L V 2014 Independent heating modules with condensing hot water boilers as heat generators as an effective alternative to centralized heat supply *Adv. Environ. Biol.* **8(13)** 89-93

[8] Bychikhin S A, Svirin M V and Trubaev P A 2019 Evaluation of energy-saving effect during the installing process of automated heating stations in educational institutions *IOP Conf. Ser.: Mater. Sci. Eng.* **552** 012019

[9] Varlamov G, Romanova K, Daschenko O, Ocheretyanko M and Kasyanchuk S 2016 The use of contact heat generators of the new generation for heat production *EasternEuropean J. Of Enterprise Technologies* **1** 52-8

[10] Grinenko G P, Kozhevnikov V P, Kuleshov M I and Pogonin A A 2012 Perspektivy' razvitiya ry'nka kondensacionny'h kotlov v Rossii [Prospects of the market development of condensation boilers in Russia] *Vestnik BGTU im. V.G. Shukhova* [*Bulletin of BSTU named after V G Shukhov*] **3** 145-9 [In Russian]

[11] Che D, Liu Ya and Gao Ch 2004 Evaluation of retrofitting a conventional natural gas fired boiler into a condensing boiler *Energy Conversion and Management* **45** 3251-66

[12] Gubarev A V and Lozovoj N M 2019 Construction and variants of the modernization of the condensing hot water boiler *IOP Conf. Ser.: Mater. Sci. Eng.* **552** 012004

[13] Kozhevnikov V P, Kuleshov M I, Gubarev A V, Trubaev P A, Pogonin A AMochalin A A and Fejgelman M O 2012 Stend I nekotory'e rezultaty' ispy'tanij toplivosberegayuschego kondensacionnogo vodogrejnogo kotla [Stand and some test results of fuel-saving hot water boiler]*Vestnik BGTU im. V.G. Shukhova* [*Bulletin of BSTU named after V G Shukhov*] **3** 182-4 [In Russian]

[14] Kozhevnikov V P, Kuleshov M I, Gubarev A V, Trubaev P A and Andreev A B 2014 Condensing Hot Water Boiler: Applicability, Design, Research *Adv. Environ. Biol.* **8(13)** 122-7

[15] Todorović M N, Živković D S, Mančić G S and Ilić G S 2014 Application of energy and exergy analysis to increase efficiency of a hot water gas fired boiler *Chem. Ind. Chem. Eng. Q.* **20(4)** 511-21

[16] Pan Zh, Niu J, Lu Ch, Zhang Zh and Zhu J2010 Study on thermal efficiency of gas-fired vacuum hot water boiler with installing flue gas condensing waste-heat exchanger *4th Int. Conf. on Bioinformatics and Biomedical Engineering* (Chengdu: IEEE) 11495956

Preparation of the forecasted fuel and energy balance of a boiler in an inadequate data condition

S V Guzhov[1], E G Gasho[2] and V A Shepel[1]

[1]Department of Heat and Mass Transfer Processes and Installations, National Research University MPEI, Krasnokazarmennaya St., 14, Moscow, 111250, Russia
[2]Department of Industrial Heat Energy Systems, National Research University MPEI, Krasnokazarmennaya St., 14, Moscow, 111250, Russia

E-mail: GuzhovSV@yandex.ru

Abstract. The task of fuel and energy balance preparation for heat energy sources in inadequate data conditions is an urgent issue, especially for the outdated boilers. It is difficult to use formulas for physically worn-out systems description, mainly because thermotechnical characteristics of the insulating components have significant deviations from the standard values. Boiler`s equipment physical depreciation increases greenhouse gas emissions into the atmosphere. Therefore, dealing with modernization funding feasibility of the heat energy sources, and also for periodic adjustment works that help to reduce harmful substance emissions into the atmosphere, requires the use of a fairly complete mathematical model of a boiler and a heat network. However, the lack of data makes it very difficult to formulate a reliable model. Status analysis of every element in the system is a labor-intensive and expensive task. Another extremely labor-intensive issue is a determination of the forecasted thermal energy demand of the municipal consumers and sector of housing and public utilities. Physically worn-out buildings have low thermal resistance, which increases heat demand. Calculation of the necessary amount of heat generation by the worn-out boilers for the city districts with worn-out buildings is difficult due to lack of a wide range of data about the boiler's condition, network condition, and heated buildings condition. The paper presents the predictive energy balance preparation method using the statistical method of information processing only.
Keywords: modernization, flue gas, heat supply system, forecasted heat energy demand, mathematical modeling, regression analysis, multicollinearity, heat energy consumption forecasting.

1. Introduction

Implementation of the modernization policy for the outdated boilers requires a preliminary assessment of technical and economic effects from modernization, but this is not enough. The complexity of such calculations is that existing outdated boilers very often work in modes that are far from the designed values. Let`s take a look at how closely the operation of these boilers on the fuel oil influence the current ecological situation and also effect from heat source transition to modern technologies. Additional immediacy of the problem is in the environmental impact reduction. Boiler`s flue gas releases into the atmosphere large amount of hard and gaseous pollutants, such as fuel ash, oxides of carbon, sulfur, and nitrogen. Aside from that, a large amount of carbon dioxide and water vapor get into the atmosphere. The forecast with an engineering accuracy of technical and ecological effects from such heat

IOP Conf. Series: Materials Science and Engineering **791** (2020) 012012 doi:10.1088/1757-899X/791/1/012012

sources is a very difficult problem, especially in conditions with substantially inadequate and partially unreliable initial information.

Reliable calculation of the modernization feasibility will allow increasing heat supply quality, system reliability, heat energy tariff growth rate reduction, improvement of the existing ecological situation.

The objective of this paper is the calculation with given accuracy of the technical-and-economic effect from modernization of the existing obsolete and worn-out heat supply system of a city district.

2. Current ecological situation

Implementation of the modernization policy for the outdated boilers requires a preliminary assessment of technical and economic effects from modernization and not just that. Environmental impact reduction is the top priority. The complexity of such calculations is that the existing outdated boilers very often work in modes that are far from the designed values. Let`s take a look at how closely the operation of these boilers on fuel oil influences the current ecological situation and also effect from heat source transition to modern technologies. Boiler`s flue gas releases into the atmosphere large amount of hard and gaseous pollutants, such as fuel ash, oxides of carbon, sulfur, and nitrogen. Aside from that, a large amount of carbon dioxide and water vapor get into the atmosphere.

Sulfur dioxide or sulfurous anhydrite (sulfurous gas) leads to a tree dwarfism and leaf chlorosis (yellowing or discoloration). Humans suffer from upper air passages exasperation caused by this gas.

Nitrogen oxide (NO) and dioxide (NO_2) take part in photochemical smog formation in the surface layer of the atmosphere. Nitrogen oxide dissolution in the atmosphere vapors leads to the acid formation, which is the main cause of the "acid" rains, that are very harmful to flora and fauna. Humans develop susceptibility to acute and chronic respiratory diseases. In the upper atmosphere, nitrogen oxide triggers a lot of reactions, that make the ozone layer thinner, which can cause skin cancer in humans [1].

Solid waste, such as ash and slag, is partially responsible for environmental damage [2]. In spite of filters use, which can be in non-serviceable conditions on the worn-out equipment, every year several million tons of fine aerosol is being released into the atmosphere from the powerplant emissions, which are condensation nucleus for the water vapor and an additional cause of a precipitation formation. Its elevated concentration creates different respiratory diseases in humans.

Soot particles don`t interact with the atmosphere`s oxygen, but participate in a smog formation and promote skin cancer.

Carbon oxides (CO and CO_2 – carbon dioxide) practically don`t interact with the other substances in the atmosphere and its lifetime is almost unlimited. Its overload in the air decreases the ability of the hemoglobin in red blood cells to carry oxygen, resulting in slow human responses, weak perception, headache appearance, drowsiness, nausea. High concentration can lead to fainting and even coma.

Product of the fuel oil combustion is a fuel oil ash, which consists of vanadium oxide (VO, V_2O_5), and also metal oxides (SiO_2, MgO, Al_2O_3, Fe_2O_3, Ni_2O_3, and other oxides). The most toxic agent - vanadium pentoxide was taken as a controlling factor. Air quality standards define the allowable limits of harmful substances content for the industrial area, and also for the city`s residential area.

This paper`s objective is the problem statement of the technical-and-economic effect evaluation of the outdated heat supply system modernization. Modernization will allow improvement of quality and reliability of the heat supply system, keep existing tariffs on heat energy, and also substantially improve the existing environmental situation. The main prerequisites for modernization:

- heavy wear out of the heat supply system, which is in use since 1970;
- high degree of physical wear out of the existing heat energy equipment (more than 70%);
- designed service life expiration and low operating efficiency of the main equipment;
- connection hourly heat load of the consumers is less than 20% of the designed equipment power, which has a negative impact on the effectiveness and safe operation of the boiler`s equipment;

IOP Conf. Series: Materials Science and Engineering **791** (2020) 012012 doi:10.1088/1757-899X/791/1/012012

- heavy harmful substances emission into the atmosphere due to fuel oil use as a fuel.

3. The existing heat supply system description

Administrative center`s square footage within a municipal district is 835.7 thousand ha (43% of the MD “Tersky District” territory), where lands of the inhabited localities – 1.47 thousand ha, including being studied residential location – 1.41 thousand ha. The number of the residential population as of January 1, 2018, is 4578 residents and has a gradual decline tendency (Table 1).

Table 1. Dynamic of the population change

Year	1970	1979	1989	2002	2009	2013	2016	2017	2018
Population, ppl.	7 268	7699	8 309	6 497	5 442	5 170	4 772	4 658	4578

The considered territory is in the subarctic area, because its climate is being formed by the influence of the White Sea and the continental part of the Kola Peninsula, and characterized as near sea climate. The warmest month – July, with a daily mean temperature of 14.3°C. Absolut maximum – +32°C. Absolut minimum – minus 40°C. The average duration of the frost-free season is 85-90 days. Territory climatic conditions fall into the construction climatic zone II-A. Air temperature of the coldest five-day period, with the reliability of 0.92, is assumed as minus 29°C, and air temperature with the reliability of 0.94 – minus 17°C. Heating season duration, on average, is 263 days. The average annual air temperature is 0.5°C. The coldest month is February, with a daily mean temperature of 11.9°C (Table 2).

Table 2. Monthly average temperature values

Month	Jan.	Feb.	March	April	May	Sept.	Oct.	Nov.	Dec.
Monthly average temperature, °C	-11.5	-11.9	-6.4	-1.5	4.3	7.5	1.4	-4.4	-8.3

Heat supply network length is 13.2 km (considering double lines). For the moment, the residential area has five boilers in use with the total power output of 31.5 MWt.

Authorized temperature charts of the heat sources and heat supply networks operation have the following values – 95/70°C and 130/70°C. However, mathematical modeling has shown that the boiler`s actual temperature charts differ from the authorized temperature chart of the heat supply network 95/70°C (Figure 1) due to several reasons, including the following:

- fuel economy;
- break up possibility of the equipment of the heat supply source and networks, due to substantial wear;
- manual control of the heat energy production process.

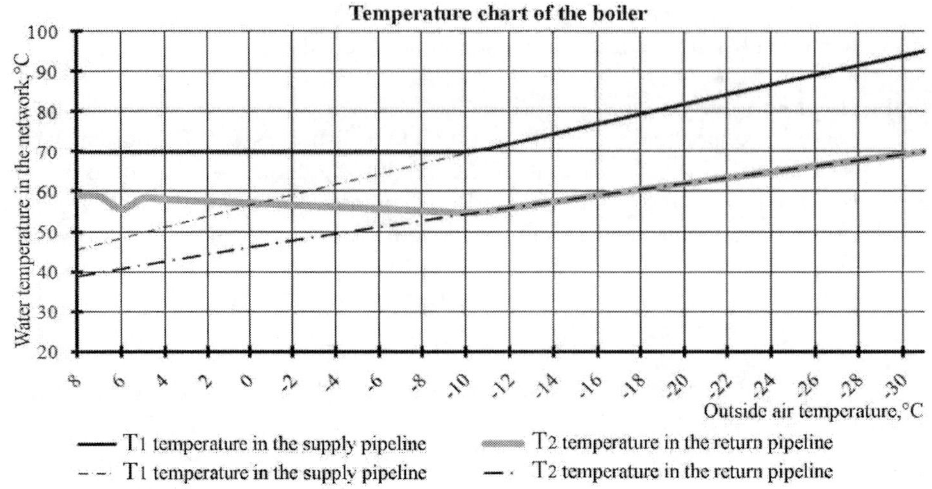

Figure 1. Calculated actual chart of heat energy delivery by the boiler

Heat supply system – closed double pipeline. Boiler`s outlet pipes – D219 mm (delivery), D219 mm (return) provide heating load and DHW 2.011 Gcal/hr. Pipelines are under and above ground. Available head is 24 m, return pipeline pressure – 30 m. Thermal power flow rate for auxiliary needs 0.599 Gcal/hr. Thermal power losses in the networks 1.593 Gcal/hr. During the 2018/19 heating season load received 22.956 thousand Gcal. The fuel oil was used as fuel (Table 3). The estimated amount of emissions [3] into the atmosphere is around 5.8 million tons per year.

Table 3. Schedule of the fuel oil supply to the boiler in the 2018-19 heating season, MT/mo

July	Aug.	Sept.	Oct.	Nov.	Dec.	Jan.	Feb.	Marc.	Apr.	May	June
34	46.5	78	125	146	172	191	170	158	140	129	64

The examination discovered the following drawbacks of the technology:

1. Fuel oil unload requires the use of steam, which accompanied by the emission of the fuel oil light fractions into the atmosphere. Substantial fuel oil quality decrease creates a smell problem, and, therefore, light hydrocarbon emissions during fuel oil unloading. When fuel oil is being heated using direct steam, the water content in the fuel oil rises sharply, which leads to the loss of steam and condensate, but also fuel oil quality degrades. As a result, the boiler`s efficiency and performance reliability decrease. Water in the fuel oil, during heating, induces foaming.

2. The boiler has an inefficient flue gas cleaning system. Several high toxic components are being released into the city`s atmosphere – benzopyrene, fuel oil ash, nitrogen oxides. The gas cleaning system is not an option due to the highly corrosive activity of the exhaust gas.

3. The fuel oil with a sulfur content of more than 2% is being used. The fuel oil contains sulfur in active and passive forms. Active sulfur causes corrosion in pipes, heating units installed in tanks, heat-exchange units and back-end surfaces at the temperature of a steel wall lower than the dew point.

Complexity of the stated problem is in the following aspects:

I. impossibility of using formula-based models for forecast, because the 3D model of the heat supply network was composed at the point of technical-and-economic feasibility study [4];

II. lack of the reliable information about wear out degree of each element in the main and supporting equipment of heat supply source, main pipelines, CHP units and consumers` network [5-7];

III. characteristics of the heated building enclosing structures and degree of wear out are unknown, which doesn`t allow to estimate the demand in different climatic conditions with acceptable engineering precision [8, 9];

IV. impossible to rely fully on the source`s heat delivery alone, due to the increased self-consumption and losses, and also due to the boiler`s manual control errors [10, 11];

V. complexity of the heat supply source analysis, because its operational schedule of heat energy delivery differs from the recommended schedule [12], etc.

The described above factors undoubtedly disallow using formula-based models. Therefore, it is necessary to accept the considered boiler system as a «black box» and examine it using statistical methods of information analysis.

4. Selection of the data analysis method for construction of the heat consumption regression model

Let`s examine several main methods and models, which at present are being used in a statistical approach for loads planning in heat and power systems:

- extrapolation method – statistical method, which uses the trend identification and description for forecasting;
- correlation-and-regression model – statistical model, which conducts the forecast using the mathematical description of the existing dependencies between variables:
- neural network model – nonlinear model, forecast based on it results in the model of heat energy consumption dependence on external factors, and also on retrospective data [13].

Energy consumption forecast of the considered city is done by the correlation-and-regression analysis in connection with the following aspects:

a) method is based on the consideration of the correlation (statistical) relationships between loads on separate parts, caused by the presence of several factors [14];
b) inadequacy of the extrapolation method of boiler`s operation modes due to lack of the annual and monthly similarity of climatic factors [15];
c) forecast accuracy far exceeds the forecast accuracy of the other models;
d) neural network model-based forecast is a complex problem that requires BigData analysis, which is absent in this case.

The chosen correlation and regression analysis are based on the test of the hypothesis of the extent of the correlation ratio between input and output parameters. For the checking purpose most frequently used Pearson`s linear correlation coefficient or Spearman`s and Kendall`s rank correlation coefficients. Pearson`s linear coefficient – a measure of the linear correlation ratio between two quantitative attributes x and y. It is based on assumption that when characteristics of x and y completely independent, value deviations of factorial feature from the average ($\bar{x}-x_i$) have random nature and should randomly match with different deviations ($\bar{y}-y_i$). In the case of the substantial advantage of matching or mismatching of these deviations, it is assumed that there is a relationship between x and y.

Rank correlation coefficients are based not on the correlation of the correlated characteristic values itself but on their ranks – numerical orders, which are assigned separately to each individual value x and y in the ranked series. Calculation formula of Kendall`s rank correlation coefficient (τ) for connected ranks:

$$\tau = \frac{S}{\sqrt{\left(\frac{n\cdot(n-1)}{2}-U_x\right)\cdot\left(\frac{n\cdot(n-1)}{2}-U_y\right)}}. \quad (1)$$

where S – actual total score sum when estimator +1 of each pairs of ranks with the same change order and when estimator −1 of each pairs of ranks with the opposite change order;
$U_x = U_y = 0.5 \cdot \sum[t \cdot (t-1)]$ – number of scores, which correct (decrease) maximum score sum at the expense of repetitions (grouping) of t ranks in each series.

Using the examined coefficients, it is possible to prove the presence of connection, establish its direction and relationship degree, but impossible to determine the significance of the being examined factor among the other factors.

It is worthwhile to evaluate the degree of conformity of the forecast model and the object`s actual behavior by using the determination coefficient. When R^2=1, it is possible to conclude that the func-

IOP Conf. Series: Materials Science and Engineering **791** (2020) 012012 doi:10.1088/1757-899X/791/1/012012

tional relationship between input and output variables exist, when R^2=0 – that it is non-existent. If 0<R^2>1, then the closer R^2 to 1, the better quality of the function approximation with initial data. Value R^2 increases when one more regressor is added, therefore, when choosing between several regression equations one shouldn`t rely on determination coefficient R^2 alone. To eliminate this, the following formula can be used:

$$R^2 = 1 - \frac{ESS/(\mathrm{N-k-1})}{NSS/(N-1)}, \quad (2)$$

where numerator – unbiased estimator of errors dispersion, denominator – unbiased estimator of dispersion Y; TSS (Total Sum of Squares) - all dispersion or variation Y, which characterizes the degree of random value spread of regression function around mean value Y; ESS (Error Sum of Squares) - sum of squares of regression residuals.

The singularity of the suggested method is in the introduction of the «idealized heat consumption» factor into an initial data array, which is determined by using classical formula-based mathematical methods. This step allows both increase accuracy of the statistical calculation and clearly determine deviations in the object`s actual operational mode from the recommended.

5. Heat energy consumption forecast

As a test set for model construction was used heat energy consumption data for every 3 hours during the period from 01.06.2018 to 31.05.2019. The being composed regression model doesn`t consider the summer period when boilers deliver heat in the DHW mode. Goal seek and best model selection were done in the EXEL environment. The process of actual heat consumption forecast includes the following stages:

a) analysis of the initial information and filtering off of the unreliable data such as single unexplainable discharge or drop of heat consumption; dimension errors; lack of metering and monitoring readings, etc.;
b) determination of the actual and «idealized» production of heat energy for the heat supply system under study during the selected period of time;
c) determination of the set of factors, that is, based on the expert's opinion, having an impact on heat consumption values and filtering off some of them using collinearity and multicollinearity criteria;
d) formulation of the forecast mathematical function or group of predictive rules;
e) determination of the values of all input factors at the forecast time period, or range of their values using principle: «optimistic», «neutral», «pessimistic» forecasts;
f) determination of the forecast heat consumption values considering all selected factors and periods.

As a factorial features were taken: outside air temperature (X_1); atmospheric pressure at the station level (X_2); relative humidity at the altitude of 2m, % (X_3); wind direction, rhumb (X_4); wind speed, m/s (X_5); total cloud cover (X_6); sum of all clouds in view (X_7); dew point temperature at the altitude of 2m (X_8); accumulated precipitation quantity, mm (X_9). Total heat consumption is taken as an effective feature.

Formula based calculation of the «idealized» heat energy delivery can be done using formula:

$$Q_{form} = G_{suppl} \cdot C_P \cdot (t_{suppl} - t_{ret}) \quad (3)$$

where C_P=1,001kcal/kg·R, G_{suppl} - heat transfer medium consumption in the supply pipeline kg/hr; t_{suppl} and t_{ret} – heat transfer medium temperature in the supply and return pipeline.

Calculation of the paired correlation coefficients can be done by the Student`s t-criterion:

$$t_{obs} = \frac{r\cdot\sqrt{n-2}}{\sqrt{1-r^2}}. \quad (4)$$

where r – paired correlation coefficient, calculation based on the selected for research data, n - number of observations.

To find t_{criter} using Student`s table the value of significance level was taken as α=5% (confidence probability 95). To test on statistical significance, let`s compare t_{criter}= -1.646 and paired correlation coefficients, calculated by EXEL. Factors X_6, X_9 – statistically are not significant and can be eliminated from further analysis.

Test of the input factors multicollinearity assumption using the Farrar-Glauber method on first type statistical criteria:

$$FG = -\left[n - 1 - \frac{2k+5}{6}\right]\ln(\det[R]), \quad (5)$$

where n=1976 – number of observations; k=9 – number of factors.

At v=k/2(k-1)=1 degrees of freedom and significance level α=0.05, the number of degrees of freedom d=9·(9-1)/2=36. Tabulated criterion value equals to 50.998.

Calculated value:

$$FG = -\left[1976 - 1 - \frac{18+5}{6}\right]\ln(0.00148623) = 6975.$$

Considering that the calculated value is bigger than the tabulated then with the assumed reliability one can say that in the array of factors exists multicollinearity with the probability of 95%.

Let`s identify a multicollinearity of initial factors using statistical criteria of the second type (Fisher`s criterion):

$$F_k = (d_{kk} - 1)\frac{n-k}{k-1}, \quad (6)$$

where d_{kk}-diagonal elements of the matrix n=1976 – number of observations; k=9 – number of factors.

At v_1=n-k and v_2=k-1 degrees of freedom and significance level α: d_1=1976-9=1967; d_2=9-1=8; then F_{table} corresponds to value 2.071.

If $F_{calcul} < F_{table}$ then there is no reason to decline the hypothesis of multicollinearity absence. Based on the conducted analysis a relationship between factors X_1, X_8 was identified and followed by the elimination of X_8 from the calculation.

Test of the variables on multicollinearity on statistical criteria of the third type (Student`s criterion) demonstrated the increase of residual dispersion from 0.70 for Q=F($X_1$) to $R^2$= 0,762 Q=F ($X_1$; $X_4$; $X_5$; $X_2$; $X_3$; $X_7$). To test the validity of the factor's selection, let`s introduce into calculation factors X_6, X_8, X_9 that have been excluded before. Considering that the determination coefficient value will not change R^2=0.762, then predictors selection was adequate.

Carry out of the linear regression coefficients calculation determines the type of the function:

$$Q=3.11-0.0936\cdot X_1-0.0010\cdot X_4+0.0203\cdot X_5-0.0010\cdot X_2+0.0042\cdot X_3+0.0038\cdot X_7.$$

Based on the obtained mathematical dependence and calculated the actual schedule of heat energy delivery the data about heat energy delivery for the next heating season was obtained. It was assumed that the climatic factors for future and present calendar years are similar.

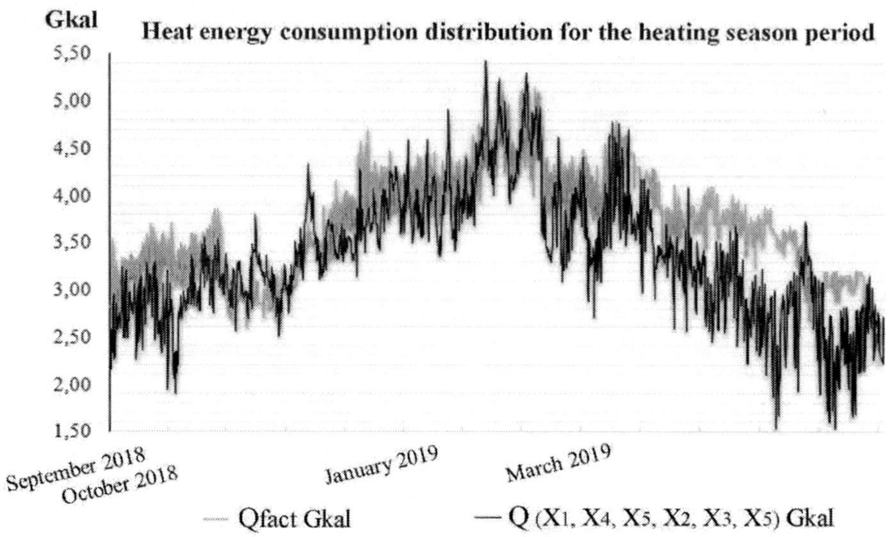

Figure 2. Forecasted value of heat energy consumption by boiler`s load during heating season with sampling period 3 hr.

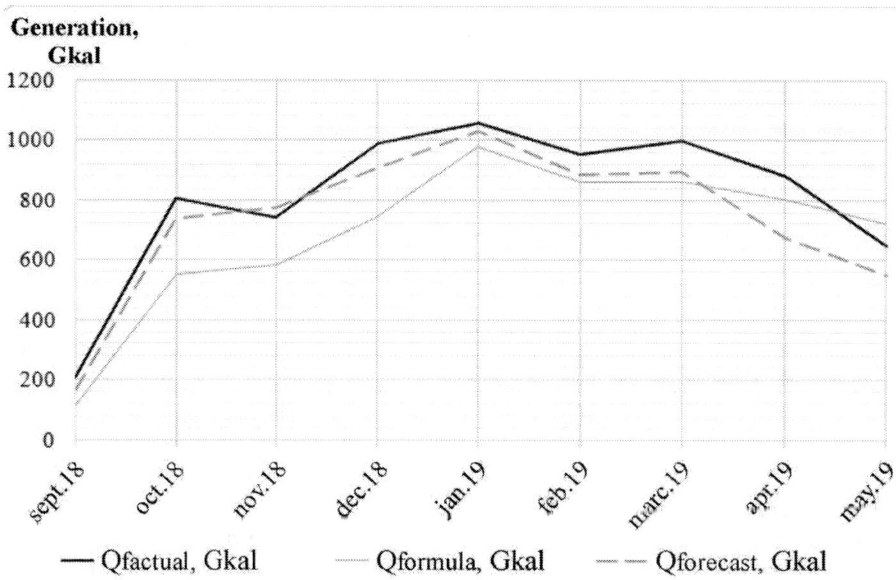

Figure 3. Forecasted value of heat energy consumption by boiler`s load during heating season with monthly sampling period

A decrease in the forecasted delivery of heat energy was observed during the whole heating season. September and November are the exceptions because the heat source intentionally delivers more heat to warm up buildings.

6. Conclusion

The total decrease in the forecast with confidence probability 95% and significance level 5% is 529 Gcal, which is 13.3% and equivalent to the emissions decrease by 0.77 million tons per year. This effect can be achieved by the exclusive use of the heat source operational modes optimization.

The paper presents the construction method of the forecast function of heat consumption by the heat supply system in a substantial lack of information about the network`s structure and condition. Affecting factors set justification has been conducted. The formulation of the forecast mathematical function of heat consumption by the city`s district has been conducted.

It has been shown that formula-based models, which can be used with the existing level of data, deliver big differences from the actual and cannot be used.

The results of the process analysis of different factors influence on the heat energy consumption amount for the object of research have been presented. It has been discovered, during research, that the amount of heat consumption decreased in comparison to the previous period due to several parametric factors, which confirms the necessity of both additional modes tuning up and heat source modernization. The presented mathematical methods allow us to determine the degree of climatic factors influence on the total heat consumption amount. It should be noted, that discovered dependence has strictly individual character, which is unique to the system under analysis only. Test for collinearity of different system with similar characteristics, and as a result, the general form of the function Q=f (Xi), can be different from the obtained in the paper.

The results of the model calculation can also be used for calculation of the expected productive delivery during the next periods, economical estimation of the worn-out heat supply system further modernization, pricing policy formulation, estimation of the budget and borrowed fund investment efficiency, etc.

Acknowledgments
This paper was prepared using the results of the project, which has been conducted with the support of the Russian Science Fund (unique identifier of the grant RSF № 16-19-20568).

References
[1] Gao J, Kovats S, Vardoulakis S, Wilkinson P and Liu Q 2018 Public health co-benefits of greenhousegas emissions reduction: A systematic review *Science of The Total Environment* **627** 388-402
[2] Cho J, Kim J, Lee S and Koo J 2016 A bi-directional systematic design approach to energy optimization for energy-efficient buildings *Energy and Buildings* **120** 135-44
[3] Zheng X, Streimikiene D, Balezentis T, Mardani A and Liao H 2019 A review of greenhousegasemission profiles, dynamics, and climate change mitigation efforts across the key climate change players *J. of Cleaner Production* **234** 1113-33
[4] Huang X and Wang Y 2019 Investigating the effects of 3D urban morphology on the surface urban heat island effect in urban functional zones by using high-resolution remote sensing data: A case study of Wuhan, Central China *ISPRS J. of Photogrammetry and Remote Sensing* **152** 119-31
[5] Gasho E G, Guzhov S V and Krolin A A 2019 On the forecasting of trends of specific mixed loads of electric consumers of microdistricts and cities change in the conditions of inadequate data *IOP Conf. Ser.: Mater. Sci. Eng.* **552** 012013
[6] Larsen S, Hongn M, Castro N and González S 2019 Comparison of four in-situ methods for the determination of wallsthermal resistance in free-running buildings with alternating heat flux in different seasons *Construction and Building Materials* **224** 455-73
[7] Rasooli A and Itard L 2019 In-situ rapid determination of walls' thermal conductivity, volumetric heat capacity, and thermal resistance, using response factors *Applied Energy* **253** 113539
[8] Deconinck A and Roels S 2015 A Maximum Likelihood Estimation of the Thermal Resistance of a Cavity Wall from On-site Measurements *Energy Procedia* **78** 3276-81
[9] Rasooli A, Itard L and Ferreira C 2016. A response factor-based method for the rapid in-situ determination of wall'sthermal resistance in existing buildings *Energy and Buildings* **119** 51-61

[10] Al-Sanea S and Zedan M 2012 Effect of thermal bridges on transmission loads and thermal resistance of building walls under dynamic conditions *Applied Energy* **98** 584-93

[11] Ostrowski P, Szelejewski F and Zymelka P 2019 Assessment of improvement in the energy and exergy efficiency of the gas heat plant after the exhaust gas cooled down below the dew point and the use of recovered heat *Energy* 116179

[12] Brøgger M, Bacher P and Wittchen K 2019 A hybrid modelling method for improving estimates of the average energy-savingpotential of a building stock *Energy and Buildings* **199** 287-96

[13] Huang S, Tao B, Li J and Yin Z 2019 On-line heat flux estimation of a nonlinear heat conduction system with complex geometry using a sequential inverse method and artificial neural network *Int. J. of Heat and Mass Transfer* **143** 118491

[14] Nouvel R, Mastrucci A, Leopold U, Baume O and Eicker U 2015 Combining GIS-based statistical and engineering urban heatconsumption models: Towards a new framework for multi-scale policy support *Energy and Buildings* **107** 204-12

[15] Amasyali K and El-Gohary N 2018 A review of data-driven building energy consumption *Renewable and Sustainable Energy Reviews* **81(1)** 1192-205

Development of an algorithm for analyzing the causes of increased oxygen content of condensate

T V Iglina, A I Popov, N P Krasnova, J I Rakhimova and A S Gorshenin

Heat and Power Department, Samara State Technical University, 244, Molodogvardeyskaya St., Samara, 443100, Russia

E-mail: 89615663193@mail.ru

Abstract. To increase the efficiency of condensers, one of the most important methods is to constantly maintain a high hydraulic density, since malfunctions of the condensing unit lead to a significant reduction in electricity generation. The increased air content in the wet vapor in the condenser and the air suction into the turbine, lead to exceeding the standard values of the condensate oxygen content and therefore have a significant impact on the uninterrupted operation of the TPP, causing the pipeline to corrode from the condensation unit to the deaerating unit. In addition, the increased oxygen content makes heat transfer difficult, since an insulating layer is formed between the condensate film on the heated surface and the steam. Currently, there are a number of methods for thermal calculation of a condensation unit. In this work, the most accurate Iglin-Shepelev technique was used. Using the developed algorithm, the work of the Kirov CHPP-4 and Bezymyanskaya CHPP was analyzed. The experimental data with the calculated characteristics plotted on them are also clearly shown at various temperatures of cooling water with an average flow rate of circulating water, and the reasons for the increased oxygen content of the condensing units of the stations under consideration are identified. Based on the calculation results, measures were proposed to eliminate the increased oxygen content.

1. Introduction

The condensation device of the steam turbine installation (STP) has a significant impact on the efficiency, continuity and reliability of the thermal power plant (TPP). The efficiency of the condenser is affected by: the contaminants presence on the heat exchange surfaces; the presence air suction in the condensation system of the turbine; vacuum depth in the condensation unit (including the efficiency of the ejectors), etc.

Any negative changes in the operation of the condenser lead to a decrease in the depth of vacuum, a decrease in the ejector efficiency, a decrease in the thermal power plant efficiency, and, as a result, an increase in fuel consumption.

According to studies [1, 2], a change in the pressure in the condenser by 1 kPa (1% vacuum) leads to a change in the power of the TPP turbine with an initial superheated steam pressure of 13-24 MPa by about 0.8-0.9% of the nominal power, data for various the turbines are shown in Figure 1. The deepening of the vacuum directly depends on the air extraction devices operation (ejectors).

In addition, the increased air content in wet steam in the condensing unit and the air suction into the turbine, lead to exceeding the standard values for the oxygen content of the condensate (20 μg/ kg) and, as a result, have a significant impact on the uninterrupted operation of the TPP, causing the pipeline to corrode [3, 4] from the condensing unit before deaerating installation [5, 6].

IOP Conf. Series: Materials Science and Engineering **791** (2020) 012013 doi:10.1088/1757-899X/791/1/012013

Air and other non-condensable gases entering the vapor significantly impede heat transfer, since they form an insulating layer between the condensate membrane on the heated surface with steam (Figure 2). The presence of oxygen or air has a negative effect not only in condensers [7].

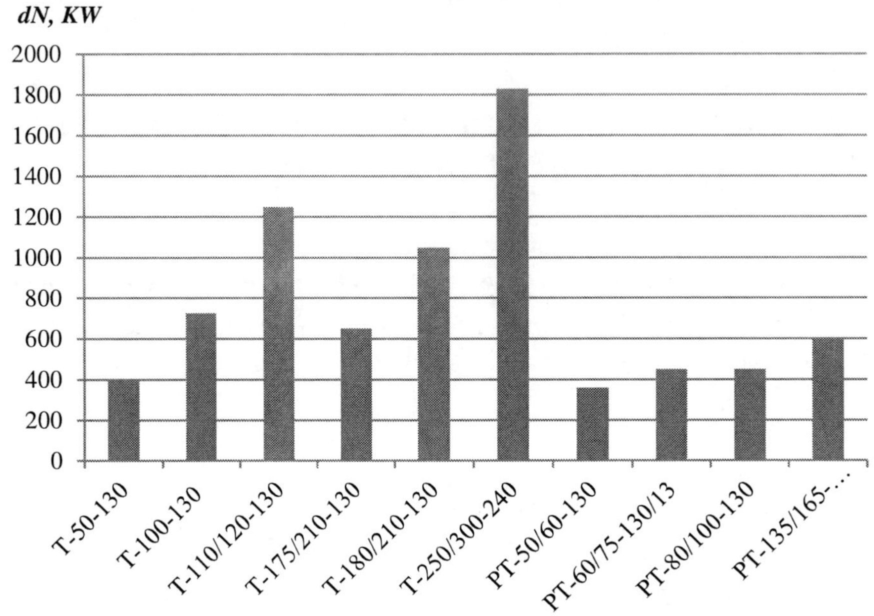

Figure 1. Change in turbine power with a pressure change in heating turbines condensers 1 kPa, operating in condensation mode.

With a decrease in the vapor load of the condenser, especially in the modes with increased air suction, the intensity of the heat and mass transfer process changes, the zone of intense vapor condensation decreases, the average heat transfer coefficient decreases, and the condensate supercooling increases.

In [8], it was shown that brass tubes corrosion in a condenser can also be caused by ammonia intense concentration in the apparatus zones. For example, the tests at one of the units of the Severnaya CHPP (St. Petersburg) showed that the ammonia concentration in the local damage zones of network heaters exceeds the standard value (1000 μg / l) by 50-60 times.

Thus, for the efficient operation of condenser, one of the most important methods is the high hydraulic density constant maintenance, since malfunctions of the condensing unit lead to a significant reduction in electricity generation. However, in order to develop these measures, it is necessary to create an effective condenser diagnostic system [9], which would make it possible to identify the cause of the condenser malfunction (decrease in the quality of its operation). Such a system should have a module for recording and storing all condenser operation parameters: steam flow rate, flow rate and temperature of the main condensate, recirculation condensate flow rate, flow rate and temperature of cooling water. Particular attention must be paid to the automatic fixation of the oxygen content of the main condensate. Most stations measure oxygen twice a shift, this is not enough to quickly eliminate problems and accidents in the event of their occurrence, the consequences of such accidents can lead to corrosion of equipment, and therefore to high costs. In addition, the diagnostic system must have an information analysis module, and this requires mathematical models [10-12] of all the processes occurring in the condenser, which will become the basis for the software.

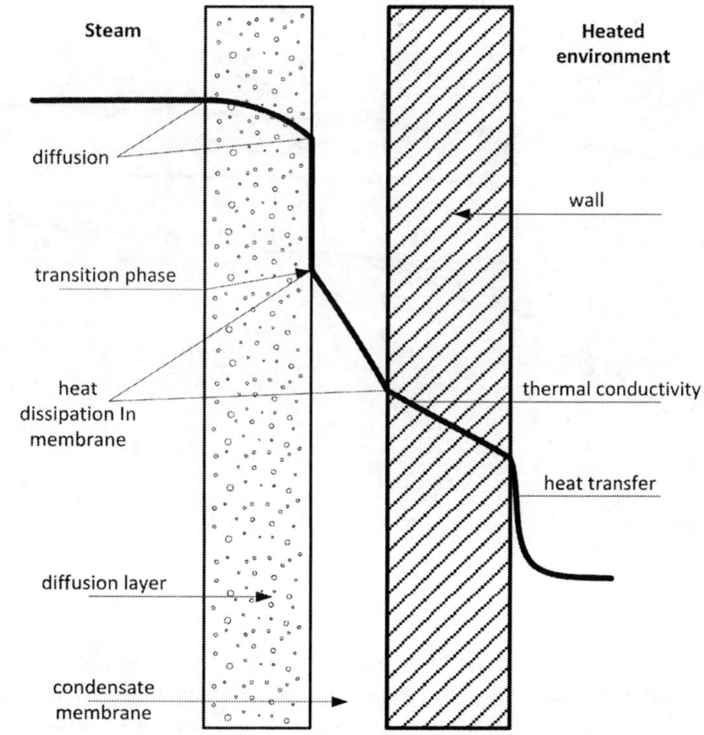

Figure 2. The effect of the air gap and the condensate membrane on the temperature regime.

It is necessary to develop an algorithm for analyzing the condenser reliability, which could be used to assess the reasons for the increase in oxygen content in the main condensate based on the available operational parameters.

2. Materials and Methods

To create an algorithm for the condenser reliability, it is necessary to choose its mathematical model as a basis. The difficulty in creating an accurate condenser calculation system is to find the heat transfer coefficient k. Currently, there are a number of different methods for calculating the heat transfer coefficient of a condensing unit [13, 14].

The US Institute of Heat Transfer [15] does not take air suction into the condenser and the dependence of the heat transfer coefficient on the steam load of the installation. In addition, it is still being finalized.

Also, there is a Russian method for calculating the heat transfer coefficient of a condenser. It belongs to the Kaluga Turbine Plant (KTP technique). In it, in addition to the speed of the coolant, the initial temperature and the contamination of the condenser tubes, air suction is also taken into account. However, during the condensation unit operation, the measurement of the pollution layer thermal conductivity coefficient λ in the condenser tubes is an almost impossible task. The KTP technique also does not take into account the nature of the condenser and ejector joint operation.

P V Iglin, A G Shempelev [16] proposed a different methodology for calculating the heat transfer coefficient in the condenser, which is in good agreement with experimental and regulatory data.

A comparison of the methods described above is shown in Figure 3.

As shown in Figure 3, when the condenser loads are less than half of the nominal, a large discrepancy in the results of the calculated and standard values of the heat transfer coefficient begins, associated with the influence of the ejector on the operation of the condenser and air suction.

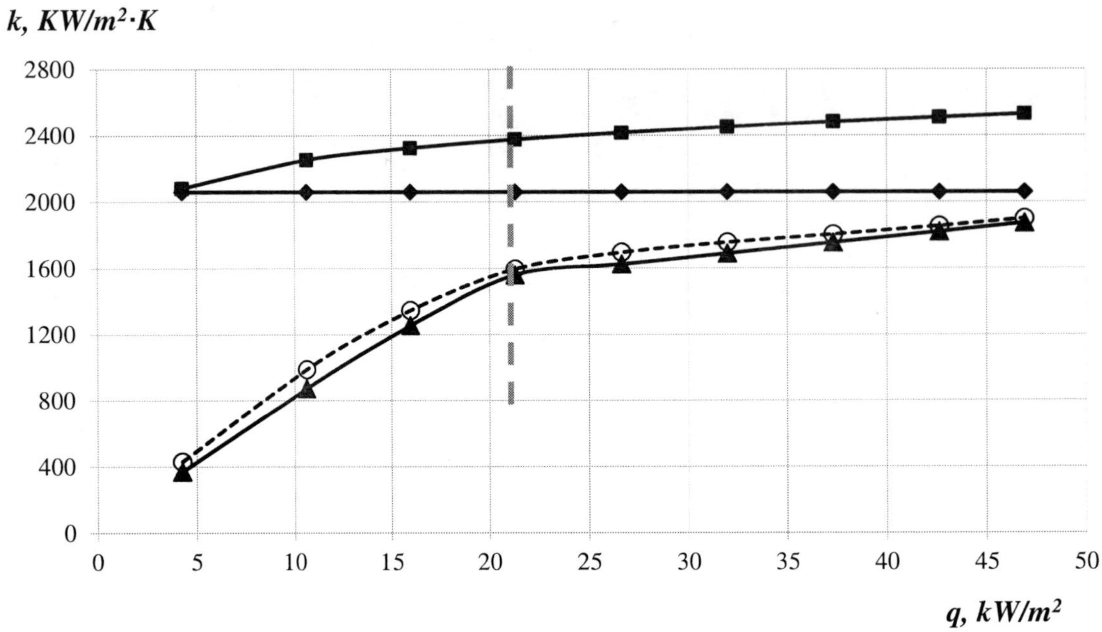

Figure 3. Comparison of the heat transfer coefficient using the example of a 100 MW turbine unit condenser and heat transfer area 3000 m^2: ▲– calculation according to the Iglin-Shempelev technique, ■ – KTP calculation, ♦ – calculation according to the ITO method, ○ – normative characteristic.

In the work, methods of numerical calculations using the mathematical model of the Iglin-Shempelev condenser were used, as well as regression analysis as applied to the experimental data for the Kirov CHPP-4 and Bezymyanskaya CHPP (Samara).

3. Results

The following algorithm was proposed to analyze the increased oxygen content causes in the main condensate after the condenser:

1) The necessary parameters of the condenser operation are fixed: the steam flow into the condenser neck D_c, the pressure in the condenser pc, the temperature of the main condensate at the outlet of their condenser t_c, the flow rate of cooling water W, the temperature of cooling water at the inlet t_{1w} and at the outlet t_{2w}, air suction in the condenser (or turbo) G_{air}, the actual oxygen content in the main C_{O2} real condensate. To do this, it is necessary to equip the condenser with measuring instruments at various points in the installation, which should record and take measurements and transfer them to a computer for storage and processing.

2) The measured parameters are transferred to the mathematical model of the condenser, in which the KTP methods are used to determine the characteristics of the condenser during condensation of pure steam, and the Iglin-Shepelev method is used to determine the characteristics of the condenser during steam-air mixture condensation in it. Under the characteristic of the capacitor is understood the dependence of the saturation temperature (pressure) in the capacitor on its specific heat load. For quick calculations, the algorithm can be written in Python.

3) The oxygen content of the main condensate is determined on the basis of Henry's law with air suction into the condenser corresponding to the measured (actual) $C_{O2\,fact}$ and standard $C_{O2\,norm}$.

4) The calculated values are compared with the actual $C_{O2\,real}$ and the maximum allowable (20 μg/kg):

• $C_{O2\,real} \leq 20$ μg/kg - the oxygen content in the main condensate is normal, no control actions are required;

IOP Conf. Series: Materials Science and Engineering **791** (2020) 012013 doi:10.1088/1757-899X/791/1/012013

• C_{O2_real} > 20 μg/kg, to determine the cause of the increase, it is necessary to compare the value of the actual G_{air} air suction with the standard value determined by the formula $G_{air.norm} = 8 + 0.065 \cdot N$ (where N is the nominal turbine power, MW). If $G_{air} > G_{air.norm}$, then the excess of oxygen content is caused by large suction of air into the turbine. In this case, it is necessary to increase the hydraulic density of vocational schools. If $G_{air} \leq G_{air.norm}$, then the increased oxygen content is associated with the supply of flows to the condenser, in addition to the steam stream from the turbine with high oxygen content, for example, recirculation condensates, condensates of regenerative heaters, ejector coolers, etc. In this case, the following measures can be taken to reduce the oxygen content: installing a deaerating device in the condensate collector (if it is absent), excluding the supply of drains under the level of condensate in the condensate collector.

For clarity, in Figure 5 shows a block diagram of the algorithm.

Using the algorithm above, the work of the Kirov CHPP-4 and Bezymyanskaya CHPP (Samara) was analyzed.

At Kirov CHPP-4, a turbine No. 2 with a condenser of 60 MW was considered during its operation for one calendar year.

During this period, the temperature of the cooling water changed from 9°C to 32°C, and the flow rate of cooling water from 500 t/h to 4000 t/h. Using the above algorithms, it was found that this installation works with air suction from standard (12 kg/h) to high (60 kg/h). Supercritical oxygen values above 400 μg/mg were also observed. Such an excess is explained by the discharge into the condensate collector of the drainage contaminated with the oxygen flow, while the maximum value was 10,000 μg/mg. In addition, it is worth noting that oxygen is measured at the Kirov Thermal Power Plant-4 by the dissolved oxygen analyzer MARK-302T. This device has a low threshold of an acceptable measured value - 40 μg/mg, therefore, for large values, the indigo carmine method is used. In the indigo carmine method, the oxygen content in water is determined by comparing the colored test sample with the scale. This method is very subjective, as it depends on the person color perception.

Figure 4 shows the experimental data with the calculated characteristics plotted on them, at various temperatures of cooling water with a flow rate of circulating water W = 2000 t/h. The calculated data revealed the reasons for the increased oxygen content of the installation.

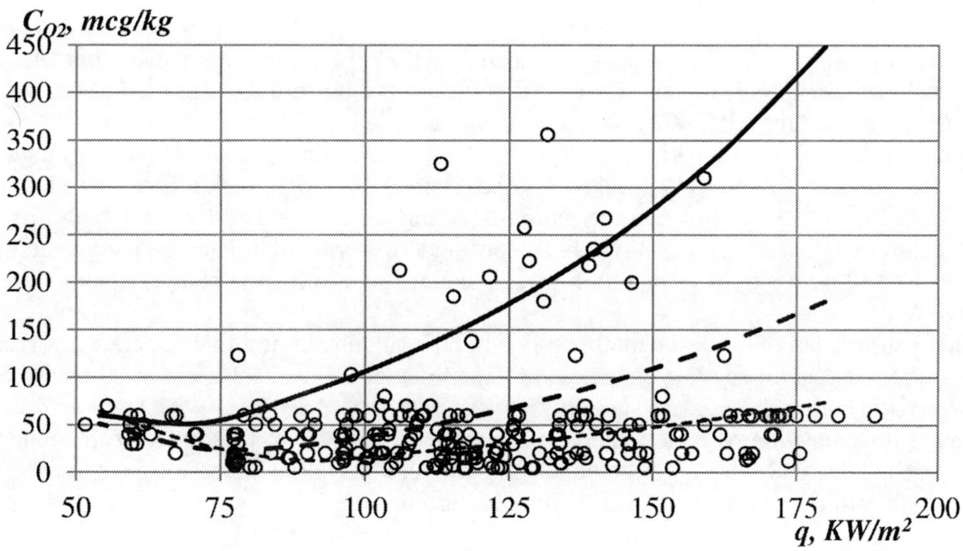

Figure 4. The dependence of the oxygen content on the specific heat load of the installation: ——— – t_{1w} = 30°C; — — — – t_{1w} = 20°C; -------- - t_{1w} = 10°C.

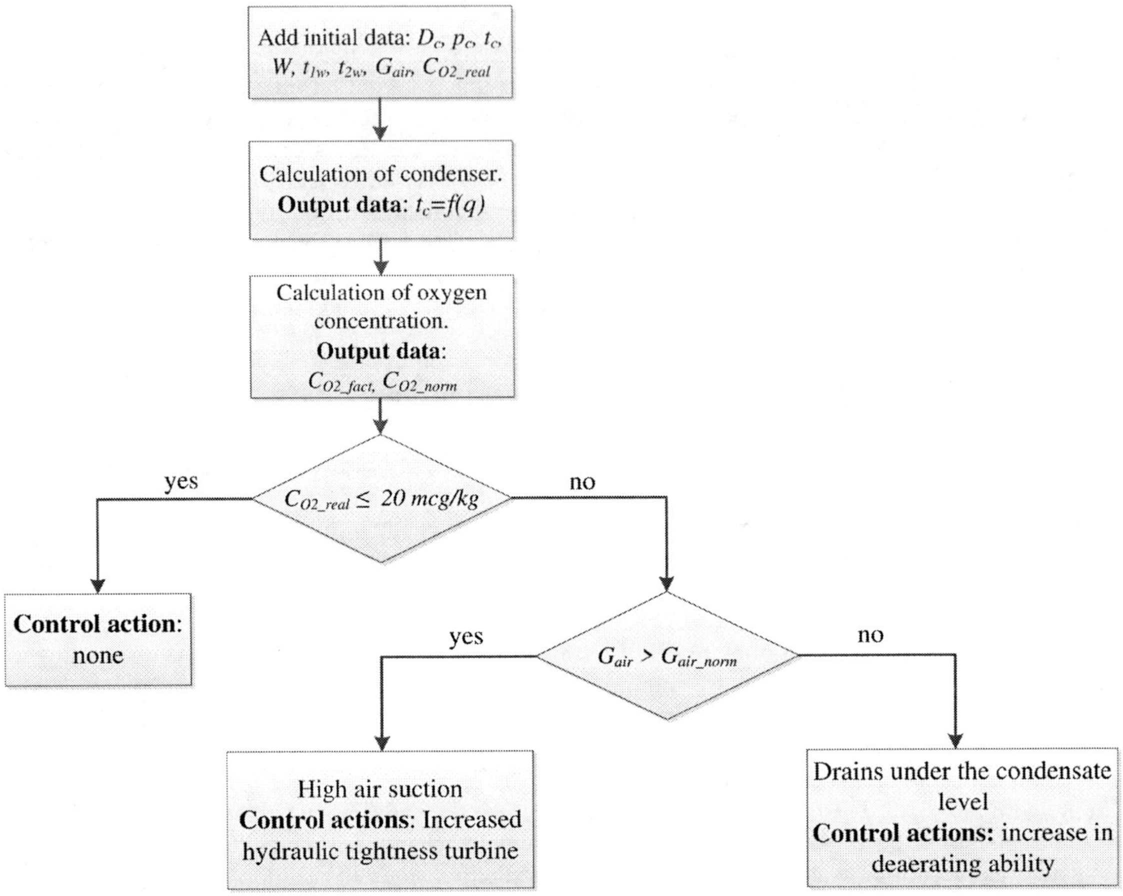

Figure 5. The flowchart of the analysis of the causes of the increased oxygen content in the main condensate after the condenser.

Further, an analysis of data on Bezymyanskaya CHPP in Samara was carried out. According to Bezymyanskaya CHPP, calculations were carried out according to the experimental data of a turbine unit No.6 with a capacity of 50 MW, with a week work period.

For the considered period of time, the specific load of the condenser q was in the range of 10–25 kW/m^2, the pressure of the condenser p_k varied from 3.101 kPa to 4.335 kPa. Using the Iglin-Shempelev technique, the average temperature of cooling water was determined, which was 14°C and the air suction into the turbine was 44 kg/h. According to experimental data, the oxygen content of the condenser was 12-40 μg/kg. As a result of the calculations, the following characteristics were obtained (Figure 6, 7).

Figure 4 shows that the experimental points lie in the calculated design characteristics area. In connection with increased air suction, an excess of experimental values is also observed over the calculated characteristic built with standard suction.

Figure 6 illustrates the oxygen content in the main condensate at the outlet of the condenser, depending on the heat load. Thus, we can conclude that there is a slight excess of oxygen content, which is caused by a 5-fold increase in air suction into the turbine unit.

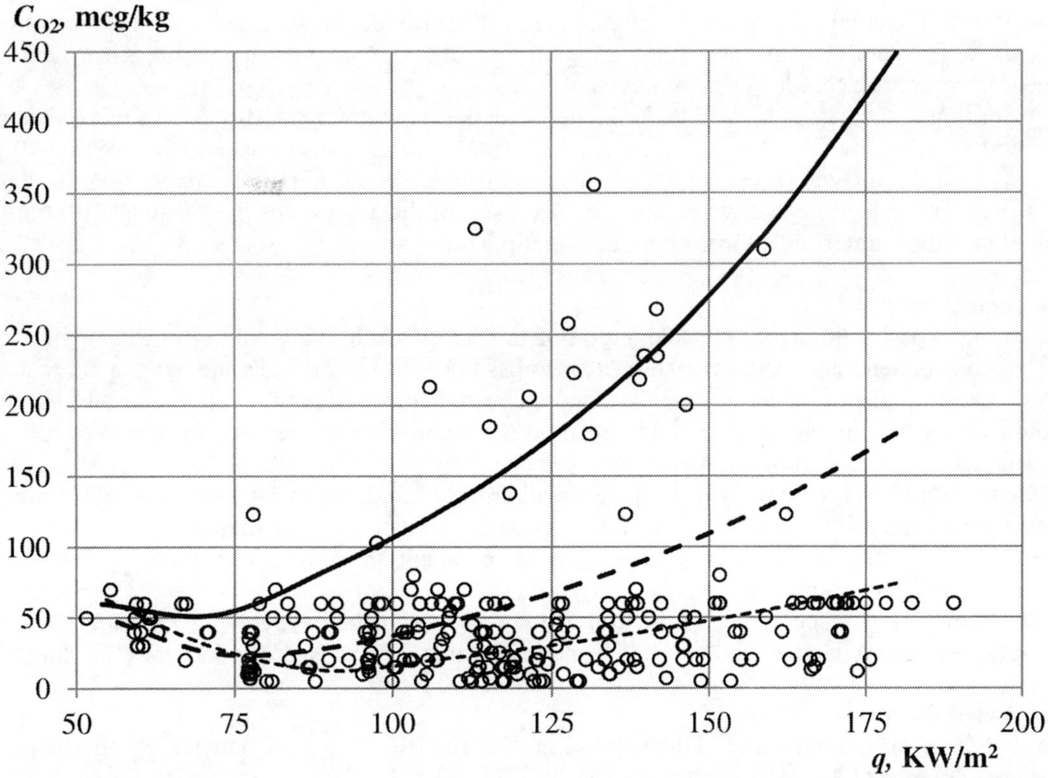

Figure 6. Characteristics of the condenser: ——–at G_{air} = 11.2 kg/h; – – – – at G_{air} = 44 kg/h.

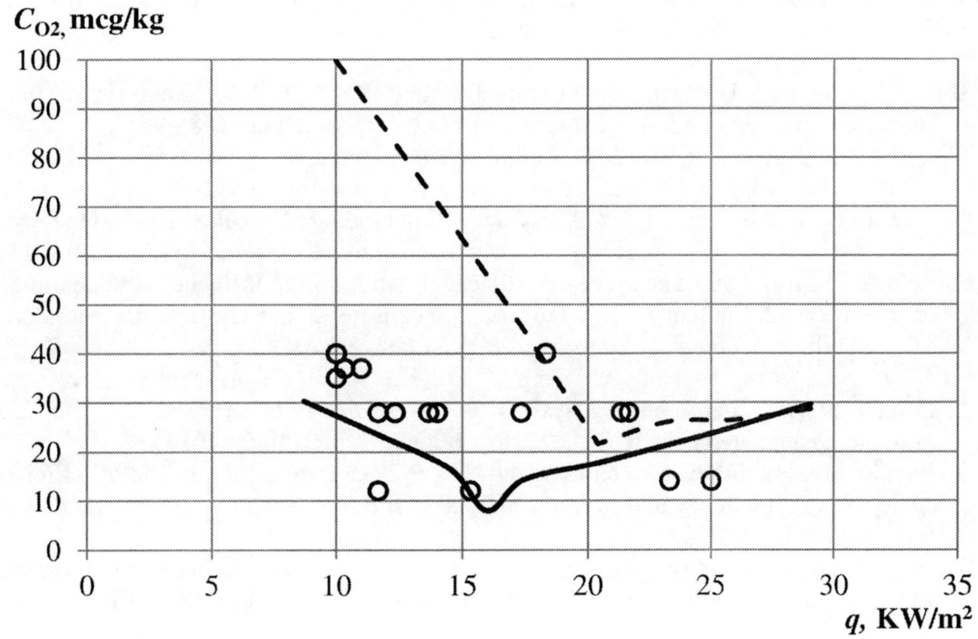

Figure 7. Oxygen content of the condenser: ——–at G_{air} = 11.2 kg/h; – – – – at G_{air}= 44 kg/h.

4. Discussion

Based on the analysis of the oxygen content of the condensers of the Kirov CHPP-4 and Bezymyanskaya CHPP, it can be judged that the employees do not provide the necessary care for the equipment, since the oxygen content at both stations is higher than the standard ones, and at the Kirov CHPP-4 the standard values are exceeded even hundreds of times.

According to the calculated and experimental data, Bezymyanskaya CHPP works more reliably and efficiently than Kirov CHPP-4. The maximum oxygen content at Bezymyanskaya CHPP is 40 μg/kg, at Kirov CHPP-4 this value reaches 10,000 μg/kg, thus there is a difference in exceeding the standards 250 times. The indicators that were obtained as a result of the analysis of the Kirov CHPP-4 are unacceptable for the uninterrupted operation of the equipment.

5. Conclusion

Due to the experimental data, the calculation of indicators according to the Iglin-Shempelev method and the constructed visual characteristics, it was possible not only to detect the oxygen content and air suction excess in the turbine unit over the standard ones, but also to put forward the closest assumption of the reasons for the excess values. The method of establishing the reasons for the excess is clearly described in the flowchart of Figure 5.

For the Kirov CHPP-4, the reason for exceeding the excess values by more than 500 times is the discharge into the condensate collector of drains that are contaminated with oxygen.

For Bezymyanskaya CHPP, as the algorithm passed result, the reason for the excess was air suction in the turbine low-pressure cylinder.

However, a more accurate determination of the reasons for the oxygen content and air suction excess requires a more detailed and deep instrumental examination of each condenser at the stations.

Acknowledgments

The study was carried out with the financial support of the RF President Council on grants as part of the research project MK-2614.2019.8.

References

[1] Iglin P V, Suvorov D M and Shempelev A G 2018 Efficiency of Using Built-In Bundles of Cogeneration Steam Turbine Condensers for Make-up Water Heating *Problemele Energeticii Regionale* **3** 36-51

[2] Shin J Y, Jeon Y J, Maeng D J, Kim J S and Ro S T 2002 Analysis of the dynamic characteristics of a combined-cycle power plant *Energy* **27(12)** 1085-98

[3] Wang Y, Wang T and Sun M 2018 Failure Analysis on Leakage of Brass Condenser Tube in Thermal Power Plant *IOP Conf. Ser.: Mater. Sci. Eng* **439(5)** 052005

[4] Faes W, Lecompte S, Ahmed Z Y, Van Bael J, Salenbien R, Verbeken K and De Paepe M 2019 Corrosion and corrosion prevention in heat exchangers *Corrosion Reviews* **37(2)** 131-55

[5] Sushchikh V, Iglin P and Shempelev A 2017 Estimation of the Influence of Operational Factors on the Oxygen Content of the Turbine Condensate at the Outlet from the Condenser of Steam Turbine *Problemele Energeticii Regionale* **2** 82-90

[6] Fu Q, Tian L, Wang R, Ning Y, Yang Y and Xie X 2015 Corrosion and Protection of the Condenser Seawater Cooling System *In 3rd Int. Conf. on Material, Mechanical and Manufacturing Engineering, IC3ME 2015* (Guangzhou, China) pp 1753-7

[7] Gorshenin A S, Rakhimova J I and Krasnova N P 2019 Increasing the energy efficiency of the cycle of heat treatment of aluminum ingots *In IOP Conf. Series: Earth and Environmental Science* **288(1)** 012090

[8] Syafei N S 2019 Events of corrosion phenomena on carbon steel pipes in environment of sea water and ammonia solutions due to the presence of sweet gas *EKSAKTA: Berkala Ilmiah Bidang MIPA* **20(1)** 86-99

[9] Cheng S Y, Xue R J, Peng M J, Gong C and Zhao Q 2010 Study on control system of a vapor

condenser of nuclear power plant *In 2010 Int. Conf. on Machine Learning and Cybernetics* (Qingdao, China) vol 3 pp 1239-43

[10] Yan D I N G, Chang G U and Qiong F A N G 2004 The Dynamical Mathematical Model and Simulation of Condenser in 300MW Unit *Turbine Technology* **5**

[11] Zhu D B, Ma J, Jiao X L, Qiao Z X and Zhao H 2010 Condenser Modeling Research for Ship Steam Power Plants *Energy Conservation Technology* **6**

[12] Zhou L X, Zhou Y and Cao Z J 2012 Computation of Condenser Heat Transfer Area of 600MW Direct Air Cooling Unit *Turbine Technology* **2**

[13] Li S X and Wang J S 2015 Dynamic modeling of steam condenser and design of PI controller based on grey wolf optimizer *Mathematical Problems in Engineering* **2015** 120975

[14] Nebot E, Casanueva J F, Casanueva T and Sales D 2007 Model for fouling deposition on power plant steam condensers cooled with seawater: Effect of water velocity and tube material *Int. J. of Heat and Mass Transfer* **50** 3351-8.

[15] Zeng H, Meng J A and Li Z 2012 Numerical study of a power plant condenser tube arrangement *Applied Thermal Engineering* **40** 294-303

[16] Shempelev A, Iglin P and Tatarinova N 2017 On condenser mathematical model method introduction into steam turbine unit mathematical model. *In 2017 Int. Conf. on Industrial Engineering, Applications and Manufacturing, ICIEAM* (St. Petersburg, Russia) 1-4

IOP Conf. Series: Materials Science and Engineering **791** (2020) 012014 doi:10.1088/1757-899X/791/1/012014

Development of technologies to increase efficiency and reliability of combined cycle power plant with double-pressure heat recovery steam generator

A A Kudinov, S K Ziganshina, K R Khusainov and Yu E Demina

Department of Thermal Electrical Station, Samara State Technical University, Molodogvardeyskaya St., 244, Samara, 443100, Russia

E-mail: zakharova_yu@bk.ru

Abstract. To increase the efficiency of a combined cycle power plant, it was suggested in a double-pressure heat recovery steam generator to carry out a reheat of water steam that had been spent in a high pressure cylinder of a steam turbine. The researchers introduce a mathematical algorithm and calculation methodology of the waste-heat boiler. It is proved that after re-superheat of water vapour in a double-circuit waste-heat boiler after the high-pressure superheater (with vapour already used in the high-pressure cylinder of a condensing steam turbine), a CHP-plant efficiency and reliability increase. Calculations for the combined cycle gas turbine unit PGU-450 were performed. It is shown that when the reheater is installed in the recovery steam generator, its efficiency increases by 2.15%. At the same time the efficiency of CCGT for power generation is increased by 3.36%, the degree of dry water steam used in the turbine is increased by 0.082. Consequently, the specific consumption of conventional fuel for electric power generation is reduced by 6.1%. Re-superheating of the total flow of water vapour allows to increase CGP-450 steam-gas unit efficiency by increasing the parameters of the steam entering the low-pressure cylinder of the steam turbine.

1. Introduction

Thermal efficiency of the combined cycle power plant depends on the degree of perfection of the gas-turbine installation, steam turbine and recovery steam generator [1]. Significant reserves in increasing the efficiency of the recycling CCGTU are laid down in the improvement of the steam cycle, which, as is known, provides about 1/3 of the power of the CCGTU [2]. The efficiency of the CCGTU-U is greatly influenced by the efficiency of the recovery steam generator, which depends not only on the temperature of the exhaust gases at the outlet of its convoy shaft, but also on the temperature of the ambient air [3]. One of the ways to increase efficiency of the recovery steam generator and economy of the recovery-type CCGTU is to increase the number of convective surfaces of RSG heating in order to reduce the rate of exhaust gases of GTI.

The purpose of the work is to develop technologies to increase the efficiency [4] and reliability [5] of the combined cycle power plant with a double-pressure heat recovery steam generator due to the use of intermediate reheater of water steam used in the high-pressure cylinder of the steam turbine [6].

IOP Conf. Series: Materials Science and Engineering **791** (2020) 012014 doi:10.1088/1757-899X/791/1/012014

2. Materials and Methods

In order to increase the economy and reliability of the combined cycle gas turbine unit operation PGU-450 is proposed to place the reheater of water steam [7] partially used in the high-pressure cylinder of the steam turbine (figure 1) in a double-pressure heat recovery steam generator.

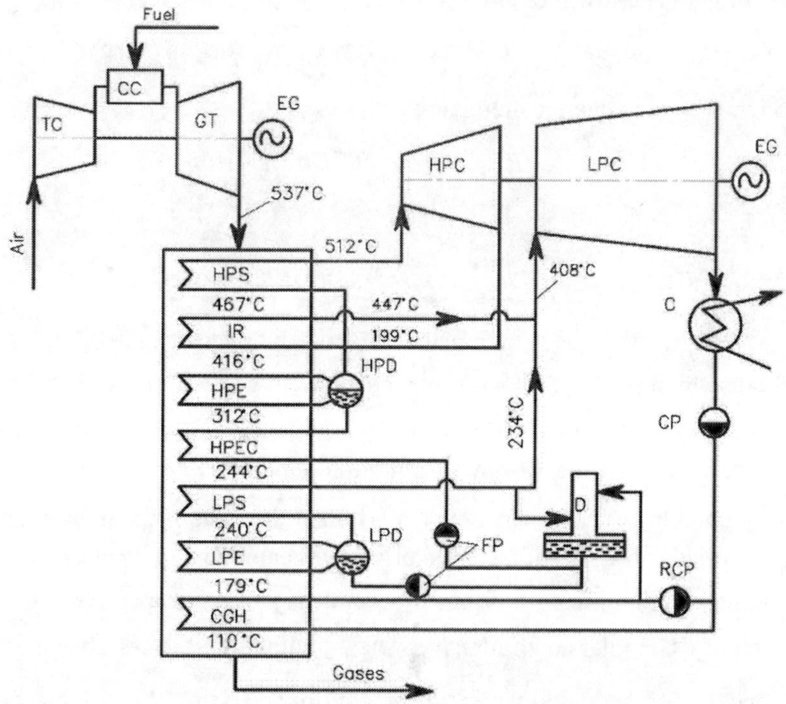

Figure 1. Thermal schematic diagram of CCGTU-U with reheat of steam in recovery steam generator.

The considered unit of the PGU-450 includes: Siemens V94.2 gas turbine installation with nominal capacity of 155 MW, two-pressure recovery steam generator P-90 with forced circus of working medium in high and low pressure evaporation circuits and steam turbine condensing plant KT-150-8. Design electric power of PGU-450 in nominal mode at operation without reheat of water steam is 460.44 MW, and its efficiency is 51.76%.

To determine efficiency of using secondary superheat [8] of water steam in CCGT with double-pressure heat recovery steam generator, single reheat is considered with arrangement of heating surface of reheater in zone of high temperatures of recovery steam generator downstream of main superheater of high pressure [9]. The double-pressure heat recovery steam generator P-90 includes the following heat exchange surfaces: high and low pressure superheaters (HPS, LPS), intermediate reheater (IR); high and low pressure evaporator (HPE and LPE); high pressure economizer (HPEC); condensate gas heater (CGH).

The heat scheme of the PGU-450 was calculated and the economic parameters of its operation were determined [10]. Calculations are performed for two components: 1) during CCGTU operation without intermediate reheat; 2) during CCGTU operation with reheat of water steam. The methodology described in is adopted as the basis. The results of the calculations are shown in figure 2, 3.

From the equations of thermal balance for the heating surfaces of the heat recovery steam generator, the enthalpies of gases after the superheaters (HPS, LPS, IR), evaporator (HPE and LPE), economizer (HPEC) and condensate gas heater (CGH) were calculated.

Enthalpies of gases:

IOP Conf. Series: Materials Science and Engineering **791** (2020) 012014 doi:10.1088/1757-899X/791/1/012014

after superheaters of high and low pressure

$$h''_{hps} = h'_{rsg} - D_0^{hp}(h_0^{hp} - h''_{s(hp)})/G_g\ ; \tag{1}$$

$$h''_{lps} = h''_{hpec} - D_0^{lp}(h_0^{lp} - h''_{s(lp)})/G_g\ ; \tag{2}$$

after superheaters of high and low pressure

$$h''_{ir} = h''_{hps} - D_0^{hp}(h''_{ir} - h'_{ir})/G_g\ ; \tag{3}$$

downstream of high pressure water economizer

$$h''_{hpec} = h''_{hpe} - D_0^{hp}(h_1 - h_d^{hp})/G_g\ ; \tag{4}$$

at condensate gas heater outlet

$$h''_{rsg} = h''_{lpe} - G_{cgh}(h''_c - h'_c)/G_g\ , \tag{5}$$

where h''_{hps}, h''_{lps}, h''_{ir}, h''_{hpec}, h''_{rsg}, h'_{rsg} – enthalpy of gases after high and low pressure superheaters, intermediate reheater, high pressure economizer, condensate gas heater, enthalpy of gases at heat recovery steam generator inlet respectively; h_0^{hp}, h_0^{lp} – high and low pressure superheated steam enthalpy; $h''_{s(hp)}$, $h''_{s(lp)}$ – enthalpy of water steam at saturation temperature in the drum of you-well and low pressure; h'_{ss}, h''_{ss} – enthalpy of superheated water steam at the inlet of the intermediate reheater and at the outlet of it; G_g – mass flow rate of gases entering the heat recovery steam generator; h_1 – enthalpy of feedwater entering the high pressure drum; h_d^{hp} – enthalpy of feedwater entering the HPEC; h'_c, h''_c – enthalpy of water at CGH inlet and outlet; h''_{hpe}, h''_{lpe} – enthalpy of gases after high and low pressure evaporator; D_0^{hp} and D_0^{lp} – high and low pressure steam flow rates; $G_{cgh} = D_0^{hp} + D_0^{lp} - D_d + G_r$ – water flow rate through condensate gas heater; D_d and G_r – steam consumption for deaerator and water for recirculation.

From the thermal balance equations for each evaporator, high and low pressure steam flow rates were determined:

$$D_0^{hp} = \frac{G_g(h''_{ir} - h''_{hpe})}{h''_{s(hp)} - h_1}; \tag{6}$$

$$D_0^{lp} = \frac{G_g(h''_{lps} - h''_{lpe})}{h''_{s(lp)} - h_d^{lp}}. \tag{7}$$

h_d^{lp} – low pressure circuit feedwater enthalpy.

Steam flow to deaerator:

$$D_d = (D_0^{hp} + D_0^{lp})(h_d - h''_c)/(h_0^{lp} - h''_c) \tag{8}$$

h_d – enthalpy of water after deaerator.

Condensate flow for recirculation to increase water temperature before CGH:

$$G_r = (D_0^{hp} + D_0^{lp} - D_d)(h'_c - h_{cp})/(h''_c - h'_c) \tag{9}$$

h_{cp} – condensate enthalpy downstream of condensate pump.

Mass flow rate of exhaust gases of GTI $G_g = 506.15$ kg/s, air excess factor $\alpha = 3.33$ was calculated according to formulas given in. By enthalpy of gases at condensate gas heater outlet temperature of gases at heat recovery steam generator t''_{rsg}.outlet is determined.

The efficiency of the heat recovery steam generator is determined by the ratio of the difference of enthalpies of gases at the inlet of the control unit and at the outlet of its convective shaft to the difference of enthalpies of gases at the inlet of the control unit and enthalpies of gases discharged into the environment, cooled in the atmosphere to the ambient air temperature:

$$\eta_{rsg} = \left(h'_{rsg} - h''_{rsg}\right) / \left(h'_{rsg} - h_{at}\right). \tag{10}$$

To find of gas temperature after high and low pressure evaporator, the values of temperature pressures at pinch points of HPE and LPE - $\delta t_s^{hp} = 8$ °C, $\delta t_s^{lp} = 7$ °C were accepted. The temperature of the gases after each evaporator was determined from the saturation temperature in the drum and the accepted temperature head:

$$\theta_s^{hp} = t_s^{hp} + \delta t_s^{hp}; \theta_s^{lp} = t_s^{lp} + \delta t_s^{lp}. \tag{11}$$

In figure 2, the process of superheating in the intermediate reheater installed after the superheater with a high pressure line.

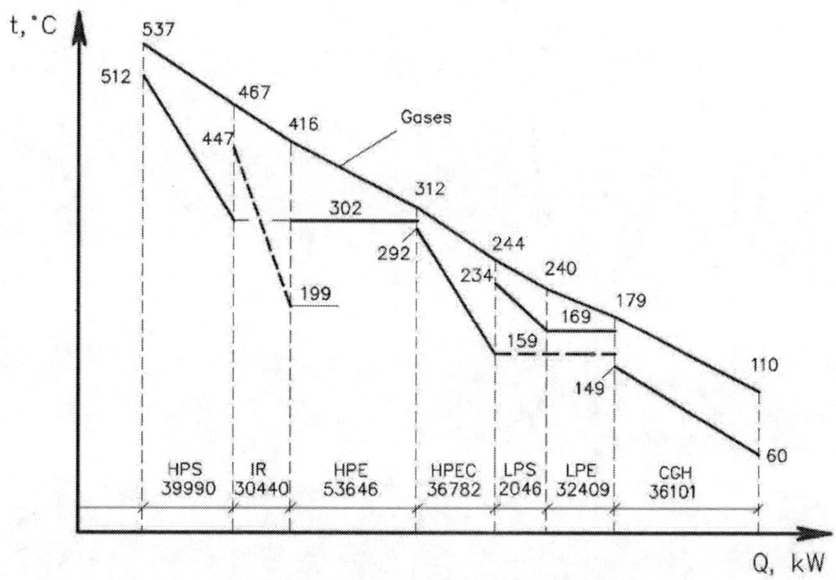

Figure 2. Heat diagram of the heat recovery steam generator P-90 with arrangement of reheater after the superheater of high pressure.

Figure 3 shows the process of water steam expansion in a steam non-reheat turbine. Figure 4 shows the process of water steam expansion in steam turbine at single intermediate reheat of water steam in heat recovery steam generator with arrangement of intermediate reheater downstream of high-pressure superheater along the flow of outgoing combustion products [11].

Steam enthalpy in the mixing camera at implementation of intermediate reheat of water steam in a zone of high temperatures of the heat recovery steam generator was determined by a formula:

$$h_{mix} = \left[D_0^{hp} h''_{ss} + \left(D_0^{lp} - D_d\right) h_0^{lp}\right] / \left(D_0^{hp} + D_0^{lp} - D_d\right) \tag{12}$$

IOP Conf. Series: Materials Science and Engineering **791** (2020) 012014 doi:10.1088/1757-899X/791/1/012014

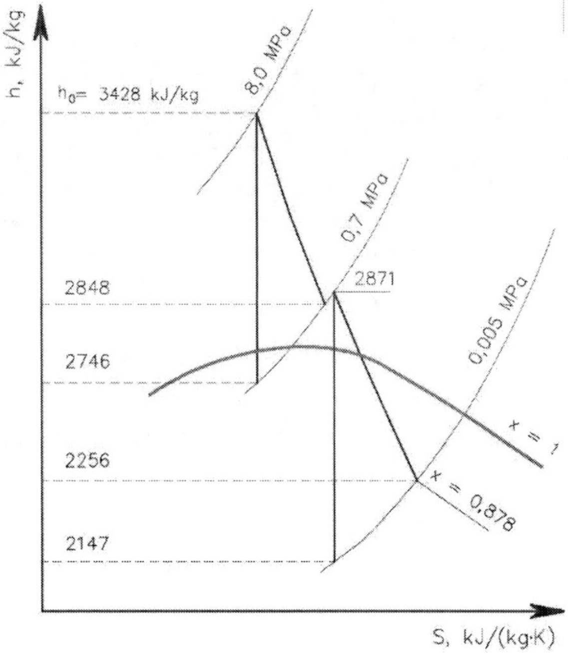

Figure 3. Steam expansion process for turbine KT-150-8 in case of single-stage intermediate reheat in heat recovery steam generator.

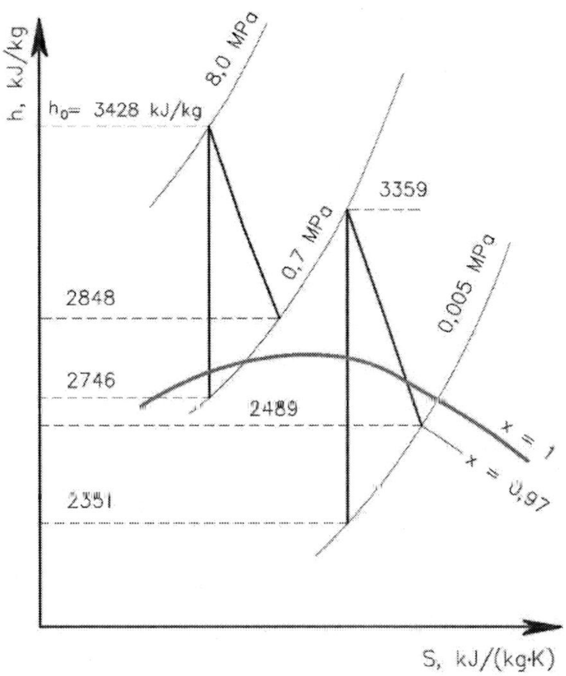

Figure 4. Process of expansion of water steam for turbine KT-150-8 at single reheat of steam in heat recovery steam generator.

3. Results

Calculations for the technical specifications of the PGU-450 were performed with the following initial data: p_0 = 8.0 MPa; Pc = 5 kPa; Rd = 0.6 MPa; Ne = 155.3 MW; steam pressure of a contour of low pressure of – 0.7 MPa; electric efficiency of GTI – 0.344; gas temperature on an entrance to a heat recovery steam generator – 537 °C. The calculation results of the main technical and economic indica-

IOP Conf. Series: Materials Science and Engineering **791** (2020) 012014 doi:10.1088/1757-899X/791/1/012014

tors of CGP-450 with the use of water vapour re-superheat in the waste-heat boiler and without its use are given in Table 1.

Table 1. The calculation results of PGU-450 operation with the use of water vapour resuperheat in the waste-heat boiler and without its use.

Parameter	Parameter point	
	Without intermediate reheat	With reheat of water steam
Electric power of GTI, MW	155.27	155.27
Electric efficiency of GTI, %	34.4	34.4
Electric power of STP, MW	149.9	178.86
Electric efficiency of STP, %	33.21	38.59
Efficiency RSG, %	79.71	81.86
Electric power of CCGTU, MW	460.44	489.4
Electric efficiency of CCGTU, %	51.76	55.12
Steam properties high pressure loop		
– pressure, MPa	8.0	8.0
– temperature, °C	512	512
– flow, t/h	198.29	198.29
Steam properties after intermediate reheater:		
– pressure, MPa	–	0.6
– temperature, °C	–	464
– flow, t/h	–	198.29
Steam properties in the steam turbine mixing chamber:		
– pressure, MPa	0.7	0.7
– temperature, °C	203	408
– flow, t/h	396.58	396.58
Steam properties low pressure loop:		
– pressure, MPa	0.7	0.7
– temperature, °C	245.0	234.0
– flow, t/h	64.43	64.43
Gas temperature of the heating surfaces of the heat recovery steam generator. °C:		
–after GTU	537	537
–after HPS (base and intermediate)	467	467
–after intermediate reheater	–	416
–after HPE	327	312
–after HPEC	265	244
–after LPS	259	240
–after LPE	194	179
–after CGH	122	110
Specific consumption of conventional fuel, g/kWh	237.64	223.15

4. Conclusion

The heat scheme of the PGU-450 was calculated for two versions of operation: without performing intermediate reheat and with performing intermediate reheat of steam used in the high-pressure cylinder of the steam turbine [12]. When performing the intermediate reheat of the water steam in the heat recovery steam generator. the main characteristics of the CCGTU increase: electric power from 460.44 to 489.4 MW (by 28.96 MW). efficiency of power generation from 51.76 to 55.12% (by 3.36%) and the degree of dryness of the waste water steam in the turbine from 0.877 to 0.959 (by 0.082). Accordingly. the specific consumption of conventional fuel on the electric power supply is reduced from

IOP Conf. Series: Materials Science and Engineering **791** (2020) 012014 doi:10.1088/1757-899X/791/1/012014

237.64 to 223.15 g/kWh (by 6.1%) [13]. At the same time. saving of conventional fuel for PGU-450 at its operating time of 4500 hours per year and cost of conventional fuel of 3700 rubles/t. In monetary terms is 118.72 million rubles/year. Analysis [14] of the calculation results indicates that when the intermediate reheater is installed in the heat recovery steam generator after the superheater of high pressure. the efficiency of the CCGTU operation is substantially increased [15].

References

[1] Kuznetsov V A and Trubaev P A 2018 Resources and Problems of the Mathematical Simulating Thermo-Technological Processes *J. Phys.: Conf. Ser* **1066** 012024

[2] Truini A D 2013 *Parogazovye ustanovki elektrostantsii* [Combined-cycle plants of power stations] (Moscow: Izdatel'stvo MEI) [In Russian]

[3] Martínez González A, Lesme Jaén R and Silva Lora E 2020 Thermodynamic assessment of the integrated gasification-power plant operating in the sawmill industry: An energy and exergy analysis *J. Renewable Energy* **147(1)** 1151-63

[4] Sultanov M M. Boldyrev I A and Gorban Yu A 2019 Research of dynamic characteristics of signals for calculation of technical and economic indicators and increase in power efficiency of the equipment of thermal power plant *IOP Conf. Ser.: Mater. Sci. Eng.* **552** 012007

[5] Sultanov M M. Truhanov V M and Gorban Yu A 2019 Parametric model for predicting the reliability of heat power equipment of TPP *IOP Conf. Ser.: Mater. Sci. Eng.* **552** 012006

[6] Liu Z. Ren X. Yan Z. Zhu H. Zhang T. Zhu W and Li X 2019 Effect of inlet air heating on gas turbine efficiency under partial load *J. Energies* **12 (17)** 3327

[7] Kudinov A A. Ziganshina S K and Khusainov K R 2018 Feedwaterheating in wasteheat boiler with exhaust gases from gas-turbine power unit *J. Phys.: Conf. Ser* **1111 (1)** 012044

[8] Wang S. Liu Z. Cordtz R. Imran M and Fu Z 2019 Performance prediction of the combined cycle power plant with inlet air heating under part load conditions *J. Energy Conversion and Management* **200** 112063

[9] Maheshwari M and Singh O 2019 Thermodynamic study of different configurations of gas-steam combined cycles employing intercooling and different means of cooling in topping cycle *J. Applied Thermal Eng.* **162** 114249

[10] Hofmann R. Panuschka S and Beck A 2019 A simultaneous optimization approach for efficiency measures regarding design and operation of industrial energy systems *J. Computers & Chemical Eng.* **128** 246-60

[11] Alexandrov A A 2006 *Termodinamicheskie osnovy tsyklov teploelektricheskikh ustanovok* [Thermodynamic bases of thermal power plant cycles] (Moscow: Izdatel'stvo MEI) [In Russian]

[12] Kindra V O. Rogalev N D. Lisin E M and Khudyakova V P 2017 Razrabotka i tekhniko-ekonomicheskii analiz teplovyh skhem otopitelnykh gazoturbinnykh elektrostantsyi [Development and technical and economic analysis of thermal schemes of heating gas turbine power plants] *Novoe v Rossiiskoi Elektroenergetike* **3** 6-20 [In Russian]

[13] Ziganshina S K and Kudinov A A 2016 Thermal gain of CHP steam generator plants and heat supply systems *J. Thermal Eng.* **63 (8)** 587-91

[14] Kudinov A A and Ziganshina S K 2018 Analysis of the efficiency of using the heat of natural gas combustion products to preheat forced air supplied to the boiler *J. Power Technology and Eng.* **52(3)** 319-24

[15] Derevyanov M Yu. Pleshivtseva Yu E. Kordyukova L S and Rapoport A V 2019 Using DEA models to measure the efficiency of energy saving projects *IOP Conf. Ser.: Mater. Sci. Eng.* **552** 012020

IOP Conf. Series: Materials Science and Engineering **791** (2020) 012015 doi:10.1088/1757-899X/791/1/012015

An elaborated diffusion mathematic model of radiative transfer in an extinction medium

V A Kuznetsov, P A Trubaev and O A Ryazancev

Department of Energy Engineering of Heat Technologie, Belgorod State Technological University named after V.G. Shukhov, Kostyukov St., 46, Belgorod, 308012, Russia

E-mail: kousnezow@mail.ru

Abstract. The computational method of diffusion approach, used sometimes for simulating the radiative heat transfer, differs in simplicity of its algorithm, but can not take ac-count of radiation anisotropy. The reasons of its insufficient accuracy have been re-vealed; it is shown that inadequate boundary con-ditions produce unpredictable er-rors. Ways for eliminating shortcomings of the diffusion method have been devel-oped. The accurate diffusion model has been theoretically elaborated for radiative heat transfer in dust-laden gases. To eliminate singularity of temperature curves on the surface of confining walls, the en-closed extinction medium was hypothetically considered as an unbounded one. The radiation intensity has been expanded in a se-ries and integrated over the spherical solid angle, what has allowed to obtain more precise differential equation for the resulting radiation flux both in the unbounded and enclosed extinction medium. As a result, shortcomings peculiar to the previous method of diffusion approach have been eliminated, and correct Neumann and Robin boundary conditions have been formulated. As a result, the elaborated diffu-sion model got the increased accuracy in simulating radiative heat transfer, and con-currently it offers an advantage of simplicity and universality of the diffusion ap-proach method.
Key words: radiative heat transfer, selective gases, scattering dust particles, bounda-ry conditions, diffusion mathematic model.

1. Introduction

The efficiency of many high-temperature technological plants can be improved with the help of numerical researching the radiative heat transfer in dust-laden gases. To ensure adequate simulation results, it is necessary to admit that triatomic gases emit and absorb the radiant energy selectively, generally in lines of the spectrum [1], while particles of the technological dust or ash usually form the gray extinction medium, which can scatter rays. An elongated scattering indicatrix can be approximately replaced by an isotropic one [2], considering that exactly forward scattering is equivalent to no scattering at all [3].

This assumption allows to write down the conservation differential equation for the radiation intensity I_l in any ray direction:

$$\frac{\partial I_l}{\partial l} = \alpha_{\mathrm{p}} I_{\mathrm{o}} - k I_l + \beta \bar{I} \, , \qquad (1)$$

IOP Conf. Series: Materials Science and Engineering **791** (2020) 012015 doi:10.1088/1757-899X/791/1/012015

where I_o is the total intensity of equilibrium radiation, W/(m^2·sr); α_p is the mean Planck absorption coefficient, characterizing the common emitting capacity of the medium, m^{-1}; k is its local extinction coefficient, m^{-1},

$$k = \alpha + \beta ;$$

α is the local absorption coefficient of the medium, m^{-1}; β is the isotropic-scattering coefficient of dust particles, m^{-1}; l is a distance along the ray, m; $\bar{I}$ is a local radiation intensity, averaged in the spherical solid angle, W/(m^2·sr),

$$\bar{I} \equiv \frac{1}{4\pi} \int_{(4\pi)} I_l \, \mathrm{d}\omega = \frac{\sigma}{\pi} T_{\mathrm{rad}}^4 ;$$

σ is the Stefan–Boltzmann constant, W/(m^2·K^4); T_{rad} is the radiant temperature [4], K,

$$T_{\mathrm{rad}}^4 \equiv \frac{1}{4\sigma} \int_{(4\pi)} I_l \, \mathrm{d}\omega ; \tag{2}$$

ω is the solid angle, sr, defined by means of the plain polar angle θ and azimuthal one ψ,

$$\mathrm{d}\omega = \sin\theta \, \mathrm{d}\theta \, \mathrm{d}\psi .$$

Integrating the equation (1) over spherical solid angle gives the conservation differential equation for radiation flux:

$$\mathrm{div}\mathbf{q}_{\mathrm{rad}} = 4\sigma \left(\alpha_{\mathrm{p}} T^4 - \alpha T_{\mathrm{rad}}^4 \right), \tag{3}$$

where $\mathbf{q}_{\mathrm{rad}}$ is a vector of the resulting radiation flux, W/m^2; T is the thermodynamic temperature, K.

In scientific publications, much attention is given at present time to problems of adequate absorption coefficients of selective gases [5, 6]. Not less important is an aim to elaborate some easy algorithm for solving differential equations of the radiative transfer [7].

A main problem ensues from that there is not any distinct differential equation for interrelating the vector of the resulting radiation flux $\mathbf{q}_{\mathrm{rad}}$ with radiation parameters. Therefore it is usually found numerically by means of angular integrating projections of the radiation intensity:

$$q_{\mathrm{rad}} = \int_{(4\pi)} I_l \cos\theta \, \mathrm{d}\omega = \int_0^{2\pi} \mathrm{d}\psi \int_0^{\pi} I_l \cos\theta \sin\theta \, \mathrm{d}\theta , \tag{4}$$

such an algorithm requires the radiation intensity be computed previously at many points along numerous rays by solving the equation of radiative heat transfer (1).

Different simulating methods have been applied for this purpose. There are methods of discrete transfer [8], discrete ordinates [9], finite volumes [10], finite elements [11], natural elements [12], and a zonal one [13]. Despite their variety, all these algorithms display excess complicacy and too high time of numerical solving tasks.

A specific place is held by the method of spherical harmonics [14], which higher orders of approximation allow for spatial anisotropy of the radiation intensity with the help of numerical solving a sequence of differential equations. Comparison of various numerical models of heat transfer in a high-temperature flame [7] showed that the approach P-3 of this method yields results of high precision at smaller expense of the computer time. The shortest computation resides in the approach P-1 of the method of spherical harmonics. However, it does not consider the anisotropy of radiation, what leads to some raised error of computational results.

The approach P-1 essentially coincides with the method of diffusion approach, which has been considered in details in [15]. Its advantage is an easy compatibility with simulating algorithms of the

IOP Conf. Series: Materials Science and Engineering **791** (2020) 012015 doi:10.1088/1757-899X/791/1/012015

convective heat transfer. Therefore, it would be worthwhile to adapt the method of diffusion approach to the anisotropic radiative transfer in the extinction medium. The aim of this report is to represent a more perfect diffusion method for numeric solving the radiative heat-transfer equations in dust-laden gases.

2. Methods

An obvious source of anisotropic radiation intensity is the surface of the walls confining a volume of dusty gases. To exclude it temporarily from consideration, a concept of an unbounded medium has to be introduced. This hypothetical medium ought to have the same radiation properties and temperature, as the dust-laden gases in the confined volume. The hypothetical absence of the confining walls simplifies substantiating equations of radiative heat transfer, since the temperature curve in the unbounded medium can be presented by continuous smooth functions without any singularity in points that coincide with the surface of the confining walls.

An expression for the radiation intensity in the unbounded medium follows from the equation (1):

$$I_l = \frac{1}{k}\left(\alpha_p I_o + \beta \bar{I} - \frac{\partial I_l}{\partial l}\right).$$

It is significant to mention that such a representing form of the radiation intensity has sense only in respect to the energy of spectrum that would be absorbed in the extinction medium. As a consequence, it requires some new modes of computing the absorption coefficient for its using in the radiative heat transfer equation.

This mathematical form can be repeatedly differentiated along length of the ray, supposing that lumped radiation sources are absent there, and the extinction coefficient would be constant. Higher derivatives in right-hand sides of each of obtained equalities have to be replaced with expressions, represented by their next equalities. As a result, the radiation intensity will be expanded in a series:

$$I_l = \frac{\alpha_p}{k}\left(I_o - \frac{\partial I_o}{k\partial l} + \frac{\partial^2 I_o}{k^2\partial l^2} - \frac{\partial^3 I_o}{k^3\partial l^3} + \frac{\partial^4 I_o}{k^4\partial l^4} - \frac{\partial^5 I_o}{k^5\partial l^5} + \ldots\right) + \\ + \frac{\beta}{k}\left(\bar{I} - \frac{\partial \bar{I}}{k\partial l} + \frac{\partial^2 \bar{I}}{k^2\partial l^2} - \frac{\partial^3 \bar{I}}{k^3\partial l^3} + \frac{\partial^4 \bar{I}}{k^4\partial l^4} - \frac{\partial^5 \bar{I}}{k^5\partial l^5} + \ldots\right). \quad (5)$$

Equality (5) can be written more compactly as follows:

$$I_l = \frac{1}{k}\left[\left(\alpha_p I_o + \beta \bar{I}\right) + \sum_{n=1}^{\infty}(-1)^n \frac{\partial^n\left(\alpha_p I_o + \beta \bar{I}\right)}{k^n \partial l^n}\right].$$

Hereinafter, a simplified one-dimensional consideration of spatial transformations will be applied, what, however, would not depreciate generality of the achieved mathematic results. Thus, it is assumed that a distance l along any ray and coordinate y in direction of the radiation-flux vector are interrelated by the ratio:

$$\mathrm{d}l = \mathrm{d}y/\cos\theta.$$

The series (5) is convergent under certain conditions and, consequently, it admits mathematical transformations. For example, its integrating over the solid angle in accordance with formulae (2) and (4) gives series of the radiant temperature and the resulting radiation-flux density in the unbounded medium:

$$T_{\mathrm{rad}}^4=\frac{1}{k}\left[\left(\alpha_{\mathrm{p}}T^4+\beta T_{\mathrm{rad}}^4\right)+\frac{\pi}{\sigma}\sum_{n=1}^{\infty}\frac{\partial^{2n}\left(\alpha_{\mathrm{p}}I_{\mathrm{o}}+\beta\bar{I}\right)}{(2n+1)k^{2n}\partial l^{2n}}\right], \tag{6}$$

$$q_{\mathrm{rad}}^{\infty}=-\frac{4\sigma}{k}\left[\frac{\partial\left(\alpha_{\mathrm{p}}T^4+\beta T_{л}^4\right)}{3k\partial y}+\frac{\pi}{\sigma}\sum_{n=1}^{\infty}\frac{\partial^{2n+1}\left(\alpha_{\mathrm{p}}I_{\mathrm{o}}+\beta\bar{I}\right)}{(2n+3)k^{2n+1}\partial l^{2n+1}}\right]. \tag{7}$$

The series (6) can be written in the following form:

$$\alpha_{\mathrm{p}}T^4+\beta T_{\mathrm{rad}}^4=kT_{\mathrm{rad}}^4-\frac{\pi}{\sigma}\sum_{n=1}^{\infty}\frac{\partial^{2n}\left(\alpha_{\mathrm{p}}I_{\mathrm{o}}+\beta\bar{I}\right)}{(2n+1)k^{2n}\partial l^{2n}}.$$

If the right-side part of this expression would be introduced under the sign of the first derivative in the series (7), then a more accurate expression will be obtained for the resulting radiation-flux density in the unbounded extinction medium:

$$\begin{aligned}q_{\mathrm{rad}}^{\infty}=&-\frac{4\sigma}{3k}\frac{\partial T_{\mathrm{rad}}^4}{\partial y}-4\pi\frac{\alpha_{\mathrm{p}}}{k}\left(\frac{4}{45}\frac{\partial^3 I_{\mathrm{o}}}{k^3\partial y^3}+\frac{8}{105}\frac{\partial^5 I_{\mathrm{o}}}{k^5\partial y^5}+\frac{4}{63}\frac{\partial^7 I_{\mathrm{o}}}{k^7\partial y^7}+\ldots\right)-\\&-4\pi\frac{\beta}{k}\left(\frac{4}{45}\frac{\partial^3\bar{I}}{k^3\partial y^3}+\frac{8}{105}\frac{\partial^5\bar{I}}{k^5\partial y^5}+\frac{4}{63}\frac{\partial^7\bar{I}}{k^7\partial y^7}+\ldots\right).\end{aligned} \tag{8}$$

It can be produced in the follow generalized form:

$$q_{\mathrm{rad}}^{\infty}=-\frac{4\sigma}{k}\left[\frac{\partial\left(\alpha_{\mathrm{p}}T^4+\beta T_{л}^4\right)}{3k\partial y}-\frac{\pi}{\sigma}\sum_{n=1}^{\infty}\frac{4n}{3(2n+1)(2n+3)}\frac{\partial^{2n+1}\left(\alpha_{\mathrm{p}}I_{\mathrm{o}}+\beta\bar{I}\right)}{k^{2n+1}\partial l^{2n+1}}\right].$$

The sum of items with higher partial derivatives can be replaced here with a simpler expression that has been got from the following approximate equality:

$$\begin{aligned}\frac{\partial}{k\partial y}\int_{(4\pi)}I_l\cos^8\theta\,d\omega=&4\pi\frac{\alpha_{\mathrm{p}}}{k}\left[\frac{\sigma}{\pi}\frac{\partial T^4}{9k\partial y}+\left(\frac{\partial^3 I_{\mathrm{o}}}{11k^3\partial y^3}+\frac{\partial^5 I_{\mathrm{o}}}{13k^5\partial y^5}+\ldots\right)\right]+\\&+4\pi\frac{\beta}{k}\left[\frac{\sigma}{\pi}\frac{\partial T_{\mathrm{rad}}^4}{9k\partial y}+\left(\frac{\partial^3\bar{I}}{11k^3\partial y^3}+\frac{\partial^5\bar{I}}{13k^5\partial y^5}+\ldots\right)\right]\approx\frac{4\sigma}{9}\frac{\partial T_{\mathrm{rad}}^4}{k\partial y}.\end{aligned}$$

Indeed, the items with higher derivatives contain in this equality numerical coefficients that differ little from those in corresponding items of the right-hand side of the equation (8). As a final result, an easy and precise enough formula would substitute for the series (7):

$$q_{\mathrm{rad}}^{\infty}\approx-\frac{4\sigma}{3k}\frac{\partial T_{\mathrm{rad}}^4}{\partial y}-\frac{4\sigma}{9k}\left(\frac{\alpha}{k}\frac{\partial T_{\mathrm{rad}}^4}{\partial y}-\frac{\alpha_{\mathrm{p}}}{k}\frac{\partial T^4}{\partial y}\right).$$

It allows for anisotropy of the radiation in the hypothetical unbounded extinction medium.

IOP Conf. Series: Materials Science and Engineering **791** (2020) 012015 doi:10.1088/1757-899X/791/1/012015

3. Discussions

Radiative heat transfer in an unbounded extinction medium

To simplify equation forms, it would be advantageous to introduce the notion of the determining temperature T_Σ, which accounts for the summary influence of the radiant and thermodynamic temperatures on transferring the radiant energy in the unbounded extinction medium:

$$T_\Sigma^4 \equiv T_{\text{rad}}^4 + \frac{1}{3k}\left(\alpha T_{\text{rad}}^4 - \alpha_p T^4\right). \tag{9}$$

Following this, the specified differential expression of the resulting radiation-flux vector in the unbounded extinction medium takes a gradient form that is similar to the formula of the diffusion-approach method:

$$\mathbf{q}_{\text{rad}}^{\infty} = -\frac{4\sigma}{3k}\,\text{grad}\,T_\Sigma^4. \tag{10}$$

Following this, the specified differential expression of the resulting radiation-flux vector in the unbounded extinction medium takes a gradient form that is similar to the formula of the diffusion-approach method:

$$\text{div}\mathbf{q}_{\text{rad}}^{\infty} = \frac{4\sigma}{1+\alpha/(3k)}\left(\alpha_p T^4 - \alpha T_\Sigma^4\right). \tag{11}$$

If the gradient expression (10) for the resulting radiation-flux vector would be introduced in the last equality, a specified transport equation for the radiation-heat transfer in the hypothetical unbounded medium can be obtain. It takes the following form in Cartesian coordinates x, y, z:

$$\frac{\partial}{\partial x}\left(\frac{\partial T_\Sigma^4}{k\partial x}\right) + \frac{\partial}{\partial y}\left(\frac{\partial T_\Sigma^4}{k\partial y}\right) + \frac{\partial}{\partial z}\left(\frac{\partial T_\Sigma^4}{k\partial z}\right) + \frac{9}{3+\alpha/k}\left(\alpha_p T^4 - \alpha T_\Sigma^4\right) = 0. \tag{12}$$

This equation differs from the differential equation of the diffusion-approach method both in a functional variable and in a numerical coefficient of its item.

The temperature of the hypothetical unbounded medium is not strictly determined outside the confined gaseous space. Hence, it makes sense to accept a physically reliable assumption that the temperature curves have inflection points that coincide with the surface of enclosing walls. According to this postulate, all the derivatives of even orders are equal to zero at these points.

At the flexion points of the temperature curve, the series (6) establishes the following relation between the radiant and thermodynamic temperatures:

$$T_{\text{rad.bn}}^4 = \frac{1}{k}\left(\alpha_p T_{\text{w}}^4 + \beta T_{\text{rad.bn}}^4\right),$$

where T_{w} is the wall thermodynamic temperature; $T_{\text{rad.bn}}$ is a radiant temperature at points of the unbounded extinction medium that coincide with the wall surface.

This relation can be reduced to explicit form by means of identical transformations:

$$\alpha T_{\text{rad.bn}}^4 = \alpha_p T_{\text{w}}^4.$$

Finally, boundary conditions of the differential equation (12) would be found with the help of the formula (9) for determining temperature:

$$\alpha T_{\Sigma.\text{bn}}^4 = \alpha_p T_{\text{w}}^4, \tag{13}$$

were $T_{\Sigma.\text{bn}}$ is a determining temperature at points of the unbounded extinction medium that coincide with the wall surface.

Radiative heat transfer in the enclosed medium

A difference between the effective radiation flux, propagating from confining walls, and the hypothetical radiation flux outside of the unbounded extinction medium forms an anisotropic additional radiation flux. Separate calculating absorption of the additional radiation would not only complicate the computing algorithm, but also would introduce new errors into the computer simulation. It is more appropriate to find conditions allowing to apply the equations (9) – (12) to the summary radiation transferred. With the defining temperature in enclosed space denoted as T_s, they receive following forms:

$$T_{\text{s}}^4 \equiv T_{\text{rad}}^4 + \frac{1}{3k}\left(\alpha T_{\text{rad}}^4 - \alpha_{\text{p}} T^4\right), \tag{14}$$

$$\mathbf{q}_{\text{rad}} \approx -\frac{4\sigma}{3k}\,\text{grad}\, T_{\text{s}}^4, \tag{15}$$

$$\text{div}\mathbf{q}_{\text{rad}} = \frac{4\sigma}{1+\alpha/(3k)}\left(\alpha_p T^4 - \alpha T_s^4\right), \tag{16}$$

$$\frac{\partial}{\partial x}\left(\frac{\partial T_{\text{s}}^4}{k\partial x}\right)+\frac{\partial}{\partial y}\left(\frac{\partial T_{\text{s}}^4}{k\partial y}\right)+\frac{\partial}{\partial z}\left(\frac{\partial T_{\text{s}}^4}{k\partial z}\right)+\frac{9}{3+\alpha/k}\left(\alpha_{\text{p}}T^4-\alpha T_{\text{s}}^4\right)=0. \tag{17}$$

Despite formal identity, expression (14) differs in values of its radiant temperature T_{rad} from the analogue (9). Also, the differential equations (12) and (17) might produce different results being solved with distinct boundary conditions.

The radiation fluxes that are required for stating the boundary conditions on surface of the walls are depicted in Fig. 1. It is accepted here that the effective radiation of walls is diffusive, i.e. it is emitted and reflected with equal radiation intensity in all the directions within the hemispherical solid angle.

When the confined dust-laden gaseous volume has a fairly large optical thickness, the additional radiation propagated from the opposite wall is expected to be absorbed completely by the medium. Thus, it may be assumed, that the incident radiation flux on the enclosing walls practically equals in this case to such of a flux produced by the hypothetical unbounded medium with the same temperature field. This radiation flux can be determined by integrating the series (5) of the radiation intensity, multiplied by $\cos\theta$, over the hemispherical solid angle. Since the even temperature derivatives at the bounds of the domain are supposed to be equal to zero, the surface density of the incident radiation flux would be represented as follows:

$$E_{\text{inc}} = \frac{\alpha_{\text{p}}}{\alpha}\sigma T_{\text{w}}^4 - 2\frac{\pi}{k}\sum_{n=0}^{\infty}\left[\frac{\partial^{2n+1}\left(\alpha_{\text{p}} I_{\text{o}} + \beta\bar{I}\right)}{(2n+3)k^{2n+1}\partial l^{2n+1}}\right]_{\text{bn}}$$

where subscript "bn" denotes derivatives of the radiation intensity found at the points of the unbounded medium that coincide with the wall surface.

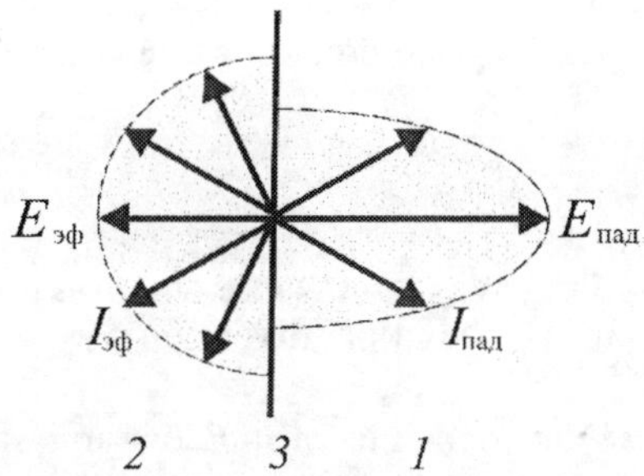

Figure 1. A scheme of radiation fluxes on surface of the walls: *1* – incident radiation *2* – effective radiation; *3* – wall surface.

This equality would be presented more obviously in a wide form:

$$E_{\text{inc}} = \frac{\alpha_{\text{p}}}{\alpha}\sigma T_{\text{w}}^{4} - 2\pi\frac{\alpha_{\text{p}}}{k}\left[\frac{1}{3k}\left(\frac{\partial I_{\text{o}}}{\partial y}\right)_{\text{bn}} + \frac{1}{5k^{3}}\left(\frac{\partial^{3} I_{\text{o}}}{\partial y^{3}}\right)_{\text{bn}} + \frac{1}{7k^{5}}\left(\frac{\partial^{5} I_{\text{o}}}{\partial y^{5}}\right)_{\text{bn}} + \ldots\right] -$$

$$-2\pi\frac{\beta}{k}\left[\frac{1}{3k}\left(\frac{\partial \bar{I}}{\partial y}\right)_{\text{bn}} + \frac{1}{5k^{3}}\left(\frac{\partial^{3} \bar{I}}{\partial y^{3}}\right)_{\text{bn}} + \frac{1}{7k^{5}}\left(\frac{\partial^{5} \bar{I}}{\partial y^{5}}\right)_{\text{bn}} + \ldots\right].$$

From here, in view of the series (7), a simple expression will be obtained for the surface density of the incident radiation flux on the walls that enclose the extinction medium:

$$E_{\text{inc}} \approx \frac{\alpha_{\text{p}}}{\alpha}\sigma T_{\text{w}}^{4} + \frac{1}{2} q_{\text{rad.w}}^{\infty} \tag{18}$$

The surface density of the effective flux of radiation emitted and reflected by the confining walls can be determined with the well-known formula [4]:

$$E_{\text{eff}} = \frac{\alpha_{\text{p}}}{\alpha}\sigma T_{\text{w}}^{4} + \left(\frac{1}{\varepsilon_{\text{w}}} - 1\right) q_{\text{rad.w}}, \tag{19}$$

where ε_{w} is the emissivity of the wall surface; $q_{\text{rad.w}}$ is the density of the resulting radiation flux from the extinction medium to the wall. The ratio of absorption coefficients had been introduced in formula (18), in order to match more accurately the radiation fluxes of the walls and those of a volume of the extinction medium in accordance with the equality (13).

The difference between the incident (18) and effective (19) radiation fluxes gives a formula that connects on the wall surface the resulting radiation fluxes in confined and unbounded extinction media:

$$q_{\text{rad.w}} = q_{\text{rad.w}}^{\infty}\, \varepsilon_{\text{w}}/2. \tag{20}$$

Boundary conditions

In the diffusion approach method, the boundary conditions on the wall surface are inherently approximate. Besides, the incident radiation is assumed, as a rule, to be diffusive. Such a rough stating leads to unpredictable mistakes in numerical simulations.

In the elaborated diffusion model, Neumann more accurate boundary conditions for the differential equation (17) directly follow from a ratio (20) in view of the gradient expressions (10) and (15):

$$\left(\partial T_{\text{s}}^{4}/\partial y\right)_{\text{w}} = \frac{\varepsilon_{\text{w}}}{2}\left(\partial T_{\Sigma}^{4}/\partial y\right)_{\text{w}}. \tag{21}$$

The right-side part of this equality has to be computed in each of iteration after solving the differential equation (12) of radiative heat transfer in the unbounded medium.

Robin boundary conditions can be derived from the balance equation that interrelates the radiant temperature and surface fluxes of radiation [16]:

$$4\sigma T_{\text{rad.w}}^4 = m_{\text{eff}} E_{\text{eff}} + m_{\text{inc}} E_{\text{inc}},$$

where $T_{\text{rad.w}}$ is the radiant temperature, calculated on the wall surface; m_{eff} and m_{inc} are ratios of hemispherical integrals that determine volumetric and surface radiation fluxes on the wall. It can be assumed m_{eff}=2 for the diffused radiation of wall.

If to take into consideration formulae (17) and (18) for densities of incident and effective radiation fluxes, the preceding equality would be reduced to an expression that contains only coefficient m in its last item:

$$4\sigma T_{\text{rad.w}}^4 = 4\frac{\alpha_{\text{n}}}{\alpha}\sigma T_{\text{w}}^4 + (1-\varepsilon_{\text{w}})q_{\text{rad.w}}^{\infty} + \frac{m}{2}q_{\text{rad.w}}^{\infty}. \quad (22)$$

This coefficient is defined here as follows:

$$m = \left[\int_{(2\pi)} I_l \,\text{d}\omega - 2\frac{\sigma}{k}\left(\alpha_{\text{p}} T_{\text{w}}^4 + \beta T_{\text{rad.bn}}^4\right)\right] \Big/ \left(\frac{1}{2}q_{\text{rad.w}}^{\infty}\right).$$

The denominator of this ratio corresponds to the series (7). Its numerator ought to be found by means of integrating termwise the series (5) and admitting the even derivatives to be equal zero:

$$\int_{(2\pi)} I_l \,\text{d}\omega - 2\frac{\sigma}{k}\left(\alpha_{\text{p}} T_{\text{w}}^4 + \beta T_{\text{rad.bn}}^4\right) =$$

$$-2\pi\frac{\alpha_{\text{p}}}{k}\left[\frac{1}{2k}\left(\frac{\partial I_{\text{o}}}{\partial y}\right)_{\text{bn}} + \frac{1}{3k^3}\left(\frac{\partial^3 I_{\text{o}}}{\partial y^3}\right)_{\text{bn}} + \frac{1}{5k^5}\left(\frac{\partial^5 I_{\text{o}}}{\partial y^5}\right)_{\text{bn}} + \ldots\right] -$$

$$-2\pi\frac{\beta}{k}\left[\frac{1}{2k}\left(\frac{\partial \bar{I}}{\partial y}\right)_{\text{bn}} + \frac{1}{3k^3}\left(\frac{\partial^3 \bar{I}}{\partial y^3}\right)_{\text{bn}} + \frac{1}{5k^5}\left(\frac{\partial^5 \bar{I}}{\partial y^5}\right)_{\text{bn}} + \ldots\right]$$

Comparing numerical coefficients before the first derivatives in both numerator and denominator gives us a rough estimation: $m \approx 3/2$. Now replacing the determining temperature $T_{\text{s.w}}$ in the equation (22) with the radiant one in accordance with formula (14) results in a computational expression for the Robin boundary conditions:

$$\sigma T_{\text{s.w}}^4 \approx \sigma T_{\Sigma.\text{bn}}^4 + \frac{3k+\alpha}{12k}(1{,}75 - \varepsilon_{\text{w}})q_{\text{rad.w}}^{\infty}. \quad (23)$$

This expression includes the resulting radiation-flux density in the unbounded medium. Thus, in order to set correctly the boundary conditions for the differential equation (17), one has to solve previously the differential equation (12) with the boundary condition (13).

Owing to such an algorithm, the precision of the whole numerical method rises [16]. At some distance from walls, where the dust-particle absorption nullifies the wall additional radiation over the total spectrum, the formula (15) becomes accurate. If, in addition, the resulting radiation flux had been determined by means of the relation (20) with acceptable accuracy, some potential faults of the gradient formula (15) might be revealed only restrictedly in a narrow field beside the walls [17], and they would not be capable to yield any essential inadequacy to computational results.

IOP Conf. Series: Materials Science and Engineering **791** (2020) 012015 doi:10.1088/1757-899X/791/1/012015

4. Conclusion

The accurate diffusion model has been theoretically elaborated for radiative heat transfer in dust-laden gases. To eliminate singularity of temperature curves on the surface of confining walls, the enclosed extinction medium was hypothetically considered as an unbounded one. The radiation intensity has been expanded in a series and integrated over the spherical solid angle, what has allowed to obtain more precise differential equation for the resulting radiation flux both in the unbounded and enclosed extinction medium. As a result, shortcomings peculiar to the previous method of diffusion approach have been eliminated, and correct Neumann and Robin boundary conditions have been formulated.

The elaborated diffusion model has an easy algorithm, minimum requirements to computer resources and small time of its numerical realization.

Acknowledgments

The article was prepared within development program of the Flagship Regional University on the basis of Belgorod State Technological University named after V.G. Shukhov.

References

[1] Siegel R and Howell J R 1972 *Thermal radiation heat transfer* (N.Y.: McGraw-Hill)

[2] Huang Zh, Cheng Q and Lou Ch 2018 A direct solution for radiative intensity with high directional resolution in isotropically scattering media *Int. J. Heat and Mass Transfer* **117** 296-302

[3] Goody R M and Yung Y L 1995 *Atmospheric Radiation* (N.Y.: Oxford Univers. Press)

[4] Shorin S N 1964 *Teploperedacha* [Heat Transfer] (Moscow: Vysshaya Shkola) [in Russian]

[5] Cai J and Modest M F 2014 Improved full-spectrum k-distribution implementation for inhomogeneous media using a narrow-band database *J. Quant. Spectrosc. Rad. Transfer* **141** 65-72

[6] Rothman L S, Gordon I E, Barber R J, Dothe H, Gamache R R, Goldman A, Perevalov V I and Tashkun S A 2010 HITEMP, the high-temperature molecular spectroscopic database *J. Quant. Spectr. Rad. Transfer* **111** 2139-50

[7] Pal G, Gupta A, Modest M F and Haworth D C 2015 Comparison of accuracy and computational expense of radiation models in simulation of non-premixed turbulent jet flames *Combustion and Flame* **162** 2487-5

[8] Feldheim V and Lybaert P 2004 Solution of radiative heat transfer problems with the discrete transfer method applied to triangular meshes *J. Comp. Appl. Math.* **168** 179-90

[9] Ozen G and Selçuk N 2013 Performance of DOM and IDA with different angular discretization methods in 3-D absorbing–emitting–scattering media *Int. J. Therm. Sci.* **65** 104-10

[10] Raithby G D and Chui E H 1990 A finite-volume method for predicting a radiant heat transfer in enclosures with participating media *J. Heat Transfer* **112** 415-23

[11] An W, Ruan L M, Qi H and Liu L H 2005 Finite element method for radiative heat transfer in absorbing and anisotropic scattering media *J. Quant. Spectr. Rad. Transfer* **96** 409-22

[12] Zhang Y, Yi H-L and Tan H-P 2014 Natural element method analysis for coupled radiative and conductive heat transfer in semitransparent medium with irregular geometries *Int. J. Therm. Sci.* **76** 30-42

[13] Hottel H C and Sarofim A F 1967 *Radiative Transfer* (N.Y.: McGraw-Hill)

[14] Ou Szu-C S and Liou Kuo-N 1982 Generalization of the spherical harmonic method to radiative transfer in multi-dimensional space *J. Quant. Spectrosc. Rad. Transfer* **28** 271-88

[15] Chetverushkin B N 1985 *Zadachi matematicheskogo modelirovaniya dinamiki izluchayushchego gaza* [Mathematic simulating problems of the radiating-gas dynamics] (Moscow: Nauka) [in Russian]

[16] Kuznetsov V A and Trubaev P A 2018 Resources and Problems of the Mathematical Simulating Thermo-Technological Processes *J. Phys.: Conf. Ser.* **1066** 012024

[17] Kuznetsov V A 2017 Mathematical Model of the Radiative Heat Exchange in the Selective Gases of a Diffusion Flame *J. of Eng. Physics and Thermophysics* **90(2)** 357-65

IOP Conf. Series: Materials Science and Engineering **791** (2020) 012016 doi:10.1088/1757-899X/791/1/012016

Determination of the physicochemical properties of boric acid for VVER emergency modes

A V Morozov, A R Sahipgareev, A S Shlepkin and A S Soshkina

State Scientific Centre of the Russian Federation – Institute for Physics and Power Engineering named after A.I. Leypunsky, 1, Bondarenko Sq., Obninsk, 249033, Russia

E-mail: sas@ippe.ru

Abstract. The results of experimental research of physicochemical properties of water solutions of H_3BO_3 are presented in this article. The literature review about the influence of hydrogen ion exponent of H_3BO_3 solutions on its heat and mass transfer, and corrosiveness of an acid in the VVER reactor core in is given. It is found that the existing data does not cover the entire range of parameters (acid concentration, temperature and pressure) characteristic of a possible Loss-of-Coolant Accident (LOCA) at nuclear power plant with a VVER. The problem of the accumulation and crystallization of boric acid in case of LOCA and the operation of passive safety systems (such as hydro accumulators of the first, second and third stages, and a passive heat removal system) is formulated. Test equipment and techniques of research are described. The generalizing equations for the pH and acidity constant of hydrous solutions of H_3BO_3 in range of the concentration (5-100 gkg^{-1} H_2O) and temperature (25-50 °C) has been obtained.

1. Introduction

Within the concept of the safety of the VVER-TOI project, emphasis is paid to issues related to the accumulation and crystallization of H_3BO_3 in the core during Loss-of-Coolant Accident (LOCA) within 72 hours. As is known, the reactor autonomy during an emergency process is determined by the operation of the passive safety systems (PSS). PSS provide the core cooling by feed into the reactor of the water solution of boric acid from the hydro accumulators of the passive core flooding system (figure 1). As well as it is necessary to take into account the flow of condensate from the steam generators (SG) of three non-emergency loops, which operating into condensing mode due to the passive heat removal system [1-3].

Based on this, it is postulated that further mass transfer of boric acid in the core can be carried out in two ways. With a sufficient flow rate of the water solution of boric acid from the system of hydro accumulators of second and third stage (HA-2 and HA-3 systems), mixing of the flows occurs: the incoming solution, the condensate from the SG and the coolant remaining in the lower part of the core. Subsequently, a part of boric acid is carrying out into the break (figure 1, position 4). If the mixing of the condensate flows and the boric acid solution from HA-2 and HA-3 does not occur, then due to the higher density, the boric acid solution drops down to the lower part of the core. Above this layer, condensate accumulates with a low acid concentration. Further excess condensate from the steam generators carry away into the break. Thus, conditions are created for increasing the concentration of H_3BO_3 in the lower part of the core and for achieving the conditions for its crystallization.

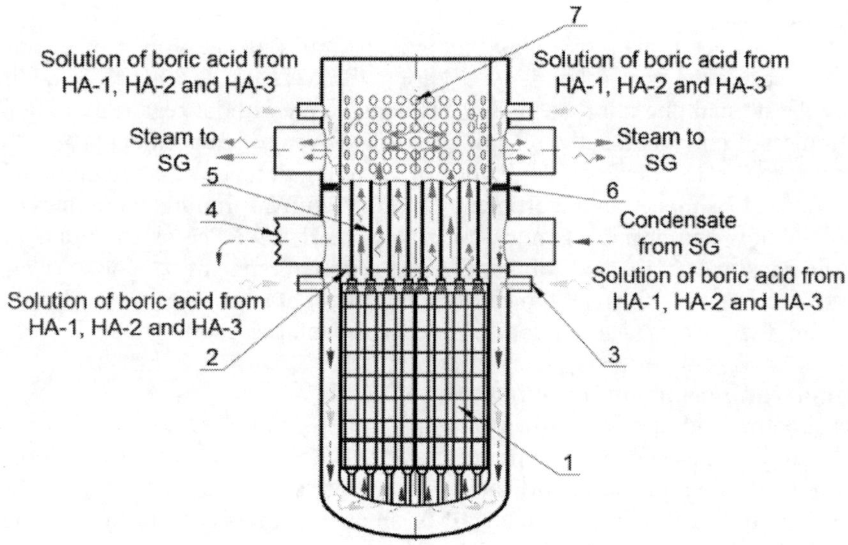

Figure 1. Mass transfer of H_3BO_3 solution in the VVER in LOCA. 1 – core; 2 – evaporation surface; 3 – nozzle of the emergency core cooling system; 4 – primary circuit break; 5 – protective tube unit; 6 – separation collar; 7 – the perforation of reactor barrel

In the research of the accumulation and crystallization of boric acid in the VVER core in the event of an accident conducted by the authors of this paper, it has been found that exceeding the maximum concentration of boric acid starts after 45 hours of the emergency process [5]. It should be noted that in the framework of these studies it was supposed that the only form of the existence of boric acid is orthoboric acid. However, in inorganic chemistry [6–8] there is a known tendency for the formation of polyboric acids with increasing concentration of orthoboric acid (H_3BO_3).

Orthoboric acid has very weak acidic properties, due to addition of a hydroxyl anion rather than to removal of a proton of hydrogen. The structure and composition of polyboric acids resulting from this process depend on the initial concentration of orthoboric acid and temperature.

The effect of polyborate complexes on the boric acid crystallization process has not been studied yet. The solubility of these particles may be higher than the solubility of orthoboric acid. In this case, their formation will favorably affect the processes of heat removal from the reactor core in case of an accident, because it will reduce the risk of sludge formation on the reactor internal surfaces. But, we can't exclude the possibility that the solubility of the formed polymer boron-containing particles will be lower than the solubility of orthoboric acid. This, in turn, can complicate the process of heat removal from the core. The intensity of the formation of polymer particles depends on the temperature and *pH* of the solution.

At the present, data on the effect of boric acid concentration on the *pH* of the solution are very limited due to the lack of experimental results of the studying of highly concentrated H_3BO_3 solutions. Boric acid in this form has very weak acidic properties and does not act as a proton donor, accordingly, its corrosion activity is extremely small. At the same time, in [9] the information is given that with increasing concentration of H_3BO_3 acidity of the environment increases and the formation of polymer particles, the structure of which is described above, occurs. In addition, as the degree of acidity increases, the corrosivity of the solution also increases. Under normal operating conditions of VVER, the acidity of the medium due to the presence of an orthoboric acid solution with a concentration of 4-8 gl^{-1} in the reactor primary circuit is compensated by the addition of potassium hydroxide. In the event of LB LOCA, the entire mass of coolant that is in the reactor at the time of the accident is taken out to the containment. Thus, subsequently, in the reactor core there is only a solution of boric acid with a gradually increasing concentration (due to the operation of HA-2 and HA-3 systems) without any alkalizing reagents.

As noted above, currently there are only theoretical studies of the properties of polyboric acids in assessing their effect on the water chemistry at NPPs with VVER reactors [9]. In addition, in the calculations, only the normal operation of the NPP is considered without regard to a significant increase in the concentration of boric acid in the reactor core in the event of an accident [10].

Thus, to improve the accuracy of calculations of accumulation and crystallization of boric acid and taking into account their impact on the efficiency of heat removal from the core in case of emergency, it is necessary to know and physicochemical properties of H_3BO_3. The existing data on the degree of acidity (*pH*) and density of boric acid solutions cover a limited range of parameters (temperature, pressure, acid concentration), which is not typical for an accident at a nuclear power plant with VVER [11, 12]. In this regard, there was a need for additional experimental research.

2. Experimental equipment and techniques

At Joint Stock Company «State Scientific Centre of the Russian Federation – Institute for Physics and Power Engineering named after A.I. Leypunsky» (IPPE) the experimental research of physicochemical properties of water solutions orthoboric acid with the parameters characteristic of the first circuit of VVER reactor facility in 24 hours after the start of an accident has been made [13]. The experiments were conducted in two stages. At the first stage, the density of boric acid solutions with a concentration of 2.5-400 gkg^{-1} H_2O at a temperature of 25-130 °C in the pressure range of 0.1-0.4 MPa were measured in a specially designed test facility [14].

At the second stage of research, the experiments for determining the degree of acidity of aqueous solutions of boric acid depending on the concentration (5-100 gkg^{-1} H_2O) and temperature (25-50 °C) were conducted. pH-meter MARK-901 – a universal device designed to measure the activity of hydrogen ions (*pH*) was used for measurements. The measurement error of the activity of hydrogen ions for this gauge is ±0.1 *pH*.

The experiments were carried out according to the following technique. Before starting the measurements, the pH meter was calibrated in a special buffer solution with a *pH* of 7.0. Further, in accordance with the given program of experiments, the degrees of acidity of distilled water and boric acid solutions with different concentrations were measured. On laboratory scales BP2100 (maximum weight – m_{max}=2100±0.1 g) the required mass of distilled water (1 kg) is weighed at room temperature. Then this volume of water is poured into four conical flasks of 250 ml each. These vessels are then sequentially placed in a thermostat filled with distillate and equipped with a heating device for warming up the liquid up to the required temperature (25, 40 and 50 °C). When the specified temperature is reached, the heater is switched off and a pH-meter and an electronic thermometer are placed inside the bulb (figure 2). Readings from both devices are recorded within 10 minutes. Then the flask with distilled water is removed from the thermostat. The next vessel with the test liquid is placed inside the device and the procedure is repeated according to the technique described above.

After determining the degree of acidity of distilled water, a solution with a concentration of 5 gkg^{-1} H_2O was prepared. For this purpose, 5 g of dry orthoboric acid, pre-weighted on laboratory scales Adventurer Pro AV412C (maximum weight – m_{max}=410±0.01 g), is added to 1 kg of distillate placed in a beaker. Then, with the help of magnetic stirrer MSH-300 a complete dissolution of H_3BO_3 crystals is achieved. For the preparation of the solutions with concentrations exceeding the limit of solubility of boric acid at room temperature, a heater is switched at the stirring device. Further procedures are performed according to the method described above.

Figure 2. Test facility for hydrogen ion exponent measurement of water solutions of H_3BO_3.

3. Results of experimental research

On the test facility twenty one series of experiments was carried out, each of which included 4 measurements by the method described above. The obtained experimental data on temperature and *pH* were averaged and presented in table 1. Dependences of *pH* of boric acid water solution on temperature at various concentrations are shown in figures 3 and 4.

Table 1. The experimental data.

No. exp.	Concentration H_3BO_3 (gkg^{-1} H_2O)	Temperature (°C)	*pH*	No. exp.	Concentration H_3BO_3 (gkg^{-1} H_2O)	Temperature (°C)	*pH*
1	0	25.7	6.10	12	10	50.0	4.67
2	0	39.9	5.93	13	25	25.3	4.44
3	0	49.8	5.77	14	25	39.7	4.16
4	2.5	25.2	5.72	15	25	49.6	4.03
5	2.5	39.7	5.59	16	50	25.2	3.81
6	2.5	49.9	5.26	17	50	39.8	3.49
7	5	25.1	5.52	18	50	49.6	3.38
8	5	40.0	5.31	19	75	39.8	3.04
9	5	49.8	4.99	20	75	49.9	2.85
10	10	25.3	5.13	21	100	49.6	2.53
11	10	39.9	4.92				

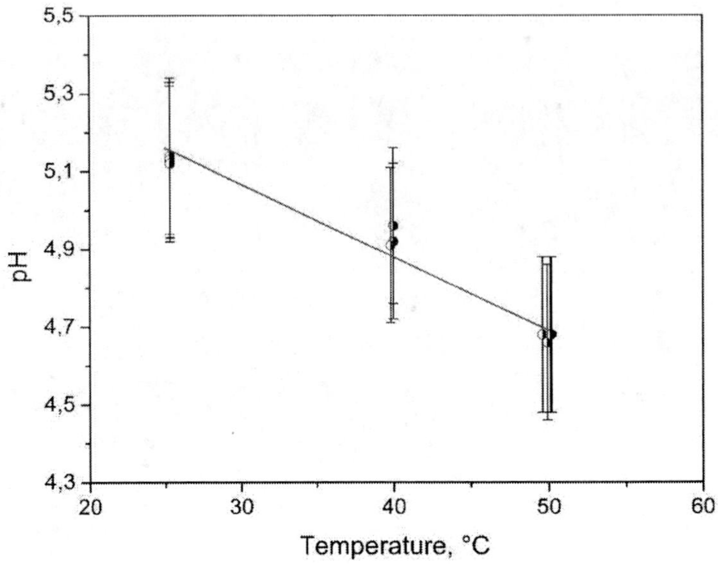

Figure 3. The change of the hydrogen ion exponent of boric acid as a function of temperature at a concentration of 10 gkg^{-1} H_2O.

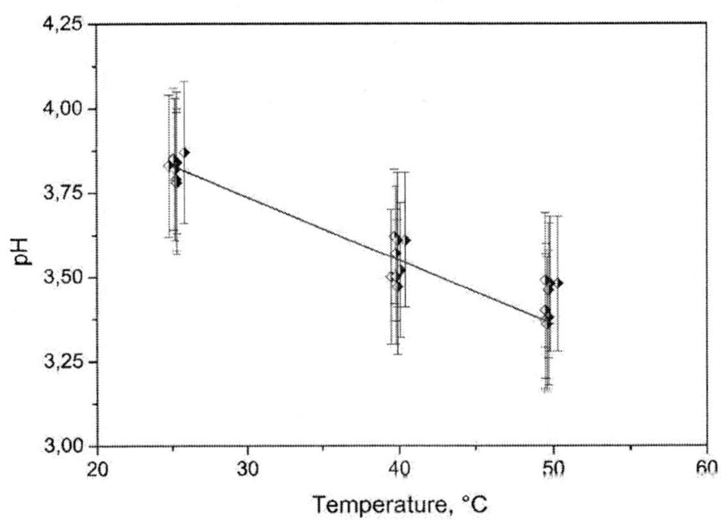

Figure 4. The change of the hydrogen ion exponent of boric acid as a function of temperature at a concentration of 50 gkg^{-1} H_2O.

The experimental data shown in figures 3 and 4 are well approximated by the following dependence:

$$pH_{sol} = pH_{dist}(T_{sol}) - (12.72C_{H_3BO_3}{}^{0.59} + 0.49\times10^{-2}T_{sol}), \quad (1)$$

where the hydrogen ion concentration of distilled water is:

$$pH_{dist}(T_{sol}) = 6.46 - 1.36\times10^{-2}T_{sol}, \quad (2)$$

where T_{sol} is the temperature of the H_3BO_3 solution, °C; and $C_{H_3BO_3}$ is the concentration of boric acid in mixture, gg^{-1}. The discrepancy between the experimental points and the approximating curves is less than 10%.

The obtained experimental data on the hydrogen ion concentration of boric acid were compared with the known literature data. It has been established that, for example, in chemistry manuals [15], the *pH* of H_3BO_3 is given at a temperature of 20 °C and a concentration of less than 2 gkg^{-1} H_2O. The data obtained in experiments conducted by Westinghouse lie outside the range under study [17]. Also, in [16], an equation for the estimated calculation of the hydrogen ion exponent of weak acids is presented:

$$pH = \frac{1}{2} pK_a - \frac{1}{2} \lg(C_M), \tag{3}$$

where pK_a is acidity constant; and C_M is the molar concentration of the water solution of H_3BO_3, $moll^{-1}$.

The molar concentration can be determined taking into account the known empirical dependency for calculating the solution density of boric acid [4] according to the following equation:

$$C_M = \frac{m_{H_3BO_3} \rho_{sol}}{M_{sol}(1000 + m_{H_3BO_3})}, \tag{4}$$

where $m_{H_3BO_3}$ is the mass of H_3BO_3, g; M_{sol} is the molar mass of the solution, $gmol^{-1}$; and

$$\rho_{sol} = 1141 - 0.48(T_{sol} + 273.15) + 39217(T_{sol} + 273.15)^{-0.843} C_{H_3BO_3}, \tag{5}$$

is the density of the boric acid in solution, gl^{-1}; and $C_{H_3BO_3}$ is the concentration of boric acid in solution, gg^{-1}.

In addition, to calculate the concentration of hydrogen ions according to equation (3), it is required to know the acid constant pK_a. Known literature data on the acidity constant are limited [15], and in the formulas for its determination the influence exerted by the concentration of H_3BO_3 is not taken into account [12].

In this regard, it is required to determine the dependence of the acidity constant of boric acid on its concentration in solution. The equation for calculating of acidity constant of H_3BO_3 using empirical dependency hydrogen ion exponent (1) and molar concentration of boric acid (4) was obtained and the following form:

$$pK_a = 2(pH_{dist}(T_{sol}) - (12.72 C_{H_3BO_3}{}^{0.59} + 0.49 \times 10^{-2} T_{sol})) + \lg(C_M), \tag{6}$$

the terms of the equation are described above.

Figure 5 presents a dependence of *pH* of boric acid aqueous solutions on various concentrations.

As shown in figure 5, not only temperature, but also concentration affects the change in *pH* of boric acid.

Equations (1) and (6) for calculating the hydrogen ion concentration and acidity constants presented in this paper can be used in the following parameter interval of boric acid solution:

- temperature of solution T_{sol} =25-50 °C;
- boric acid concentration in water solution – $C_{H_3BO_3}$ = 0-100 gkg^{-1} H_2O.

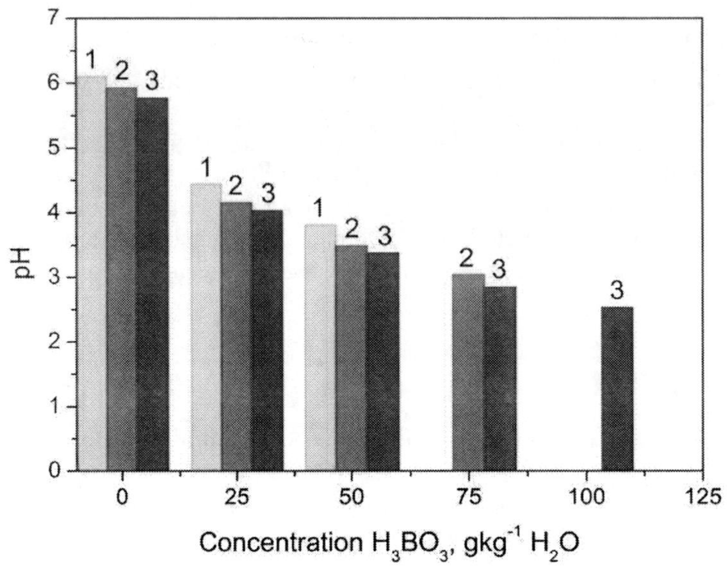

Figure 5. Comparison of the hydrogen ion exponent of boric acid at the change its concentration in the range of 0-100 gkg^{-1} H_2O: 1 – 25 °C; 2 – 40 °C; 3 – 50 °C.

4. Conclusion

As a result of research at JSC «SSC RF – IPPE», experimental data on the physicochemical properties of highly concentrated H_3BO_3 solutions at the parameters typical for LOCA on NPPs with VVER have been obtained. These results can significantly expand the range of known properties of H_3BO_3 solutions.

The results of experiment are of profound interest for NPPs with VVER reactors equipped with passive safety systems. The obtained data can be used to assess the effect of boric acid on the water chemistry parameters of the VVER primary circuit, both in normal operation and in the case of an accident, and to clarify the results of calculations of emergency heat removal processes in the reactor, carried out using computer codes.

Acknowledgments

The study has been financed by grant of the Russian Science Foundation (project 16-19-10649).

References

[1] Berkovich V M, Korshunov A S, Taranov G S, Kalyakin S G, Morozov A V and Remizov O V 2006 *Atomic Energy* **100(1)** 14-9
[2] Morozov A V and Remizov O V 2012 *Thermal Eng.* **59(5)** 365-70
[3] Morozov A V, Remizov O V and Tsyganok A A 2010 *Trans. American Nuclear Society 2010 ANS Annual Meeting* (San Diego) 676-7
[4] Morozov A V, Pityk A V, Ragulin S V, Sahipgareev A R and Shlepkin A S 2017 *Proc. Int. Cong. on Advances in NPP (Fukui and Kyoto)* 186-92
[5] Morozov A V, Pityk A V, Ragulin S V, Sahipgareev A R, Soskina A S and Shepkin A S 2017 *Journal of Physics: Conference Series* **899(9)** 092010
[6] Bates R G 1973 *Determination of pH. Theory and practice* (NY: John Wiley & Sons)
[7] Karapet'yants M Kh and Drakin S I 1981 *Obshchaya i neorganicheskaya khimiya* [General and inorganic chemistry] (Moscow: Khimiya) [In Russian]
[8] Cotton F, Wilkinson G and Gaus P 1987 *Basic inorganic chemistry* (NY: John Wiley & Sons)

[9] Gavrilov A V, Kritsky V G, Rodionov Yu A and Berezina I G 2012 *Thermal Eng.* **60(7)** 482-6

[10] Morozov A V, Sorokin A P, Ragulin S V, Pityk A V, Sahipgareev A R, Soshkina A S and Shlepkin A S 2017 *Thermal Eng.* **64(7)** 490–5

[11] Azizov N D and Akhundov T S 1996 *High temperature* **34(5)** 788-91

[12] Manov G G, DeLollis N J, Lindvall Ph W and Acree S F 1946 *J. of Research of the National Bureau of Standards* **36** 543-58

[13] Morozov A V, Sorokin A P, Pityk A V, Sahipgareev A R, Soshkina A S and Shlepkin A S 2018 *Atomic Energy* **125(3)** 178-84

[14] Morozov A V, Pityk A V, Sahipgareev A R and Shepkin A S 2018 *J. of Physics: Con. Series* **1128** 012103

[15] Simanova S A 2004 *Novyi spravochnik khimika i tekhnologa* [New Handbook of chemist and technologist] (St. Petersburg: ANO NPO "Professional") [In Russian]

[16] Glinka N L 1981 *Obshhaya ximiya* [*General chemistry*] (Moscow: Mir publishing) [In Russian]

[17] Byers W A, Brown W L, Kellerman B E, Shearer K S and Fink D J 2010 Summary of Tests to Determine the Physical Properties of Buffered and Un-buffered Boric Acid Solutions *WCAP-17021-NP Revision 1*

IOP Conf. Series: Materials Science and Engineering **791** (2020) 012017 doi:10.1088/1757-899X/791/1/012017

Study of temperature fields in a ball bearings bodies under boundary conditions of the third kind

A I Popov, P V Iglin, K V Gubareva, T V Iglina and A S Doronin

Department of Industrial Heat Power Engineering, Samara State Technical University, 244, Molodogvardeyskaya St., Samara, 443100, Russia

E-mail: pixinot@icloud.com

Abstract. The reliability of the rolling bearing body directly affects the continuity and duration of operation of industrial equipment. Ball bearings are often used in centrifugal pumps of thermal power plants and water supply systems. Overheating and subsequent failure of such bearings lead to failure of pumping equipment. To avoid such breakdowns, it is necessary to study the thermal processes inside the rolling elements, which can lead to their deformation and destruction. In this paper, the problems of non-stationary heat transfer inside rolling bodies (spherical bodies) are considered. An analytical solution of the problem in dimensionless values of temperatures and coordinates is obtained on the basis of Fourier and Bubnov-Galerkin methods under boundary conditions of the third kind. The solution of such a problem requires the use of an extensive mathematical apparatus, so special software tools such as Mathcad and MatLab were used in the work. So solving the problem analytically in dimensionless quantities is convenient because it allows you to scale the results to all such objects without reference to a specific rolling body. As a result, the solution is obtained in the form of a power polynomial that does not contain special functions.

1. Introduction

Ball bearings are widely used in modern industry (from agriculture to aerospace) and in the energy industry. The reliability of most technological equipment depends on it condition. Ball bearings are everywhere used in centrifugal pumps at thermal power plants, as well as in step-up pumping stations of heat supply and water supply systems. The analysis of actual works showed [1-5] that the most part of breakdowns of the pumping equipment is connected with failure of bearings. For example, [1] bearing defects at Nile pumping stations resulted in an 18% reduction in pump efficiency.

Therefore, the study of the influence of various factors (external and internal) on the characteristics of centrifugal pump bearings is of particular interest [6]. In the course of its operation, the rolling bearing body is heated because of friction forces, especially when there is insufficient or complete absence of lubricants. This can cause deformation of the rolling body, jamming and destruction of the bearing. Many studies [7-9] of thermal processes in bearings have shown a negative effect of overheating of the rolling body on the operation of the bearing and all equipment (centrifugal pump).

There are several works [10-16], which are aimed at studying the thermal processes in the rolling bodies of bearings. However, the majority of researchers consider questions of thermal interaction of rolling bodies with bearing lubrication, without paying due attention to internal thermal processes of a rolling body. These processes can cause structural damage inside the bearings. Thus, the actual problem is to study the processes of thermal conductivity inside the rolling bearings bodies (spherical bodies). The results of this study will allow to better analyze the work of centrifugal pump bearings at thermal stations and can be used in the system of diagnostics and automatic control of pumping equipment of the station. Moreover, the results of the study should be scaled to bearings of different sizes under different external conditions (ambient temperature), so the task must be solved in dimensionless values of temperatures and coordinates.

Statement of the problem: the boundary value problem of thermal conductivity for a sphere under symmetric boundary conditions of the third kind will have the following form:

$$\frac{\partial t(x,\tau)}{\partial \tau} = a\left(\frac{\partial^2 t(x,\tau)}{\partial x^2} + \frac{2}{x}\frac{\partial t(x,\tau)}{\partial x}\right) \quad (\tau > 0;\ \ 0 \le x \le r); \tag{1}$$

$$t(x,0) = t_0; \tag{2}$$

$$\partial t(0,\tau)/\partial x = 0; \tag{3}$$

$$-\lambda\frac{\partial t(r,\tau)}{\partial x} = \alpha(t(r,\tau) - t_{\text{cp}}), \tag{4}$$

where t – temperature; x – coordinate; τ – time; t_0 – initial temperature; $t_{\text{ст}}$ – wall temperature; a – thermal diffusivity; r – radius of a sphere.

Using dimensionless parameters, we reduce the problem (1) – (4) to a dimensionless form

$$\frac{\partial\Theta(\xi,\text{Fo})}{\partial\text{Fo}} = \frac{\partial^2\Theta(\xi,\text{Fo})}{\partial\xi^2} + \frac{2}{\xi}\frac{\partial\Theta(\xi,\text{Fo})}{\partial\xi} \quad (\text{Fo} > 0;\ \ 0 < \xi < 1); \tag{5}$$

$$\Theta(\xi,0) = 1; \tag{6}$$

$$\frac{\partial\Theta(0,\text{Fo})}{\partial\xi} = 0; \tag{7}$$

$$\frac{\partial\Theta(1,\text{Fo})}{\partial\xi} + \text{Bi}\Theta(1,\text{Fo}) = 0, \tag{8}$$

where $\Theta = (t - t_{\text{cp}})/(t_0 - t_{\text{cp}})$ – dimensionless temperature; $\xi = x/r$ – dimensionless coordinate; $\text{Fo} = (a\tau)/r^2$ – Fourier criterion (dimensionless time).

2. Method of solution

2.1. Analytical solution

The approximate analytical solution is based on the joint application of the method of separation of variables and orthogonal methods of weighted residuals. Obtaining a high-precision approximate solution of the thermal conductivity problem is achieved by directly satisfying the differential equation of the Sturm-Liouville boundary value problem in a given number of points of the spatial variable [17].

Obtaining a solution is limited only by the possibility of separating the variables in the original differential equation. At the same time, there are practically no restrictions on the form of the differential equation of The Sturm-Liouville boundary value problem obtained after the separation of variables.

The solution of the problem (5-8), following the method of separation of variables, is found in the form

$$\Theta(\xi,\text{Fo}) = \phi(\text{Fo})\psi(\xi). \tag{9}$$

Substituting (9) in (5), we find

$$\frac{d\phi(Fo)}{dFo}+\nu\phi(Fo)=0; \tag{10}$$

$$\frac{d^2\psi(\xi)}{d\xi^2}+\frac{d\psi(\xi)}{d\xi}\frac{2}{\xi}+\nu\psi(\xi)=0, \tag{11}$$

where ν – some constant.

The solution of equation (10) is known and has the form

$$\phi(\mathrm{Fo})=A\exp(-\nu\mathrm{Fo}), \tag{12}$$

where A – unknown coefficient.

Substituting (9) into (7), (8) we obtain

$$d\psi(0)/d\xi=0; \tag{13}$$

$$\frac{\partial\psi(1)}{\partial\xi}+\mathrm{Bi}\cdot\psi(1)=0. \tag{14}$$

Solving the Sturm-Liouville boundary value problem (11), (13), (14) accepted as

$$\psi(\xi)=B_0+\sum_{i=1}^{r}B_i\xi^{i+1}, \tag{15}$$

where $B_i\ (i=\overline{1,\ r})$ – unknown coefficients. Note that the relation (15) satisfies the boundary condition (13).

The relation (13) allows to enter one more boundary condition

$$\psi(0)=const=1. \tag{16}$$

Substituting (15) in (16), we find $B_0=1$. Demand that the relation (15) satisfies the boundary condition (14) and the equation (11) at the points $\xi=0;1/3;2/3;1$. Substituting (15), limited to five terms of the series, in the ratio (14) and equation (11), with respect to points $\xi=0;1/3;2/3;1$ with respect to unknown coefficients $B_i\quad(i=\overline{1,5})$ we obtain a system of five algebraic linear equations. Then from the solution of this system at $Bi=1$ we find the coefficients $B_i\quad(i=\overline{1,5})$:

$$B_1=-\frac{\nu}{6};$$

$$B_2=\frac{13\nu^4-2943\nu^3+179250\nu^2-2906280\nu+6123600}{3c};$$

$$B_3=-\frac{29\nu^4-7782\nu^3+563580\nu^2-9520740\nu+20207880}{3c};$$

$$B_4=\frac{9\nu^4-2601\nu^3+197892\nu^2-3450600\nu+7348320}{c};$$

$$B_5=-\frac{3\nu^4-900\nu^3+69390\nu^2-1229580\nu+2624400}{c};$$

$$c = 4\nu^3 - 512\nu^2 + 1620\nu - 204120.$$

Find the integral of the weighted residual of equation (11)

$$\int_0^1 \left[\frac{\partial^2}{\partial \xi^2} \left(1 + \sum_{i=1}^{r} B_i \xi^{i+1} \right) + \frac{2}{\xi} \frac{\partial}{\partial \xi} \left(1 + \sum_{i=1}^{r} B_i \xi^{i+1} \right) + \nu \left(1 + \sum_{i=1}^{r} B_i \xi^{i+1} \right) \right] d\xi = 0. \quad (17)$$

Calculating the integrals in (17), taking into account the found values of the coefficients $B_i \quad (i = \overline{1,5})$ with respect to the eigenvalues ν_k, we obtain an algebraic equation of the fifth degree

$$\nu^5 - 1161\nu^4 + 246360\nu^3 - 14911200\nu^2 + 243777600\nu - 514382400 = 0. \quad (18)$$

From the solution of equation (18) we obtain five eigenvalues $\nu_1 = 907.242263$; $\nu_2 = 166.248673$; $\nu_3 = 63.157269$; $\nu_4 = 21.884347$; $\nu_5 = 2.467446$.

Substituting (12), (15) in (9), for each eigenvalue we will have partial solutions of the form

$$\Theta_k(\xi, Fo) = A_k \exp(-\nu_k \mathrm{Fo}) \left(1 + \sum_{i=1}^{r} B_i(\nu_k) \xi^{i+1} \right). \quad (19)$$

Each particular solution of (19) exactly satisfies the boundary conditions (7), (8) and approximately (in the third approximation) – equation (5). However, none of them, including their sum

$$\Theta(\xi, Fo) = \sum_{k=1}^{n} \left[A_k \exp(-\nu_k \mathrm{Fo}) \left(1 + \sum_{i=1}^{r} B_i(\nu_k) \xi^{i+1} \right) \right], \quad (20)$$

do not satisfy the initial condition (6). To fulfill the initial condition, its residual is made and the orthogonality of the residual to each eigenfunction is required, i.e.

$$\int_0^1 \left\{ \sum_{k=1}^{5} \left[A_k \left(1 + \sum_{i=1}^{r} B_i(\nu_k) \xi^{i+1} \right) \right] - 1 \right\} \psi_j(\nu_j, \xi) d\xi = 0. \quad (j = 1, 2, 3; \ r = 5) \quad (21)$$

By calculating the integrals in (21) to find $A_k \quad (k = \overline{1,5})$ we obtain a system of five algebraic linear equations. Its decision:

$$A_1 = 0.018464; \ A_2 = -0.108447; \ A_3 = 0.228104; \ A_4 = -0.398377; \ A_5 = 1.265116.$$

The results of calculations by the formula (20) in the fourth approximation are presented in Fig. 1. Their analysis allows us to conclude that in the range of numbers $0{,}01 \le \mathrm{Fo} < \infty$ the difference between the obtained solution and the numerical one does not exceed 1.5%.

To improve the accuracy of the solution, it is necessary to increase the number of members of the series (15). To obtain additional equations in order to determine the unknown coefficients of B_i we will increase the number of points along the ξ, coordinate in which equation (11) should be performed. And, in particular, taking the number of such points equal to 8 (with a step $\Delta\xi = 1/7$, starting from point $\xi = 0$), we obtain 9 equations with respect to unknown coefficients B_i (another equation is added as a result of the boundary condition (14)). After solving the system of these equations and finding the coefficients $B_i (i = \overline{1,10})$ the further course of the solution is repeated.

In Figure 1 in addition to the numerical solution and the solution in the 4-th approximation, as represented by decision in the 8th approximation and from the analysis of the temperature distribution can be judged to increase the solution accuracy by increasing the number of approximations. In addition, the discrepancy between the eighth approximation and the numerical solution in the range of numbers $0.01 \le \text{Fo} < \infty$ does not exceed 0.5%.

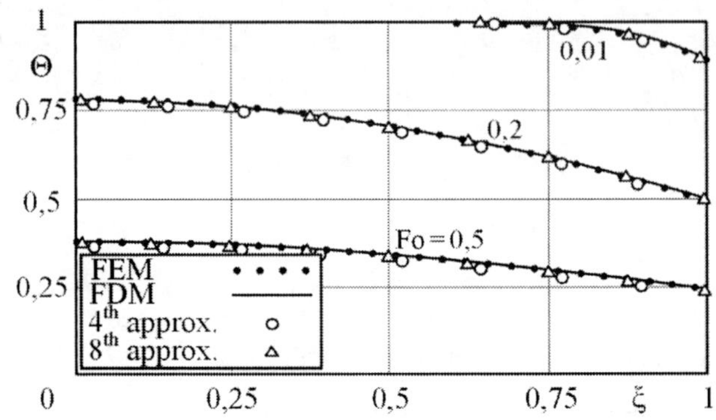

Figure 1. Graphs of temperature changes in the ball.

Figure 2 shows the residual of differential equation (5) in the 4th, 6th and 8th approximations at $\text{Fo} = 0.01$, the analysis of which allows us to conclude that the above method converges.

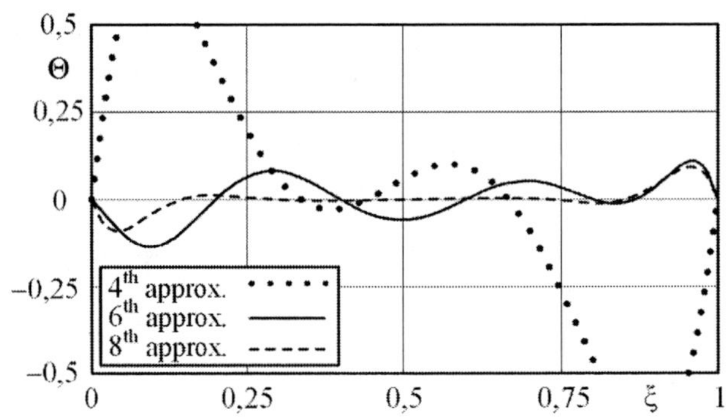

Figure 2. Residual of the differential equation.

2.2. Solution in ANSYS

In the framework of this work, the problem of thermal conductivity for the ball was also solved in the ANSYS. In order to compare the solution obtained in ANSYS and dimensionless analytical solution, the corresponding properties and parameters presented in table 1 were set.

Table 1. Properties and parameters.

Property	Value	Unit of measurement
Density ρ	1	kg/m^3
Specific heat C	1	J/kg·°C
Thermal conductivity λ	1	W/m·°C
Ambient temperature t_{cp}	0	°C
Initial temperature t_0	1	°C
Heat transfer coefficient α	1	W/m^2°C
The radius of the ball r	1	m

Figure 3. The original geometry.

The solution obtained as a result of simulation of the studied process in ANSYS allows to estimate the body temperature at any point and at any time, as well as to derive graphical and tabular temperature distributions. Figure 4 shows the temperature distribution at $Fo = 0{,}01$. The obtained data clearly illustrate the process of thermal conductivity in the direction from the center of the ball to the outside. Also, the inverse dependence of the heat conduction process of the ball will have the same form.

Table 2 shows the specific temperature values imported from the ANSYS solution at different points of the globe at different times, and Figure 1 they can be compared with numerical and analytical solution.

Figure 4. Temperature distribution.

3. Conclusion

As a result of joint application of a method of separation of variables and orthogonal methods of the weighted residuals the approximate analytical solution allowing to estimate thermal processes in a sphere (bearing) has been received. The obtained dependences can be presented in a simple and logical form, which will allow to estimate the influence of thermal conductivity processes on the rolling bodies of bearings in different temperature conditions of the environment, also these mathematical dependences can be easily converted into a program code that can be used in automation systems, diagnostics systems and control systems of the centrifugal pump.

A variant of the solution of this problem in the ANSYS program was also described, the results of which correspond to the analytical solution. However, the solution of such tasks in the program ANSYS require the installation of this software and, in addition, there is a need for high computing power of the PC.

Table 2. Temperature values imported from the ANSYS.

ξ	Fo		
	0.01	0.2	0.5
0	0.99990	0.78271	0.41581
0.05	0.99990	0.78209	0.41538
0.10	0.99989	0.78023	0.41412
0.15	0.99987	0.77712	0.41202
0.20	0.99983	0.77275	0.40909
0.25	0.99978	0.76710	0.40534
0.30	0.99969	0.76015	0.40079
0.35	0.99957	0.75188	0.39544
0.40	0.99938	0.74226	0.38931
0.45	0.99909	0.73128	0.38243
0.50	0.99865	0.71890	0.37481
0.55	0.99798	0.70511	0.36649
0.60	0.99694	0.68990	0.35750
0.65	0.99535	0.67326	0.34785
0.70	0.99289	0.65521	0.33760
0.75	0.98906	0.63576	0.32676
0.80	0.98309	0.61495	0.31539
0.85	0.97375	0.59284	0.30352
0.90	0.95913	0.56950	0.29120
0.95	0.93615	0.54502	0.27846
1	0.90001	0.51952	0.26536

4. Acknowledgment

The study was carried out with the financial support of the Russian National Science Foundation within the framework of the research project No. 18–79–00171 and the RF President Council on grants as part of the research project MK-2614.2019.8.

References

[1] Abu-Zeid MA and Abdel-Rahman S M 2013 Bearing problems effects on the dynamic performance of pumping stations *Alexandria Engineering J.* **52(3)** 241-8

[2] Vartha V, Kumar A, Mathew S, Aneesh R, Bejoy J, Thomas T K and Shajimon A C 2015 Failure analysis of ball-bearing of turbo-pump used in Liquid Rocket Engine *Materials Science Forum* **830** pp 709-12

[3] Bari H M, Deshpande A A, Jalkote P S and Patil S S 2012 Pump Coupling & Motor bearing damage detection using Condition Monitoring at DTPS *J. of Physics: Conf. Series* **364(1)** 12-22

[4] Feng Y, Li X, Ke Z, Chen Z and Tao M 2018 Pump Bearing Fault Detection Based on EMD and SVM *26th Int. Conf. on Nuclear Engineering* (London, England) ICONE26-81584, V001T01A008

[5] Jagtap H P and Bewoor A K 2017 Development of an algorithm for identification and confirmation of fault in thermal power plant equipment using condition monitoring technique *Procedia Engineering* **181** 690-7

[6] Korolev A V, Korolev A A and Krehe R 2017 Mathematical simulation and analysis of rolling contact fatigue damage in rolling bearings *The Int. J. of Advanced Manufacturing Technology* **89(1-4)** 661-4

[7] Takabi Jafar and Khonsari M M 2013 Experimental testing and thermal analysis of ball bearings *Tribology Int.* **60** 93-103

[8] Mizuta K, Inoue T, Takahashi Y, Huang S, Ueda K and Omokawa H 2003 Heat transfer characteristics between inner and outer rings of an angular ball bearing *Heat Transfer-Asian Research* **32(1)** 42-57

[9] Yan K, Wang Y, Zhu Y, Hong J and Zhai Q 2016 Investigation on heat dissipation characteristic of ball bearing cage and inside cavity at ultra high rotation speed *Tribology Int.* **93** 470-81

[10] De-xing Z, Weifang C and Miaomiao L 2018 An optimized thermal network model to estimate thermal performances on a pair of angular contact ball bearings under oil-air lubrication *Applied Thermal Engineering* **131** 328-39

[11] Than V T and Huang J H 2016 Nonlinear thermal effects on high-speed spindle bearings subjected to preload *Tribology Int.* **96** 361-72

[12] Zheng D and Chen W 2017 Thermal performances on angular contact ball bearing of high-speed spindle considering structural constraints under oil-air lubrication *Tribology Int.* **109** 593-601

[13] Yan K, Wang Y, Zhu Y and Hong J 2017 Investigation on the effect of sealing condition on the internal flow pattern of high-speed ball bearing *Tribology Int.* **105** 85-93

[14] Yan K, Wang N, Zhai Q, Zhu Y, Zhang J and Niu Q 2015 Theoretical and experimental investigation on the thermal characteristics of double-row tapered roller bearings of high speed locomotive *Int. J. of Heat and Mass Transfer* **84** 1119-30

[15] Bian W, Wang Z, Yuan J and Xu W 2016 Thermo-mechanical analysis of angular contact ball bearing *J. of Mechanical Science and Technology* **30(1)** 297-306

[16] Liu Y, Li D, Tang Z, Deng Y and Wu D 2017 Thermodynamic modeling, simulation and experiments of a water hydraulic piston pump in water hydraulic variable ballast system *Ocean Engineering* **138** 35-44

[17] Eremin A V, Stefanyuk E V, Rassypnov A Y and Kuznetsova A E 2013 Nestacionarny`j teploobmen v cilindricheskom kanale pri laminarnom techenii zhidkosti [Non-stationary heat exchange in cylindrical channel at laminar flow of fluids] *Vestnik SGTU Ser. Fiz.-Mat. nauki* **4(33)** 122-30 [In Russian]

The efficiency of gas combustion in TGM-84B boilers with a different number of burners

M A Taymarov[1], R V Akhmetova[2], E A Akhmetov[3] and R G Sungatullin[4]

[1]Department of Power Engineering, Kazan state power engineering university, Krasnoselskaya Str., 51, Kazan, 420066, Russia
[2]Department of Electric station, Kazan state power engineering university, Krasnoselskaya Str., 51, Kazan, 420066, Russia
[3]Department of Industrial power system and systems of heat supply, Kazan state power engineering university, Krasnoselskaya Str., 51, Kazan, 420066, Russia
[4]Department of Power supply of enterprises and energy resource saving technologies, Kazan state power engineering university, Krasnoselskaya Str., 51, Kazan, 420066, Russia

E-mail: ahmetova_rv@bk.ru

Abstract. The efficiency of fuel combustion in energy boilers of thermal power plants largely depends on the layout of the burners in the combustion furnace, on their number and unit capacity. An important condition for the cost-efficiency of the selected combustion mode is the absence of a torch projecting from the combustion products onto the screen surfaces of the heating of the furnace. To prevent the formation of large quantities of harmful emissions of nitrogen oxides into the atmosphere, the necessary requirement for fuel combustion is to reduce the local values of the torch temperature in the volume of the furnace. In addition, the intensity of the radiation of the torch from the furnace into the horizontal gas duct on the pipe surface of the heating of the superheater should not exceed the permissible values for the thermal resistance of the metal pipes. With a frontal arrangement, the use of burners with a low unit thermal power with an increase in their number allows to reduce the length of the torch and prevent the torch from projecting onto the rear screen. However, this reduces the coefficient of efficiency of the boiler, as it increases the loss of heat with flue gases due to an increase in their temperature.
Keywords. Boiler, combustion, gas, furnace, burner, twisting of air, torch, temperature, radiation, combustion products, coefficient of efficiency, emissions, nitrogen oxides.

1. Introduction

The choice of unit power and the number of gas-oil burners in the design of energy steam boilers for thermal power plants depends on many of its design features and operating conditions [1-6]. The main ones are the size of the furnace and its cross-sectional area, the location of the superheater elements, and the design of the lowering gas duct. As a rule, steam power boilers have balanced traction and are operated at small and large steam loads. In this regard, duringoperation at low loads, part of the burners is turned off or transferred to the mode of operation with low fuel consumption. The efficiency of fuel combustion depends on the method of controlling the combustion mode for a given number of working burners. For the experimental research of the efficiency of fuel combustion, three TGM-84 boilers with a nominal capacity of 420 t/h and the number of burners equal to 6.8 and 18 were selected. During studying the efficiency of burning natural gas from Urengoy field with lower heat of com-

IOP Conf. Series: Materials Science and Engineering **791** (2020) 012018 doi:10.1088/1757-899X/791/1/012018

bustion LHV=34257 kJ/m^3. The efficiency of fuel combustion is characterized by the coefficient of efficiency of the boiler. Another important indicator is the concentration of nitrogen oxides in the products of combustion. These indicators depend on the intensity of the torch radiation and the load of the boiler. Therefore, the planned volume of experiments included measurements of the intensity of torch radiation while measuring the concentration of nitrogen oxides at different loads.

2. Materials and Methods

The layout schemes of the burners on the investigated TGM-84B boilers of Nizhnekamsk TPP-1 and Ufa TPP-4 are shown in Figure 1.

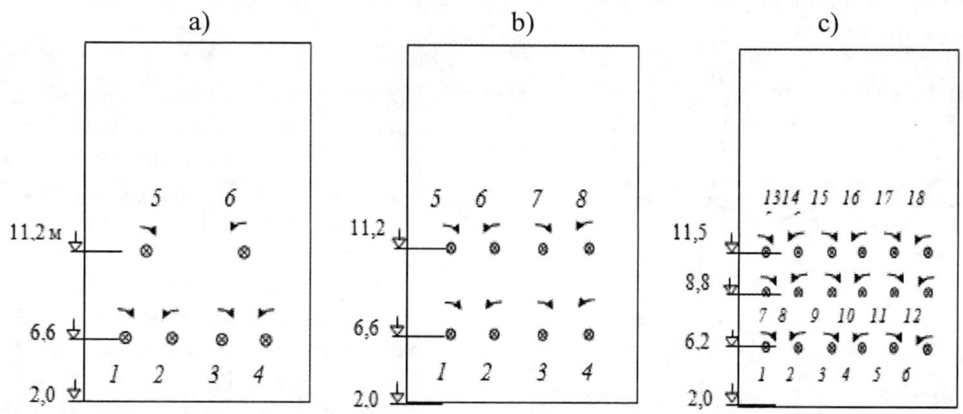

Figure 1. The layout schemes of the numbers and directions of twisting of air revolutions for the burners on the investigated TGM-84B boilers: a) TGM-84B boiler of Nizhnekamsk TPP-1 (station No. 6) with the number of burners equal to 6; b) TGM-84B boiler of Ufa TPP-4 (station No. 14) with the number of burners equal to 8; c) TGM-84B boiler of Nizhnekamsk TPP-1 (station No. 1) with the number of burners equal to 18.

The productivity of TGM-84B boiler is 420 t/h at a superheated steam pressure of 14 MPa and temperature of 560°C. The burners of Taganrog boiler plant construction are located on the front wall of the furnaces of TGM-84B boilers at elevations ranging from 6.2 m to 11.5 m.

The furnaces have a two-light screen that divides the furnace volume in height into two half-furnaces. The sizes of the furnaces are: width 16 m, depth 7.5 m, height 24 m. For TGM-84B boilers of Nizhnekamsk TPP-1 the unit capacity of each of the six burners is 50 MW, each of the 18 burners is 17 MW, for TGM-84B boiler of Ufa TPP - 4 - 38 MW. The burners have a central gas distribution through a coaxial conical nozzle 1 with three rows of gas supply holes of various diameters (Figure 2).

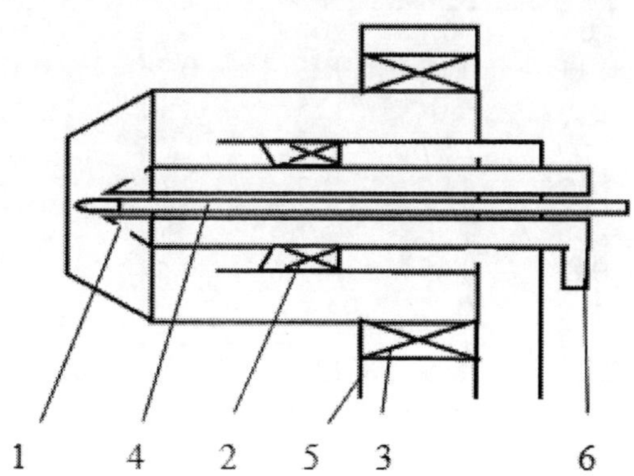

Figure 2. Diagram of a two-flow vortex burner for TGM-84B boilers: 1 - conical gas nozzle, 2 - blades of an axial internal air twisting, 3 - blades of a peripheral tangential air twisting, 4 - nozzle, 5 - air supply duct, 6 - gas supply.

IOP Conf. Series: Materials Science and Engineering **791** (2020) 012018 doi:10.1088/1757-899X/791/1/012018

The air twisting in the burners is rotated by axial blade 2 and tangential peripheral 3 air swirlers, which are the main ones. During burning gas,the nozzle 4 is removed from the burner and placed on a rack for intermediate storage.

Measurements of the parameters of Urengoy gas combustion process were carried out in the operational sections of the boilers in the cut of the water economizer and in the balance sections after the smoke exhaust. The test procedure is given in [7-11]. Coefficient of efficiency of boilers was determined by the reverse balance. During the tests, the indications of standard station instruments were used.

A radiometer with a calibration of RK-15 through the hatches on the walls of the furnaces measured the intensity of the radiation of the torch from the furnace volume at various loads. The concentration of nitrogen oxides NO_x in the combustion products in the operating section was measured using the DAG-500 gas analyzer. The error in determining the coefficient of efficiency was ±70.71%. Figure 3 shows the measurement scheme of the operating parameters of the operation of the TGM-84B boilers during experiments.

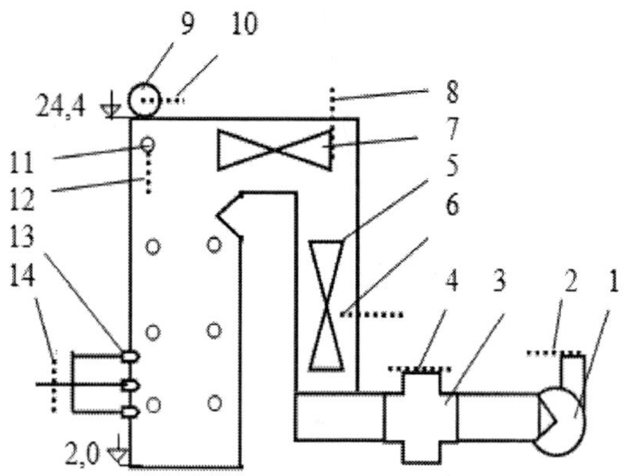

Figure 3. The measurements scheme of the operating parameters of the operation of TGM-84B boilers during the experiments: 1 - smoke exhaust, 2 - balance section, 3 - regenerative air heater, 4 - air parameters after the regenerative air heater, 5 - economizer, 6 - operating section, 7 - superheater, 8 - parameters of superheated steam, 9 - cylinder, 10 - parameters of the steam-water mixture in the boiler cylinder, 11 - inspection hatches, 12 - measurement of the temperature and intensity of the torch radiation, 13 - burners, 14 - measurement of the parameters of air and fuel supplied to the combustion.

3. The received results and their discussion

Tables 1-3 show experimental data on the operation parameters of TGM-84B boilers (station No. 6 and No. 1) with the number of burners 6 and 18 for Nizhnekamsk TPP-1 and for Ufa TPP-4 (station No. 14) with the number of burners 8 for various steam loads.

Table 1. The operation parameters of TGM-84B boiler (station No. 6) of Nizhnekamsk TPP-1 with the number of burners equal to 6.

Parameter	Load, t/h				
	210	260	320	380	420
Number of operational burners, pc.	6	6	6	6	6
Gas pressure in front of the burners, kg/cm^2	0.11	0.12	0.17	0.24	0.28
Air pressure in front of the burners, kg/cm^2	50	80	125	180	225
Coefficient of sir excess	1.09	1.07	1.06	1,05	1.05
Temperature of flue gases,°C	105	113	123	131	136
Heat losses with flue gases, %	4.55	4.58	4.75	5	5.21
Heat losses in the environment, %	0.8	0.65	0,5	0,44	0.4
Boiler gross coefficient of efficiency, %	94.65	94.77	94.75	94.56	94.39
NO_xcontent in flue gasesreduced to air excess 1.4, mg/m^3	105	125	150	175	190

The air temperature in front of the regenerative air heaters was 30°C, the temperature of the feed-water was 230°C.

Table 2. The operation parameters of TGM-84B boiler (station No. 1) of Nizhnekamsk TPP-1 with the number of burners equal to 18.

Parameter	Load, t/h			
	240	300	360	420
Number of operational burners, pc.	10	12	16	18
Gas pressure in front of the burners, kg/cm^2	0.32	0.34	0.36	0.41
Air pressure in front of the burners, kg/cm^2	60	75	100	125
Coefficient of sir excess	1.07	1.07	1.06	1.05
Temperature of flue gases, °C	138	143	148	154
Heat losses with flue gases, %	7	7.05	6.9	6.85
Heat losses in the environment, %	0.7	0.56	0.47	0.4
Boiler gross coefficient of efficiency, %	92.3	92.39	92.63	92.75
NO_x content in flue gases reduced to air excess 1,4, mg/m^3	190	240	290	345

Table 3. The operation parameters of TGM-84B boiler (station No. 14) of Ufa CHPP-4 with the number of burners equal to 8

Parameter	Load, t/h					
	220	260	300	340	380	420
Number of operational burners, pc.	8	8	8	8	8	8
Gas pressure in front of the burners, kg/cm^2	0.15	0.21	0.27	0.36	0.43	0.52
Air pressure in front of the burners, kg/cm^2	35	45	65	85	110	145
Coefficient of sir excess	1.05	1.047	1.045	1.042	1.04	1.04
Temperature of flue gases, °C	132	134	137	141	145	149
Heat losses with flue gases, %	5.58	5.63	5.76	5.93	6.09	6.27
Heat losses in the environment, %	0.76	0,65	0.56	0.49	0.44	0.4
Boiler gross coefficient of efficiency, %	93.06	93.72	93.68	93.58	93.47	93.38
NO_x content in flue gases reduced to air excess 1,4, mg/m^3	165	175	182	189	197	205

As can be seen from the table. 1-3, the highest coefficient of efficiency at all loads has TGM-84B boiler (station No. 6) with six burners at Nizhnekamsk TPP-1. The lowest values of the coefficient of efficiency has TGM-84B boiler (station No. 1) with the number of burners 18. The decrease in the coefficient of efficiency with an increase in the number of burners is associated with an increase in the temperature of the flue gases. The concentration of NO_x in flue gases increases with an increase in the number of burners. During the experiments it was found that a change in the temperature of the feed-water by ±10°C leads to a change in the temperature of the flue gases by ±2.5°C, and a change in the temperature of the cold air by ±10°C leads to a change in the temperature of the flue gases by ±5°C.

As can be seen from figure 4, with an increase in the number of burners n, a decrease in the intensity of the torch radiation q_f near the rear screen of the furnace occurs due to a decrease in the length of the torch. However, in this case, the radiation intensity and the temperature of the central area of the torch increase. This leads to an increase in the formation of thermal nitrogen oxides NO_x with an increase in the number of burners. The data of this research are consistent with the experimental results [11] obtained by burning gas for TGM-84B boiler with the number of burners n=4.

Results of a pilot study on influence of number of torches on regime indicators at joint combustion of gas and fuel oil (tab. 4-6) are received during the work of coppers at air temperature in front of regenerative air heaters 70°C and at a temperature of feed water of 230°C. Heating of air in front of regenerative airheaters to 70°C at combustion of fuel oil is carried out in order that will increase temperature of products of combustion at the exit from a fire chamber in front of boiler superheaters.

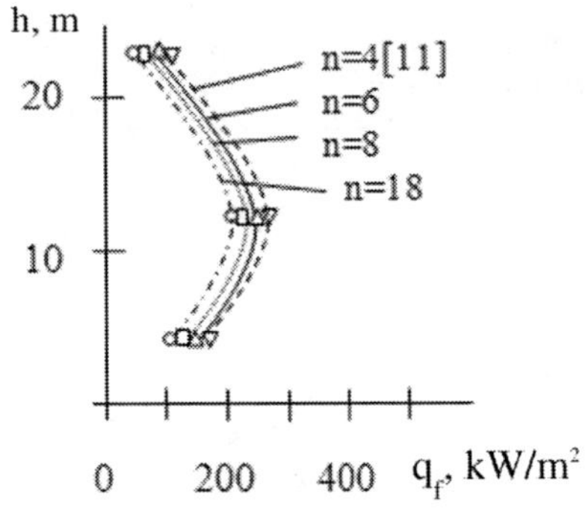

Figure 4. Dependence of torch radiation intensity q_f on the height of the furnace h for the furnaces of TGM-84B boilers with different number of burners n at a load of 320 t/h for burning gas with a heat combustion of 34257 kJ/m^3 as measured through the side hatches of the right wall at the rear screen in comparison with the research data [11].

Table 4. The operation parameters of TGM-84B boiler (station No. 6) of Nizhnekamsk TPP-1 with the number of burners equal to 6 for combined combustion of gas and fuel oil.

Parameter	Load, t/h			
	210	270	350	420
Gas burner numbers	5, 6			
Fuel oil burner numbers	1, 2, 3, 4			
Gas pressure in front of the burners, kg/cm^2	0.10	0.11	0.18	0.28
Fuel oil pressure in front of the burners, kg/cm^2	10	14.5	17.5	25
Air pressure in front of the burners, kg/cm^2	75	105	170	245
Gas flow rate per boiler, nm^3/h×10^3	6.5	7.2	9	11.5
Fuel oil consumption per boiler, t/h	9.5	11	15.5	18.8
Coefficient of sir excess	1.13	1.12	1.09	1.08
Temperature of flue gases, °C	136	141	148	153
Heat losses with flue gases, %	6.4	6.51	6.65	6,72
Boiler gross coefficient of efficiency, %	92.8	92.85	92.87	92.88
NO_x content in flue gases reduced to air excess 1,4, mg/m^3	148	183	225	283

Table 5. The operation parameters of TGM-84B boiler (station No. 1) of Nizhnekamsk TPP-1 with the number of burners equal to 18 for combined combustion of gas and fuel oil.

Parameter	Load, t/h		
	300	360	420
Gas burner numbers	8,11,13-18		
Fuel oil burner numbers	2-5, 7, 12		1-7,12
Gas pressure in front of the burners, kg/cm^2	0.18	0.24	0.4
Fuel oil pressure in front of the burners, kg/cm^2	28.5	31	35
Air pressure in front of the burners, kg/cm^2	65	80	120
Gas flow rate per boiler, nm^3/h×10^3	12	15	18.5
Fuel oil consumption per boiler, t/h	7.2	12	13
Coefficient of sir excess	1.06	1.06	1.06
Temperature of flue gases, °C	137	143	156
Heat losses with flue gases, %	6.56	6.88	7.59
Boiler gross coefficient of efficiency, %	92.80	92.65	92.01
NO_x content in flue gases reduced to air excess 1,4, mg/m^3	178	210	298

IOP Conf. Series: Materials Science and Engineering **791** (2020) 012018 doi:10.1088/1757-899X/791/1/012018

Table 6. The operation parameters of TGM-84B boiler (station No. 14) of Ufa CHPP-4 with the number of burners equal to 8 for combined combustion of gas and fuel oil.

Parameter	Load, t/h		
	260	340	420
Gas burner numbers		5-8	
Fuel oil burner numbers		1-4	
Gas pressure in front of the burners, kg/cm^2	0.21	0.36	0.52
Fuel oil pressure in front of the burners, kg/cm^2	12	17	23
Air pressure in front of the burners, kg/cm^2	70	110	160
Gas flow rate per boiler, $nm^3/h\times10^3$	1.065	1.058	1.048
Fuel oil consumption per boiler, t/h	150	160	166
Coefficient of sir excess	6.50	6.80	6.94
Temperature of flue gases, °C	92.82	92.75	92.66
Heat losses with flue gases, %	280	320	360
Boiler gross coefficient of efficiency, %	0.21	0.36	0.52
NO_x content in flue gases reduced to air excess 1,4, mg/m^3	12	17	23

As can be seen from tables 4-6, when gas and fuel oil are combusted together, the efficiency decreases with increasing number of burners at all boiler loads.

4. Conclusions

1. An increase in the number of gas-oil burners in TGM-84B boilers from 6 to 18 leads to a decrease in the coefficient of efficiency due to an increase in heat losses with flue gases

2. Due to the decrease in the torch length with an increase in the number of burners, the radiation intensity and temperature in the central area of the torch increase, that leads to an increase in the formation of thermal nitrogen oxides.

References

[1] Taymarov M A, Lavirko Yu V and Belyaeva E A 2015 Intensivnost` luchistogo teploobmena v topke kotla pri izmenenii parovoj nagruzki [The intensity of radiant heat transfer in the boiler furnace with a change in steam load] *Izvestiya vy`sshix uchebny`x zavedenij. Problemy` e`nergetiki* [*University News. Energy problems*] **7-8** 1-4 [In Russian]

[2] Zhang Ji, Yuan Han, Zhao Jian and Mei Ning 2016 Viscosity estimation and component identification for an oil-water emulsion with the inversion method *J. of Petroleum Science and Engineering* **111** 759-67

[3] Gelderen L, Malmquist M and James G 2017 Vaporization order and burning efficiency of crude oils during in situ burning on water *Fuel* **191** 528-37

[4] Kadota T and Yamasaki H 2002 Recent advances in the combustion of water fuel emulsion *Progress in Energy and Combustion Science* **28** 385-404

[5] Ahrenfeldt J, Thomsen T, Henriksen U and Clausen L 2013 Biomass gasification cogeneration *Applied Thermal Engineering* **50** 1407-17

[6] Sanchez A and Williams F 2014 Recent advances in understanding of flammability characteristics of hydrogen *Progress in Energy and Combustion Science* **41** 1-55

[7] Taymarov M A, Akhmetova R V, Chiklyaev E G and Sungatullin R G 2017 Pokazateli ekspluatatsionnykh parametrov kotlov TGM-84B pri szhiganii v nikh metano-vodorodnykh fraktsiy [Indicators of operational parameters of TGM-84B boilers during methane-hydrogen fractions are burned in them] *Vestnik KSPEU* **33** 58-63 [In Russian]

[8] Taymarov M A, Kuvshinov N E, Akhmetova R V, Sungatullin R G and Chiklyaev D E 2016 Issledovaniye khimicheskikh protsessov obrazovaniya oksidov azota pri szhiganii gaza i mazuta [Research of the chemical processes of the formation of nitrogen oxides during the combustion of gas and fuel oil] *Vestnik KSPEU* **19** 80-3 [In Russian]

[9] Taymarov M A, Akhmetova R V, Safin R G, Lavirko Yu V, Sadykov R A and Medvedeva G A 2016 Determination of the dependence of the burn-up time of a drop of fuel oil on the intensity of flame radiation *Research J. of Applied Sciences Year* **11** 1660-5

[10] Taymarov M A, Akhmetova R V and Akhmetov E A 2019 Efficiency of application of various layout arrangements of oil-gas burners in thermal power plant boilers *IOP Conf. Ser.: Mater. Sci. Eng* **552** 012008

[11] Abryutin A A and Vyazova S K 2011 Investigation of heat transfer in a boiler furnace and a screen superheater during the combustion of fuel oil and gas with small excesses of air *Thermal Engineering* **2** 16-21

[12] Taymarov M A, Akhmetova R V, Margulis S M and Saltanayeva Ye A 2017 *Teplovoe izluchenie v topkax kotlov* [Thermal radiation in boiler furnaces] (Kazan`: IPK Brig) [In Russian]

[13] Taymarov M A, Akhmetova R V, Lavirko Y V, Sungatullin R G and Zheltuhina Ye S 2017 Snizhenie vredny`x vy`brosov v atmosferu oksidov azota kotlami TE`S [Decrease in harmful emissions in the atmosphere of nitrogen oxides by boilers of thermal power plant] *Izvestiya Kazanskogo gosudarstvennogo arxitekturno-stroitel`nogo universiteta* **1(39)** 180-8

[14] Roslyakov P V and Yegorova L Ye 1996 Influence of the main characteristics of a zone of active burning on nitrogen oxides output *Power System* **9** 22-6

[15] Taymarov M A 2013 *Razrabotka metodov snizheniya vy`brosov okislov azota kotlami TE`S* [Development of methods of decrease in emissions of oxides of nitrogen by boilers of thermal power plant] (Kazan`: Kazanskij gosudarstvenny`j e`nergeticheskij universitet)

[16] Shimizu K, Hibi A, Koshi M, Morii Y and Tsuboi N 2011 Updated Kinetic Mechanism for High-Pressure Hydrogen Combustion *J. of Propulsion and Power* **27** 383-95

[17] John C C 2004 Transient radiative transfer in 2D cylindrical medium with collimated pulseirradiation *J. of Quantitative Spectroscopy and Radiative* 299-313

[18] Mishra S C and Rath P 2003 Transient radiative heat transfer in participating media of Numerical heat transfer *Int. J. of Heat and Mass* 746-52

[19] Abdallah P B 2000 Thermal Emission of Semitransparent slab with variable spatial refractive index *J. of Quantitative Spectroscopy and Radiative* 185-98

[20] Krishna N A and Subhash C 2006 Discrete transfer method applied to radiative transfer in a variable refractive Index semi-transparent medium *J. of Quantitative Spectroscopy and Radiative* 432-40

[21] Burke M P, Chaos M, Ju Y, Dryer F L and Klippenstein S J 2012 Comprehensive H2/O2 kinetic model for high-pressure combustion *Int. J. of Chemical Kineticsvol* **44** 444-74

[22] Lu TJ and Jin JM 2017 Electrical-thermal co-simulation for analysis of high-power rf/microwave components *IEEE Trans. Electromagn. Compat*. 93-102

[23] Skoplaki E and Palyvos J A 2009 On the temperature dependence of photovoltaic module electrical performance *Solar Energy* **63** 614-24

[24] Sánchez AL, Williams FA 2014 Recent advances in understanding of flammability characteristics of hydrogen *Progress in Energy and Combustion Science* **41** 1-55

[25] Rui Lou and Shu Bin Wu 2011 Products properties from fast pyrolysis of enzymatic/mild acidolysis lignin *Applied Energy* **88** 316-22

IOP Conf. Series: Materials Science and Engineering **791** (2020) 012019 doi:10.1088/1757-899X/791/1/012019

Method for determining the characteristics of a gas-steam chp with a steam turbo drive of a compressor

V K Tian[1], S A Gulina[1], L P Sheludko[2], V V Biryuk[3] and I Y Goryunova[4]

[1]Department of Pipeline Trans, Samara State Technical University, Molodogvardeyskaya St., 244, Samara, 443100, Russia

[2]Department of Management and System Analysis of Thermal Power and Social Engineering Complexes, Samara State Technical University, Molodogvardeyskaya St., 244, Samara, 443100, Russia

[3]Department of Heat Engineering and Heat Engines, Samara University, Moskovskoye Shosse 34, Samara, 443086, Russia

[4]Department of Turbines and Engines, Ural Federal University, Mira Str., 19, Yekaterinburg, 620000, Russia

E-mail: kr_oeg@mail.ru

Abstract. A new type of maneuverable combined-cycle plant with a steam-turbine drive of a low-pressure compressor is considered. The analysis of its technological processes in the generation of electric and thermal energy in heating and non-heating periods of the year is carried out. It is shown that in heating conditions, the burning of additional fuel in the afterburner between the stages of the evaporator of the utilizer boiler provides an increase in its steam production, maneuverability, thermal and electric power of the CCGT-CHP plant. The influence of the chemical composition and thermophysical properties of natural gas on obtaining the required excess air in the combustion chamber of a combined cycle gas turbine is considered. A mathematical model of technological processes is developed. The influence of deep utilization of the heat of the flue gases of the utilizer boiler on increasing the efficiency and improving the environmental characteristics of the gas-steam plant of a thermal power plant was recorded.
Key words: electric and thermal power, waste heat boiler, backpressure steam turbine, gas turbine installation, combined cycle gas turbine, compressor steam drive.

1. Introduction

One of the urgent problems of the energy sector is the creation of cogeneration CCGT-CHP with a high level of cogeneration generation of heat and electric energy [1]. To increase the thermal and electric power in the heating modes of their operation, in the patent [2] it was proposed to use a two-stage evaporator with a fuel afterburning chamber (*AC*) between its steps in a waste heat boiler (*HB*), to supply this *CCGT-CHP* with two counter-pressure turbines with network heaters. This allows, due to the combustion of fuel in the boiler house, to increase the production of steam in the boiler unit and significantly increase the electric and thermal power [3]. These measures contribute to improving the maneuverability of the combined cycle power plant. In the patent [4], a scheme is proposed for a gas-steam heat and power plant in which expanded steam in a counter-pressure steam turbine (*ST*) is mixed in the combustion chamber (*CC*) with the products of fuel combustion. The resulting mixture is expanded in a combined cycle gas turbine (*SGT*) driving an electric generator. The installation pro-

IOP Conf. Series: Materials Science and Engineering **791** (2020) 012019 doi:10.1088/1757-899X/791/1/012019

vides contact condensation of the vapor component of the gas-vapor mixture. The article [5] compares the characteristics of *CCGT* with steam and gas turbine drives of compressors. It was found that the steam drive helps to reduce the degree of increase in air pressure in the compressor and increase the gas temperature in front of the compressor unit. In the application for the invention [6], the authors proposed a new type of gas-steam *CCGT-CHP*, which has high mobility when working in both non-heating and heating periods of the year and increased thermal efficiency. *CCGT-CHP* (Fig. 1) contains a gas turbine unit with (gas turbine), a waste heat boiler with a two-stage evaporator and a chamber for afterburning fuel, and two counter-pressure transformer substations.

2. The principle of operation of the installation

The principle of its operation is as follows: atmospheric air is compressed in a low-pressure compressor (LPC) driven from the main counterpressure steam turbine PT1. The air is compressed in a high-pressure compressor (HPC) and then fed into the combustion chamber (CC). To increase the proportion of steam in the gas-vapor mixture, the combustion process in the CC is carried out at a small coefficient of excess air $\alpha \approx 1.2$. This ratio of fuel and air will ensure complete combustion at a low air flow through the compressor. In the mixing chamber CC serves steam expanded in PT1.

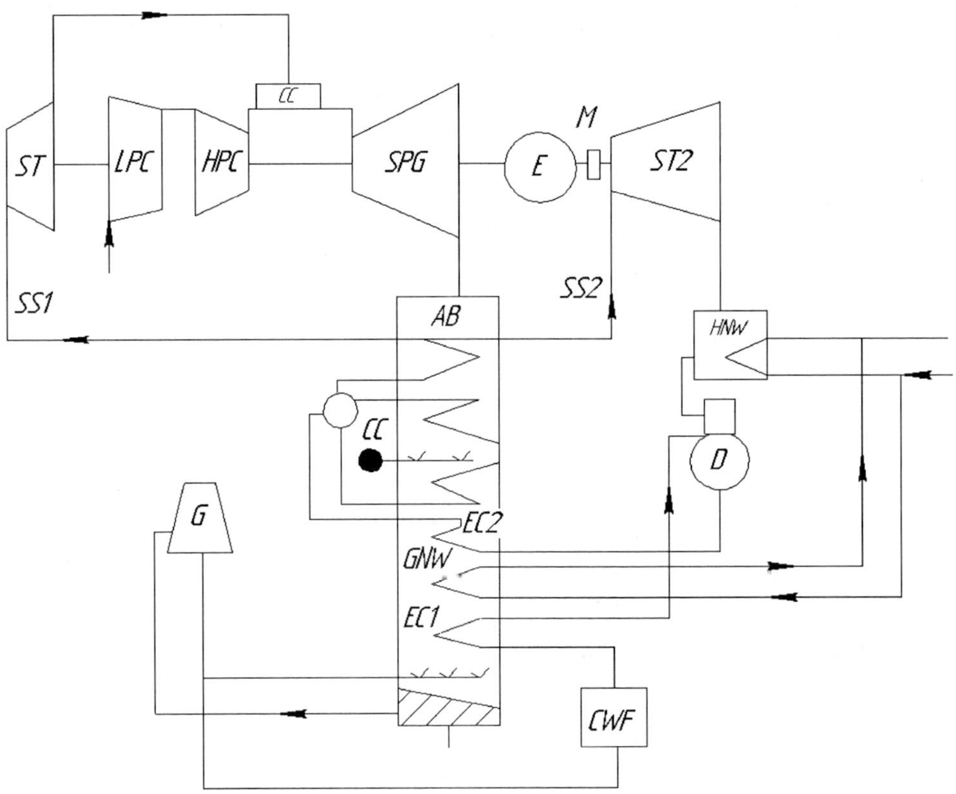

Figure 1. Schematic diagram of maneuverable gas and steam combined heat and power plants with steam compressor (*ST* - counterpressure steam turbine LPC drive, *ST2* - energy counterpressure steam turbine, *LPC* - low pressure compressor, *HPC* - high pressure compressor, *CC* - combustion chamber, *SGT* - gas and steam turbine, *E* - electric generator with flexible coupler, *SS1* - steam supply steam line to the main anti-pressure *ST*, *AB* - waste heat boiler, *SS2* - steam supply pipe for energy *ST2*, *HNW* - mains water heater, *CC* - afterburner with fan, *G* - cooling tower, *ECI* - economizer of the first stage, *EC2* - economizer of the second steps, *C* - steam condensate separator, *CWF* - chemical water treatment, *GWH* - gas-water heater, *M* - electric coupler).

The resulting gas-vapor mixture is expanded in a gas-vapor turbine (*GVT*). Her work is used to drive *HPC* and an electric generator (*E*). The steam-gas mixture expanded in the PGT is sent to a waste heat boiler (*HB*), where its heat is used: for generating superheated high-pressure steam in the economizer of the second stage (*EcII*), in the evaporator and steam superheater, for heating the heating network water, in gas and water heater (*GWH*) and for heating of feed water in the economizer of the first stage (*EcI*). As in [1], to increase the maneuverability of the unit during the heating period, additional fuel is burned in the fuel boiler and the steam production in the boiler is increased. This steam is fed to *PT2* to increase the production of electricity and heat network water in (*HNW*). When the temperature of the outside air changes, the steam capacity of the *HB* is regulated by changing the fuel consumption in the boiler.

A feature of this scheme is the low-pressure switch from the steam turbine *ST1*, and the high-pressure switch from the turbine. Introducing steam into the *CC* mixing zone allows increasing the work of expanding the gas-vapor mixture and increasing its power. *HB* provides for the use of heat of condensation of water vapor from a gas-vapor mixture and the return of condensate to the cycle. Its high potential heat is used to generate steam in the *HB*, and low potential heat is used in the hot water supply to heat the network water of the heating system.

3. Materials and Methods

Natural gas (*NG*) is used as a fuel for CCGT, which differs in its chemical composition in the content of hydrocarbons and other gases depending on the field. Therefore, the *NG* has different values of the lower heat of combustion of the fuel Hu and the stoichiometric coefficient L_0. In [7, 8], an analysis was made of the influence of the composition of the working fluid on the parameters of *GTU* efficiency. Failure to take into account the composition of the fuel gas for a mixture with $\alpha=1...4$ reduces the accuracy of engineering thermodynamic calculations, then it decreases from 7% to 3% [9]. When compiling a mathematical model for calculating the parameters of combined cycle steam turbine units with *LPC* steam drives, the air humidity at the inlet to the gas turbine and the humidity of the fuel gas were taken into account. The coefficient of excess air in the compressor station is determined taking into account the properties of the components of the gas mixture: air and combustion products of gaseous fuels N_2, CO_2, O_2, H_2O, depending on the parameters of the gas turbine based on the heat balance equation:

$$\alpha = \frac{\eta_G \cdot Hu - h^*_{гG}\left(1-\frac{1}{L_0}\right) + h^*_{вC}}{h^*_{вG} - h^*_{вC}}, \tag{1}$$

where: η_{CC} – the coefficient of completeness of fuel combustion in the CC, $h^*_{вG} = V^O C^в_p T^*_G$ - is the enthalpy of air at the theoretically necessary amount for combustion at a temperature T^*_G in front of the turbine, $h^*_{гG} = \left(V^0_{CO_2} \cdot C^{CO_2}_p + V^0_{N_2} \cdot C^{N_2}_p + V^0_{H_2O} \cdot C^{H_2O}_p\right) T^*_G$ is the enthalpy of the products of combustion with the coefficient of excess air $\alpha=1$ and temperature T^*_G, $h^*_{вC} = V^O C^в_p T^*_C$ - enthalpy of air at a theoretically necessary amount at temperature T^*_C, $C^{CO_2}_p, C^{N_2}_p, C^{H_2O}_p, C^в_p$ - mass heat capacities of carbon dioxide, nitrogen, water vapor and air, respectively, at the temperature of the combustion products T^*_G and air at the entrance to the combustion chamber T^*_C; $V^0_{RO_2}$ - is the volume of triatomic gases of combustion products (without water vapor at $\alpha=1$), V_{H_2O} - is the volume of water vapor taking into account the moisture content in fuel and in air, V_{R_2} - is the volume of diatomic gases of combustion products (without water vapor at $\alpha=1$): $V^O = V^0_{RO_2} + V_{R_2} + V_{H_2O}$ - is the total volume of combustion products.

The coefficient of excess air in the combustion chamber α depends on the enthalpy of gas in front of the turbine h_G and the enthalpy of air behind the compressor h_C, which is determined by the degree of increase in pressure in the compressor π_C. For the temperature range in front of the turbine from *1373* K to *1673* K, the dependence of α on the degree of pressure increase in the compressor π_C.

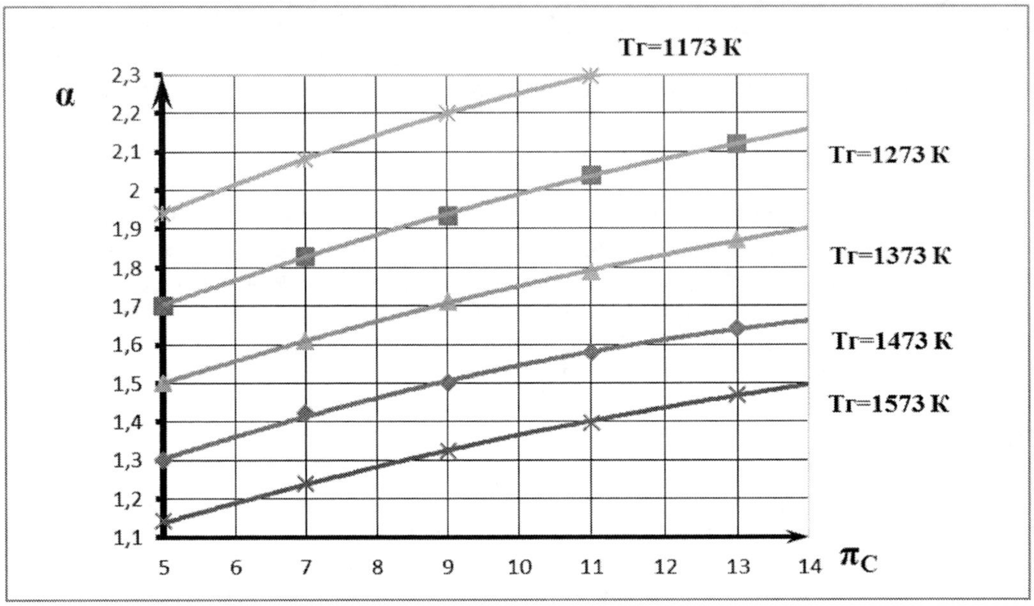

Figure 2. The dependence of α on the degree of pressure increase in the compressor π_C.

Based on the mathematical model, a graphical dependence of the coefficient of excess air in the compressor station on the parameters of the gas turbine unit was constructed (Figure 2), which allows one to estimate the range of the choice of the degree of increase in pressure π_C and the gas temperature in front of the turbine T_T for a mixture with an excess air coefficient $\alpha \approx 1.15 ... 1.25$.

From the choice of the total degree of increase in pressure in the compressor $\pi_{\Sigma C} = \pi_{LPC} \cdot \pi_{HPC}$ depends on the amount of useful work of the cycle of gas turbines Le and the obtaining in the *CC* value α at the level of *1.15 .... 1.25*. The power consumed by the compressor N_C was calculated, depending on the air flow rate G_V through it at different $\pi_{\Sigma C}$. The compressor power increases with increasing $\pi_{\Sigma C}$ and the flow rate of compressed air G_V. The steam parameters at the *CC* inlet are set taking into account the steam parameters in front of typical backpressure turbines and the level of hydraulic losses in the steam path. Accepted: $R_{S.S.} = 9.0$ *to* 13.0 *MPa*; $M_{S.} = 818$ *K*; $R_{S.S.} = 1.6$ *MPa*; $T_{S.S.} = 513$ *K*.

4. Calculation method

Steam flow through the PT is determined by a given G_V and π_{LPC} [10, 11]. In Figure 3 shows the dependence of the relative steam flow rate through PT1 on the distribution of πK over the compressor stages, where:

$$\overline{\pi_{LPC}} = \frac{\pi_{LPC}}{\pi_{\Sigma C}} \qquad (2)$$

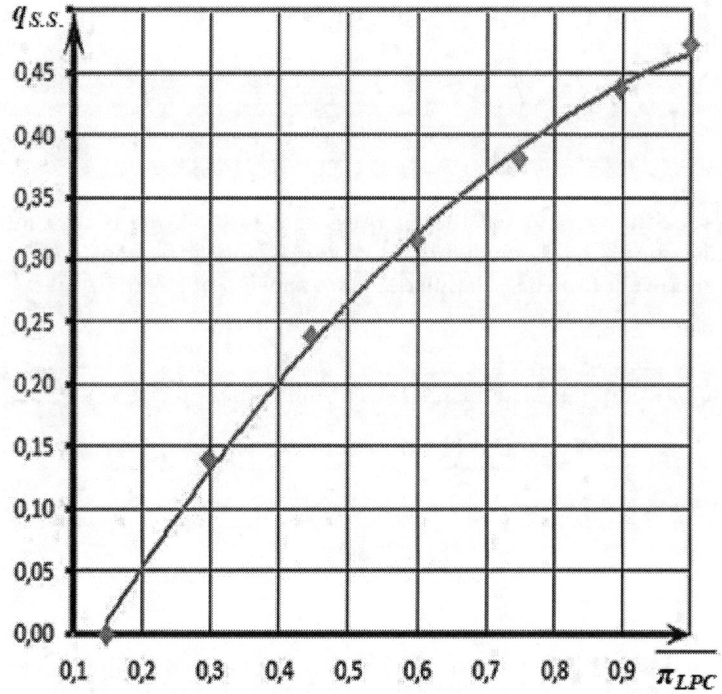

Figure 3. The dependence of $q_{S.S.}$ on the relative $\overline{\pi_{LPC}}$.

The use of low pressure π_{LPC} in low pressure steam pumps allows to reduce the consumption of superheated steam and to reduce the power of the drive *PT1* and the steam productivity of *HB* in non-heating operating modes of the *CCGT-CHP*. The characteristics of the gas-vapor mixture in the *SGT* and in the *HB* — the flow rate, pressure, and enthalpies of the mixture were determined taking into account the shares of the products of combustion and steam [12]. The electric power of the gas-steam *CCGT-CHP* in non-heating modes of its operation is determined:

$$N_{E\Sigma} = G_V\left[\left(1-q_C+q_T+q_{ST}\right)L_{SGT}\cdot\eta_M - L_{HPC}\right] \tag{3}$$

where: L_{SGT} is the specific operation of a combined cycle gas turbine, L_{HPC} is the specific operation of *HPC*, η_M is the mechanical efficiency, G_V is the mass flow rate of gas turbine units, q_i is the relative flow rate of the working fluid (cooling air, fuel, saturated steam). Enthalpy of gas-vapor mixture at the outlet of the steam generator *HB*:

$$h_{outAB} = \frac{\left(1-q_C+q_T\right)h^G_{SGT} - q_{ST}\cdot h^P_{SGT}}{\left(1-q_C+q_T+q_{ST}\right)} \tag{4}$$

where: h^{G}_{SGT} - is the enthalpy of the vapor-gas mixture, h^P_{SGT} - s the enthalpy of the vapor-gas mixture at the outlet of the steam section of the *HB*.
Steam production *HB*:

$$G_{P\Sigma} = \frac{\left(1-q_C+q_T\right)h^G_{SGT} - q_{NP}\cdot h^P_{SGT}}{\left(h_{outAB} - h_{PV}\right)}. \tag{5}$$

In heating modes, additional fuel is burned in the boiler to increase steam production in the boiler for *ST2*, its operation is used to drive the electric generator. The total electric power of *SGT* and *ST2* in heating modes of the gas-steam *CCGT-CHP*:

$$N_{E\Sigma} = G_{SGmix} L_{SGT} + G_{SS} L_{ST2} - G_V L_{HPC}; \tag{6}$$

$$L_{ST2} = \left(h_{SS} - h_{BP}\right)\eta_{SGs} \tag{7}$$

where: L_{ST2} is the specific operation of the energy steam turbine, h_{SS} is the enthalpy of superheated steam, h_{BP} is the adiabatic back pressure enthalpy, and η_{STs} is the efficiency of the steam turbine 2. Additional electrical power of the installation due to expansion of steam in *ST2*:

$$N_{EST2} = \left(G_{V\Sigma} - G_{ST1}\right) L_{ST2}. \tag{8}$$

The total electric power of gas-steam CCGT-CHP in heating modes:

$$N_{E\Sigma} = G_V \left[\left(1 - q_C + q_T + q_{SS}\right) L_{SGT} \cdot \eta_M - L_{HPC}\right]. \tag{9}$$

Relative fuel consumption in AB:

$$q_{fAB} = \frac{G_{fAB}}{G_{V2}} = \frac{h^*_{inAB} - h^*_{outAB} - h_T + h_V}{\eta_{AB} \cdot H - h^*_7 + h_T}, \tag{10}$$

where: h^*_{inAB}, h^*_{outAB} are the enthalpies of the mixture in front of and behind the afterburner, G_{fAB} is the fuel consumption in AB. h_T is the enthalpy of fuel gas, h_V is the enthalpy of air.

Thermal power of the installation in heating conditions:

$$Q_\Sigma = Q_{NW} - Q_{GWH}, \tag{11}$$

$$Q_{NW} = G_{P\Sigma} \cdot r, \tag{12}$$

$$Q_{GWH} = \left(G_{P\Sigma} + G_{AB}\right)\left(h_{2HB} - h_{CU}\right)\eta_{GWH} \tag{13}$$

where: Q_{NW} - heat generated in the *HNW*; - specific heat of vaporization, Q_{GWH} - thermal energy of G_{WH}.

Using the proposed mathematical model, an analysis of the characteristics of the gas-steam CCGT-CHPP was carried out during its operation in the non-heating and heating periods of the year according to recommendations [14]. In the calculations, the following were taken: air flow through compressors G_A=25 kg/s, the degree of increase in pressure in the low-pressure switch and in the high-pressure switch $\pi_{\Sigma C}$=6.4; excess air coefficient α = 1.2. The temperature of the mixture before *SGT* T_{SG}=1100K. The parameters of superheated steam produced in the *HB* - 13 MPa, 540 [11]. The regulation of the characteristics of the units is carried out in accordance with the operating conditions of the CCGT unit by changing the fuel consumption in *CC* and *AB*. In fig. Figures 3 and 4 show the effect of fuel consumption in *AB* on *HB* steam production and on the generation of additional electric power in ST2. The use of the afterburner with an additional fan gives a double increase in the electric capacity of the CCGT unit.

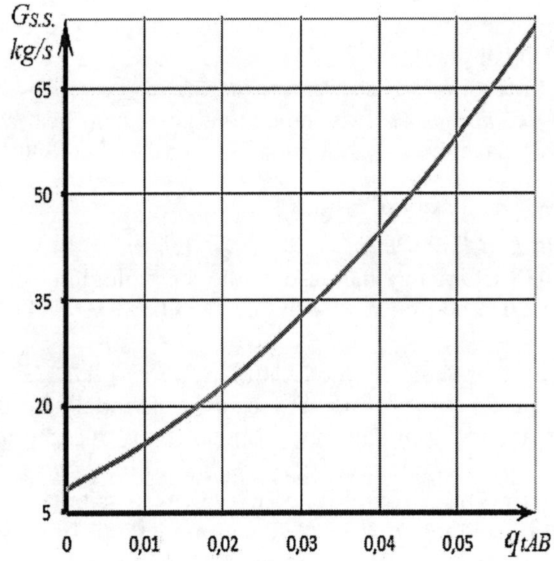

Figure 4. The dependence of the *HB* steam productivity on the relative consumption of fuel gas in the gas compressor $G_{S.S.}=f(q_{tAB})$

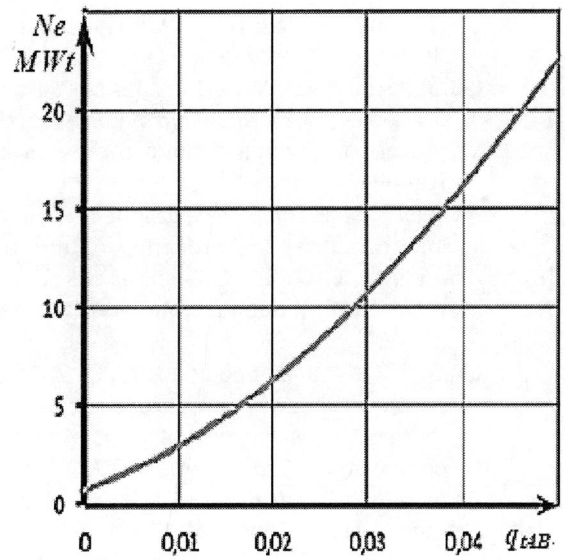

Figure 5. The dependence of the additional electric power N_E on the relative fuel consumption q_{tAB} in *AB*.

5. Conclusion.

The proposed scheme of a gas-steam installation with a low-pressure steam drive with an excess air coefficient α=1.2 in the compressor station and supplying steam from the *ST1* to the mixing chamber makes it possible to reduce the metal consumption of the design of compressors and *ST1* at a low level of G_V and $\pi_{\Sigma C}$. The expansion of the gas-vapor mixture in the gas-turbine *SPT* contributes to an increase in temperature before the *CC* increases its steam production. The installation between the stages of the *KD* evaporator with a fan for supplying air to it makes it possible to increase the combined heat and power capacity of the *CCGT* unit with an increase in the indicators of joint cogeneration generation of heat and electric energy [15]. The technological scheme of the gas-steam *CCGT-CHP* is relatively simple and has lower capital costs compared to a binary cycle *CCGT*. According to recommendations [16], the introduction of steam into the gas turbine combustion chamber improves the environmental performance of the *CCGT-CHP*

References

[1] Veselov F V, Erokhina I V, Makarova A S and Khorshev A A 2017 Comprehensive Assessment of the Effective Scope of Modernization of Thermal Power Plants to Substantiate the Rational Structure of the Generating Capacities for the Future until 2035 *Thermal Eng.* **64(3)**

[2] Biryuk V V, Kirsanov Yu G, Livshits M Yu, Tsapkova A B and Sheludko L P 2016 *Teplofikacionnaya parogazovaya ustanovka* [Combined-cycle heat and gas installation] Patent RF No 2650232 [In Russian]

[3] Teplov B D and Radin Yu A 2019 Improving Flexibility and Economic Efficiency of CCGT Units Operation in the Conditions of the Wholesale. Electricity Market *J. of Thermal Eng.* **66(5)** 323-30

[4] Nikishin V A, Peshkov L I, Ryzhinsky I N and Sheludko L P 2017 *Parogazovaya ustanovka s paroturbinny`m privodom kompressora i regenerativnoj gazovoj turbinoj* [Combined-cycle plant with a steam-turbine compressor drive and a regenerative gas turbine] Patent RF No 2409746 [In Russian]

[5] Zaryankin A E, Zaryankin V A, Mager A S and Noskova M A 2015 Parogazovy`e ustanovki s dopolnitel`ny`m e`nergeticheskim kotlom i paroturbinny`m privodom kompressora [Combined-cycle plants with an additional energy boiler and a steam-turbine compressor drive] *Ga-zoturbinny`e texnologii* **3** 40-45[In Russian]

[6] Gulina S A, Tyan V K, Livshits M Yu and Sheludko L P 2019 *Manevrennaya gazoparovaya teploe`lektrocentral` s parovy`m privodom kompressora* [Maneuverable gas-steam heat and power plant with a steam compressor drive] Zayavka na izobretenie RF No 2019127680 [In Russian]

[7] Avdeev K A, Agureev I E, Elagin M Y and Hmelev R N 2017 Determination of thermodynamic and thermophysical properties of alter fuels *J. of Eng. Physics and Thermophysics* **90(6)**

[8] Sheludko L P, Gulina S A and Biruk V V 2018 Efficiency increase of the technologic electric plant on oil-pumping stations of the main oil pipe line *J. of Physics: Conf. Series* **1111(1)** 012064

[9] Gulina S A and Sheludko L P 2016 Modelirovanie termodinamicheskogo cikla GTD, rabotayushhego na gazoobraznom toplive proizvol`nogo sostava [Modeling the thermodynamic cycle of a gas turbine engine running on gaseous fuel of arbitrary composition] *Sbornik trudov XXIX mezhdunarodnoj nauchnoj konferencii «Matematicheskie metody` v texnike i texnologiyax MMTT – 29»* Vol 12 (Saratov) pp 67-73 [In Russian]

[10] Zamaleev M M, Gubin I V, Sharapov V I and Bushuev E N 2018 Calculation methods of power efficiency of combined heat and power plant at change of equipment operating modes and thermal schemes *J. of Physics: Conf. Series* **1111(1)** 012039

[11] Astvatsaturova A A, Zorin V M and Trukhnii A D 2015 Assessment of steam work efficiency as applied to a turbine being designed *J. of Thermal Eng.* **62(1)** 26-33

[12] Biryuk V V, Shimanov A A, Larin E A, Livshits M Y and Sheludko L P 2018 Unitized Power and Heat Generating Steam–Gas Plant *J. of Eng. Physics and Thermophysics* **91(4)** 1029-37

[13] Shase D L 2001 Combined-Cycle Development Evolution and Future GER-4206 (GE Power Systems) URL https://www.ge.com/content/dam/gepower-pgdp/global/en_US/documents/technical/ger/ger-4206-combined-cycle-development-evolution-future.pdf

[14] Ankudinova M S, Larin E A and Sandalova L A 2018 Influence of ambiguity of reliability indices information on configuration shaping of combined cycle plant *J. of Physics: Conf. Series* **1111(1)** 012011

[15] Mikhailov V E, Khomenok L A and Kovalev I A 2018 Development of High-Powered Steam Turbines by OAO NPO Central Research and Design Institute for Boilers and Turbines *Thermal Engineering* **65** 1-10

[16] De D K, IkonoU I and Akinmeji S O 2018 Health, Environmental Effects, the Control of Emission from Power Plants and the Need for a New Emission Capture Technology *J. of Physics: Conf. Series* **1072(1)** 012019

IOP Conf. Series: Materials Science and Engineering **791** (2020) 012020 doi:10.1088/1757-899X/791/1/012020

Structural and operational optimization of the phase transition accumulator during operation as part of a nuclear power plant based on long-term durability

V E Yurin and A B Moskalenko

Department of Energy Problems, Saratov Scientific Center of the Russian Academy of Sciences, Rabochaya St., 24,Saratov, 410028, Russian Federation

E-mail: oepran@inbox.ru

Abstract. The purpose of the article is the operational optimization of the heat accumulation system taking into account the requirements of the energy market, daily differentiated tariffs and the influence of a specific operating regime on the main equipment performance using the example of a phase transition accumulator installed at a nuclear power plant. There was developed a three-dimensional model for the non-statutory process of flowing fluids and heat-accumulating material hardening which comes as a result of the heat transfer and the stress field exchange between them. The service life of 3D models with various collector wall thicknesses was calculated. There was performed the calculation of the economic effect and payback period for options being compared based on of phase transition accumulator operating cycle, his performance with various collector metal wall thickness. The performance change in the economic analyses is based on the useful lives change while the change in daily operating time is based on the power generation change. As a result, the research shows that the most profitable option for accumulated net income is to use a collector with an accumulator wall thickness of 40 mm and the following operation mode: 5 hours charge + 16 hours discharge.

1. Introduction

Covering irregularities in the electrical loads schedule of the Unified Energy System of Russia is a high priority of today. Combining nuclear power plants with heat energy accumulation systems is one of the possible problem solutions. Accumulation due to the phenomenon of phase transition is a promising direction [1-3]. The combination of nuclear power plants with a heat accumulator will make it possible to ensure the operation of reactor plants with the same installed capacity during recession hours of electric loads by reducing the generated electricity during accumulation process. It will also provide additional electricity during peak and half-peak hours due to the use of accumulated night energy [4].

$LiNO_3$, LiF, $NaNO_3$, LiCl are widespread as heat accumulation materials for accumulators [5, 6]. However, the melting point of the presented materials does not allow their use in pure form in the considered energy complex. It is possible to use an alloy consisting of $LiNO_3$ and NaCl to obtain the required operating parameters of the accumulator. [7-9].

The new design of the heat accumulator with phase transition material was developed to provide the possibility of use the heat accumulation system in the required conditions (patent of the Russian Federation No. 179855) (figure 1).

IOP Conf. Series: Materials Science and Engineering **791** (2020) 012020 doi:10.1088/1757-899X/791/1/012020

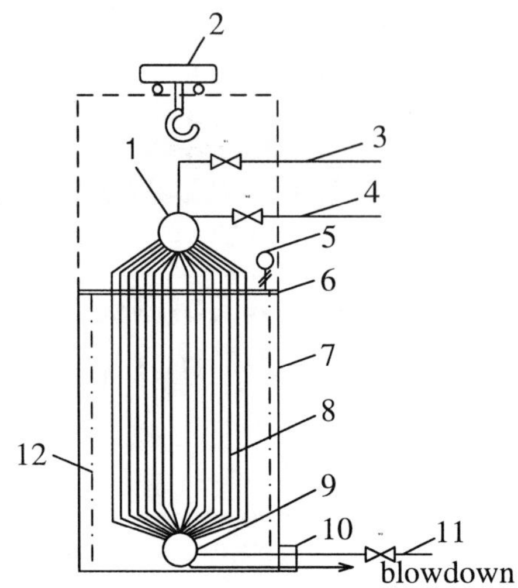

Figure 1. Design of phase transition accumulator. 1 – overhead collector; 2 – overhead traveling crane; 3 – steam conductor; 4, 11 – pipeline; 5 – breathing hole; 6 – tube sheet; 7 – jar; 8 – metal heat exchange tube bundle; 9 – bottom collector; 10 – downtake tube; 12 – glass tubes, in which the parameters control sensors of the heat-accumulating material are located.

Main steam is directed through steam conductor 3 to the overhead collector 1 during charging. Then it is distributed through the metal heat exchange tube bundle 8 and gives off heat to the accumulation material. After that, the steam condenses, merges into the bottom collector 9 and is directed into the main cycle through the pipeline 11.

During the discharge process, feed water is supplied to the bottom collector 9 through the pipeline 11. Then, it is evenly distributed over the heat exchange metal tubes 8. Heated feed water through a pipeline 4 from the overhead collector 1 is directed into the main feed water line in front of the steam generators.

The operating conditions of the phase transition accumulator are characterized by intense thermal power cyclic effects on its structural elements. These effects arise due to changes in pressure and temperature of the coolant in transient operation modes. Damage in the form of fatigue cracks can occur during prolonged exposure to stresses in the equipment structural material. Besides, corrosion processes can intensify, which can lead to a loss in its availability. These processes are non-stationary character and lead to changes in the temperature of the metal structural elements. These effects also have a negative impact on the lifetime of the equipment and, therefore, on the lifetime of the accumulation system as a whole.

A large of cycles amount of exposure to excess pressure and heating-cooling occur in the heat exchange tube bundle as the phase transition accumulator is used. These processes lead to the gradual destruction of heat transfer tubes. Despite a large number of studies of thermal accumulators, a full-fledged study of the unsteady process of heat transfer between a heat-accumulating substance and a heat carrier in a phase transition accumulator (used with NPP) absents. Only the thermodynamic efficiency of the accumulator is investigated, and the design life of the accumulator itself is not considered. The task of lifetime researching of the accumulator structural elements, taking into account the unsteady heat transfer process, is urgent and needs to be solved. Research of the lifetime will optimize the design of the accumulator and make more precise the capital investment in the installation. This will have a direct impact on the annual operating costs: repair, depreciation and others. The results of these researches will allow to fully evaluate the economic efficiency of combining nuclear power plants with phase transition accumulators.

The modern technology for studying fatigue life of equipment under thermocycling influences is based on three-dimensional numerical modelling using computational fluid dynamics programs, stress-strain state calculation programs in combination with mathematical models and durability calculation programs. Based on this technology, modelling of the unsteady thermal state of the heat transfer tubes of the phase transition accumulator is carried out.

2. Description of the computation model

The presence of excess pressure inside the collectors and tubes of the phase transition accumulator leads to the occurrence of mechanical stresses in them. Also, cyclic temperature changes are followed by thermal stresses. The following conditions are accepted for calculating the lifetime of phase transition accumulator:

The temperature of the heat accumulating material, water in the tubes, and phase transition accumulator structural elements after charging is equal to 272 °C. During discharge, water having a temperature of 225 °C is supplied to the down collector of the phase transition accumulator, is heated in the phase transition accumulator and at the outlet has a temperature of at least 245 °C. Thus, in the bottom collector, at the beginning of the discharge process, a difference of 47 °C is observed between the water temperature and the metal temperature.

During charging, steam with a temperature of about 275 °C is supplied to the overhead collector of phase transition accumulator, which has a temperature of about 247°C, which leads to a temperature unevenness of 28 °C. Thus, the analysis of the bottom collector during the discharge of the phase transition accumulator is since the difference between the temperatures of the water and the metal at this element is the largest and that leads to greater thermal stresses.

The study of the bottom collector completely (i.e., along its entire length) is not advisable, because along its length all geometric dimensions are constant and the result of the calculations will be similar at any point and the studing only part of the bottom collector allows to reduce the required computing resources and the calculation time.

3 variants of 3D models of the bottom collector part of the phase transition accumulator with the tube bundle part, which differs from each other by the collector wall thickness were built. The stress calculation for the three variants will determine the optimal wall thickness of the collectors in terms of its lifetime. The geometric dimensions of the collector part with the tube bundle part are presented in table 1. For example, figure 2 shows a 3D model of the collector part with tube bundle part.

Table 1. The geometric dimensions of the part of the collector with part of the tube bundle.

	Value
The inner diameter of the heat exchange tubes, D_t^{in} (m)	0.021
Wall thickness of the heat exchange tubes, δt (m)	0.002
The distance between the axes of the tubes of the heat exchange surface in the longitudinal and transverse directions of the collector, δd (m)	0.126
The inner diameter of the collector, D_c^{in} (m)	0.8
Collector wall thickness, δc (m)	0.03 (a); 0.04 (b); 0.05 (c)
The number of tubes in the row of the collector, in the transverse direction, n (pcs)	18

Figure 2. 3D model of the collector and the heat exchange surface of the phase transition accumulator.

In the simulation adopted steel of grade 10. Its physical properties are shown in Table 2.

Table 2. The physical properties of steel of grade 10 depending on temperature.

	Temperature, °C	
	200	300
Density, ρ (kg/m3)	7800	7765
Elastic modulus, $E \cdot 10^{-5}$ (MPa)	1.99	1.9
Coefficient of thermal expansion, $\alpha \cdot 10^{6}$ (1/°C)	13.2	13.9
Thermal conductivity, λ (W/(m · K))	53	49.6
Heat capacity, cp (J/(kg · K))	532	565

Next, an unsteady process of fluid flow and solidification of TAM was simulated. The temperature of the collector metal depending on time was determined.

3. Results and discussion

Based on the data obtained, stresses from temperature non-uniformity and the influence of internal pressure (6.91 MPa) at any point of the calculation model depending on time were determined (figure 3). Figure 4 shows the maximum stresses for all three variants at the corresponding time points. The moment of pressure increases from zero was not simulated, because during this process maximum stresses cannot be achieved.

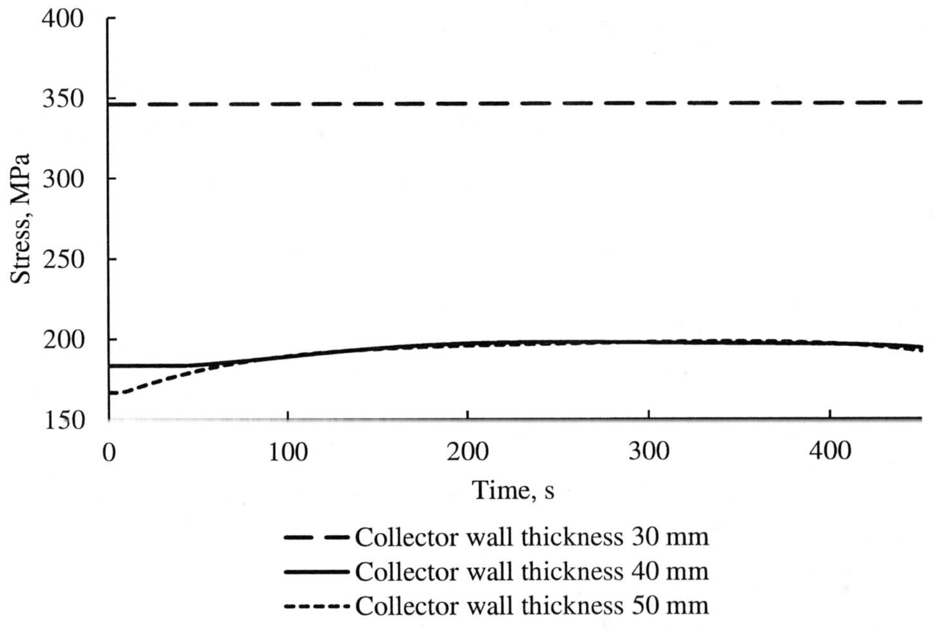

Figure 3. The dependence of the maximum stresses on the time.

IOP Conf. Series: Materials Science and Engineering **791** (2020) 012020 doi:10.1088/1757-899X/791/1/012020

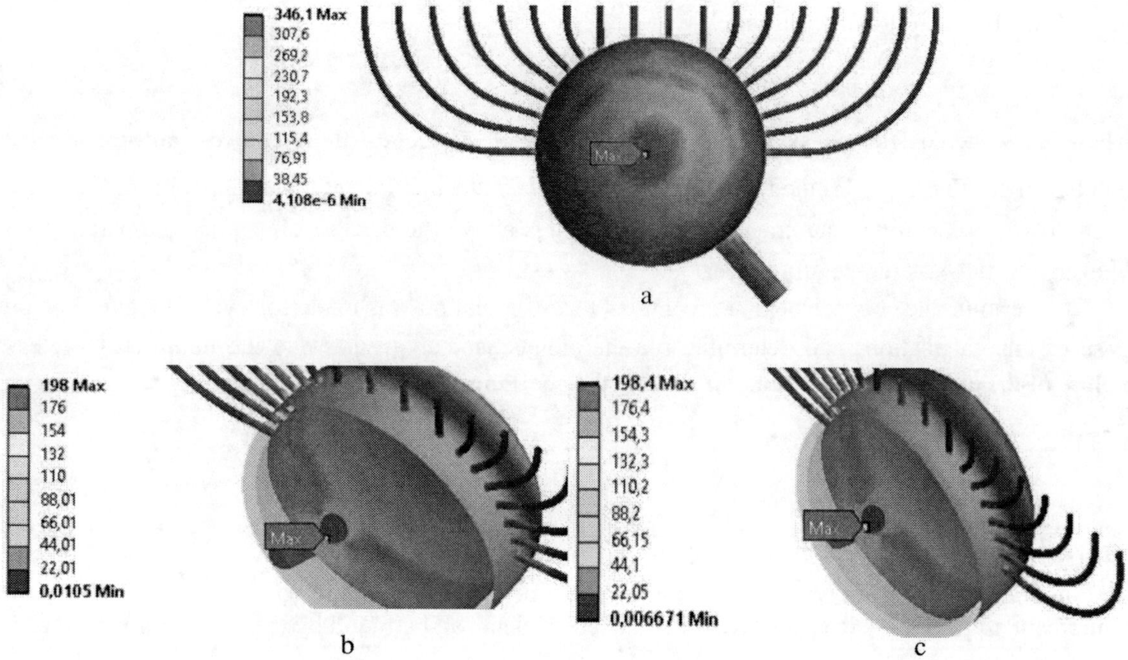

Figure 4. Stress field in 3D models with different collector wall thicknesses, MPa: a – 30 mm; b – 40 mm; c – 50 mm.

Based on the previously obtained data, figure 5 presents a graph of the dependence of the maximum stresses in the calculation model per cycle on its thickness.

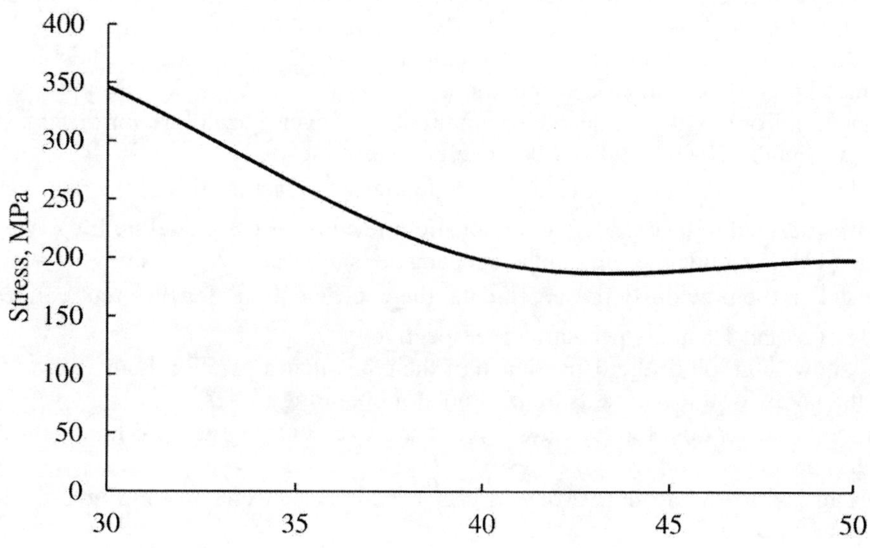

Figure 5. The dependence of the maximum stresses in the calculation model per cycle on its thickness.

IOP Conf. Series: Materials Science and Engineering **791** (2020) 012020 doi:10.1088/1757-899X/791/1/012020

It is proposed to use the methodology for calculating equivalent operating hours to estimate the lifetime of the tube bundle of the phase transition accumulator [10]. Equivalent operation hours are determined by the following formula:

$$T_{eq} = a \cdot n + \tau_{sum} + b \cdot \tau_p \tag{1}$$

where a is the coefficient taking into account low-cycle fatigue of the metal for start-up; n is the number of start-ups; τ_{sum} is the total operating time; b is the coefficient taking into account long-term strength when operating equipment with temperatures above the base level; τ_p is the operating time in excess of the base temperature level.

To determine the coefficient a in formula (1), a calculation was made for cyclic fatigue. The purpose of this calculation is to determine the thermomechanical stresses and the number of cycles to failure of the element. The formula for finding the coefficient a is as follows:

$$a = \frac{T_{life}}{N_f} \tag{2}$$

where T_{life} is the lifetime of the element; N_f is the number of cycles to failure.

Finding the coefficient b involves the calculation of long-term strength during operation of the equipment above the basic characteristics. This coefficient can be found by the following formula:

$$b = \frac{T_{life}}{\tau_f} \tag{3}$$

where τ_f is the time to failure of the element during operation of the equipment in excess of the base temperature, which can be defined by the long-term strength equation.

According to the strength calculation standards for stationary boilers and steam and hot water pipelines (RD 10-249-98), the maximum stress amplitudes at wall thicknesses of 30, 40, and 50 mm are 173.1 MPa, 99 MPa, and 99.2 MPa, respectively. Figure 6 shows the dependence of the number of cycles to failure on the stress amplitudes at a temperature of 272 °C, which is the maximum temperature of the metal in the cycle under consideration.

As can be seen from Figure 6, the maximum number of cycles to failure for different collector wall thicknesses is 14,600, 170,000, and 169,000 cycles, respectively.

To calculate the coefficient a , it is necessary to know the factory life of the phase transition accumulator. It was decided to take the accumulator lifetime equal to the guideline life of heat exchangers. According to [11], the guideline life of heat exchangers is 30 years.

Thus, based on the previously presented data, the coefficients a for different collector wall thicknesses are 18, 1.5, and 1.6 hours per start-up, respectively.

Figure 7 shows a graph of the dependence of the equivalent operation hours for a year on the collector wall thickness for three variants of accumulator operating cycles:

- Variant 1 - Operation for 292 days a year for 2 cycles (1 charge for 5 hours + 1 discharge for 6 hours) per day.
- Variant 2 - Operation for 292 days a year for 2 cycles (1 charge for 5 hours + 1 discharge for 12 hours) per day.
- Variant 3 - Operation for 292 days a year for 2 cycles (1 charge for 5 hours + 1 discharge for 16 hours) per day.

IOP Conf. Series: Materials Science and Engineering **791** (2020) 012020 doi:10.1088/1757-899X/791/1/012020

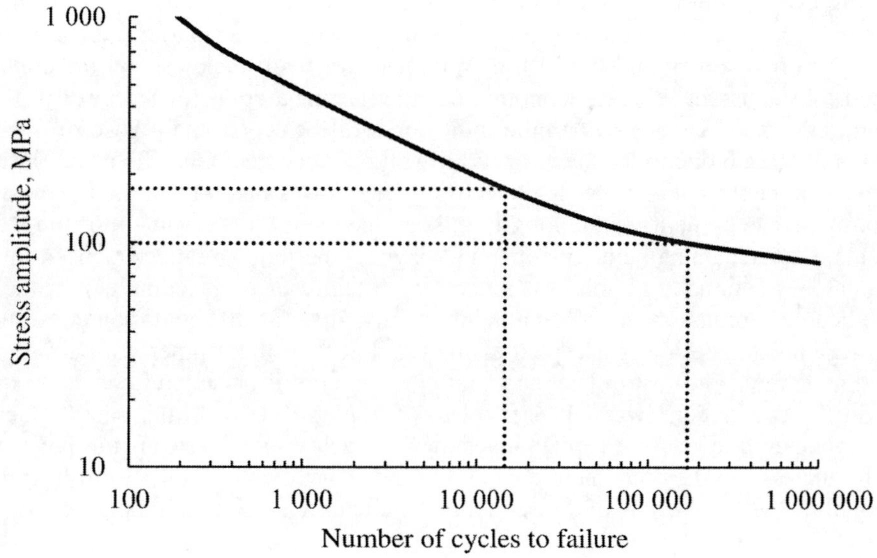

Figure 6. Dependence of the number of cycles to failure on the stress amplitude for carbon steels.

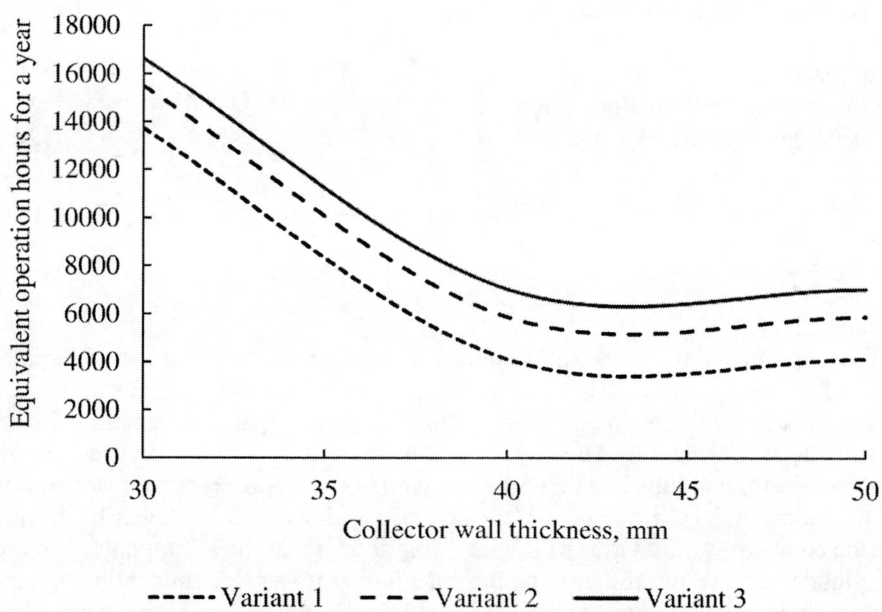

Figure 7. Dependence of the equivalent operation hours for a year on the collector wall thickness and duration of operation per day.

It should be noted that an increase in the collector wall thickness leads to an increase in temperature non-uniformity inside the collector wall. This appear in an increase in stresses in the cycle (figures 3-5), and leads to an increase in the equivalent operation hours for a year, but this begins to appear when the collector wall thickness is more than 40 mm (figure 7).

IOP Conf. Series: Materials Science and Engineering **791** (2020) 012020 doi:10.1088/1757-899X/791/1/012020

The economic effect and payback period were calculated to compare different variants for accumulator operating cycles and different accumulator lifetime at different collector wall thicknesses. The need to compare and search for the optimal solution is since with a smaller thickness of the collector wall, there is less investment and useful life. With a larger wall thickness, on the contrary, more investment and longer useful life. An economic calculation for a collector with a wall thickness of 50 mm is not necessary. Obviously, economic indicators will be worse than those of a collector with a wall thickness of 40 mm due to the lower useful life and greater steel intensity of the structure.

The increase in nuclear fuel prices and electricity tariffs was taken from the Forecast of the Ministry of Economic Development for the long-term socio-economic development of the Russian Federation to calculate the accumulated net income and payback period. The investment in the phase transition accumulator is estimated as follows. Since the structure of the accumulator without taking into account the heat accumulation material contains mainly pipes of different diameters and lengths, the cost of the accumulator design is accepted as the retail cost of these pipes with a margin of 10% for delivery and assembly. The cost of heat accumulation material is accepted to be 3.1 $/kg [12-15]. The dollar rate in the calculations was taken equal to 65 rubles. Based on the data of the TURBOPAR group of companies, the relative capital investment in a low-power steam turbine plant is about $ 640/kW. The increase in the equipment cost and spends was adopted in the form of inflation, for calculations equal to 3%. The results of the economic calculation are shown in table 3.

Table 3. The results of the calculation of accumulated net income and payback period for different variants for using the phase transition accumulator.

	Variant 1		Variant 2		Variant 3	
Collector wall thickness, (mm)	30	40	30	40	30	40
Capital investment in accumulator design, ($ million)	0.4	0.5	0.4	0.5	0.4	0.5
Useful life, (years)	19	64	17	45	16	37
Amortization for the accumulator design (in the first year of operation), (thousand $)	22.5	7.7	24.6	10.8	26.2	13.8
Accumulated net income (time horizon – 60 years), ($ million)	73.1	76.1	217.0	225.9	312.9	325.7
Payback period, (years)	11.9	12.0	9.1	9.2	8.3	8.4

4. Conclusion

The operational optimization of the heat accumulation system was carried out taking into account the requirements of the energy market, daily differentiated tariffs and the influence of a specific operating regime on the lifetime of the main equipment using the phase transition accumulator installed at nuclear power plants as an example. The paper assesses the effect of cyclic and long-term loads on the lifetime of the collector and tube bundle of the accumulator with different collector wall thicknesses as a result of thermomechanical influences. As calculations showed, a thickness of 40 mm is the most optimal for the collector wall in terms of lifetime. It is shown that the accumulator lifetime depends on its operating mode. As a result of the economic calculation, it was determined that the most profitable option is to use collectors with a wall thickness of 40 mm in the design of the accumulator and an operating mode: 5 hours charge + 16 hours discharge in the half-peak part of the electric load graph.

Acknowledgments

The given research was funded by RFBR according to the research project № 18-38-00881.

IOP Conf. Series: Materials Science and Engineering **791** (2020) 012020 doi:10.1088/1757-899X/791/1/012020

References

[1] Mehling H and Cabeza L F 2008 *Heat and cold storage with PCM : An up to date introduction into basics and applications* (Berlin: Springer-Verlag Berlin and Heidelberg GmbH & Co. KG)

[2] Beckmann G and Gilli P V 1984 *Thermal Energy Storage* (Vienna: Springer-Verlag Vienna)

[3] Raoux S and Wuttig M 2010 *Phase Change Materials: Science and Applications* (New York: Springer Science and Business Media)

[4] Aminov R Z, Yurin V E and Egorov A N 2018 *Kombinirovanie AE`S s mnogofunkcional`ny`mi e`nergeticheskimi ustanovkami* [Combining NPPs with multifunctional power installations] (Moscow: Nauka) [In Russian]

[5] Fultz B 2014 *Phase Transitions in Materials* (Cambridge: Cambridge University Press)

[6] Redkin A, Zaikov Yu, Tkacheva O and Kumkov S 2016 Molar thermal conductivity of molten salts *Int. J. of Ionics The Science and Technology of Ionic Motion* **22** 143-9

[7] Shokouhmand H and Kamkari B 2013 Experimental investigation on melting heat transfer characteristics of lauric acid in a rectangular thermal storage unit *Experimental Thermal and Fluid Science* **50** 201-12

[8] Kenisarin M M 2010 High-temperature phase change materials for thermal energy storage *Renewable and Sustainable Energy Reviews* **14** 955-70

[9] Mozgovoj A G, Shpilrain E E, Dibirov M A, Bochkov M M, Levina L N and Kenisarin M M 1990 *Teplofizicheskie svojstva teploakkumuliruyushhix materialov. Kristallogidraty`.* [Thermophysical properties of heat storage materials. Crystal hydrates] (Moscow: IVTAN AN SSSR) [In Russian]

[10] PJM Manual 15: Cost Development Guidelines URL www.pjm.com/~/media/documents/manuals/m15.ashx

[11] Aronson K E, Blinkov S N, Brezgin V I, Brodov Yu M, Kuptsov V K, Larionov I D, Nirenshteyn M A, Plotnikov P N, Ryabchikov A Yu and Khayet S I 2015 *Teploobmenniki e`nergeticheskix ustanovok* [Power plant heat exchangers] (Yekaterinburg: UrFU) [In Russian]

[12] Kenisarin M and Mahkamov K 2016 Salt hydrates as latent heat storage materials:Thermophysical properties and costs *Solar Energy Materials and Solar Cells* **145** 255-86

[13] Zhou D and Eames P 2016 Thermal characterisation of binary sodium/lithium nitrate salts for latent heat storage at medium temperatures *Solar Energy Materials and Solar Cells* **157** 1019-25

[14] Infante Ferreira C A 2016 *Advances in Solar Heating and Cooling* ed R Z Wang and T S Ge (Amsterdam: Elsevier)

[15] Fernandez A G and Gomez-Vidal J C 2017 Thermophysical properties of low cost lithium nitrate salts produced in northern Chile for thermal energy storage *Renewable Energy* **101** 120-5

IOP Conf. Series: Materials Science and Engineering **791** (2020) 012021 doi:10.1088/1757-899X/791/1/012021

Application of phase portraits for estimation of reactive power of electric circuits

D P Andrianov, N P Badalyan and G P Kolesnik

Department of power engineering and electrical engineering, Vladimir State University named after A.G. and N.G. Stoletovs, Gorky St., 87, Vladimir, 600000, Russia

E-mail: andrianov@vlsi.ru

Abstract. In this paper we consider the principal possibility of using the estimation of the reactive power of electric circuits by phase portraits of electric current oscillations. The shift of the phase angle of current and AC voltage oscillations relative to each other is displayed on the phase portrait of oscillations by bifurcation of the phase trajectory. The region described by the phase trajectory corresponds to the reactive power. Reactive power must be kept to a minimum to produce an energy efficient power supply. As an optimization criterion it is proposed to use the size of the phase portrait area of stationary oscillations of alternating current. Simulation of the optimization process was carried out on the basis of numerical solution of the system of differential equations using dynamic programming methods. Typical functions of numerical solution of systems of differential equations and optimization of the target function, available in mathematical packages, were used.

1. Introduction

The safety and cost-effectiveness of power systems depend to a large extent on the availability of reactive power. Management and optimization of reactive power is currently receiving increased attention.

In [1] attention is drawn to the fact that the nature of voltage stability in the power grid can be analyzed by studying the generation, transmission and consumption of reactive power.

A number of researches are devoted to methods of evolutionary calculations at optimization of reactive power [2], algorithms of control of tension and reactive power [3], [4], adaptive algorithms [5], [6].

The issues of multi-purpose optimization of active power losses, voltage deviations, static voltage reserve and reactive power compensation are considered [7]. Separately, it should be noted the issues related to the dispatching of reactive power [8].

At the hardware level, the issues of reactive power compensation using thyristor compensators [9], SAPF compensators [10], in GRID-systems [11] are considered.

2. Materials and Methods

The peculiarity of AC electric circuits is the shift of current and voltage fluctuations relative to each other, introduced by inertial links-inductances and capacitances. The divergence of oscillations is characterized by a phase shift, the magnitude of which depends on the reactive power.

Of interest is the consideration of changes in current and voltage without taking into account time, that is, obtaining a phase portrait of oscillations-a picture in which the change of parameters relative to each other is represented, and not relative to time [12].

IOP Conf. Series: Materials Science and Engineering **791** (2020) 012021 doi:10.1088/1757-899X/791/1/012021

In the steady-state mode, the oscillations of the parameters of the electric circuits of alternating current are strictly periodic, repetitive, which greatly simplifies the obtaining of a phase portrait. It should be noted that currently existing mathematical packages (MATLAB, MathCAD) allow to obtain acceptable solutions with minimal costs.

It is generally accepted to consider phase portraits of dynamic objects in relation to the stability analysis of their regime [13].

Phase trajectories, as components of the phase portrait, make it possible to simply and visually represent the presence of reactive power in the operation of AC electrical circuits.

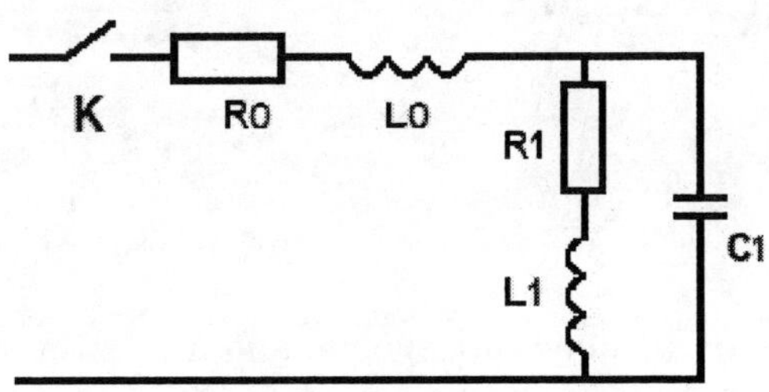

Figure 1. Diagram of a typical electrical circuit.

A typical electrical circuit (Figure 1) contains a power branch with key K, resistance R0 and inductance L0, as well as a load of resistance R1, inductance L1 and capacitance C1. The scheme corresponds to a system of three differential equations, of which two equations describe the dynamics of the current at the input to the circuit and in the load branch with resistance R1 and inductance L1, and the third equation-the voltage change at the capacitance C1. The input of the circuit is supplied with a voltage varying according to the sinusoidal law.

Typical functions of numerical solution of systems of differential equations in the form of Cauchy (Mathcad package) were used for modeling. The system of equations was transformed to the matrix form with the allocation of derivatives of the first order. The result of the numerical solution of the system of differential equations is a table of values describing the graphs of the behavior of the parameters (Figure 2).

The above parameters R0, R1, inductance L0, L1 and capacitance C1 determine the nature of the transition process that occurs in the circuit when switching the key K.

Under zero initial conditions, the change in the parameters of the alternating current acquires the character shown in Figure 2.

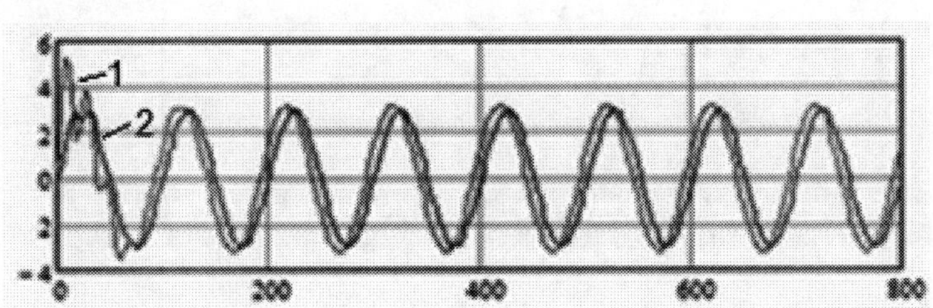

Figure 2. Transient in the presence of reactive power: 1 – the current in the initial branch; 2 – the voltage between nodes.

The discrepancy between the current and voltage fluctuations on the phase shift is clearly traced. During one period, the fluctuations take a steady character.

IOP Conf. Series: Materials Science and Engineering **791** (2020) 012021 doi:10.1088/1757-899X/791/1/012021

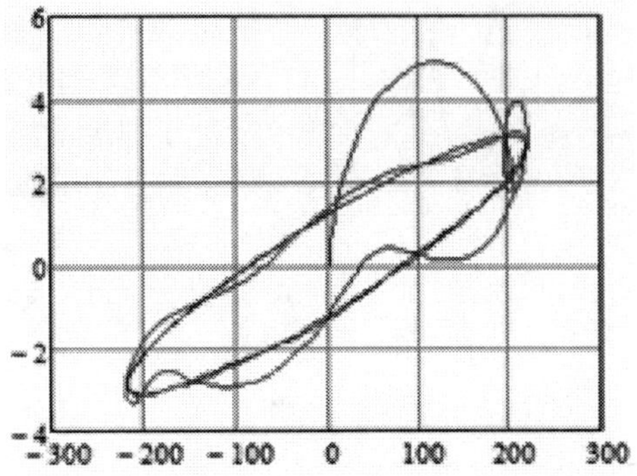

Figure 3. Phase portrait of current fluctuations taking into account the transition process.

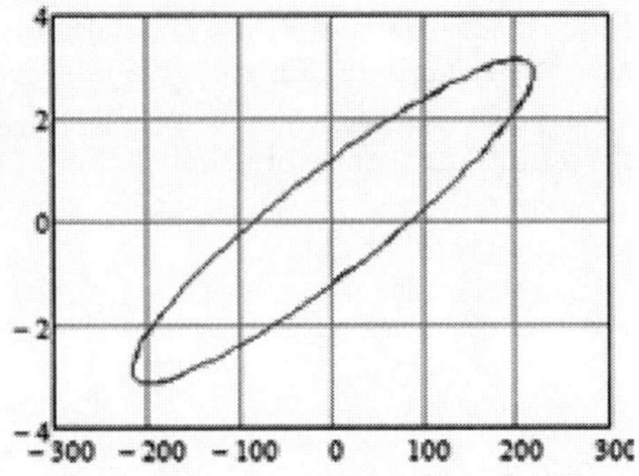

Figure 4. Phase portrait of current fluctuations without taking into account the transition process.

The phase portrait of the oscillations corresponding to the transition process (Figure 2) is shown in Figure 3. The abscissa axis of the phase portrait corresponds to a change in voltage, the ordinate axis-a change in current.

If the transition process corresponding to the switching of the key is excluded from the phase portrait, the phase portrait of steady – state oscillations will be obtained-Figure 4. The reference point in time for steady-state oscillations is the mark «200» on the horizontal axis of the Figure 2 diagram.

The phase portrait for steady-state oscillations is determined by a closed phase trajectory in the form of an ellipse whose axes depend on the phase angle between the current and voltage oscillations.

The Diagram of Figure 5 is a voltage characteristic that displays the dynamics of the behavior of the current and voltage relative to each other. The inclined line «a-b» corresponds to the phase portrait without reactive power, that is, with synchronous oscillation of current and voltage: the ellipse «acbd» does not have a small half-axis OS and degenerates into a segment of the straight line «ab». The area of the «acbd» figure corresponds to the reactive power due to the phase shift of the current and voltage oscillations.

-1×10^4

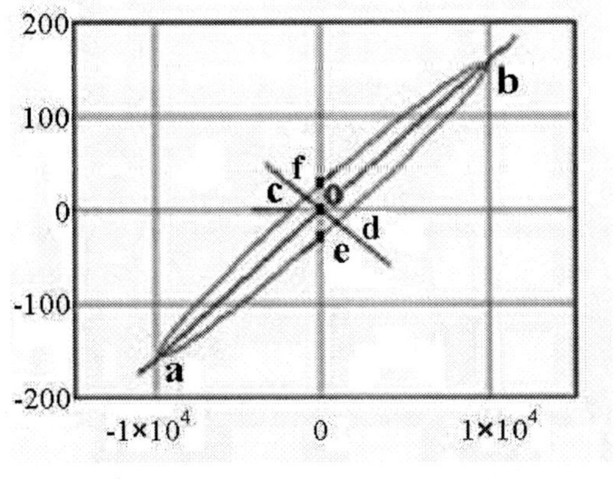

Figure 5. Phase portrait of steady-state oscillations.

For the replacement circuit (Figure 1), you can choose the value of the capacitance C so that the current and voltage fluctuations become common-mode (Figure 6).

IOP Conf. Series: Materials Science and Engineering **791** (2020) 012021 doi:10.1088/1757-899X/791/1/012021

This combination of parameters of the electric circuit will correspond to the transition process with the phase portrait in Figure 7, and for steady-state oscillations-the phase portrait in Figure 8.

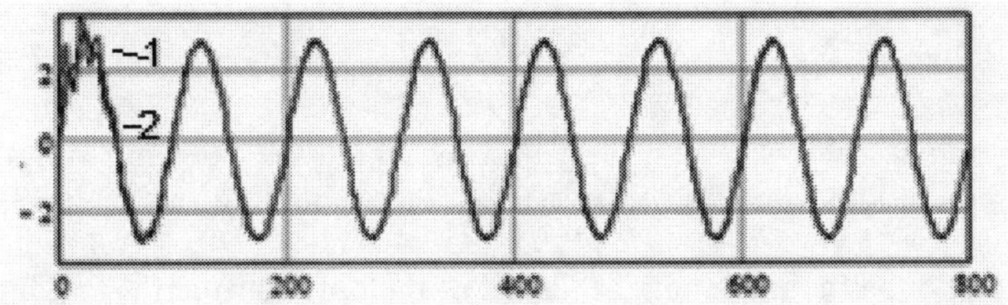

Figure 6. The transition process in reactive power compensation: 1 – the current in the initial branch; 2 – the voltage between nodes.

3. Results

It is of interest to optimize the parameters of the power supply system to achieve a minimum reactive power. The solution of the problem is carried out on the basis of numerical simulation of the system of differential equations using dynamic programming methods [14], [15].

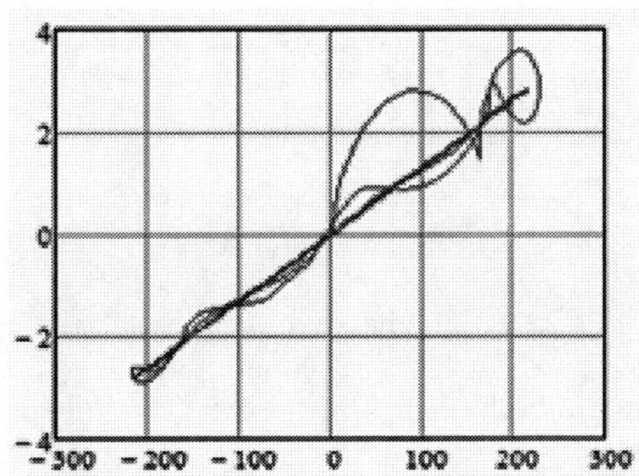

Figure 7. Phase portrait of current fluctuations taking into account the transition process.

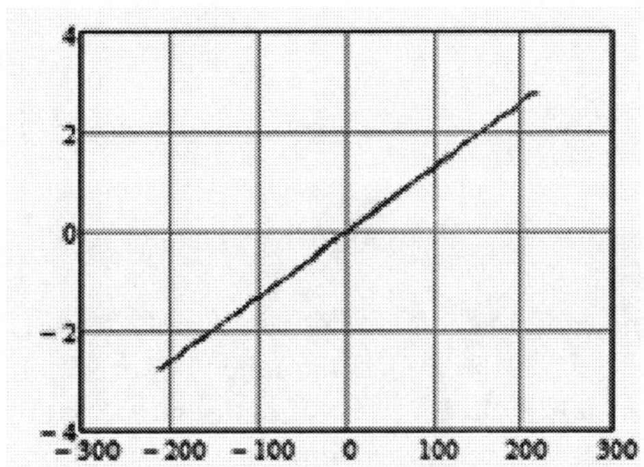

Figure 8. Phase portrait of current fluctuations without taking into account the transition process.

The algorithm for solving the problem consists of several steps:

1. For a fixed combination of parameters a system of differential equations is solved and the reactive power is determined.
2. On the basis of the search algorithm of the optimization function the parameters of the electric circuit are changed.
3. Step 1 for new parameters is executed.
4. The calculated variants are compared according to the selected criterion and the parameters of the optimal variant are fixed. If the compared variants differ from each other by an amount less than the allowable value, the calculation is terminated, otherwise paragraph 1 is performed.

The software implementation of the algorithm is based on combining the matrix equation with the optimization function of minimum search [16].

Typical algorithms and functions of Math Cad package were used. The matrix solution of the system of differential equations was formed as a function that returns the value of the parameters processed by the target function. When calling the optimization function of the Math Cad package, the

IOP Conf. Series: Materials Science and Engineering **791** (2020) 012021 doi:10.1088/1757-899X/791/1/012021

reference to the function of solving the system of differential equations was passed as the first parameter.

The objective function, as a mathematical record of the optimality criterion, has the form

$$Z(C1, C2, C3) \rightarrow \min \tag{1}$$

where C1, C2, C3 – the desired variables, the values of which are determined in the process of solving the problem, the combination of values of these parameters should ensure the achievement of the minimum reactive power.

The objective function, as a mathematical record of the optimality criterion, has the form.

Taking into account the fact that the search is conducted for the ranges of possible changes in the parameters, the problem is reduced to conditional optimization.

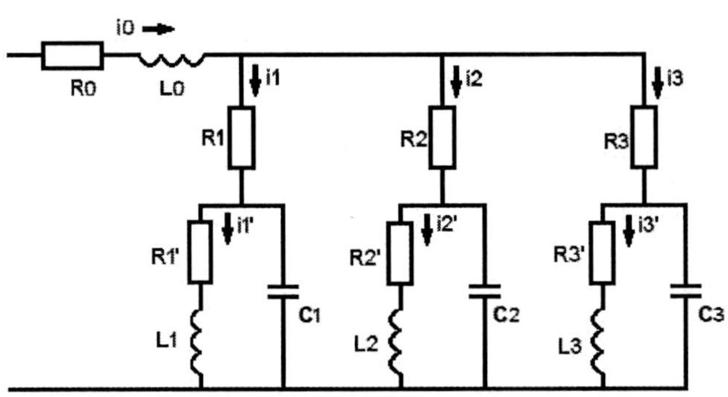

Figure 9. Design scheme for radial power supply scheme.

The radial scheme of power supply was modeled (Figure 9). The behavior of the electric current in the circuit of three load branches is described by a system of differential equations of the seventh order, which includes three equations of current changes in the branches of the load, three equations of voltage changes in the branches of capacitive reactive power compensators, and the equations of current changes at the input to the circuit.

The system of differential equations was reduced to the Cauchy form and solved by the 4th order Runge-Kutta numerical method with automatic step selection. The transient process arising as a result of connection to the circuit of an alternating voltage source was simulated. The initial conditions for the variables of the considered system were assumed to be zero.

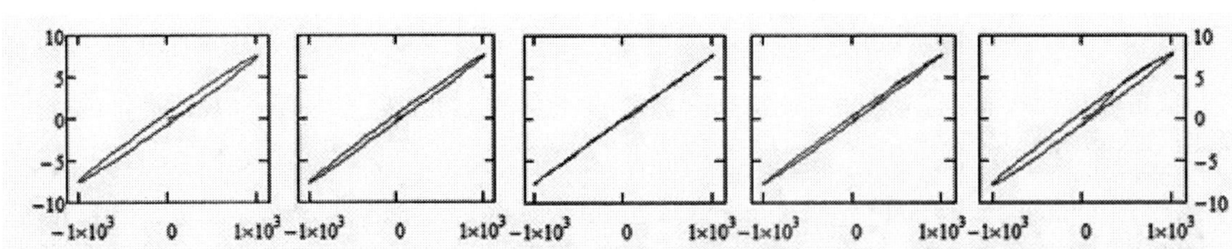

Figure 10. Evolution of the phase portrait of electric current oscillations with successive increase of the capacitance value.

On the basis of the obtained solutions (the change of the current at the input to the circuit and in the branches of the load, as well as the voltage at the terminals of the capacitors in time), the phase portrait "input current – input voltage».

The influence of capacitance of capacitors on reactive power was analyzed. Figure 10 shows the evolution of the phase portrait of the electric current oscillations with a sequential increase in the capacitance of the capacitor. It is obvious that there is a certain capacitance value at which the area of the figure described by the phase trajectory is zero.

The combination of calculation blocks implementing the solution of the system of differential equations and optimization search allowed to compose the program of optimization of the considered scheme on reactive power. As the target function, the value of the characteristic size of the phase portrait (Figure 5) «ef» was taken, as the optimization parameters – the value of the capacitances of reactive power compensators.

4. Discussion

The above approach is justified when taking into account the nonlinearities of electrical circuits, such as relay switching functions, hysteresis of magnetic circuits, mutual inductance of branches, the influence of temperature on the resistance of conductors, etc. It is of particular importance in the analysis of the stability of nonlinear dynamical systems.

In the case of considering the dynamics of nonlinear systems the application of generally accepted methods of the theory of automatic control requires a careful approach to the linearization of nonlinearities. On the other hand, the use of phase portraits as a research tool implies the use of specific mathematical methods that require the researcher to have in-depth knowledge of higher mathematics.

From the point of view of reactive power optimization in power supply systems, it should be noted that the consumer may have specific requirements for the minimum possible level of reactive power, this should be taken into account in the form of additional restrictions in the optimization search.

From the point of view of hardware, it should be noted that at present there are no great difficulties in the manufacture of devices that allow to obtain phase portraits of electric current oscillations.

5. Conclusion

This article considers the possibility of using phase portraits of alternating current oscillations to assess the reactive power and organization of optimization search of electric circuit parameters based on the criterion of minimizing the area of the phase portrait of stationary electric current oscillations. The proposed approach makes it possible to analyze the process of generation, transmission and consumption of reactive power, as well as the possibility of controlling it in order to improve the energy efficiency of power supply systems.

References

[1] Kumar S and Goyal S K 2014 A Particle Swarm Optimization for Reactive Power Optimization *Int. J. of Computational Engineering Research (IJCER)* **4** 40-6

[2] Sahu G and Swarnkar K 2014 Review of Reactive Power Optimization Using Evolutionary Computation Techniques *Advance in Electronic and Electric Engineering* **4** 73-82

[3] Varadarajan M and Swarup K 2008 Differential evolutionary algorithm for optimal reactive power dispatch *Electrical Power and Energy Systems* **30** 435–41

[4] Wu P and Zhang J 2013 Power system reactive power optimization based on global differential evolution algorithm *25th Chinese Control and Decision Conf., CCDC* (Guiyang, China) pp 1019-22

[5] Jikeng L, Xudong W and Weihong Z 2008 Reactive Power Optimization Based on Adaptive Immune Algorithm *Proc. of the Int. Multi Conf. of Engineers and Computer Scientists, IMECS 2008* (Hong Kong) **2** pp 1499-503

[6] Mandal B 2013 Optimal reactive power dispatch using quasi-oppositional teaching learning based optimization *Electrical Power and Energy Systems* **53** 123–34

[7] Li S, Zeng L, Luo X, Li Y. and He Z 2011 Reactive power optimization of power system based on multiobjective concordance evolutionary algorithm *Int. Conf. on Electrical and Electronic Eng.* pp 4878-88

[8] Patel V and Patel R. 2019 Reactive Power Management through Dispatch Analysis using Evolutionary Algorithms *Int. J. of Innovative Technology and Exploring Eng. (IJITEE)* **8** 759-65

[9] Dixon J, Moran L, Rodríguez J and Domke R 2005 Reactive power compensation technologies, state-of-the-art review *Proceedings of the IEEE* **93(12)** 2144-64

[10] Mujawar I, Patil D and Mujawar I 2013 Reactive Power Compensation and Harmonic Mitigation of Distribution System using SAPF Compensator *Int. J. of Eng. Science and Innovative Technology (IJESIT)* **2** 494-02

[11] Patidar R D and Singh S P 2009 Active and reactive power control and quality management in dg-grid interfaced systems. *J. of Engineering and Applied Sciences (ARPN)* **4** 81

[12] Magnus K 1982 *Kolebaniya Vvedenie v issledovanie kolebatel`ny`kh sistem* [Oscillation. Introduction to the study of vibrational systems] (Moscow: Mir) [In Russian]

[13] Songhao Y, Masahide H and Baohui Z 2018 A phase-plane trajectory vector-based method for real-time identification of critical machines. *Transactions on Electrical and Electronic Engineering (IEEJ)* **13** 1-8

[14] Bertele U and Brioshi F 1972 *Nonserial dynamic programming* (New York: Academic Press)

[15] Sundaram L R 1996 *A First Course in Optimisation Theory* (Cambridg: Cambridge University Press)

[16] Andrianov D P and Badalyan N P 2017 K voprosu ob optimizacii parametrov dinamicheskix sistem [On the optimization of parameters of dynamic systems] *Energy system: Proceedings of the II Int. Scientific and Technical Conf. BSTU named after VG Shukhov* (Belgorod: BSTU) 134-7 [In Russian]

IOP Conf. Series: Materials Science and Engineering **791** (2020) 012022 doi:10.1088/1757-899X/791/1/012022

Development of an algorithm for searching optimal solutions for compensation of reactive power

E A Bekirov and S N Voskresenskaya

Institute of physics and technology, V.I. Vernadsky Crimean Federal University, Kievskaya St., 181, Simferopol, 295004, Russia

E-mail: Kaf_energo@cfuv.ru

Abstract. The problem of reactive power compensation, which includes many tasks, was substantiated in the work. Taking into account a number of specific problems of reactive power compensation and analysis of various electrical systems, an effective method of the calculation of the steady-state operation of the electrical network, on the basis of which the program "KRM_2021V7" was developed, is proposed. The method of nodal voltage, which was accompanied by the compilation of the system of equations and schematic matrices, among which the matrix of nodal conductivities and nodal loads were distinguished, is considered. An algorithm for finding optimal solutions for reactive power compensation is formulated and developed. Also, the work explains the use of genetic algorithms of mutation and population to find optimal solutions. The result of the calculation of losses in power lines for the considered electrical network is given.

1. Introduction

Reactive power causes additional power and electric energy losses in the elements of the electrical network with active resistance [1]. The var flows decreases the capacity of power lines and transformers, and loads the electrical power supply system of enterprises. The value of var flows can be reduced by means of compensating devices, which are most often used as capacitor banks [2] installed directly in areas of reactive power consumption. The problem of reactive power compensation includes a number of technical and economic tasks.

To consider specific problems of reactive power compensation, it is necessary to conduct an analysis of electrical systems using the concept of the balance of active and reactive power and electrical energy. Increasing of var flow or reactive power factor leads to the increase of total current; it is causes the additional voltage losses and tends to the number of negative events. For the calculation and installation of compensating devices, it is necessary to conduct an analysis of electrical networks of power supply systems. Power supply systems for industrial enterprises, although they can have the same electrical schemes, but at the same time designed to power various consumers of electrical energy and can include elements with different electrical parameters. If part of the electrical network also includes residential buildings as consumers, then for them the voltage can vary significantly with changing operating modes of industrial enterprises.

Based on the above, it is necessary to have energy parameters of the power supply system for more efficient use of compensating devices. Therefore, the methodology and program for conducting research and calculating of the steady-state operating mode of the electrical network is proposed.

2. Materials and Methods

Analysis of electrical systems, their parameters and characteristics is an integral part of the process of power supply to consumers. At the moment, they are trying to carry it out using various methods and tools, programs oriented to a quick [3] and effective [4-6] state assessment. Moreover, genetic algorithms are often used for predicting situations in complex ramified power systems [7 - 10]. They are based on assigning different values to a data array.

Programs and methods of analysis are aimed either at changing the parameters of the electrical network itself, or at optimizing operating modes, including administrative methods with the introduction of penalties for enterprises for violations [11].

Reactive power generation can be estimated using nonlinear programming [12], often the load is represented as a random matrix [13] and a semblance of load is used on certain days during the considered time period. In this case, the economic component [14] should be taken into account, since the location of capacitor banks at different points and nodes leads to change losses in electrical network. The reliability of power supply to consumers depends on reactive power [15].

In connection with the foregoing, this article uses the method of genetic analysis and computer simulation with the compilation of load matrices and the possibility of their subsequent correction in accordance with real data. The program is designed in such a way that it is possible to calculate the optimal location of the capacitor banks to compensate for reactive power in terms of electrical network losses and the cost of electricity transmission.

3. Results

3.1. Calculation of the steady-state operating mode of the electrical network

Calculations of complex and meshed electrical networks were carried out using the electronic computing system and the nodal voltage method. This allowed building a universal calculation algorithm, which is convenient to apply to any scheme of the electrical networks or enter corrections, when changing networks parameters. Thus, equivalent scheme conversion can be avoided.

For calculation by nodal voltage method, it is necessary to form and find solution a system of equations.

For the searching networks scheme (Figure 1), number the nodes. The node 0 – it is the basic node, the voltage in which is equal to the known value U_b.

Then, compose for this scheme the matrix of complex nodal conductivities.

$$\dot{Y} = \begin{vmatrix} \dot{Y}_{11} & \dot{Y}_{11} & \dot{Y}_{11} & \dot{Y}_{11} & \dot{Y}_{11} \\ \dot{Y}_{21} & \dot{Y}_{22} & \dot{Y}_{23} & \dot{Y}_{24} & \dot{Y}_{25} \\ \dot{Y}_{31} & \dot{Y}_{32} & \dot{Y}_{33} & \dot{Y}_{34} & \dot{Y}_{35} \\ \dot{Y}_{41} & \dot{Y}_{42} & \dot{Y}_{43} & \dot{Y}_{44} & \dot{Y}_{45} \\ \dot{Y}_{51} & \dot{Y}_{52} & \dot{Y}_{53} & \dot{Y}_{54} & \dot{Y}_{55} \end{vmatrix}, \quad (1)$$

where $\dot{Y}_{ij}$ – are the complex conductivities between the node i and the node j, and in the case $i = j$ – it is the sum of the complex conductivities of the given node with all the connected nodes.

In accordance with the expression (1), the matrix of nodal conductivities for the electrical networks scheme with electric parameters, shown in Figure 1, has the form:

$$\dot{Y} = \begin{vmatrix} 0.234 - j0.368 & 0 & -0.062 + j0.108 & -0.12 + j0.16 & 0 \\ 0 & 0.247 - j0.365 & -0.195 + j0.244 & 0 & 0 \\ -0.062 + j0.108 & -0.195 + j0.244 & 0.351 - j0.486 & 0 & -0.094 + j0.134 \\ -0.12 + j0.16 & 0 & 0 & 0.22 + j0.108 & -0.1 + j0.3 \\ 0 & 0 & -0.094 + j0.134 & -0.1 + j0.3 & 0.194 - j0.434 \end{vmatrix}$$

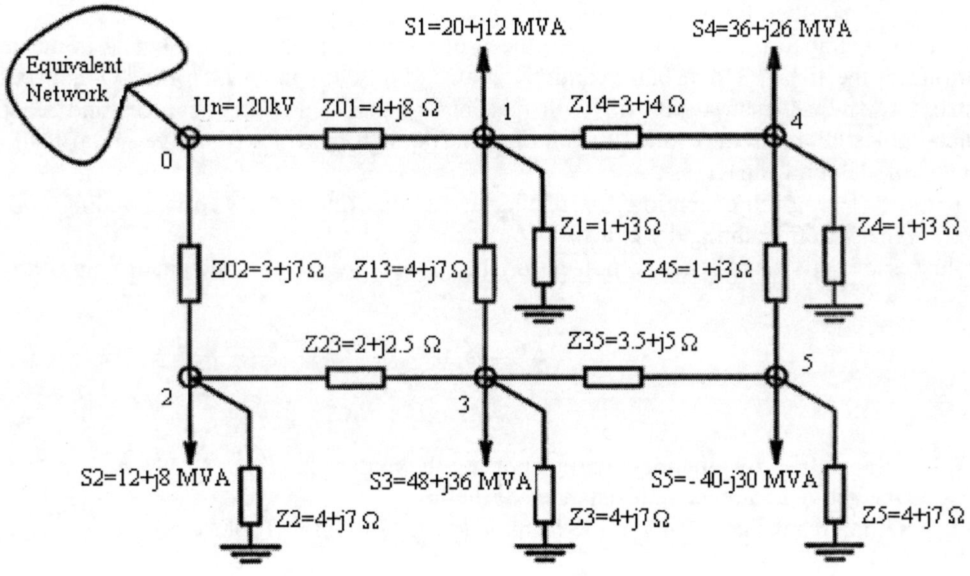

Figure 1. Test electrical network.

The number of columns and rows is determined by the number of nodes in the electrical network, except for the basis node.

Create a matrix of capacities of the consumers, connected to the corresponding nodes of the scheme:

$$\dot{S} = \begin{vmatrix} \dot{S_1} \\ \dot{S_2} \\ \dot{S_3} \\ \dot{S_4} \\ \dot{S_5} \end{vmatrix}, \tag{2}$$

where $\dot{s_i}$ – a capacity of the consumer connected to the *i*-th node.

In accordance with the expression (2), the matrix of nodal loads for the electrical network scheme showing in Figure 1 has the form:

$$\dot{S} = \begin{vmatrix} 20 + j12 \\ 12 + j8 \\ 48 + j36 \\ 36 + j26 \\ -40 - j30 \end{vmatrix}.$$

The values of the voltage at the nodes can be determined from the matrix formula:

$$\dot{U} = nU_b - \dot{Z}\left(\dot{U}_{\dot{D}}\right)^{-1}\dot{S}, \tag{3}$$

where n – is the matrix column consisting of units, the number of rows of which corresponds to the number of rows of the nodal conductivities matrix;

$\dot{Z}$ – is the matrix of nodal complex resistances equal to $\dot{Y}^{-1}$;

$\left(\dot{U}_{\dot{D}}\right)^{-1}$ – is the inverse diagonal matrix of nodal voltages, the elements of which complex interfaced.

In formula (3), the nodes voltages enter both the left and right parts of the equation. Therefore, it is possible to solve this equation only by an iterative method. For this it is necessary to set the initial ap-

IOP Conf. Series: Materials Science and Engineering **791** (2020) 012022 doi:10.1088/1757-899X/791/1/012022

proximation of nodal voltages. Also at each stage of the calculation in the right part of equation should insert the values obtained from the previous iteration.

Calculation using formula (3) is made in the programming language c ++. The input data for the calculation are the files "G_matrix.txt" and "X_matrix_txt", "P_matrix.txt" and "Q_matrix.txt". Files "G_matrix.txt" and "X_matrix_txt" fill with matrices of active nodal resistances and reactive nodal resistances. Files "P_matrix.txt" and "Q_matrix.txt" fill with matrices of active nodal load capacities and reactive nodal load capacities.

As a result of program execution, the results of calculations for the corresponding nodes will be displayed in the system command window.

The losses of active and reactive power for the power lines were calculated using the following formulas:

$$\Delta\dot{P} = \dot{S}_t(\dot{U}_{\dot{D}})^{-2}\dot{R}\dot{S},$$
$$\Delta\dot{Q} = \dot{S}_t(\dot{U}_{\dot{D}})^{-2}\dot{X}\dot{S}, \quad (4)$$

where $\dot{S}_t$ – is the matrix row obtained by transporting the matrix $\dot{S}$;
R and X – is the active and reactive resistances of the line.

The calculation results according to the formula (4) are shown in Table 1.

Table 1. Results of power loss calculation.

Node number	Power at the beginning of the line, MVA	Power at the end of the line, MVA	Power losses for the line, MVA	Power compensating device in the node, MVAr
01	37.9+j23.9	37.3+j23.5	0.6+j1.3	0
02	39.5+j27.6	37.0+j27.1	0.5+j1.2	0.05
13	27.0+j19.1	6.1+j6.9	0.02+j0.04	0.1
14	11.2+j5.2	11.2+j5.5	0.03+j0.	0.3
35	15.1+j9.6	15.0+j10.0	0.1+j0.1	0
45	24.9+j20.4	24.8+j20.5	0.1+j0.2	0.06

3.2. Search conditions for optimal solutions of reactive power compensation

To solve the problem of optimization of reactive power compensation, two criteria are adopted:

- minimization of costs for generation and transmission of reactive power;
- regulation of voltage in the set limits from U_{min} to U_{max}.

The objective size of the costs will be determined by the formula:

$$C = C_l + C_g, \quad (5)$$

where C_l – is the costs for reactive power losses;
C_g – is the costs for generation of reactive power by capacitor banks.

Costs for reactive power losses:

$$C_l = c_0\Delta P,$$

where c_0 – is the specific costs of active power losses;
ΔP – is the active power losses.

Costs for capacitor banks:

$$C_g = (E_n + a_{op})(K_c + K_{tr} + K_m) + c_0\Delta P_{CB},$$

where E_n – is the norm coefficient of efficiency of the investments, accepted for calculations in the electric energy industry equal to 0.125;
a_{op} – is the norm coefficient of amortization for maintenance and repair equal to 0.05 for electrotechnical installations;

K_c – is the initial costs for the acquisition capacitor banks;
K_{tr} – is the costs for transportation of capacitor banks to the destination;
K_m – is the cost of mounting capacitor banks;
ΔP_{CB} – is the specific power loss in a capacitor bank.

Fixed costs independent from the capacity of capacitor banks are taken equal to:

$$\mathrm{K_0} = K_{tr} + K_{\mathrm{m}}.$$

The optimal voltage must satisfy the requirement:

$$U_{min} \leq U \leq U_{max}.$$

In this case, the reactive power factor must satisfy the condition:

$$tan\phi > 0.1.$$

The search for optimal solutions is based on genetic algorithms. The multiple values that need to be found in this case will correspond to the multiple power values of the compensating devices. Matrix of the multiple power values of the compensating device Q_{CB} consists from n rows and m columns, where the columns are the schema nodes and the rows are the values sought.

The mutation of the population is a random variation in one of the elements of the Q_{CB} matrix. The value of the mutable element takes a new value, which is randomly selected from desired range. The range of possible values of the Q_{CB} elements is determined based on the maximum value of the reactive power of the load Q_{max} in the researching scheme of the electrical network.

The application of the genetic algorithm makes it possible to account the discreteness of the scale of the produced capacitor devices, which makes it possible to increase the calculation speed by tens of times. For this, it is necessary that the individuals of the population (solutions) are taken the discrete values.

The searching of optimal solution is suggested to determine not by the annual number of hours of the maximum losses, because this indicator does not give an accurate estimate of the power losses for the year, but according to the graphs of the daily loads of consumers. Thus, it is proposed to use an information model that takes into account the network operation mode for each individual period of per day, which will allow obtaining more accurate results.

For the generated population (the range of Q_{CB} values), adaptions is calculated. The adaptions function will estimate the value of the total annual costs for generating and transmitting reactive power and matching the voltage to the standardized values. The most adapted solutions will fall into the new population, with that crossing with other relevant solutions. There may be a large number of possible solutions. Let's show by example one of the possible variants of crossing, which is performed in the decimal system of calculation:

$$Q_{CB\ k} = \frac{Q_{CB\,[k-1][i+1]} - Q_{CB\,[k-1][i]}}{2},$$

where k – is the generation number;
i – is a number of the group of individuals.

When taking into account the discreteness of the scale of capacitor devices, the Q_{CB} values are also discrete, in which case the operation of crossing solutions is simplified or completely eliminated, and the diversity of the population (solutions) will be provided by other types of crossover (random increment of solutions). The optimal for the developed algorithm will be randomly changing one of the individuals or all individuals to a random increment Q_{CB}. When accounting the discreteness, population is replaced by a multiple of integer values, each of which corresponds to the technical and economic parameters of one of the capacitor devices. In this case, a crossover based on the displacement of any random individuals will be $\Delta N = 1$.

When calculating the discrete values Q_{CB}, formula (5) takes the form:

$$C = K_0 + C_K + C_O + c_0\,\Delta P_{CB} + c_0 \Delta P,$$

IOP Conf. Series: Materials Science and Engineering **791** (2020) 012022 doi:10.1088/1757-899X/791/1/012022

where C_K – cost of the capacitor bank of the certain nominal;
C_0 – costs for servicing a capacitor bank of a certain brand and nominal for the considered period of time.

A new generation of the solutions or a new population will be a new Q_{CB} matrix, the individuals of which will consist of a certain number of individuals (multiple of solutions) of the previous generation (the previous Q_{CB} matrix), a certain number of individuals that subjected of a little change (crossover) and new individuals consisting of randomly generated (mutated) Q_{CB} values. To prevent premature convergence to the local imaginary cost function, and to ensure acceptable value, it is necessary to keep a certain balance between individuals undergoing mutation and crossover. When writing the developed program "KRM_2021V7", the relationships between individuals, subjected to a crossover (small increment) and a mutation (random change), were determined experimentally. The optimal number of individuals in the population was determined to be 500. That is, the number of rows of the Q_{CB} matrix will be 500, and the number of columns is determined by the number of nodes in the electrical network.

The more generations of solutions will be grown, the more accurate the solution will be. The recommended number of generations (iterations) for the developed methodology varies depending on the complexity of the calculated scheme. But it should be noted that with each increase in the scheme per node, the complexity of the calculations increases according to the parabolic law.

Figure 2 shows a block diagram of the operation of the developed algorithm.

Subprograms (functions) are not presented in the form of a block diagram because of their volume, but their work can be used with the source code of all functions in the language c ++ / 11.

A technique for theoretical analysis of the determination of power and nodes locations has been developed, where it is necessary to introduce compensating devices to compensate reactive power.

4. Discussion

The program is written in the programming language with ++ / 11, she use only the standard language libraries. This program was compiled in the operating system "Windows 10" by the "min GW" compiler. Also compilation of the program by compilers MS VC ++, Intel c ++, Borland c ++ is possible.

The program does not contain its own graphic interface and uses the console host to interact with the user. To set the initial data, it is need to create files with the names: "G_matrix.txt" and "X_matrix_txt", "P_matrix.txt" and "Q_matrix.txt", "G_base_matrix.txt" and "X_base_matrix.txt". The results are displayed in the console window.

Also for each load, it is need to create a matrix of the diagram of daily load, the rows of which will correspond to the nodes, and the time-of-day columns with an interval of one hour. These matrices need to fill in the files "P_T.txt" (for active loads) and "Q_T.txt" (for reactive loads).

Data on the costs of losses and capacitor banks, as well as data on the nominal values of the produced compensating devices, are in the program itself.

5. Conclusion

Thus, from the above, we can conclude:

- Mathematical models of optimization tasks are analyzed and an optimal model is chosen for solving the problem of reactive power compensation. It was revealed, that the optimal for the solving problem is a model based on genetic algorithms, which has an advantage over other methods both in the calculation speed and in the reliability of the results obtained.
- A program has been developed in the programming language c ++, which solves the problem of optimal power and location of compensating devices using genetic algorithms. The calculation is based on the analysis of the diagrams of the daily consumption of active and reactive power by electric loads.
- Investigations have been carried out on a fragment of the urban electrical network, which prove the effectiveness of the reactive power compensation method developed in the work.

IOP Conf. Series: Materials Science and Engineering **791** (2020) 012022 doi:10.1088/1757-899X/791/1/012022

- The algorithm developed in the work can be used by both energy-saving enterprises to optimize the transfer of reactive power, and large energy consumers to reduce the electricity costs.

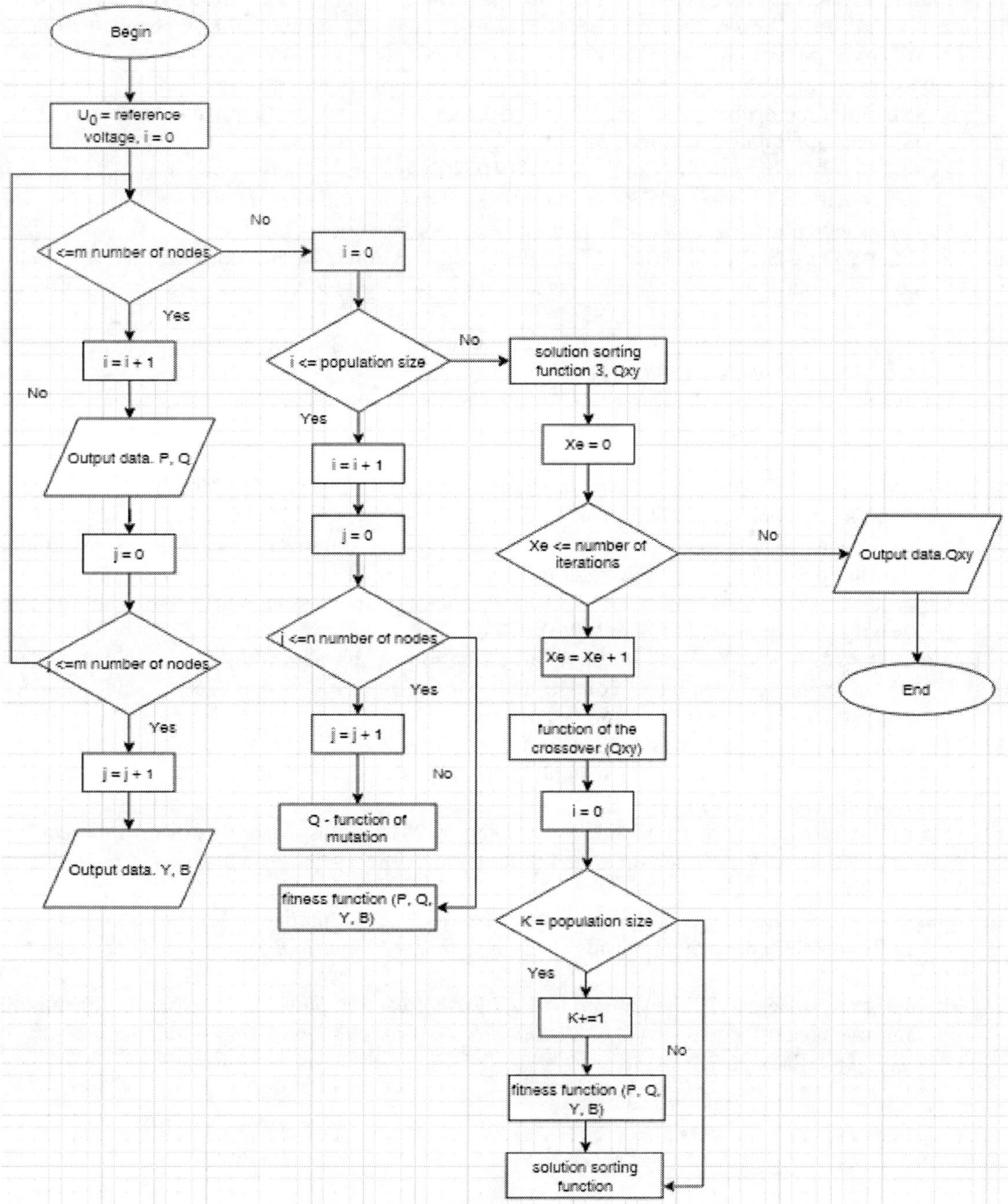

Figure 2. Program "KRM_2021V7" operation algorithm.

References

[1] Zhelezko U, Artemyev A and Savchenko O 2004 *Raschet, analiz i normirovaniye poter' elektroenergii v elektricheskikh setyakh* [Calculation, analysis and rationing of electricity losses in electrical networks] (Moscow: NC ENAS) [In Russian]

[2] Ersan Kabalci 2017 Reactive Power Compensation in AC Power Systems *Reactive Power Control in AC Power Systems* ed N M Tabatabaei, A J Aghbolaghi, N Bizon, F Blaabjerg (Cham: Springer) pp 275-315

[3] Abdolkarimzadeh M and Aghdam F H 2019 A novel and efficient power system state estimation algorithm based on Weighted Least Square (WLS) approach service *Journal of power technologies* **99(1)** 15–24

[4] Zeng S, Yu W, Hong X and Cheng C-K 2010 Efficient power network analysis with modeling of inductive effects *IEICE Transactions on Fundamentals of Electronics Communications and Computer Sciences* **93-A(6)** 1196-203

[5] Zhang W, Zeng S, Yu W, Wang J and Hong X 2003 Analysis and optimization of structured power/ground networks *IEEE Transactions on Computer-Aided Design of Integrated Circuits and Systems* **22(11)** 1533-44

[6] Li H 2019 A novel reactive power optimization method for distributed power system using PSO *J. of Phys.: Conf. Ser.* **1303** 012107

[7] Charishma D, Upadyay P, Kumar D R and Haritha U 2015 Reactive power compensation in power system using evolutionary algorithm *Int. J. of Latest Research in Science and Technology* **4(4)** 110-3

[8] Niu M and Xu Z 2014 Efficiency ranking-based evolutionary algorithm for power system planning and operation *IEEE Transactions on Power Systems* **29(3)** 1437-38

[9] Akbari S and Amooshahi M K 2009 Power system stabilizer design using evolutionary algorithms *Int. Review of Electrical Engineering* **4(5)** 925-31

[10] Chiou J-P, Chang C-F and Li C-J 2017 A backtracking evolutionary algorithm for power systems *MATEC Web of Conf.* **119(3)** 01046

[11] Volovikov A A and Yudin A A 2018 Investigation of the algorithm optimization of electric power regimes according to voltage and reactive power with the use of penalty functions method *J. Phys.: Conf. Ser.* **1066** 012016

[12] Faraji D, Rabiei A, Mohammadi B and Hoseynpoor M 2011 Reactive power generation management to improve voltage stability margin of power systems Australian *J. of Basic and Applied Sciences* **5(6)** 957-63

[13] Sheng W, Liu K, Pei H, Li Y, Jia D and Diao Y 2016 A fast reactive power optimization in distribution network based on large random matrix theory and data analysis *Applied Sciences* **6(6)** 158

[14] Saxena N K and Kumar A 2016 Cost based reactive power participation for voltage control in multi units based isolated hybrid power system *J. of Electrical Systems and Inf. Technology* **101** 12-24

[15] Benidris M, Sulaeman S, Tian Y and Mitra J 2016 Reactive power compensation for reliability improvement of power systems *IEEE/PES Transmission and Dist. Conf. and Exposition (T&D)* (Dallas) pp 7519910

IOP Conf. Series: Materials Science and Engineering **791** (2020) 012023 doi:10.1088/1757-899X/791/1/012023

Use of the PMU Infrastructure to Determine the Location of Short-circuit Power Lines

N Buryanina[1], Yu Korolyuk[2], M Koryakina[2], K Suslov[3], N Solonina[3] and E Lesnykh[4]

[1]Chukotka Branch of North-Eastern Federal University named after M.K. Ammosova, Studencheskaya St., 3, Anadyr, 698000, Russia
[2]Department of General Disciplines, Chukotka Branch of North-Eastern Federal University named after M.K. Ammosova, Studencheskaya St., 3, Anadyr, 698000, Russia
[3]Department of Power Supply and electrical engineering, Irkutsk National Research Technical University, Lermontova St., 83, Irkutsk, 664074, Russia
[4]Siberian Transport University, Dusi Kovalchuk St., 191, Novosibirsk, 630049, Russia

E-mail: bns2005_56@mail.ru

Abstract. One of the serious problems of the normal operation of power systems is a violation of efficiency of overhead power lines. Short circuits are especially dangerous. The probability of occurrence of faults is reduced, but not reduced to zero, with an increase in the quality of installation, reliability, insulators and conductive materials. The proposed method is based on the fact that the short circuit current at the beginning of the transmission line of power electric system depends on the distance (location) to the fault point. This article discusses the following topics: theoretical possibility of fault location on the time of arrival of the responses to the beginning and end of the line, the development of the algorithm processing of information, the development of the block diagram of additional devices that are not included in the phasor measurement units (PMU). The article gives an example of the application of this technique.

1. Introduction

Reduction in losses for the economic entities by the improvement in reliability and power supply quality is a topical task. At present some measures are taken to meet the requirements on reliability and quality of power supply. A serious problem facing electric power systems is failure of overhead and cable transmission lines [1-4]. Short circuits in transmission lines represent a particular danger. The probability of short circuits decreases but does not disappear with an increase in the quality of installation, reliability of insulators and conducting materials of transmission lines. Short-circuits cause disconnection of individual consumers and entire regions. This imposes high requirements on reliability and fast operation of relay protection. However, if a short circuit has occurred and relay protection has successfully operated there a remains the task to promptly and where possible accurately detects coordinates of the short circuit. This, in the final analysis, allows us to quickly restore the transmission line and place it into operation which in turn minimizes economic losses.

2. The state of the research

Recently, power electric systems are one of the most dynamic sectors which employ the advances of fundamental and applied knowledge. Right now the opportunity has appeared to pay special attention

IOP Conf. Series: Materials Science and Engineering **791** (2020) 012023 doi:10.1088/1757-899X/791/1/012023

to the global system of universal time. Even two decades ago the universal time was the province of special organizations dealing with space, industry, defense, astrophysics, etc. The adoption of the Global positioning system has made it possible to receive the standard time signals at an individual substation and use this precise time for solving a wide range of problems. Based on these technologies it has become possible to develop and implement adopt intelligent power electric systems, Smart Grids and PMU.

In the electric power industry there are two possible types of research: active and passive experiments [5-7]. As an active experiment, it is possible to study the error of the transmission line with the help of short current pulses and an estimate of the response time. This time depends on the distance between the source end of the transmission line and a short circuit point. A disadvantage of active experiments is the necessity to apply special equipment and the time to get prepared to the experiment. There is a great variety of methods for detection of overhead and cable line fault locations [5-12, etc]. The pulse method is based on measuring time intervals between the moment of transmitting a probe pulse of alternating current and the moment of receiving a reflected pulse from the fault location. To make measurements by the method of oscillation discharge the voltage supplied to the faulted cable conductor is gradually raised to the voltage of cable fault. The loop method is based on measuring resistances by the direct current bridge. The capacitance method suggests measuring capacitance of a broken conductor by measuring bridges. The acoustic method supposes creation of a spark discharge at the fault location and listening to sound vibrations that occur above the fault point. There is also the induction method and others [5].

Passive experiment however makes it possible to use modern high speed digital technologies, and obtain data directly from the signals associated with the processes that occur in transmission lines after a short circuit, i.e. in real time. Such data can be the instant values of current and voltage, voltage and current phases. One of the possible solutions to this problem is related to the dependence of currents at the source end of a line on the distance to the short circuit point [13-17]. However, there are reasons that decrease the accuracy of the fault point detection, namely: variations in the effective values of voltage at the transmission line connection point, as well as the dependence of current amplitude on voltage phase at the time of short circuit. We suggest the use of time factors related to the final velocity (vF) of power (electric signal) transmission along the considered line. It is obvious that there will be a response (echo) spreading in both directions along the line, that will have the form of a front of increasing or declining voltage or current and the time of the response arrival at the source end or load end of the line can be recorded with a high accuracy.

This paper focuses on the following issues:

- Consideration of possibility of determining the fault place on the basis of time when responses come to the source end and load end of the transmission line;
- Determination of the possibility of using the available infrastructure of PMU to determine the above time instants and transfer the data to the processing center;
- Development of an algorithm for primary data processing, obtained from primary measuring devices;
- Development of a block diagram of additional devices which are not envisaged within PMU.

3. The main principles of the approach

We will consider the idea of the suggested method on the example of a transmission line without branches with one-way supply.

Figure 1 shows the calculation scheme for the determination of the short circuit place, taking into consideration the time of signal arrival at the source and load ends of the line.

Let the short circuit occur at time t_{sc} at point K (Figure 1) and a transient process start. For the sake of simplification we make the following assumptions:

- voltage at point K drops to zero;
- length of the considered line is much shorter than the length of the incident current wave λ ($\lambda \approx 6000$ km at the frequency of 50 Hz);

- the instant values of current and voltage in the steady state are constant along the whole line.

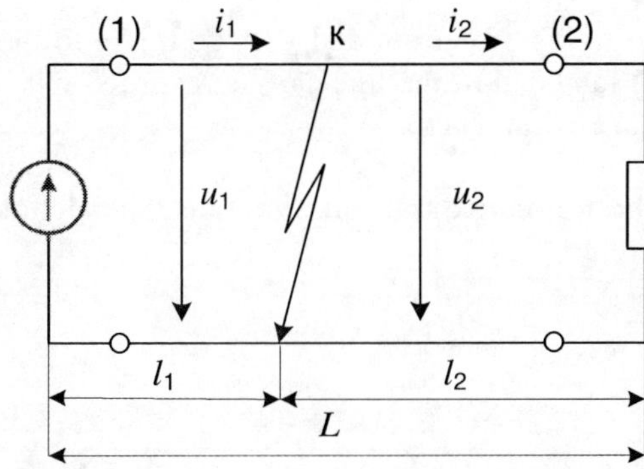

Figure 1. Design scheme of a short circuit in a line.

In Figure 1: L-length of the transmission line; l_1, l_2 – distances from the short circuit point to the source and load ends of the transmission line, respectively; i_1, i_2, u_1, u_2 – current and voltage of the first and second sections of the line, respectively.

Figure 2 presents a model of the line with lumped parameters for the calculation of transient process in the line with distributed parameters.

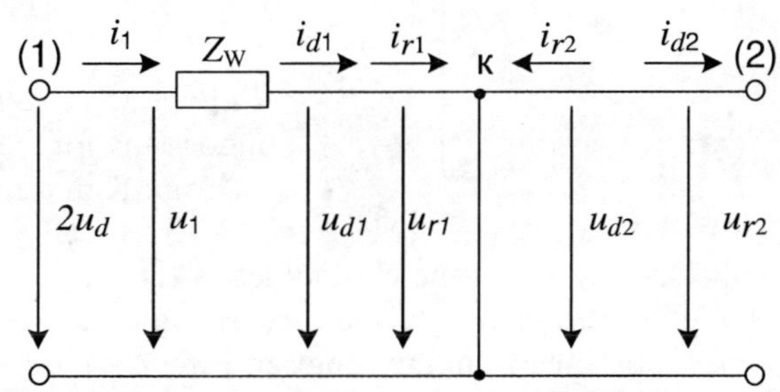

Figure 2. Model of the transmission line for calculation of transient process.

In Figure 2: Z_w- wave impedance of the line; i_{d1} - incident current wave of the first section, i.e. from the source end to the short circuit point; i_{r1} – reflected current wave of the first section; u_{d1}– incident voltage wave of the first section; u_{r1} – reflected voltage wave of the first section; i_{d2} - incident current wave of the second section, i.e. from the short circuit point to the load end of the line; i_{r2} – reflected current wave of the second section; u_{d2} – incident voltage wave of the second section; u_{r2} – reflected voltage wave of the second section.

It is easy to see from the model that at time t_{sc}:

$$i_{r1} = i_{d1} \text{ and } u_{r1} = -u_{d1}$$

Thus, a positive current wave front and a negative voltage wave front travel at velocity v_F in the direction from point K to the source end of the line. The fronts can be detected with the aid of current and voltage sensors. Time t_1 of their arrival at the source end of the line can be recorded with high accuracy thanks to the clock showing the time synchronized with the universal time.

At the second section of the line there is also a transient process. At time t_{sc} current i_2 and voltage u_2 at the beginning of the second section vanish which means that

IOP Conf. Series: Materials Science and Engineering **791** (2020) 012023 doi:10.1088/1757-899X/791/1/012023

$$i_{d2} + i_{r2} = 0\,,\; i_{r2} = -i_{d2}$$

$$u_{d2} + u_{r2} = 0\,,\, u_{r2} = -u_{d2}$$

Hence, from point K negative fronts of current and voltage travel to the load end of the line. Time t_2, when these fronts arrive at the load end of the line can be recorded with the aid of current and voltage sensors as well as with high precision clock.

Let us show that knowing t_1 and t_2, we can detect the place of short circuit. To this end we will consider a geometric model of the short-circuited transmission line. The technique of identifying the short circuit place is explained in Figure 3.

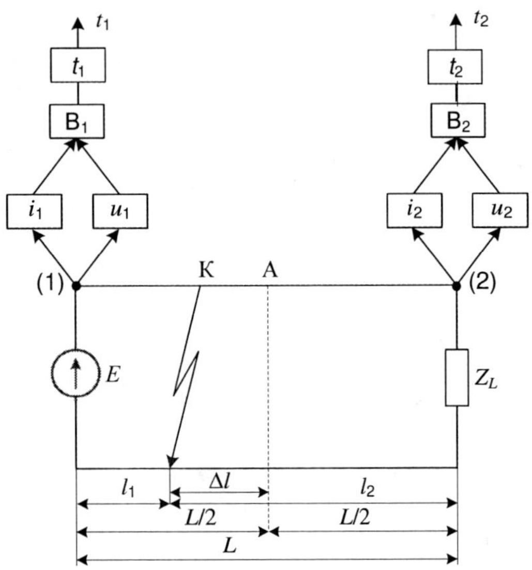

Figure 3. A scheme of devising an algorithm for determining the place of point K in transmission line.

In Figure 3: (1), (2) are the sites at which the chronometers (t) and primary current (i) and voltage (u) sensors are installed; A – geometric center of the line; B – modules for processing the data from current and voltage sensors; Δl – distance from the center of Line A to the short circuit point K.

Using this scheme we determine Δt_1 - the time of wave propagation from the short circuit point to the source end of the line:

$$\Delta t_1 = \frac{l_1}{V_\Phi}$$

In the same way we find Δt_2 - the time of wave propagation from the short circuit point to the load end of the line:

$$\Delta t_2 = \frac{l_2}{V_\Phi}$$

Express time t_1 of the signal (response) arrival at the source end of the line through the short circuit time:

$$t_1 = t_{sc} + \Delta t_1$$

Similarly find time t_2 of the response arrival at the load end of the line:

$$t_2 = t_{sc} + \Delta t_2$$

IOP Conf. Series: Materials Science and Engineering **791** (2020) 012023 doi:10.1088/1757-899X/791/1/012023

Determine the difference between the time of response arrival at the source end of the line and the time of response arrival at its load end:

$$t_1 - t_2 = \Delta t_1 - \Delta t_2 = \frac{l_1}{V_\Phi} - \frac{l_2}{V_\Phi} = -\frac{2\Delta l}{V_\Phi}$$

and finally determine Δl

$$\Delta l = -\frac{(t_1 - t_2)\,V_\Phi}{2} \tag{1}$$

Knowing Δl, we find l_1 and l_2 by the equations

$$l_1 = \frac{l}{2} - \Delta l, \tag{2}$$

$$l_2 = \frac{l}{2} - \Delta l \tag{3}$$

If point K is closer to the source end of the line (to the left of point A), then $\Delta l > 0$. If point K is closer to the load end of the line, then $\Delta l < 0$.

When the time aspect is taken into consideration the issue of accurate determination of the time instants t_1 and t_2 is particularly important.

We suggest using the available infrastructure of PMU (Figure 4). PMU make it possible to take phasor measurements of currents and voltages at the given points of the power system [18, 19].

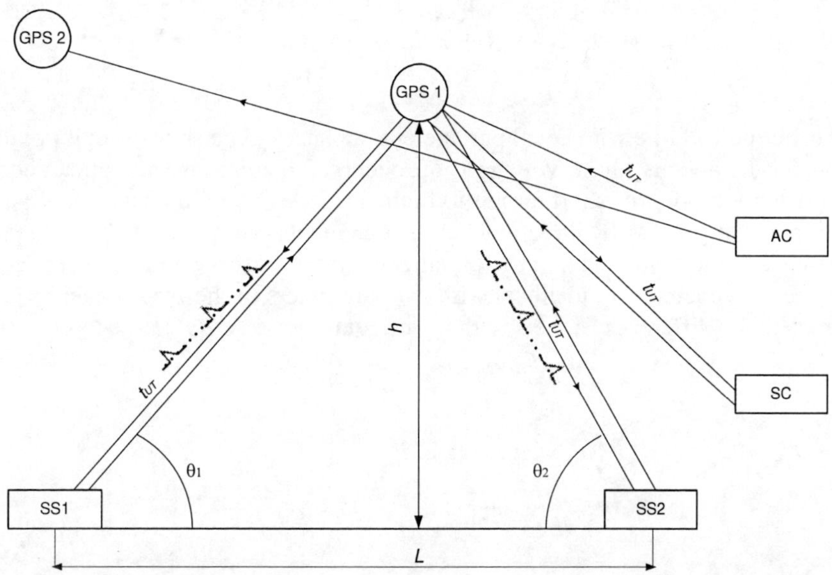

Figure 4. PMU infrastructure. (where SS1- feeding substation; SS2- receiving substation; L– length of the line; AC- atomic clock; GPS1, GPS2 – satellites sending time signals; SC – control centre; SS1, SS2 – network substations; t_{UT} – time pulse of the atomic clock; h – height of the satellite above the Earth in the area, where the substations are located; Q_1, Q_2 – angles at which the satellite is seen from SS1 and SS2, respectively.

Phasor measurement implies simultaneous measurement of both the effective value and the phase of current and voltage. These parameters allow us to calculate current values of transmitted power, voltage drops in the sections of the transmission line, power loss in the transmission line, etc. Meas-

IOP Conf. Series: Materials Science and Engineering **791** (2020) 012023 doi:10.1088/1757-899X/791/1/012023

urement of effective values and phases of current and voltage, although possible, have not become widespread, since the control of the system under dynamic operating conditions requires that the data be definitely connected with the universal time. For instance, in order to determine losses in the line we should simultaneously measure active power at the source and load ends of the line precisely at the same time.

The time measurement resolution of PMU is not sufficient to accurately determine the time of the event. Therefore, at the measurement points we should form our own time (count) pulses with a short time interval of, for example, 10^{-6} seconds, using additional devices.

Consider the use of the PMU and additional devices for accurate determination of time t_1 and t_2. These pulses are formed by the pulse generators installed in modules B_1 and B_2, and received at the input of the pulse counters located in modules t. The pulses are generated with the same frequency. At the outputs of modules t we obtain t_1 and t_2, respectively, in the following form:

$$t_1 = t_{gps} + ndt \tag{4}$$

$$t_2 = t_{gps} + mdt \tag{5}$$

where t_{gps} – a synchronizing pulse from GPS satellite, that contains complete information about the universal time, namely: year, month, day, hour, minute, second, milliseconds;
n, m – number of count pulses from the arrival of t_{gps} to the moment, when the response to the short circuit is received at the source end and load end of the line, respectively.

We can determine the difference between the time the responses arrive at the source end of the transmission line and the time they arrive at its load end, using expressions (4) and (5):

$$t_1 - t_2 = (n - m)dt \tag{6}$$

Knowing t_1 and t_2, we find Δl, l_1, l_2 according to expressions (1), (2), (3).

4. Results

Consider a short-circuit of line with equal parameters in phases. At a short-circuit point, in the event of a ground fault, instantaneous phase voltage drop occurs. This causes the appearance of rectangular waves moving from the point of short circuit to both sides. A line with phase equal parameters can be described in a transient process by three independent channels – α, β and 0. And the parameters of the line in the channels α and β are equal to the parameters in the positive and negative sequences, and the parameters in the zero channel are identical with the parameters of the zero sequence.

Line parameters for different channels with a particular phase A are defined as

$$u_0 = \frac{u_a + u_b + u_c}{3}, \quad i_0 = \frac{i_a + i_b + i_c}{3} \tag{7}$$

$$u_\alpha = u_a - u_0, \quad i_\alpha = i_a - i_0 \tag{8}$$

$$u_\beta = \frac{u_b - u_c}{\sqrt{3}}, \quad i_\beta = \frac{i_b - i_c}{\sqrt{3}} \tag{9}$$

The proposal consists in fixing the difference in the appearance times at the installation site of protection through the channels α and β. Consider a specific 220 kV line with AS-120 wire.

$$Z_\alpha = j0{,}43\ \Omega/\text{km}, \quad Y_\alpha = 2{,}65 \cdot 10^{-6}\ \text{S/km}$$

$$Z_0 = j1{,}2\ \Omega/\text{km}, \quad Y_0 = 1{,}94 \cdot 10^{-6}\ \text{S/km}$$

The constants for distribution channels are:

$$\beta_\alpha = \sqrt{x_\alpha b_\alpha} = 1{,}067 \cdot 10^{-3} \text{ rad/km}$$

$$\beta_0 = \sqrt{x_0 b_0} = 1{,}526 \cdot 10^{-3} \text{rad/km}$$

Velocities of wave motion along channels:

$$V_\alpha = \frac{\omega}{\beta_\alpha} = 2{,}943 \cdot 10^{-6} \text{ km/s}$$

$$V_0 = \frac{\omega}{\beta_0} = 2{,}059 \cdot 10^{-6} \text{ km/s}$$

One kilometer the wave on the zero channel moves for a time - 4.86 ms, on channel α - for a time - 3.4 ms. The distance from the place of measurement of the time of appearance of wave fronts is equal to:

$$L_{cs} = \frac{t_\alpha - t_0}{4{,}86 - 3{,}4}$$

Thus, it is possible to determine the point of a short circuit on the line with one-sided fixation of waves.

5. Conclusion

Other The proposed method is very promising, since it mainly uses PMU devices, and the costs related to the development and use of additional devices are insufficient. This is due to the fact that it is proposed to use the existing infrastructure of power systems.

It is suggested that the moments of time and information transmission to the processing center should be determined with the help of PMU, up-to-date tools of digital communication and UT system.

We propose that the structural schemes of auxiliary devices, which are not foreseen in PMU, should be added to the available tools.

This method allows to determine the location of the short-circuit in real time. Studies have shown that this method has very high accuracy.

Acknowledgments

The study was carried out with the financial support of the Russian Foundation for Basic Research and the Subject of the Russian Federation - the Republic of Sakha (Yakutia) № 18-48-140 010.

References

[1] Yang Z, Zhong H, Xia Q and Kang C 2016 Optimal transmission switching with short-circuit current limitation constraints *IEEE Transactions on Power Systems* **31(2)** 1278-88

[2] Chai Y, Jiang C, K Zhang and S Xu 2016 Safe operation improvement of an electrical power system by superconducting fault current limiters *Chinese Control and Decision Conf., CCDC* (Yinchuan, China) pp 28-30

[3] Gao F, Ai S, Ding R and Huang Y 2015 Development and experiment of fast breaker-type fault current limiter *5th Int. Conf. on Electric Utility Deregulation and Restructuring and Power Technologies, DRPT* (Changsha, China) pp. 1790–1794

[4] Phadke A, Wu Z and Zora L 2015 Simultaneous transmission line parameter and PMU measurement calibration *IEEE Power & Energy Society General Meeting* (Denver, USA) pp 7286115

[5] Ferreira V H, Zanghi R, Fortes M Z, Sotelo G G, Silva R B M, Souza J C S, Guimarães C H C and Gomes Jr S 2016 A survey on intelligent system application to fault diagnosis in electric

power system transmission lines *Electric Power Systems Research* **136** 135-53

[6] Mohamed E A and Talaat H A 2010 Khamis Fault diagnosis system for tapped power transmission lines *Electric Power Systems Research* **80** 599-613

[7] Suslov K V, Solonina N N and Smirnov A S 2011 Smart Grid: A new way of receiving primary information on electric power system state *2nd IEEE PES Int. Conf. and Exhibition on Innovative Smart Grid Technologie* (Manchester, UK) pp 6162654

[8] Bahmanyar A, Jamali S, Estebsari A and Bompard E 2017 A comparison framework for distribution system outage and fault location methods *Electric Power Systems Research* **145** 19-34

[9] Rafinia A and Moshtagh J 2014 A new approach to fault location in three-phase underground distribution system using combination of wavelet analysis with ANN and FLS *Int. J. of Electrical Power & Energy Systems* **55** 261-74

[10] Gururajapathy S S, Mokhlis H and Illias H A 2017 Fault location and detection techniques in power distribution systems with distributed generation: A review *Renewable and Sustainable Energy Reviews* **74** 949-58

[11] Gazzana D S, Ferreira G D, Bretas A S, Bettiol A L, Carniato A, Passos L F N, Ferreira A H and Silva J E M 2014 An integrated technique for fault location and section identification in distribution systems *Electric Power Systems Research* **115** 65-73

[12] Liao Y 2011 Generalized fault-location methods for overhead electric distribution systems *Trans. Power Deliv* **26** 53–64

[13] Gilany M, El Din E S T, Aziz M M A and Ibrahim D K 2005 An accurate scheme for fault location in combined overhead line with underground power cable *IEEE Power Eng. Society General Meeting* (San Francisco, USA) pp 489308

[14] Hajjar A A, Mansour M M and Talaat H A 2004 High-phase order power transmission lines relaying approach based on the wavelet analysis of the fault generated traveling waves *39th Int. Universities Power Eng. Conf. UPEC 2004* (Bristol, UK) vol 2 pp 847660

[15] Ferreira G D, Gazzana D S, Bretas A S and Netto A S 2012 A unified impedance-based fault location method for generalized distribution systems *2012 IEEE Power and Energy Society General Meeting* (San Diego, CA, USA)

[16] Wang C, Zhang B, Li G, Yan L, and Zhao J 2011 One practical method of fault location for transmission lines based on dual-voltage fault components *Int. Conf. on Electrical and Control Eng.* (Yichang, China) pp 6057787

[17] van Erp F T G, van Riet M J M, van Rasing J F G, Provoost F and van Deursen A P J 2005 Modelling and live measurements of step and touch voltages at LV customers in urban areas caused by MV faults *CIRED 18th Int. Conf. and Exhibition on Electricity Distribution* (Turin, Italy) pp 0203

[18] Suslov K V, Stepanov V S and Solonina N N 2014 PMU for detection of short-circuit point in the transmission line *3rd Int. Conf. on Smart Grids and Green IT Systems SMARTGREENS Barselona, Spain)* pp 63-7

[19] Suslov K V, Stepanov VS and Solonina N N 2012 Improving the reliability of operation Microgrids *IEEE Int. Energy Conf. and Exhibition (ENERGYCON)* (Florence, Italy) pp 6348275

IOP Conf. Series: Materials Science and Engineering **791** (2020) 012024 doi:10.1088/1757-899X/791/1/012024

The process of charge carrier generation in photosensitive elements based on heterojunction of monocrystalline and amorphous silicon

M A Bykov[1], S A Zuev[2] and E A Bekirov [1]

[1]Department of Electrical Energy and Electrical Engineering, V I Vernadsky Crimean Federal University, Academician Vernadsky Ave., 4, Simferopol, 295007, Russia
[2]Department of Radiophysics and Electronics, V I Vernadsky Crimean Federal University, Academician Vernadsky Ave., 4, Simferopol, 295007, Russia

E-mail: misha23b@yandex.ru

Abstract. In this article, to describe the process of charge carrier generation and current transfer, we consider a model for calculating the characteristics of photovoltaic cells based on heterojunctions of amorphous - monocrystalline silicon in the kinetic approximation. Developed the main provisions of a numerical model for the study, based on the statistics of the broken bonds, and physics of photogeneration and sign conclusion nonequilibrium carriers, and the experimental verification of the obtained simulation results. The use of a simplified model of the density of states in the mobility gap of hydrogenated amorphous silicon is substantiated. This model allows to carry out researches of influence of temperature, degree of alloying and the geometrical sizes on characteristics of photoconverters. The generation of current carriers is characterized by the rate of optical generation and for the interpretation of photoconductivity, the following models of recombination are used such as zone-tail, which occurs when a nonequilibrium carrier is captured from the free zone to the tail state of the opposite zone, which has captured the carrier of another sign and acts as a recombination center. The number of tail states increases as the temperature decreases. Tail - broken link recombination occurs when captured carriers are tunneled from the tail zone state to the broken link state. The dependence of the photoconductivity on the temperature of these structures is calculated.
Key words: monocrystalline silicon, amorphous silicon, heterojunction, photoconductivity.

1. Introduction

The end of the twentieth and early twenty-first centuries were marked by large-scale achievements in the development, creation, optimization of optical converters, their electrical parameters and the study of physical phenomena in semiconductor structures based on heterojunctions of amorphous-monocrystalline silicon (a-Si:H) - (C-Si), widely used as materials for the creation of solar photovoltaic energy. The study of such structures is located on the border of different scientific areas, and there are many scientific publications devoted to them. These are papers of foreign scientists: Martin A. Green, Jianhua Zhao, A. I. Shkrebtii, C. B. Honsberg; scientists from Russia: J. I. Alferov, V. G. Budagyan, O. A. Golikova, A. G. Kazansky, M. M. Kazanin; scientists from Ukraine: V. G. Litovchenko, A. M. Gorban, D. I. Levinzon, N. I. Kluy, V. P. Kostylev, V. N. Borschev, G. S. Khripunova, A. V. Sachenko, V. A. Skrichevskiy. It is necessary to develop new semiconductor materials in order to improve the parameters of photovoltaic converters. At the same time, semiconductor structures with high

IOP Conf. Series: Materials Science and Engineering **791** (2020) 012024 doi:10.1088/1757-899X/791/1/012024

photoelectric conversion capacity are needed along with low production cost. The importance and timeliness of using such structures is directly related to the increasing energy consumption and trends in the transition from traditional to renewable energy resources, including solar energy. Certain hopes in solving this problem are pinned on the development of technology for photosensitive elements production on the basis of heterostructures amorphous-monocrystal silicon (GAMS). They combine a good photoelectric conversion in amorphous silicon with the possibility of rapid withdrawal of carriers at the heterojunction. The considered semiconductor GAMS structures have promising photo convector parameters in combination with acceptable technological production costs. Insufficiently developed in terms of theory, experiment and technology are the issues related to the physics of processes that affect the formation of broken bonds and the possibility of their compensation, with the withdrawal of non-equilibrium carriers that arose as a result of photogeneration from the amorphous silicon area before their recombination [1, 2].

Must be considered that lack of an in-depth understanding of the physics of the process occurring in the photopreforming structures based on GAMS makes it necessary to conduct numerical and analytical modeling to study the physics of photogeneration and current transfer processes on the basis of broken connections statistics with the conclusion of non-equilibrium carriers, and to conduct experimental verification of the simulation results.

In works [3, 5-7] the dependence of p-type photoconductivity α -Si:H on the doping level and hydrogen concentration with argon-hydrogen plasma and defects in the structure was studied. The studies were carried out using both a numerical model and experimental samples. An energy model of amorphous silicon photoconductivity is proposed. In work [5], the properties and structure of α -Si:H films obtained by the method of silane decomposition (silicon-hydrogen compound) in a high-frequency glow discharge from external influences, including the annealing temperature, were studied. In works [4, 5] a simpler and more cost-effective magnetron sputtering method for thin films of amorphous hydrogenated silicon with similar characteristics is proposed.

The effectiveness of photopleasure equally depends on the characteristics of the film and geometry of the transducers.

The purpose of this work is to develop the basic principles of the numerical model to study the dependence of the photoconductivity of heterojunctions α -Si:H – c-Si and its approbation.

2. Model of current transfer carriers

To solve this problem it is advisable to carry out the calculation in the kinetic approximation. The model proposed in [6, 7, 10-12] supplemented with representations from [4, 5] can be used to describe the processes of carrier transfer in silicon.

Carrier transport in amorphous silicon is described in terms of hopping conductivity for an unordered system with a random field of randomly located charged point centers. The conductivity process is described by media transitions between different localized positions.

The correlation function of an additional random field is represented as [8]:

$$\phi(r) = \frac{2\pi n_t e^4}{\varepsilon^2} r_0 \exp\left(-\frac{r}{r_0}\right), \tag{1}$$

when $r_0 = \left(\frac{4\pi n_0 e^2}{\varepsilon T}\right) = \alpha^{-1}$ is the radius of screening, n_t is a concentration of centers, e is the electron charge, ε is the dielectric constant of the exemplar, and $r = r_1 - r_2$ is a distance between two points in the exemplar.

The kinetic equation for amorphous silicon in a stationary positions can be written in the form proposed in [8], rewriting the right part taking into account all the main processes of generation and recombination of current carriers.

$$\frac{\partial f_\lambda}{\partial t} = -\sum_{\lambda'}\left\{W_{\lambda\lambda'} f_\lambda\left(1-f_{\lambda'}\right) - W_{\lambda\lambda'} f_{\lambda'}\left(1-f_\lambda\right)\right\},$$

when f_λ - the diagonal element is non-equilibrium one-particle density matrix, $W_{\lambda\lambda'}$ - probability of transition between positions λ and λ'. The probability of scattering corresponding to different types of transitions is determined through the scattering cross sections.

This equation describes the evolution of the diagonal part of the density matrix, and in the localized positions it describes the balance of electronic transitions between different positions. The probability of jumps between different pairs of centers varies randomly and in a wide range. As a result of changes in population conditions, the easiest directions of jumps can be blocked.

Current carrier generation is characterized by the rate of optical generation. The following models of recombination are used for the interpretation of photoconductivity:

1) recombination of "zone - tail" occurs when the capture of nonequilibrium media of the free zone on the condition of the tail opposite the zone, seized the front of this carrier other character, and acting as a recombination center. The number of tail positions increases as the temperature decreases;

2) recombination "tail - dangling bonds" occurs when the tunneling of the captured carriers from the condition of the tailings areas in the state of dangling bonds;

3) tail -to-tail recombination is a tunneling recombination between electrons trapped in the tail positions of the conduction band and holes trapped in the tail states of the valence band.

In [8] it is shown that in reality it is necessary to take into account all possible mechanisms of recombination and to allocate from them the basic depending on a spectrum of carriers positions, type and concentration of defects, external conditions (temperature, light intensity, etc.).

When describing the processes of carriers recombination and their capture to the appropriate traps, you can use the model proposed in [8, 9], supplementing it with the concepts described in [8].

The density of positions in disordered semiconductor at $|E| >> \sqrt{\psi_1}$ [9]:

$$\rho(E) = \frac{\psi_1 \exp\left(-\frac{E^2}{2\psi_1}\right)}{2\pi^2|E|^{3/2}} + \frac{\alpha\cdot\exp\left(-\frac{E^2}{2\psi_1}\right)}{8\sqrt{2}\pi^{3/2}\sqrt{\psi_1}}\left(E - \frac{\alpha|E|^{3/2}}{\sqrt{2\pi\psi_1}}\right), \quad (2)$$

when $\psi_1 = 2\pi n_t e^4 r_0/\varepsilon^2$, E - energy of positions, $\alpha^{-1} = r_0$ - characteristic length, the role of which is played by the radius of the shield.

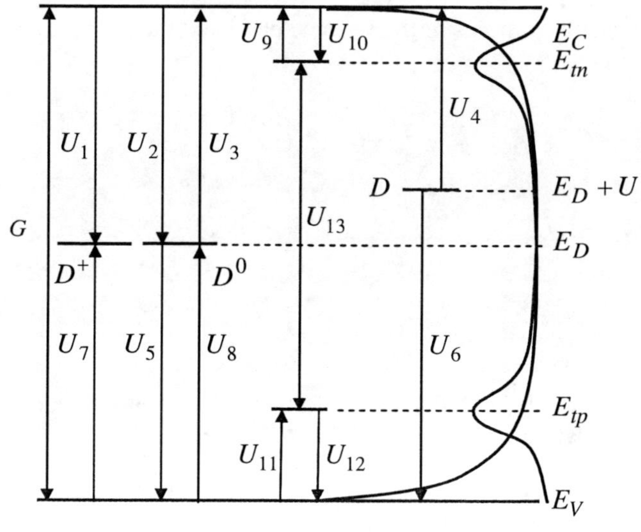

Figure 1. Schematic representation of electronic transitions for a simplified state density model and a view of the state density in the mobility gap.

Calculations are carried out for a simplified model of the density of positions in the mobility gap, consisting of four discrete levels. Levels E_{tp} and E_{tn} with density of positions N_{tp} and N_{tn} - effective tail levels of the valence band and conduction band respectively. Levels E_D and $E_D + U$ - energy levels of positions of dangling bonds.

We consider the positions of the zone tails as small traps in thermodynamic equilibrium with the nearest zone, i.e. "traps for holes" and "traps for electrons". This simplified representation of the density of positions in the mobility gap, as shown in [5], does not lead to qualitative changes in the results of numerical calculation.

Determining the probability of finding the D-center in the state D^+, D^0 and D^-how , f^+, f^0 and f^- accordingly. Because the D-center can only be in three positions $f^+ + f^0 + f^- = 1$. You can record on the basis of the formalism of Shockley-Reeda[8]:

$$U_1 = nN_D f^+ c_n^+,\ U_2 = nN_D f^0 c_n^0,\ U_3 = N_D f^0 e_n^0,\ U_4 = N_D f^- e_n^-,\ U_5 = pN_D f^0 c_p^0,$$
$$U_6 = pN_D f^- c_p^-,\ U_7 = N_D f^+ e_p^+,\ U_8 = N_D f^0 e_p^0,\ U_9 = nN_{tn}(1 - f_{tn})c_{tn}, \tag{3}$$
$$U_{10} = N_{tn} f_{tn} e_{tn},\ U_{11} = pN_{tp} f_{tp} c_{tp},\ U_{12} = N_{tp}(1 - f_{tp})e_{tp},\ U_{13} = N_{tp} f_{tp} N_{tn} f_{tn} T_{np},$$

when c_n^0 and c_n^+ - the coefficients of electron capture on D^0 - and D^+ - position; c_p^0 and c_p^- - the coefficients of capture of holes on D^0 - and D^- - position; e_n^0, e_n^-, e_p^0 and e_p^+ - coefficients of thermal emission of electrons from D^0 - and D^- - position and holes of D^0 - and D^+ - position; f_{tn}, f_{tp} - the probability of filling of traps for electrons and holes; c_{tn}, c_{tp} and e_{tn}, e_{tp} - the capture coefficients of electrons and holes at the respective traps, and heat release of these; N_D - concentration D-centers; T_{np} - the ratio of the tunneling recombination «tail - tail».

The increase in electron density in the conduction band is due to optical generation G and transitions U_3, U_4, U_{10}. The decrease of the density due to the transitions U_1, U_2 and U_9. It can be written that the change in the concentration of conduction electrons:

$$\frac{\partial n}{\partial t} = G - U_1 - U_2 + U_3 + U_4 - U_9 + U_{10}. \tag{4}$$

Similarly, it can be written for the density of holes in the valence band and the filling density of the levels: E_{tn}, E_{tp} and D:

$$\frac{\partial p}{\partial t} = G - U_5 - U_6 + U_7 + U_8 - U_{11} + U_{12},$$
$$\frac{\partial [N_{tn} f_{tn}]}{\partial t} = U_9 - U_{10} - U_{13},\qquad \frac{\partial [N_{tp} f_{tp}]}{\partial t} = U_{11} - U_{12} - U_{13},$$
$$\frac{\partial [N_D f^+]}{\partial t} = U_3 + U_5 - U_1 - U_7. \tag{5}$$

These relations allow us to determine the probability of generation and recombination of carriers included in the right part of the kinetic equation.

The kinetic equation in the nonstationary state can be written for the distribution function of carriers in the following f form:

$$\frac{\partial f_\lambda}{\partial t}+\left(\vec{v}\nabla_r f_\lambda\right)+\frac{1}{\hbar}\left(\vec{F}_i\nabla_k f_\lambda\right)=\left(\frac{\partial f}{\partial t}\right)_i-\sum_{\lambda'}\left\{W_{\lambda\lambda'}f_\lambda\left(1-f_{\lambda'}\right)-W_{\lambda\lambda'}f_{\lambda'}\left(1-f_\lambda\right)\right\}$$

when $\vec{v}_i$ - charge carrier speed e_i, $\vec{F}_i$ - force, acting on the charge from the external electromagnetic field and other charge carriers, $\left(\frac{\partial f}{\partial t}\right)_i$ – collision integral, ρ – bulk charge density of ions in the sample, $\sum_i e_i \int f_i d\vec{v}_i$ - density of the volume charge generated by free current carriers.

Force $\vec{F}_i$ is electric and can be defined as the field gradient which is the solution of the Poisson equation:

$$\Delta\phi=-\frac{1}{\varepsilon}\left(\rho(\vec{r})-\sum_i e_i\int f_i d\vec{v}_i\right)$$

with the addition of a random field (1).

The method of solving the above system of equations is described in detail in the works [6, 7].

3. Results of numerical calculations

To compare the efficiency of GAMS and amorphous silicon structures, such characteristics as VAC were used. The developed software package made it possible to obtain numerically the VAC of the studied samples. In the course of numerical experiments using the proposed method, light load VAC for structures based on GAMS (Fig.2, curve 2) when illuminated by monochromatic light with a quantum energy of 1.85 eV (0.67 μm)and intensity of 1017 $sm^{-2}\cdot s^{-1}$ and structures of amorphous silicon grown on glass with the same irradiation parameters (Fig.2, curve 1). In the case of a lighted transition (Fig.2, curve 2) at negative voltage, the current is significantly different from zero. The increase in the back branch of the VAC in the case of GAMS is due to the retraction of the photoconductivity electrons in the p-region of monocrystalline silicon.

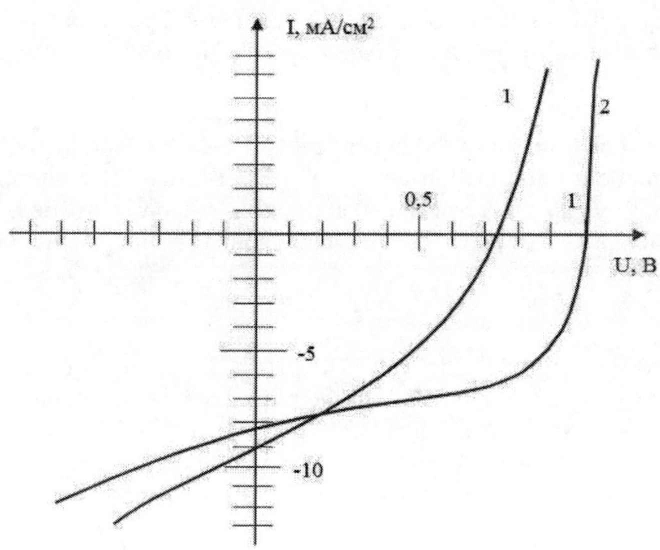

Figure 2. Light load VAC of amorphous structure (1) and GAMS-based structure (2).

IOP Conf. Series: Materials Science and Engineering **791** (2020) 012024 doi:10.1088/1757-899X/791/1/012024

Comparative analysis of these curves in Figure 2 allows you to set the share of generated carriers in the total current. Using numerically obtained VAC, the efficiency was calculated in GAMS, while it was possible to achieve an increase in efficiency up to 15%. From the curve comparison in Fig.2 shows that the GAMS photogrammetrists better performance amorphous structure.

The model parameters were chosen in accordance with the literature data and data obtained from experimental studies [4, 5]. The calculation was carried out for amorphous silicon with different degree of hydrogenation. The width of the mobility slit was taken to be 1.8 eV. The position of the Fermi level relative to the ceiling of the valence band was set to 1.05 eV. For the parameters of broken links the following values were taken: $E_D = 0.9\,\text{eV}$, $U = 0.3\,\text{eV}$. The rate of photogeneration was taken to be equal to $G = 10^{19}$ sm^{-3}s^{-1}, $c_n^0 = c_p^0 = 3 \cdot 10^{-15}$sm^{-2} and $c_n^+ = c_p^- = 1.5 \cdot 10^{-13}$sm^{-2}.

The calculation was carried out for conductivity in the region of low external field stresses up to 3 kV/cm. For fig.3 the results of numerical simulation of thermal conductivity for different values of hydrogen concentration are presented.

The results of modeling the specific conductivity dependences are in good agreement with the experimental data.

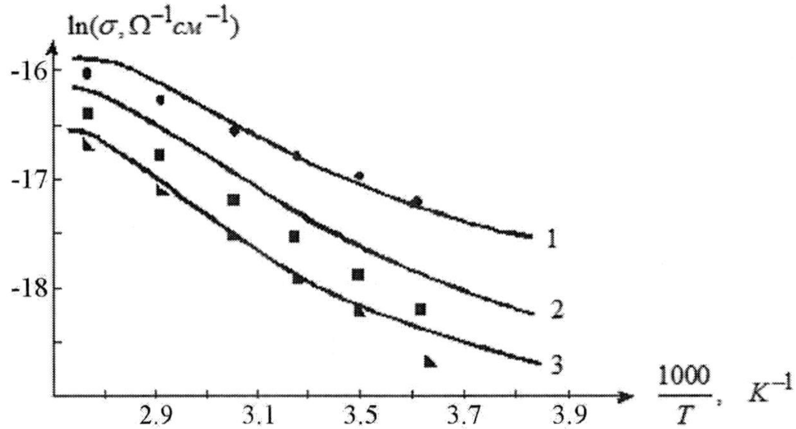

Figure 3. The thermal dependence of the calculated (solid lines) and experimental (points) conductivity a-Si films obtained by magnetron sputtering method at a concentration of hydrogen: 1 - 10%, 2 – 30%, 3- 40%.

4. Conclusion

In this paper, the basic principles of the model to study the dependence of the photoconductivity of heterojunctions α-Si:H – c-Si in the kinetic approximation are developed. The use of a simplified model of the positions density in the mobility gap of hydrogenated amorphous silicon is justified. This model allows us to study the effect of temperature, doping degree and geometric dimensions on the characteristics of the photoconverters.

References

[1] Bekirov E, Asanov M, Voskresenskaya S and Alkaata A 2017 Real time processing of the phase shift and the frequency by voltage signal conversion into the sequence of rectangular pulses *Signal Processing Symposium, SPSympo* (Debe, Poland) 220-5

[2] Abibullaev A, Bekirov E, Asanov M and Kudria S 2015 Generating system operation real-time processing under unsteady conditions *Signal Processing Symposium, SPSympo* (Debe, Poland) 232-6

[3] Zuev S A et al. 2016 Dependence of the conductivity on the active - region thickness in GaAs

thin-film Schottky diodes *Semiconductors* **50** 810

[4] Kilessa V G, Asanov E E, Zuev S A, Starostenko V I and Slipchenko M I 2014 Selection of kinetic parameters for the transfer current in GaAs *Telecommunications and Radio Engineering* **73** 1273-92

[5] Bykov M A, Zuyev S A, Slipchenko M I and Shadrin A A 2013 A description of carrier current flow transport model in thin photoconverting a-SI:H film *Telecommunications and Radio Engineering* **72** 1881-92

[6] Arsenichev P., Bendeberya N, Grigor'ev Ye, Zuev S, Slipchenko M, Starostenko V, Taran Ye. 2014 Experimental research of diffraction properties of thin conducting films in a waveguide *Telecommunications and Radio Engineering* **73** 1241-50

[7] Kuznecov S V 2001 Chislenny`j raschet temperaturny`x zavisimostej fotoprovodimosti α-Si: H p-tipa [Numerical calculation of temperature dependences of photoconductivity α-Si:H p-type] *Fizika I Tehnika Poluprovodnikov* **35(10)** 1244-9 [In Russian]

[8] Zuyev S A, Starostenko V V, Tereshenko V Y, Churumov G I and Shadrin A A 2004 Model PTSH submicronnih razmerov na kremnii [Model MESFET submicron size on silicon] *Radioelectronika I informatika* **3** 47–53 [In Russian]

[9] Gordienko Y E, Zuyev S A, Starostenko V V, Tereshenko V Y and Shadrin A A 2006 Osobenosti lavinnogo proboya v kremnievih PTSH po rezyltatam chislennogo modelirovaniya [Features of avalanche breakdown in silicon MESFET on the results of numerical simulation] *Radiotehnika, Vseukrainskiy majvedomstvenniy nauchno-tehnicheskiy sbornik* **121** 146-52 [In Russian]

[10] Bykov M A and Mazinov A S 2006 Optical spectral characteristics of thin-film constructions on the basis of hydrogenated amorphous silicon *Proc. of SPIE* (Alushta, Ukraine) **6023**

[11] Slipchenko N I, Bykov M A and Mazinov A S 2008 Research of spectral description films of amorphous hydrogenated silicon received by a magnetron method with the use of an ionic source *Proc. of SPIE* **7008**

[12] Bykov M A, Mazinov A S and Bykov A M 2005 Structural and optical features of state-of-the-art industrial solar cells and units on their basis *Technologies For Future Generations Int. Home of Scientists «F.J.Curie» «St. St. Constantine and Helena»* (Varna) 309-12

[13] HOMER Energy URL https://www.homerenergy.com/

[14] Natural Resources Canada URL https://www.nrcan.gc.ca/energy/retscreen/7465

[15] PVSYST Photovoltaic Software URL https://www.pvsyst.com/

[16] Solar Design Software URL https://www.accasoftware.com/en/solar-design-software

IOP Conf. Series: Materials Science and Engineering **791** (2020) 012025 doi:10.1088/1757-899X/791/1/012025

The alternator and the ancillary components continuous drive from the gearbox

R Yu Dobretsov, G P Porshnev, A G Semenov, I A Komarov and DE Telyatnikov

Higher School of Transport, Institute of Mechanical Engineering, Materials and Transport, Peter the Great St. Petersburg Polytechnic University, Politekhnicheskaya St., 29, Saint-Petersburg, 195251, Russia

E-mail: komarov@eering.ru

Abstract. This paper considered implementation variations of the alternator and/or other units (systems) continuous mechanical drive from summing reduction unit, combined power flows from input and output gearbox valves. The three-link planetary mechanism is used in the simplest case as a reduction unit. These principals could be used in the hybrid cars transmission. The main purpose of this concept is to provide an uninterrupted power supply to the vehicle electrical onboard system from the alternator, including cases of the engine off. Describing principles could be used in transmissions of the wheel traction transport vehicles and tracked vehicles with the hybrid propulsion system. Variants of double way transmission kinematics schemes are provided. The differential characteristics of the schemes are movement and steering abilities with the inoperable internal combustion or electric engines. It worth noting that such engineering solution have extra advantages. The steering control level is increasing, electromechanical drive could replace hydrostatic steering control transmission as a more chip, reliable and technical.

1. Introduction

The number of an electric energy consumers and their power in transport vehicles are always increasing. The units operational integrity, which providing vehicle movement and driving control, more often depends on the on-board electrical network. The oil pump with an electric drive, the power distribution mechanism with an electromechanical drive of a control elements [1,2], the traction electrical motor as a part of the hybrid propulsion system [3,4], the hybrid power distribution mechanism [5], steering device, the electric power steering, etc. have already implemented in the automotive industry. The battery capacity without working alternator is not enough to ensure the all consumers operation. But the alternator does not work, when the engine is not running.

The alternator or the reversible electric machine is a propulsion system part of almost any modern heat engine vehicle. The traction electrical motor, which could be made as the reversible electric machine, in the parallel/series type hybrid propulsion system could have a function of an alternator and a starter. However, this solution requires using the complex reduction gearbox with an electronic control device in the transmission for the implementation of all operation modes of the propulsion system.

The operation of the alternator should be ensured for a vehicle towing, for example. Therefore, the reserve alternator driven by the driven shaft of the transforming mechanism could be used for cases of this kind. The low power alternator would provide systems for vehicle mobility in critical situations.

The alternator drive from a planetary gear set is alternative to drive from the heat engine. The planetary gear set provides summing rotation speed of the engine drive shaft and the output shaft of the reduction unit. So, we could use all control systems during the towing operation. This solution helps to solve the problem with "the bump start".

The planetary gearbox oil pump is also should be driven, when the engine is not running. Nowadays the most part of the vehicle manufacturers provide the oil pump driving from the torque converter pump wheel, which is part of the hydro mechanical transmission. The oil pump driving is indirectly provided from the heat engine output shaft. In this case it is necessary to limit the towed vehicle speed and the towing distance [6]. The towing is generally prohibited sometimes, and evacuation is recommended by spec lift or flatbed recovery truck, which is unacceptable in many cases.

2. Results

The principles of continuous mechanical drive Focus on the options of the electric alternator and the oil pump drive providing from the transforming mechanism. The connection with the ancillary component is carried out through the link G. The output of the link G could be from the driving or driven link of the transforming mechanism, depending on the transmission layout. The latter case options shown in Figure 1.

Figure 1, a: the plan is universal to the transforming mechanism type. The link G is an ring gear of a simple planetary gear mechanism which isn't include into the kinematic scheme of the actual transforming mechanism. The rotation speed plan of the proposed planetary gear mechanism is shown in Figure 2.

The changing of a direction of link G rotation in mechanism(when reversing) should be considered when using this plan, so it is necessary to provide a method for automatically reversing the driven alternator and pump.

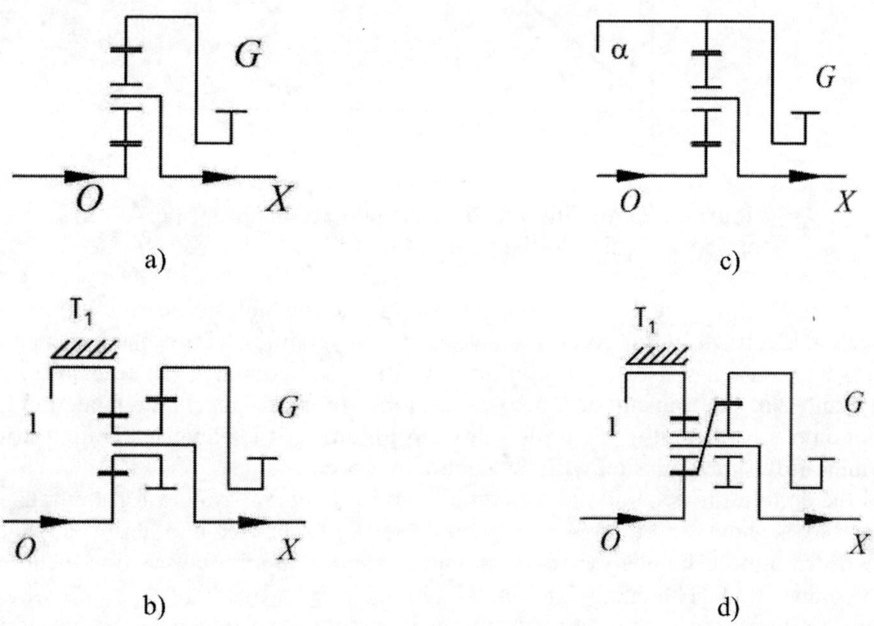

Figure 1. Variants of the power take-off transforming mechanism(TM) organization: O - is the driving link, X - is the driven link and G - is driving link of the alternator and / or the oil pump; α- is the connecting link1- isan arbitrary link of the planetary gearbox (special case TM) and T1- is the control element associated with an arbitrary link.

Figure 1, b: the plan for the planetary gearbox with more than 3 degrees of freedom. The link αis a connecting link which isn't concerned with the gearbox controls. The enforcing conditions $\text{sign}(\omega_\alpha) = \text{const}$ is theoretically possible.

The rotation speed of the link αis ω_α. The alternator and pump aren't need in the reverse gear in these conditions. The plan involves using the one of the planetary gearbox rows. Such an approach makes it possible to reduce the transforming mechanism axial dimension in comparison with version 1, a.

Figure 1, c and 1, d: the link G is included in the complex four-link mechanism (optionc - low-tech stepped satellites are used, option d - the plan with the concatenated satellites is used).The link 1 is one of planetary gearbox arbitrary main links connected to the control element T1. This simplifies the gearbox plan synthesis while retaining the other advantages of option 1,b.

It should be noted that the kinematic plans options in Figure 1 are also applicable to solve the specific task of a parallel power flow driving of the high-speed tracked machine steering mechanism [7-11].However, in this case, attention is being given to another drive feature. It becomes possible to provide a hyperbolic law of the design turning radius changing to the vehicle speed, this relation in traditional mechanisms is linear, which limits the average vehicle speed and causes heat engine underutilization.

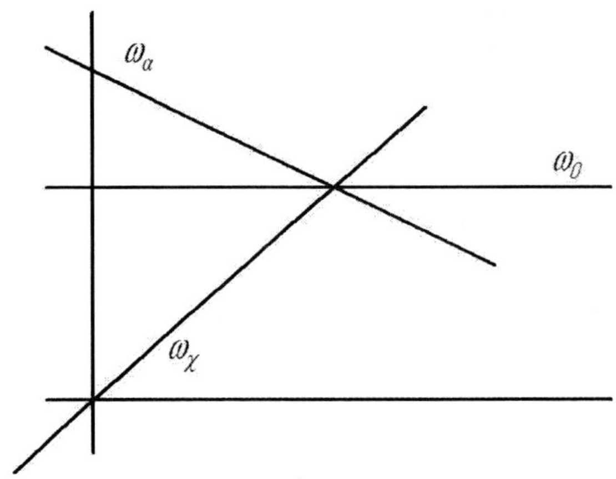

Figure 2. Three-link mechanism velocity diagram, performed according to the plan in Figure 1,a.

The principle shown in Figure 1, b-d, could be used in the transmission of the hybrid power plant vehicle.The traction electrical motor could be made as the reversible electric machine and discard the alternator. Refusing to use a starter isn't advisable, as itis quite compact, the transmission plan gets easier and the number of the transmission layout solutions are increasing.The connection host of heat engine and traction electrical motor engine is shown in Figure 3. It is named "parallel/series type hybrid" for thedouble-flows transmission with the central gearbox.

The plan of the double flows parallel/series type hybrid propulsion system for the transport vehicle with central gearbox is shown in Figure 4,a. You could see that it is used another principlein the parallel flow power distribution. The energy of the storage system 8 supply the traction electric motor for power input to parallel flow. The energy storage system is charged by the alternator 9 which is driven from the gearbox 3. Thus, the traction electric motor is a part of the tracked vehicle steering mechanism.

The double flows mechanisms are used in the tracked vehicles. The designing and the power, kinematic aspects analyzing issues are considered in the fundamental literature sources [4, 6, 8, 12]. However, the problem of the transmission use for the hybrid propulsion system isn't considered to this moment.

IOP Conf. Series: Materials Science and Engineering **791** (2020) 012025 doi:10.1088/1757-899X/791/1/012025

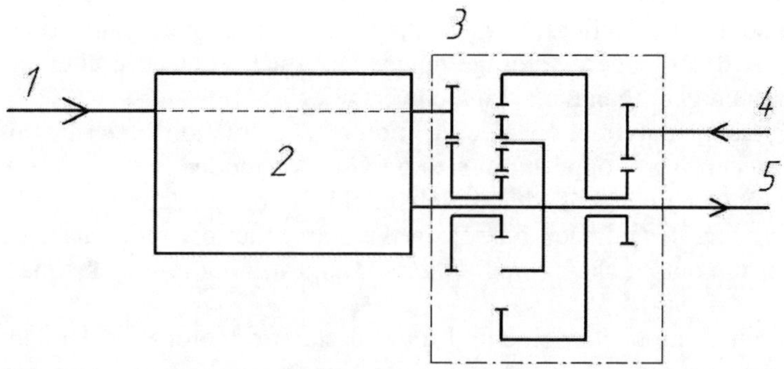

Figure 3. The transmission plane variants for the parallel/series type hybrid propulsion system of the transport vehicle with central gearbox (see the Fig.1,a): 1 – power input from the heat engine; 2 – converting mechanism (for example: gearbox); 3 – reduction gearbox; 4 – power input from the traction electric motor; 5 – power output to the final gear.

The mechanism operates as a hybrid mechanism of the power distribution [13]. The steering device friction C_R is on for the linear motion. The brake T_R is on for the power redistribution between to the turn. At the same time the control element isn't worked on the slipping mode [14], all types of turnings are provided by the traction electric motor. Two-steps reduction unit is providing size reduction of the traction electric motor. Considering transmission designing principle isapplicable for high-speed tracked vehicles weighting 8-30 tons.

The plan of the double flows parallel/series type hybrid propulsion system for the transport vehicle with on-board gearboxes is shown in Fig. 4, b [15]. It commonly used for the special transport technological vehicle built on the main battle tank (military engineering vehicle, vehicle-launched bridge, vehicle to work in the radiation electrochemical reaction areasetc.).

In this scheme the central shaft should be stopped by the brake T_0 to provide neutral turn. At this moment the friction C_0 should be turned off. We note that in this casethe traction electric motor could be used as a heat engine starter. This starting mode is recommended only for the emergency conditions as it leads to the extra power loss.

The operating modes

Rectilinear motion using only the heat engine. The heat engine is connecting with input shaft of the transmission and traction electric alternator main power way. The economic operating mode is preferable. There is no mechanical connection between the engine and the transmission parallel power way. The traction electric motor is connected to the parallel power way by the reduction unit P. The reduction unit transmission ratio is selected for the traction electric motor according to provide the whole range of the turning radiuses. Symmetrical spur-gear differential with two control elements (brake T_R and locking clutch C_R) is set up in the parallel power way. During the rectilinear motion traction electric motor is turned off, the parallel power way isn't used: the brake T_R and clutch C_R are turned on (the planetary gear set is locked).

Rectilinear motion is stable. The traction electric alternator output energy is accumulated in the energy storage device. It could be used to other consumers supply connected to high-voltage network. As the traction electric motor of domestically made cars should be used next to 50% of the vehicle movement time, the traction electric alternator power could be less than traction electric motor power by 20-30%.

The turning using the heat engine and traction electric motor. The clutch C_R is turned off, brake T_R (transmission ratio of the planetary gear set D is (-1)). The traction electric motor turns on (the rotation direction is chosen by the vehicle turning direction). The same gears are selected for the both on-board gearboxes (the using of low range gear forlow speed track gearbox could be considered). According to the kinematics of the double flows steering device, the speed of the high-speed track increases and the speed of the low-speed track declines, because the boards differential connection. The traction electric motor supply is provided by energy storage device and traction electric alternator, therefore the sum wheel power could be higher than the engine powerfor a short time.

Neutral turning using onlytraction electric motor. The traction electric motor maximum power should be selected according to opportunities to provide this mode.

When the main power flow fails, the vehicle could use only traction electric motor for rectilinear motion when failing engine (fallback mode). In that case, the low range gear of the low speed track gearbox turns on or the output shaft stops. The minimum turning radius is equal to a half of the track width.

Rectilinear motion using heat engine and tracked electric motor (accelerated mode). In that case, the both power ways are used. The mode is applied to achieve the maximum speed. The parallel flow clutch C_R turns on.

Rectilinear motion using only tracked electric motor (fallback mode). When the main power flow fails (the main power flow shafts are stopped, for example, by the dynamic braking modes) the traction electric motor could be used to move the vehicle for a short time. The parallel power flow clutch C_R turns on. The stopping on-board brakes are used for the steering control system because the low vehicle speed. The energy storage system supplies the movement. The development of the SPBTU specialist could be implemented in such transmission design [16-18].

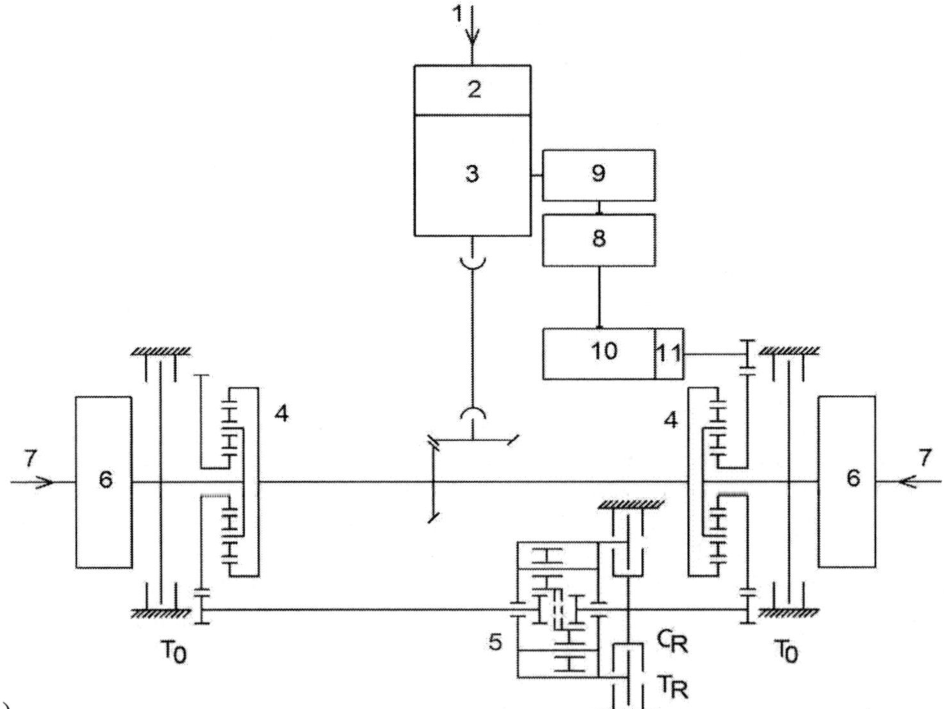

a)

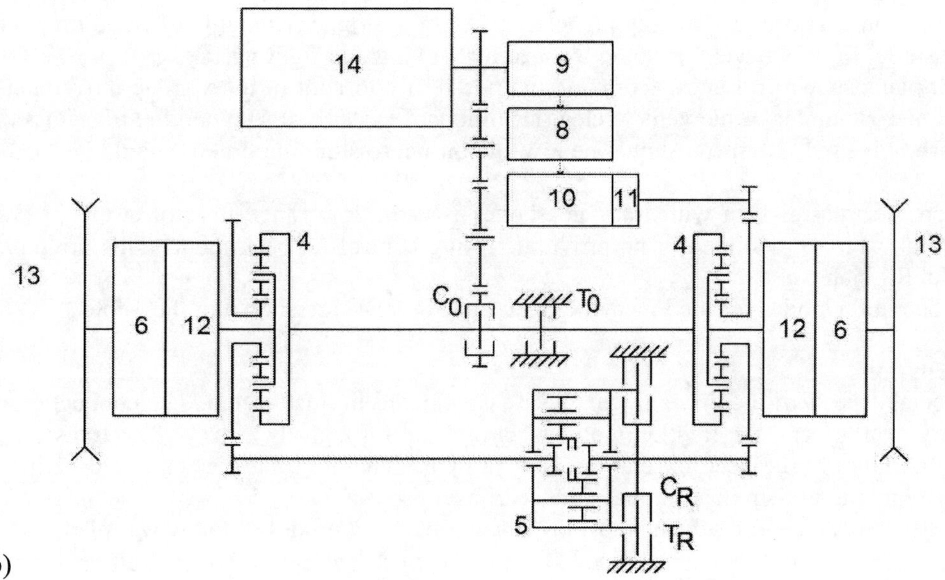

b)

Figure 4. Kinematic diagram of double-flows transmission: a –with central gearbox; b –with on-board gearboxes; 1 –power supply from engine; 2 – coupling mechanism; 3 – gearbox; 4 – summing planetary gear set; 5 – differential with clutches C_R and T_R; 6 – track gear reducer; 7 –power transfer to the drive wheel; 8 – energy storage device; 9 – traction electric alternator; 10 – tracked electricmotor; 11 – parallel flow reverse gear; 12 – tracked gearbox; 13 – drive wheel; 14 – internal combustion engine; C_0 – central clutch; T_0 –brake.

Table 1. Vehicle operation modes.

Operation modes	**With the central gearbox**		**With the on-board track gear-boxes**	
	Tracked electric motor	Activated control elements	Tracked electric motor	Activated control elements
Rectilinear motion using only the heat engine	Off	T_R and C_R	Off	T_R, C_0 and C_R
The turning using the heat engine and traction electric motor	Engine-on mode with a given rotation direction	T_R	Engine-on mode with a given rotation direction	C_0 and T_R
Neutral turning using only traction electric motor	Engine-on mode with a given rotation direction	T_R	Engine-on mode with a given rotation direction	T_0 and T_R
Rectilinear motion using heat engine and tracked electric motor (Accelerated mode)	Engine-on mode	C_R	Engine-on mode	C_R
Rectilinear motion using only tracked electric motor (fallback mode).	Engine-on mode	C_R	Engine-on mode	C_R

The turning using only heat engine (fallback mode). When the parallel power flow fails, tracked gearboxes or on-board stopping brakes is used to control vehicle movement. The brake T_R and the clutch C_R (sun gear of the summing planetary gear set is stopped) should be turned on in both variants of the parallel flow. The vehicle could be used for moving and fighting.

Both plans shown in Figure 4 provide the possibility to control the vehicle movement in case the traction electric motor or the vehicle electrical onboard system fails. When the transmission with central gearbox is used, the speed should be low and on-board stopping brakes should be used for steering control.

When the transmission with track gearboxes is used, low range gear or stopping brake for low speed track gearbox is turned on to provide steering control fallback mode. This principle is used in the serial Russian tanks.

All operation modes for the kinematic diagram (Figure 4) are shown in the Table 1.

3. Discussion

The special aspects of the drive kinematicsare considered in this section. The connection between the planetary gearbox and the traction motor is carried out through the link G. The transforming mechanism in addition to the gearbox may includethe hydrodynamic transmission could be included in the transforming mechanism in addition to the gearbox.

The gearbox kinematic scheme must provideall operating modes of the power plant:

- operation of the heat engine (The traction electrical motor works in alternator mode);
- multiple operation of the heat engine and traction electrical motor;
- operation of the traction electrical motor;
- neutral;
- parking.

Such principle of the connection between the gearbox and traction electrical motor allows to eliminate the use of the complex reduction gearbox that reconciles the operation of the heat engine and traction electrical motor isn't not necessaryif we use describing principle.

The drive principle described above could be used in tracked transport vehicle transmission that provides the principle of a parallel/series type hybrid, the same as such wheeled vehicles transmissions with tracked electric motor as part of the power distribution mechanism [19,20,21,22]. The lubrication system pump of the planetary gearboxes, for example.

The inventive drive method isuniversal, and it could be used for all gearbox types. But an additional summing planetary gear set is necessary in a helical or spur gearboxes, it complicates the transmission that isn't always economically acceptable. Such function in the planetary gearbox can be performed by one of the planetary gears set with minimal complication in the construction.

4. Conclusion

According to the results of work, Russian patents [23-25] were obtained.

1. The possibility of organizing the alternator and / or an oil pump drive using a differential mechanism which combines the input and output links of the transforming mechanism was shown.
2. The direction changing of the leading link in such drivesshould be, therefore it is necessary to provide the pump or alternator reverse.
3. This principle of communication with the transforming mechanism could be use in new plan development of the implementing hybrid transmission.

References

[1] Ushiroda Yu, Sawase K, Takahashi N, Suzuki K and Manabe K 2003 Development of Super AYC *Technical Review* **20** 73-6
[2] Bulyashov A H Automotive Handbook URL http://www.irs.kit.edu/131213_GC_TorqueVectoring_ZF_Handout.pdf

IOP Conf. Series: Materials Science and Engineering **791** (2020) 012025 doi:10.1088/1757-899X/791/1/012025

[3] Dobretsov R Yu, Lozin A V and Semenov A G 2018 *Gibridnyj mekhanizm raspredeleniya moshchnosti v transmissii transportnoj mashiny* [The hybrid power distribution mechanism in the transport vehicle transmission] Patent RU No 2656938 [In Russian]
[4] Wong J Y 2001 *Theory of ground vehicles* (USA: John Wiley & Sons)
[5] Grigoryev I V, Dobretsov R Yu and Gazizov A M 2017 Puti uluchsheniya upravlyaemosti lesny`x i transportny`x gusenichny`x mashin [Ways to improve controllability of logging and transport tracked vehicles] *Vestnik Bashkirskogo gosudarstvennogo agrarnogo universiteta* **23** 97-106
[6] Fischer R, Küçükay F, Jürgens G, Najork R and Pollak B 2015 *The Automotive Transmission Book* (Switzerland: Springer International Publishing)
[7] Lozin A V 2016 *Dvuhpotochnaya transmissiya transportnoj mashiny s bortovym sposobom povorota* [The double-flows transmission of the transport tracked vehicles with skid steering] Patent RU No 2599855 [In Russian]
[8] Ogorkiewicz R M 1991 *Technology of Tanks* Vol 1 (UK: Jane's)
[9] Wang G G, Wang S H and Chen C W 1990 Design of turning control for a tracked vehicle *IEEE Control Systems Magazine* **10** 122–5
[10] Demidov N N 2019 The drivetrain and the steering mechanism for the twin engines tracked vehicle *J. of Phys.: Conf. Ser.* **1177** 012010
[11] Lozin A V 2018 Hyperbolic steering for tracked vehicles *Proceedings of the 4th Int. Conf. on Industrial Eng., ICIE 2018* (Cham: Springer) pp 2367-74
[12] Shelomov V B 2013 *Teoriya dvizheniya mnogocelevyh gusenichnyh i kolesnyh mashin. Tyagovyj raschet krivolinejnogo dvizheniya* [*The theory of multi-purpose tracked and wheeled vehiclesmovement. Traction calculation of curvilinear motion*] (Saint-Petersburg: SPbPolyTechU) [In Russian]
[13] Didikov R A 2017 Power distribution control in perspective wheeled tractor transmission *Procedia Engineering* **206** 1735-40
[14] Uvakina D 2018 Performance improvement of Arctic tracked vehicles *MATEC Web of Conferences* **245** 17001
[15] Didikov R A 2018 Power distribution control in the transmission of the perspective wheeled tractor with automated Gearbox *Advances in Intelligent Systems and Computing* (Cham: Springer) 192-200
[16] Belko V O, Glivenko D E, Emelyanov O A, Ivanov I D and Plotnikov A A 2017 Current pulse polarity effect on metallized film capacitors failure *IEEE Transactions on Plasma Science* **45** 1020-5
[17] Belko V O and Emelyanov O A 2016 Self-healing in segmented metallized film capacitors: experimental and theoretical investigations for engineering design *J. of Applied Physics* **119** 024509
[18] Belko V O, Petrenya Y K, Andreev A M, Kosteliov A M and Roitgarz M B 2019 Numerical simulation of discharge activity in hv rotating machine insulation *Proc. of the 2019 IEEE Conf. of Russian Young Researchers in Electrical and Electronic Eng.* (Saint-Petersburg: IEEE) pp 800-2
[19] Dobretsov R Yu, Gubachev M, Naletov I and Vasiliev A 2018 On the way to driverless road-train: digital technologies in modeling of movement, calculation and design of a road-train with hybrid propulsion unit *SHS Web of Conf.* **44** 00030
[20] Didikov R A 2018 Transmission of the perspective wheel tractor with automatic gearbox: management of the power distribution mechanism *Int. Review of Mechanical Eng.* **12** 790-6
[21] Kondakov S V, Kharlapanov D V and Vansovich E I 2016 Models of the turn resistance for high – speed caterpillar vehicles *Russian Engineering Research* **36** 1-5
[22] Bukashkin A Yu 2017 Split transmission of tractor with automatic gearbox *Procedia Engineering* **206** 1728-34

IOP Conf. Series: Materials Science and Engineering **791** (2020) 012025 doi:10.1088/1757-899X/791/1/012025

[23] Dobretsov R Yu, Lozin A V and Semenov A G 2018 *Privod elektricheskogo generatora v sostave samohodnogo nazemnogo transportnogo sredstva* [The drive electric alternator as a part of an automotive surface vehicle] Patent RU No 265694 [In Russian]

[24] Dobretsov R Yu, Lozin A V and Semenov A G 2018 *Bortovaya nasosnaya stanciya sistem smazki i gidravlicheskogo upravleniya nazemnogo transportnogo sredstva* [The onboard pump for lubrication and hydraulic control systems of surface vehicle] Patent RU No 2656956 [In Russian]

[25] Izotov V Z, Pyatkov VA, Starovoytov V S and Suslov AA 1976 *Mekhanizm povorota gusenichnoj mashin* [The tracked vehicle steering device] Patent RU No 521174 [In Russian]

IOP Conf. Series: Materials Science and Engineering **791** (2020) 012026 doi:10.1088/1757-899X/791/1/012026

Analysis of applicability of conventional units in power grid companies in the Russian Federation

Y V Dronova[1], N A Chernenko[2], D A Tikhonova[3] and D I Shinkaryuk[3]

[1]Department of Production Management and Power Economics, Novosibirsk State Technical University, Karl Marx ave., 20, Novosibirsk 630073, Russia
[2]Department of High Voltage Engineering and Electrophysics, Novosibirsk State Technical University, Karl Marx ave., 20, Novosibirsk 630073, Russia
[3]Department of Production Management and Economics of Energy, Novosibirsk State Technical University, Karl Marx ave., 20, Novosibirsk, 630073 Russia

E-mail: N.Chernenko@corp.nstu.ru

Abstract. Value formation for energy transmission services in the Russian Federation is carried out under strict state control, basing on the widespread practice of benchmarking or "best practice" methods. The basis for calculating some of the best indicators is the concept of "unit credit". (c.u.). Over the past few years, the issue of the "unjustified" application of the CU and the need to review it has been raised several times. The analysis of the current normative documentation has shown that use of different documents allows changing the results of calculation of the basic items of the expenses using c.u. as the basic indicator, for example, the expenses for maintenance service and repair (MRO) regarding a labor payment. This situation arises in the process of transferring the actual labor intensity of maintenance and repair works in c.u. The value of deviation of labor inputs determined by the difference in periodicity and labor intensity leads to significant distortion of result when forming costing and defining the optimum or the most effective values. Thus, these differences can become a tool for manipulation by energy companies when costing making use of benchmarking.

1. Introduction

The concept of "conventional unit" (c.u.) of equipment is used to calculate a large number of grid companies' performance indicators. Appearance of this indicator is due to the necessity to calculate the number of personnel, expenses for maintenance of grid equipment, and it was impossible to use as a basis the capacity or length of power lines due to significant differences in the structure of equipment of individual enterprises of power grids [1-10]. The modern pricing system also relies on this concept in terms of calculations. However, the majority of the regulatory and legal framework defining the use of conventional units became obsolete now or has the status of "inactive documents". This leads to the possibility of changing the calculated values in accordance with the company's goals.

2. Materials and Methods

A conventional unit is estimating certain labor intensity by a dimensionless coefficient (unit). The first documents defining the concept of conditional unit are attributed to 1960. For 1 c.u. were taken labor costs for the operation of 1 km of single-circuit 110 - 154 kV overhead line on metal and reinforced concrete supports. The structure of service volumes in conditional units for power grid facilities (pow-

IOP Conf. Series: Materials Science and Engineering **791** (2020) 012026 doi:10.1088/1757-899X/791/1/012026

er lines and substations) corresponded to the structure of operating personnel, and the number of conditional units per 1 person of production personnel (specific load for power grids of all voltage classes) was approximately the same [1, 3].

In course of time, the standards have been repeatedly corrected due to the regular analysis of the actual number of staff in the electricity grid. After having restructured the industry, such an analysis was practically discontinued, and the companies gained partial independence to determine the structure of personnel, and all these aspects resulted by appearance of "gaps" between the actual labor intensity of maintenance of electrical equipment of the networks and the standard number.

This problem is mentioned in the article by F.L. Golberg, where it is proposed to adopt the "bulk forming units" developed by ORGRES as an alternative to the existing conventional units [10]. In 2008, the FSK Company accepts these units for calculation by the Order No. 162. However, other power grid companies continue to work with conventional units.

Currently, the Russian Federation does not have a legal document that would enshrine this concept. The following definitions can be found in the literature:

- the number of conditional units per 1 person of production personnel is a specific load on the objects of electric networks of all voltage classes 1 c.u. = 40 h-hour [10];
- the number of conventional units determines the labor costs for one repair of this type on the accepted meter [1];
- the number of conventional units is an indicator characterizing the cost of live labor, expressed in working time spent for producing products (services) [3].

In terms of application of C.O.U. today use is made of various legal acts, as follows:

- The order of the State Building Committee of the Russian Federation from 03-04-2000 № 68, defining calculation of number of the personnel for repair and maintenance service;
- Order of the Federal Tariff Service of August 6, 2004 № 20-e/2;
- Regulatory document RD 39-0148311-601-88;
- Methodical guidelines for developing process charts and projects for maintenance and repair of overhead power lines, STO 56947007-29.240.55.168-2014.

Due to the transition of the pricing system to a new procedure of forming the gross revenue required (GRR) basing on the Methodology for comprehensive determination of the technical and economic condition of power facilities, including the indicators of physical wear and tear and energy efficiency of power grid facilities (approved by Resolution of the Government of the Russian Federation No. 1401 of December 19, 2016), grid companies are forced to use the benchmarking system or the "best practice method".

To realize this approach, the Ministry of Energy initiated data collection since 2016. The list of these indicators is quite extensive, and some of them are calculated considering the number of c.u. at the power grid facility, and some of them take into account this number indirectly. The indicators calculated on the basis of the number of conventional units include, first of all, maintenance and repair costs, which are the most significant expenses of any grid company. Fig. 1 shows the rating of power grid companies by specific maintenance and repair costs.

Effective companies are highlighted in gray, and ineffective companies are highlighted in black. The middle line shows the value that will be set as the target for all companies. However, the issue of calculating maintenance and repair costs remains methodologically undeveloped.

3. Results

3.1. Credibility analysis of c.u. indicators

To assess the reliability of the calculation of key performance indicators of power grid companies, we compare different values used to calculate the labor intensity of some types of electrical equipment. The basis for calculating the labor intensity in accordance with the current normative and technical documentation is the technical documents developed within the organization that define the conditions and content of work on maintenance of electrical equipment.

IOP Conf. Series: Materials Science and Engineering **791** (2020) 012026 doi:10.1088/1757-899X/791/1/012026

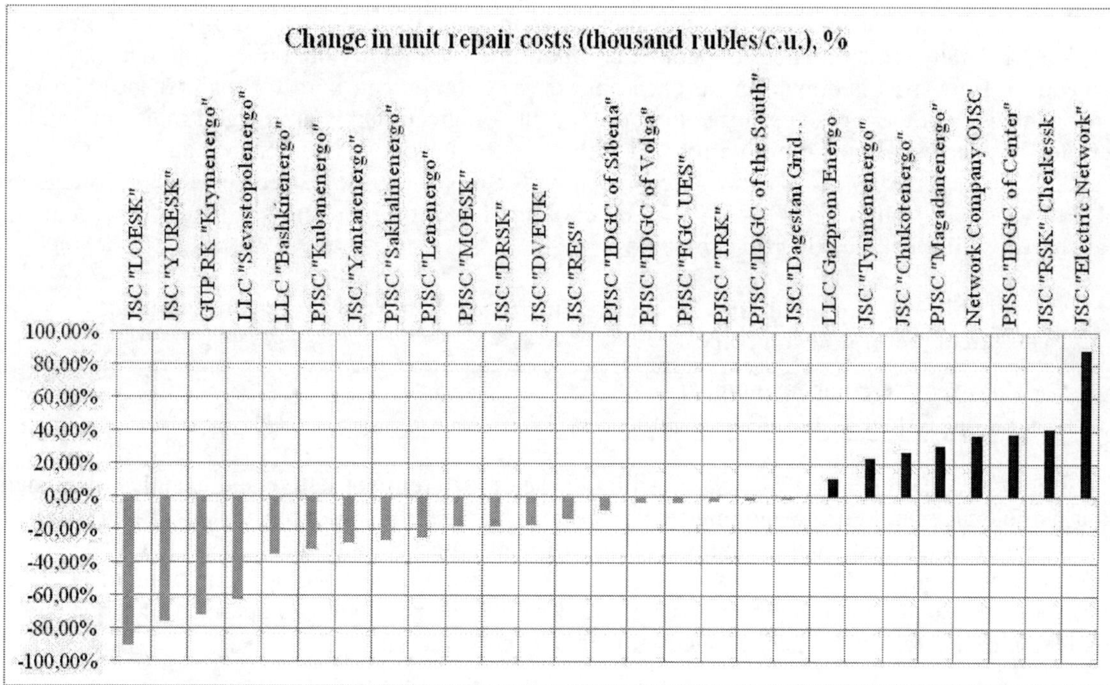

Figure 1. Rating of grid companies by unit maintenance costs in 2018 (according to the Ministry of Energy).

In accordance with the "Rules of Organization of Maintenance and Repair of Electric Power Facilities" approved under the Order of the Ministry of Energy of the Russian Federation dated 25.10.2017 No. 1013 for the purpose of increasing reliability in the operation of electric power facilities, as well as for the purpose of increasing the duration of overhaul periods and ensuring the quality of maintenance and repair work, electric power industry entities develop and approve the Standards for Maintenance and Repair of Electric Power Facilities or pieces of equipment of electric power industry. It is envisaged to carry out both planned preventive maintenance and technical repair. The standard is developed in strict accordance with current legislation, technical requirements of regulatory documents, and local acts of the organization. The main form of technical documentation reflecting the entire process of maintenance or repair of an electric power industry facility is the technological chart for maintenance and repair. Technical standards and requirements, methods and recommendations (taken into account in the process of drawing up the technological charts) should reflect the main provisions of the technical policy of the organization, long-term experience of maintenance and repairs, both in its organization and in other organizations, to take into account the features of the operating equipment, service life, technical condition and the remaining life of the equipment, as well as individual elements and units, the requirements of the design documentation of manufacturers, as well as occupational health and safety requirements.

The flow chart contains the following sections:

- measures and conditions of safe work performance (hazardous production factors are reflected at the workplace, fire safety measures are listed, and environmental requirements to work performance are specified);
- work organization (the name and quantity of protective equipment and clothes, tools and devices, materials and spare parts, mechanisms of necessary mechanisms are listed, as well as the composition of the team with indication of position, electric safety group, number of people and their qualification features);

IOP Conf. Series: Materials Science and Engineering **791** (2020) 012026 doi:10.1088/1757-899X/791/1/012026

- work technology (the procedure of work performance is given, which includes both organizational (beginning and end of work) and technological operations (inspections, checks, main maintenance or repair works, tests and measurements) with indication of the performer, electric safety group, number of people, norm of time for an individual operation in h·h).

Thus, a flow chart is drawn up for each item of grid equipment, both for maintenance and repair. Labor intensity of the works performed under the technological map is an integral indicator and is determined by the list of operations performed within the framework of maintenance and repair.

Table 1 presents the norms of labor intensity (h-h) for repair of power transformer for voltage class 6(10) kV with capacity of 1000 kV-A with oil cooling of TM-1000/6(10) type, taken from technological charts of different electric grid companies.

Table 1. Labor input rates for the ongoing repair of the power transformer in different power grid companies.

Power grid company number	1	2	3	4
Transformer TM-1000/6(10), h·h	19,7	42	24	17

Significant differences are associated with the use of different normative and technical documentation as guidance ones for developing technological charts. As an example, Table 2 shows the labor intensity norms on maintenance and repair for different types of equipment according to several methodical documents. Table 3 shows the frequency of repair impacts on the same units of equipment indicating the source of these indicators. Table 4 shows the results of comparison of labor costs on maintenance and repair and their recalculation in conventional units.

Table 2. Norms of labor intensity of maintenance and repair according to various methodical documents for current repair.

Type of electrical equipment	In accordance with STO 56947007-29.240.55.168-2014, h·h	According to the Handbook [10], h·h	According to RD-39-0148311-601-88, h·h
Air lines up to 1000 V on metal and reinforced concrete supports	6	8	6
Cable lines up to 10 kV	23	21	23
Three-phase double-winding oil transformers up to 10 kV	18	-	5.2
Oil switch 35-110 kV	7	-	7.2

Table 3. Maintenance and Repair Frequency Rates according to various methodical documents for repair.

Type of electrical equipment	In accordance with STO 56947007-29.240.55.168-2014, month.	According to the Handbook [10], month	According to RD -39-0148311-601-88, month
Air lines up to 1000 V on metal and reinforced concrete supports	36	36	72
Cable lines up to 10 kV	24	12	72
Three-phase double-winding oil transformers up to 10 kV	36	-	24
Oil switch 35-110 kV	12	-	24

Table 4. Labor intensity standards of maintenance and repair according to different methodical documents in c.u. and h·h.

Type of electrical equipment	Recalculation in c.u. in accordance with the order of the Federal Tariff Service No. 20-e/2, c.u.	Full labor intensity of maintenance and repair in accordance with STO 56947007-29.240.55.168-2014, h·h	Complete labor-intensive maintenance and repair in accordance with RD-39-0148311-601-88, h·h	Maximum difference, h·h
Air lines up to 1000 V on metal and reinforced concrete supports	200	216	432	232
Cable lines up to 10 kV	350	529	1656	1306
Three-phase double-winding oil transformers up to 10 kV	-	648	124,8	523.2
Oil switch 35-110 kV	23	84	172,8	149.8

3.2. Modeling calculation of efficiency indicators using various methodical documents.
To illustrate possible distortion of the calculation results key indicators through the use of various methods for determining the cost of maintenance and repair, calculations were made for the real section of the power grid. Table 5 shows the list of electric equipment of the real section of electric grids and defines annual labor costs both in h·h and in c.u.

Table 5. Labor intensity of maintenance by type of equipment for the power grid section.

Type of electrical equipment	U, kV	Quantity, pcs. (unit)	Labor costs per year in accordance with STO 56947007-29.240.55.168-2014, h·h	Labor costs per year in accordance with FTS Order No. 20-e/2, c.u.
Complete transformer substation (2 transformer substation)	110/35/6	3	72	9
Complete transformer substation (2 transformer substation)	35/6	19	380	57
Complete transformer substation	35/6	5	80	15
Air switches	35	50	460	550
Air switches	6(10)	555	2608.5	3052.5
Oil switches	35	47	352.5	300.8
Oil switches	6(10)	254	2997.2	787.4
Power oil transformer	110/35/6	6	376.8	46.8
Power oil transformer	35/6(10)	43	1982.3	90.3
Power oil transformer	6(10)/0.4	685	15549.5	658
Separator with short-circuit breaker	110	10	177	47
Separator with short-circuit breaker	35	68	1067.6	646
IN TOTAL		1745	26103.4	6259.8

Calculations showed that the recalculation of the total labor intensity of one unit of equipment varies from 1 h-hour to 11 hhour. The range of deviations is shown in Figure 2.

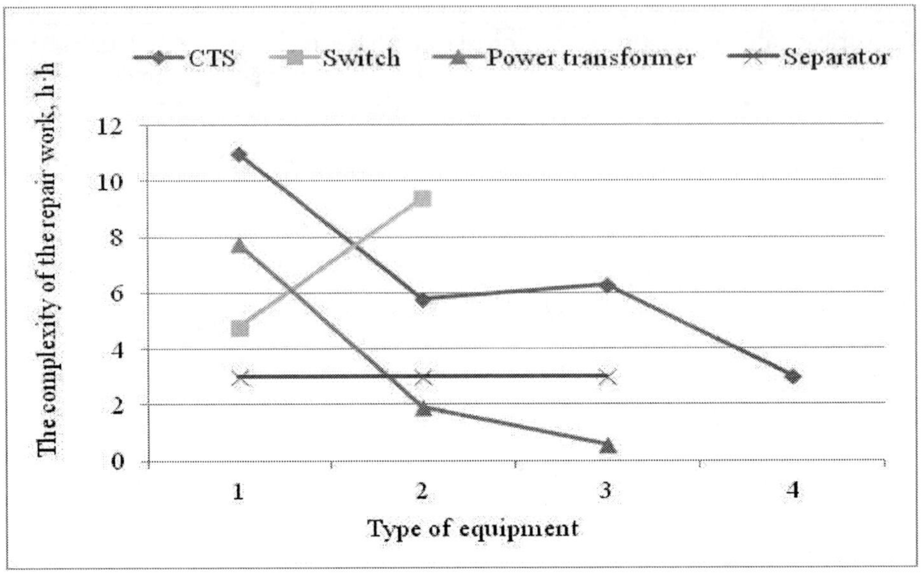

Figure 2. Calculation of h•h in 1 c.u. by groups of electrical equipment.

4. Discussion

Application of various methodical approaches for calculating maintenance and repair costs and estimation of unit costs in c.u. differs significantly depending on the type of the document used. This allows the energy company (in accordance with its objectives) to manipulate the calculations, overestimate or underestimate the absolute value of costs and ROE, and influence the place of companies in the efficiency rating. Benchmarking can be used only in case of availability of accurate and uniform guidelines, and if they are not available, KPI comparisons are not reliable [11-15].

Currently, power grid companies, especially those that can use industry guidelines (energy service companies of the oil and gas industry) can use internal industry guidelines or guidelines of the Ministry of Energy when calculating maintenance and repair costs. Differences in calculations can reach 10 times the value, which will eventually lead to an increase/decrease in RET, and eventually the tariff for energy transmission services.

It is necessary to develop new normative-methodical documents to determine the labor intensity and frequency of maintenance and repair. Using the existing regulatory framework does not allow performing reliable implementation of benchmarking when assessing the performance of power grid companies.

Acknowledgments

The research is executed within the research contract No AAAA-B19-219012990011-9.

References

[1] Afanasyev N A and Yusipov M A 1989 *Sistema tekhnicheskogo obsluzhivaniya i remonta oborudovaniya energokhozyaystv promyshlennykh predpriyatiy (sistema TOR EO)* [System for maintenance and repair of equipment of energy facilities of industrial enterprises] (Moscow: Energoatomizdat) [In Russian]

[2] Digitalization & Energy/International Energy Agency URL: https://www.connaissancedesenergies.org/sites/default/files/pdf-actualites/ digitalizationandenergy3.pdf

[3] Golberg F L 2008 Sistema ob``emoobrazuyushhix pokazatelej i ee primenenie v ENE`S

E`nergetik [The system of volume-forming indicators and its application in the Unified National Electric Grid] *Energetik* **4** pp 7-10 [In Russian]

[4] Kassakian J G et al 2011 *In The Future of the Electric Grid: An Interdisciplinary MIT Study* (Massachusetts: MIT Press)

[5] Khuntia S R, Rueda J L, Bouwman S and van der Meijden M A M M 2016 A literature survey on asset management in electrical power [transmission and distribution] system *Int. Transactions on Electrical Energy Systems* **26(10)** 2123-33

[6] Lam T C and Yeh R H 1994 Optimal Maintenance-Policies For Deteriorating Systems Under Various Maintenance Strategies *IEEE Transactions on Reliability* **43** 423-30

[7] Mißfeldt S A 2012 *Comparative analysis between the Dutch and German electricity distribution network industry with regard to regulation and efficiency* (Deutchland: Aarhus University)

[8] Natti S and Kezunovic M A 2008 A Risk-Based Decision Approach for Maintenance Scheduling Strategies for Transmission System Equipment *Proceedings of the 10th Int. Conf. on Probabilistic Methods Applied to Power Systems* (Puerto Rico, USA) 1-6

[9] Schneider J, Gaul A J, Neumann C, Hogräfer J, Wellßow W, Schwand M and Schnettlere A 2006 Asset management techniques *Int. J. Electrical Power and Energy Systems* **28(9)** 643-54

[10] Yashchura A I 2006 *Sistema tekhnicheskogo obsluzhivaniya i remonta energeticheskogo oborudovaniya* [System for maintenance and repair of power equipment] (Moscow: Publishing House NTs ENAS) [In Russian]

[11] Agle B R, Donaldson T, Freeman R E, Jensen M C, Mitchell R K and Wood D J 2008 Dialogue: Toward Superior Stakeholder Theory *Business Ethics Quarterly* **18(2)** 153-90

[12] Balmer J and Greyser S A 2002 Managing the Multiple Identities of the Corporation *California Management Review* **44(3)** 72-86

[13] Farsi M, Filippini M and Greene W 2005 Efficiency Measurement in Network Industries: Application to the Swiss Railway Companies *J. of Regulatory Economics* **28(1)** 69-90

[14] Klein P, Mahoney J, MeGahan A and Pctelis C. 2012 Who is in charge? A property rights perspective on stakeholder governance *Strategic Organization* **10(3)** pp 304-15

[15] Machek O 2011 Regulatory Benchmarking in Central Europe Current Practice and Possibilities ot Development for the Energy Sector *The annals of the university of Oradea. Economic Sciences Series* **20(1)** 80-6

Increase in accuracy forecast for electrical energy consumption of the WEM subjects using fuzzy neural networks

Ya P Fedorov[1], V L Burkovsky[2] and A L Rutskov[3]

[1]Corporate finance department and corporate management of the Financial University under the Government of the Russian Federation, Leningradsky prospect, 49, Moscow, 125993, Russia
[2]Department of electric drive, automation and control in an engineering system, Voronezh State Technical University, Moscow prospect, 179, Voronezh, 394066, Russia
[3]Forecasting department of electricity balance, power and loss analysis, branch office OJSC «IDGC Centre» - «Voronezhenergo», Arzamaskaya 2, Voronezh, 394033, Russia

E-mail: alex_8_90@mail.ru

Abstract. This paper covers in depth functioning essence of the balancing electricity market in the Russian Federation in accordance with the current economic stimuli. Significant role of the prediction process for energy resources consumption has been noted, which has a direct influence on a pricing policy within the marked trade area. Analysis of different factors, that impact execution accuracy of the one day in advance model, was conducted. Problem of the weakly defined and unknown parameters, that influence the performance of the balancing electricity market, is suggested to resolve by means of fuzzy neural controller. The paper presents its structure and settings, and possibility to scale it to use not only as an analyzer but as an executive element in the subsystems of electric power systems. High accuracy performance of the fuzzy neural controller was shown - around (0,5-3) %, that is much higher than methods already in use (for instance, based on a coefficient of decline/growth of power consumption).
Keywords: wholesale electricity market; economic stimuli; electric energy consumption forecast; fuzzy neural networks; simulation modelling.

1. Introduction

Wholesale electricity market (WEM) in the Russian Federation is a technical and economic platform that stands for competitive environment for production, transportation and consumption of an energy resources within 75 federal subjects. Subjects of this system are sellers and buyers of electricity and power that meet requirements mentioned in the RF Government Regulation dated 27.10.2010 №1172 (changed 31.07.2019) «On the establishment of the Rules of Wholesale Electricity and Capacity Market».

WEM consists of the following elements (figure 1):

- electrical energy market (consists of three segments: regulated contracts (RC); «day-ahead market» (DAM); and also balancing market (BM)) - electrical power production and sale;
- capacity market – bidding process to deliver connected capacity reserve;

IOP Conf. Series: Materials Science and Engineering **791** (2020) 012027 doi:10.1088/1757-899X/791/1/012027

- system services market – provide qualitative and quantitative characteristics of the energy resources;
- system operator of the unified energy system (SO «UES») - technical market regulator;
- non-commercial partnership trading system administrator (NP «TSA») - economic governing organ;
- federal antimonopoly service (tariff regulation that concerns PC); Ministry of Energy of the Russian Federation (legislative support, supervisory responsibilities) - not shown on the picture.

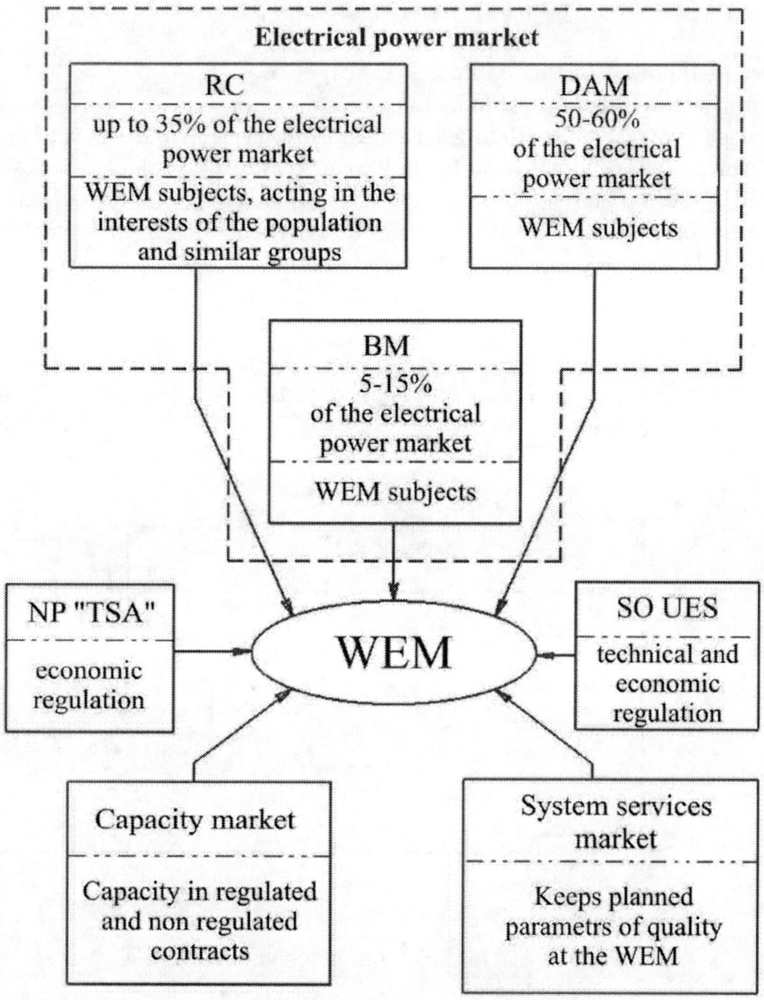

Figure 1. WEM function chart.

2. Materials and Methods

Specific nature of energy resources doesn`t allow to accumulate it (at least for now), therefore, balance between production and consumption must be mantained in real time mode. Last named circumstance is a major technical factor which is the necessity of ensuring electricity consumption forecast error minimization. It can be explained in a way that electricity production is a complicated technological process during which all of the generated product must be consumed (under conditions of trasportation losses optimization).

Therefore, setting up WEM creates an economic stimuli to icrease forecast accuracy, meaning that all electricity consumption deviations from planed values (preliminarily requests submitted by subjects during forming of DAM) are purchased or sold using BM prices. Specifics of the balancing market makes dynamics of pricing policy for purchase/sale unsufficient/excessive energy resources volume multidirectional.

$$C_{BM\_pur} = \max\,(C_{DAM};\, C_{BM}) \quad (1)$$

$$C_{BM\_sale} = \min\,(C_{DAM};\, C_{BM}) \quad (2)$$

where C_{DAM} – electrical power cost by volume at DAM;

C_{BM} – electrical power cost by volume at the BM.

Way of forming DAM assumes finding an equilibrium price (figure 2) and volumes of delivery (consumption) - chosen bids, that has been taken into account for this electrical power market segment the day before billing period starts. Volumes that are not collected should be traded at BM or should be excluded by the subject of market on its own initiative (cancellation of production/consumption of electrical power).

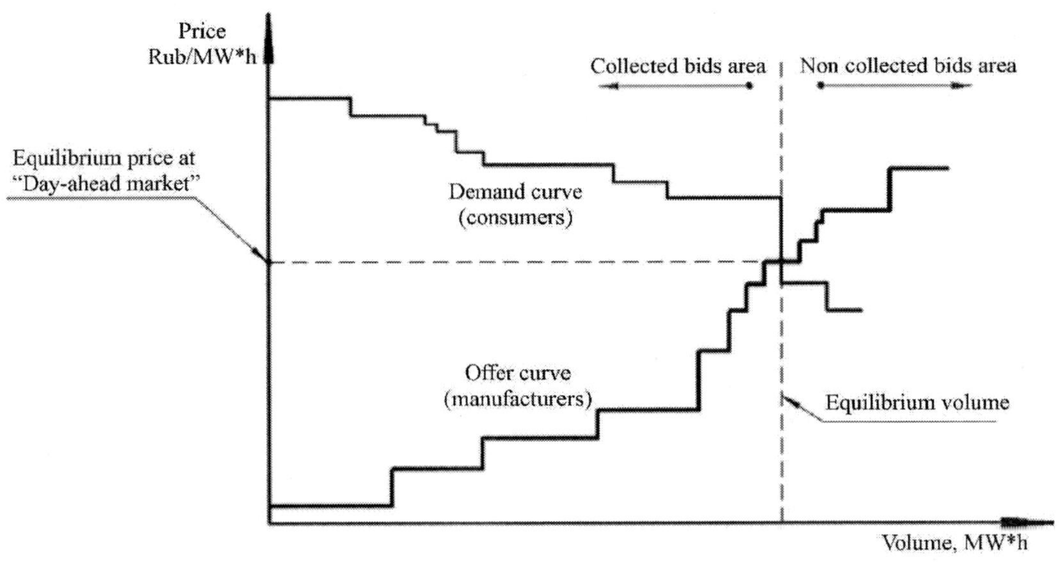

Figure 2. Bids selection mechanism at DAM.

Therefore, deviation from forecasted numbers of bids, chosen for DAM, as part of the actual energy resources consumption within design day, creates losses for the subject who allowed such deviations, if it leads to breaking planed schedule [1, 2]. For instance, during supply shortage (produced volumes less than current requests from energy receiving devices) any WEM participant, that needs to increase purchases, according to expression (1), will purchase electricity at BM. That is the way to implement a kind of extra charge penalty which compensates expenses of manufacturers and infrastructure participants for unexpected operating modes (characterized by increasing expenses). Note that in case of supply shortage, during sale of energy resources to WEM consumers, this initiative is put into action for the price – C_{DAM} (because it helps to restore balance, moving it closer to equilibrium volume).

Data analysis at the WEM website [3] demonstrates that mean value of absolute deviations C_{DAM} from C_{BM} over the period 6 months in 2019 for the first pricing zone of the RF is in range ± (17-35) %; averaged volume of balancing market over the same period - 47,468 billion KWh (or 12,25 % of the electrical power market).

Decrease of the BM portion is possible by increasing accuracy of the forecast for electricity consumption. For this purpose, proved way is to use methods based on fuzzy neural network (FNN) [2, 4] and fuzzy neural controller (FNC).

Figure 3 presents a generalized function chart for FNC that use assignment vector current value at the i-th moment of time – $g_{a.}(i)$ - input data to evaluate electricity consumption forecast for deferred expenses; and also exit of the object`s reference model – $y_2(i+k)$ - forecasted value. Reference model based on ANN (which is the reference model for the subject to control (SC), in this case – WEM electrical power market) describes dynamics of the mentioned system within corresponding timeframes, trained on forecasted error $e_2(i+k) = y_2(i+k) - y_{SC}(i)$, and control signal – $y_1(i+k)$. Setting vectors W_1, W_2 give the ability to exercise parameterization of the input terms (for FNN) and activation functions (for ANN).

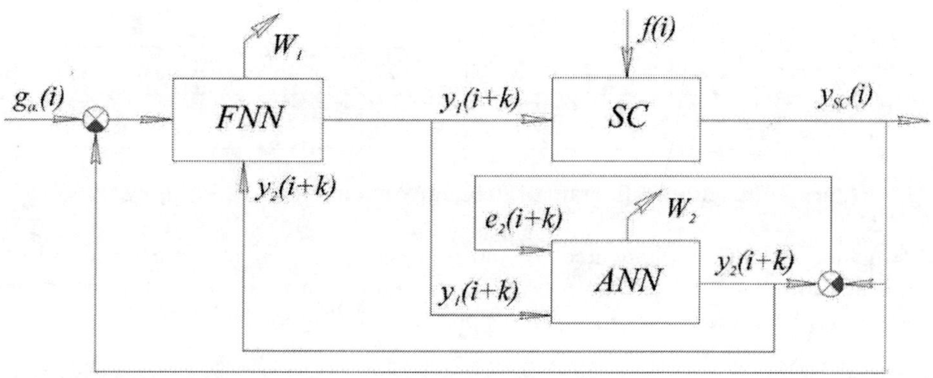

Figure 3. FNC function chart for electricity consumption of the WEM subjects forecast.

Vectors $g_{a.}(i)$ and $y_2(i+k)$ take the following data bases (DB) as an input value:

- consumption volume distribution of the WEM subject in the prior period $P(i)$;
- cyclic and random factors (environmental influence, modes of operation that connected with the calendar cycles) – $M(i)$;
- planned structure of the energy receiving devices – $S(i)$.

3. Results

By applying settings mentioned above, FNC uses inverse model SC and becomes the process controller with high adaptive characteristics, driven by undetermined factors that are fully taken into consideration.

Specifics of such an arrangement should be noted:

- fuzzy neural network (FNN), as part of the fuzzy controller, has freely scalable structure, and therefore, possibility to use it on unstructured subjects of WEM;
- artificial neural network (ANN), by using reference model SC, increases the effectiveness of decision making for current status assessment of the energy receiving devices.

Best results within the framework of studies [15] demonstrated P-shaped and Gaussian membership functions of terms for input variables. Exit of FNC is executed as linearized value.

Structure of the reference model (ANN) is presented with the help of the multilayered perceptron, trained on fast backpropagation of error algorithm. Structure of the network is variable and depends on SC. Type of the activation function – sigmoid.

FNC functional diagram in the ANFIS system is presented on the figure 4. Implementation method – Sugeno.

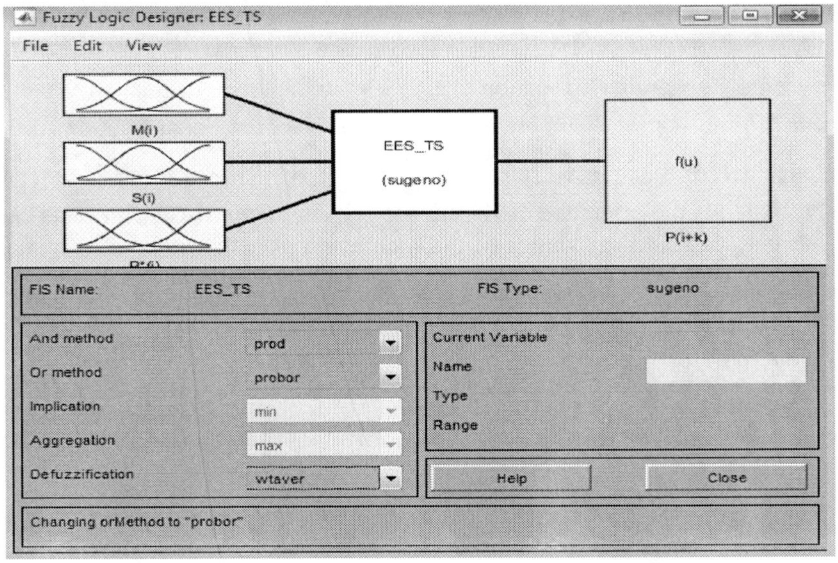

Figure 4. Functional diagram of FNC implementation in ANFIS system.

General structure of FNC implemented for the process control of the forecasted WEM subject's electricity consumption dynamics for short-term and medium-term periods (for the case of input terms assignment – $9\times3\times3$: $P(i)\times M(i)\times S(i)$) presented on figure 5.

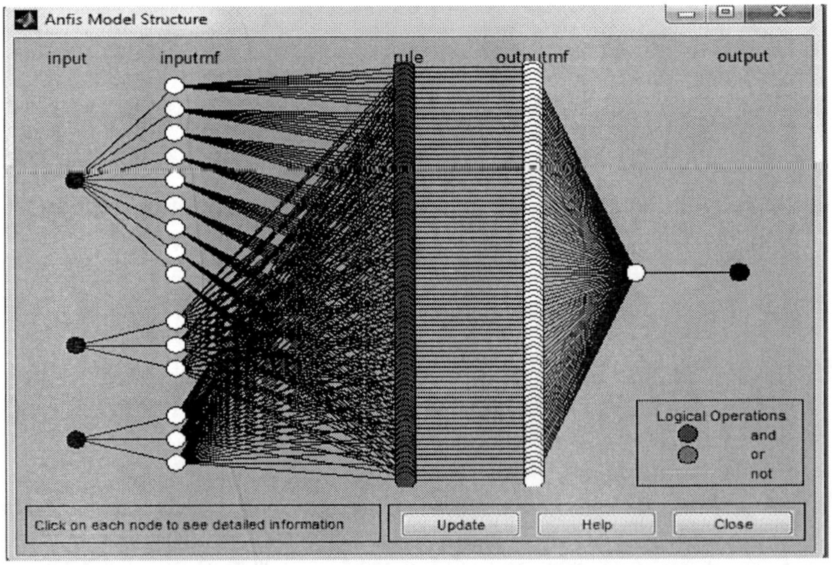

Figure 5. Basic FNN structure within the framework of the developed FNC.

IOP Conf. Series: Materials Science and Engineering **791** (2020) 012027 doi:10.1088/1757-899X/791/1/012027

Combination of the least squares method and reverse gradient descent method (hybrid) was utilized as the learning algorithm.

Table 1 presents settings for FNN, training by the combination of the least squares method and reverse gradient descent method along with variations of structure types of membership functions of input membership terms for implementation of Sugeno.

Table 2 presents simulation modeling qualitative results based on [4-13] in relation to the accuracy of functioning FNC forecast for electricity consumption of WEM subjects [14].

Table 1. Settings for FNN in compliance with implementation of Sugeno.

1. Netork type – FNN (Sugeno)			
2. Topology and implementation of fuzzy input-output data			
Membership terms assignment ($M(i)\times S(i)\times P^{*}(i)$)	Number of learning rules, ea.	Transfer function types for input terms	Transfer function types for output terms
$3\times3\times3$; $6\times3\times3$; $9\times3\times3$	**176**	**trapmf; dsigmf; psigmf**	**constant**
3. Topology and ANN learning (multilayered perceptron)			
Total number of layers, ea.	Number of hidden layers, ea.	Number of elements in training data set, ea.	Number of elements in test data set, ea.
4	**2**	**17 850**	**324**
Stopping conditions of learning – **increase of root mean square error**			
Method of weight renewal in an artificial neural network – **epoch** (scale free time period)			

Table 2. Comparative analysis of models implementation based on FNC.

FNN learning algorithm – Sugeno (hybrid). Processor type - Intel Core i3-7350				
Input membership terms type (configuration $3\times3\times3$)				
	Term	trapmf	psigmf	dsigmf
Estimated parameters	ε_n	0,271	0,228	0,228
	r	0,981	0,989	0,987
	$\varepsilon_{\%}$	3,145	2,012	2,005
Number of training epochs		500,0	6,0	6,0
Learning time, sec		67,46	1,03	1,05
FNN learning algorithm – Sugeno (hybrid). Processor type - Intel Core i3-7350				
Input membership terms type (configuration $6\times3\times3$)				
Estimated parameters	ε_n	0,090	0,053	0,061
	r	0,980	0,987	0,988
	$\varepsilon_{\%}$	2,954	1,903	1,985
Number of training epochs		500,0	16,0	17,0
Learning time, sec		68,74	2,33	2,48
FNN learning algorithm – Sugeno (hybrid). Processor type - Intel Core i3-7350				
input membership terms type (configuration $9\times3\times3$)				
Estimated parameters	ε_n	0,021	0,012	0,012
	r	0,983	0,991	0,993
	$\varepsilon_{\%}$	2,901	1,853	1,751
Number of training epochs		500,0	51,0	53,0
Learning time, sec		69,41	7,03	7,36

IOP Conf. Series: Materials Science and Engineering **791** (2020) 012027 doi:10.1088/1757-899X/791/1/012027

Best results in the analyzed data set demonstrated the implementation of FNC based on Sugeno algorithm using input terms configuration such as $(9\times3\times3)$ and membership function – dsigmf. Final results of the training, marked by FNC implementation on test data set, presented on the figure 6.

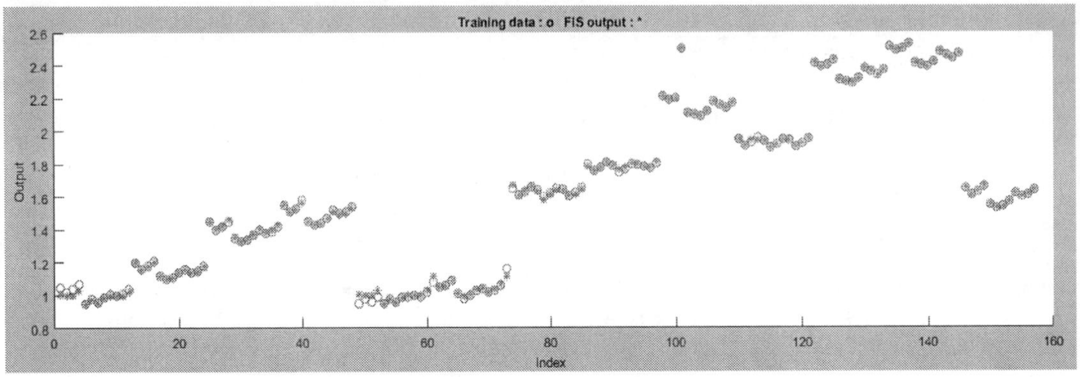

Figure 6. Results of FNC training on test data set.

Analysis of the functioning FNC forecast for electricity consumption in WEM allows to mention that its characteristics has high accuracy. Based on evaluation of the test data set, error is no more than 3,2%. With that in mind, and also using evaluation of the BM volumes and weighted mean value of deviation P_{DAM} from P_{BM} over a period of 6 month in 2019, one can observe qualitative estimation of potential savings, from utilization of FNC by WEM subjects, which is about 4,231 billion KWh.

As an expert recommended guideline, authors [1] point out that it is worthwhile to correct the consumption forecast graph by a value that doesn`t exceed forecast error on technology factors. However, in reality, due to lack of (significant lacking) data on competitive companies, it is more beneficial to increase accuracy of its own consumption volume with ability to correct this value after receiving high degree of confidence information about adjacent consumers.

4. Conclusion

1. Usage of models that control dynamic of power flow distribution as part of ADCS based on FNN (FNC) allows to obtain error within (0,5-3) %, which is vastly superior to already used in existing systems method of coefficient of decline/growth of power consumption with error (3-6) %.

2. Pointed out decrease in error control of power flow dynamic in operational and mid-term periods in models based on FNN (FNC) is possible as a result of complex use of both outside (weather conditions, cyclical pattern, factory operation) and inside (change of operating mode) undefined factors that influence elements of EPS.

3. Developed model of control power dynamic based on FNN (FNC) is easily scalable which allows to apply it both for operational usage within planner and for mid-term analysis of energy resources distribution.

4. It should be noted that possibility of development of optimal strategy in accordance with approaches, mentioned in this paper, is useful for generating companies and energy resources market as a whole.

References

[1] Solov`eva I A and Dzyuba A P 2013 Prognozirovanie potrebleniya e`lektroe`nergii v sootvetstvii s texnologicheskoj i ry`nochnoj sredoj [Electricity consumption forecasting in accordance with technological and market environment] *Nauchny`j dialog: E`konomika Pravo Politologiya* **7(19)** 97-113 [In Russian]

[2] Kry`sanov V N, Ruczkov A L and Shukur O M 2015 Optimizatsiya sistemy ekonomicheskikh faktorov funktsionirovaniya sistemnogo operatora v Voronezhskoy elektroenergeticheskoy

sisteme [Economic factors system optimization of system operator functioning in Voronezh electric power system] *E`lektrotexnicheskie Kompleksy` i Sistemy` Upravleniya* **3** 62-7 [In Russian]

[3] Wholesale electricity market data URL https://br.so-ups.ru/

[4] Alanis A Y, Ricalde L J, Simetti C and Odone F 2013 Neural model with particle swarm optimization kalman learning for forecasting in smart grids *Mathematical Problems in Engineering* **2013** 197690

[5] Jiang X, Ling H, Yan J, Li B and Li Z 2013 Forecasting electrical energy consumption of equipment maintenance using neural network and particle swarm optimization *Mathematical Problems in Engineering* **2013** 194730

[6] Shi B, Yu-Xia L I and Xin-Hua Y U 2009 Short-term load forecast based on modified particle swarm optimizer and back propagation neural network model *J. of Computer Applications* **29(4)** 1036-9

[7] Monteleoni C, Schmidt G A, Saroha S and Asplund E 2011 Tracking climate models *Statistical Analysis and Data Mining* **4(4)** 372-92

[8] Çevik H H and Çunkaş M 2015 Short-term load forecasting using fuzzy logic and ANFIS *Neural Computing and Applications* **26(6)** 1355-67

[9] Li P, Li Y, Xiong Q, Chai Y and Zhang Y 2014 Application of a hybrid quantized elman neural network in short-term load forecasting *Int. J. of Electrical Power & Energy Systems* **55** 749-59

[10] Nedellec R, Cugliari J and Goude Y 2014 Gefcom2012: Electric load forecasting and backcasting with semi-parametric models *Int. J. of Forecasting* **30(2)** 375-81

[11] Zadeh L A 1974 Outline to a new approach to the analysis complex systems and decision processes *IEEE Transactions on Systems, Man and Cybernetics* vol 3 pp 28-44

[12] Cheng Z and Juncheng T 2015 Adaptive combination forecasting model for china's logistics freight volume based on an improved PSO-BP neural network *Kybernetes* **44(4)**

[13] Devaine M, Gaillard P, Goude Y and Stoltz G Forecasting electricity consumption by aggregating specialized experts *Machine Learning* **90(2)** 231-60

[14] Burkovskij V L, Kry`sanov V N and Ruczkov A L 2016 Realizaciya programmnogo kompleksa prognozirovaniya urovnya regional`nogo e`nergopotrebleniya [Implementation of software system to forcast regional energy consumption demand] *Vestnik Voronezhskogo gosudarstvennogo texnicheskogo universiteta* **12(3)** 41-7 [In Russian]

[15] Kry`sanov V N, Gamburg K S and Ruczkov A L 2014 Prognozirovanie potrebleniya e`lektroe`nergii territorial`ny`mi setevy`mi organizaciyami s ispol`zovaniem metodov nejro-nechyotkix setej [Prediction of electricity consumption by regional network organizations using neuro-fuzzy network methods] *E`lektrotexnicheskie kompleksy` i sistemy` upravleniya* **2** 40-6 [In Russian]

IOP Conf. Series: Materials Science and Engineering **791** (2020) 012028 doi:10.1088/1757-899X/791/1/012028

The method of power transformers and autotransformers windings condition determination

E R Gayfutdinova, V M Bulatova and A R Ilkevich

Department of Electric Stations named after V. K. Shibanov, Institute of Electric Power and Electronics, Kazan State Power Engineering University, Krasnoselskaya Str., 51, Kazan, Republic of Tatarstan, 420066, Russia

E-mail: mirag_mg@list.ru

Abstract. In article advantages and shortcomings of a method of the frequency analysis of a condition of windings of transformers are considered. This control method of a condition of windings of the transformer is based on comparison of the amplitude-frequency characteristics (AFC) of the diagnosed transformer with its amplitude-frequency characteristics in good repair. It is reported that in various frequency ranges it is possible to find manifestations of various defects and bruises of windings of power transformers and autotransformers. Using a pulse frequency analysis method, the amplitude frequency response spectrum is examined in frequency ranges from zero to 2 MHz. Types and places of defects of transformer winding are determined from value and direction of frequency offset of amplitude-frequency characteristic lines. Such types of defects can include changes in the state of windings, violations of the pressing mechanism, displacement of winding turns relative to each other. In order to determine the type and magnitude of the defect, this method creates a mathematical model of the transformer. An important task in the simulation of transformer windings is the connection between strain types and transformer elements.
Keywords: method of the frequency analysis, amplitude-frequency characteristic, transformer winding, defect of a winding.

1. Introduction

Electric equipment of electrical power systems and networks has to be in operating state throughout all serviceable life. The power transformers (PT) are one of electrical power systems and networks major elements, and reliability of power supply of consumers depends on their correct work. In electrical networks of any tension transformers and autotransformers of single-phase and three-phase execution are used. According to GOST 11677-85 and GOST 52719 2007 the manufactured transformers till 2008 have standard endurance of 25 years and after 2008 – 30 years. Now, about 50% of the transformers which are on stream have endurance more than 30 years and developed the standard resource. [1-3].

For life extension of transformers methods of their monitoring and diagnostics are constantly improved. Early detection of defects in the transformer allows to reveal possibility of a contingency situation and to make the decision on a planned stop without interruption of power supply of consumers. On the basis of results of comprehensive examination of the transformer recommendations about the mode of its further operation and volume of the repair work necessary for life extension of the transformer up to 40 years and more are developed. [1].

According to the conducted researches power transformers at power plants and substations are affected by following hazards:

IOP Conf. Series: Materials Science and Engineering **791** (2020) 012028 doi:10.1088/1757-899X/791/1/012028

- turn–on magnetization currents at inclusion of which lead to damage of windings because of development of electric and mechanical transition phenomena;
- the storm and switching overvoltages causing damages of the major and coil insulation at small stocks of electric strength;
- increases in a running voltage because of the noncompensated charging power of power lines which lead to saturation of cores of power transformers;
- short-circuit currents making mechanical impacts on windings. External short-circuits have also larger danger. They can lead to deformation of windings at their insufficient dynamic firmness;
- an overload of power transformers which leads to additional heating and as a result, to decrease in endurance because of an isolation aging;
- the geomagnetic and induced currents arising at the strong geomagnetic storms. These currents proceed on the grounding windings of power transformers and cause saturation of cores. It leads to an increase in nonsinusoidal magnetization currents, to increase in heating of structural elements of power transformers. In this case there is a malfunctioning of relay protection and automatic equipment and indexes of quality of electric energy go down. On extended high-voltage air-lines the magnitude of geomagnetic and induced currents reaches the working values of currents in lines [4-6].

The main methods of diagnosing and the revealed defects of power transformers are given in table 1 [7].

Table 1. The main methods of diagnosing and the revealed defects.

№	Diagnosing method	The revealed defects
1	Measurement of insulation resistance	The strong humidification, pollution
2	Measurement of a complex admittance, dielectric losses and capacity of isolation	Humidification, local destruction by categories, deterioration in characteristics of oil
3	Definition of the absorptive characteristics of isolation	Humidification
4	Definition of physicochemical characteristics of oil	Humidification, aging, overheating, pollution, thermal decomposition of materials
5	The analysis of the gases dissolved in oil	Thermal and electric breakdown of structural elements
6	Measurement of partial categories	Local defects (inclusions), change of distribution of tension on designs, electric destruction
7	Measurement of resistance of current carrying parts	Damage of current carrying elements and switches of devices of regulations of tension
8	Measurement of losses of a no-load operation	Violation of isolation of elements of a magnetic circuit
9	Measurement of resistance of a short-circuit	Deformations of windings, interturn short circuits
10	Measurement of the frequency characteristics of windings	Deformations of windings

2. Materials and Methods

For decrease in damage from the emergency shutdowns of transformers and autotransformers, also for decrease in economic damage from errors of measurements of transformers of tension and current, it is developed and the system of various control of their state is exercised. One of the modern and effec-

IOP Conf. Series: Materials Science and Engineering **791** (2020) 012028 doi:10.1088/1757-899X/791/1/012028

tive control methods of a condition of windings of transformers is the continuous and pulse methods of the frequency analysis (MFA). In various frequency ranges, as a rule, from 10 kHz to 1 MHz it is possible to find manifestation of defects and bruises of windings of power transformers and autotransformers. The changes of the inductive and capacitive characteristics of a winding arising at emergence of various defects of a winding significantly influence an amplitude-frequency characteristic. Modeling of defects is performed if change of winding dimensions and accordingly change of winding electrical parameters are known. The connection between strain types and transformer elements is an important task in the simulation of transformer windings. The main defects of power transformer windings include the following.

1. Winding displacement - When the winding is displaced, the distance between the winding coils changes. At the same time the natural inductance of coils does not change, but their mutual inductance changes. Longitudinal capacity between coils increases as distance decreases and decreases as distance between coils increases.

2. Increase in a winding - leads to increase in length of a winding at the size ΔL that leads to reduction of inductance and longitudinal capacities.

3. Compression of part of turns - reduces the distance between turns or coils, if the winding is completely affected, the whole height of the winding decreases. This results in an increase in inductance values and longitudinal capacitances.

4. Local deformation (crease) of turns to winding axis changes the diameter of turns, which leads to reduction of inductance values and increase of transverse capacitance values.

5. Local deformation (increase) of rounds from a winding axis - the mentioned part of a winding takes the ellipse form, diameter increases in one party and decreases in another, it leads to increase in inductance and reduction of transverse capacitance.

6. Winding part twisting - leads to reduction of diameter of affected winding part, winding size by height does not change, its radial size also remains unchanged. Such a defect results in a decrease in the respective inductance values and an increase in the respective transverse capacitance values.

7. Winding short circuits between turns are the end result of destruction of paper insulation, loss of electrodynamic resistance of windings under the action of operational factors and, first of all, short circuit currents with further breakdown of insulation in the place of deformation. The winding voltage is reduced and the transformation coefficient is increased, as well as the temperature is increased and the winding overheats, the number of turns in the winding is reduced and the ohmic and inductive resistance of the turn is accordingly reduced and the longitudinal capacitance is increased.

8. Ground turn closure - the affected turn acquires zero potential and the ground capacity becomes zero.

9. When the winding is bent, the transverse capacitance and inductance of the substitution circuit are changed by 20%, as a result, it is obtained that resonances are shifted towards higher frequencies, with subsequent appearance of a difference between amplitude-frequency characteristics.

10. Damage of magnetic circuit is simulated by change of successive elements, and curve of amplitude-frequency characteristic is shifted towards higher frequencies.

11. Oil leakage leads to changes in parallel capacity and conductivity, amplitude-frequency response changes towards decreasing frequencies. The effect of the tank and oil on the amplitude-frequency response is more noticeable at a frequency of 10 kHz and more.

The method of the frequency analysis of windings is based on comparison of the amplitude-frequency characteristics of the diagnosed transformer with the amplitude-frequency characteristics removed in good repair. But for successful application of a method of the frequency analysis for the purpose of monitoring of a condition of windings it is necessary to have transformer amplitude-frequency characteristics in operating state. All newly produced power transformers and autotransformers have factory (reference) amplitude-frequency characteristics. For transformers released earlier than 2005, such factory characteristics are largely absent. For primary monitoring of the state of power transformer windings, which do not yet have reference amplitude-frequency characteristics, a number of techniques have been developed: a) comparison of amplitude-frequency characteristics of different phases of the same transformer; b) comparison of amplitude-frequency characteristics of identical phases of different single-type transformers. The method of diagnostics consists in definition of inter-

IOP Conf. Series: Materials Science and Engineering **791** (2020) 012028 doi:10.1088/1757-899X/791/1/012028

vals of frequencies of separate spectral components of an amplitude-frequency characteristic of a winding at accidental variations of electric parameters of efficient power transformers and autotransformers.

The method of the frequency analysis originally called method of low-voltage impulses (LVI) was offered in 1966 (LechandTuminski) for diagnostics of deformation of windings of power transformers [8]. The method of the frequency analysis is more sensing and exact method in comparison with other methods of diagnostics. In 50 - the 60th years the frequency method was generalized on lumped parameter systems of management. In 1978, the Disk and Egvin were among the first who used a method of the frequency analysis for detection of deformation in windings of the transformer [9]. The method of the frequency analysis was included in the standard of the Republic of China in 2004 [Frequency response analysis on winding deformation of power transformers // The electric power industry standard of people's Republic of China. Promulgated on December 14th, 2004, executed from June1st, 2005]. The principle of a method of the frequency analysis is based on use of Impluls-9 installation which includes the laptop with the plateaus established in it for record and processing of signals and a square wave-form oscillator. The software under the name IFRA allows to make calibration, testing of a metering circuit, record of signals, filtration, preservation in the database, comparison of oscillograms, means of their processing and the analysis. A conclusion about existence or lack of deformations in windings is accepted on the basis of the criteria put in the program.

In a method of the frequency analysis two approaches are used: the continuous and pulse [10-12]. Depending on the scheme and group of connection of windings of a power transformer, measurement by method of the frequency analysis is taken in the different ways. Generally windings of low tension are connected in a one-piece triangle. At such connection of windings the source of the testing signal is connected to one of winding conclusions, and the response is removed from other conclusion. At such scheme of connection, all three phases of a winding are excited. In certain cases, at removal of a tank of the transformer it is possible to take measurements at which windings of low tension are disunited [13]. In this case each phase is excited separately. The way of measurements with the disunited winding of low tension demands opening of the transformer that is a heavy work and takes a lot of time. At connection of a winding in a star each phase is excited separately, and the testing signal is given on an entrance of one phase, and the response is photographed from a generic neutral point.

The pulse method of the frequency analysis (IFRA – Impulse Frequency Response Analysis) is that on one of windings of the transformer the short square-wave probing pulse of low tension about 100 ÷ of 600 V and lasting about 1 microsec of (figure 1) moves.

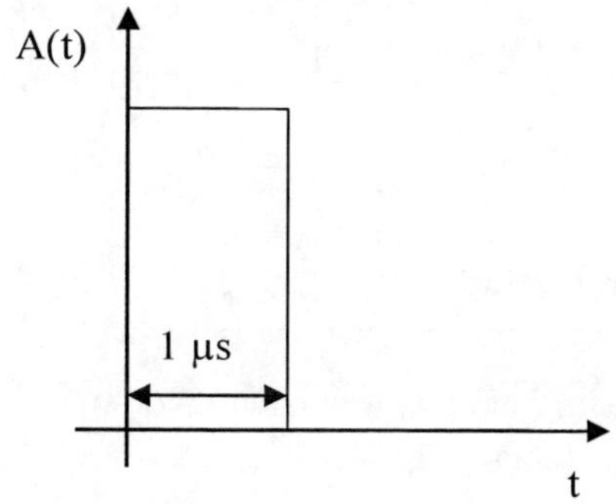

Figure 1. Form of a signal of the Impulse-9 device given on the studied winding.

Impulse-9 device is used in frequency analysis method. Transient process is excited in examined winding. Digitization of amplitudes of input and output signals with different sampling rate is performed (in Impulse-9 device maximum sampling rate is 200 MHz). The ratio of the output signal and the input signal gives a time-dependent value of the transmission coefficient of the winding K(t). Impulse-9 has built-in Fourier transform function of signals depending on frequency [10]. Fourier transform of output signal produces pulse amplitude-frequency response of winding. Since the amplitude of the input signal in the frequency range depends on the frequency, the ratio of the output signal to the input signal, that is, the value of the transmission coefficient, is determined.

IOP Conf. Series: Materials Science and Engineering **791** (2020) 012028 doi:10.1088/1757-899X/791/1/012028

Dependence of transmission coefficient obtained in pulse mode using Impulse-9 device coincides with amplitude-frequency characteristic of winding obtained at smooth continuous change of frequency of input sinusoidal signal of fixed amplitude. The thus obtained amplitude-frequency characteristic of the winding of the analysed transformer is compared with the amplitude-frequency characteristic of the same winding of the transformer, previously recorded in its serviceable condition. Such amplitude-frequency characteristic recorded at the factory or during operation of the transformer prior to occurrence of the defect is called a normogram. The amplitude-frequency characteristic recorded after the possible occurrence of a defect is called a defectogram. If the correlation coefficient between the normogram and the defectogram is greater than 0.96, the transformer is considered usable. If the correlation coefficient is less than 0.96, it is considered that the defects in the transformer require immediate elimination, and the transformer must be taken out for repair. The type and number of defects within the frequency analysis method are not determined. In order to link the defect in the winding to the change in the amplitude-frequency spectra, it is necessary to create a theoretical model of the winding with distributed parameters (type of long line model).

Recently, works have appeared in which attempts have been made to apply standard spectroscopy methods to the analysis of amplitude-frequency characteristics of transformers in order to increase the information value of the frequency analysis method and to obtain information on the type of defect, its value and position in the winding before the beginning of repair with opening of the transformer tank.

The procedure of measurement by a pulse method is explained by the key diagram represented in figure 2 and the scheme of connection of devices in figure 3. After start-up of the program, control of parameters of measurements and selection of the command "Start" is launched the generator of the probing impulses via the control unit connected to an entrance of the Start-up generator. At this Analog-to-digital converter (ADC) it is automatically transferred to a waiting mode of record of signals. From the output of the Impulse – 9 generator the square-wave probing pulse passes on the internal connector and a measuring cable on an entrance of the coordinating (connecting) block installed on a transformer tank cover.

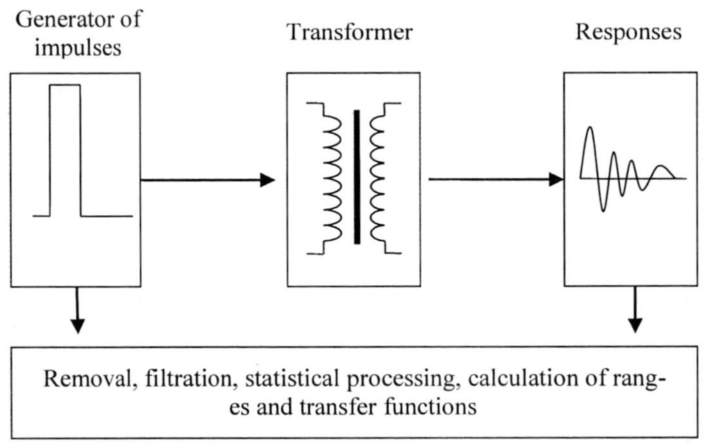

Figure 2. The basic scheme of measurement by a pulse method of the frequency analysis.

Further, from the output of the coordinating block via the external connector connected to one of inputs of the transformer (input of a neutral, input of a winding or connection of inputs of windings of one tension) by means of a clip, the probing impulse enters on an object winding. From the same input the signal via other external connector coordinating the block, a measuring cable, the internal connector and a divider of K1 (1:20) enters on the 1st ADC channel. Reaction of windings (responses) to influence of the square-wave probing pulse from inputs of three phases sequentially enter via the external connector coordinating the block and a measuring cable on the 2nd ADC channel – through a divider of K2 (1:5) or immediately on ADC.

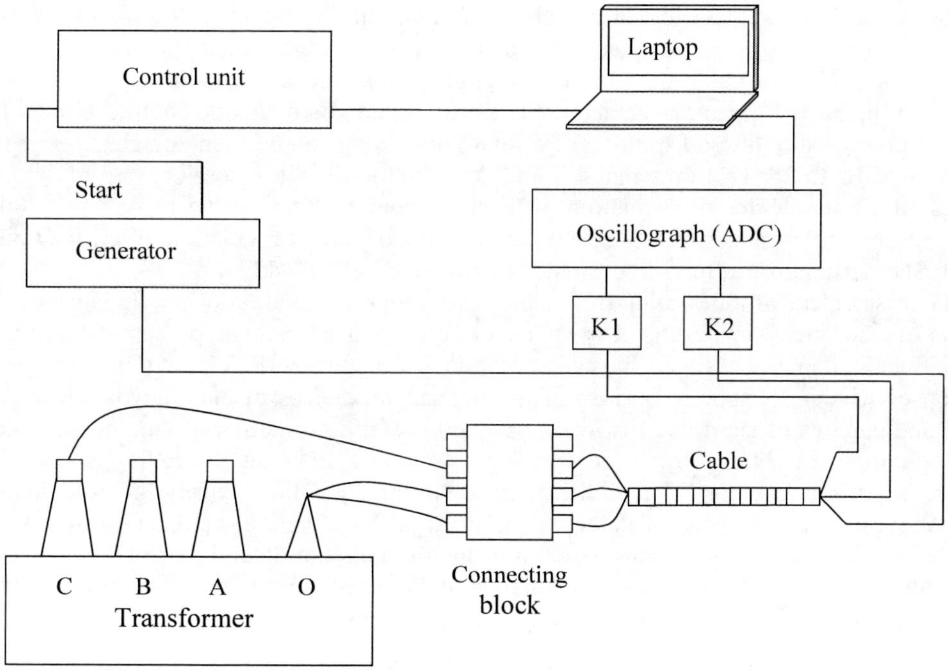

Figure 3. Scheme of connection of devices at measurement by a pulse method of the frequency analysis.

The Impulse-9 installation represented in figure 3 includes:

- square-wave pulses generator;
- laptop;
- to ADC plateau;
- control unit;
- the software (IFRA) providing the procedure of measurements, processing, the analysis and assessment of results of diagnostics.

Also are included in the installation package the calibrator, dividers, cables, connectors coordinating the block, the instruction (user's guide).

3. Results

Failure of windings of transformers makes in the sum about 44% of total number of damageability. Ranges of amplitude-frequency characteristics change depending on the changes of parameters of inductance and capacity of a winding caused by change of sizes of windings of the transformer. It does a method of the frequency analysis sensitive to internal parameters of a winding. Model operation of defects is made if change of the sizes of a winding and respectively change of electric parameters of winding dimensions is known. Communication between types of deformation and elements of the transformer is an important task at model operation of windings of the transformer. For example: – at a winding bend, the cross capacity and inductance of the equivalent circuit for 20% changes, as a result it turns out that resonances are displaced towards the highest frequencies, with the subsequent emergence of a difference between amplitude-frequency characteristics; – damage of a magnetic circuit is modelled by change of serial elements, and the curve of an amplitude-frequency characteristic is displaced towards the highest frequencies; – oil leak, as shown in [10, 14], leads to changes of parallel capacity and conduction, the amplitude-frequency characteristic changes towards decrease of frequencies. Influence of a tank and oil on an amplitude-frequency characteristic more considerably with a frequency of 10 kHz and more.

IOP Conf. Series: Materials Science and Engineering **791** (2020) 012028 doi:10.1088/1757-899X/791/1/012028

4. Discussion

According to authors [15], vertical displacement of a winding is found with a frequency of 100 kHz, and radial deformation – with a frequency up to 10 KHz and is higher than 100 kHz.

Authors of work [16] claim that different types of defects lead to change in different ranges of a range of an amplitude-frequency characteristic. Deformation in a magnetic circuit, residual magnetization, interturn short circuits and splitting of chain of a winding can be found on changes in a frequency range from 20 Hz to 2 kHz. In the range from 2 kHz to 400 kHz it is possible to find deformations in windings and in the device of regulation of tension under loading. In the range from 400 kHz to 2 MHz the shift and clamping of wires of a winding in the device of regulation of tension under loading is found. Such assumption allows to estimate location of defect quickly.

In [17] assessment of a deviation of resonances through calculation of coefficient of a percentage deviation of resonances for finding of the place of defect is used. Such approach of localization of defect is defined in the larger ranges that can cause monitoring errors. It is known that destruction or deformation of windings happens because of mechanical influences of electrodynamic forces at rated currents and currents of short-circuits. It is possible to refer deformation of windings to internal damages of windings of the transformer (compression, drowning, twisting, a swelling, shift etc), interturn short circuits, partial categories, humidification, pollution by solid impurity, gaseous inclusions and others. Above-mentioned types of defects come to light by means of application of various control systems of a state. In work [18] consideration of longitudinal and lateral deformations is given in the transformer winding conductor, caused by short circuit current, and the useful ratios allowing to estimate lengthening and shift of conductors in a winding at the known current.

5. Conclusion

It is known that the method of the frequency analysis, as a rule, is the comparative method applied to detection of defects of windings of transformers. In case of lack of input datas of the transformer, results of measurement of an amplitude-frequency characteristic are compared to data of other phases (interphase comparison) or with data of the same transformers. A number of shortcomings, such as distinction of amplitude-frequency characteristics of extreme windings and the winding which is in the middle of a magnetic circuit, various serviceable lifes of the same transformers, or distinction of issue dates has this approach. These distinctions influence ranges of amplitude-frequency characteristics and complicate applicability of a method of the frequency analysis. To lower the listed above shortcomings, it is necessary to develop techniques of definition of frequency ranges which can serve as reference data for the same transformers. In work [10] the spectral method of processing of observed datas which demands initial model operation of the experimental range and finding of characteristics of lines is used. For increase in effectiveness of monitoring of a condition of windings of transformers the method of the frequency analysis in [10] offered a technique of decomposition of a range of an amplitude-frequency characteristic on separate lines of a range. Such technique allows to determine types and places of defects of a winding by the size and the direction of shift of frequencies of lines. To determine type and size of defect by this technique it is necessary to create mathematical model of the transformer. As each type of the transformer has an individual amplitude-frequency characteristic, the model has to be made for each of the existing types of transformers. Usually, when using a pulse method of the frequency analysis the range of an amplitude-frequency characteristic is investigated in frequency ranges from zero to 2 MHz. Some researchers work in higher frequencies (up to 10 MHz) [19].

In work it is reported about a method of definition of a condition of windings of power transformers and autotransformers on the basis of amplitude-frequency characteristics of these transformers. Windings of a power transformer are presented to models by the long line with various number of RLC elements depending on the frequency of a signal and length of a winding. On shift of resonance lines of a range it is possible to judge type and size of defect of a winding of the transformer. It is possible to refer changes of geometry of windings as a result of influence to such types of defects round short circuits, violations of the mechanism of pressing, drowning of rounds, shift of rounds of a winding relatively each other which can lead to loss of dynamic stability and emergency subsequently.

References

[1] L`vov M Yu 2010 Analiz povrezhdaemosti silovy`x transformatorov napryazheniem 110 kV i vy`she [Analysis of damageability of power transformers of 110 kV and above] *E`lektrichestvo* **2** 27-31 [In Russian]

[2] Abdallah J 2009 Using Frequency Response analysis (FRA) in transformers internal fault detection *WSEAS Transactions on Power Systems Issue* **9(4)** 297-306

[3] Wang M and Vandermaar A J 2002 Review of condition assessment of power transformers in service *IEEE Electrical Insulation Magazine* **6** 12-25

[4] Elovaara J 2007 Finnish experience with grid effect of gic's *Space Weather* Ed J Lilensten (Dordrecht, Netherlands: Springer) pp 311-26

[5] Kappenman J G 2010 *Geomagnetic storms and Their Impact on the U.S. power grid* (Goleta, CA: Metatech Corporation)

[6] Pulkkinen A et al 2010 Solar shield: Forecasting and mitigating space weather effects on high-voltage power transmission systems *Natural Hazards* **53(2)** 333-45

[7] Sanchez J 2011 *Assistant to the diagnosis of defects of the transformers of power* [Thesis to get the rank of Docteur, supported publicly on June 21st, 2011] (Grenoble: University of Grenoble)

[8] Lech W and Tuminski L 1966 Detecting transformer winding damage by the low voltage impulse method *Electrical review* **179(21)** 768-72

[9] Al-Amar E, Karady G G and Hevia O P 2007 Improved technique for fault detection sensitivity in transformer maintenance test *IEEE Power Eng. Society, General Meeting* (Tampa) pp 505-8

[10] Il`darxanov R G 2011 *Sovershenstvovanie metoda ocenki sostoyaniya obmotok silovy`x transformatorov na osnove ix chastotny`x xarakteristik* [Perfecting of a method of assessment of a condition of windings of power transformers on the basis of their frequency characteristics] (Kazan: KGEU) [In Russian]

[11] Islam A, Khan S I and Hoque A 2011 Detection of mechanical deformation in old aged power transformer using cross correlation co-efficient analysis method *Energy and Power Eng.* **3(4)** 585-91

[12] Wang M, Vandermaar A J and Srivastava K D 2004 Transformer winding movement monitoring in service – Key factors affecting FRA measurements *IEEE Electrical Insulation Magazine* **5** 5-12

[13] Vikash M, Sharlinprija K and Ilampoornan M K 2012 Transformer fault detection by frequency response analysis *Int. J. of Computer Science and Electrical Eng.* **2** 26-31

[14] Kalpana P, Narottam D, Abu-siada A and Syed I 2013 Power transformer winding fault analysis using transfer function *Australasian Universities Power Eng. Conf.* (Hobart, Australia) pp 1-3

[15] Hidalen H K, Brede A P and Lundgaard L 2001 Transformer winding diagnosis using the transfer function approach *Nordic Insulation Symposium Stockholm* (Stockholm, Sweden) pp 113-120

[16] Gonzalez A A, Jacobo J C and Ruiz S S 2012 Incipient failure detection in power transformers through *Frequency Sweep Response Proceedings of the World Congress on Eng. and Computer Science* vol 2 (USA, San Francisco) pp 4971-8

[17] Barzegaran M R, Mirzaie M and Shayegani Akmal A 2009 Frequency response analysis in power transformer for detection of winding short-circuit using *Quasi-static Finite Element and Circuit-based Method World Applied Sciences J.* **7(8)** 1006-15

[18] Isaev Yu N, Elgina G A and Lavrinovich V A 2013 Opredelenie deformacij provoda obmotok transformatora sobstvenny`m magnitny`m polem [Definition of deformations of a wire of windings of the transformer internal magnetic field] *Izvestiya Tomskogo Politexnicheskogo Universiteta* **4** 176-9 [In Russian]

[19] Ilampoornan M K and Vikash M, 2012 Transformer fault detection by frequency response analysis *IOSR Journal of Electrical and Electronics Engineering* **1(4)** 27-32

IOP Conf. Series: Materials Science and Engineering **791** (2020) 012029 doi:10.1088/1757-899X/791/1/012029

Two-point measurement approach to diagnostics of partly damaged insulator chains in high-voltage overhead lines

V E Kachesov[1], A Ju Shutovich[1] and E T Kitova[2]

[1]Department of High voltage equipment and electrophysics, Novosibirsk State Technical University, 20 Prospekt K. Marksa,, Novosibirsk, 630073, Russia
[2]Department of Foreign languages, Novosibirsk State Technical University, 20 Prospekt K. Marksa,, Novosibirsk, 630073, Russia

E-mail: shutovich.a@gmail.com

Abstract. A system for location of fault insulator strings is proposed. Insulation failure in a defective isolator results in the occurrence of a significant current discharge in the circuit, which achieves some amperes and whose main own oscillation frequency is in the range of 15-30 MHz. The exact location of the defective string of insulators can be determined by aerial inspection with two small aircraft equipped with synchronized oscillographs which record high-frequency electromagnetic pulses. Processing of the recorded waveforms with geo tags makes it possible to locate a defective insulator string. The selection of the useful signal from the recorded waveforms is performed using wavelet analysis based on the main antenna characteristics. It was found that the accuracy of the location depends to a large extent on the characteristics of the instruments of the satellite exact time system.

1. Introduction

The modern world puts particularly high reliability requirements to the industries that are of great importance for the normal industry functioning and human life. The power industry is undoubtedly the basic industry which should be highly reliable. Insufficient electricity supply resulted from the disruption of the high-voltage overhead lines (HV OHL) can lead to large economic losses in industry. The analysis of OHL failures carried out in [1,2] shows that line isolation faults accounts for a significant part of all failures, therefore, timely and accurate location of the defects is an urgent problem. The physical reason for the occurrence of such a defect is a damaged glass insulator. Due to the influence of external factors such an insulator can lose its insulating properties, however, not completely. In practice, it is customary to call such insulators zero insulators, assuming that the remaining electric strength of this element is much less than the initial one. But the remaining reduced electric strength in the situation when the voltage distribution across the string of insulators is uneven, creates conditions when the damaged element can be blocked by a discharge along the surface of the remaining insulating part.

Currently, methods for inspecting overhead lines are divided into several groups [3-8]:

- visual and video inspection;
- thermal imaging control;
- acoustic monitoring;
- control by ultraviolet radiation;
- laser scanning of the line;
- analysis of electromagnetic radiation in the high frequency range.

IOP Conf. Series: Materials Science and Engineering **791** (2020) 012029 doi:10.1088/1757-899X/791/1/012029

In this work we propose a method based on the last group of methods - synchronous measurement and processing of high-frequency electromagnetic radiation at two points during a flight over an overhead power line using various small aircraft. For example, measurement information collection tools may be placed on board unmanned aerial vehicles (UAV).

2. Materials and Methods

2.1. Discharge processes caused by a defective insulator

When the remaining electrical strength of the damaged insulator is small, the applied voltage is sufficient to cause a discharge along the surface of the remaining dielectric. This overlap is characterized by a high rate of current change, the frequency of which is determined directly by the current loop. To obtain an accurate measurement of the amplitude of the discharge current, the OHL support must be represented as line sections with different wave impedances for given frequencies, but such a solution of the problem entails large and complex calculations by means of the programs for calculating electromagnetic transient processes such as EMTP [9] or SPICE [10], which is not always justified. In this case we can use the engineering methodology for assessing the amplitude and frequency of the emerging current. It will be shown below that such an approach satisfies the accuracy of the process prediction at the moment of discharge. In this approach, the OHL support is substituted for a lumped inductance (L_{tow}), proportional to the sum of its element lengths - traverses and racks. The discharge current is limited by the phase wave impedance of the line at high frequency, the characteristic impedance of the discharge circuit and the tower footing resistance at high frequency. The current flow path and the equivalent circuit of the power line and discharge circuit are shown in Figure 1.

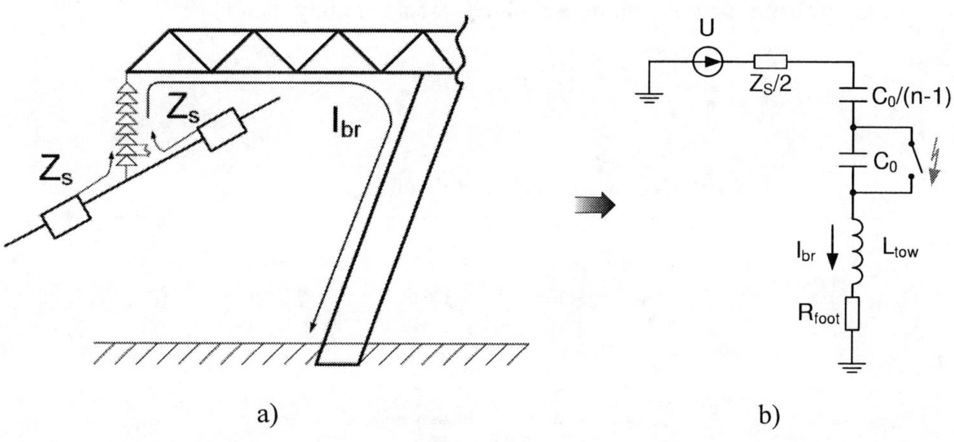

Figure 1. The path of the HF discharge current (a) and the equivalent circuit for estimating the discharge current (b) (the support is represented by a lumped inductance).

At high frequencies, due to the weak penetration of the electromagnetic field into the earth (and, therefore, assuming that the soil resistance is $\rho \to 0$), the surge impedance of the power line conductor can be approximately calculated as

$$Z_S \cong 60 \ln\left(\frac{2H}{r}\right), \quad r = \begin{cases} r_0, n=1 \\ \left(r_0 n r_b^{n-1}\right)^{1/n}, n \geq 2 \end{cases}, \quad r_b = \frac{a}{2\sin(\frac{\pi}{n})},$$

where H - the average conductor suspension height, r - the radius of the conductor (r_0 when n = 1), or the radius of the equivalent conductor in case of a bundle conductor ($n \geq 2$); a - separation between sub-conductors; n - a sub-conductors number.

IOP Conf. Series: Materials Science and Engineering **791** (2020) 012029 doi:10.1088/1757-899X/791/1/012029

The amplitude of the high-frequency discharge current, taking such a simplified model, is defined as

$$I_{\text{br.m}} = \frac{U_{\text{br}}}{\beta L_{\text{tow}}} e^{-\delta\pi/(2\beta)} \quad (1)$$

where $\delta = \dfrac{\dfrac{Z_C}{2} + R_{\text{foot}}}{2L_{\text{tow}}}$ and $\beta = \sqrt{\beta_0^2 - \delta^2}$ – oscillations decay factor and angular frequency, $R_{foot} \cong 100$ Om – tower footing resistance at a high frequency, $\beta_0 = \left(\dfrac{L_{\text{tow}} C_0}{N_{\text{ins}} - 1}\right)^{-0.5}$.

On HV OHL and especially EHV OHL, the discharge current is limited, to a greater extent, by the characteristic impedance of the discharge circuit (Z_C), which, in the general, is determined by the formula:

$$Z_C = \sqrt{\frac{L_{\text{tow}}(N_{\text{ins}} - 1)}{C_0}}.$$

With increasing operating voltage of the power transmission line, the number of insulators in the string (N_{ins}) increases, which leads to an increase in the characteristic impedance of the circuit Z_C. The high-frequency line impedance (Z_S) and the tower footing resistance at high frequency (R_{foot}) have a smaller effect on the current amplitude. Due to the relatively small attenuation ($Z_C \gg (Z_S/2 + R_{\text{foot}})$), the amplitude of the discharge current can be estimated less accurately than (1) as

$$I_{\text{br.m}} \cong \frac{U_{\text{br}}}{Z_C} e^{-\delta\pi/(2\beta)}. \quad (2)$$

Table.1 Characteristics of discharge circuit

U_{rated}, kV	110	220	330	500	750
Z_S, Ohm	450	460	360	330	315
Z_C, Ohm	1800	2800	3700	4600	6500
N_{ins}	7	13	19	26	38
L_{tow}, μH	19	23	26*	30*	40*
$I'_{br.m}$, A	4.8	3.3	2.6	2.1	1.5
F, MHz	15	19	22	25	26
$I'_{\text{br.m}} F$ (A·MHz)	79	65	58	50	37

*supports of the portal type (portal structure), the conductor is split; $r_0 = 0.0104$ m; $U_{br} = 10$ kV

Table 1 shows the basic input data for the HV OHL, the calculated frequencies of natural oscillations and the approximate amplitudes of the discharge current, depending on the class of line voltage. In the calculations it was assumed that the intrinsic capacitance of the insulator is $C_0 = 35$ pF, and the linear inductance of the support, depending on its design, is 0.8 - 1 $\mu H/m$ (a lower value for the portal type supports).

When analyzing the results presented in the table, it can be noted that the product of the initial discharge current ($I_{br.m}$) and the frequency $F = \beta/(2\pi)$ differs relatively little (approximately twice) from a change in the line operating voltage. This suggests that the EMF induced in the electromagnetic (EM) antenna of the device fixing the HF radiation will not depend heavily on the operating voltage of the high-voltage line.

IOP Conf. Series: Materials Science and Engineering **791** (2020) 012029 doi:10.1088/1757-899X/791/1/012029

2.2. Location of the damaged string

The methodology for determining a damaged element (an insulator in a string) is described in [11-14] and consists in recording EM radiation during a HV OHL inspection by aerial aircraft. The method involves the use of two registrars located at a certain distance from each other. When performing aerial inspection, they synchronously (through time synchronization receivers) register signals coming from the defective section (string of insulators) of the high-voltage power transmission line. During the flight, the additional data on their location is collected and registered, and for each captured waveform there is a reference to the place - a geographical tag. Next, the signal is processed using software that extracts pairs of waveforms taken at one moment in time from the data collected. When having the geo data for each waveform, it is possible to find out the distance between the two registrars and then analyze the recorded signals in order to accurately locate the defective string.

Table 1 shows the expected signal frequency (F), which provides useful information about the presence of a fault. When the emitted frequency is known as well as the characteristics of the receiving antenna and by using the methods of representing the signal in the frequency and time domain (window Fourier transform, wavelet transform), it is possible to find a signal that meets the search criteria. Such a criterion is the sum of the pulse occurrence times (Δt_1 and Δt_2) from the moment of simultaneous start of horizontal oscilloscope scans. This sum is the travel time of the electromagnetic wave to cover the distance between the two sensors - τ. It should be equal to the calculated wave travel time by using geotags. If this condition is met, then we can say that the defect is between two sensors and its coordinate is calculated by the expression

$$x_d = x_1 - \Delta t_1 v, \quad (3)$$

where x_1 is the "x" coordinate for the first sensor (i.e., the UAV), v is the propagation velocity of the EM wave (for air, $v \cong 300$ m/us). An example of such a waveform in a simplified form is shown in Figure 2, *b*. It shows that there are two such signals caused by the breakdown of a zero/damaged insulator. By the delay time Δt_1=1.7 μs the coordinate of the fault on the line is determined.

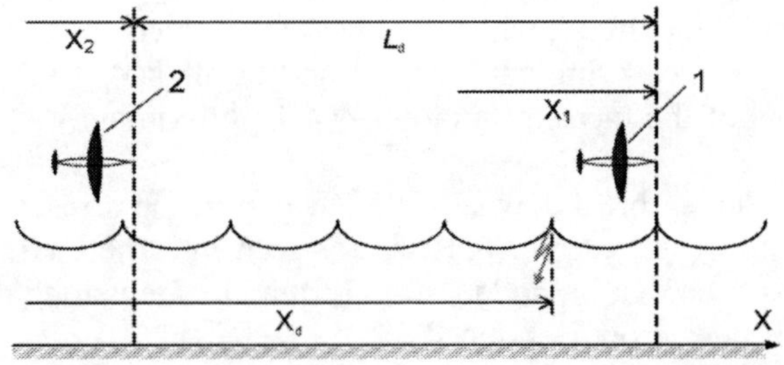

a)

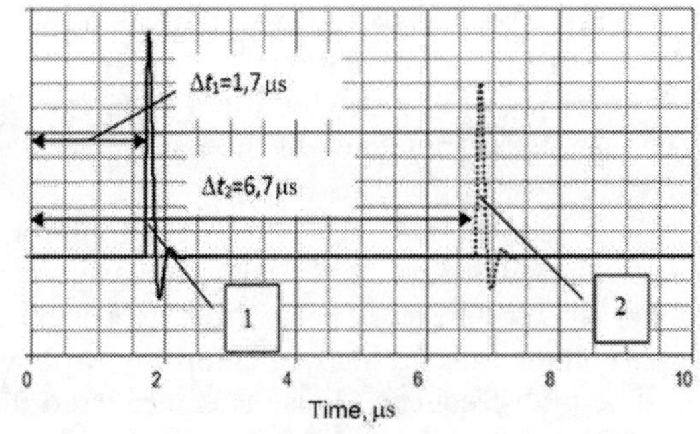

b)

Figure 2. The position of the aircraft during the aerial OHL inspection (a) and the pulses of electromagnetic sensors (b).

3. Results

To confirm the operability of this method, an experiment was conducted on the active 10 kV overhead line, on which an HF discharge was initiated. For these purposes, a cap insulator was suspended on a phase conductor, and an air gap with a breakdown voltage of 6 to 8 kV was connected to it in series (Figure 3). It acted as a defective insulator and, during breakdowns, generated high-frequency current pulses into a power line.

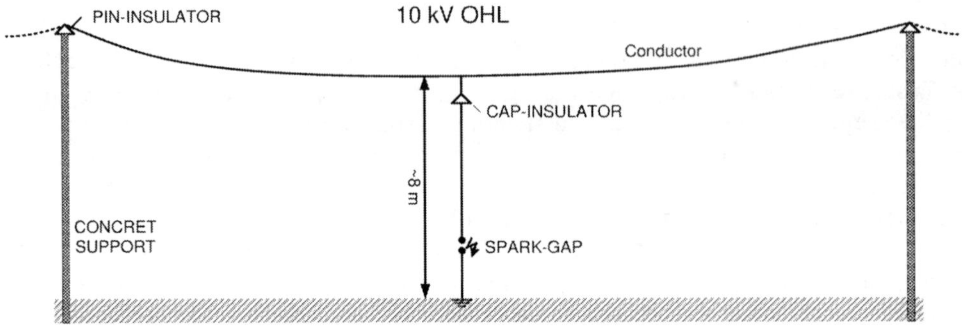

Figure 3. The scheme of the experiment.

Assessment of the current amplitude (I_m) can be performed by the expression (1), in which, instead of the support inductance, the inductance of the conductor connecting the linear wire to the ground electrode through an insulator and a spark gap should be taken. According to (1), for a conductor 8 m long, $C_0 = 35$pF, $H = 8$ m, $R_{foot} \cong 200$ Om and $r_0 = 6.9$ mm, the current is in the range $I_m = 6.4$ - 8.5 A. Due to the absence of high-quality artificial ground electrode in the middle span, the resistance R_{foot} is a priori assumed to be of the specified value.

A current pulse flowing into the OHL conductor, as it spreads to both sides of the spark gap, induces a high-frequency electromagnetic field. During the pulses generation, an electromagnetic sensor was moving along the OHL, in the form of a square loop antenna with a side of 200 mm, located at a 1 m height from the surface of the earth. The signal induced in the sensor was recorded by a digital storage oscilloscope with a bandwidth of 100 MHz. The signal was recorded at several points: 200, 800, and 1000 meters from the HF radiation source. In addition, to simulate the real distance between the overhead power line and the aircraft, the sensor with the recording oscilloscope was separated by 35 meters from the line. The received signals were processed to determine the possibility of identifying points in time corresponding to the arrival of signals caused by breakdowns in the arrester. Waveform processing was carried out using wavelet analysis.

Analysis of the recorded signals using wavelets made it possible to accurately identify the main frequencies in the oscillations that occur in the measuring circuit, which included the inductance of the loop antenna (L_{loop}) and the input capacitance of the measuring cable (C_{cab}) and oscilloscope (C_{osc}). The calculated frequency of the natural oscillations of the measuring circuit was approximately = 16 MHz.

The frequency of the main oscillations formed during breakdowns of the air gap in the arrester is relatively small - about 8 MHz. However, the front of the current pulse introduced into the OHL is rather steep (~ 15 ns). Therefore, it is possible to assume that in the measuring frame located along the high-voltage line, an EMF of complex shape with a steep front is induced.

The high-frequency pulse was measured during a half-period of industrial frequency (10 ms) during which a breakdown of the discharge gap occurred. The total recording length was 10 million points, the time sampling step was 1 ns. Figures 4 shows fragment of oscillogram recorded at 200 m distance from the radiation source, and its spectrogram obtained by wavelet analysis [15]. Oscillations with a frequency of natural oscillations f_0 have the highest energy, they are exactly determined through wavelet analysis. The removal of the electromagnetic sensor at 35 m from the axis of the OHL path led to a weakening of the signal by approximately 10 dB. Without special signal processing algorithms for oscillograms it can be determined that the arriving time of the measuring pulse can approximately be considered the time of the first oscillation period occurrence with the fundamental frequency f_0. Based on this fact, the accuracy of fixing the time of pulse arrival to an electromagnetic sensor can be estimated at the period level of this frequency, i.e. $\sigma_{\Delta t} \cong 1/f_0$. For the loop antenna which was used $\sigma_{\Delta t} \cong 60$ ns, i.e. the geometric error is about 18 m. The average distance between adjacent HV OHL supports (span length) is about ~ 300 m, the travel time by the electromagnetic wave for this distance, respectively, is τ_{span}=1 *μs*. The estimated time error $\sigma_{\Delta t}$ is much less than the travel time ($\sigma_{\Delta t} << \tau_{span}$). Therefore, the assignment of a defective string of insulators to a specific HV OHL/EHV OHL support is carried out without error. However, it is necessary to take into account the limited accuracy of synchronization of the oscillograph system built on a satellite exact time system. The accuracy of the time synchronization system is about 100 ns, which results in the real maximum error in the location of defective insulation increase up to ~ 50 m.

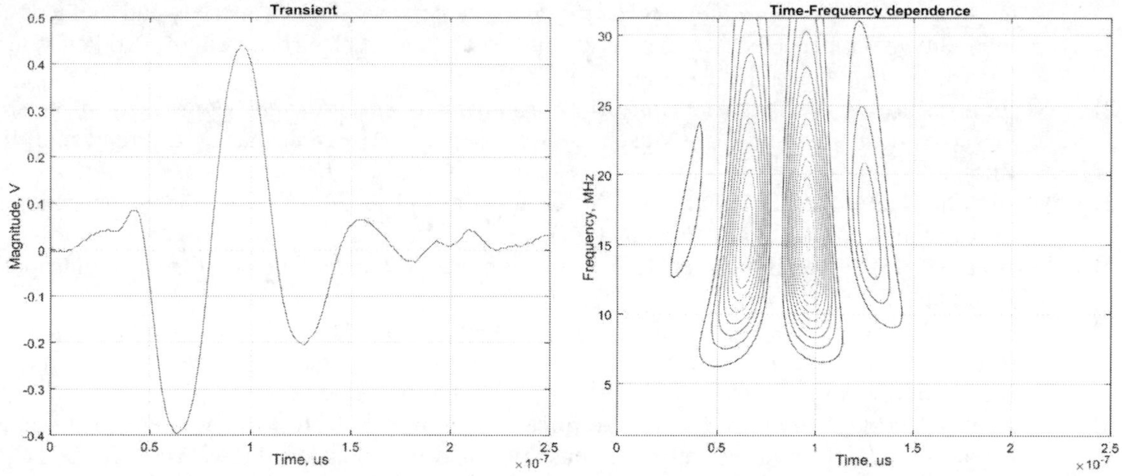

Figure 4. A waveform fragment for the voltage induced in the electromagnetic sensor (a) and the spectrogram obtained by using wavelet analysis (the distance from the sensor to the arrester is 200 meters, the location of the sensor is under the high-voltage line.

4. Discussion

It should be noted that the object where full-scale experiments on the discharge pulses generation were conducted and their registration at some distance from the radiation source (on an overhead power transmission line 10 kV) differs from the planned field of application - HV and EHV overhead power lines. The difference in the high-frequency signal propagation conditions can be explained by the fact that the 10 kV OHL supports are often located at small distances from each other - the distance between them is approximately 60 m. When high frequency pulse comes to each support with lumped capacity insulator on the reinforced concrete support having finite inductivity and earth resistance the pulse refraction occurs. On the HV and EHV lines, the distance between the supports is considerably greater (up to 300-350 m) and the signal refraction at the nodes is expected to be smaller. This has a positive effect on the signal propagation conditions and the possibility for its measuring. To estimate the signal propagation range of a similar (high-frequency / pulse) origin, it is planned to conduct an experiment on an overhead transmission line of 110 kV. It should also be noted that the developed method for locating defective strings can be used in any vehicles moving along the overhead lines routes.

5. Conclusion

1. It has been theoretically shown that during the damaged insulators breakdowns, a sufficiently large high-frequency current of Amperes (1.5 ... 5 A) is introduced into the line depending on the operating voltage of the high-voltage line.

2. Experimentally, on the active 10 kV line, the possibility of detecting high-frequency electromagnetic radiation similar to breakdowns of defective insulators is shown. The current level of measuring equipment development is quite sufficient for recording electromagnetic signals from defective insula-

IOP Conf. Series: Materials Science and Engineering **791** (2020) 012029 doi:10.1088/1757-899X/791/1/012029

tors at a distance of ~ 1 km. When using specialized low-noise amplifiers, this distance can be increased.

3. Determination of the useful signal associated with the moment of the defective insulator breakdown from the recorded waveforms is reliably performed using wavelet analysis.

4. The error in the location of defective insulation using the method described depends mainly on the characteristics of the exact time devices installed on aircraft.

References

[1] Vasil'ev A P 2014 *Metody i sredstva obespechenija nadezhnost'ju i bezopasnost'ju jelektricheskih setej I ustanovok jelektrojenergeticheskih sistem* [Materials and methods for reliability assurance, electrical networks, and equipment of electrical system] (St. Petersburg: Izd-vo Politehn. Un-ta) [In Russian]

[2] Nepomnyashhij V A 2013 *Nadezhnost' oborudovaniya energosistem* [*Reliability of electric power systems equipment*] (Moscow: Elektroenergiya. Peredacha i Raspredelenie) [In Russian]

[3] Whitworth C C, Jones D I, Duller A W G and Earp G K 2001 Aerial video inspection of overhead power lines *Power Eng. J.* **15(1)** 25–32

[4] Marshall M 1999 Aerial infrared line inspection *Rural Electric Power Conf.* (Indianapolis, USA) pp A3-1-2

[5] El-Hag A, Mukhopadhyay S, Al-Ali K and Al-Saleh A 2017 An intelligent system for acoustic inspection of outdoor insulators *3rd Int. Conf. on Condition Assessment Techniques in Electrical Systems, CATCON 2017* (Rupnagar, India) pp 122-5

[6] Arbusov R and Ovsyannikov A 2000 Application of optical-electronic flaw detector to detect damaged and defective insulators in substation and transmission lines *Proc. CIGRE SC 33 Conf.*

[7] Golinelli E, Musazzi S, Perini U and Pirovano G 2009 Laser Based Scanning System for High Voltage Power Lines Conductors Monitoring *20th International Conference and Exhibition on Electricity Distribution, CIRED 2009* (Prague, Czech Republic) pp 784

[8] Khawaja A H, Huang Q and Khan Z H 2017 Monitoring of Overhead Transmission Lines: A Review from the Perspective of Contactless Technologies *Sensing and Imaging* **18** 24

[9] Dommel H 1986 *Electromagnetic Transients Program Reference Manual* (Portland, United States: Bonneville Power Administration)

[10] Mahseredjian J, Dennetière S, Dubé L,Khodabakhchian B and Gérin-Lajoie L 2005 On a new approach for the simulation of transients in power systems *Electric Power Systems Research* **77(11)** 1514-20

[11] Gole A M, Nayak O B, Sidhu T S and Sachdev M S A 1996 Graphical Electromagnetic Simulation Laboratory for Power System Engineering Programs *IEEE Trans. on Power Systems* **11(2)** 599-606

[12] Banzhaf W 1989 *Computer aided circuit analysis using SPICE* (Upper Saddle River, United States: Printice Hall)

[13] Shutovich A Yu and Kachesov V E 2016 Lokaciya defektov v linejnoj izolyacii vozdushny`x linij po dvum izmereniyam s pomoshh`yu letatel`ny`x apparatov [Location of defects in overhead transmission lines insulation by two-point measurement made by aircraft] *Elektroenergetika Glazami Molodezhi-2016* (Kazan: KGEU) vol **1** pp 262-5 [In Russian]

[14] Messina A R 2009 *Inter-area Oscillations in Power Systems* (New York: Springer-Ferlag)

[15] Sadowsky J 1994 The continuous wavelet transform: A tool for signal investigation and understanding *Johns Hopkins APL Tech. Dig.* **15(4)** 306-18

IOP Conf. Series: Materials Science and Engineering **791** (2020) 012030 doi:10.1088/1757-899X/791/1/012030

Electric Grids Technical Evaluation Method based on their Failure Probability

B S Kompaneets and D A Boyarkov

Department of Production and Household Electrification, Polzunov Altai State Technical University, 46 Lenin pr., Barnaul, 656038, Russia

E-mail: dmitrij.bojarkov@gmail.com

Abstract. The article provides the substantiation of the necessity to develop a mechanism for determining the electric grid element with the lowest reliability taking into account the equipment operation peculiarities in the Russian Federation. Critical analysis of existing approaches to identifying dilapidated electrical equipment, as well as recommendations of the Ministry of Energy of the Russian Federation on assessing technical condition of electric networks has been carried out. It is proposed to determine the technical condition of electric grids by the value of their failure probability. Failure probabilities mathematical models for large groups of emergency modes are given: insulation failure, failure of current-carrying elements and failure due to mechanical damage, as well as for all electrical equipment on the whole. Criteria and an algorithm for mechanical defects of electric grids elements expert evaluation are presented. The calculation of the failure probabilities for two elements of real electric grid is made. The conclusions is made on the feasibility of replacing one of the considered elements.

1. Introduction

Today the situation in the Russian electric power industry is that a large part of the electric grid infrastructure is very outdated. Moreover, more than 70% of power lines and transmission substations have worked out their physical and moral resources to the utmost. Such electrical grids are characterized by frequent breakdowns resulting in an interruption in power supply which leads to huge material losses, both for electricity consumers and electric grid companies.

In order to reduce dangerous situations occurrence the electricity companies should timely repair electrical equipment. However it is impossible to timely repair all the equipment because of the limited financial resources coming from the electricity tariff [1-2]. Thus, in order to reduce the occurrence of emergency conditions and to minimize their consequences the grid companies must clearly identify the most unsatisfactory technical condition of the equipment [3-8].

At present in Russia the most common method of identifying unreliable electric grids equipment involves carrying out electrical parameters control measurements and their subsequent comparison with threshold regulatory values given in specialized scientific and technical documentation. However this approach to evaluating technical condition of electric grids is difficult to consider completely correct. For example, according to the rules of electrical installations the insulation resistance of cables with voltage up to 1 kV during control measurements should not be less than 0.5 MΩ, otherwise its further use is prohibited. But the existing experience in operating cable lines shows that the cable does not fail for a long time. after this threshold value is reached.

2. Materials and Methods

To evaluate the technical condition of electric networks special automated systems [9-14], based on the expert evaluation method, have been developed. This technique is based on the opinion of highly qualified specialists, the experts in the field of operating electrical networks and the real technical condition of the equipment under study. However, according to the studies [9], with the widespread introduction of such systems, employees of electric grid companies conducting the corresponding evaluation have different qualifications and experience in this field. So the final assessment of the technical condition of electric grids will obviously be subjective and inaccurate. Therefore identifying unreliable equipment based on such estimates will not be possible.

The main body regulating the electric power industry in Russia, the Ministry of Energy of the Russian Federation, has expressed its interest and support for this topic. The Ministry of Energy made recommendations on the further research development direction, namely, that the technical condition of electrical equipment should be determined on the basis of technical diagnostics, statistics of defects and failures in order to assess the equipment wear level and the failures probability, what is not implemented in existing methods. Guided by this recommendation, as an indicator describing the electric grid technical condition we will take the probability of electric grid infrastructure elements failure.

As previously mentioned, in order to evaluate technical condition of the electric grid, electricity companies make control measurements of some indicators specific to the type of electrical installation under consideration [15-20]. In this regard the construction of mathematical models of electrical networks failure probabilities should be carried out on the basis of these values. This will allow to rely on the accumulated statistical base and will not require additional costs for creating new approaches to electrical equipment testing.

Electrical grids elements failure can occur due to a huge number of reasons which, in turn, leads to various options for emergency development [16-19]. However, after analyzing the essence of the physical processes leading to emergency, as well as the measured control electrical values of electrical equipment, three large groups of emergency conditions were identified:

- current-conducting elements failure;
- insulation failure;
- electric network element failure an due to mechanical damage.

Let us consider mathematical models of a separate emergency mode for each of the indicated types and simulate failure probability of live parts of the electric network elements.

3. Results

As an example, let us consider a contact of arbitrary shape, for which P_{break} is the probability of contact breaking. The contact zone has a rather complex shape (Figure 1).

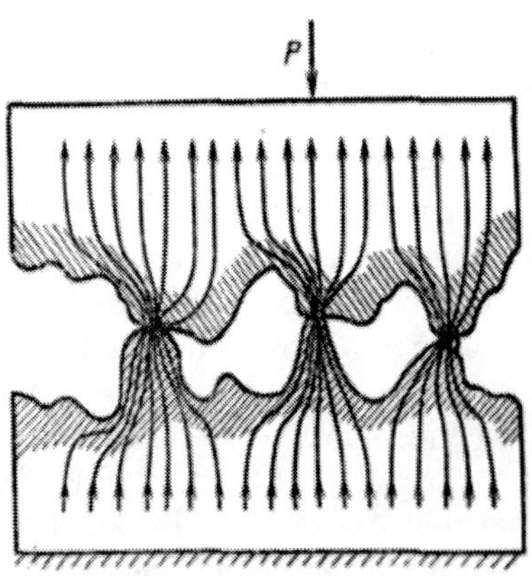

Figure 1. Contact zone form.

IOP Conf. Series: Materials Science and Engineering **791** (2020) 012030 doi:10.1088/1757-899X/791/1/012030

Let us divide the entire contact area into small areas. Then the contact break probability for the considered zone will be determined by the formula:

$$P_{break} = (1 - p_{work})^n. \tag{1}$$

As a result of dividing the contact zone into small areas, we can imagine that the total resistance (R_{cont}) will be composed of the resistances of its parts (r_{cont}) connected in parallel. In this case:

$$R_{cont} = r_{cont} / n \Rightarrow n = r_{cont} / R_{cont} . \tag{2}$$

Let us consider the failure probability of one such part contact. The probability of the area operation capability is that lower, the higher the contact resistance is [18]. The nature of such relationship is very difficult to evaluate. However, any dependence can be represented in the form of a polynomial, and, therefore, the following expression will be true:

$$P_{work} \sim \frac{1}{\left(a_0 r_{cont}^m + a_1 r_{cont}^{m-1} + \ldots + a_m r_{cont}^0\right)}, \tag{3}$$

where m is polynomial degree of the.

Let us consider the probability of contact operability for the area allocated in a certain contact zone. It will depend on a large number of factors: the material state (surface roughness, films formation, the presence of cracks and indentations), the operating conditions of the electric grid (overload, overvoltage, high currents, mechanical stress), and environmental conditions (temperature, air humidity, the presence of aggressive media), etc [21]. All these factors in one or another way affect the probability of contact breaking. And since the research is conducted for some fixed state, we have the opportunity to express this set of parameters in terms of some constant (C).

In this case the expression will be true:

$$P_{work} = \frac{C}{\left(a_0 r_{cont}^m + a_1 r_{cont}^{m-1} + \ldots + a_m r_{cont}^0\right)}. \tag{4}$$

Then the probability of contact breaking will be calculated by the formula:

$$P_{break} = \left(1 - \frac{C}{\left(a_0 r_{cont}^m + a_1 r_{cont}^{m-1} + \ldots + a_m r_{cont}^0\right)}\right)^{\frac{r_{cont}}{R_{cont}}}. \tag{5}$$

We take out the value r_{cont}^m outside the polynomial and replace $r_{cont} = -x$, then we get:

$$P_{break} = \frac{1}{\left(1 + \dfrac{C}{x \cdot r_{cont}^{m-1}\left(\dfrac{a_0 r_{cont}^m + a_1 r_{cont}^{m-1} + \ldots + a_m r_{cont}^0}{r_{cont}^m}\right)}\right)^{\frac{x}{R_{cont}}}}. \tag{6}$$

As the number of partitions increases to infinity, the resistance of the part r_{cont} will tend to infinity, therefore x will also tend to infinity, and the quantity $\left(1 + \frac{1}{x}\right)^x$, when x approaches infinity, will equal

IOP Conf. Series: Materials Science and Engineering **791** (2020) 012030 doi:10.1088/1757-899X/791/1/012030

the exponent «*e*» (the second special limit). Ratio limit $\frac{a_0 r_{cont}^{m} + a_1 r_{cont}^{m-1} + \ldots + a_m r_{cont}^{0}}{r_{cont}^{m}}$ for r_{cont} tending to infinity will be determined by the ratio of the coefficients of the polynomial highest degrees and is equal to a_0. We replace the ratio of two constants $\frac{C}{a_0} = A_{cont}$.

Thus, the expression will be true:

$$P_{\mathrm{break}} = \frac{1}{R_{cont}\sqrt{e^{\frac{A_{cont}}{r_{cont}^{m-1}}}}}. \tag{7}$$

Due to the fact that the probability value should not depend on the method of the contact surface partitions, i.e. it should not depend on the method of the part discretization (how small considered parts are) and it should not depend on r_{cont} of discrete part, therefore, the value r_{cont}^{m-1} should be a constant value, and therefore $m = 1$. Then finally:

$$P_{\mathrm{break}} = \frac{1}{R_{cont}\sqrt{e^{A_{cont}}}}. \tag{8}$$

Note that the derived formula will be valid not only for the contact connections failure, but also for all current conducting parts as a whole, since the nature of the physical phenomena occurring in these objects is identical.

In order to unify the symbols in the constructed mathematical models we shall indicate the contact breaking probability P_{break} as P_{cont} by this value we will generally understand the probability of current conducting parts failure including electrical contact connections.

We will transfer the contact resistance from absolute values to relative onees:

$$R_{cont} = \frac{R_{\mathrm{cont.meas}}}{R_{\mathrm{cont.norm}}}, \tag{9}$$

where $R_{cont\cdot meas}$ is the value of contact resistance measured during the test of electrical equipment;

$R_{cont\cdot norm}$ – the standard value of the contact resistance regulated by Electric Installation Code.

Then formula (8) will have the following form:

$$P_{cont} = \frac{1}{\frac{R_{\mathrm{cont.meas}}}{R_{\mathrm{cont.norm}}}\sqrt{e^{A_{cont}}}}. \tag{10}$$

In this case, the constant A_{cont} is determined as follows:

$$A_{cont} = \ln\left(\frac{1}{\omega \cdot \alpha_{cont}}\right) \cdot \frac{R_{\mathrm{cont.meas}}}{R_{\mathrm{cont.norm}}}, \tag{11}$$

where ω is the failure flow parameter for the considered type of electric grid equipment;

α_{cont} – the proportion of the investigated electrical equipment of a particular type with which an accident occurred due to current conducting parts failure.

The conclusion of the formula for the probability of failure (breakdown) of electric grid element insulation is similar to the conclusion made for current conducting parts. Therefore, we present a mathematical model of insulation breakdown without proof:

$$P_{insul} = \frac{1}{\frac{R_{\text{insul.meas}}}{R_{\text{insul.norm}}}\sqrt{e^{A_{insul}}}}. \tag{12}$$

The constant A_{insul} is defined as follows:

$$A_{insul} = \ln\left(\frac{1}{\omega \cdot \alpha_{insul}}\right) \cdot \frac{R_{\text{insul.meas}}}{R_{\text{insul.norm}}}, \tag{13}$$

where α_{insul} is a statistical indicator showing the share of electrical equipment that has failed due to insulation breakdown.

Various regulatory and technical documents do not regulate the measurement of electrical equipment control parameters to determine mechanical damage. Moreover their very presence, nature and intensity can indirectly indicate a certain technical condition of the electric grid element [22-23]. Since mechanical damage cannot be formalized in the form of electrical quantities (resistance or current) it is possible and advisable to use the method of expert evaluation in this case.

In the conditions of Russian electric grids operation the experts are often operational personnel who can evaluate the technical condition of the same element in different ways or incorrectly. To reduce subjectivity and increase evaluation accuracy a rigid scale of generalized expert evaluation of electrical equipment was developed. An expert from the operating personnel clearly defines the actual state of the electrical equipment by the description given in the developed scale and sets a certain point rating corresponding to the scale. The rating scale has a gradation of 1 to 10 points, where "1" is the worst condition, "10" is the best condition.

We note that we will set three qualitative states of the electric network elements:

- "Good" is the state of the least failure probability;
- "Satisfactory" - the state when failure probability is possible;
- "Unsatisfactory" - the state with high failure probability.

On the basis of the foregoing we will compile generalized criteria and a scale for the expert evaluation of mechanical defects in power grid equipment (Table 1).

A universal algorithm was developed for the expert evaluation of mechanical defects in power grid equipment according to which the operating personnel of the power grid company should make a decision on assessing the technical condition of the considered electrical installation.

Evaluation must be carried out starting with a description of "10". Highest rating this description is a set of the element qualitative characteristics, which can be estimated at 10 points and determine the condition as the best. If the qualitative description of the element state does not match to any of the presented items it is necessary to lower the maximum rating:

- by 1 point if the element state corresponds to a similar paragraph in the description of "6" expert evaluation;
- by 2 points if the element state of corresponds to a similar item in the description of "2" expert evaluation.

After conducting expert evaluation of electrical equipment mechanical defects it is necessary to transform the result of expert evaluation into probabilistic form. In the course of research it was determined that the mathematical model of the probability of these types of damage would be:

$$P_{\text{mech.def}} = A_{mech} \cdot \frac{\sqrt[10]{10.3} - \sqrt[10]{B_{EA}}}{\sqrt[10]{10.3}}, \tag{14}$$

IOP Conf. Series: Materials Science and Engineering **791** (2020) 012030 doi:10.1088/1757-899X/791/1/012030

where $P_{mech.def}$ is the of power grid equipment failure probability due to mechanical defects (damage);
B_{EA} is an expert assessment of the electric grid company operating personnel;
A_{mech} is a coefficient taking into account the statistics of defects in electrical equipment in case of mechanical damage.

Table 1. Generalized criteria and scale of expert evaluation of mechanical defects in power grid equipment

Evaluation Scale of Electric Network Element	Qualitative Description of Electric Network Element Technical Condition Evaluation
10 "good"	1) The fastening of electrical equipment is strong and reliable without changing its position from the original; 2) The surface of the insulators is clean and smooth, has no defects (chips, cracks); 3) In the places of current conducting elements connection there are no traces of corrosion, deformation, mobility; 4) Other elements do not have chips, cracks, deformations, traces of corrosion.
6 "satisfactory"	1) The fastening of electrical equipment is dense, a slight change in position from the original one is possible due to mechanical stress; 2) The surface of the insulators has insignificant chips, cracks, discoloration of part of the surface (at least one sign); 3) At the junction of current conducting elements slight traces of corrosion are observed while there are no other defects; 4) Other elements have insignificant chips, cracks, deformations, traces of corrosion (one sign at least).
2 "unsatisfactory"	1) The fastening of electrical equipment is weak, the housing is movable; 2) The surface of the insulators has significant chips, cracks, there are traces of an electric arc (one sign at least); 3) Blackening, significant deformations are observed at the junction points of current conducting elements, there is mobility in the joints (one sign at least); 4) Other elements have significant chips, cracks, deformations, traces of corrosion (one sign at least).

The failure of electric network element will occur on the whole if at least one of the above types of emergency conditions comes fully out. In the theory of power supply reliability this situation is regarded as buck connection of electric circuit elements. Then the mathematical model of the failure probability of electric network element as a whole will have the following form.

$$P = 1 - (1 - P_{insul})\cdot(1 - P_{cont})\cdot(1 - P_{mech.def}). \quad (15)$$

4. Discussion

As an example of using this technique to evaluate technical condition of the electric network elements and on this base to select the most unreliable element for its repair let us consider two real electric network elements of the same type, for example, two disconnectors. Table 2 shows the results of control measurements of electrical parameters and expert evaluation of mechanical defects of these disconnectors.

Table 2. Information on conducting control measurements of electrical parameters and expert assessment of mechanical defects of disconnectors

Disconnector Name	Measured / estimated parameter	Parameter value
Disconnector # 1	Insulation Resistance, MΩ	4350
	DC Resistance, μOhm	115
	External defects evaluation	«5»
Disconnector # 2	Insulation resistance, MΩ	13600
	DC Resistance, μOhm	107
	External defects assessment	8

We will calculate the probability of failure for Disconnector # 1.
Breakdown probability:

$$P_{\text{insul.discon.1}} = 1 - \frac{1}{\frac{R_{\text{insul.discon.1}}}{R_{\text{insul.discon.norm}}}\sqrt{e^{A_{\text{insul.discon.}}}}} = 1 - \frac{1}{\frac{4350}{1000}\sqrt{e^{0,04846}}} = 0.01108.$$

The failure probability of current conducting parts:

$$P_{\text{cont.discon.1}} = \frac{1}{\frac{R_{\text{cont.discon.1}}}{R_{\text{cont.discon.norm}}}\sqrt{e^{A_{\text{cont.discon.}}}}} = \frac{1}{\frac{115}{175}\sqrt{e^{2.941399}}} = 0.01137.$$

The probability of failure of Disconnector # 1 due to external defects:

$$P_{\text{mech.discon.1}} = A_{mech}\frac{\sqrt[10]{10.3} - \sqrt[10]{B_{\text{def.discon.1}}}}{\sqrt[10]{10.3}} = 0.015552 \cdot \frac{\sqrt[10]{10.3} - \sqrt[10]{5}}{\sqrt[10]{10.3}} = 0.001084.$$

General probability of Disconnector # 1 failure:

$$P = 1 - (1 - P_{insul.discon.1}) \cdot (1 - P_{cont.\ discon.1}) \cdot (1 - P_{mech\ discon.1}) = 0.023384.$$

Let us make a similar calculation for Disconnector # 2.
Breakdown probability:

$$P_{\text{insul.discon.2}} = 1 - \frac{1}{\frac{R_{\text{insul.discon.2}}}{R_{\text{insul.discon.norm}}}\sqrt{e^{A_{\text{insul.discon}}}}} = 1 - \frac{1}{\frac{13600}{1000}\sqrt{e^{0,04846}}} = 0.003556.$$

The failure probability of current conducting parts:

$$P_{\text{cont.discon.2}} = \frac{1}{\frac{R_{\text{cont.discon.2}}}{R_{\text{cont.discon.norm}}}\sqrt{e^{A_{\text{cont.discon}}}}} = \frac{1}{\frac{107}{175}\sqrt{e^{2.941399}}} = 0.008142.$$

The failure probability of Disconnector # 2 due to mechanical damage:

$$P_{\text{mech.discon.2}} = A_{\text{mech.discon}}\frac{\sqrt[10]{10.3} - \sqrt[10]{B_{\text{def.discon.2}}}}{\sqrt[10]{10.3}} = 0.015552 \cdot \frac{\sqrt[10]{10.3} - \sqrt[10]{8}}{\sqrt[10]{10.3}} = 0.000388.$$

General probability of Disconnector # 2 failure:

$$P = 1 - (1 - P_{insul.discon.2}) \cdot (1 - P_{cont.\ discon.2}) \cdot (1 - P_{mech\ discon.2}) = 0.012053.$$

Comparing the values of disconnectors failure probabilities it is clear that the value is greater for Disconnector # 1. Therefore Disconnector # 1 should be repaired.

5. Conclusion

Thus, taking into account the conditions of limited financial resources of electric grid companies the methodology presented in this article allows to determine the priority of replacing dilapidated electrical equipment by failure probability value of the electric network elements, what in its turn will significantly increase the reliability of power supply to consumers and reduce the level of material losses arising from emergency situation. In addition, the developed methodology for determining failure probability of electric grids completely eliminates the critical comments of the Ministry of Energy of the Russian Federation to existing methods.

Acknowledgments
The authors express their gratitude to Polzunov Altai State Technical University Administration, the department staff and mark the special role of the international cooperation department staff in the work on the presented materials.

References

[1] Brown R E and Marshall M 2000 *IEEE Transactions on Power Systems* **15** 887-92
[2] Alvehag K and Awodele K 2014 *IEEE Transactions on Power Systems* **29** 6594877
[3] Bertling L, Allan R and Eriksson R 2005 *IEEE Transactions on Power Systems* **20** 75-82
[4] Xu P, Fang H, Wang H and Su Y 2018 *Power System Technology* **42** 3024-32
[5] Huo M-L, Liu Y and Yang L 2014 *Dianli Xitong Baohu yu Kongzhi/Power System Protection and Control* **42** 100-6
[6] Selvik J T and Terje A 2011 *Reliability Engineering & System Safety* **96** 324-31
[7] Da Silva A M L, Cassula A M, Billinton R and Manso L A F 2002 *IEE Proceedings: Generation, Transmission and Distribution* **149** 1-6
[8] Jurgensen J H, Nordstrom L and Hilber P 2019 *IEEE Transactions on Power Delivery* **34** 8701475
[9] Zhang X, Yue S and Zha X 2018 *IET Generation, Transmission and Distribution* **12** 295-302
[10] Park Y M, Kim G-W and Sohn J-M 1997 *IEEE Transactions on Power Systems* **12** 363-9
[11] Lin X, Ke S, Li Z, Weng H and Han X 2010 *IEEE Transactions on Power Delivery* **25** 5446314
[12] Sun J, Qin S-Y and Song Y-H 2004 *IEEE Transactions on Power Systems* **19** 2053-9
[13] Gupta P, Lin C-T, Mehlawat M K and Grover N 2016 *IEEE Transactions on Systems, Man, and Cybernetics Part A: Systems and Humans* **46** 7287778
[14] Xu B, Yin X, Zhang Z and Li X 2018 *Power System Technology* **42** 3241-8
[15] Dehghanian P, Fotuhi-Firuzabad M, Aminifar F and Billinton R 2013 *IEEE Transactions on Power Delivery* **28** 6468146
[16] Dehghanian P, Fotuhi-Firuzabad M, Aminifar F and Billinton R 2013 *IEEE Transactions on Power Delivery* **28** 6463471
[17] Brown R E, Frimpong G and Willis H L 2004 *IEEE Transactions on Power Systems* **19** 782-87
[18] Bompard E, Huang T, Wu Y and Cremenescu M 2013 *Int. J. of Electrical Power and Energy Systems* **50** 50-64
[19] Ndawula M B, Djokic S Z and Hernando-Gil I 2019 *Energies* **12** 531
[20] Shpiganovich A N, Zatsepina V I, Shpiganovich A A and Stepanov V M 2018 *EAI Endorsed Transactions on Energy Web* **18** e10
[21] Xiang Z and Gockenbach E 2006 *IEEE Transactions on reliability* **55** 361-68
[22] Allan R and Billinton R 2000 *Proceedings of the IEEE* **88** 140-62
[23] Jiang X and Sheng G 2018 *Gaodianya Jishu/High Voltage Engineering* **44** 1041-50

IOP Conf. Series: Materials Science and Engineering **791** (2020) 012031 doi:10.1088/1757-899X/791/1/012031

Research of Thyristor Voltage Regulator Effect on Power Flows Control in Distribution Electric Networks

A A Kralin, R Sh Bedretdinov, E V Kryukov and O A Pronina

Department of Electric Power Engineering, Power Supply and Power Electronics, Nizhny Novgorod State Technical University n.a. R.E. Alekseev, 24 Minin St., Nizhny Novgorod, 603950, Russia

E-mail: kryukov@nntu.ru

Abstract. The article is devoted to the power flows redistribution process research in distribution electric networks (DEN) with thyristor voltage regulator (TVR). TVR is intended to regulate power flows and voltage stabilization in the medium-voltage network of 6-20 kV. The problems in the medium-voltage network of 6-20 kV were analyzed and solutions were determined. It was established that the key development trend of DEN is the engineering and adoption of devices based on D-FACTS technologies. In addition, we considered the features, technical characteristics and control system of TVR. The research of power flow distribution in electric system was based on a simulation model of DEN with TVR. TVR allows to regulate power flows by changing the amplitude and phase angle of the voltage, with the help of longitudinal-transverse regulation. The results have shown that the TVR allows to align load schedules, to stabilize the voltage level in complex networks with unevenly loaded nodes and to realize optimal distribution of power flows in the electrical network.

1. Introduction

One of the actual problems of modern electric power industry is the development of power flow controlling devices. This is especially true for electric distribution networks (DEN) of 6-20 kV [1]. These networks are characterized by: high length, unbalanced load of transmission lines, high electric power dissipation and continuous growth of loads. Power flows occur in loop and ring circuits with multilateral power supply they causing suboptimal distribution of energy flows in the DEN. It is not always advisable replacing electrical equipment with more powerful and increasing the cross-section of power lines. Therefore, the crucial solution of the problems of DEN is the introduction of technologies D-FACTS (distribution flexible alternative current transmission systems) [2]. Devices implementing D-FACTS technologies include unified power flow controller – UPFC [3], interline power flow controller – IPFC [4], distributed static series compensator – DSSC [5], thyristor controlled phase angle regulator – TCPAR [6], thyristor switched series capacitor – TSSC [7], smart transformers – SmartTrafo [8]. A comparative characteristic of these devices is presented in Table 1.

Devices based on D-FACTS technologies allow to realize the following functions:

• generation or consumption of active and/or reactive power to change the voltage at the connection point. This leads to an increase of carrying capacity of transmission lines and a decrease of power losses in the electrical networks;

• inclusion of a controllable voltage source or a controllable resistance in the network to regulate power flows [9].

D-FACTS technologies allow to switch from an electric network with a "passive" structure of electric energy transmission to an "active" one, which realizes the functions of operating modes control [10]. However, D-FACTS devices are still under development [11].

Table 1. Comparative characteristic of devices based on D-FACTS technologies.

Functions	Devices D-FACTS					
	UPFC	IPFC	DSSC	TCPAR	TSSC	SmartTrafo
Power flow control	+	+	+	-	-	-
Active-power generation	+	+	-	-	-	-
Reactive power compensation	+	-	+	-	+	-
Voltage amplitude regulation	+	-	+	+	-	+
Phase angle regulation	+	-	-	+	-	-
Integration of renewable energy sources into power network	-	+	+	-	-	-

Scientists of NNSTU are developing an experimental prototype of the thyristor voltage regulator (TVR), representing an automatic thyristor regulator of the value and phase of the voltage with parallel and serial transformers. This electrical installation is designed for intelligent DEN 6-20 kV. TVR belongs to the class of controlled devices that reduce electrical losses and increase the carrying capacity of power lines. Scientific and technical solutions for the development of TVR are based on modern technologies of power electronics [12].

The aim of the work is to study the process of redistribution of power flows in the medium voltage distribution network with TVR.

2. Materials and methods

2.1. TVR Description

TVR is a high-speed 6 kV power semiconductor device based on transformer-thyristor voltage regulators [13]. It is intended to control power flows and voltage regulation in medium voltage DEN. A simplified TVR circuit is shown in Figure1 [14].

The principle of TVR operation is based on the shared use of longitudinal (change the magnitude) and transverse (change the phase) voltage regulation. The TVR power circuit contains a parallel transformer TV1 and a serial transformer TV2. The primary windings of the TV1 transformer are connected to the input terminals of the TVR (6-20 kV network). The secondary windings of TV1 are connected to the primary windings of the TV2 transformer via the transverse and longitudinal control modules. The secondary windings of the transformer TV2 connected between the input (A, B, C) and output (A2, B2, C2) TVR terminals (circuit break), making a longitudinal-transverse component of the voltage regulated by the thyristor switches in phase and magnitude.

The transverse control module is based on thyristor switches VS1-VS8. This module generates 90° shifted out-of-phase voltages at the terminals of the secondary windings of the TV1 transformer. This allows to implement amplitude control of the TVR output voltage.

The longitudinal control module is based on thyristor switches VS9-VS16. This module introduces a variable inphase or antiphase with the phase voltage of the source EMF value. This allows you to realize the regulation of the value of the output voltage TVR.

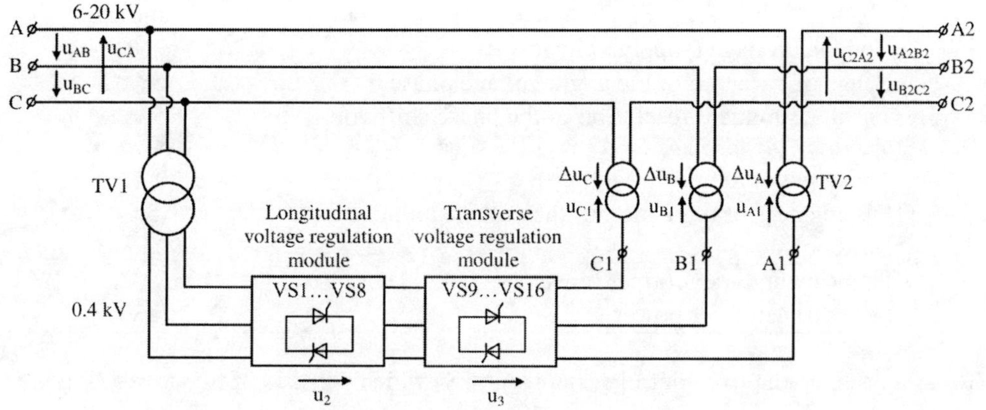

Figure 1. Simplified TVR circuit.

Both modules can be switched on or off from the power supply circuit of the primary windings of the TV2 transformer using the adopted circuit connections of thyristor switches. The combined use of transverse and longitudinal control modules allows for longitudinal and transverse regulation of the TVR output voltage. Vector diagrams of the input and output line voltages when the longitudinal and transverse control modules work together are in Figure 2.

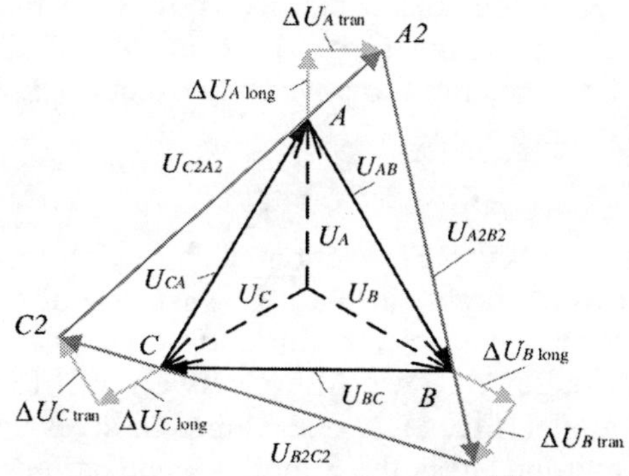

Figure 2. Vector diagram of longitudinal-transverse voltage control.

Summation of the longitudinal and transverse components of the voltage provides a wide range of regulation of the EMF boost. This allows the output voltage vector to be shifted relative to the input voltage by ±5 electrical degrees and ±10% amplitude. Thus, in the longitudinal-transverse mode of TVR there is a change in the magnitude and phase of the voltage, which allows influencing the flows of active and reactive power:

$$P = \frac{U_1 \cdot U_2}{x_{TDL}} \cdot \cos\delta; \tag{1}$$

$$Q_{1(2)} = \frac{1}{x_{TDL}} \cdot \left(U_{1(2)}^2 - U_1 \cdot U_2 \cdot \cos\delta\right); \tag{2}$$

where U_1 and U_2 – power transmission line voltages; x_{TDL} – power transmission line reactance; δ - angle between U_1 and U_2 voltages vectors.

Technical specifications of the TVR experimental prototype are shown in the Table 2.

Table 2. Technical specifications of the experimental TVR sample.

Parameter	Value
Nominal voltage (supply voltage)	6 kV ±10%;
Range of variation of the angle voltage phase α	± 5°
The measurement resolution of the phase shift voltage α	1.5°
The range of regulation voltage	± 10%;
Discreteness of regulation of the voltage amplitude	≤ 1.5%
Load power	≤ 1000 kVA
Shunt-wound transformer power	106 kVA
Series transformer power	3x28 kVA

TVR has an active-adaptive control system (AACS) which consists of two levels (Fig. 3).

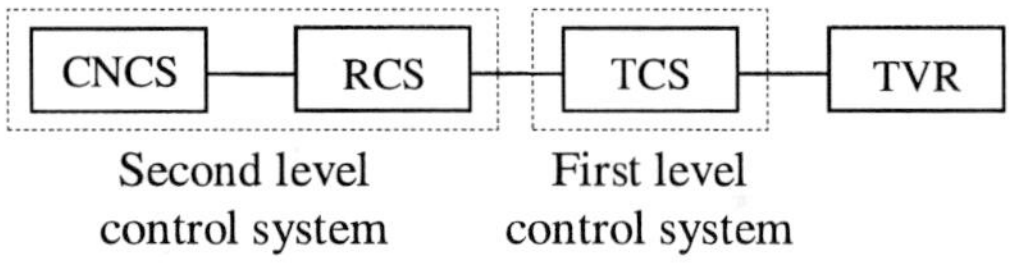

Figure 3. TVR AACS structure.

The first level of AACS consists of a centralized network control system (CNCS) and a regulator control system (RCS). The CNCS provides centralize control of the TVR group, and the RCS performs monitoring and controlling tasks the TVR. The second level of AACS includes a technological control system (TCS). The TCS is required to physically execute the TVR control commands coming from the first level control system.

2.2. Simulation model of DEN with TVR

The object of the research is the 6 kV DEN section. This section of the electric network has an ring configuration with several power sources. In the Fig. 4 is a simplified scheme of a fragment of the DEN 6 kV with TVR in one of the lines. Power comes to the ring network from three feeders from the substation "Priokskaya". Distribution of power flows between distribution centers (DC) is carried out via cable lines. A simulation model (Fig. 5) was developed to investigate flow distribution in the DEN with TVR. The simulation environment is the Simulink application of the Matlab software package. Simulink has the ability to simulate a wide range of ready-made electrical devices. The TVR model is a development of the authors [15].

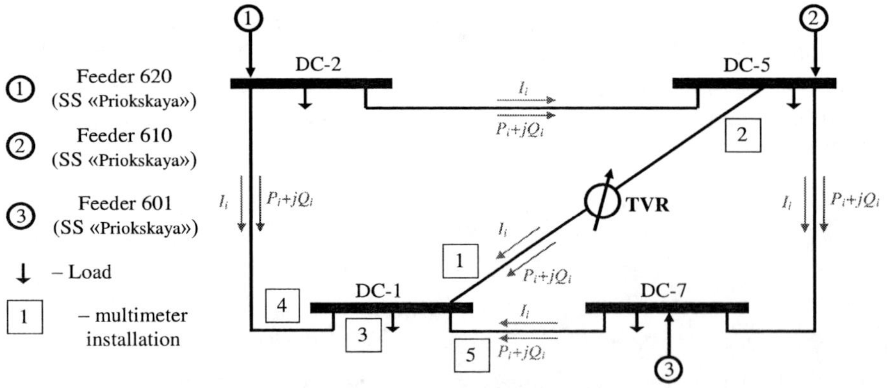

Figure 4. The fragment of the DEN 6 kV with TVR.

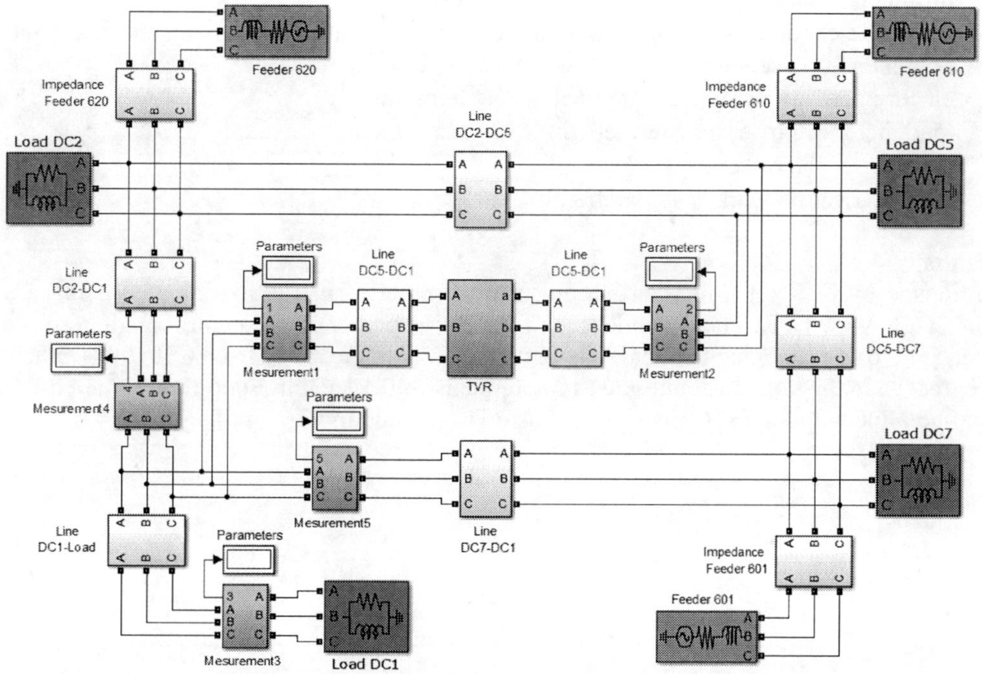

Figure 5. Appearance of the simulation model of the DEN with TVR (Initial conditions).

To measure the value of power flows in the simulation model, measurement units (multimeters) are provided.

The source data for the simulation are: the type and length of cable lines, the nature and nominal power of the electrical load, circuitry and parameters of TVR (Table 2, 3, 4).

Table 3. Parameters of cable lines of DEN with TVR.

Line name	Mark and cross section of the line	Line length, km	The resistivity of the line, Om/km	
			active	reactive
Feeder 620	AAB-6 2(3×185)	2.5	0.064	0.040
Feeder 610	ASB-6 2(3×185)	2.5	0.064	0.040
Feeder 601	AAB-6 2(3×240)	2.5	0.049	0.039
DC-2-DC-1	ASB-6 2(3×150)	0.8	0.080	0.041
DC-2-DC-5	ASB-6 2(3×120)	1.0	0.098	0.042
DC-5-DC-7	ASB-6 2(3×240)	0.9	0.049	0.039
DC-5-DC-1	AAB-6 3×150	1.2	0.159	0.081
DC-7-DC-1	AAB-6 2(3×120)	1.1	0.098	0.042
DC1-Load	AABl-6 3×35	0.7	0.890	0.087

Table 4. Parameters of electrical loads in DEN with TVR.

Consumers	Voltage, kV	Frequency, HZ	Active power, kW	Reactive power, kA
DC-1	6	50	809.2	142.8
DC-2			4154.4	733.1
DC-5			3655.8	645.2
DC-7			2622.7	462.8

IOP Conf. Series: Materials Science and Engineering **791** (2020) 012031 doi:10.1088/1757-899X/791/1/012031

The following assumptions are made during the developing a simulation model of DEN with TVR:

- infinite power sources are used as supply lines;
- the active conductivity of the insulation of the electrical network is insignificant and therefore not taken into account;
- electric network parameters are unchangeable in time;
- load factor of power transformers 6/0.4 kV is assumed to be 0.85;
- load power factor cosφ = 0.85;
- the electrical network is symmetrical.

3. Results

The influence of TVR on flow distribution in a close network was investigated using a simulation model of DEN with TVR. The dependences of the active power on the control angle α of the TVR are obtained (Fig. 6) by changing the phase angle of the TVR in the range from 0° to 90° (in increments of 1.5° degrees) and the voltage amplitude (in increments of 0.1 kV). In Fig.6 represented the flow distribution in the line without TVR (line from DC2 to DC-1), and in Fig.7 in line with TVR.

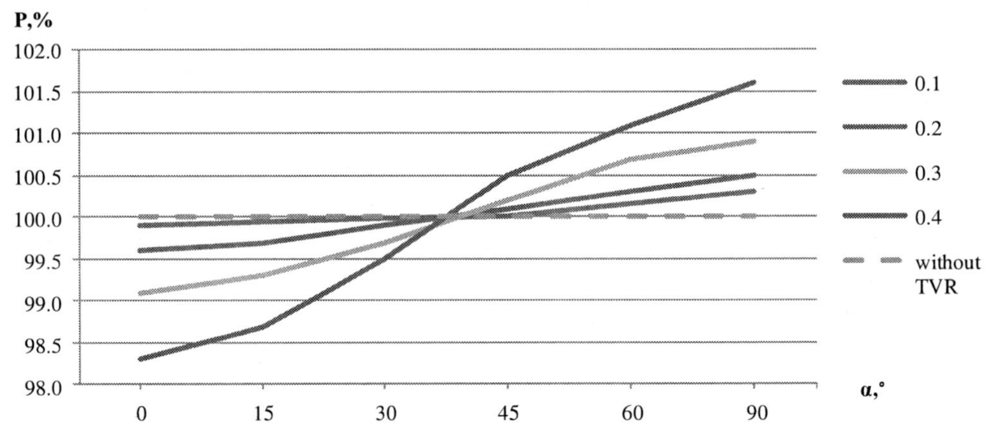

Figure 6. Dependence of the active power on the phase angle of the output voltage TVR in the line without TVR.

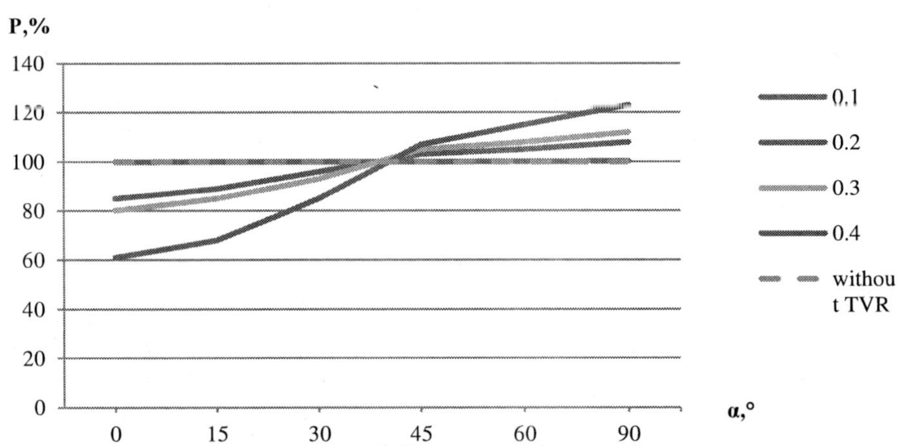

Figure 7. Dependence of the active power on the phase angle of the output voltage TVR in the line with TVR.

The research of the redistribution of power flows in DEN with unevenly loaded DC was performed similarly. In the simulation model, the TVR phase angle was changed in the range from 0° to 120°. Graphs of the dependence of the active power on the phase angle of the TVR control were formed based on the readings of the measurement units (Figure 8) for the following lines:

- from DC-5 to TVR;
- from TVR to DC-1;
- from DC-1 to the load;
- from DC-2 to DC-1;
- from DC-7 to DC-1.

Measurement points 1, 2, 3, 4, 5 correspond to the installation locations of measurement units (multimeters).

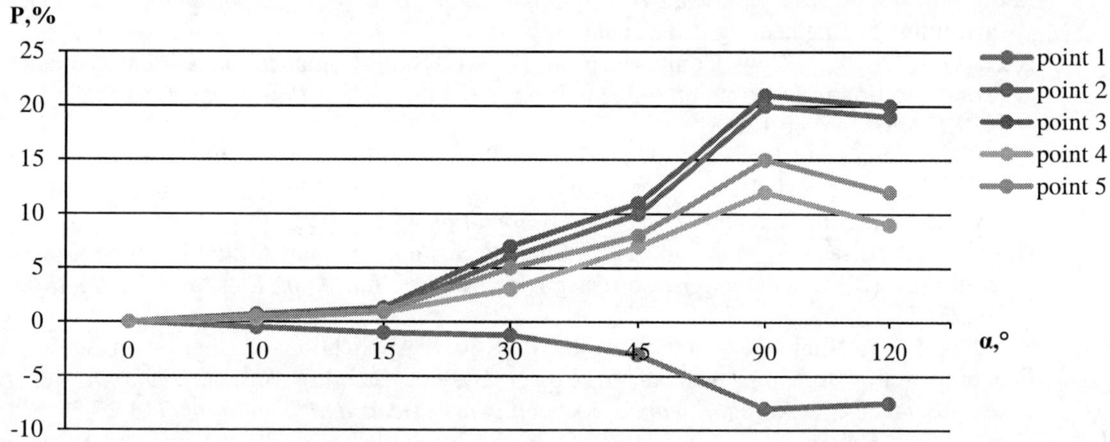

Figure 8. Dependence of the active power on the phase angle of the output voltage TVR in unevenly loaded DEN.

The construction of line dependencies between DC2 and DC5, DC5 and DC-7 was not done due to a minor change in the capacity of these lines.

4. Discussion

Analyzing dependence of the active power on the angle of the output TVR voltage in the line with TVR and without TVR we can find that the highest percentage of regulation occurs in lines close to TVR. This is because the network has a long length, and the TVR power is much less than the total power of the power supplies. In lines remote from the TVR, the change in power flows is negligible, about 0.2-1.5%.

Analysis of the researching results of uneven load distribution centers (DCs) showed that the installation of TVR near the overloaded DC allows you to achieve the uniform load DEN in common. The location of the TVR is chosen in such way that the DC1 is powered by all the feeders supplying the ring complex-closed network with a power shortfall on DC1.

5. Conclusion

The development and implementation of devices based on D-FACTS technologies is one of the crucial factors in the transition to DEN with an "active" energy transmission structure. This approach best meets the requirements of the electric network of complex closed configuration with multi-power supply. The problem of redistribution of power flows between unevenly loaded DEN nodes is acute in such networks.

The influence of TVR on the DEN parameters regime has been studied with the help of simulation modeling. The results of the research prove the effectiveness of TVR application for solving the problem of DEN power flow regulation. TVR allows to regulate power flows by changing the amplitude and phase angle of the voltage, with the help of longitudinal-transverse regulation. TVR allows you to align load schedules and stabilize the voltage level in complex networks with unevenly loaded nodes.

Thus, TVR can be successfully used in 6-20 kV DEN, implementing flexible flow distribution in networks with multi-circuit configuration and multiple power supplies.

Acknowledgments

This work is supported by the Ministry of Science and Higher Education of the Russian Federation (agreement №14.577.21.0242 for a grant on September 26, 2017). The unique project identifier is RFMEFI57717X0242.

References

[1] Radosavljevic J 2018 *Metaheuristic Optimization in Power Engineering* (London: The Institution of Engineering and Technology)

[2] Bach Q K 2019 A New Comparison of D-FACTS Performance on Global Voltage Sag Compensation in Distribution System *Proc. 2019 IEEE Asia Power and Energy Eng. Conf., APEEC* (Macao) pp 126-31

[3] Rakhi K, Prerna and Chitrangada R 2017 Modelling of UPFC (Unified Power Flow Control) to Improve Stability of Power System by Real and Reactive Power Control of Transmission Line *Advances in Systems, Control and Automation* **442** 647-55

[4] Hussein N A, Eisa A A, Mahmoud H M, Shehata S A and Othman A 2018Interline power flow controller (IPFC) characterization in power systems *Int. J. of Eng. end Technology* **7(3)** 1656-65

[5] Kumbhare P P, Nimje A A and Sawarkar P R 2018 Application of Distributed Static Series Compensator for Improvement of Power System Stability *Silicon Photonics & High Performance Computing. Advances in Intelligent Systems and Computing* **718** 27-34

[6] Hardas A V, Rajderkar V, Chandrakar D V K and Hardas V D 2018 Optimum Location of Thyristor Controlled Phase Angle Regulator Based on Performance Index *Proc. 2018 Int. Conf. on Smart Electric Drives and Power System (ICSEDPS)* 132-6

[7] Kathal P and Bhandakkar A 2013 Power Flow Control In Power System Using FACT Device Thyristor Controlled Series Capacitor (TCSC): A Review *Int. J. for Research and Development in Eng., IJRDE* **1(3)** 82-91

[8] Graaff R 2010 *Flexible distribution systems through the application of multi back-to-back converters: Concept, implementation and experimental verification* (Eindhoven: Technische Universiteit Eindhoven)

[9] Corredor A, Beleed H, Johnson B K and Hess H L 2018 D-FACTS for Improved Reliability of the Transmission System During Contingencies *Proc. 2018 North American Power Symp., NAPS*, (Fargo, USA) pp 1-6

[10] Rogers K M and Overbye T J 2009 Power flow control with Distributed Flexible AC Transmission System (D-FACTS) devices *Proc. 41st North American Power Symp. NAPS*, (Starkville, USA) pp 1-6

[11] Akram Z 2018 A comprehensive review on D-FACTS devices *Int. J. of Trend in Scientific Research and Development (IJTSRD)* **2(3)** 1811-5

[12] Kralin A, Kryukov E and Eremenko V 2019 Thyristor Voltage Regulator Application in Radial Electrical Networks for Power Flow Control *Proc. 2019 Int. Ural Conf. on Electrical Power Eng., UralCon* (Chelyabinsk, Russian Federation) pp 81-5

[13] Sosnina E, Asabin A, Kralin A and Kryukov Eeditor 2019 Voltage Control with Thyristor-Regulated Booster Transformer *Proc. 6th IEEE Int. Conf. on Smart Grid, icSmartGrids 2018* (Nagasaki, Japan) pp 202-7

[14] Sosnina E, Kralin A, Bedretdinov R and Kryukov E 2018 Harmonic analysis of the thyristor regulator output voltage *Proc. of the 2018 IEEE PES Transmission and Distribution Conf. and Exhibition - Latin America, T and D-LA 2018* (Lima, Peru) pp 339-43

[15] Sosnina E N, Asabin A A, Kralin A A and Kryukov E V 2018 Research of TRBVT regulation characteristics *Proc. of the 2018 IEEE PES Innovative Smart Grid Techn. Conf. Europe, ISGT-Europe 2018* (Sarajevo, Bosnia and Herzegovina) pp 8571591

Calculation and analysis of the arcs efficiency of small and large capacity arc steel-melting furnaces

A N Makarov, V V Okuneva and Yu M Pavlova

Departament of Electrical Supply and Electrical Engineering, Tver State Technical University, nab. Af Nikitina, 22, Tver, 170026

E-mail: tgtu_kafedra_ese@mail.ru

Abstract. In the course of the research, the calculation and performed analysis of the arc efficiency of furnaces ASF-5 and ASF-100 power furnaces were performed and the average arc efficiency for melting was determined. The reasons of the increased specific consumption of electric power on charge melting by arc steel-melting furnaces of small capacity in comparison with heavy-load furnaces are revealed. Efficiency of arcs of low-tonnage furnaces is much less than the same indicator of modern heavy-duty arc steel-melting furnaces (ASF). Low efficiency of arcs of light-duty furnaces leads to a significant specific power consumption for smelting, comprising 700-750 kWh/t, that is 1.9-2 times higher than the specific energy consumption in modern furnaces ASF-120, amounting to 375 kWh/t. With the use of devices for foaming slag in furnaces ASF-5 the average melting efficiency of furnaces can be increased, and the specific electricity consumption for melting will be reduced to 440-480 kWh/t.
Keywords: electric arc, furnace, steel, efficiency, thermal radiation

1. Introduction

Modern advances in steel electrometallurgy are determined by a new generation of energy-saving arc steel-melting furnaces (ASF). Electric furnaces are considered as steelmaking units, providing a breakthrough in modern metallurgy. The results and prospects for the development of furnaces and electric steel production show the complex path of the innovation process from the emergence of a new technical idea to its implementation and entry into the market of modern technologies and equipment. New technological solutions lead to radical innovative changes in the design of furnaces and in the design solutions of electric steelmaking shops and complexes.

In the process of reconstruction, it is important to choose the direction of technical re-equipment of furnaces to increase productivity, reduce energy consumption and environmental impact. Furnaces of small capacity, these include furnaces with a capacity of 1.5 to 20 tons of metal, are characterized by increased specific electricity consumption for melting the charge and for melting as a whole, which is, respectively, 700-750 kWh/t for 5-ton ASF-5 furnaces. In modern high-power high-capacity furnaces with a capacity of 80 to 180 tons, which are melting units, the specific electricity consumption for melting is 375 kWh/t. Figure 1 the filling of the charge in a small capacity ASF-5 (a) and a large capacity furnaces-120 (b) is shown.

From the comparison of the specific energy consumption for melting in small and large capacity furnaces, it follows that small capacity furnaces consume 35-45% more electricity for melting one ton of steel compared to large capacity furnaces. Specific power consumption is 20-35% higher both in old and modern low-tonnage furnaces equipped with modern automated control systems (ACS), devices for slag foaming and other devices for smelting intensification [1, 9, 13].

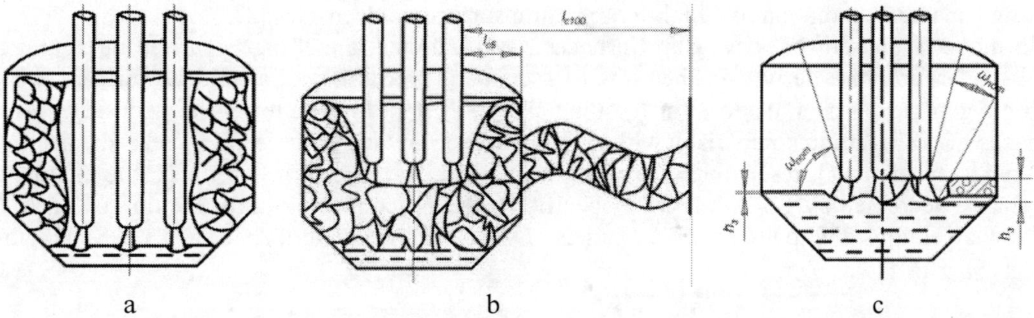

Figure 1. Cutting wells and melting the charge in the furnace ASF-120 (a), ASF-100, ASF-5 (b) and deepening the arc in the metal and slag in the furnace ASF-120 (right side figure c), ASF-5 (left side figure c).

It is necessary to solve the problem of determining the causes of high power consumption for melting in low-capacity furnaces compared to the same indicator of high-capacity furnaces. To do this, we reviewed the development of low-tonnage and heavy-duty furnaces, and calculated the efficiency of arcs in the process of melting the charge and analyzed the changes in the efficiency of arcs in furnaces.

2. Analysis of ASF of small and large capacity

We will analyze the development of small and large capacity furnaces and changes in the efficiency of arcs in the process of melting steel in furnaces. According to the method developed by the author [2, 3], the efficiency of arcs and other energy parameters of steel melting in low-tonnage and heavy-load furnaces were calculated. We reviewed and compared the technical characteristics of the ASF-5 and ASF-100 furnaces, which the furnaces had in the 1980s and in the 2010s. The method of calculation and analysis of energy parameters of melting and efficiency of furnace arcs is described in [2, 3, 12].

In the 2000s, transformers with a capacity of 5 MVA were installed on the ASF-5 furnace. The specific capacity of modern ASF-5 furnaces is 1000 kVA/t, it exceeded the specific capacity of modern ASF-100 furnaces by 20%. However, by increasing the capacity of the furnace ASF-5, without the use of other methods of intensification of melting, breakthrough technical and economic results of the furnace could not be obtained [15, 16].

We found out that equipping a furnace ASF-5 with a transformer of 1.8 times more power, without the equipment of gas-oxygen burner (GOB) and devices for slag foaming (DSF), led to higher furnace productivity by 1.8 times, but did not reduce the specific consumption of electricity for melting and for melting. This approach was not effective. Furnaces ASF-5, ASF-100 worked in the 1980s in equal conditions. We compared the technical characteristics of the furnaces ASF-5, capacity of 2.8 MVA, and furnaces ASF-100, capacity of 45 MVA, working in the 1980s with a full cycle of melting. In ASF-100 furnaces the maximum of thermal radiation of arcs is opposite to arcs in the middle of height of walls from level of a bath of metal [4]. On the lining in the middle of the height of the walls formed hot spots from thermal radiation arcs, causing reflow lining. In the lower and upper parts of the walls refractories had wear not exceeding 20-30% [4]. However, since the refractories in the central part of the height of the walls had a wear of 90-95%, it was necessary to stop the furnace for repair, destroy the masonry of the old walls, install new masonry walls. The walls lined with refractory brick in furnaces ASF-100 had a small resistance to an average of 200 heats.

In the ASF-5 furnace, the arc current is 6.3 kA, it does not have significant radial and axial forces, does not cause a deepening in the metal bath, the arc deviation is 15-20° to the electrode axis. As a result of this position of the arcs in the ASF-5 furnace, the maximum thermal radiation of the arcs and the wear of the walls are concentrated mainly in the lower part of the walls near the slopes [4]. After each melting, the walls are restored with refractory powder. In ASF-100 furnaces it is impossible to restore walls as filling powder rolls down from a vertical surface of walls. Resistance of lined vaults of furnaces ASF-5, ASF-100 differs slightly as the reason of wear of lining of arches in furnaces of small and big capacity is radiation of a high-temperature surface of electrodes [4].

In modern high-power heavy-duty furnaces ASF-120 charge melting is carried out, other technological operations are derived in the units of secondary processing of steel. The furnace is equipped with oxygen-gas burners, devices for foaming slag [5,11,19]. For the thirty-year period at heavy-load furnaces of ASF-120, in comparison with ASF-100 the following parameters increased: specific power in 2 times, tension and, respectively, arc length in 2-3 times, current in 1,5 times; the melting time of the charge decreased by 1.5-2 times, the specific electricity consumption for melting by 15%, the furnaces were equipped with water-cooled panels of walls, vaults, modern ACS, GOB, DSF (Figure 2).

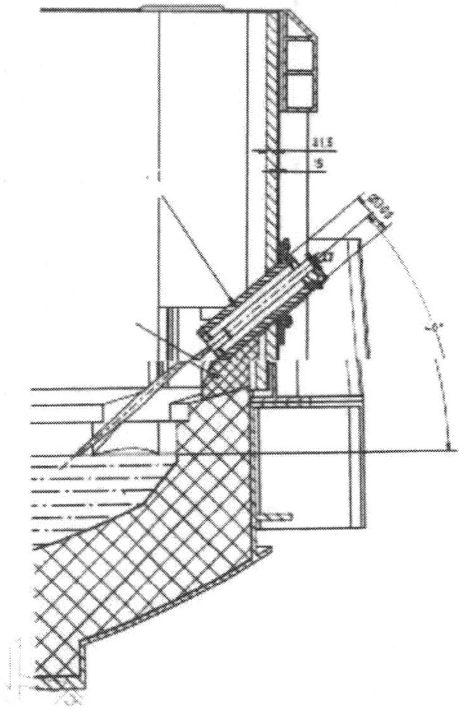

Figure 2. Combined use of the manipulator for foaming slag and oxygen burner in the furnace

3. Calculation of the arcs efficiency of ASF of small and large capacity

Calculations revealed that, depending on the capacity of the furnaces, the average efficiency of arcs increases from the minimum values of 0.55-0.57 in low-tonnage furnaces without the use of GOB, DSF to 0.78-0.80 in heavy-duty furnaces in which GOB, DSF are used. The specific power consumption for charge melting, on the contrary, has the highest value of 475-500 kWh / t in low-tonnage furnaces with the lowest efficiency, decreases depending on the increase in capacity and efficiency of arcs, reaching the lowest value of 360-375 kWh/t in heavy-duty furnaces with the highest efficiency of arcs [5,6].

In the ASF-5 furnace, the average thermal efficiency has the lowest value $\eta_h = 0.64$, the highest $\eta_h = 0.75$ in the ASF-100 furnace, in the ASF-120 furnace $\eta_h = 0.68$. In the furnace ASF-100 energy to melt the charge is introduced by electric arcs, the efficiency of which $\eta_{av} = 0.67$, and the use of the energy of the metal oxidation reaction C efficiency equal to one. In the furnace ASF-120 energy is introduced using arcs with $\eta_{av} = 0.78$, GOB with an efficiency of 0.45-0.55 and the oxidation reaction of coke and electrons with an efficiency of 0.35-0.45 [3], so the average thermal efficiency of the furnace η_h = 0.68. Input of energy into the furnace by means of electric arcs is the most effective way with the maximum efficiency, heat of GOB, coke and electrodes is less effectively used. However, the use of coke powder in DSF allows to increase the height of the slag layer and to bury the arcs in the slag, increasing their efficiency. In furnaces ASF-100 without DSF slag layer height h_{sl} = 100 mm and η_a = 0.67, in furnaces ASF-120, h_{sl} = 240-360 mm and η_a = 0.78. In ASF–5 furnaces, the height of the slag layer h_{sl} = 35 mm, the arc length during the melting period is 115-135 mm, 60-70% of the arc length is not covered by slag and the arc radiates 2/3 of the power to the lining of the walls and arch, which corresponds to the efficiency of the arc at the end of the melting η_a = 0.35-0.4 [4]. At the beginning of the melt when burning arcs in the well in the charge in the furnace ASF–5 η_a = 0.77, and the average for the melting efficiency is equal η_{av} = 0.57. Based on the fact that the average efficiency of arcs ASF-5 η_{av} = 0.57 it follows that during the melting period about 43% of the thermal radiation flux of arcs falls not on the charge, metal bath, slag, but on the lining of the walls, the arch, causing their heating, melting, wear of the lining. Due to the fact that in the ASF-5 furnaces the arcs are at a distance of less than 1 m from the walls, the radiation of the arcs reaches the walls and is uselessly spent on heating and melting the lining. The efficiency of arcs in ASF-5 furnaces is reduced compared to ASF-120 furnaces, in which the arc radiation does not reach the lining and is useful for heating and melting the charge.

The average fusion efficiency of arcs of the furnace ASF-5 34% less than the average for the period of melting and smelting on the whole efficiency of arcs η_{av} = 0.78 furnace ASF-120 and 16% less than η_{av} = 0.67 furnace ASF-100. The low average melting efficiency of low-tonnage furnace arcs is the reason for the increased specific energy consumption in the furnaces. The average melting efficiency of arcs in low-tonnage ASF-5 η_{av} = 0.58 (Figure 3, c), which corresponds to the specific consumption of electricity for melting 700-790 kWh/t. The average melting efficiency of arcs in heavy-duty furnaces, in ASF-100, which worked in the 1980s under the same conditions as low - tonnage furnaces with a full melting cycle, is η_{av} =0.67-0.72, and the specific electricity consumption is 420-460 kWh/t [4,5]. The ARC efficiency increases with the depth height h_d of the arc penetration into the metal and slag. In ASF-5 furnaces, the arc depth height h_d =37-40 mm, the average arc efficiency during the melting period η_{av} = 0.57, the specific electricity consumption for melting is 475 kWh/t. In ASF-100 furnaces, the depth height of the arc h_d = 140-150 mm, the average efficiency of the arc η_{av} = 0.67, the specific power consumption for melting is 425 kWh / t.

In furnaces of ASF-120 height of deepening of arcs h_d= 240-410 mm, average efficiency of arcs η_{av} = 0.78 (Figure 1,b) specific energy consumption for melting of 375 kWh/t. Thus, the larger the height of the slag layer , penetration of the arc into the metal and slag, the greater efficiency of the arc and the less the specific energy consumption in furnaces. By foaming slag in low-tonnage furnaces up to 100-150 mm, it is possible to bury the arc in the slag and increase their efficiency during the melting period and for melting as a whole to η_{av} = 0.72-0.74, and the specific electricity consumption for melting and melting as a whole is reduced to 440-480 kWh/t, reduce the melting time, increase the productivity of furnaces.

The results of energy balances in small and large capacity furnaces [6-8], as well as the results of arc efficiency calculations show that it is impossible to achieve efficiency at the beginning of the melting period in small capacity furnaces greater than η_a = 0.77 (Figure 3, a). For the first 20 mins of the melting period, the average efficiency of arcs furnace ASF-5 amounted η_{av} = 0.77, analyzed over the first 20 minutes of the period, the melting efficiency of the arcs in furnaces ASF-120, ASF-100 made up η_{av} = 0.83 (Figure 3, b) or 8% more compared to low-tonnage furnaces. Since the specific consumption of electricity in the furnaces is directly proportional to the efficiency of arcs, the specific consumption of electricity for the first 20 minutes. Melting in the ASF-5 furnaces will be 8-10% more than in the ASF-120, ASF-100 furnaces.

Figure 1 shows a section of the working space during the cutting of wells and melting of the charge in furnaces ASF-120 (figure 1, a), in furnaces ASF-5, ASF-100 (figure 1, b) and the penetration of arcs into metal and slag in the liquid periods of melting in the furnace ASF-120 (right part figure 1,c), in the furnace ASF-5 (left part figure 1,c). Lining of the hearth, slopes, walls, arch, water-cooled pan-

IOP Conf. Series: Materials Science and Engineering **791** (2020) 012032 doi:10.1088/1757-899X/791/1/012032

els are not shown. On the figure 1 different scales of construction of furnaces are accepted, therefore furnaces ASF-120, ASF-100, ASF-5 are represented in identical dimensions. In the furnace ASF-120 arcs at the beginning of the melting period introduced a large power, about 70 MW, arcs for 6 min. cut one common well in the charge and burn inside the well (figure 1, a). At such position of arcs their efficiency is equal to $\eta_a = 0.93$ and 93% of power of arcs is radiated by them on a metal bath and a firm charge and it is useful spent on their heating and melting. Arc radiation power loss is 7%.

a

b

Figure 3. Charge loading in small-capacity chipboard DSP-5 (a) and large-capacity chipboard DSP-120 (b)

In the ASF-5 of power of arcs at the beginning of melting is 2.7 MW, this power is not enough for cutting a common well in the charge for 6 minutes. In the ASF-5 of melting the charge, the arcs create three shallow wells, 1/3 the height of the charge, which after 10-12 minutes of melting are combined into one common well (figure 1, b). Cutting wells in the furnace ASF-5 is slow, there are collapses of the charge and short circuits of the arc, the molten metal flows down into the bath (figure 1, b). Part of the thermal radiation of the arcs comes out of the well and falls on the central part of the arch of the furnace, the efficiency of the arcs is reduced. The depth of wells with the melting of the charge in a furnace ASF-5 lasts 20 minutes. Through 17-20 shadowing electrodes and arc in furnace ASF-5 are lowered until liquid metal baths, their length expands in 8-10 times and, so as wall on 1/2 - 2/3 its heights freed from charge, then arc radiate 60-65% capacity on vault and wall and 35-40% on charge and bath metal, efficiency arcs declines until 0.35-0.40. At the 20th minute of melting, the voltage stages are switched, the voltage, current, length, power of the arcs and reduced efficiency of the arcs is increased to 0.44. Similarly, during all periods of melting, oxidizing, refining, reduce the voltage, length, current, arc power, increasing the efficiency arc within 0.5-0.76. However, such maneuverable operation of the ASF-5 furnace does not lead to a significant increase in the efficiency of the furnace arcs during the melting period, which is $\eta_{av} = 0.57$.

Compared with the ASF-5 furnace ASF-100 worked in a similar mode of melting charge, the power of the arcs was 25-26 MW, which is not enough to cut through the well for 5-6 minutes the well cutting in the furnace ASF-100, similar to the furnace ASF-5, lasted 20 minutes, after which the arcs burned on a liquid metal bath. However, the average efficiency of arcs in the furnace ASF-100 for the melting period $\eta_{av} = 0.67$, and in the furnace ASF-5 the same $\eta_{av} = 0.57$. Therefore, the reason for the lower efficiency of arcs during the melting period in the ASF-5 furnace compared to the same indicator in the ASF-100 furnace is not the slow cutting of wells, but another, let's find it.

In the furnaces ASF-100, ASF-120 the distance between the arcs and the refractory or water cooled panels of the walls is more than 3m and the thermal radiation of arcs passing between adjacent arcs voids between the pieces of the charge, gets into pieces following radial layers of the charge and does not reach the walls, it is useful spending on heating and melting of the charge. During the next melting in the open periods, the temperature of the lining of the walls in the ASF-100 furnace reaches 1500°C. After filling the charge and the beginning of melting, cutting wells in the charge, the lining temperature begins to decrease. The decrease in the temperature of the lining of the walls of the furnace ASF-100 is due to the fact that the thermal radiation of the arcs does not reach the lining of the walls and the lining, which has a temperature of 1500°C, begins to give thermal energy to a colder charge. The process of transferring thermal energy to the charge by lining the walls continues during the entire period of melting of the charge [17, 18].

On the whole, for melting the efficiency of arcs in furnaces ASF-100 η_{ASF} = 0.68-0.72 and the specific consumption of electricity for melting 475-500 kWh/t, which is 35-50% less than in low-tonnage furnaces ASF-5, where Q_e = 640-750 kWh/t and the arc is half immersed in slag [20]. The above is illustrated in figure 1,where the right side of the figure shows schematically thermal performance of the arc furnace ASF-100, in the left part of the figure thermal performance of the arc furnace ASF-5, and the solid angle of the loss ω_{con} demonstrates the loss of thermal radiation of the arc, thermal radiation arcs on the walls and roof of the furnace, the solid angle of the loss of thermal radiation of the arc ω_{con} minimum in heavy furnaces ASF-100, ASF-120 and the maximum in light-duty furnaces of ASF-5.

4. Conclusion.

The reasons of the increased specific consumption of electric power on charge melting by arc steel-melting furnaces of small capacity in comparison with heavy-load furnaces are revealed. The compared furnaces ASF-5, ASF-100 worked in equal conditions without gas-oxygen burners, devices for foaming slag with a full cycle of melting. The performed calculations helped to find out that in low-tonnage furnaces compared with heavy-duty furnaces, the efficiency of arcs is less important, which affects the increased specific consumption of electricity in low-capacity furnaces compared to heavy-duty furnaces. To increase the efficiency of arcs and reduce the specific power consumption in low-tonnage furnaces, it is necessary to increase the slag height by using slag foaming devices.

Acknowledgments.
The study was carried out with the financial support of RFBR within the framework of the scientific project № 18-33-00511.

References

[1] Bershiczkij I M and Protasov A V 2015 Nekotory`e osobennosti i perspektivy` razvitiya malotonnoj e`lektrometallurgii [Some features and prospects of development of low-tonnage electrometallurgy] *Elektrometallurgiya* **10** 28-35 [In Russian]

[2] Kuznetsov M S et al 2010 Vliyanie massy metalloshihty I tolshiny shlaka na tekhnologiyu vyplavki stali v dugovoj pechi [Influence of mass of a metal charge and thickness of slag on technology of melt of steel in the arc furnace] *Electrometallurgy* **2** 2-6 [In Russian]

[3] Shakirov Z, Shumakov A M, Uryupin G P, Zinurov I Y and Zapolsky A A 2013 Osobennosti tekhnologii vyplavki stali v elektropechi DSP-120 Consteel i puti ee sovershenstvovaniya [Peculiarities of the technology of smelting steel in an electric furnace DSP-120 Consteel and ways of its improvement] *Elektrometallurgiya* **9** 2-7 [In Russian]

[4] Nikolsky L E, Smolyarenko V D and Kuznetsov L N 1981 Teplovaya rabota dugovykh staleplavilnykh pechej [Thermal operation of arc steel-melting furnaces] (Moscow: Metallurgiya) [In Russian]

[5] Makarov A N 2019 Raschet i analiz vzaimosvyazi KPD i raspolozheniya dug s e`lektropotrebleniem v dugovy`x staleplavil`ny`x pechax maloj i bol`shoj vmestimosti. Chast` I. Raschet i analiz vzaimosvyazi KPD dug i e`lektropotrebleniya [Calculation and analysis of the relationship between efficiency and location of arcs with power consumption in arc steel furnaces of small and large capacity. Part I. Calculation and analysis of the relationship between arc efficiency and power consumption] *Metallurg* **4** 29-35 [In Russian]

IOP Conf. Series: Materials Science and Engineering **791** (2020) 012032 doi:10.1088/1757-899X/791/1/012032

[6] Samways N 2005 Revitalized with a new continuous steelmaking process *Iron and Steel Technology* **6** 29-45
[7] Nikol`skij L E, Smolyarenko V D and Kuzneczov L N 1981 *Teplovaya rabota dugovy`x staleplavil`ny`x pechej* (Moscow: Metallurgiya) [In Russian]
[8] Evstratov V G, Kiselev A D , Zinurov I Yu , Shaxirov Z X , Mamenko Yu F , Shumakov A N and Gindullin M G 2012 Osobennosti teplovoj raboty` e`lektropechi DSP-120 Consteel na Ashinskom metallurgicheskom zavode [Features of thermal operation of electric furnace ASF-120 Consteel at Ashinsky metallurgical plant] *Elektrometallurgiya* **8** 2-6 [In Russian]
[9] Gudim Yu A, Zinurov I Yu, Kiselev A D and Shumakov A M 2005 Racional`ny`e sposoby` intensifikacii plavki v sovremenny`x dugovy`x staleplavil`ny`x pechax [Rational methods of intensification of melting in modern arc steel furnaces] *Elektrometallurgiya* **9** 2-6 [In Russian]
[10] Makarov A N, Rybakova V V and Galicheva M K 2014 Electromagnetic and the Arc Efficiency of Electric Arc Steel Melting Furnaces *J. of Electromagnetic Analysis and Applications* **6** 184-92
[11] Makarov A N 1994 Raschet izlucheniya dugi zaglublennogo plazmotrona v plazmenno-dugovoj pechi [Calculation of arc radiation of a buried plasmatrona in a plasma-arc furnace] *Izvestiya vuzov. Chernaya metallurgiya* **6** 14-6 [In Russian]
[12] Makarov A N 2016 Calculations of heat transfer in torch furnaces by gas volume radiation haws *Word J. of Eng. and Technology* **4** 488-503
[13] Krivandin V A 2001 E`nergosberezhenie kak rezul`tat neprery`vnogo sovershenstvovaniya teplovoj raboty` i konstrukcii nagrevatel`ny`x ustrojstv [Energy saving as a result of continuous improvement of thermal operation and design of heating devices] *Izvestiya vuzov. Chernaya Metallurgiya* **7** 48-54 [In Russian]
[14] Kuzneczov L N, Pirogov N A and Egorov A V 1981 Raschet parametrov dugovy`x staleplavil`ny`x pechej dlya plavki metallizirovanny`x materialov [Calculation of parameters of arc steel furnaces for melting metallized materials] *Issledovanie v Oblasti Promyshlennogo Elektronagreva: sb. tr. VNIIE`TO* 88-97 [In Russian]
[15] Borovinskih S V 2007 V Rossiyu prihodit innovacionnyj elektrostaleplavilnyj process [Russia comes innovative steelmaking process] *Elektrometallurgiya* **9** 13-6 [In Russian]
[16] Bourne K and Granderat R 2013 Potentsialnye vozmozhnosti I problemy utilizatsii tepla v staleplavilnykh pechah [Potential opportunities and problems of utilization of heat in steelmelting furnaces] *Metallurgicheskoye Proizvodstvo i Tekhnologii* **2** 8-19 [In Russian]
[17] Timofeev E S, Timofeeva A S and Fedina V V 2009 Issledovanie e`nergotexnologicheskogo rezhima DSP-150 pri plavke stali s primeneniem v shixte goryachebriketirovannogo zheleza [Investigation of energy-technological regime of ASF-150 during steel melting with application of hot-briquetted iron in charge] *Elektrometallurgiya* **8** 6-9 [In Russian]
[18] Makarov A N 2017 Izmenenie KPD dug v processe plavki v dugovy`x staleplavil`ny`x pechax [Change of efficiency of arcs in the process of melting in arc steel-melting furnaces] *Metallurg* **4** 55-8 [In Russian]
[19] Argenta P and Corbella M 2006 EAF integration info the blast furnace route at Wheeling Pittsburgh *MPI International* **2** 42-7
[20] Makarov A N 2014 Theory of radiative heat exchange in furnaces, fire boxes, combustion chamber is replenished by four new laws *Science Discovery* **2** 34–42

IOP Conf. Series: Materials Science and Engineering **791** (2020) 012033 doi:10.1088/1757-899X/791/1/012033

The magnetic field of a three-phase power line with an ordinary and triangular arrangement of wires

O V Oleshuk[1], A V Sysun[2], A A Tikhomirov[1] and N V Sobolev[1]

[1]Department of Electric Power Engineering and electronic, Petrozavodsk state university, Lenina str., 33, Petrozavodsk, 185910, Russia
[2]Department of applied mathematics and cybernetic, Petrozavodsk state university, Lenina str., 33, Petrozavodsk, 185910, Russia

E-mail: sasha.82@mail.ru

Abstract. Power lines are sources of magnetic fields, which requires attention to ensure safety for people and to consider electromagnetic compatibility. In this article the nondimensional form expressions for calculation of magnetic field strength of a three-phase line with an ordinary and triangular arrangement of wires were obtained what makes it possible to calculate the necessary distance to the core of the line if the maximum permissible level of magnetic field strength is provided. The approximate expression of magnetic field strength at large distance from the line was considered. It is valid with the measurement accuracy less than 1%, which shows the proportion between magnetic field strength and distances between wires and inverse proportion of square distance to the wire core. It was determined that the triangular arrangement of three-phase line wires has lower magnetic field strength than the arrangement of wires in row due to more compact arrangement of wires.

1. Introduction

High current cables of a power system are a source of electromagnetic field strength [1-6]. It requires calculations to estimate electromagnetic safety and electromagnetic compatibility in system designing [7-9].

Electromagnetic fields of 0.4-10 kV cables can usually be neglected, but industrial frequency magnetic fields can exceed the critical value, especially in case of lines with an ordinary and triangular arrangement of wires.

A magnetic field of one wire or two wires single-phase line is calculated and can be written in expressions [10-12].

For a three-phase line, vectors of a magnetic field are not equal in amplitude (due to different distances to the point of observation), shifted in angle in space (due to the mismatch of directions of vectors) and shifted in phase. It makes calculations of a magnetic field more difficult. Approximate values and graphics methods are usually used, which leads to a large measurement error [11-15].

This paper considers universal nondimensional expressions for magnetic field strength of a three-phase line. There are three cases of wire arrangement: horizontal row, vertical row and triangular.

2. Materials and Methods

In figure 1 three-phase line with arrangement of wires in horizontal row is shown. Coordinate axes are taken along the line connecting wires instead of the case where axes are connected with position vectors and changed with shift of the observation point.

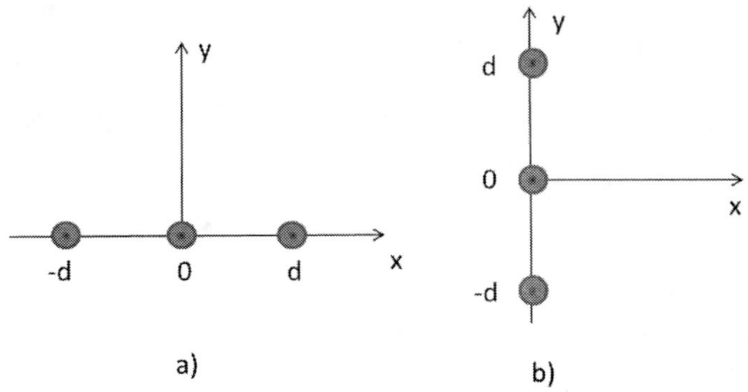

Figure 1. Three-phase line with arrangement of wires in row:
a) horizontal arrangement; b) vertical arrangement.

Figure 2 shows the location of magnetic vectors of each of the wires of figure 1(option a). Vectors of magnetic field act along lines H_A, H_B, H_C, which are perpendicular to position vectors of the corresponding wire core.

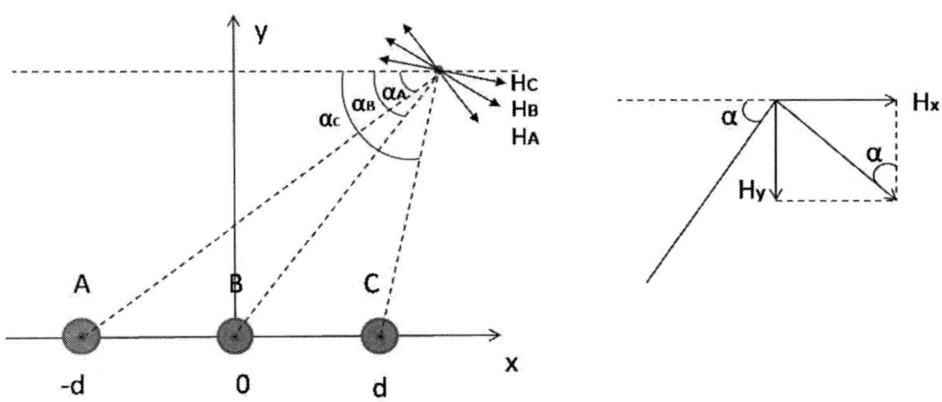

Figure 2. Scheme of location of magnetic field strength vectors for horizontal arrangement Magnetic field strength of a straight wire with current I(t) is expressed as.

$$H(t) = \frac{I(t)}{2\pi r}$$

where r is the distance from the observation point to the wire core. Projections of magnetic field are equal: $H_x(t) = H(t)\cos\alpha$, $H_y(t) = H(t)\sin\alpha$.

Then, in our case, according to Figure 2, the projections of the vectors at the point with coordinates xy from the wire core:

$$H_{Ax} = \frac{I_0 sin(\omega t - 120°)}{2\pi(y^2 + (x+d)^2)^{\frac{1}{2}}} \cdot \frac{y}{(y^2 + (x+d)^2)^{\frac{1}{2}}} H_{Ay} = -\frac{I_0 sin(\omega t - 120°)}{2\pi(y^2 + (x+d)^2)^{\frac{1}{2}}} \cdot \frac{(x+d)}{(y^2 + (x+d)^2)^{\frac{1}{2}}}$$

IOP Conf. Series: Materials Science and Engineering **791** (2020) 012033 doi:10.1088/1757-899X/791/1/012033

$$H_{Cx} = \frac{I_0 \sin(\omega t + 120°)}{2\pi(y^2 + (x-d)^2)^{\frac{1}{2}}} \cdot \frac{y}{(y^2 + (x-d)^2)^{\frac{1}{2}}} H_{Cy} = = -\frac{I_0 \sin(\omega t + 120°)}{2\pi(y^2 + (x-d)^2)^{\frac{1}{2}}} \cdot \frac{(x-d)}{(y^2 + (x-d)^2)^{\frac{1}{2}}} \quad (1)$$

where I_0 – current amplitude. Projections of the magnetic field are obtained by summing up the projections of individual cores:

$$H_x = H_{Ax} + H_{Bx} + H_{Cx}, \qquad H_y = H_{Ay} + H_{By} + H_{Cy} \quad (2)$$

The square of the effective value of the magnetic field is determined from the expression:

$$H_{RMS}^2 = \frac{1}{2\pi}\int_0^{2\pi} (H_x + H_y)^2 d\omega t \quad (3)$$

Expressions (1) and (2) are summed up, and then they are squared:

$$H_x^2 = \frac{I_0^2}{4\pi^2}\left(\frac{y^2 sin^2(\omega t - 120°)}{(y^2 + (x+d)^2)^2} + \frac{y^2 sin^2(\omega t)}{(y^2 + x^2)^2} + \frac{y^2 sin^2(\omega t + 120°)}{(y^2 + (x-d)^2)^2} + \frac{2y^2 sin(\omega t) sin(\omega t - 120°)}{(y^2 + (x+d)^2)(y^2 + x^2)} + \frac{2y^2 sin(\omega t) sin(\omega t + 120°)}{(y^2 + (x-d)^2)(y^2 + x^2)} + \frac{2y^2 sin(\omega t - 120°) sin(\omega t + 120°)}{(y^2 + (x+d)^2)(y^2 + (x-d)^2)}\right)$$

$$H_y^2 = \frac{I_0^2}{4\pi^2}\left(\frac{(x+d)^2 sin^2(\omega t - 120°)}{(y^2 + (x+d)^2)^2} + \frac{x^2 sin^2(\omega t)}{(y^2 + x^2)^2} + \frac{(x-d)^2 sin^2(\omega t + 120°)}{(y^2 + (x-d)^2)^2} + \frac{2x(x+d) sin(\omega t) sin(\omega t - 120°)}{(y^2 + (x+d)^2)(y^2 + x^2)} + \frac{2x(x-d) sin(\omega t) sin(\omega t + 120°)}{(y^2 + (x-d)^2)(y^2 + x^2)} + \frac{2(x-d)(x+d){I_0}^2 sin(\omega t - 120°) sin(\omega t + 120°)}{(y^2 + (x+d)^2)(y^2 + (x-d)^2)}\right) \quad (4)$$

The obtained expressions are substituted in (3) and integrated over the period:

$$H_{RMS}^2 = \frac{I_0^2}{8\pi^2}\left(\frac{1}{y^2 + (x+d)^2} + \frac{1}{y^2 + x^2} + \frac{1}{(y^2 + (x-d)^2)^2} + \frac{y^2 + x(x+d)}{(y^2 + (x+d)^2)(y^2 + x^2)} - \frac{y^2 + x(x-d)}{(y^2 + (x-d)^2)(y^2 + x^2)} + \frac{y^2 + (x-d)(x+d)}{(y^2 + (x+d)^2)(y^2 + (x-d)^2)}\right) \quad (5)$$

The following expressions were used during the integration:

$$\overline{sin^2(\omega t)} = \frac{1}{2}$$

$$\overline{sin^2(\omega t \pm 120°)} = \frac{1}{2}$$

$$\overline{sin(\omega t) sin(\omega t \pm 120°)} = \frac{1}{2}\left(cos(\pm 120°) - cos(2\omega t \pm 120°)\right) = -\frac{1}{4}$$

$$\overline{sin(\omega t + 120° \omega t) sin(\omega t - 120°)} = \frac{1}{2}\left(cos(240°) - cos(2\omega t)\right) = -\frac{1}{4}$$

The expression (5) is led to the common denominator:

$$H_{RMS}=\frac{I_{RMS}}{2\pi}\left(\frac{d^2[3(y^2+x^2)+d^2]}{(y^2+(x-d)^2)(y^2+(x+d)^2)(y^2+x^2)}\right)^{\frac{1}{2}} \quad (6)$$

The maximum value of magnetic field strength is directly over the vertical line at x=0:

$$H_{RMS}=\frac{I_{RMS}}{2\pi}\frac{d(3h^2+d^2)^{\frac{1}{2}}}{h(h^2+d^2)} \quad (7)$$

At $\frac{h}{d}\geq 10$ with accuracy better that 1% the approximate expression is applicable

$$H_{RMS}=\frac{I_{RMS}}{2\pi}\frac{d\sqrt{3}}{h^2}=0{,}276 I_{RMS}\frac{d}{h^2} \quad (8)$$

Approximation accuracy (8) referred to (7) is shown in table 1.

Table 1. Approximation accuracy (8) referred to (7).

$\frac{h}{d}$	Accuracy (%)
5	3.2
10	0.8
20	0.2

According to the expression (6), the exposure radius to the value $0.5H_{RMSmax}$ is approximately equal to $x \approx h$.

2.1. A three-phase line with the vertical arrangement of wires in row

The case of vertical arrangement of wires in row is obtained by shifting the system to 90 and replacing coordinates x and y in the expression (6)

$$H_{RMS}=\frac{I_{RMS}}{2\pi}\left(\frac{d[3(x^2+y^2)+d^2]}{(x^2+(y-d)^2)(x^2+(y+d)^2)(x^2+y^2)}\right)^{\frac{1}{2}}$$

The maximum value of magnetic field strength would be above the line x=0. In case y>d

$$H_{RMS}=\frac{\sqrt{3}I_{RMS}d}{2\pi}\frac{\left(y^2+\frac{1}{3}d^2\right)^{\frac{1}{2}}}{y(y^2-d^2)}$$

The approximate expression (8) is applied for large distances h. The approximate expression is also applied for any direction from the line and any distance h>>d.

$$H_{RMS}=0{,}276\frac{I_{RMS}d}{(h+d)^2}$$

2.2. A three-phase line with the triangular arrangement of wires

A three-phase line with the triangular arrangement of wires is shown in figure 3.

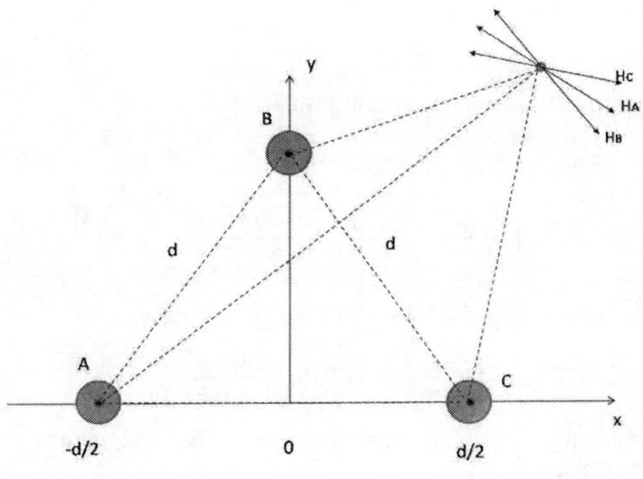

Figure 3. Three-phase line with arrangement of wires in triangle.

The origin of axes is placed in the middle of the lower side of an equilateral triangle. The distance between the wires is "d". We have obtained expressions for effective magnetic field strength by applying the same methodology as we did for the line with the arrangement of wires in row.

$$H_{RMS} = \frac{I_{RMS}\mathrm{d}}{2\pi}\left(\frac{\frac{3}{2}(x^2+y^2)-\frac{\sqrt{3}}{2}y\mathrm{d}+\frac{5}{8}\mathrm{d}^2}{\left(x^2+(y-\frac{\sqrt{3}}{2}\mathrm{d})^2\right)\left(\left(x-\frac{\mathrm{d}}{2}\right)^2+y^2\right)\left(\left(x+\frac{\mathrm{d}}{2}\right)^2+y^2\right)}\right)^{\frac{1}{2}} \quad (9)$$

Figure 6 shows graphs of magnetic field strength which were calculated by using the expression (9) for the different ratio of h/d. In this case, h is also counted from the top of the triangle.

2.3. Special cases are considered for this method of laying the wire core
Case of x=0:

$$H_{RMS} = \frac{I_{RMS}\mathrm{d}\left(\frac{3}{2}y^2-\frac{\sqrt{3}}{2}y\mathrm{d}+\frac{5}{8}\mathrm{d}^2\right)^{\frac{1}{2}}}{2\pi(y-\frac{\sqrt{3}}{2}\mathrm{d})\left(y^2+\frac{\mathrm{d}^2}{4}\right)} = \frac{0{,}195\frac{\mathrm{d}}{l}\left(1+\frac{1}{2}\left(\frac{\mathrm{d}}{l}\right)^2\right)^{\frac{1}{2}}}{l\left(1-\frac{1}{\sqrt{3}}\frac{\mathrm{d}}{l}\right)^2\left(1-\sqrt{3}\frac{\mathrm{d}}{l}+\frac{1}{3}\left(\frac{\mathrm{d}}{l}\right)^2\right)} \quad (10)$$

where

$$l = \left(x^2+\left(y-\frac{\sqrt{3}}{6}x\right)^2\right)^{\frac{1}{2}}$$

distance from the centre point of the line to the observation point.

Case of y=0, x>d/2:

$$H_{RMS} = \frac{I_{RMS}\mathrm{d}}{2\pi\left(x^2-\frac{\mathrm{d}^2}{4}\right)}\left(\frac{\frac{3}{2}x^2+\frac{5}{8}\mathrm{d}^2}{x^2+\frac{3}{4}\mathrm{d}^2}\right)^{\frac{1}{2}}$$

IOP Conf. Series: Materials Science and Engineering **791** (2020) 012033 doi:10.1088/1757-899X/791/1/012033

The appoximate expression is applicable for large distances h/d:

$$H_{RMS} \approx 0.195 \frac{I_{RMS} d}{h^2} \tag{11}$$

For a two-wire single-phase line we have the following expression:

$$H_{RMS} = \frac{I_{RMS} d}{2\pi \left(\left(y^2 + \left(x + \frac{d}{2} \right)^2 \right) \left(y^2 - \left(x - \frac{d}{2} \right)^2 \right) \right)^{\frac{1}{2}}} \tag{12}$$

If d<<h we have

$$H_{RMS} = \frac{I_{RMS} d}{2\pi h^2} = 0.159 \frac{I_{RMS} d}{(h + 0{,}86d)^2}$$

According to the graphs, the triangular arrangement has lower magnetic field strength due to more compact arrangement of wires.

It means a lower value than a three phase line at the same distance between wire cores.

Using expressions (8) and (12), it is possible to calculate the necessary distance to the centre of the line if the maximum permissible level of magnetic field strength 50 Hz [1-3] is provided.

3. Results

Figure 4 shows graphs of distribution of magnetic field strength for a three-phase line in horizontal arrangement at distance *h*.

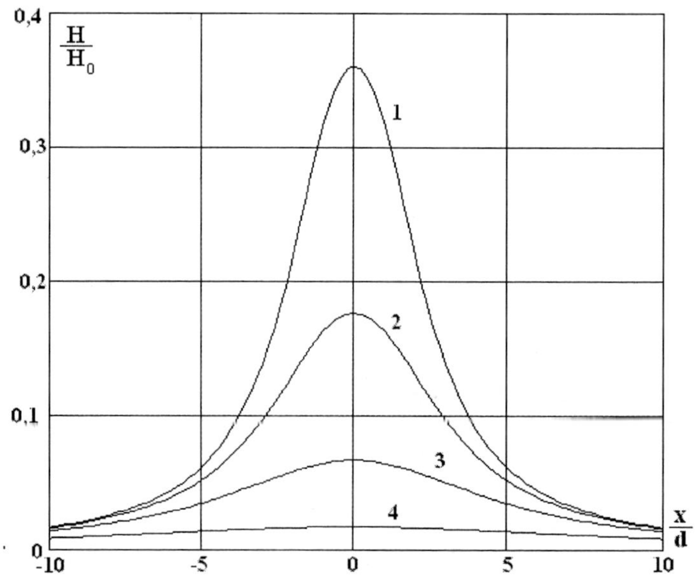

Figure 4. Distribution of nondimensional root mean square value of magnetic field strength in horizontal direction for cases of different distance from horizontal arrangement of row wires $H_0 = \frac{I_{RMS}}{2\pi d}$,

$$1 - \frac{h}{d} = 10, 2 - \frac{h}{d} = 5, 3 - \frac{h}{d} = 3, 4 - \frac{h}{d} = 2.$$

Figure 5 shows the distribution of magnetic field strength for different h/d. The distance h in this case is calculated from the upper wire.

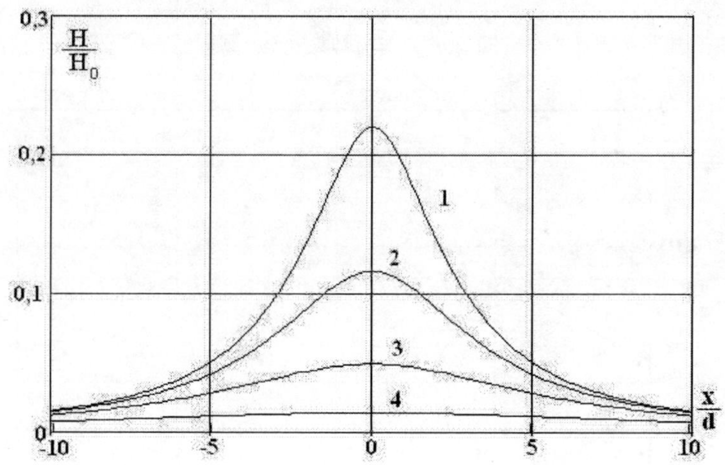

Figure 5. Distribution of nondimensional root mean square value of magnetic field strength in horizontal direction for different distances from the upper wire $H_0 = \frac{I_{RMS}}{2\pi d}$,

$$1 - \frac{h}{d} = 10, 2 - \frac{h}{d} = 5, 3 - \frac{h}{d} = 3, 4 - \frac{h}{d} = 2.$$

Figure 6 shows graphs of magnetic field strength which were calculated by using the expression (9) for the different ratio of h/d. In this case, h is also counted from the top of the triangle.

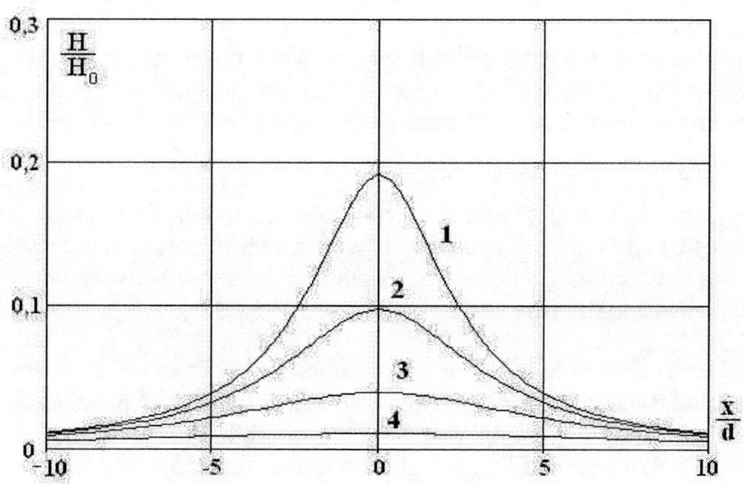

Figure 6. Distribution of nondimensional root mean square value of magnetic field strength in horizontal direction for different distances from the upper wire $H_0 = \frac{I_{RMS}}{2\pi d}$

IOP Conf. Series: Materials Science and Engineering **791** (2020) 012033 doi:10.1088/1757-899X/791/1/012033

Figure 7 shows graphs of distribution of magnetic field strength in horizontal plane for 3 options of underground arrangement of wires and h/d=10. (h is counted from the ground level to upper wires). It is less than in case of the arrangement of wires in row.

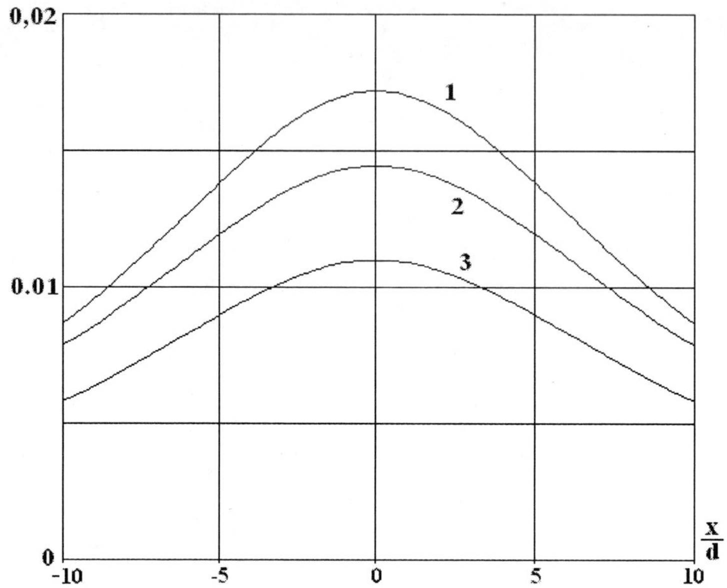

Figure 7. Distribution of nondimensional root mean square value of magnetic field strength in horizontal direction for different arrangement of wires: 1 – horizontal arrangement, 2 – vertical arrangement, 3 – triangular arrangement. $H_0 = \frac{I_{RMS}}{2\pi h}$.

4. Discussion

According to the graphs, the triangular arrangement has lower magnetic field strength due to more compact arrangement of wires.

It means a lower value than a three phase line at the same distance between wire cores.

Using expressions (8) and (12), it is possible to calculate the necessary distance to the centre of the line if the maximum permissible level of magnetic field strength 50 Hz [1-3] is provided.

5. Conclusion

1. Accurate expressions for calculation of magnetic field strength of three-phase line were obtained.
2. Approximations (8,11,12) are valid for d<<h with a measurement accuracy less than 1%, which show the proportion between magnetic field strength and distances between wires (d) and inverse proportion of square distance to the wire core.

References

[1] Swanson J and Renew DC 1994 Power frequency fields and people *Engineering Science & Education J. Institution of Eng. and Technology* (IET) **3(2)** 71-9
[2] Maddock B J 1992 Overhead line design in relation to electric and magnetic field limits *Power Eng. J.* **6 (5)** 217-24
[3] Merchant CJ, Renew DC and Swanson J 1994 Exposures to power-frequency magnetic fields in the home *J. of Radiological Protection* **14(1)** 77-87
[4] Bracken TD 2002 Assessing compliance with power-frequency magnetic-field guidelines *Health Phys* **83(3)** 409–16

[5] Stuchly M and Shengkai Z 1996 Magnetic field-induced currents in the human body in proximity of power lines *IEEE Transactions on Power Delivery* **11(1)** 102-9
[6] Barn F 1990 The Effect of Time Varying Magnetic Field on Biological Materials *IEEE Truns Magnetics* **26(5)** 2092-7
[7] Baishya MJ, Kishore NK and Bhuyan S, 2014 Calculation of electric and magnetic field safety limits under *UHV AC Transmission Lines Power Systems Conf. (NPSC) Eighteenth National* (India: IEEE) pp 1-6
[8] Budnik K and Machczyński W 2006 Contribution to studies on calculation of the magnetic field under power lines *European Transactions on Electrical Power* **16** 345
[9] Dawalibi F. 1986 *IEEE Transactions on Power Delivery PWRD* **1** 30-62
[10] Phan Tu V and Tlusty J 2003 The induced magnetic field calculation of three phase overhead transmission lines above a lossy ground as a frequency-dependent complex function *Power Eng. 2003 Large Eng. Systems Conf.* (Canada) pp 154-8
[11] Rozov V, Grinchenko V, Tkachenko O and Yerisov A 2018 Analytical calculation of magnetic field shielding factor for cable line with two-point bonded shields *IEEE 17th Int. Conf. on Mathematical Methods in Electromagnetic Theory (MMET)* (Kiev: IEEE) pp 358-61
[12] Dein A 2009 Magnetic-Field Calculation Under EHV Transmission Lines for More Realistic Cases *IEEE Transactions on Power Delivery* **24(4)** 2214-22
[13] Hernández VJ, Castronuovo ED and Sánchez I 2018 Optimal statistical calculation of power cables disposition in tunnels, for reducing magnetic fields and costs *Int. J. of Electrical Power and Energy Systems* **103** 360-8
[14] Memari AR and Janischewskyj W 1996 Mitigation of magnetic field near power lines *IEEE Transactions on Power Delivery* **11(3)** 1577-86
[15] Hanafy M Ismail 2009 Characteristics of the magnetic field under hybrid ac/dc high voltage transmission lines *Electric Power Systems Research* **79(1)**

Elements realization of software-algorithmic system optimization of technological processes based on adaptive methods

A L Rutskov[1], E V Sidorenko[2], V L Burkovsky[3] and Ya P Fedorov[4]

[1]Forecasting department of electricity balance, power and loss analysis, branch office OJSC «IDGC Centre» - «Voronezhenergo», Arzamaskaya 2, Voronezh, 394033, Russia
[2]Novovoronezh representation of NR Corporation SA ESKM, Moskovsky prospect, 179, office building 3, Voronezh, 394066, Russia
[3]Department of electric drive, automation and control in an engineering system, Vornezh State Technical University, Moscow prospect, 179, Voronezh, 394066, Russia
[4]Corporate finance department and corporate management of the Financial University under the Government of the Russian Federation, Leningradsky prospect, 49, Mocow, 125993, Russia

E-mail: alex_8_90@mail.ru

Abstract. The paper examines a possibility to optimize complex technological processes of energy resources transfer and consumption. Notes the importance of unidentified and weakly formalized factors consideration, which influence quality of the described object classes functioning. Suggests use of fuzzy neural networks and controllers on their basis for implementation of technological processes optimization of energy resources transfer and consumption at the expense of its dynamic accuracy forecast improvement. Presents algorithmic realization, that later on configured in the MATLAB simulated environment. Quantitative and qualitative parameters, which characterize fuzzy neural controller's usage quality, were received. Later, software environment elements are being described, which constructed on the basis of the suggested adaptive approach for complex technological processes optimization. Points out a possibility to conduct comparative study, and also simulation modeling implementation by way of presented solutions.
Keywords: optimization of complex technological processes, electrical power systems, electrical energy loss minimization, fuzzy neural networks, accuracy of forecast, elements of a software system.

1. Introduction

Optimization of complex technological processes in electrical power systems (TP EPS), in the most common case, is implemented with the criterion of minimum capacity loss and can be presented as follows:

IOP Conf. Series: Materials Science and Engineering **791** (2020) 012034 doi:10.1088/1757-899X/791/1/012034

$$F_j=\sum_{i=1}^{k} W_i=\sum_{i=1}^{k}\left[\frac{\left(\sqrt{P_i(U)^2+Q_i(U)^2}\right)\cdot R_{Ei}}{U_i^2}\cdot\left(\frac{1}{\left(1-\frac{\Delta U_{\%i}}{100}\right)^2}-1\right)+\right. \tag{1}$$

$$\left.+\Delta P_{xxi}\cdot T_{pi}\cdot\left(\frac{U_i}{U_{nom\,i}}\right)^2+\Delta p_{cor\,i}\cdot L_i\cdot k_{Ucor\,i}\right]\rightarrow\min,$$

where W – total losses of electrical power in TP EPS;
ΔR_E, L – equivalent active resistance and length of elements for energy power transmission;
ΔU, % – change of the input voltage value as a percentage of the nominal value;
$\Delta P_{xx\,i}$ – no-load power loss in TP EPS elements;
$T_{p\,i}$ – number of TP EPS hours of work;
$\Delta p_{cor\,i}$ – mean specific losses due to corona in the electrical power transmission elements in TP EPS;
$k_{Ucor\,i}$ – corona loss coefficient.

In addition, the following constrains should be met

$$\begin{cases}\left|P_{int\,i}+P_{out\,i}\right|>0;\\ \left|Q_{int\,i}+Q_{out\,i}\right|>0;\\ i=1,\ldots,k;\end{cases} \tag{2}$$

where $P_{int\,i}$, $Q_{int\,i}$ – values of internal real and reactive power flow of elements in TP EPS;
$P_{out\,i}$, $Q_{out\,i}$ – values of outer real and reactive power flow of elements in TP EPS.

2. Materials and Methods

TP EPS performance indicators optimization is often performed on the basis of the Lagrangian method applied to the expression (1) with constrains (2). Weakness of this solution is an actual neglect of undetermined components, which have strong influence on the electrical power flow distribution [1, 2-7].

Gradient methods with fixed and variable step size are taken as an alternative variant (in particular, on the basis of the Newton-Raphson algorithm).

For TP EPS performance indicators optimization with consideration of unidentified components it is worthwhile to use a modified Newton-Raphson algorithm with fuzzy neuron controllers (FNC) inclusion in its structure [8, 9]. Realization of the mentioned approach have sequence of steps, which presented in the algorithm shown in Figure 1 [9].

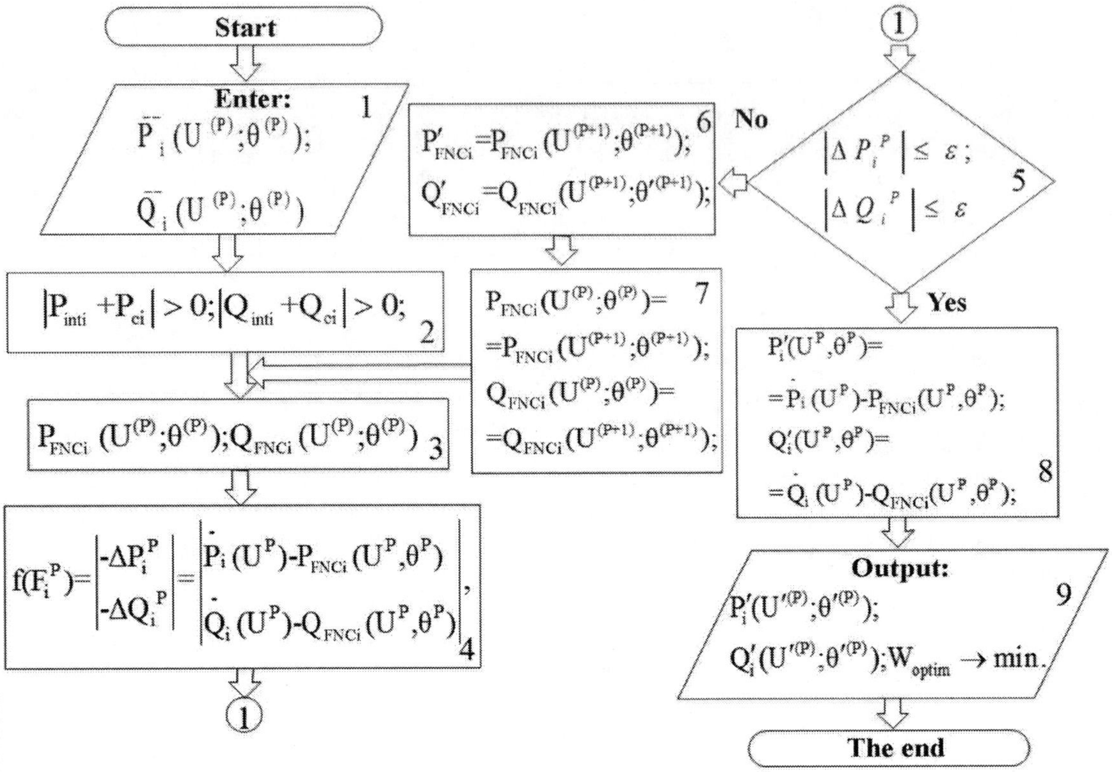

Figure 1. Modified Newton-Raphson algorithm for TP EPS optimization with FNC inclusion

Figure 1 presents the following notations:

$P_{FNCi}(U^{(P)};\theta^{(P)})$, $Q_{FNCi}(U^{(P)};\theta^{(P)})$ – forecast of real and reactive power with use of FNC;

$f(F^P)$ – unbalanced vector on n - th iteration;

$\bar{P}(U^P), \bar{Q}(U^P)$ – assignment of real and reactive power in the nodal of TP EPS;

ΔP^P, ΔQ^P – real and reactive power residuals;

P'_{FNCi}, Q'_{FNCi} – adjusted vectors of real and reactive power;

$U^{(P+1)}, \theta^{(P+1)}$ – voltage amplitude and phase lag during the iterations.

Table 1 presents comparative study results of TP EPS control system functioning using modified Newton-Raphson algorithm, shown in figure 1. Realization implemented within the ANFIS structure of MATLAB software system framework.

IOP Conf. Series: Materials Science and Engineering **791** (2020) 012034 doi:10.1088/1757-899X/791/1/012034

Table 1. Comparative study of based on fuzzy neural network models realizations.

FNN learning algorithm – Sugeno (hybrid)					FNN learning algorithm – Mamdani (hybrid)		
Input membership term types (configuration $3\times3\times3$)							
	term	trapmf	psigmf	dsigmf	trapmf	psigmf	dsigmf
Estimated prameters	ε_n	0.271	0.228	0.228	0.152	0.061	0,056
	r	0.981	0.989	0.987	0.972	0.989	0,974
	$\varepsilon_{\%}$	3.145	2.012	2.005	3.478	2.054	2,159
Nuber of training epochs		500	6	6	500	30	27
Learning time, sec.		67.46	1.03	1,05	67.44	4.27	3.85
FNN learning algorithm – Sugeno (hybrid)					FNN learning algorithm – Mamdani (hybrid)		
Input membership term types (configuration $6\times3\times3$)							
	term	trapmf	psigmf	dsigmf	trapmf	psigmf	dsigmf
Estimated prameters	ε_n	0.090	0.053	0.061	0.087	0.028	0,032
	r	0.980	0.987	0.988	0.979	0.987	0,988
	$\varepsilon_{\%}$	2.954	1.903	1.985	2.903	1.874	1,889
Number of training epochs		500	16	17	200	40	33
Learning time, sec.		68,74	2.33	2.48	27.01	5.47	4.68
FNN learning algorithm – Sugeno (hybrid)					FNN learning algorithm – Mamdani (hybrid)		
Input membership term types (configuration $9\times3\times3$)							
Estimated pa-rameters	term	trapmf	psigmf	dsigmf	trapmf	psigmf	dsigmf
	ε_n	0.021	0.012	0.012	0.073	0.011	0,013
	r	0.983	0.991	0.993	0.985	0.997	0,994
	$\varepsilon_{\%}$	2.901	1.853	1.751	2.101	1.609	1,645
Number of training epochs		500	51	53	100	38	34
Learning time, sec.		69,41	7.03	7.36	13.88	5.24	4.69
Processor type - Intel Core i3-7350 – Core Processor (4.2 GHz)							

Best results from the analyzed sample demonstrated FNN realization on the basis of Mamdani algorithm using input terms configuration and membership function - psigmf. figure 2 presents results of fuzzy neural network training in the context of this variant realization. Test sample deviations from the exit results is an indication of high accuracy functioning of given system.

IOP Conf. Series: Materials Science and Engineering **791** (2020) 012034 doi:10.1088/1757-899X/791/1/012034

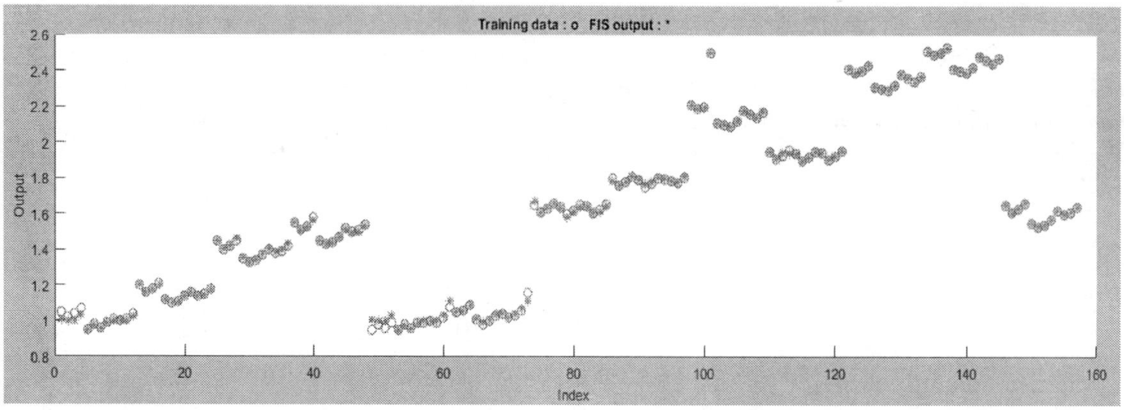

Figure 2. Fuzzy neural network training using an input membership term function of psigmf type.

3. Results

Received during functioning of mentioned TP EPS algorithm optimization results are realized in the form of software system elements [9]. User, in addition, is provided with a possibility to perform both on-line (up to 1 day) and medium-term parameters control (weekly, monthly variations) of technological cycle of conversion, transfer and consumption of energy resources. To construct necessary dependences, one should choose a period of interest, that needs analysis of power flow in TP EPS. Realized software system also allows to conduct collaborative calculations for normative method of energy resources distribution and perform comparative study with system functioning based on FNC.

Graph of the collaborative analysis results for normative (classic) process of energy resources control and FNN system is shown in figure 3. It can be seen that in the presented block of software system was also realized a function «Compare control forecast with real values», which gives a possibility in graphical format to compute combined graphs of forecast control samples and real values for the same periods.

Interface of the developed software elements contains interrelated forms, first of which has an area for automatic or manual fill-out of user`s information.

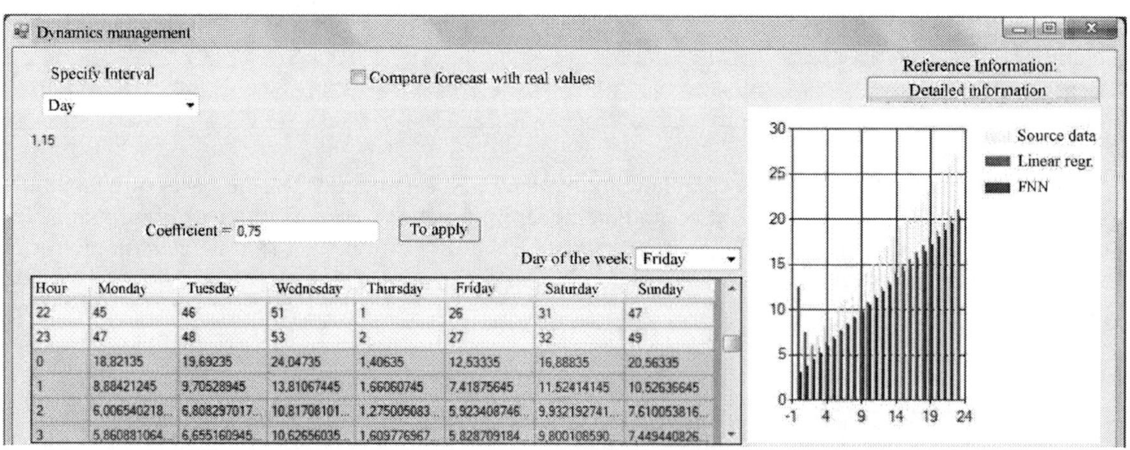

Figure 3. Collaborative analysis bloc of real and controlled measures for on-line and medium-term periods.

Upon completion of the corresponding data bases generation appears an option to fill-out parameter settings both classic method and realization on the basis of FNN (FNC) figure 4. It is worth noting,

IOP Conf. Series: Materials Science and Engineering **791** (2020) 012034 doi:10.1088/1757-899X/791/1/012034

that the considered approach framework gives a possibility to analyze data not only in the context of power but also transferred electricity volume.

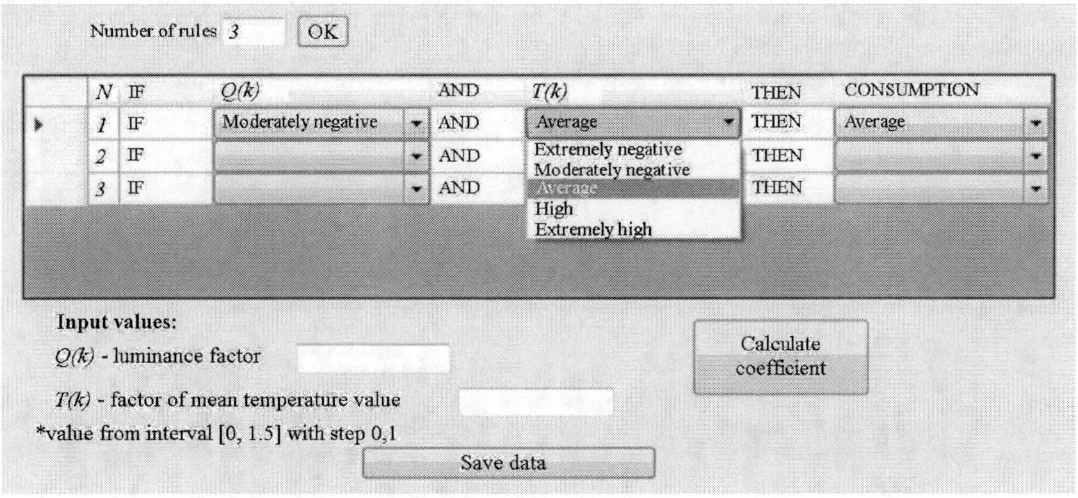

Figure 4. Generation of the functioning rules for FNN (FNC).

In the software system elements is also organized an option to preview reference data about algorithms [10], which are used to control flow of energy resources in TP EPS [11-14].

The developed solution also provides input data validity check at different stages, among them, for the data bases in the previous periods. So, before TP EPS optimization algorithm begins to work there is a check for correctness and completeness of the input data, which can be updated when necessary.

The developed interface of the software system is user friendly both for automatic and for manual mode use, and that was confirmed by the testing results.

Hardware and software tools for modeling and analysis of alternative structures in the control system that handles technological processes of energy resources conversion, transfer and consumption realized by means of the host-program ODP.exe, layout of which is presented in Figure 5 for the modeling subsystem of «Generation/CHP» type [15].

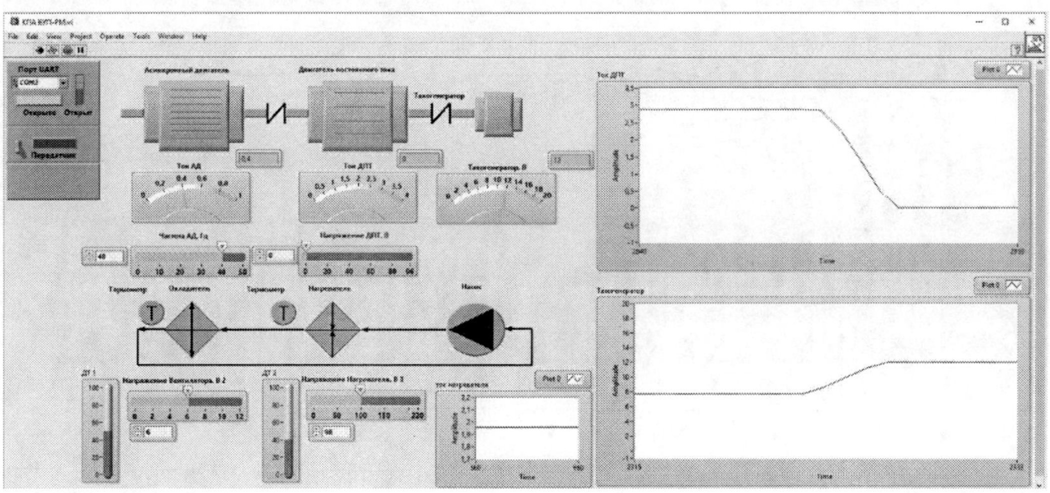

Figure 5. User`s interface for the modeling subsystem of «Generation/CHP» type.

To establish connection to a hardware part there are switch and window to choose a port to communicate by data transmission virtual port. To assign work mode (DAC control) there are sliders and windows for data digital writing. Both dial indicators and oscillogram graphs of transient processes are used for the information output.

Analisis of the TP EPS optimization model functioning using measure of the energy resources loss at its minimum, which has been constructed on the FNC basis, allows to say about its high accuracy characteristics, that are much more accurate than existing classic realizations (around 3-6%), which is proved by the data in figures 6,7.

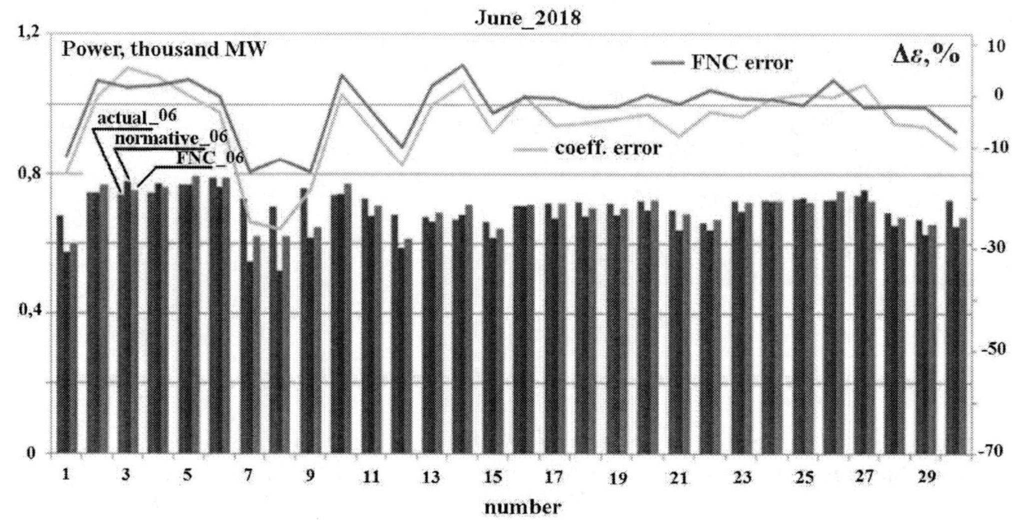

Figure 6. Optimization forecast for the TP EPS load center of Voronezh city (June, 2018): actual, normative, FNC – thousand MW; forecast systems functioning error based on the normative method and FNN usage, %.

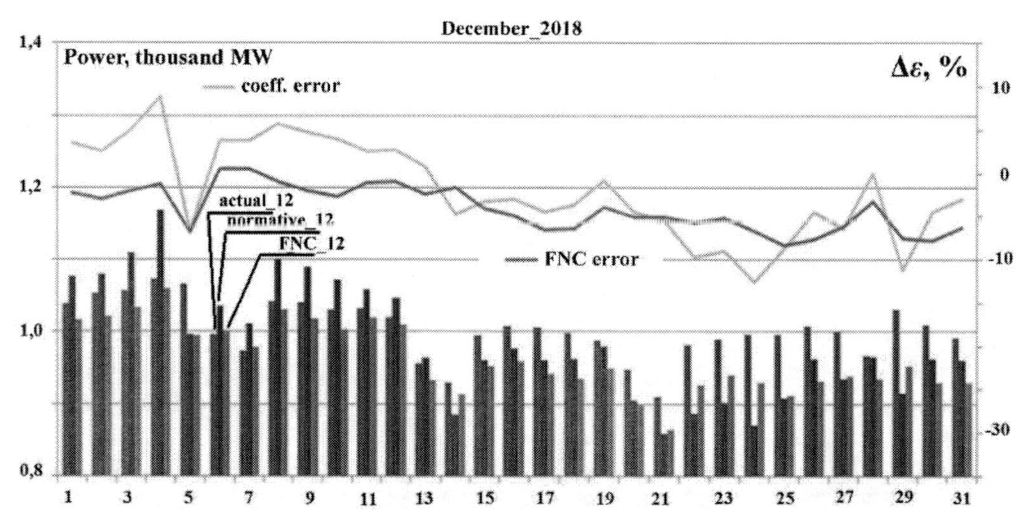

Figure 7. Optimization forecast for the TP EPS load center of Voronezh city (December, 2018): actual, normative, FNC – thousand MW; forecast systems functioning error based on the normative method and FNN usage, %.

4. Conclusion

1. Conducted optimization of complex TP EPS with consideration of weakly formalized factors by using fuzzy neural networks and controllers on their basis. Presented algorithm which allows realization of the control parameters correction in real time mode.

2. Forecast error in on-line and medium-term period is in the range of 1,6% to 3,5%, that allows to say about high potential usage of the described approach in the existing energy load centers (in particular in power generating industry).

3. Developed software system elements for TP EPS optimization allow to carry out automated monitoring and control of load parameters based on evaluation of actual and forecast graphs of power flows distribution dynamics.

4. There is a possibility to organize feedback from TP EPS actuator mechanisms and their optimization on the basis of the performed in advance simulation modeling.

5. Received within the framework of the developed software system elements results of TP EPS Voronezh load center functioning prove reliability of conclusions, with regard to relevance usage of FNN (FNC) as an adaptive structure, which allow to take into consideration weakly formalized and also unidentified factors.

References

[1] Kry`sanov V N, Sharapov YuV and Ruczkov A L 2015 Analiz programmno- apparatnogo obespecheniya setej Smart Grid [Hardware and software analisis of Smart Grid networks] *Vestnik VGTU* **11(4)** 97-101 [In Russian]

[2] Migliore M et al. 2006 Parallel network simulations with NEURON *J. of Computational Neuroscience* **21(2)** 119-29

[3] Berenji H R 1992 A Reinforcement Learning- Based Architecture for Fuzzy Logic Control *Int. J. of Approximate Reasoning* **6** 267-92

[4] Wilamowski B M, Iplikci S, Kayank M O and Efe M O 2001 An Algorithm for Fast Convergence in Training Neural Networks *Int. Joint Conf. on Neural Networks (IJCNN'01)* (Washington) pp 1778-82

[5] Xie T T, Yu H and Wilamowski B M 2010 Replacing fuzzy systems with neural networks *Proc. IEEE Conf. on Human System Interaction* (Rzeszow, Poland) pp 189-93

[6] Yu H and Wilamowski B M 2009 Efficient and reliable training of neural networks *Proc. 2nd IEEE Human System Interaction* (Catania, Italy) pp 109-15

[7] Borkowski L S 2010 Response of a Hodgkin-Huxley neuron to a high-frequency input *Phys. Rev. E* **80(5)** 051914

[8] Nechaev Yu I et al. 2007 Strength and vibration multimode control for ship, moving in the ice condition *Proceedings of the 10th Int. Symp. on Practical Design of Ships and Other Floating Structures* (Houston, Texas USA) vol 2 1295-1302

[9] Kry`sanov V N, Ruczkov A L and Shukur O Ma 2015 Sistemnaya optimizaciya e`konomicheskix pokazatelej funkcionirovaniya sistemnogo operatora v Voronezhskoj e`nergosisteme [System optimization of economic function indicators of the system operator in Voronezh power system] *E`lektrotexnicheskie Kompleksy` i Sistemy` Upravleniya* **3** 62-7 [In Russian]

[10] Burkovskij V L, Kry`sanov V N and Ruczkov A L 2016 Realizaciya programmnogo kompleksa prognozirovaniya urovnya regional`nogo e`nergopotrebleniya [Realization of software system for regional energy consumption level forecast] *Vestnik VGTU* **12(3)** 41-7 [In Russian]

[11] Davison A.P. et al. 2008 PyNN: A Common Interface for Neuronal Network Simulators *Frontiers in Neuroinformatics* **2** 11

[12] Miśkowicz M 2007 Latency characteristics of event-driven task scheduler embedded in neuron chip *IJCSNS Int. J. of Computer Science and Network Security* **7(12)** 132-49

[13] Brovko A V, Murphy E K and Yakovlev V V 2009 Waveguide microwave imaging: neural

network reconstruction of functional 2-D permittivity profiles *IEEE Transactions on Microwave Theory and Techniques* **57(2)** 406-14

[14] Sultan Alneyadi, Elankayer Sithirasenan and Vallipuram Muthukkumarasamy 2016 A survey on data leakage prevention systems *J. of Network and Computer Appl.* **6(62)** 137-52

[15] Burkovskij V L, Kry`sanov V N, Ruczkov A L and Shukur O M 2016 Realizaciya e`lementov programmnogo kompleksa prognozirovaniya regional`nogo e`nergopotrebleniya na baze nejronnoj seti [Realization of software system elements for energy consumption level forecast on the basis of neural network] *Sovremenny`e Metody` Prikladnoj Matematiki, Teorii Upravleniya i Komp`yuterny`x Texnologij (PMUKT-2016) Sbornik Trudov IV Mezhdunarodnoj Konf.* (Voronezh) pp 59-63 [In Russian]

IOP Conf. Series: Materials Science and Engineering **791** (2020) 012035 doi:10.1088/1757-899X/791/1/012035

The selection of a priority package of measures for technical re-equipment and reconstruction of the district of electrical grids

N G Semenova[1] and A D Chernova[2]

[1]Department of Automatic Electric Drive, Electromechanics and Electrotechniques, Orenburg State University, Pobedy Av., 13, Orenburg, 460018, Russia
[2]Department of Heat Technology and Power Engineering, Orenburg State University, Prospect Pobedy, 13, Orenburg, 460018, Russia

E-mail: fiara@inbox.ru

Abstract. Making decisions on the technical re-equipment and reconstruction of electrical grids needs analyzing a large amount of information, comparing a variety of options, evaluating the impact of a decision in the long term. Therefore, the process of formation of the package of measures is complicated. In this regard, it is necessary to develop an automated decision support system for selecting a priority package of measures for the technical re-equipment and reconstruction of the electrical grid district. The article presents the structure of the decision-making process for the technical re-equipment and reconstruction of the district of electrical grids, the model of formation and the method for selecting a priority package of measures. It is proposed to use graph theory for compiling a decision tree and methods of combinatorial optimization for the formation and selection of a priority package of measures. A two-level structural scheme of the decision-making process for the selection of a priority package of measures for the technical re-equipment and reconstruction of the electrical grids at the local and global levels has been developed.

1. Introduction

The variety of features of the technical re-equipment and reconstruction (TR&R) of electrical grids: many goals of functioning with the uncertainty of growth of load; increase in the number of estimated parameters; modern socio-economic requirements, leads to an increase in the number of development options and the amount of information processed. This necessitates not only the automation of information processes but also the introduction of intelligent decision support systems (DSS)[1]. Currently, DSS are being actively developed for analyzing, modeling, planning, operating and servicing customers in the electricity distribution system from alternative sources [2-3], creating microgrids and connecting them to the Smart Grid [4-5], optimizing investment management for power grid companies [6], projects on the repair of electrical grids [7], the reconfiguration of the energy distribution system [8], the regulation of the hydro-power plant's operating mode [9], managing of the generating company [10]. However, there is no modern DSS research in the field of decision making on the technical re-equipment and reconstruction of electrical grids; therefore, the topic of paper is relevant and significant.

For electrical grids characterized by a hierarchical structure, multiply connected and nested objects. As noted in [11], "the emergence of a hierarchical structure was due to the increasing complexity of

IOP Conf. Series: Materials Science and Engineering **791** (2020) 012035 doi:10.1088/1757-899X/791/1/012035

the technology of controlled objects, which creates great difficulties for centralized management. Therefore, it has become necessary to divide the entire decision-making process into such a number of levels so that the solution to the optimization problem at each of them is not difficult."

In the article district of electrical grids is a set of technologically interconnected objects, located on the same territory and served by one structural unit. An electrical grid object (EGO) is a collection of substation and power lines that supplying it. The study proposed to consider a two-tier decision-making structure: at the first (local) level, the selection of preferred alternatives for the technical re-equipment and reconstruction of electrical grid facilities is made, and at the second (global) level – the selection of a priority package of measures for the technical re-equipment and reconstruction of the district of electrical grids, formed from the preferred alternatives for EGO.

2. Two-Level Decision-Making Structure, Mathematical Model of the Formation and the Method of Selecting a Priority Package of Measures for the Technical Re-Equipment and Reconstruction of Electrical Grid District

2.1. Two-level decision-making structure for the technical re-equipment and reconstruction of electrical grid district

Analysis of the literature in the field of decision-making on the technical re-equipment and reconstruction has revealed the following main stages of decision-making [12]: the formation of alternatives; evaluation of alternatives; selection of a preferred alternative. Based on the proposed two-level decision-making structure, the authors adapted the classical decision making steps in accordance with the purpose of the study.

Stages of the decision-making process for the technical re-equipment and reconstruction of electrical grid object at the local level:

1 Determining the need for a solution: formalization of the problem of making a decision on the technical re-equipment and reconstruction of the EGO by assessment the technical state of the equipment;

2 Generation of possible solutions: the formation of alternatives depending on the type of nomenclature work in the framework of technical re-equipment or reconstruction of EGO;

3 Selection of decision criteria: justification and development of mathematical models of criteria for evaluating alternatives for the technical re-equipment and reconstruction of EGO;

4 Comparison and selection of preferred EGO technical re-equipment and reconstruction alternatives;

5 Evaluation of results and feedback.

Stages of the decision-making process for the technical re-equipment and reconstruction of electrical grid district at the global level:

1 The selection of the district of the electrical grids, limiting the number of objects, limiting the number alternatives;

2 Generation of solutions: the formation of the possible package of measures from the preferred alternatives for the technical re-equipment and reconstruction of electrical grid objects;

3 Evaluation of alternative package of measures by decision criteria;

4 Ranking and selection of a priority package of measures for the technical re-equipment and reconstruction of the district of electrical grids;

5 Evaluation of results and feedback.

Based on the above, the authors have developed a structural decision-making model for the technical re-equipment and reconstruction of the district of electrical grids. It is presented in Figure 1.

The local level of decision making was considered in detail by the authors in [13]. This article describes the second stage of decision making at the global level. At this stage, it was proposed to select the priority package of measures for the technical re-equipment and reconstruction of the district of electrical grids as a solution to the optimization problem.

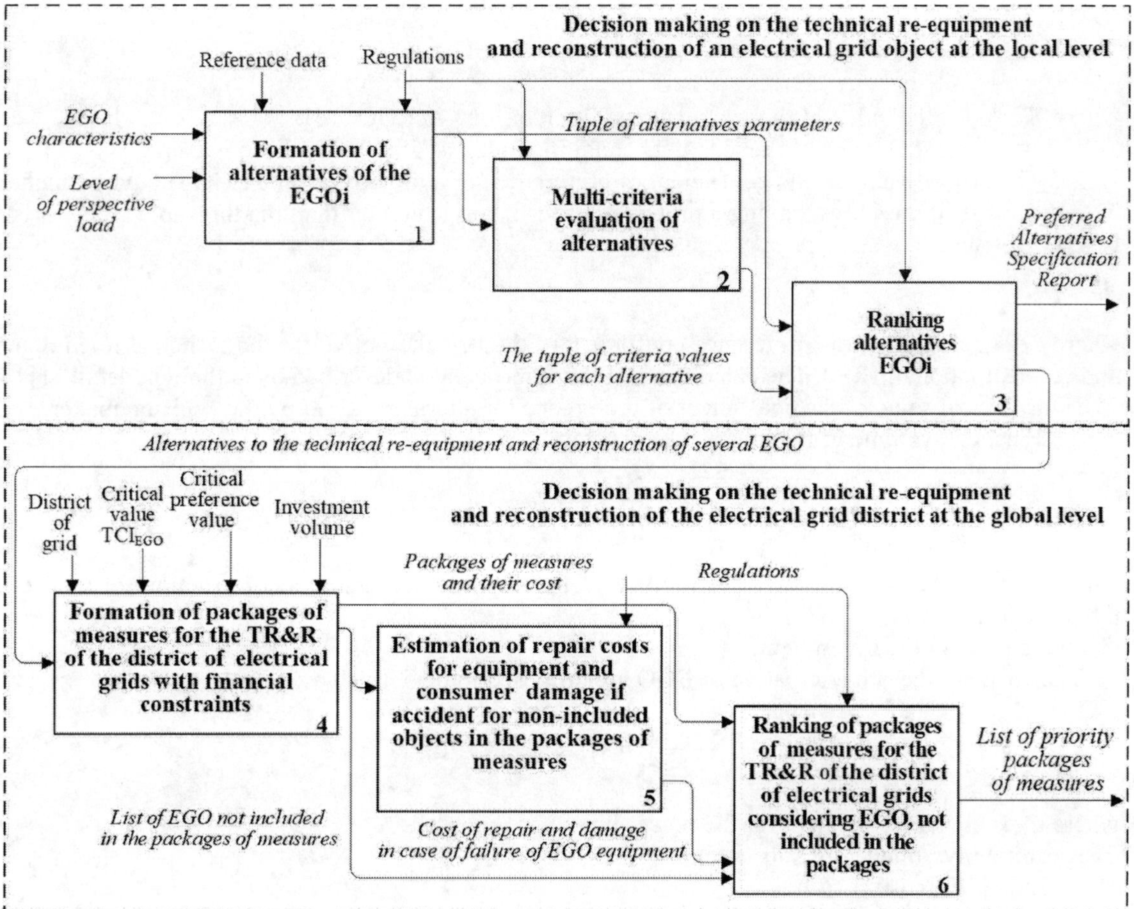

Figure 1. Structural model of decision-making on the selection of the priority package of measures for the technical re-equipment and reconstruction of the district of electrical grids.

2.2. Mathematical model of the formation of a package of measures for the technical re-equipment and reconstruction of the district of electrical grids

The selection of a package of measures for the technical re-equipment and reconstruction of the district of electrical grids is the problem of combinatorial optimization. To solve it, the authors identified and formalized:

1 Objective function: minimization of repair costs and consumer damage from interruptions in power supply if the equipment fails:

$$\sum_{j=1}^{n}(M_j + F_j) \rightarrow \min \tag{1}$$

where M_j is operating costs for the refurbishment of the j-th equipment that is not included in the equipment replacement program;

F is damage from undersupply of electricity in case of failure of worn-out equipment of the *j-th* EGO.

2 Constraints:

IOP Conf. Series: Materials Science and Engineering **791** (2020) 012035 doi:10.1088/1757-899X/791/1/012035

2.1 Critical value of the technical condition index. The EGO should be included in the package of measures, having a technical state index less than the threshold value. It is set by the decision maker (DM):

$$TCI_{EGOi} \le TCI_{cr} \tag{2}$$

where TCI_{EGOi} is the index of the technical condition of the *i-th* EGO [14];
TCI_{cr} is the threshold value of the technical condition index, set by the decision maker.

2.2 The critical value of the coefficient of preference for alternatives. The EGO should be included in the package of measures, having a preference coefficient is greater than the threshold value. It is set by DM:

$$T_i \ge T_{cr} \tag{3}$$

where T_i is the value of the preference coefficient of the *i-th* alternative for the technical re-equipment and reconstruction of EGO. The calculation of this coefficient is described by authors in detail in [13];
T_{cr} is the critical value of the coefficient of preference for alternatives, set by the decision maker.

2.3 Financial volume of the investment program:

$$\sum_{i=1}^{n} C^{EGOi} \le C_{\max} \tag{4}$$

where C^{EGOi} is the costs of new construction, technical re-equipment or reconstruction for the i-th alternative of EGO;
C_{max} is the size of the investment.

The costs for the i-th alternative of EGO are proposed to be defined as:

$$C^{EGO} = \sum_{t=0}^{3} (C_{Lt} + C_{St} + C_{dt} - C_{rest}) \cdot (1 + E_н)^{-t} \tag{5}$$

where C_L is capital investment for the power line;
C_S is capital investments for substations;
C_d is dismantling costs;
C_{res} is residual value of the equipment.

3 Boundary conditions: the considered quantities must be positive, and the sequence numbers of the objects must belong to the natural numbers:

$$\begin{aligned} &M_j, F_j, C_i, C_{\max}, TCI_{EGOi}, TCI_{cr}, T_i, T_{cr}, > 0 \\ &\mathrm{i, j} = 1, 2 \ldots n \end{aligned} \tag{6}$$

2.3. The method of selecting a priority package of measures for the technical re-equipment and reconstruction of the electrical grids

All measures should use modern technical solutions and a comparative analysis of options, ensure the achievement of strategic development goals and fulfill key performance indicators, but exclude redundant technical solutions. If the volume of the investment program is not enough for the implementation of the whole package of measures, the objects are ranked according to the degree of prioritization. So, it is necessary to search for combinations of alternatives for the technical re-equipment and reconstruction of objects corresponding to typical technological solutions with financial constraints. The authors developed a method with the following stages for the formation of a package of measures for the technical re-equipment and reconstruction of the district the electrical grids.

Stage 1. Reduction of the number of elements used in the formation of a package of measures The reduction is carried out by considering the EGO, located on the territory of only one of the districts of electrical grids, as well as the constraints defined by (2-4).

Stage 2. Determining the priority of adding the EGO to the package of measures for the technical re-equipment and reconstruction of the districts of electrical grids. An analysis of internal regulations of electrical grid companies on investment programs was conducted. On its basis, the division of the EGO into three groups of prioritization depending on the technical condition, level of prospective load was developed and implemented in software.

The rationales for the first group of objects included implementation were:

- restoration of the power supply scheme and the achievement of the standard parameters of the damaged object as a result of the accident;
- the need to reduce the risks of a power failure caused by weather conditions;
- the need to comply with the requirements of supervisory and regulatory authorities;
- the need to replace objects in pre-emergency condition.

The second group of objects has measures to fulfill obligations under concluded and forward-looking contracts for technological connection.

The third group of objects has measures, which are necessary for technical re-equipment: an increase in the capacity of the electrical grid and the capacity of substations.

Stage 3. Formation of a preliminary package of measures for the technical re-equipment and reconstruction of the district the electrical grids. To find all the possible combinations of measures generated decision tree [15]. Its nodes contain combinations of EGO numbers and alternatives for each of them. The tree has a root – a zero node without designation.

Pre-determined the sequence of adding EGO_i in the branches of the tree according to their group prioritization. First, an EGO from the first group is added to the branch, then an EGO from the second group is added, and at the end, an EC EGO from the third group is added. The authors explain the process of forming a decision tree by the example of three objects (Figure 2).

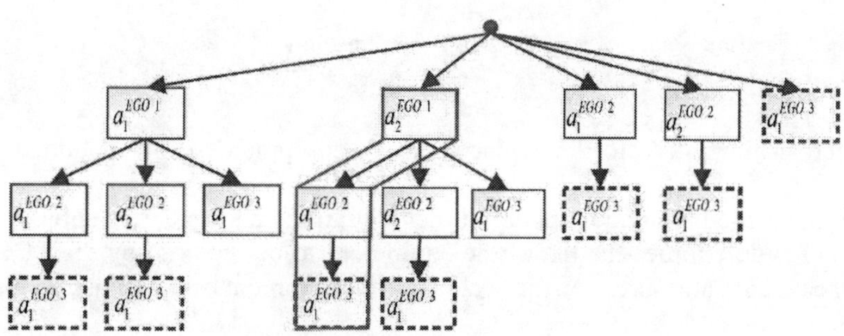

Figure 2. Decision Tree for 3 EGO.

Possible solutions for the technical re-equipment and reconstruction of the district of electrical grids are the branches of this tree. For example, the solution may contain for the first object a second alternative A_2^{EGO1}, for the second object – the first alternative A_1^{EGO2}, for the third object – the first alternative A_1^{EGO3}. In Figure 2, an example of the solution is highlighted with a broken line.

To find all potentially realizable combinations, the tree is traversed in direct order by a depth-first search. The tree is traversed down as much as possible on each child, before moving on to the next related branch. The tree is traversed in direct order: each of the parent nodes is viewed before its child.

The transition to the child node is carried out if the total cost of alternatives for all EGO in the branch is less than the volume of the investment program (4). If the volume of the investment program for $i + 1$ node is exceeded:

$$\sum_{i=0}^{n} C^{EGO_{i+1}} > C_{\max} \tag{7}$$

IOP Conf. Series: Materials Science and Engineering **791** (2020) 012035 doi:10.1088/1757-899X/791/1/012035

this node-parent i (with respect to the considering node $i + 1$) was the final vertex of the branch - a leaf.

At this stage, the formation of the branch is completed. All EGO and their alternatives a_m^{EGOi} in the branch nodes were added to the list of the preliminary package of measures for the technical re-equipment and reconstruction of the electrical grids:

$$A^{EGO_i} = \{a_1^{EGO_1}, a_3^{EGO_2} ..., a_m^{EGO_i}\} \quad (8)$$

4 stage. Determination of the list of point-replaceable equipment. Formation of the list of equipment of electrical grids objects not included in the package of measures. Compiled a list of equipment objects that do not fall into the branch of the decision. It was ranked by technical condition index:

$$E = \{e_1^{EGO'_1}, e_2^{EGO'_2} ... e_k^{EGO'_j}\}, TCI_{e_k} < TCI_{e_{k+1}} \quad (9)$$

where e_k is a unit of equipment installed on EGO'_j.

For each piece of equipment, a list of alternatives A^{ek} was compiled. It ranked by a coefficient of preference T_{am}:

$$A^{e_k} = \{a_1^{e_k}, a_2^{e_k} ..., a_m^{e_k}\}, T_{a_m} > T_{a_{m+1}} \quad (10)$$

The unspent funds of the investment program were determined for each package of measures:

$$C_{res} = C_{\max} - \sum_{i=1}^{n} C^{EGO_i} \quad (11)$$

If the unspent funds of the investment program were sufficient for a point replacement of one or several units of worn-out equipment:

$$C_{res} - \sum_{m=1}^{k} C_{a_m^{e_k}}^{EGO'_j} \to 0 \quad (12)$$

then a cost-effective alternative to the replacement of equipment a_m^{ek} was added to the package of measures.

5 Determination of the list of equipment not included in the package, and subject to repair.

The equation (12) determines the list of the equipment to be repaired by the EGO, which is not included in any package of measures for the technical re-equipment and reconstruction of the district of electrical grids:

$$V^{EGO'_j} = \left\{ v \middle| \; v \notin A^{EGO_i} \wedge v \notin A^{e_k} \right\} \quad (13)$$

The equipment included in this set cannot be replaced within the investment program. It is recommended for him to conduct repairs at the expense of operational costs.

3. Results

The result of the formation of the package of measures for the technical re-equipment and reconstruction of the district of electrical grids are:

1 Objects with equipment replacement alternatives included in the A^{EGOi} solution branch;

2 List of pointwise replaceable worn equipment a_m^{ek};

3 For each package a list of EGO'_j with equipment $V^{EGO'j}$ not included in the investment program, subject to repair at the expense of operating resources.

The selection of the priority package was carried out by ranking the packages of measures according to the value of the objective function. The objective function was determined by (1) by calculating the amount of operating costs M associated with the repair of equipment $V^{EGO'j}$ not included in the

IOP Conf. Series: Materials Science and Engineering **791** (2020) 012035 doi:10.1088/1757-899X/791/1/012035

package of measures, and the damage F from an interruption in the power supply in case of accidents. The ranking was carried out by a selection sort algorithm. As a result, an ordered sequence of packages was created by attaching one package after another in order of increasing costs. Thus, a priority package of measures was determined for the most responsible and worn-out object with limiting the investment program.

Priority package of technical re-equipment and reconstruction measures for the Central district of electrical grids without financial constraints is presented in Table 1. Financial volume of the investment program consist 653159.9 ths rubles.

Table 1. Priority package of technical re-equipment and reconstruction measures for the Central district of electrical grids without financial constraints.

Number of EGO	Power Transformer MVA	Circuit Breaker HV	Circuit Breaker MV	Circuit Breaker LV	Cost ths rubles
1	Change to 2x40	Change to SF6	Change to VCB	Change to VCB	292817.7
2	Change one to 40	No change	-	No change	84127.4
3	Change to 2x10	Change to VCB	-	Change to VCB	49743.3
4	Change to 2x25	Change to SF6	Change to VCB	Change to VCB	226471.5

Priority package of technical re-equipment and reconstruction measures for the Central district of electrical grids with financial constraints in 610000 ths rubles is presented in Table 2. Financial volume of the investment program consist 607220.3 ths rubles.

Table 2. Priority package of technical re-equipment and reconstruction measures for the Central district of electrical grids with financial constraints.

Number of EGO	Power Transformer MVA	Circuit Breaker HV	Circuit Breaker MV	Circuit Breaker LV	Cost ths rubles	Repair costs and consumer damage ths rubles
1	Change to 2x40	Change to SF6	Change to VCB	Change to VCB	292817.7	9377.1
2	Change one to 40	No change	-	No change	84127.4	
3	No change	Change to VCB	-	Change to VCB	3804.3	
4	Change to 2x25	Change to SF6	Change to VCB	Change to VCB	226471.5	

4. Discussion

Software implemented method in an automated decision support system. Testing of DSS and approbation on the example of EGO of the Central District of electrical grids of the Orenburg region was carried out. The obtained results do not contradict the proposals for the substations of the Central District of electrical grids indicated in the documents on the development of the power industry of the Orenburg region.

5. Conclusion

A two-level structural scheme of decision making on the technical re-equipment and reconstruction of the district of electrical grids has been developed. It contains the local level of decision making on the selection of preferred alternatives for the technical re-equipment and reconstruction of electrical grid object; the global level of decision-making on the selection of a priority package of measures for the technical re-equipment and reconstruction of the EC region, formed from the preferred alternatives to the technical re-equipment and reconstruction of the EGO, taking into account financial constraints.

A mathematical model for the formation of a package of measures for the technical re-equipment and reconstruction of the district of electrical grids is proposed. It consists of a set of equations:

- the objective function to minimize the cost of repairs and damage from interruptions in power supply in case of equipment failure;
- three constraints to take into account the threshold values of the technical condition index, the preference coefficient of alternatives and the financial volume of the investment program;
- the boundary condition for considering only positive values of quantities.

A method for selecting a priority package of measures for the technical re-equipment and reconstruction of the district of electrical grids has been designed. It is based on graph theory and combinatorial optimization. To implement the method proposed the following steps:

- reduction of the number of elements in the formation of a package of measures for the technical re-equipment and reconstruction of the district of electrical grids;
- determination of the sequence of adding electrical grid objects to the package of measures;
- formation of a preliminary package;
- determination of the list of point-replaceable equipment;
- definition of the list of repaired equipment, not included in the packages.

The developed model and method are implemented in a decision support system that has been tested. The results of the DSS are consistent with the schemes and energy development programs for the electrical grids of the Orenburg region.

References

[1] Ada S and Ghaffarzadeh M 2015 Decision making based on management information system and decision support system *European Researcher* **93** 260-9

[2] Barriquello C H, Garcia V J, Schmitz M, Bernardon D and Fonini J S 2017 A decision support system for planning and operation of maintenance and customer services in electric power distribution systems *System Reliability* (London: IntechOpen) pp 355-70

[3] Jones L E 2011 Strategies and decision support systems for integrative variable energy resources in control centers for reliable grid operations URL https://www.osti.gov/servlets/purl/1032531

[4] Asaleye D A, Breen M and Murphy M D 2017 A decision support tool for building integrated renewable energy microgrids connected to a smart grid *Energies* **10** 1765

[5] Mazurek M 2013 Concepts of distributed database for decision support in smart grid *Biuletyn WAT* **LXII** 115-28

[6] Yang C, Su Z, Wang X and Liu Y 2017 Design of investment management optimization system for power grid companies under new electricity reform *AIP Conf. Proceedings* **1820** 040007

[7] Ramsay B, McPherson A, Eastwood R, Ozveren C S and Glare J Oatley 1997 A decision support system for electricity distribution network refurbishment project *Electric Power Systems Research* **40** 27-35

[8] Andervazh M-R, Javadi S and Hosseini Aliabadi M 2018 Decision support system for multicriteria reconfiguration of power distribution systems using CSO and efficient graph traversal and repository management techniques *Int. Trans. Electr. Energ. Syst.* **28** e2579

[9] Liao J, He L, Yuan X and Li H 2008 Application of decision support system to three gorges cascade hydropower station *Int. Conf. on Genetic and Evolutionary Computing* **2** 500-3

[10] Moshkin B N, Myatezh T V, Sekretarev Y A and Averbukh M A 2019 Mathematical model of managing of the generating company on the criterion of the profit maximization *IOP Conf. Ser.: Mater. Sci. Eng.* **552** 012016

[11] Altunin A Y and Semukhin M V 2000 *Modeli i algoritmy` prinyatiya reshenij v nechetkix usloviyax* [Decision models and algorithms in fuzzy conditions] (Tyumen: Publishing house of Tyumen State University) [In Russian]

[12] Soloviev N A, Chernoprudova E N and Lesovoy D A 2012 *Osnovy` teorii prinyatiya reshenij dlya programmistov* [Basics of decision making theory for programmers] (Orenburg: OSU) [In Russian]

[13] Semenova N G and Chernova A D 2018 Vy`bor predpochtitel`nogo resheniya po razvitiyu e`lektricheskix setej na osnove nejrosetevy`x texnologij [Selecting the Preferred Solution on Development of Electrical Supply System Based on Technology of Neural Networks] *Vestnik Yuzhno-Ural`skogo Gosudarstvennogo Universiteta. Seriya: E`nergetika* [*Bulletin of SUSU. Series Energy*] **18** 38-43 [In Russian]

[14] Khal'yasmaa A I, Dmitriev S A and Kokin S E 2016 An automated system for taking decisions to assess the actual state of electrical equipment *Power technology and Eng.* **49** 389-92

[15] Levitin A 2011 *Introduction to the Design and Analysis of Algorithms* (Boston: Addison-Wesley)

IOP Conf. Series: Materials Science and Engineering **791** (2020) 012036 doi:10.1088/1757-899X/791/1/012036

Development of an evaluation method for electromechanical systems unification

L N Titova[1], U A Pertsev[1], S A Samodurova[2] and P O Nuhin[1]

[1]Faculty of electromechanical systems and electricity supply Voronezh State Technical University, Moscow prospect 14, Voronezh, 394026, Russia
[2]Faculty of cadastre, real estate, land management and geodesy Voronezh State Technical University, Moscow prospect 14, Voronezh, 394026, Russia

E-mail: TLN64@mail.ru

Abstract. The paper examines problematic areas of electromechanical systems unification, formulates and presents formal writing of problem for systemic unification of electromechanical systems and its solution procedure, describes general evaluation scheme of technical-and-economic effect from implementation of different variants of electromechanical systems unification. Presents results of the calculation relations that allow to carry out full and thorough formulation of technical specifications with due consideration of possible unification variants that brings out maximum result at the designing and execution project stage, operation and servicing of the electromechanical systems. Method is designed to evaluate values of technical-and-economic outcome from work that is being conducted in the interests of the suggested unification variants implementation for electromechanical system specimens at the development stage. Presented method is the primary way to construct criterial function of system unification problem. Project`s future outlook can be determined at any point of the designing stages with consideration to service life of the product that is being in development, product family, its components, which included in the electromechanical system, as a separate entity and also as a part of an entire system.

1. Introduction

As of now electromechanical systems consist of a big number of interdependent parts that are diverse by function-based tasks and structural design nature. Development of component parts (CP) for electromechanical systems should be done with unification of this systems in mind. This kind of work is critical for the electromechanical converters due to a variety of structural designs, that limits area of its application. Electromechanical systems converters assortment acknowledges insufficient unification in this industry. [1].

Electromechanical systems development and operation requires a lot of time, material and financial spending. Existing limitations on this kind of expenses makes unification work during the development of this systems very important. [2].

2. Materials and Methods

Conducted analysis of existing way and content of complicated technical systems unification work, which electromechanical systems part of, and also status analysis of this kind of work allows to make several very important conclusions that outlines directions for improvement of not just unification work itself but also methodological support for this kind of work.

IOP Conf. Series: Materials Science and Engineering **791** (2020) 012036 doi:10.1088/1757-899X/791/1/012036

In the first place, while unification work for complicated technical systems is conducted at the research and project justification stages, impact from its implementation should be evident and assessed for all stages of the life cycle. However, at present time, process of unification, continuous in terms of content, its results and subjects, practically «disintegrates» into several independent and consequent work efforts for unification at every stage of the life cycle. Those inconsistencies, in the first place, are connected to the different degree of detail consideration in these systems as a subject of unification [3]. In fact, if during research and project justification of electromechanical system subjects, within the context of strategic (program) planning, all interdependent advanced electromechanical system entirety at the related planning period considered as a subject of unification, then at the development stage of specific subject of energy system as a subject of unification, in general, considered specific subject excluding its place in the advanced entirety of such systems. Therefore, in fact, at present time, during unification work at this stage, results of the previous stage are not considered. In addition, when choosing ways of subject unification at this stage, influence of this choice on the whole advanced entirety is not considered. In other words, influence of the chosen way of unification of the subject in development on the degree of unification and other technical-and-economic and operation-and-maintenance characteristics of other system specimens, that are part of advanced entirety, is not being taken into consideration. In addition, influence of these specimens on the under development [4] is also not being taken into consideration. Disadvantage of the existing approach to organize implementation of the end-to-end unification process of electromechanical system specimens also determines disadvantages of the existing methodological support.

This allows to establish that existing methodological support is not sufficient to make well founded system-based decisions about electromechanical system unification problems, which are relevant to all levels of its disaggregation, interconnected with life cycle stages [5] and with all possible directions of unification.

In view of this, facts of research work duplication are present, especially at the CP elements level of electromechanical systems, also reduction tendency to borrow ready-made, previously developed CP, and, as a result, this creates increase in assortment of non-compatible but with similar characteristics CP, and increase of expenses for development and operation of electromechanical systems in general.

All of this suggests a necessity for improvement of methodological support for electromechanical systems unification work from the prospective of system approach. In other words, review unification work and outcome evaluation from it, in the context of all life cycle stages, and as a unification subject – all advanced entirety of one-type electromechanical systems on all levels of its disaggregation and in the context of all possible directions of unification.

From the stated above follows that during development of the problem statement for systematic unification of electromechanical systems it is necessary, as a unification subject, consider all entirety of such systems, having in mind all its structural and functional complexity and inconsistency, and also dynamics of its evolution on all life cycle stages. Estimating potentially possible outcome from electromechanical systems unification work one should consider that this outcome is achieved by implementing specific directions of unification during development of isolated samples of such systems.

In such case, during development of the problem statement for systematic unification of the advanced electromechanical systems we should consider the following model of unification.

For any direction of unification, according to the P 50-54-102-88 «Operating procedures for intersectoral, sectoral and factory unification», which included into possible variant of electromechanical system specimens unification [6], as a result of implementing one of the unification variant, some elements (subjects of unification) of the initial homogeneous elements set remain unchanged, and some, that form subsets, replaced by the others, as a rule, more advanced elements. Therefore, set of elements split up into several subsets, to be exact, subset where elements are not being replaced (not unified) and several subsets, where elements being replaced with one more advanced element. As a result of this, set (manifold) with a smaller number of elements is obtained [7]. From this kind of unification model follows that all ways of getting technical-and-economic outcome from different variants of uni-

IOP Conf. Series: Materials Science and Engineering **791** (2020) 012036 doi:10.1088/1757-899X/791/1/012036

fication implementation are connected with reduction of elements manifold – subjects of unification. Mainly, in the context of different life cycle stages of specimen, these are gains obtained by means of:

- cutback in a number of engineering designs (newly designed specimens and their CP);
- cutback in number of types, mass-produced and purchased specimens of electromechanical systems and its CP;
- cutback in an assortment of electromechanical system specimens and their CP which are on supply and in operation.

In addition, it is necessary to consider fact that for the accepted model of unification, some elements of initial manifold (set) are being substituted for more advanced elements, this creates excessive quality, and, therefore, creation of extra expenses for «servicing» this excessiveness [8].

Keeping in mind the suggested model of unification, problem of systematic unification of electromechanical systems can be formulated in the following way.

For given initial set of required specimens, and also for initial set of required CP, included in it, one should chose variant of advanced entirety unification of electromechanical systems, that provides maximum technical-and-economic outcome from its implementation for given efficiency level of electromechanical system specimens, technical level of its CP and given amount of financing for development and operation of electromechanical systems advanced entirety.

When selecting measure for technical-and-economic effect, one should consider [9] fundamental matching principle: «to what extent one state or process is better (worth) that the other state (process) ». From this perspective, using accepted model of unification, as a basis for criterial function construction of formulated problem of systemic unification is appropriate to take full impending expenses for engineering and operation of electromechanical system specimens, that has been developed using one or another variant of unification and included into the advanced entirety of electromechanical systems.

In the interests of formal writing development of formulated systemic unification problem, let us introduce the following notations.

Let us denote by:

$m = 1, 2,\ldots, M$ – work stage number, and M - its total number;

$i = 1, 2,\ldots, I$ – electromechanical systems disaggregation level number, and I - its total number;

$k = 1, 2,\ldots, K$ – homogeneous specimens group number, and K - its total number;

$n = 1, 2,\ldots, N$ – electromechanical system specimens and its CP numbers for the initial group sets of homogeneous specimen (CP) sets, respectively, and N - its total number;

$n^* = 1, 2,\ldots, N^*$ – electromechanical system specimens and its CP numbers for the resultant (unified) group sets of homogeneous specimens (CP), and N^* - its total number;

$\{O_n\}$ and $\{O^*n^*\}$ – sets of homogeneous specimens (CP) of electromechanical systems before conducting unification work (initial) and after it (resultant);

H_j – j - th direction of unification, where $j = 1, 2,\ldots, 5$;

$\{B_j\}$ – sets of unification variants of electromechanical system specimens (CP), in addition, unification variant B_j described by mapping $\{O_n\} \overset{\hat{H}_j}{\Rightarrow} \{O^*_{n^*}\}$;

H_j – operator, which transfers the initial set into the resultant on j - th direction of unification.

Then formulated problem can be written in the following way:

$$B^{m+1} = \arg\max_{j_{k_i}} \sum_{k_i, j_k=1}^{K_i, J_k} E_{k_i}\left[B^m, B^{m+1}_{j_{k_i}}(H_{j_{k_i}}, O_{j_{k_i}}, O^*_{j^*_{k_i}}), W(O^*_{j^*_{k_i}})\right] \tag{1}$$

$$F(H_{j_{k_i}}, O^*_{j^*_{k_i}}) \geq F^0_{j^*_{k_i}} \tag{2}$$

$$TY(H_{j_{k_i}}, O^*_{j^*_{k_i}}) \geq TY^0_{j^*_{k_i}} \tag{3}$$

$$C(H_{j_{k_i}}, O^*_{j^*_{k_i}}) \geq C^0 \tag{4}$$

where: W(On) – electromechanical system specimens (CP) demand function;
E – function that characterizes difference of full impending expenses for development and operation of electromechanical system specimens, which are being developed without unification and with implementation of unification variant which is under consideration;
F – function that characterizes the electromechanical system specimens effectiveness;
F^0 – required levels of the electromechanical system specimens effectiveness;
TY – function that characterizes technical excellence of electromechanical systems CP;
TY^0 – required levels of technical excellence of the electromechanical system specimens CP;
C – function that characterizes full impending expenses for the development and operation of the advanced entirety of electromechanical system specimens;
C^0 – given financing volume for the development and operation of advanced entirety of electromechanical system specimens;
k_i – group numbers of the electromechanical system homogeneous specimens (CP) on i-th level of disaggregation;
j_{k_i} – unification direction number for k_i - th group of homogeneous electromechanical system specimens (CP).

General scheme of the presented procedure which solves systemic unification problem for electromechanical systems (1)-(4), based on suggested approach, chosen principles and corresponds to its content and specific, described earlier, is presented on figure 1.

Figure 1 also presents the assortment of methods, which are necessary to solve systemic unification problem (1)-(4).

For evaluation of technical-and-economic effectiveness of work, which is being conducted in the context of the suggested unification variants [10] of electromechanical system specimens implementation as part of systemic unification of electromechanical systems and its development, procedure is presented, which is the basis for construction of criterial function of systemic unification problem.

In accordance with this and with systemic unification of electromechanical systems problem statement, value of technical-and-economic effectiveness of work to unify electromechanical systems is taken as a difference in full impending expenses for development and operation of electromechanical system specimens, which is being developed in accordance with chosen variant of unification and being developed without its unification, therefore:

$$E = C - C^u \tag{5}$$

where E – technical-and-economic effect value;
C^u, C – full impending expenses for specimens considering expenses for its unification work and without such work, respectively.

Therefore, in general case:

$$C = C_{dd} + C_n W + C_{op} W T \tag{6}$$

$$C^u = C^u_{dd} + C^u_n W^u + C^u_{op} W^u T^u, \tag{7}$$

where C^u_{dd}, C_{dd}, C^u_n, C_n C^u_{op}, C_{op} – costs of development, production and average cost of specimen operation over the year period in accordance with its unification and without it, respectively;
W^u, W – demand (requirement) for a specimen in accordance with its unification and without it, respectively;

T^u, T – useful lifetime of specimen in years in accordance with its unification and without it, respectively. Then:

$$E = (C_{dd} - C_{dd}^u) + (C_n W - C_n^u W^u) + (C_{op} WT - C_{op}^u W^u T^u) \quad (8)$$

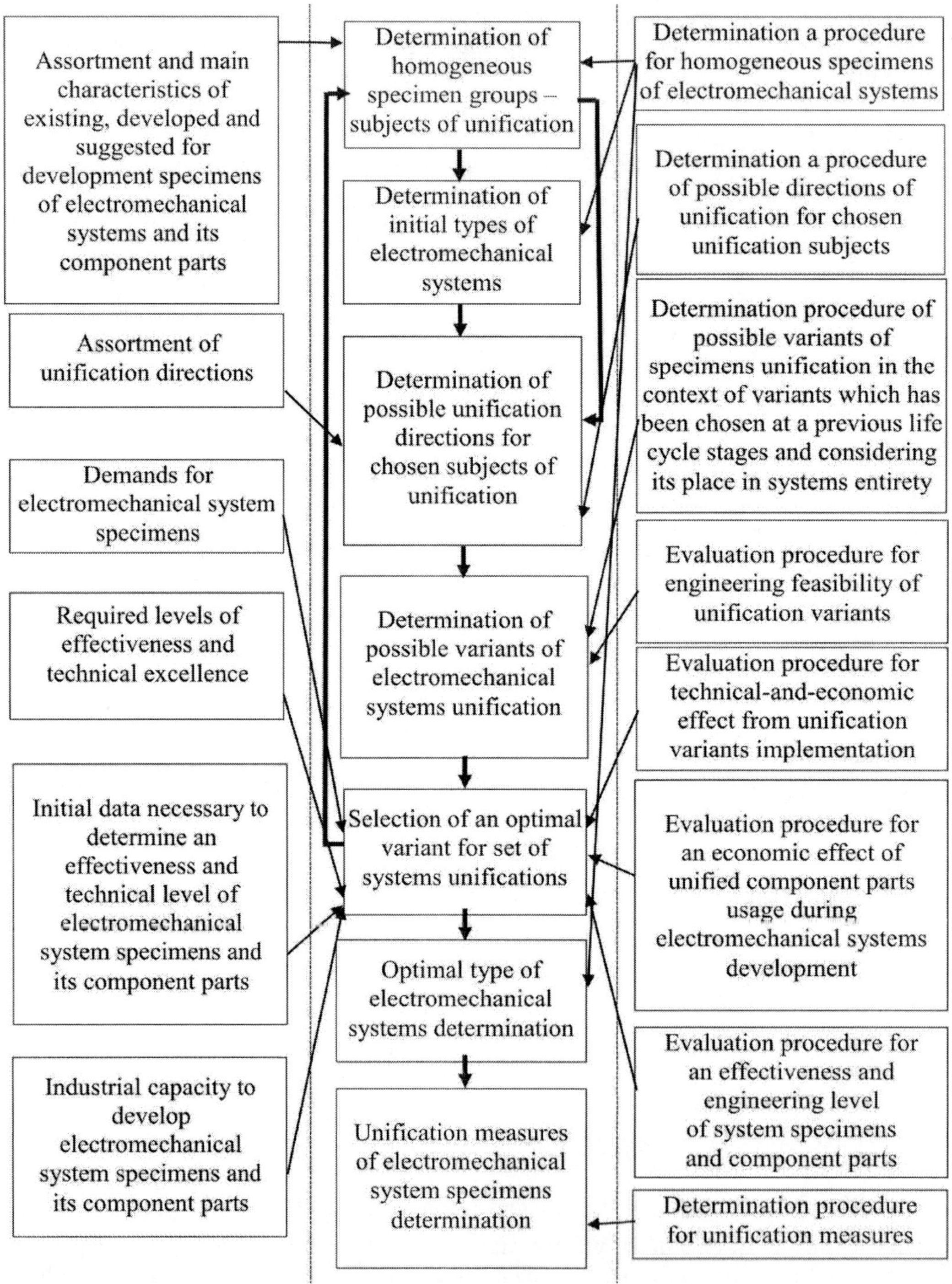

Figure 1. Problem solving procedure general scheme.

To receive calculation relations to estimate value of technical-and-economic effect of electromechanical systems unification work, on the basis of routine unification problem formulation [6], and without constrains for generality of received results, let us be guided by the following unification model [11] of advanced electromechanical systems.

Set of homogeneous electromechanical systems is given, which are part of the advanced electromechanical system specimens entirety *{Oi}*, where $i = 1, 2, \ldots, I$, basic characteristics of these specimens and demand for it are given W^i, respectively. At a later stage, during problem solving of technical-and-economic effect evaluation to determine value of demand for component parts (CP) it is necessary to consider demand, caused by inclusion of component parts into the SPTA and further use of CP during specimen operation. Also, initial set *{CPi}* of required homogeneous CP is given, which are part of the electromechanical system specimens, and its main characteristics, full impending expenses for it are given as well, in other words C^i_{dd}, C^i_n, C^i_{op}, T^i are given. It is supposed that CP, which are parts of different specimens, also different. It is assumed that CP aligned in increasing order of its quality and technical level (values of its main characteristics), and therefore, its development cost.

3. Results and Discussion

Variant of electromechanical system specimen unification has been developed (generated), which means that truncated resultant is determined after work completion for unification of CP set, which are part of electromechanical system specimens, meaning that also given a scheme to divide initial set *{CPi}* into subsets *{CPi}*$_j$, which are being replaced, on the basis of the unification work results, for one of the component part of truncated (resultant) CP set $\{CP^*_j\}$, where $j = 1, 2, \ldots, J$. Basic characteristics are also given CP^*_j.

Let us introduce notation for the scheme to replace CP of initial set for CP* of truncated (resultant) set:

$$\{CP_i : i = 1, \ldots I\} = \{CP_i : i = 1, \ldots, i_1\}_1 \cup$$

$$\{CP_i : i = i_1 + 1, \ldots i_2\}_2 \cup \{CP_i : i = i_2 + 1, \ldots, i_3\}_3 \cup \ldots \tag{9}$$

$$\{CP_i : i = i_{j-1} + 1, \ldots i_j\}_j \cup \ldots \{CP_i : i = i_{J-1} + 1, \ldots, I\}_J ,$$

where i_j can take a value from 1 to I - 1 if conditions are met:

$$\sum_{j=1}^{J} i_j = I \tag{10}$$

$$J \leq I - 1 \tag{11}$$

In addition, every subset *{CP*$_i$*}*$_j$ of the initial set is replaced for single element (component part) of the truncated (resultant) set $\{CP^*_j\}$. Let us denote full impending expenses, development, production, annual operating costs and time in service of j-th CP of truncated (resultant) set by C^{*j}, C^{*j}_{dd}, C^{*j}_n, C^{*j}_{op} and T^{*j} respectively, and by W^{*j} – demand for it.

In this case, value of technical-and-economic effect from implementation of considered unification variant of considered electromechanical system specimens can be written as follows:

$$E=\left(\sum_{i=1}^{I}C_{dd}^{i}-\sum_{j=1}^{J}C_{dd}^{*j}\right)+\left(\sum_{i=1}^{I}C_{n}^{i}W^{i}-\sum_{j=1}^{J}C_{n}^{*j}W^{*j}\right)+ \\ +\left(\sum_{i=1}^{I}C_{op}^{i}W^{i}T^{i}-\sum_{j=1}^{J}C_{op}^{*j}W^{*j}T^{*j}\right). \tag{12}$$

Having in mind the accepted replacement scheme we can write that:

$$W^{*j}=\sum_{i=i_{j-1}+1}^{i_j}W^{i}. \tag{13}$$

It is obvious, that summarized for the whole group of homogeneous electromechanical system specimens demand for CP stays unchanged, which means that:

$$\sum_{j=1}^{J}W^{*j}=\sum_{i=1}^{I}W^{i}=\sum_{j=1}^{J}\sum_{i_{j-1}+1}^{i_j}W^{i}. \tag{14}$$

Let`s describe, in addition to that, the following limitations for determining calculated relations which are perfectly acceptable for the considered problem statement and accepted scheme of advanced electromechanical systems unification, meaning that $C_{op}^{i}=C_{op}^{*j}$ and $T^{i}=T^{*j}$ for all values i and j.

Then:

$$E=\left(\sum_{i=1}^{I}C_{dd}^{i}-\sum_{j=1}^{J}C_{dd}^{*j}\right)+\left(\sum_{i=1}^{I}C_{n}^{i}W^{i}-\sum_{j=1}^{J}\sum_{i=i_{j-1}+1}^{i_j}W^{i}C_{n}^{*j}\right)= \\ =\sum_{i=1}^{I}\left(C_{dd}^{i}+W^{i}C_{n}^{i}\right)-\sum_{j=1}^{J}\left(C_{dd}^{*j}+C_{n}^{*j}\sum_{i=i_{j-1}+1}^{i_j}W^{i}\right). \tag{15}$$

Obtained expression for E is the general expression for value evaluation of the technical-and-economic effect from unification variant implementation.

To obtain much more simple relations to calculate E et us assume that:

$$C_{n}^{i}=\alpha^{i}C_{dd}^{i} \tag{16}$$

Coefficients α^{i}, which connect development cost with mass production cost [12], can be determined on the basis of available statistic support and equal to 0,1÷0,4.

In addition, let us take into account that production cost (manufacturing) is defined in the context of product cost decrease at the expense of series production increase [13], meaning increase of demand. With that the following relation is true:

$$C_{n}^{*j}=C_{n}^{i_j}\left(\frac{W^{i_j}}{\sum_{i=i_{j-1}+1}^{i_j}W^{i}}\right)^{\gamma}, \tag{17}$$

IOP Conf. Series: Materials Science and Engineering **791** (2020) 012036 doi:10.1088/1757-899X/791/1/012036

where γ - measure, which takes into consideration decrease of product cost at the expense of series production increase.

Value γ can also be determined on the basis of available statistic support and equals $\gamma = 0{,}3$.

Then:

$$E = \sum_{i=1}^{I} C_{dd}^{i}(1+\alpha^{i}W^{i}) - \sum_{j=1}^{J}\left[C_{dd}^{*j} + C_{n}^{i_j}\left(\frac{W^{i_j}}{\sum_{i=i_{j-1}+1}^{i_j} W^{i_j}}\right)^{\gamma}\right]. \quad (18)$$

At the same time, for $C_n^{i_j}$ and W^{i_j} we take a value of mass production cost and demand for the last element of *j*-th subset of initial CP.

Let us assume that *j*-th CP of truncated (resultant) set, which can replace, based on its own capability, any CP from *j*-th initial subset of CP, will possess certain excessiveness, and, therefore, its development is going to be more expensive.

Having in mind, that dependence of CP development cost on its quality and technical level (its basic characteristics) is continuous and not decreasing, we can write $C_{dd}^{*j} = \beta^{j} C_{dd}^{i_j}$, where β^{j} – coefficient, which corrects for increase (cost increase) of the CP development at the expense of its excessiveness.

Then expression (18) can be written as follows:

$$E = \sum_{i=1}^{I} C_{dd}^{i}\left(1+\alpha^{i}W^{i}\right) - \sum_{j=1}^{J} C_{dd}^{i_j}\left[\beta^{j} + \alpha^{j}\left(\frac{W^{i_j}}{\sum_{i=i_{j-1}+1}^{i_j} W^{i_j}}\right)^{\gamma}\right]. \quad (19)$$

For $C_{dd}^{i_j}$ we take development cost of the biggest by number CP in *j*-th subset of the electromechanical system specimen`s life cycle.

4. Conclusion

In summary, it is worth noting that obtained relation allows reasonably accurate prediction of the technical-and economic effect using sufficiently small volume of initial data, which allows to determine optimal values of unification measures for advanced electromechanical systems by way of problem solution of electromechanical systems systemic unification [14].

Solution of the formulated problem for electromechanical systems systemic unification, in accordance with suggested solution procedure, allows to preset, in the request for proposal for its development, optimal values of unification measures and, therefore, ensure obtaining optimal volume decrease for development and operation of advanced [15] electromechanical systems expenses.

Reference

[1] International Electrotechnical Commission (IEC) URL: http://www.iec.ch

[2] Ramani K, Cunningham R, Devanathan S, Subramaniam J and Patwardhan H 2004 Technology review of mass customization *Int. Conf. on Economic, Technical and Org. Aspects of Product Configuration Systems* pp 5-11

[3] Dennis P Hobbs 2003 *Lean Manufacturing implementation: A complete execution manual for any size manufacture* (Fort Lauderdale, FL: J. Ross Publishing)

[4] Stephen A Ruffa 2008 *Going Lean: How the best companies apply lean manufacturing principles to shatters uncertainty, drive innovation, and maximize profits* (New York: AMACOM)

[5] Iyer N, Kalyanaraman Y, Lou K, Jayanti S and Ramani K 2003 Early results with a 3D Engineering Shape Search System *Int. Symp. on Product Lifecycle Management (PLM '03), Indian Institute of Science* (Bangalore, India)

[6] Tret`yakov V M 2013 Ob unifikacii promy`shlennoj produkcii [On the unification of industrial products] *Standarty` i kachestvo* **6** 38-41

[7] Halman J I M, Hofer A P and van Vuuren W 2003 Platform-Driven Development of Product Families: Linking Theory with Practice *J. of Production Innovation Management* **20** 149-62

[8] Lou K and Prabhakar S 2004 Content-based Three-Dimensional Engineering Shape Search P*r. of the 20th Int. Conf. on Data Eng. ICDE04* (Boston, USA) pp 754-65

[9] Andersen A-L, Larsen J K, Nielsen K, Brunoe T D and Ketelsen C 2017 Exploring Barriers Toward the Development of Changeable and Reconfigurable Manufacturing Systems for Mass-Customized Products: An Industrial Survey *Customization 4.0* ed S Hankammer, K Nielsen, F T Piller, G Schuh and N Wang (Cham: Springer) pp 125-40

[10] Brunoe T D and Nielsen K 2017 Data-Driven Product Family Modeling with Feedback Survey *Customization 4.0* ed S Hankammer, K Nielsen, F T Piller, G Schuh and N Wang (Cham: Springer) pp 469-78

[11] Pandremenos J and Chryssolouris G 2009 Modular product design and customization *Proc. of the 19th CIRP Design Conf. – Competitive Design, Cranfield University* (Cranfield) pp 94

[12] Allen-Bradley White Paper, "Perspectives on future of Automation Control" URL http://www.ab.com/events/choices/

[13] Valtonen V, Parviainen A and Pyrhonen J 2006 2D-FEM Modeling of an Axial-Flux Solid-Rotor Core Induction Motor *Book of Abstracts of XVII Intern. Conf. on Electrical Machines, ICEM* (Chania, Greece)

[14] Anosov R S, Glazunov Yu M and Zelenskaya S G 2017 Metodika ocenki texniko-e`konomicheskogo e`ffekta ot realizacii variantov unifikacii e`nergeticheskix sistem [Technique for assessing the technical and economic effect of the implementation of options for the unification of energy systems] *E`nergiya – XXI vek* **4(100)** 86-91 [In Russian]

[15] Anosov R S, Glazunov Yu M and Percev Yu A 2017 Postanovka zadachi sistemnoj unifikacii e`nergeticheskix sistem [Statement of the problem of systemic unification of energy systems] *E`nergiya – XXI vek* **3(99)** 17-24 [In Russian]

IOP Conf. Series: Materials Science and Engineering **791** (2020) 012037 doi:10.1088/1757-899X/791/1/012037

Computer simulation of the operation of the device for monitoring the state of high-voltage insulators during its operation

D K Zaripov, R N Balobanov and D F Zakirov

Department of Electric Stations named after V. K. Shibanov, Institute of Electric Power and Electronics, Kazan State Power Engineering University, Krasnoselskaya Str., 51, Kazan, Republic of Tatarstan, 420066, Russia

E-mail: dzaripov@list.ru

Abstract. This article presents the design and operation principle of the high-voltage isolator state indicator of the high-voltage power transmission line. The present indicator operates without the use of a separate power supply on the principle of electric field energy accumulation. The efficiency of such a device requires that as little as possible be lost during energy storage. Therefore, the development of criteria for selection of elements of the indicator circuit with minimum leakage currents is fundamental, which was one of the first tasks set during simulation. Calculations were carried out by constructing and analyzing a three-dimensional model of the insulator and solving differential equations by finite element method in the COMSOL Multiphysics software environment. The object of the study was a linear suspended insulator with polymer insulation of LK 70/35 type, designed for insulation and attachment of wires of AC overhead power transmission lines with voltage from 35 to 500 kV. Computer simulations and all calculations were made for the said insulator. On the basis of the obtained results of simulation of the insulator operation with the optical indicator installed on it, requirements to electronic components of the developed scheme were formed and corresponding components were selected.

1. Introduction

According to the brochure "Main Results of Operation of Electric Power Facilities in 2016. The results of the autumn-winter period 2016-2017", issued by the Ministry of Energy of the Russian Federation in 2017, the number of accidents that resulted in the termination of electricity supply to consumers with a capacity of 10 MW or more due to damage to air line insulators, amounted to 12 % of the total number of accidents with such damage on overhead power transmission lines, substations and open distribution devices. If we consider only air lines, it is 22.6 %. Damage to insulators in most cases results from aging, surface contamination, manufacturing defects and accidental damage during repair work. Special place among all types of damage to isolation of overhead power transmission lines is occupied by damage to suspended polymer insulators related to their breakdown or overlap.

According to the Energy Research Institute (EPRI) in the United States, where the experience of operating polymer insulators on overhead power transmission lines is more than 30 years, over the past 10 years, US power grid enterprises have begun to experience an increasing number of polymer insulator failures on 115 and 138 kV transmission lines. EPRI studies have shown that these failures are usually associated with high electric field strength occurring near or at high voltage insulator ter-

IOP Conf. Series: Materials Science and Engineering **791** (2020) 012037 doi:10.1088/1757-899X/791/1/012037

minators. Reduction of reliability of overhead power transmission lines is caused by constant activity of electric discharges on metal terminals. Constant impact on the insulation of these corona discharges leads to cracks in the polymer shell and destruction of the end seal, so it is necessary to constantly control the condition of this type of insulators [1].

Remote isolation control methods have never been able to displace the simple visual inspection of overhead power lines and are only a supplement to it. At the same time, during visual inspection of polymer insulators it is not always possible to detect defects, as in most cases they are not visible. Consequently, there is a relevant scientific task related to the lack of effectiveness of the current methods of monitoring overhead power transmission line insulators. Its analysis revealed the need to develop and investigate new methods of insulation control with the help of devices installed on insulators, which would allow to signal their condition directly when visual inspection of overhead power transmission lines.

Installation of optical radiators signalling the presence of a defect on insulating structures in predetermined and observable places allows remote diagnosis of insulation defects under conditions of movement of the recording device. Since the position of the radiator on the insulating structure is known, algorithms for automatically detecting defects are simplified.

The principle of operation of the optical indicator is as follows: in case of violation of separate parts of the insulation structure, the distribution of the electric field along it changes. Potential difference in damaged section decreases and voltage in intact part of structure increases. Defect detection is possible by indicator illumination intensity. It depends on the drop in voltage across its electrodes and the flowing current.

Figure 1 show in a simplified way the operation of the light indicator placed in the range of the variable electric field near the insulator suspended on the metal support [2].

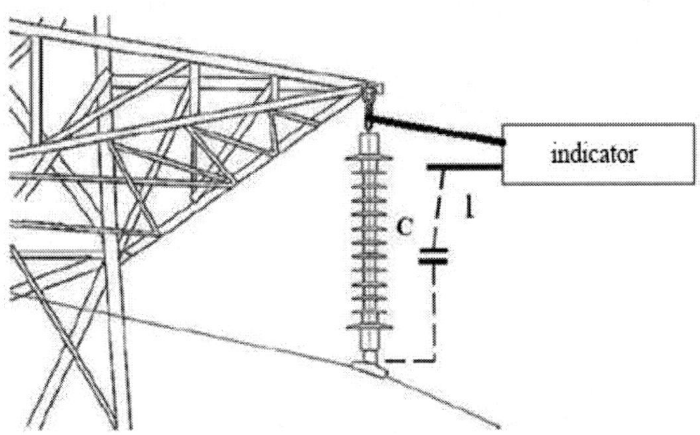

Figure 1. Indicator in the electric field near the insulator.

The figure schematically shows the electrode 1 and the electrical connection of the indicator to the metal elements of the structure. C is the capacitance of the indicator electrode relative to the high voltage portion of the structure.

The indicator diagram (figure 2) is made on the basis of dinistors [3]. A feature of this scheme is the possibility of indicating the discharges occurring in the insulation in case of damage and/or contamination. The circuit consists of two circuits. The first circuit includes the following elements: C1 capacity, R1 resistance and R2, dinistor VD6 and green LED HL1. The second circuit consists of C2 capacity, R3 resistance, VD7 dinistor and red LED HL2.

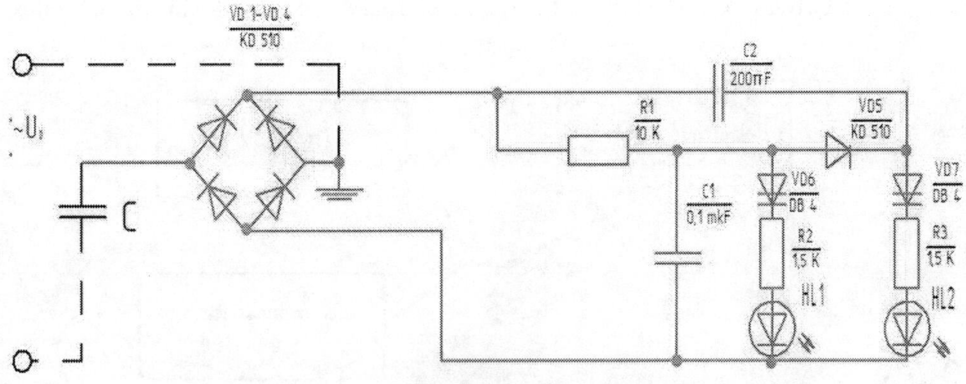

Figure 2. Dinistor-based indicator diagram.

The above-described indicator operates without using a separate power supply on the electric field energy storage principle. The efficiency of such a device requires that as little as possible be lost during energy storage. Therefore, the development of criteria for selection of elements of the indicator circuit with minimum leakage currents is fundamental. This was one of the first tasks set in the simulation.

Calculations were carried out in the COMSOL Multiphysics program [4-15] by constructing and analyzing a three-dimensional model of the LK 70/35 insulator with an optical indicator installed on it.

2. Materials and Methods

To solve this problem, a three-dimensional model of the insulator, shown in figure 3, was built in the COMSOL Multiphysics software environment.

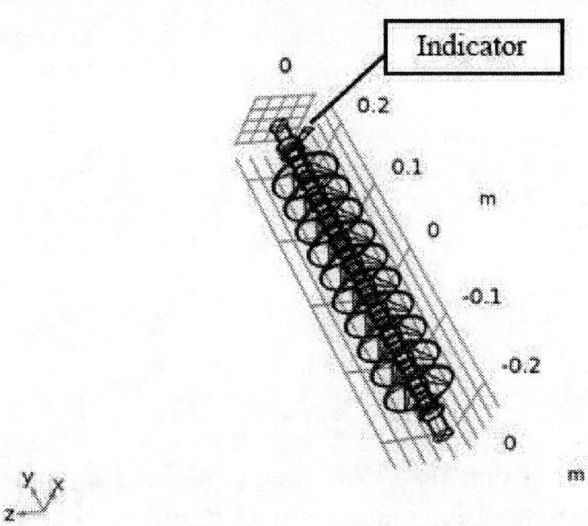

Figure 3. Geometry of three-dimensional model of LK 70/35 insulator with indicator installed at the upper end.

Figure 3 separately shows the insulator section with the indicator installed on the upper end.

A situation in which the insulator is in air was simulated, and a phase voltage of 20 kV (35 kV/1.7) is applied to both ends of the insulator, the amplitude value of which is $U_0 = 28$ kV. A parametric calculation was made where the resistance value of the R_i indicator circuit was used as a parameter. The electronic circuit of the indicator itself was modeled by a strip of semiconductor material, the ends of which abut two electrodes, one of which is fixed on the grounded end of the insulator (figure 4).

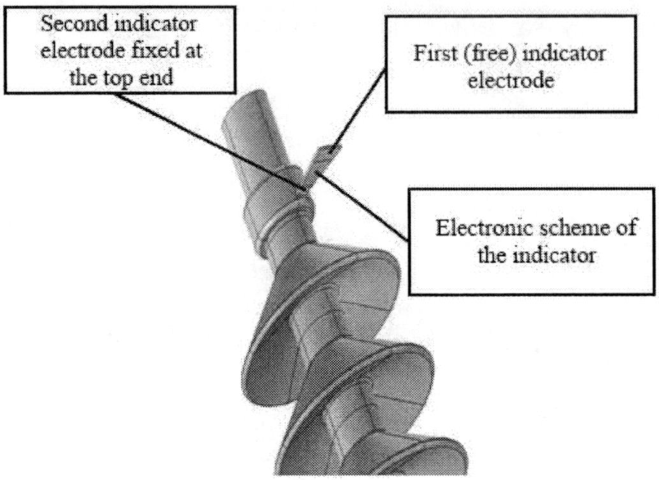

Figure 4. Indicator attached to the top end.

The variable parameter Ri was converted to the conductivity specific value G of the semiconductor strip material:

$$Ri = \frac{1}{G}\frac{l}{S} \tag{1}$$

Where l is circuit length, m; S is the section area of the strip in the middle part, m^2.

The equation by which the COMSOL Multiphysics program made calculations was derived from the continuity equation, Ohm 's law, and Gauss 's theorem. For a sinusoidal current with an angular frequency of $\omega = 2\pi f$ and an isotropic medium, these expressions will be written as:

$$J = \sigma E; \tag{2}$$

$$\nabla \cdot J = \nabla \cdot (\sigma E) = -j\omega\rho \tag{3}$$

$$\nabla \cdot D = \rho \tag{4}$$

$$E = -\nabla \cdot \mathrm{V} \tag{5}$$

$$D = \varepsilon\varepsilon_0 E \tag{6}$$

where J, E, D are vectors of current density (A/m^2), electric field intensity (V/m) and electric displacement (C/m^2), respectively; σ - specific conductivity (Cm/m); ρ – charge bulk density (C/m^3); ε – dielectric permeability of material; ε_0 - electric constant; j - imaginary unit.

After substituting the expressions for E and p into the continuity equation, the Poisson equation of the species is obtained $\nabla \cdot (-\mathrm{c}\nabla V) = \mathrm{f}$ from a set of mathematical equations solved by the COMSOL Multiphysics program:

$$\nabla \cdot (-(\sigma + j\omega\varepsilon\varepsilon_0) \cdot \nabla V) = 0 \tag{7}$$

Initial data for calculation in COMSOL Multiphysics program are given in table 1.

IOP Conf. Series: Materials Science and Engineering **791** (2020) 012037 doi:10.1088/1757-899X/791/1/012037

Table 1. Initial data for calculation in Comsol program.

Name	Value	Size	Description
l	14e-3[m]	0.014 m	Scheme length
S	1e-5[m^2]	1E−5 m²	Section area of strip (diagram) in the middle part
R	1e6[1/S]	1E6 Ω	Leakage resistance of the indicator circuit
r	R*S/l	714.29 Ω·m	Specific resistance of the indicator circuit
G	1/r	0.0014 S/m	Specific conductivity of the indicator circuit
sigma_air	0[S/m]	0 S/m	Specific conductivity of air
epsilon_air	1	1	Dielectric permeability of air
sigma_copp	5.99e7[S/m]	5.99E7 S/m	Specific conductivity of copper
epsilon_copp	1	1	Dielectric permeability of copper
sigma_al	3.77e7[S/m]	3.77E7 S/m	Specific conductivity of aluminum
epsilon_al	1	1	Dielectric permeability of aluminum
sigma_sig	1e-14[S/m]	1E−14 S/m	Specific conductivity of rubber
epsilon_sig	2.09	2.09	Dielectric permeability of rubber
sigma_gla	G	0.0014 S/m	Specific conductivity of the scheme
epsilon_gla	1	1	Dielectric permeability of the scheme
U_0	28[kV]	28000 V	Voltage amplitude at insulator
f	50[Hz]	50 Hz	Frequency

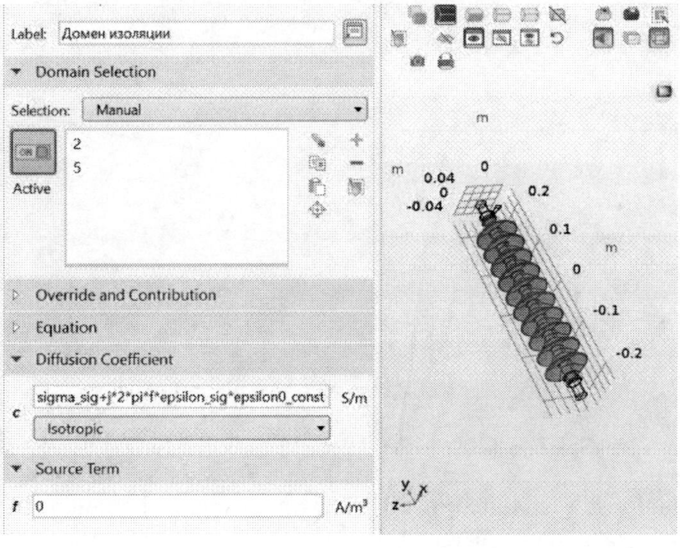

Figure 5. Set the source data for one of the domains.

IOP Conf. Series: Materials Science and Engineering **791** (2020) 012037 doi:10.1088/1757-899X/791/1/012037

The next step in simulating the operation of the insulator with an indicator installed on it for each of the regions (domains) on the example of the insulation region was to set the electrical characteristics of the materials (figure 5). Also, boundary conditions for potential, "earth" and boundary of calculation area are specified.

3. Results

Parametric calculation is made for values of leakage resistance R of electrical circuit of indicator: 1 M Ω, 10 M Ω, 100 M Ω and 500 M Ω. Figure 6 shows the voltage distribution along the insulator in pseudo-values.

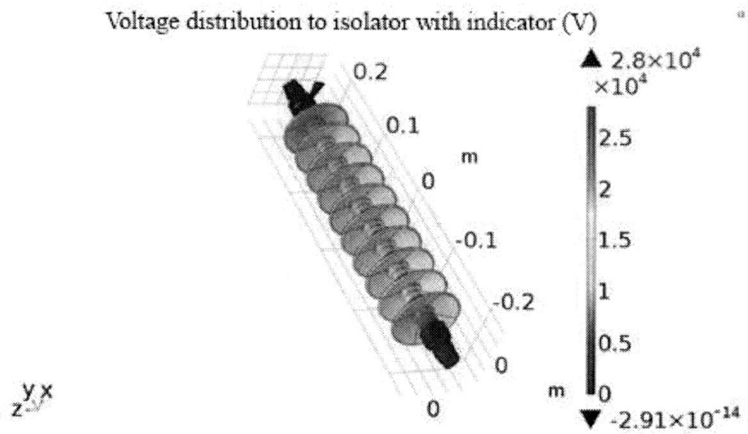

Figure 6. Voltage distribution on insulator in color.

The same graph distribution is shown in figure 7. It shows a characteristic increase in the potential gradient as it approaches the ends of the insulator.

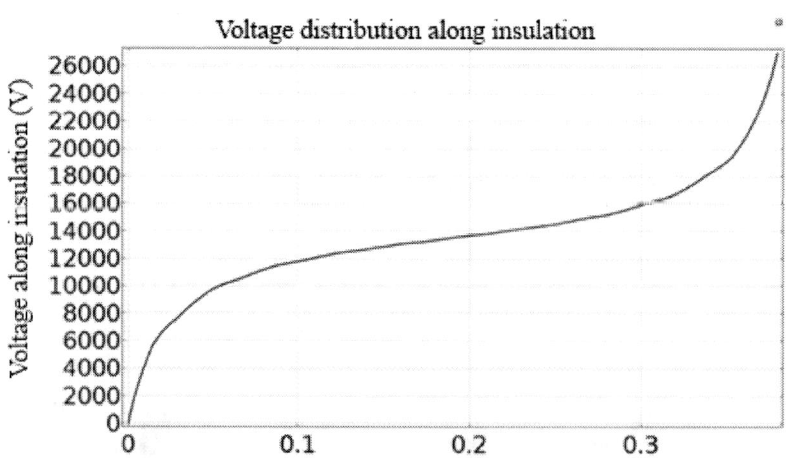

Figure 7. Voltage rise plot at insulator from grounded end to energized end.

The most important result is plotted in figure 8, which shows the potential difference between the circuit electrodes when the circuit leakage impedance changes from 1 M Ω to 500 M Ω.

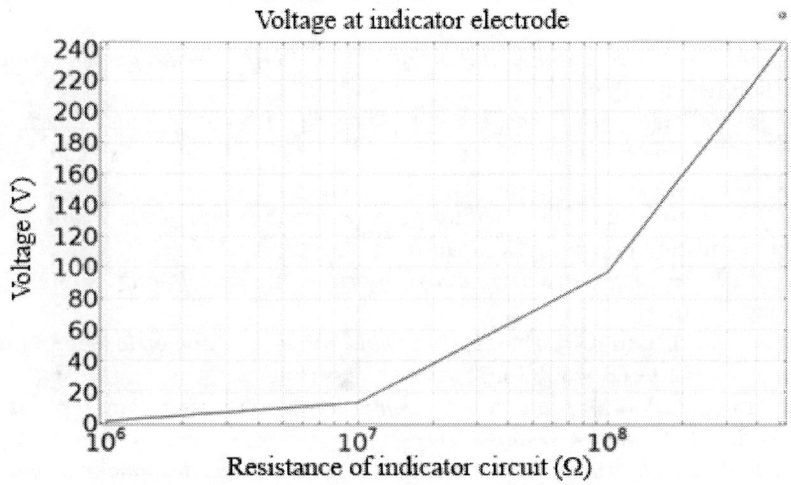

Figure 8. Dependence of amplitude value of circuit voltage on resistance of indicator circuit.

4. Discussion

From the graph shown in figure 8 it can be seen that the developed circuit based on dinistors will not work until the potential difference exceeds the voltage of dinistors actuation (~ 40 V), which can happen only if the impedance of the indicator exceeds the 20 M Ω. This important conclusion formed the basis for selecting the components and the indicator design.

5. Conclusion

In this work, first of all, the task of assessing the requirements for the design and electronic components of the indicator was solved. In addition, however, the developed model served as the basis for calculations of 35 kV insulator operation with an optical indicator, where the capabilities of the indicator were modeled and evaluated for functional purpose.

Based on the simulation results, the following components were selected: super bright LEDs (3R5 grades); dinistors (DB 4 grade), diodes (KD 510 grade) and capacitors with low leakage currents; low leakage current arrester (EC75X grade).

The most suitable emitters are ultra-bright LEDs, which have sufficient brightness to detect them from a distance of up to 50 m visually in the daytime.

References

[1] Learning from Service Experience with Composite Line Insulators URL http://www.inmr.com/learning-from-service-experience-with-composite-line-insulators/

[2] Zaripov D and Balobanov R 2016 Indikator defekta vy`sokovol`tnoj izoliruyushhej konstrukcii [High voltage insulation structure defect indicator] *E`lektrotexnika* **6** 16-21 [In Russian]

[3] Balobanov R, Zaripov D and Margulis S 2017 Ustrojstvo opticheskoj indikacii defekta vy`sokovol`tnoj izoliruyushhej konstrukcii [Device for optical indication of defect of highvoltage insulating structure] *Izvestiya vy`sshix uchebny`x zavedenij. Problemy` e`nergetiki* **4** 119-25 [In Russian]

[4] Comsol Multiphysics URL http://www.comsol.com/

[5] Bazhutin D 2015 Simulation of elastic vibrations of crane installations con-structions in comsol multiphysics package *SWVNTU* **4** 4

[6] Budko A and Vasilieva O 2014 Modelirovanie staticheskix e`lektromagnitny`x polej i raschet parametrov v comsol multiphysics [Modeling static electromagnetic fields and calculating parameters in comsol multiphysics] *Sovremennye problemy nauki i obrazovaniya* **2** 133 [In

Russian]

[7] Darrell W and Juan C 2006 *The finite element method: basic concepts and applications* (New York:CRC Press)

[8] Roger W 2009 *Multiphysics modeling using comsol: a first principles approach* (Sudbury:Jones & Bartlett Learning)

[9] Mehrzad Tabatabaian 2015 *Comsol5 for engineers* (Herndon:Mercury Learning & Information)

[10] Baker A 2012 *Finite elements: computational engineering sciences* (Chichester:John Wiley & sons LTD)

[11] William B J Zimmerman 2006 *Multiphysics modeling with finite element methods* (London: World Scientific Publishing Company)

[12] Layla S 2019 *Geometry creation and import with comsol multiphysics* (Herndon:Mercury Learning & Information)

[13] Eugenio A and David G 2015 *A practical guide to geometric regulation for distributed parameter systems* (Boca Raton:Chapman and Hall/CRC)

[14] Lennart Edsberg 2008 *Introduction to computation and modeling for differential equations* (Chichester: John Wiley & sons LTD)

[15] Darell W and Juan C 2017 *The finite element method: basic concepts and applications with matlab, maple, and comsol* (New York:CRC Press)

IOP Conf. Series: Materials Science and Engineering **791** (2020) 012038 doi:10.1088/1757-899X/791/1/012038

Optimization approach to reduce the negative impact of high harmonic components of currents and voltages in high-voltage mine networks

E V Zhilin and D A Prasol

Department of Electric Power Engineering and Automation, Belgorod State Technological University named after V.G. Shukhov, Kostyukov St., 46, Belgorod, 308012, Russia

E-mail: dapras@mail.ru

Abstract. The article presents an analysis of the work of the high-voltage mine network, which confirms the presence of higher harmonic components of currents and voltages of significant magnitude and their negative impact on the elements of the power supply system. The most pronounced higher harmonics of currents and voltages in the high-voltage mine network, as well as the most probable ranges of load currents and distortion currents are revealed. It is proved that the load change in the mine network as a function of time refers to random processes. The mathematical apparatus of the theory of random processes was used to study and predict load currents and distortion currents. The obtained quantitative values of trends of load currents and distortion currents were used in the estimation of power losses and in the calculation of parameters of filter compensating devices. An optimization approach is proposed to reduce the negative influence of higher harmonic components of currents and voltages, based on minimizing the target function of additional active power losses from the action of higher harmonics. To reduce the negative impact of higher harmonic components of currents and voltages in the high-voltage mine network, it is proposed to use two resonant passive filters tuned to the 11-th and 13-th harmonics, and a second-order broadband filter tuned to compensate harmonics, starting from the 23-rd and above.

1. Introduction

An important task in the power supply systems of ore mining enterprises in particular and in modern energy in General is to ensure electromagnetic compatibility (EMC) in electrical networks and reduce power losses and electricity due to the presence of powerful electrical receivers with nonlinear voltage characteristics. The most energy-intensive electric receivers in such systems are cage and skip lifting installations (LI) (Figure 1). As electric drives of LI the adjustable electric drive executed on system the thyristor Converter – the DC motor with independent excitation (TP-D) is used. LI, as a rule, work in repeatedly-short - term modes of rise-lowering on the set diagrams of movement. The time of movement of skips per cycle is 160 s, stands-170 s. in the cycles of LI operation, dynamic processes prevail, the duration of which is 46-50 % of the total cycle time [1].

In thyristor converters of electric drives LI used 12-pulse rectifier circuit. Electric drives of lifting units have a deep range of speed control due to changes in the voltage at the output of thyristor converters. During operation, the electric drives of LI generate higher harmonic (HG) components of currents and voltages into the supply network. Harmonics of the 11th, 13th, 23rd, 25th, 35th and 37th or-

IOP Conf. Series: Materials Science and Engineering **791** (2020) 012038 doi:10.1088/1757-899X/791/1/012038

ders appear in the supply voltage [1-3]. This leads to a decrease in the quality of electricity and violation of electromagnetic compatibility, to additional power losses in the elements of high-voltage power supply systems of mining enterprises [4-8]. Therefore, minimizing power losses due to the presence of higher harmonic components of currents and voltages is an important and urgent task for both mine electric networks and general purpose power supply systems.

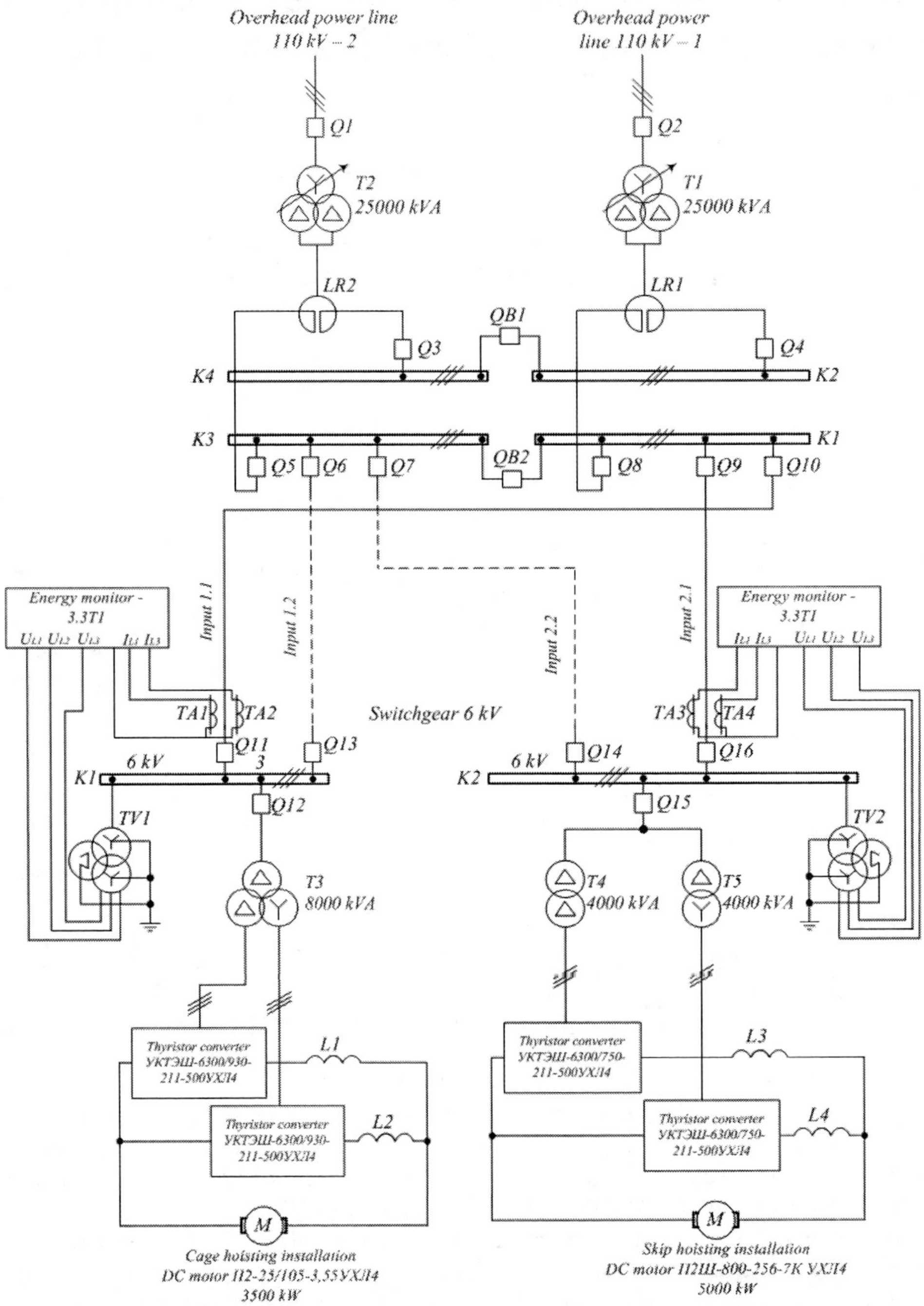

Figure 1. A fragment of the mining enterprise power supply scheme.

2. Materials and Methods

The estimation of EMC and power losses in the high-voltage mine network was carried out with the help of experimental studies, engineering calculation and simulation in the Matlab system with the Simulink extension package and the SimPowerSystems library [9, 10]. The measurements were carried out in a 6 kV switchgear on the busbar sections of a transformer substation providing power supply to the hoisting units. The location of the measurements is shown in fig. 1. Measurements are made in accordance with the requirements of GOST 30804.4.7–2013, continuously for a specified period of time for an average of three days, certified devices "Energomonitor-3.3 T1" and C. A 8335 [1, 2, 4, 9].

The fragment of results of measurements of values of parameters during operation of drives of skip and cage LI is resulted in table 1.

Table 1. Parameters of modes and indicators of EMC current.

Time	I, A	K_I, %	$I_{(1)}$, A	$K_{I(11)}$, %	$K_{I(13)}$, %	$K_{I(23)}$, %	$K_{I(25)}$, %	$K_{I(35)}$, %	$K_{I(37)}$, %
Skip LI (Energy r monitor - 3.3T1)									
17:33	311.44	8.58	308.51	5.57	4.22	2.14	2.05	1.03	1.07
17:55	332.93	8.57	329.57	5.64	4.3	2.15	2.04	1.02	1.01
18:34	323.76	9.92	269.26	5.56	4.28	2.26	1.97	1.1	0.81
18:37	370.94	8.81	321.04	5.51	4.35	2.06	2.08	0.95	1.02
18:41	311.44	8.1	369.86	5.76	4.4	2.1	2.14	0.97	1.15
Cage LI (Energy monitor - 3.3T1)									
17:43	287.03	7	288.46	5.57	1.58	2.08	0.9	0.96	0.75
17:44	447.83	9.21	439.80	7.1	2.69	2.85	1.04	1.32	0.81
17:44	435.17	9.37	438.84	7.18	2.83	2.92	1.08	1.38	0.81
17:44	408.48	9.51	409.92	7.1	2.79	2.89	1.1	1.38	0.85
17:44	310.80	8.53	323.47	6.28	2.2	2.45	1.04	1.17	0.86

As a result of measurements spectra of HG of currents and voltages in a high-voltage mine network are received. For figure 2 the spectra of harmonics of currents in the networks of 6 kV cage and skip LI of the operating ore mining enterprise are presented.

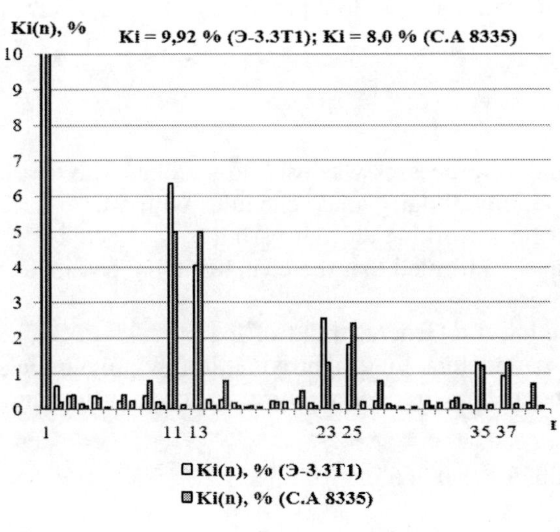

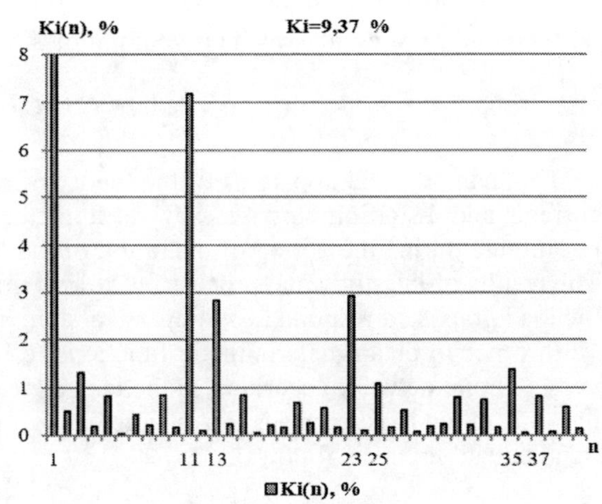

Figure 2. Spectra of current harmonics in high-voltage mine power supply system: a) – skip LI; b) – cage LI.

The results of the measurements confirm that the most pronounced are the 11, 13, 23, 25, 35 and 37 harmonics of current and voltage in the high-voltage mine power supply system. The values of the coefficients of these harmonic components of the current, respectively, were:

Energy monitor-3.3T1:

Skip LI: $K_{I(11)}$ = 5,56 %, $K_{I(13)}$= 4,28 %, $K_{I(23)}$= 2,26 %, $K_{I(25)}$= 1,97 %, $K_{I(35)}$= 1,1 %, $K_{I(37)}$= 0,81 %;
Cage LI: $K_{I(11)}$= 7,18 %, $K_{I(13)}$= 2,83 %, $K_{I(23)}$= 2,92 %, $K_{I(25)}$= 1,08 %, $K_{I(35)}$= 1,38 %, $K_{I(37)}$= 0,81 %;

C.A 8335:

Skip LI: $K_{I(11)}$= 5,0 %, $K_{I(13)}$= 5,0 %, $K_{I(23)}$= 1,3 %, $K_{I(25)}$= 2,4 %, $K_{I(35)}$= 1,2 %, $K_{I(37)}$= 1,3 %.

The maximum values of the total coefficients of the harmonic components of the current in the high-voltage mine network of skip and cage lifting units, respectively, were K_I= 9,92 % and K_I= 9,37 %. The values of the coefficients of the 11-th and 13-th harmonic components of the current have the highest level (fig. 2). The maximum values of the total coefficients of the harmonic components of the voltage were K_U= 11,32 % and K_U= 17,87 % for skip and cage LI, respectively [1, 2, 9].

The reliability of the experimental results is confirmed by the results of simulation and engineering calculations. They showed that the total coefficients of harmonic components of currents and voltages in the high-voltage mine network reach on average K_I= 8-10% and K_U= 12-17%, respectively. The reliability of the results is confirmed by a good correlation between the results of experimental studies, simulation and engineering calculations, the degree of discrepancy did not exceed 10-12 %.

The load in the high-voltage power supply system is random, which is associated with the loading of LI, with the time of day. The change in load as a function of time refers to random processes. The mathematical model of a random stationary process of load current change is given in the form of a ratio:

$$I_{(1)j} = \varphi(t_j) + \Delta_j \tag{1}$$

where $I_{(1)j}$ – values reflecting a number of observations (j = 1, 2,..., n); $\varphi(t_j)$ – some deterministic function reflecting the General trend of $I_{(1)j}$ (sometimes called "deterministic component" or "trend"); Δ_j – random deviations occurring during the process $I_{(1)j}$.

Random processes of changing the parameters of load graphs proceed uniformly and have the form of continuous random oscillations around a certain average value. Different implementations of the main frequency current load graphs have a high correlation. It follows from the results that the changes in the parameters of the load graphs correspond to stationary random processes. Often a deterministic function is represented by a polynomial of K-th degree:

$$I_{(1)}(t) = C_0 + C_1 t + C_2 t^2 + ... + C_k t^k = \sum_{j=0}^{k} C_j t^j \tag{2}$$

The mathematical apparatus of the theory of random processes was used to study and predict load currents and distortion currents. Several implementations of daily load schedules with a time interval per minute during the period of operation of the LI for a working shift are arbitrarily selected (fig. 3). The results of the study and processing of load graphs confirmed that the changes in the parameters of the load graphs correspond to stationary random processes.

In order to obtain deterministic functions reflecting the General trend of load graphs in the high-voltage mine network according to the considered parameters, their approximation by polynomials of the 6-th order is performed. For example, the fundamental frequency current and distortion current trend equations are as follows:

$$I_{(1)} = -3\cdot10^{-10}\cdot t^6 - 6\cdot10^{-8}\cdot t^5 + 3\cdot10^{-5}\cdot t^4 - 0.0025\cdot t^3 + 0.0653\cdot t^2 + 0.3552\cdot t + 255.5 \tag{3}$$

$$I_D = 6\cdot10^{-11}\cdot t^6 - 3\cdot10^{-8}\cdot t^5 + 5\cdot10^{-6}\cdot t^4 - 0.0004\cdot t^3 + 0.0107\cdot t^2 - 0.0745\cdot t + 17.657 \tag{4}$$

The values of the constant coefficients of the deterministic functions of the load graphs of the high-voltage mine network for the considered parameters are presented in table 2.

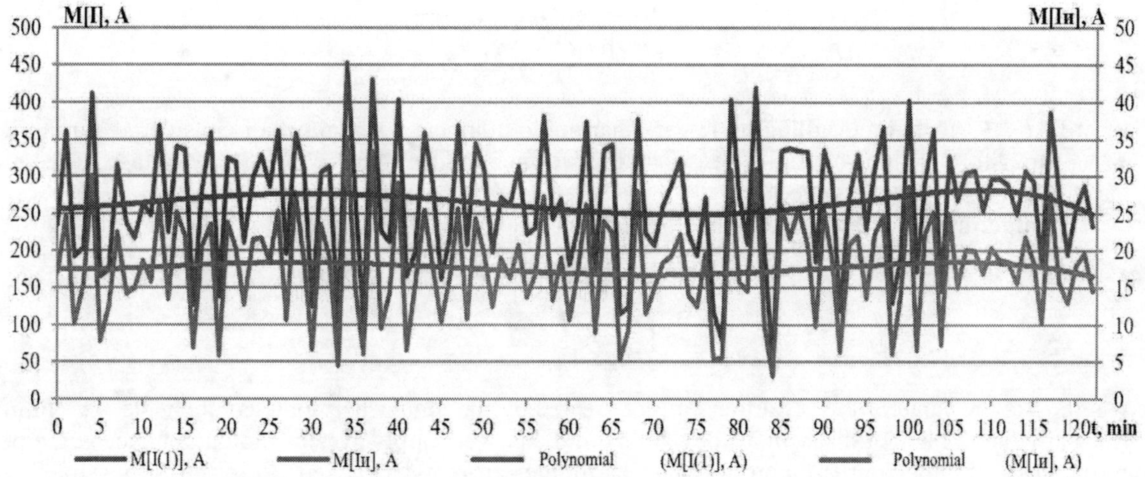

Figure 3. Graphs of current changes in the high-voltage mine power supply system and distortion current during the period of operation of the LI with graphs of deterministic functions.

Graphs of trends of the fundamental frequency currents and distortion are shown in figure 3. The obtained quantitative values of load current trends and distortion currents are used to estimate power losses. Also, these values can be used to calculate the parameters of filter-compensating devices (FCD) used to provide compensation of HG currents and voltages in the high-voltage mine network.

Table 2. The value of the constant coefficients of the deterministic function.

$I_{(1)}$, A	I_D, A	$I_{(11)}$, A	$I_{(13)}$, A	$I_{(23)}$, A	$I_{(25)}$, A	$I_{(35)}$, A	$I_{(37)}$, A
Skip LI							
255.5	17.66	12.93	9.57	4.76	4.52	2.14	2.07
Cage LI							
265.6	18.49	13.43	7.75	4.87	1.99	2.13	1.56

Additional power losses in non-sinusoidal regimes, taking into account the obtained trend values, were determined as the sum of losses in cable and overhead lines, reactors and transformers [4, 7, 8]. Thus losses from currents of the most expressed 11-th, 13-th, 23-rd and 25-th highest harmonic components were considered. To minimize active power losses in non-sinusoidal modes in high-voltage mine networks, an optimization problem based on minimizing the loss function is solved [9, 11, 12]. The target function was minimized:

$$\Delta P_{\Sigma add} = \sum_{j=1}^{p} \left(\Delta P_{PLj} + \Delta P_{Rj} + \Delta P_{Tj} \right) \rightarrow \min \qquad (5)$$

Power losses in power lines:

$$\Delta P_{PL} = 3 \cdot \sum_{n=2}^{p} I_n^2 \cdot R_{PL} \cdot k_{rn} \qquad (6)$$

where n is the number of harmonics; p – number of considered harmonics; I_n – current n-th harmonic; R_{PL} – active line resistance at the fundamental frequency; k_{rn} – coefficient of increase of resistance in transmission lines, taking into account the influence of surface effect.

Power losses from HG currents in reactors:

$$\Delta P_R = 3 \cdot \sum_{n=2}^{p} I_n^2 \cdot R_R \cdot k_{rn} \qquad (7)$$

IOP Conf. Series: Materials Science and Engineering **791** (2020) 012038 doi:10.1088/1757-899X/791/1/012038

Power losses in transformers:

$$\Delta P_T = \Delta P_{NL} \cdot \sum_{n=2}^{p} \left(\frac{U_n}{U_{nom}} \right)^2 + 0,607 \cdot \frac{\Delta P_{SC}}{u_{SC}^2} \cdot \sum_{n=2}^{p} \frac{1+0,05 \cdot n^2}{n\sqrt{n}} \cdot \left(\frac{U_n}{U_{nom}} \right)^2 \tag{8}$$

where ΔP_{NL} – transformer idling losses; n – harmonic number; p – number of considered harmonics; ΔP_{SC} – transformer short-circuit losses; u_{SC} – transformer short-circuit voltage; U_n – voltage of the n-th harmonic; U_{nom} – rated voltage.

The target function has the following limitations:

$$\begin{cases} I_{11} + I_{13} + I_{23} + I_{25} \le I_D, \\ \delta U < \delta U_{allow}, \\ K_U \le K_{U\ allow}, \end{cases} \tag{9}$$

where I_D is the value of the distortion current created by the higher harmonic components. The limitations are due to the fact that voltage deviations in the power supply system should not exceed the permissible value, and compensation of higher harmonics currents is carried out to the permissible level of non-sinusoidal supply voltage.in this case the target function has the following limitations.

3. Results

Additional power losses in the elements of the mine network in non-sinusoidal modes, taking into account the trends of HG currents, respectively amounted to: for skip LI ΔP_{add}= 9,16 kW, for cage LI ΔP_{add}= 24 kW. Total additional power losses in the high-voltage mine network amounted to $\Delta P_{\Sigma add}$= 33,16 kW. The obtained values of additional power losses have a significant value and are approximately from 16,5 to 44 % of the main losses. The optimization problem was solved by the method of indefinite Lagrange multipliers, as a result of its solution the following values of higher harmonic currents were obtained:

$$I_{11} = 6{,}476\ \text{A}; I_{13} = 5{,}957\ \text{A}; I_{23} = 4{,}479\ \text{A}; I_{25} = 4{,}296\ \text{A}.$$

In this case, the coefficients of the most pronounced harmonic components of the currents take the following values:

$$K_{I_{11}} = 2{,}319\%;\ K_{I_{13}} = 2{,}133\%;\ K_{I_{23}} = 1{,}604\%;\ K_{I_{25}} = 1{,}538\%.$$

The results of solving the optimization problem show that for minimizing of the objective function and reduce the total additional active power losses caused by higher harmonic components, it is necessary to compensate for the 11-th and 13-th harmonics and to reduce the level of 23-rd and 25-th harmonics.

An effective way to compensate for HG currents is the use of various types of passive filters (PF) [5, 6, 8, 12-15]. In this case, it can be proposed to install film-compensating devices with two resonant PF tuned to the 11-th and 13-th harmonics, and a second-order broadband filter tuned to compensate harmonics starting from the 23-rd and above [6, 8, 15]. The place of installation of FCD – on tires of 6 kV of the switchgear of lifting machines. Calculation of parameters of FCD is offered to be carried out according to the techniques offered in [6, 8, 12, 16-20].

When installing the proposed FCD on 6 kV tires of the switchgear of lifting machines and when compensating the higher harmonics of currents to the values obtained as a result of solving the optimization problem, the total loss of active power is reduced to the value $\Delta P_{\Sigma add}$= 12,59 kW, that is, more than 60 % (figure 4).

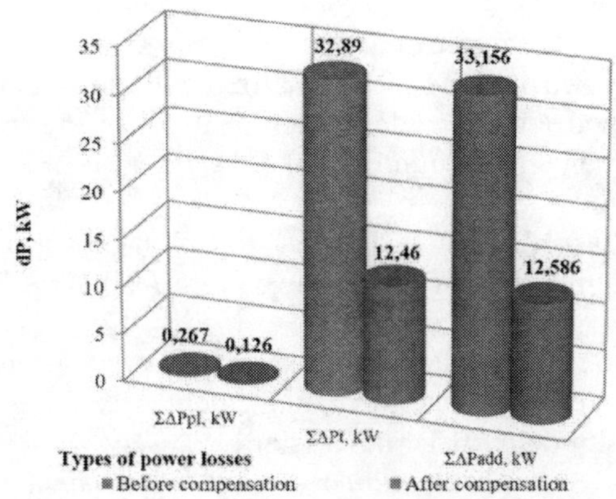

Figure 4. Histogram of additional power losses in high-voltage mine power supply system before and after compensation of HG.

4. Conclusion

The results of experimental studies, simulation and engineering calculations in the high-voltage mine network of lifting units of the ore-producing enterprise confirmed the presence of higher harmonic components of currents and voltages. At the same time, the results of measurements, simulation and engineering calculations showed good convergence. The degree of discrepancy did not exceed 10-12 %. It is confirmed that the most pronounced are the 11, 13, 23, 25, 35 and 37 harmonics of currents and voltages. The total coefficients of harmonic components of currents and voltages in the high-voltage mine network reach on average K_I= 8-10% and K_U= 12-17%, respectively.

Identification of parameters and determination of probabilistic characteristics of load currents and distortions is carried out. The identification results show that changes in load currents and distortions refer to random stationary processes. Quantitative characteristics of random processes allowed to reveal the most probable range of change of values of currents of loadings and distortions.

On the basis of the carried out identification the estimation of additional power losses in elements of high-voltage power supply systems at non-sinusoidal modes is carried out and the optimization problem of minimization of power losses is solved. At the received values of currents of the highest harmonics and taking into account the set restrictions of loss of power in mine system of power supply will be minimal. When compensating the HG to the found values, the total loss of active power in non-sinusoidal modes in the high-voltage mine network will decrease to the value $\Delta P_{\Sigma add}$= 12,59 kW, that is, more than 60 %.

It is proposed to perform compensation of HG currents by means of FCD with two resonant PF tuned to the 11-th and 13-th harmonics and a second-order broadband filter tuned to compensate harmonics starting from the 23-rd and above. Application of FCD allows to reduce losses of active power from currents of the higher harmonics and also to achieve decrease in value of total coefficient of harmonic components of tension in a high-voltage network to admissible value K_U= 4,86 %. In this case, a deeper compensation of the currents of higher harmonics is possible.

Acknowledgments

The article was prepared within development program of the Flagship Regional University on the basis of Belgorod State Technological University named after V.G. Shukhov.

References

[1] Averbukh M A, Prasol D A and Khvorostenko S V 2017 E`ksperimental`naya ocenka parametrov rezhimov v vy`sokovol`tny`x rudnichny`x setyax s moshhny`mi nelinejny`mi e`lektropriemnikami [Experimental estimation of parameters of regimes in high-voltage mining power lines with high-power non-linear consumers] *Vestnik IGTU* **21(2)** 75-84 [In Russian]

[2] Averbukh M A and Prasol D A 2018 Analiz e`lektromagnitnoj sovmestimosti seti vy`sokogo napryazheniya gornorudnogo predpriyatiya [Analysis of electromagnetic compatibility of a high voltage network of a mining enterprise] *E`nergetik* **2** 36-41 [In Russian]

[3] Semenov A S and Kuznetsov N M 2014 Analiz rezul`tatov monitoringa pokazatelej kachestva e`lektroe`nergii v podzemnom rudnike [An analysis of the results of monitoring the quality of electric power in an underground mine] *Izmeritel`naya texnika* [*Measurement Techniques*] **4** 17-20 [In Russian]

[4] Averbukh M A and Prasol D A 2018 Influence of high-power nonlinear consumers on electric energy losses in mining high-voltage power line *IOP Conf. Series: Materials Science and Eng.* **327** 052028

[5] Sharov Yu V et al. 2017 *Upravlenie kachestvom e`lektroe`nergii: uchebnoe posobie 3-e izd., pererab.* [Electric energy quality management: teaching guide. 3rd issue amended] (Moscow: Izdatel`skij dom MEI) [In Russian]

[6] Boyarskaya N P et al. 2014 *Sintez fil`trokompensiruyushhix ustrojstv dlya sistem e`lektrosnabzheniya* [Synthesis of filter compensating devices for power supply systems: collective] (Krasnoyarsk: SFU) [In Russian]

[7] Dolinger S Yu, Lyutarevich A G, Goryunov V N, Safonov D G and Cheremisin V T 2013 Ocenka dopolnitel`ny`x poter` moshhnosti ot snizheniya kachestva e`lektricheskoj e`nergii v e`lementax sistemy` e`lektrosnabzheniya [Estimation of additional power losses due to reduced electric energy quality in power supply elements] *Omskij Nauchny`j Vestnik* **2** 178-83 [In Russian]

[8] Arrillaga J and Watson N R 2003 *Power system harmonics* (Chichester: John Wiley & Sons)

[9] Averbukh M A, Pogorelov A V and Prasol D A 2018 Minimizaciya poter` moshhnosti pri nesinusoidal`ny`x rezhimax v vy`sokovol`tny`x rudnichny`x setyax [Minimization of power losses during non-sinusoidal modes in high-voltage mine networks] *Promy`shlennaya E`nergetika* **7** 38-46 [In Russian]

[10] Averbukh M A, Zhilin E V and Roschubkin P V 2017 Experimental analysis of electrical modes in a residential estate electrical power supply system *J. of Eng. and Appl. Sciences* **12** 3446-51

[11] Aleem S H E A, Zobaa A F and Aziz M M A 2012 Optimal C-type passive filter based on minimization of the voltage harmonic distortion for nonlinear loads *IEEE Transactions on Industrial Electronics* **59(1)** 281-9

[12] Kovernikova L I and Thanh N C 2012 An optimization approach to calculation of passive filter parameters based on particle swarm optimization *Int. Conf. on Renewable Energies and Power Quality* (Santiago de Compostela, Spain) vol **1**(10) pp 605-10

[13] Akagi H 2006 Modern active filters and traditional passive filters *Bulletin of The Polish Academy of Sciences Technical Sciences* **54(3)** 255-69

[14] Chou C J, Liu C W, Lee J Y and Lee K D 2000 Optimal planning of large passive-harmonic-filters set at high voltage level *IEEE Transactions on Power Systems* **15(1)** 433-41

[15] Cho Y and Cha H 2011 Single-tuned passive harmonic filter design considering variances of tuning and quality factor *J. of Int. Council on Electrical Eng.* **1(1)** 7-13

[16] Yue H, Li G, Zhou M, Wang K and Wang J 2011 Multi-objective optimal power filter planning in distribution network based on fast nondominated sorting genetic algorithms *4th Int. Conf. on Electric Utility Deregulation and Restructuring and Power Technologies, DRPT* (Weihai Shandong, China: IEEE) pp 234-40

[17] Na H, Xu D abd Huang L 2009 The application of particle swarm optimization to passive and hybrid active power filter design *IEEE Trans. on Ind. Electronics* **56(8)** 2841-51

[18] Baggini A 2008 *Handbook of power quality* (Chichester: John Wiley & Sons)

[19] Das J C 2015 *Power System Harmonics and Passive Filter Designs* (Chichester: John Wiley & Sons)

[20] Fujita H and Akagi H 1991 A practical approach to harmonic compensation in power systems series connection of passive and active filters *IEEE Trans. on Ind. Applications* **27(6)** 1020-5

IOP Conf. Series: Materials Science and Engineering **791** (2020) 012039 doi:10.1088/1757-899X/791/1/012039

Research and analysis of the low-temperature potential of heat networks

A G Batukhtin[1], M V Kobylkin[1] Yu O Rikker[2] and S G Batukhtin[1]

[1]Department of Energy, Transbaikal State University, Aleksandro-Zavodskaya St., 30, Chita, 672039, Russia
[2]Scientific Department, Transbaikal State University, Aleksandro-Zavodskaya St., 30, Chita, 672039, Russia

E-mail: mkchita@gmail.com

Abstract. The article is devoted to the use of heat pump installations in heat supply systems with TPP. The technique of the primary assessment of the efficiency of heat pumps in district heating systems is considered, on the basis of which it was concluded that it is possible to get the maximum effect from the introduction of a heat pump in the inter-heating period to compensate for the HPU load when using a heat pump at a consumer and transferring the heat and power plant to the condensation mode from the condenser to the heating system. At the same time, for greater effect, the heating system must be cooled below 20 ° C in order to use the low-temperature potential of the heating network. Based on the findings, a comprehensive heat supply system has been developed that includes TPP and HPU installed at the consumer by the universal compensation system of hot water supply load (UCS-HWSL) scheme. Modernization of the basic heat supply system to the UCS-HWSL system with a lowered temperature schedule will allow improving the technical and economic indicators of the Chita TPP-1.

1. Introduction

Energy is the basis for the development of production processes in all sectors of the economy, ensuring the development of the material and technical base of society, improving the quality of life of the population.

The basis of the uninterrupted operation of the energy industry of any state is to provide energy enterprises with fuel. However, a rough estimate of global reserves of fossil hydrocarbons indicates that by the middle of the 21st century, these reserves may be depleted. Thus, humanity at the present stage of development is on the verge of a global energy crisis.

In view of the above-mentioned global problem, the direction of the development of the country's heat supply in the context of energy conservation takes on special relevance in the energy sector. The relevance of energy saving measures is supported by the legislative base (Federal Act 261-FA On Energy Saving and on Improving Energy Efficiency and Amending Certain Legislative Acts of the Russian Federation and Related Sub-Law Acts).

Given this legislative base in Russia in recent years, interest has increased in the use of energy from renewable and secondary sources, including heat-pump unit (HPU). The emergence of demand for existing alternative technologies has expanded the existing domestic market for energy conservation, allowing consumers to use projects with HPU without the involvement of foreign experts. Regardless of the question of a full-scale transition from existing centralized heat supply systems to foreign analogues of integrated systems with district heating supply system [6-15], the use of heat pumps in modern Russian power engineering has a certain potential. This direction includes the adaptation of heat pump systems for their use in the improvement of existing district heating systems.

2. The initial assessment of the heat pump efficiency in district heating supply system

Traditionally it is believed that HPU is effective in recalculating the coefficient (COP) > 2.5. This expression is easy to verify. Supposadly, efficiency factor of condensing power plant ≈ 40%, therefore, the production of 1 kW of electricity will be spent $1/\eta=1/0.4=2.5$ kW fuel heats.

The basis of the uninterrupted operation of the energy industry of any state is to provide energy enterprises with fuel. However, a rough estimate of global reserves of fossil hydrocarbons indicates that by the middle of the 21st century, these reserves may be depleted. Thus, humanity at the present stage of development is on the verge of a global energy crisis.

Thus, the HPU seems to «return» the heat, initially spent on electricity production, and when the COP threshold is exceeded by 2.5, its work becomes energy efficient.

For TPP, the efficiency of a heat pump unit is not so obvious, since the heating, from a thermodynamic point of view, is much more efficient than separate generation [2], besides cogeneration does not allow directly estimating the efficiency of a heat pump unit through the efficiency factor of electricity generation at a heat plant.

This feature causes a lot of controversy among Russian engineers and scientists, the result of which was a lot of contradictory data about the economic feasibility or inexpediency of the use of HPU in conjunction with TPP. Most tend to the fact that the heat pump unit is notoriously inefficient in heating plants.

To obtain a specific answer to the question about the efficiency of heat supply systems in district heating systems, it is proposed to introduce the concept of extremely inefficient COP for the system in question.

This term should be understood as such COP, in which the introduction of heat pump unit into the district heating supply system of a TPP will not lead to a change in fuel consumption.

The condition for the extreme inefficiency of the HP, for isolated TPP, can be expressed as follows:

$$B_{TPP} = B_{TPP+HPU}$$

provided: BTPP,BTPP+HPU– fuel costs for TPP, respectively, in the absence and in the presence of HPU in the heating system.

Since the CHP under consideration is similar for both options, the equality of fuel consumption will be determined by the equality of steam consumption (D) for turbines:

$$D_{TPP} = D_{TPP+HPU}$$

In general, with certain assumptions, the equality of steam costs can be described as a balance:

$$\frac{N}{H_i \cdot \eta_{em}} + \frac{i_s - i_c}{H_i} \cdot \frac{Q_{rel.}}{(i_s - i_d) \cdot \eta_{h.e.}} = \frac{N + \Delta N_{HPU}}{H_i \cdot \eta_{em}}$$

provided: N – power, which must be maintained for the supply of electricity to the consumer, Hi – disposable turbine heat drop, is – steam enthalpy taken for heating, ic – steam enthalpy in condenser, id – enthalpy of drainage network water heater (NWH), Qrel. – the amount of heat released from the station to the needs of heating, ΔNHPU – necessary increase in electrical power to cover the needs of the HPU, ηem, ηh.e. – electromechanical efficiency factor of the turbine and the efficiency factor of the heat exchanger, respectively.

Assumptions:

- turbines of type T, PT;
- all the heat required by the consumer is compensated by the HP from the autonomous low-potential source of heat;

- single-stage heat line water;
- neglecting the release of power auxiliary equipment of heat networks;
- TPP has a power reserve to cover the needs of the HP.

Simplifying equality we get:

$$\frac{\Delta N_{HPU}}{Q_{rel.}} = \frac{(i_s - i_c)\cdot \eta_{em}}{(i_s - i_d)\cdot \eta_{h.e.}}.$$

Since on the left side of the equality is the value inverse COP the extremely ineffective COP for the system:

$$\mu_{ext.} = \frac{Q_{rel.}}{\Delta N_{HPU}} = \frac{(i_s - i_d)\cdot \eta_{h.e.}}{(i_s - i_c)\cdot \eta_{em}}.$$

From the obtained equation it can be seen that the efficiency of the heat pump when working together with the TPP in an isolated system depends only on the parameters of the steam selected at the NWH.

In the general case, if the released power, when closing the heat extraction samples, is sufficient only to cover the needs of a HPU, then in such a case the HPU can be considered extremely inefficient.

In particular, for the conditions of the Chita TPP-1 is extremely inefficient COP = 4,6.

Thus, the installation of HPU in the power system of the Chita TPP-1 with a COP value of up to 4.6 for the purpose of heat supply to consumers will certainly lead to excessive consumption of fuel in the system.

It is possible to reduce extremely inefficient COP and to ensure full compensation of the heat load of the TPP by using the secondary thermal resources of the TPP for heating supply to consumers equipped with heat pumps. This approach assumes the elimination of heat losses in the heat network [3, 4].

Considering the above, the maximum effect from the introduction of heat pump systems can be obtained in the inter-heating period to compensate for the load of hot water supply when using a heat pump unit at the consumer and transferring the TPP plant to the condensation mode, with the organization of heat removal from the condenser to the heating system. At the same time it is possible to cool the heating network below 20 ° C excluding the deterioration of the turbine vacuum.

At the same time, the cooling of the heat network below the ambient temperature will provide additional renewable heat, which will also reduce the extremely inefficient COP.

3. Development of complex heat supply system from TPP with HPU

A promising direction for the introduction of heat pumps at the initial stage with observance of the set conditions may be load compensation of hot water supply (HWS) in a non-heating period using a non-classical source of low-grade thermal energy [5]. In this case, the developed method of compensating for the load of HWS is accepted as a «baseline».

The hot water supply system of the building (Figure 1), modernized using the «basic» method, uses a heat carrier circulating in the closed loop of the heating system as a heat source for the heat pump. A closed loop is organized by closing the shut-off valves 8 and the three-way valve 12 along the supply line 1, and also the automation of the valves 7 is performed to reduce or completely stop the circulation of the coolant through the heat exchanger 6. The system effectively covers the load of hot water due to the heat accumulated in the building, and provides significant savings for the consumer. The method is characterized by high values of the conversion coefficient, which are characteristic of water heat pump units (HPU), with compactness and minimal implementation costs. The method also allows to utilize the excess heat of the premises during the summer period, reducing the energy consumption for air conditioning, which is characteristic only of expensive complex heat and cold supply systems [5].

IOP Conf. Series: Materials Science and Engineering **791** (2020) 012039 doi:10.1088/1757-899X/791/1/012039

If there is a «base» method on all consumers, it becomes possible to convert a source of thermal energy to work as a low-potential source of heat for the consumer according to the following method.

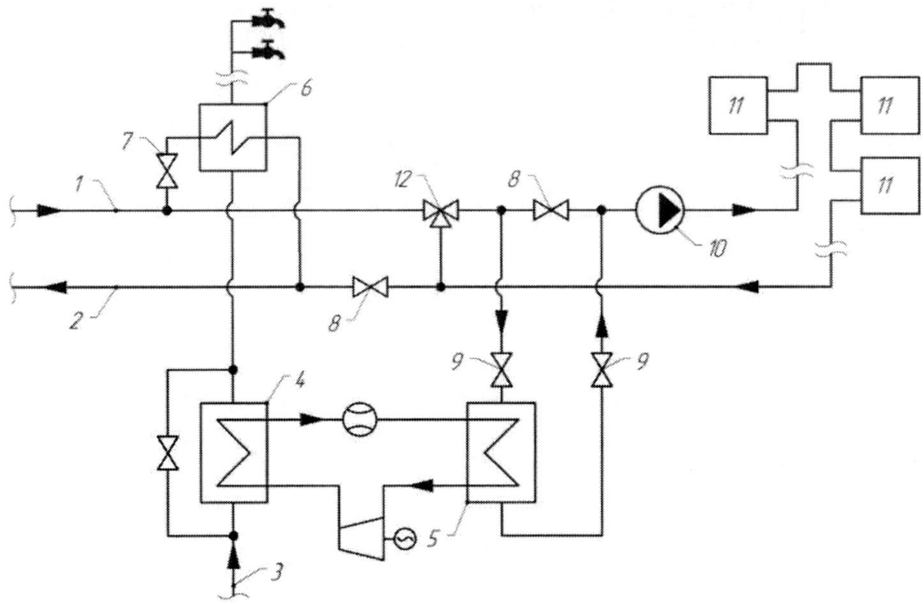

Figure 1. Hot water supply system of the building: 1 - supply pipe, 2 - return pipe, 3 - water pipe for hot water supply, 4 - heat pump condenser installation, 5 - heat pump evaporator installation, 6 - hot water heat exchanger, 7, 8, 9 - stop valves, 10 – circulation pump, 11 - heating device, 12 - three-way valve.

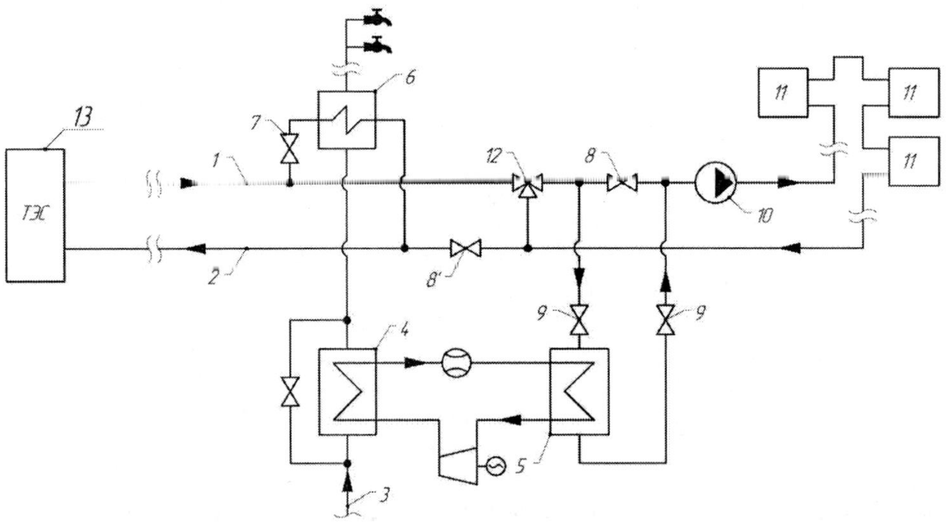

Figure 2. Integrated hot water supply system of the building (UCS-HWSL): 1 - supply pipe, 2 - return pipe, 3 - water pipe for hot water supply, 4 - heat pump installation condenser, 5 - heat pump installation evaporator, 6 - hot water heat exchanger , 7, 8, 8', 9 - stop valves, 10 - circulation pump, 11 - heating device, 12 - three-way valve, 13 - heat source.

IOP Conf. Series: Materials Science and Engineering **791** (2020) 012039 doi:10.1088/1757-899X/791/1/012039

In the transition from the «base» method to an integrated hot water system of the building, which was called universal compensation system of hot water supply load (UCS-HWSL), (Figure 2) opens valve 8' and also the three-way valve controlled by the adjusted automation of the heating system comes into operation in order to organize the supply of low potential heat. In the circuit of the heating system coolant, using a circulating pump, served in the heating system. Passing heating devices, the coolant collects the excess heat of the premises, after which it is divided into two streams, one of the streams returns to the TPP, the second stream passes the three-way valve, mixes in the required proportions with the low potential coolant from the supply pipeline and enters the HPU.

Simultaneously with the closure of the valve 8, the valve 7 is closed to shut off hot water heat exchanger.

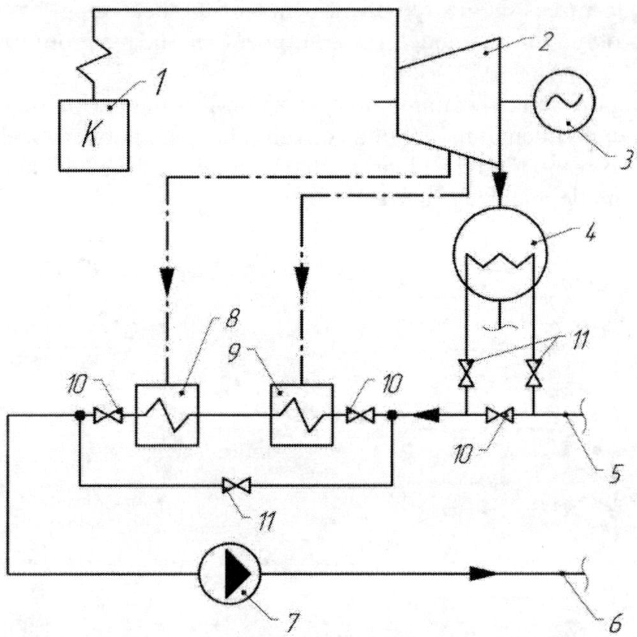

Figure 3. Schematic diagram of heating medium heating at TPP: 1 - boiler installation, 2 - turbine, 3 - generator, 4 - condenser, 5 - return pipe, 6 - supply pipe, 7 - mains pump, 8 - upper mains heater, 9 - lower mains heater, 10, 11 - stop valves.

When the coolant flow returns to TPP (Figure 3), its temperature becomes close to the ambient temperature due to heat gains in the buildings of consumers and through the insulation of heat networks. The relatively low temperature makes it possible to use it as cooling water directly in the condenser of the turbine unit when closing the shut-off valve 10 and opening the valve 11. Also, to reduce the supply high potential heat from turbine extraction, the upper and lower network heaters are disconnected.

The redistribution of steam flows from the network heaters to the condenser will provide consumers with coolant with low potential and eliminate the underproduction of power by turbines.

Transfer of consumers, heating networks and TPP to work with low potential coolant is a global expansion of the «basic» method and can significantly increase the efficiency of district heating by reducing heat losses in the condenser and increasing electricity generation based on heat consumption by reducing the pressure needed for heating. At the same time, this transition for the majority of TPP equipped with cogeneration beams is regime and does not provide for capital expenditures. In the case of CPP, there are relatively small costs for organizing the supply of coolant to the condenser.

The possibility of operating the station without shutting down the network heaters in the event of the likelihood of condensate overcooling is also considered. In this case, the mode of operation of the heating system will be similar to the standard one, but under the condition of low mains water temperature.

Thus, the gradual and controlled implementation of this technological solution will allow you to create energy-efficient systems at minimum cost and without the drawbacks of most foreign and Russian projects. In addition, the unification of plans for the development of heating systems will give impetus to the development of engineering, in particular the industry for the production of heat pump equipment, which in turn will also reduce the total cost of equipment and the entire project as a whole.

4. Assessment of the efficiency of the heat network as a source of low potential heat energy

In order to verify the presented provisions, it is necessary to simulate the heat supply system, which will make it possible to assess the degree of influence of the heat supply system on the efficiency of heat supply. As a prototype for the model, the equipment and heat supply systems of the Chita TPP-1 are adopted.

To determine heat gains in the heating system in a non-heating period based on the general model in ZuluThermo™ An experimental model of a conventional heat network with material characteristics has been created that correspond to the material characteristics of the heat networks of the Chita city, taking into account climatic features (Figure 4).

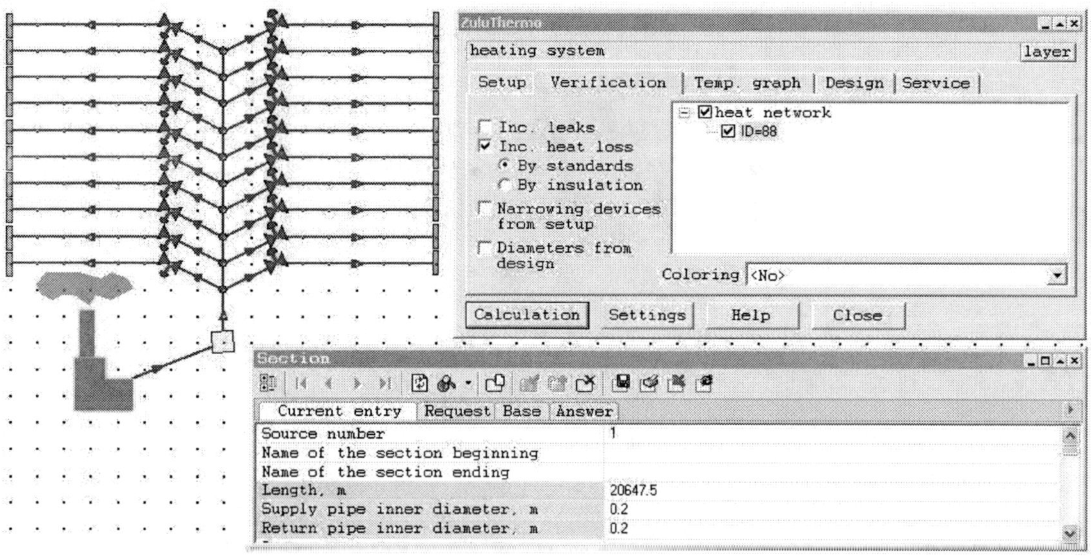

Figure 4. Thermal network topology of the experimental model

In the course of the experiment, the heat network was calculated at the average summer temperature of the soil, and the average summer air temperature for the conditions of the underground and above-ground laying. The following assumptions were made:

- the flow rate for each considered section of the heat network was taken from the condition of maintaining the flow velocity of 1 m/s.
- heat losses from the supply pipe of the heat network were excluded due to the absence of temperature pressure between the coolant and air temperature.
- leaks in the experimental system are not counted.
- thermal insulation for pipelines was calculated according to the norms of heat losses for networks not older than 2003.

According to the calculations, the highest efficiency of heat absorption can be obtained at the lowest possible coolant temperature. Since the heat pump unit is able to provide an inlet temperature of 10 ° C to the heating system, then for further calculations we will accept the heat perception, provided that the coolant temperature is minimal.

Overhead gasket, heat gain 2.27 Gcal/h.

Underground gasket, heat gain 1.8 Gcal/h.

The overall average heat absorption, taking into account the energy efficiency coefficient of the heat network and the heterogeneity of the gasket, was 2.64 Gcal/h.

This value is 2.6% of the total HWS load of the Chita TPP-1.

According to the conducted research, the UCS-HWSL is able to compensate up to 68.8% [18, 20] of the heat load of the TPP in the inter-heating period, taking into account the heat network, this value will be 71.4%.

5. Evaluation of efficiency of TPP after modernization

To obtain technical and economic indicators that determine the efficiency of the TPP-Consumer Complex with a UCS-HWSL and a lowered temperature schedule, you can use the calculation of indicators using the example of the Chita TPP-1 equipment according to the procedure outlined in [5].

In the mode without HPU, the Chita TPP-1 has the following indicators:

The specific consumption of reference fuel for the generation of thermal energy is 161.6 KgCE/Gcal, the specific consumption of reference fuel for the generation of electricity is 314.60 g / (kW•h).

Without modernization, the extremely inefficient COP for most classic projects for the implementation of heat pump units in heating systems will be 4.6.

When transferring the load of selection to the condenser and in the presence of heat gain through the heat network and taking into account the redistribution of power, the extremely inefficient COP of the system being developed was 1.71.

When introduced using the proposed HPU scheme with COP 3.2, we obtain a reduction in the specific consumption of coal equivalent for the generation of thermal energy to 138.4 KgCE /Gcal and a specific consumption of coal equivalent for generation of electric power is 303 g /(kW•h). In absolute terms, the introduction of a combined system with a heat pump system will save up to 28,670 tons of coal equivalent per inter-heating period.

6. Conclusion

1. The maximum effect of the introduction of a HPU may be obtained in the inter-heating period to compensate for the load of the HWS using the HP at the consumer and the transfer of the TPP to the condensation mode. At the same time, the heating system may be cooled below 20 ° C.

2. The most rational use of the development of TbSU with the utilization of excess heat of the premises, thereby providing trigeneration based on a standard heating system using commercially available steam-compression heat pumps.

3. It is possible to obtain the highest efficiency of heat absorption at the lowest possible coolant temperature in the heat network.

4. The overall average thermal perception of heat supply systems at low temperature, taking into account the energy efficiency coefficient of the heat network and the heterogeneity of the gasket, is 2.38 Gcal /h for the TPP under consideration.

5. The proposed system allows to reduce the extremely inefficient COP from 4.6 to 1.7.

6. Modernization of the base heat supply system to the UCS-HWSL with a lowered temperature schedule will improve technical and economic indicators, when using HPU with COP 3.2, the specific consumption of coal equivalent for the generation of thermal energy decreases from 161.6 KgCE /Gcal to 138.4 KgCE/Gcal, and the specific consumption of equivalent fuel for electricity generation from 314.60 g / (kW• h) to 303 g / (kW•h). In absolute terms, the introduction of a combined system will save up to 28,670 tons of coal equivalent per inter-heating period.

IOP Conf. Series: Materials Science and Engineering **791** (2020) 012039 doi:10.1088/1757-899X/791/1/012039

References

[1] Il'in R A and Il'in A K 2009 Kompleksnaya ocenka e`ffektivnosti kombinirovanny`x teploe`nergeticheskix ustanovok [Comprehensive assessment of the effectiveness of combined heat and power plants] *Problemy Jenergo- i Resursosberezhenija* **2** 33-41 [in Russian]
[2] Ovchinnikov Ju V 2010 *Osnovy tehnicheskoj termodinamiki* [Basics of technical thermodynamics] (Novosibirsk: NSTU) [in Russian]
[3] Rotov P V, Orlov M E and Sharapov V I 2012 O rabote sistem teplosnabzheniya bez izloma temperaturnogo grafika [About work the heating systems without thermal break schedule] *Energosberezhenie i Vodopodgotovka* **2** 12-7 [in Russian]
[4] Nikolaev Yu Ye and Baksheev A Yu 2007 Opredelenie e`ffektivnosti teplovy`x nasosov, ispol`zuyushhix teplotu obratnoj setevoj vody` TE`Cz [Definition of efficiency of thermal pumps using heat of return network water of thermal power station] *Promyshlennaya Ehnergetika* **9** 14-7 [in Russian]
[5] Batukhtin A G, Batukhtin S G, Kobylkin M V and Baranovskaia M G 2018 Application of heat pumps in a centralized heat supply as the direction of the development of district heating cogeneration *Int. Ural Conf. on Green Energy, UralCon* (Chelyabinsk, Russia) pp 45-50
[6] Guillaume R, Morten C and Pierre P 2019 Detection and characterization of domestic heat pumps *IEEE Milan PowerTech* (Milan, Italia) pp 1-6
[7] Rodrigo J, Johnny Y, Vivaldo J and Eduardo D 2019 Hybrid solar heat pump system for water heating *Engenharia Agrícola* **39** 419-25
[8] Rui F, Yiqiang J, Yang Y and Zuiliang M 2008 Theoretical study on the performance of an integrated ground-source heat pump system in a whole year *Energy* **33** 1671-9
[9] Abdeen O 2008 Ground-source heat pump systems and applications *Renewable and Sustainable Energy Reviews* **12** 344-71
[10] Pieve M and Trinchieri R 2018 Heat pump market report for Italy *HPT Magazine* **36** 14-8
[11] Blázquez C S, Borge-Diez D, Nieto I, Martín A F and González-Aguilera D 2019 Technical optimization of the energy supply in geothermal heat pumps *Geothermics* **81** 133-42
[12] Sagia Zand Rakopoulos C 2016 Alternative refrigerants for the heat pump of a ground source heat pump system *Applied Thermal Engineering* **100** 768-74
[13] Huang S, Zuo W, Lu H, Liang C and Zhang X 2019 Performance comparison of a heating tower heat pump and an air-source heat pump: A comprehensive modeling and simulation study *Energy Conversion and Management* **180** 1039-54
[14] Hundy G F, Trott A R and Welch T C 2016 Heat Pumps and Integrated Systems *Refrigeration, Air Conditioning and Heat Pumps* ed. Hundy G F, Trott A R and Welch T C (Oxford: Elsevier)
[15] Zhang X, Xiao M, He W, Qiu Z and Zhao X 2019 Heat Pump Technologies and Their Applications in Solar Systems *Advanced Energy Efficiency Technologies for Solar Heating, Cooling and Power Generation. Green Energy and Technology* ed. Zhao X and Ma X (Cham: Springer)

IOP Conf. Series: Materials Science and Engineering **791** (2020) 012040 doi:10.1088/1757-899X/791/1/012040

On the need to improve the methodology for calculating energy saving and energy efficiency of enterprises

R A Burganov, L V Maimakova, L R Urazbakhtina and L A Golitsyna

Department of Economics and Organization of Production, Kazan State Power Engineering University, 420066, Kazan, Russia. 51 Krasnoselskaya Str.,

E-mail: burraabr@gmail.com

Abstract. Currently, the development of nanotechnology and the digital economy requires a deeper study of the basic foundations of the functioning of society and the economy, among which the production, distribution and consumption of energy of various types and forms are of particular importance. The emphasis of theoreticians and practitioners on a comprehensive review of the role of energy should be reflected in the methodology for calculating energy conservation in the production process and the provision of services. This article draws attention to the need to revise the methodology for calculating the energy saving and energy efficiency indicators of a company based on the consideration of all types of energy (energy, heat, mechanical, light, human, etc.) in the company's activities. According to the author, a complete accounting of the energy used will optimize the costs of the company, maximize profits, and accordingly determine the development strategy of the company in the new business environment. In addition, the financial and economic performance of the company should also be updated and improved, taking into account the consumption of various forms and types of energy.

1. Introduction

Currently, energy as the main resource of any activity is becoming the focus of attention of both practitioners and representatives of the scientific community. And in world science, large financial and material resources are spent on finding ways to obtain and store energy in various forms [1-3]. An announcement at the RFBR of a competition for the best scientific projects of interdisciplinary basic research on the topic "Fundamentals of nature-like technologies for energy generation and consumption" indicates the relevance of a comprehensive study of the principles of energy generation and consumption even by living systems [4].

However, in the real economy, the methodology for calculating energy conservation and energy efficiency is based on the use of electricity and heat, which does not allow making the right strategic decision in the field of strategic development of economic entities, firms. In particular, in the field of energy conservation. So, in JSC "Network Company" the energy saving program of the company was implemented only in two areas: reduction of technological losses of electricity during transmission through the company's networks; optimization of electricity consumption for the company's business needs [5]. That is, the management of the enterprise does not pay due attention to the optimal use of other forms and types of energy. In such conditions, it is necessary to determine the methodology for revising the indicators of energy conservation and energy efficiency.

2. Methods

To achieve the goal of the study, both traditional and modern methods of analysis of technical, technological, financial and economic processes of the enterprise have been used. These methods included the analytical presentation and generalization of the functioning of the technical base of the enterprise through the use of various types of energy; classification of the types of energy used, including energy costs of personnel, analysis and synthesis of the calculation of

IOP Conf. Series: Materials Science and Engineering **791** (2020) 012040 doi:10.1088/1757-899X/791/1/012040

financial and economic indicators of the enterprise, etc. The interdisciplinary nature of the study made it possible to obtain more reasonable, scientifically-practical and competitive significant results in the world scientific space. Attention is paid to the use of the basic principles of implementing the energy theory of the company.

3. Results end Discussion

The aim of this work is to focus the attention of theoreticians and practitioners on a comprehensive review of the role of different types and forms of energy in the methodology for calculating energy conservation in the production process and the provision of services.

In modern times, the subject of priority tasks in economic development is the introduction of technologies based on the use of alternative energy sources in the production of goods and services. In addition, the struggle for energy is becoming a kind of driver for the development of human civilization.

The need for creating new approaches to enrego efficiency has been brought to the attention of some scientists, for example D. Bueno M. Roitman, who present a general model for energy efficiency calculation, developed with the main purpose of being capable to deal with the fast changes our society faces nowadays [6] It should be consistent with the opinion of Ali Hasanbeigi, Lynn Price, who write that industrial plants are not always aware of energy-efficiency improvement potentials [7]. One can also highlight the work of Trianni, A. et el [8].

Certain aspects of the energy saving process are highlighted in the conference materials [9-11].

The very concept of "energy" is multidimensional and abstract. From the ancient Greek language "energeia" is defined as activity or active power. Energy as a fundamental category of science is actively studied in philosophy, physics, economics, etc. Without energy, as well as without labor, capital, no actions are taken to manufacture products and provide services. However, there is still no generally accepted and fundamental scientific definition of the term "energy". So, Kogan I.Sh. notes that the concept of "energy" has acquired so many definitions and additions that it is sometimes difficult to understand what it is all about. There is also lack of clarity in metrological definitions of energy "[12]. Energy is coordinated with such concepts of the economic sphere as "the price of energy", "useful energy", "energy loss". In particular, the energy efficiency coefficient is defined as the ratio of all energy used in the household (in an established area, electrical installation, etc.) to the total amount of energy consumed in terms of its primary energy. There are also uncertainties in the definition of the terms "forms of energy" and "types of energy".

In the production process, various types of energy are used, since thermal, electric, electromagnetic, light (radiant), as well as the energy of biological resources, etc.

Mechanical energy is manifested in the movement of bodies, which has a certain direction in space (gas movement through a pipe, projectile flight, shaft rotation, etc.).

Electric energy is manifested in the latent movement of electrons along a conductor (electric current).

Thermal energy is expressed in molecular and intramolecular chaotic motion, representing the energy of the chaotic motion of atoms and molecules of matter. So, abroad, they came up with a floor covering that can generate energy when walking on it. The smart floor is made from cellulose fibers treated with a special chemical substance that allows them to generate an electric charge. According to researchers, the use of solar and wind energy has increased over the past 10 years, with an average annual increase in capacities of 51% and 21%. Although in the total production, the combined share of these types of energy accounts for only 4% [5].

All these forms of energy can be outside the field of view of management and economists, and, accordingly, are not taken into account when determining the effectiveness of the company.

The study of the use of energy of intellectual (mental) forces, energy, nerves and energy of the physical forces of employees of the enterprise is in high demand. Labor intensity is defined as the amount of energy a worker spends per unit of working time. In the anthology of economic theory, the provisions of Marxism on the need to take into account the energy expenditure of a worker to determine the level of wages are known. Malnutrition of employees does not restore the energy costs of the employee. The salary of the employee should provide opportunities for the restoration of the forces (energy) of the employee, as well as for the reproduction of labor by maintaining the viability of his family.

The process of energy consumption is based on certain principles. Under the principles understand the basic theoretical principles and rules that guide the company in its activities. The company should do only that which meets these principles, and nothing that contradicts them.

Any managerial actions of the firm's management should be taken in recognition of the leading role of various forms of energy in all production, financial, and technological operations. It is possible to have machines, machines, equipment in the balance sheet of the company, but they will not have value without connecting to an energy source. The energy balance of the company becomes the most important factor in planning the activities of the company.

Among the negative aspects of this theory is a possible manifestation of energy. The most famous and consistent theoretician of energy was the German scientist W. Ostwald. According to him, "all natural phenomena can be subordinated to the concept of energy."

The state has created special conditions for entrepreneurs. For example, when consuming renewable energy sources, tax preferences are created for real estate tax, land tax, and profit tax can be partially or completely removed.

The need for a more focused study of the role and significance of various types and forms of energy has led to the creation of the energy-consuming theory of the company. The main provisions of this theory are considered by the author in a number of scientific papers [13-14].

Existing versions of the theory of firms did not take into account the complex processes of energy consumption and energy saving, did not provide recommendations for the management of the company to achieve certain success in the field of energy efficiency, and thus did not contribute to the achievement of the energy strategy of the company and the national economy. The theory of energy consumption of the company was created based on the analysis of extensive experience in using different forms and types of energy. Solving problems in the field of optimizing energy consumption in the course of a company's activity has always been the focus of its management. This is mainly due to the desire of the company to maximize profits, an increase in the state's demand for energy conservation, and toughening competition in the production and consumption of energy resources. In modern times, this is supplemented by a decrease in renewable energy sources, a change in the operation technologies of machines, equipment, machine tools, a decrease in external energy dependence, etc. The subject of these priority tasks is the introduction of technologies using alternative energy sources.

The general scheme of the process of movement of total energy during the production of goods and the provision of services is given in figure 1.

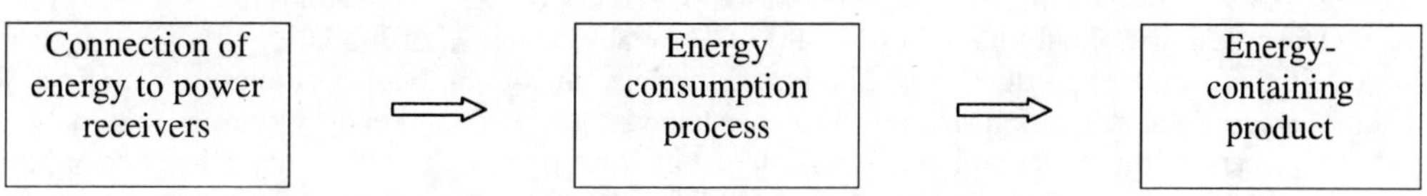

Figure 1. General scheme of the process of energy movement in a separate company

As can be seen from the figure, any product of production contains energy costs, including the energy of living organisms, that is, employees of the company.

The process of consumption of total energy is based on certain principles. Under the principles understand the basic theoretical principles and rules that guide the company in its activities. The company should do only that which meets these principles, and nothing that contradicts them.

The basic principles of the energy theory of the company include (table 1).

Each of the identified principles provides for the introduction of mandatory and (or) recommended requirements in the projects and programs of the company. Any managerial actions of the firm's management should be taken in recognition of the leading role of various forms of energy in all production, financial, and technological operations. It is possible to have machines, machines, equipment in the balance sheet of the company, but they will not have value without connecting to an energy source. The energy balance of the company becomes the most important factor in planning the activities of the company.

IOP Conf. Series: Materials Science and Engineering **791** (2020) 012040 doi:10.1088/1757-899X/791/1/012040

Table 1. The principles of energy theory of the company.

№	Name of principles	Principles
1	Principle of system	The result of the functioning of the company depends on the efficient use of all forms and types of energy, conditionally total energy
2	The principle of equal access to energy sources	All structural units of the company must be provided with energy
3	Optimality principle	Any decision in the field of energy consumption should be based on the task of optimal energy use.
4	The principle of achieving effect and energy efficiency	The use of energy should have an effect and efficiency, where there is an excess of results over costs

Compared with other versions of the theory of the company, the energy-consuming theory of the company has several advantages, at the same time, disadvantages.

The positive aspects of the energy theory of the company include universality of provisions, priority in the use of any energy (in particular, it is imperative to use the labor of a worker in production according to the factor theory of the company, and there is no way to use energy theory in energy theory), the prospect of relevance to scientific provisions.

The energy-consuming theory of the company is an integrated science, a significant place in it is occupied by physical, technical and economic concepts. This means that the researcher must possess an appropriate level of knowledge.

To determine the viability of a theory, it is necessary to consider the mechanism of its implementation in practice. The mechanism for the implementation of the theory in practice is various institutions, methods, tools that ensure the implementation of theoretical principles in the practical activities of the company. As a rule, such a mechanism includes activities of a managerial and financial and economic nature. Among the many institutions that contribute to the process of implementing the energy theory of the company into practice, we can distinguish internal company institutions involved in management decisions, mainly the management of the company and the owners who adopt regulatory documents.

Preparation of practical recommendations on energy management to increase energy and economic efficiency of energy resources based on the energy theory of the company includes the systematization of specific indicators based on the use of energy consumption.

The researcher should have in his field of view all the stages of the energy movement process in the course of production of goods and services, including the use of energy by company employees. Such scientific principles as consistency, equal access to energy sources, optimality, achievement of the effect and energy efficiency make it possible to obtain significant scientific results.In accordance with the energy theory of the company, the product of production can be measured in energy costs. Approximately, 3 conventional units of energy were expended for the production of the product. In economic science, "conventionality" is actively used. "Conventional unit of fuel", "conditional unit of production", etc. categories find their application in the economic justification of the development of the company. An important element of cost analysis is determining the marginal cost of energy consumption. Marginal cost - an increase in energy costs resulting from the production of one additional unit of output. Energy costs can be divided by type of energy consumed, cost structure, periods of energy consumption, etc. Separately, we can distinguish the calculations of the energy potential of the company, i.e. the company's capabilities in providing energy to its activities.

Each type of energy has its own price, which is created both by producers and consumers of energy (the price of the buyer and the price of the producer (seller). The price of energy is the price of the resources used in terms of the amount of energy. Hypothetically, the price of energy can be a measure of the production and exchange of goods and services, for example, one can argue that 100 units of energy were used in the production of a particular product.

According to GOST R 51541-99 "Energy Efficiency. Composition of indicators "establishes the main types of indicators of energy saving and energy efficiency introduced into regulatory

IOP Conf. Series: Materials Science and Engineering **791** (2020) 012040 doi:10.1088/1757-899X/791/1/012040

(technical, methodological) documents, technical (design, engineering, technological, operational) documentation for energy-consuming products, technological processes, works and services. Thus, the energy efficiency of a company is determined by the ratio of energy consumption to the amount of income received. When calculating energy costs, it is necessary to take into account the use of all types of energy - electrical, thermal, light, mechanical, etc., including human. Such an integrated approach requires a review of the methodology for calculating total costs. To analyze energy consumption in terms of consumption of any form of energy, four approaches can be distinguished:

- costs of connection to power grids, including technological aspects inside,
- direct energy consumption costs,
- energy distribution costs;
- costs associated with obtaining comprehensive benefits.

On the income side, it is also necessary to take into account the share of contribution to its formation of each type of energy and total energy. For example, a simple light bulb performs basically two functions - lighting a room and generating heat. Each function has a cost, respectively a price, respectively, it is necessary to find the energy value of a given product. The price of energy depends on the price of each type of energy. Of course, the problem arises of calculating such prices.

As you know, the energy potential of the company consists of technical and biological types of energy (energy of muscles, brain). The correct accounting of the processes of energy consumption of intellectual (mental) forces, energy, nerves and energy of the physical forces of the employees of the enterprise is in demand. However, in economic science there is a gap in this area of research. Therefore, the basis of the calculation can be taken to determine the energy consumption of athletes. To achieve high results, it is necessary to maintain a balance between the energy expended and its replenishment. Using the methods of determining energy consumption, it is possible to calculate how much energy is spent in preparation for the competition and during the competition, in accordance with this, draw up the athlete's mode, which allows to fully replenish the allocated energy. The main methods for determining the energy consumption of personnel include the method of determining energy consumption taking into account the coefficient of physical activity and the magnitude of the basal metabolism, which consists in multiplying the coefficient of physical activity corresponding to a certain type of activity by the magnitude of the basal metabolism. The coefficient of physical activity (CFA) is the ratio of the average daily expenditure of human energy to energy expenditure at rest. CFA can vary from 1.4 with a sedentary lifestyle to 2.4 with heavy hours of physical labor. The main metabolism is the minimum level of energy consumption, which is necessary to maintain the body's vital activity at rest. And the Aldridge method. allows you to learn about the metabolic rate or calculates the rate of energy consumption in calories. Back in Marxism, the relationship between the level of wages and the energy expended by a worker in industrial production was examined.

Digitalization of the economy requires strengthening the role and importance of financial and economic accounting for the consumption of all forms of energy in the activities of the company. In fact, the use of each type of energy should be accompanied by the conclusion of appropriate contracts, which in a generalized form goes through a number of stages, such as:

- preparatory and informational (development of a draft of measures in the field of concluding energy contracts, etc.);
- research and analysis (analysis of energy-saving measures, etc.);
- recommendatory (from the set of projects, the most optimal one, etc.);
- introduction (inclusion of the received energy in the production process);
- monitoring and corrective (monitoring the progress of the execution of concluded agreements, etc.).

Hypothetical, even the use of personnel energy should be of a contractual nature, since the energy potential of each employee of the company varies.

It should also be noted that at each stage of the use of energy, the transaction cost increases. In such a situation, it becomes necessary to combine the use of different types of energy.

Work to reduce the cost of purchasing different types of energy can be carried out in two directions: saving energy and increasing the efficiency of the production process. In

IOP Conf. Series: Materials Science and Engineering **791** (2020) 012040 doi:10.1088/1757-899X/791/1/012040

addition, there are a number of effective ways to save energy, such as using equipment that allows you to receive and use different types of energy (for example, equipment can generate heat); the use of energy-saving technologies; reduction of energy losses in power receivers and power supply systems, etc.).

In particular, the manufacturer must choose the options for providing electricity for his production. There are three possible solutions:

- purchase of electricity from outside, that is, through technological connection to energy networks;
- create their own autonomous production of electricity;
- focus on sharing the first two options.

In each case, it is necessary to evaluate the value of the alternative possibilities. In general, when making economic decisions in the field of electricity supply, all costs are of an alternative nature, i.e. depending on the options for the cost of resources, the optimal solution for their use is selected. Accounting for opportunity costs allows you to more accurately evaluate and compare the effectiveness of various economic decisions.

As you know, the explicit costs, depending on the volume of production, can be divided into fixed and variable. Constant - do not depend on the volume of production, and variable costs are not. This also applies to energy costs. So, part of such costs is aimed at constantly ensuring the working condition of technical or technological means. This can be fire or process safety.

An important element of cost analysis is the determination of marginal cost of energy consumption. Marginal cost - an increase in the cost of electricity resulting from the production of one additional unit of output. Marginal costs can be calculated on the basis of variable and fixed costs

Marginal cost of energy consumption - an increase in energy costs resulting from the production of one additional unit of output.

Marginal cost of energy consumption can be calculated on the basis of variable and fixed costs.

$$\text{Marginal cost of energy consumption} = \frac{\Delta\ \text{Gross cost of energy consumption}}{\Delta\ \text{volume of output}} \quad (1)$$

$$\text{Marginal cost of energy consumption} = \frac{\Delta\ \text{Variable cost of energy consumption}}{\Delta\ \text{volume of output}} \quad (2)$$

where Δ shows an increase in indicators.

We also find:

$$\text{Marginal energy efficiency} = \frac{\Delta\ \text{Change in energy consumption of different types}}{\Delta\ \text{volume of output}} \quad (3)$$

Any change in the data of the individual types of energy consumed is reflected in a state of energy efficiency.

The process of managing electricity costs is in the field of view of any manager. The larger the production, the more energy will be used up. The presence of a cost management program leads to lower costs, respectively, positively affects the energy efficiency of the enterprise.

Since all costs must be controlled at a certain organizational level, the enterprise must clearly determine at what level these or those costs are controlled. Optimization of energy consumption can occur due to control of energy losses and their reduction; use of secondary energy resources; optimization of technological process modes; development and coordination of a start-up schedule for energy-intensive equipment; minimize heat loss; equipment modernization

4. Conclusion

Thus, the current level of development of new technologies based on the efficient use of all types of energy, the creation of new financial and economic relations on this basis objectively requires a review of the methodology for calculating energy saving and energy efficiency indicators. In this aspect, the content of this article has a certain scientific novelty. More specific scientific results include conclusions that:

- at existing enterprises, energy conservation is achieved only by saving and optimizing the use of electricity;
- there is uncertainty in the scientific definition of the essence of "energy", its types and forms,
- enterprise management in the activity should focus on the use of all types of energy;
- a special role in obtaining energy efficiency is played by the energy of living organisms, in particular the labor resources of the enterprise personnel;
- the need to take into account all types of energy has led to the creation of an energy-consuming theory of a company that has no analogues in world economic science;
- given the substantial part of the energy theory of the company, its principles of research;
- the product of production can be measured in energy consumption, which allows to obtain more accurate data on energy efficiency indicators;
- calculation of energy costs from the side of employees allows optimizing their labor input, comparing energy costs and wage levels, as well as increasing labor productivity depending on the capabilities of each employee. The use of certain methods of accounting for energy costs of personnel depends on the maturity of company management;
- an important element of energy conservation and energy efficiency is the determination of the marginal cost of energy consumption, given the methodology for their calculation;

In general, the development trends of the innovative and digital economy, the use of nanotechnology require taking into account all factors of energy efficiency growth. More specific factors include the increased use of energy of various kinds and forms as a result of digital changes; increase in the cost of energy in determining the cost of exchange operations; development of automation and control tools based on digital technologies that allow energy measurements; control and planning of consumption parameters of all types of energy.

The very need to take into account all forms and types of energy in the methodology for calculating energy saving and energy efficiency indicators poses new challenges for the leadership of the enterprise, industry and government. And economic, technological, social phenomena or processes associated with energy consumption must be measurable. The researcher and practitioner should have in their field of vision all the stages of the energy movement process in the production of goods and services, including the use of energy by employees of the enterprise. Also, in accordance with the energy theory of the company, indicators of the financial and economic activities of the enterprise should be reviewed taking into account data on the costs of various types and forms of energy.

Acknowledgments
The reported study was funded by RFBR according to the research project № 17-02-00102-OGN-A

References

[1] Vishnever V Ya, Burganov R A and Nusratullin V 2019 Institutional Transformations In The Field Of Services In Digital Economy *The European Proceedings of Social & Behavioural Sciences EpSBS. Published by the Future Academy GCPMED 2018 – Int. Scientific Conf. "Global Challenges and Prospects of the Modern Economic Development"* (Samara, Russia) vol LVII pp 1759-68

[2] Burganov R A and Zalyalieva S G 2014 Problemy` institucionalizacii v e`lektroe`nergetike [Study of the development of institutionalism in the electric-power industry] *E`konomika regiona* [*Economy of Region*] **3(39)** 158-66 [In Russian]

[3] Terebova S V and Borisov V N 2019 The development of small innovative business in the industrial, scientific and educational sector in Russia *Economic and Social Changes: Facts, Trends, Forecast* **12(3)** 55-76

[4] Fundamental`ny`e osnovy` prirodopodobny`x texnologii i generacii i potrebleniya e`nergii (26-823) [The fundamental foundations of nature-like technology and the generation and consumption of energy (26-823)] URL:

IOP Conf. Series: Materials Science and Engineering **791** (2020) 012040 doi:10.1088/1757-899X/791/1/012040

http://www.rfbr.ru/rffi/ru/contest/o_2074257 [In Russian]

[5] Energosberezheniye. OAO "Setevaya kompaniya [Energy saving. The site of JSC "Network Company"] URL http://gridcom-rt.ru/o-kompanii/napravlenie-deyatelnosti/energosberezhenie/ [In Russian]

[6] Bueno D and Roitman M 2016 A general model for energy efficiency calculation *Conf.: IEEE Conf. on Technologies for Sustainability, SusTech* (Phoenix, USA) pp 7897151

[7] Hasanbeigi A and Price L 2010 *Industrial Energy Audit Guidebook: Guidelines for Conducting an Energy Audit in Industrial Facilities* (Berkeley: LBNL)

[8] Triann A, Cagno E, Bertolotti M, Thollander P and Andersson E 2019 Energy management: Apractice-based assessment model. *Appl. Energy* **235** 1614-36

[9] Derevyanov M Yu, Pleshivtseva Yu E, Kordyukova L S and Rapoport A V 2019 Using DEA models to measure the efficiency of energy saving projects *IOP Conf. Ser.: Mater. Sci. Eng.* **552** 012020

[10] Vasilenko N A Parametric optimization of an architectural object's form as a method to improve its energy efficiency *IOP Conf. Ser.: Mater. Sci. Eng.* **552** 012027

[11] Sivachenko L A, Sevostyanov V S and Il′inaT N 2018 Problem Tasks in the Field of Resource and Energy Saving Technologie *J. Phys.: Conf.* **1066** 012021

[12] Kogan I Sh 2017 Sistematizaciya i klassifikaciya opredelenij i dopolnenij k ponyatiyu "e`nergiya" [Systematization and classification of definitions and additions to the concept of "energy"] URL http://www.sciteclibrary.ru/rus/catalog/pages/8784.html [In Russian]

[13] Burganov R A and Yudina N A 2018 To the question of the creation of the energy consumer firm theory *J. of Entrepreneurship Education* **21** 3

[14] Burganov R A, Altynbaeva I R,, Maimakova L M, Dolonina R A and Urazbakhtina L R 2019 The Concept of Information Asymmetry in the Energy Consumer Theory of Firms *J. of Eng. and Applied Sciences* **14** 8099-108

IOP Conf. Series: Materials Science and Engineering **791** (2020) 012041 doi:10.1088/1757-899X/791/1/012041

Features of the functional reliability of 0.4-10 kV distribution networks

F L Byk, L S Myshkina and I A Sabadash

Department of Automated Electrical Power Systems, Novosibirsk State Technical Universality, 20 Prospekt K. Marksa, Novosibirsk, 630073, Russia

E-mail: Lsmyshkina@gmail.com

Abstract. The article is devoted to the study of the influence of regime restrictions on the functional reliability of distribution networks of 0.4-10 kV. This level of the network is largely determining the values of indicative indicators of reliability of power supply. In the work, by the example of the analysis of the reliability of power supply of two energy regions, the significance of the influence of the regime factor on the uninterrupted power supply is proved. Due to the complexity of the topology and the variability of the flow distribution, it is proposed to use the decomposition method for calculating the reliability and recoverability of network nodes. The expediency of calculating indicative indicators of the reliability of electricity supply at the stage of network design is shown, since today, in Russia, the economic efficiency of the functioning of territorial network companies is largely determined by the level of reliability due to the existing relationship between reliability indicators and the tariff for the transmission of electric energy.

1. Introduction

The mission of the electric grid complex is the long-term provision of reliable, high-quality and affordable power supply to consumers. Distribution electric network is an important and extended link, to a greater extent determining the level of reliability and quality of power supply. Failures in networks of 6 - 10 kV are the cause of about 70% of all power supply disturbances of consumers [1], [2] which reduces the technical efficiency of these networks [3], which actualizes the tasks of studying the reliability of networks.

At the same time, the level of the tariff for the provision of services for the transmission of electric energy by network organizations in Russia depends, inter alia, on the level of functional reliability of electricity supply, estimated by indicative indicators of reliability of electricity supply. These indicators, according to the current legislation (Government Decision of the Russian Federation 2002, Dec. 31 №1220 "On the determination of the reliability and quality indicators of the delivered goods and services provided when establishing long-term tariff"), are used in Russia to stimulate an increase in the level of reliability and quality of services provided by network organizations, and are used in the mechanism of setting tariffs for electric power transmission. Upon reaching certain values of the indicators, it is possible to increase the tariff for the transfer of electric energy to 2%, and decrease to -3%.

However, when designing electric networks of low and medium voltage classes, when deciding on the maintenance and repair, reconstruction and re-equipment, calculation of indicative indicators of reliability is not provided. However, the economic efficiency of network companies will subsequently depend on their values. What explains the relevance of research and the scientific novelty of the work.

IOP Conf. Series: Materials Science and Engineering **791** (2020) 012041 doi:10.1088/1757-899X/791/1/012041

The goal is to assess the impact of operational restrictions on the functional reliability of the distribution network. Achieving this goal contributes to the solution of the formulated tasks:

- to analyze and evaluate functional reliability using the example of a distribution network section without taking into account operational restrictions on the performance of the mutual reservation function;
- to analyze and evaluate the functional reliability of the distribution network, taking into account the identified operational restrictions on the performance of the mutual reservation function.

2. Materials and Methods

2.1. Relation of reliability and tariff

In Russia, according to Government Decision № 1220 and the relevant guidelines, the level of reliability of services for the transmission of electric energy to consumers provided by territorial grid organizations that have networks not related to the Unified National Electric Grid is determined by two indicative indicators calculated based on the duration and frequency of interruptions of power supply, the number of connection points of consumers:

- The average duration of the termination of the transfer of electrical energy to consumers at the point of delivery (SAIDI);
- The average frequency of the cessation of the transfer of electrical energy to consumers at the point of delivery (SAIFI).

$$SAIDI = \frac{\sum_{j=1}^{J} T_j \cdot N_j}{N_t} \tag{1}$$

$$SAIFI = \frac{\sum_{j=1}^{J} N_j}{N_t} \tag{2}$$

where:

T_j - the duration of the j-th termination of the transmission of electric energy in relation to points of supply to consumers of services of a network organization within the framework of a technological violation, hour;

N_j - the number of points of supply to consumers of the services of the network organization, in relation to which the j-th cessation of transmission of electric energy occurred within the framework of a technological violation, pcs;

N_t - the maximum number of points of supply for consumers of services of a network organization for the t-th settlement period of regulation for a year, pcs.

These indicators largely coincide with some used abroad, but the set of indicators used there is much wider [4, 5].

Average Service Availability Index, Energy Not Supplied and Average Energy Not Supplied [6] are widely used, which are used for demand management tasks [7].

2.2. Calculation of indicators based on the structure and topology of the network

The calculation of the expectation of indicative reliability indicators can be based on information about:

- network structure,
- topology,
- individual reliability indicators of the main network equipment,
- presence of redundancy,
- automation,
- the number of connected consumers.

IOP Conf. Series: Materials Science and Engineering **791** (2020) 012041 doi:10.1088/1757-899X/791/1/012041

Using the methodology of reliability analysis of a regional electric network [8], equivalent reliability indicators of load nodes are calculated and then indicative indicators of reliability of power supply of the analyzed network are calculated. The following are investigated in the work:

$$SAIDI = \frac{\sum_{i=1}^{m} \omega_i t_i N_i}{\sum_{i=1}^{m} N_i} \tag{3}$$

$$SAIFI = \frac{\sum_{i=1}^{m} \omega_i N_i}{\sum_{i=1}^{m} N_i} \tag{4}$$

where:

N_i - the number of points of supply to consumers of the services of the network organization, in respect of which there was an i-th termination of the transmission of electric energy as part of a technological violation, pcs.;

m is the number of load nodes in the analyzed network;

ω_i, - failure rate of power supply for node i of the network;

t_i - average recovery time of power supply for node i of the network.

The assumption of the methodology [8, 9], which is true for 110-220 kV supply networks, is that power supply interruption occurs only when all branches (equipment) that come to the node fail due to a significant amount of redundancy.

In 0.4-10 kV networks, mode studies conducted by the authors confirm the insolvency of this restriction as applied to distribution networks. To determine the reliability indicators of power supply to consumers using the above methodology, it is proposed to determine the nodes, disconnecting the load in which allows for complete mutual redundancy between the energy regions of the network.

It is worth noting that some criteria could be considered in defining the network structure, such as reliability improvement [10], [11] as well as minimization of power losses [12], [13] When choosing nodes that, based on the conditions of the regime, will be turned off under certain conditions (to keep the rest of the supply points in working condition).

We will be guided by the principle of turning off the required power (determined by operating restrictions for inter reservation), with a minimum number of disconnected consumers:

$$P \quad \rightarrow \min N_j \tag{5}$$

where N_j - is the number of points of supply to consumers of the services of a network organization that are disconnected for inter-reservation of energy areas.

Therefore, for the selected nodes when calculating the average duration of the termination of the transfer of electric energy to the supply point, the average frequency of the termination of the transfer of electric energy to the delivery point, two components are necessary: reflecting the level of structural and functional reliability.

As a scientific novelty of this work, it should be noted the proposal to use the decomposition method for calculating the reliability indicators of nodes in energy regions, taking into account the existing functional redundancy in the distribution network of 0.4-10 kV.

The essence of the method is to separately calculate the indicators for each energy region, taking into account possible redundancy from a neighboring power center, followed by superimposing the results of two calculations to determine the indicators of the structural and functional reliability of the load nodes. For this, in accordance with the methodology for reliability analysis of the regional electric network [8], matrices are compiled for the distribution network section in normal operation, as well as in the case of reserving the section supplying power from the neighboring energy region. Next, the matrix is superimposed on each other. Figure 1 shows an overlay of matrices reflecting:

- equipment failure rate;
- equipment recovery time;
- possible power distribution between network nodes.

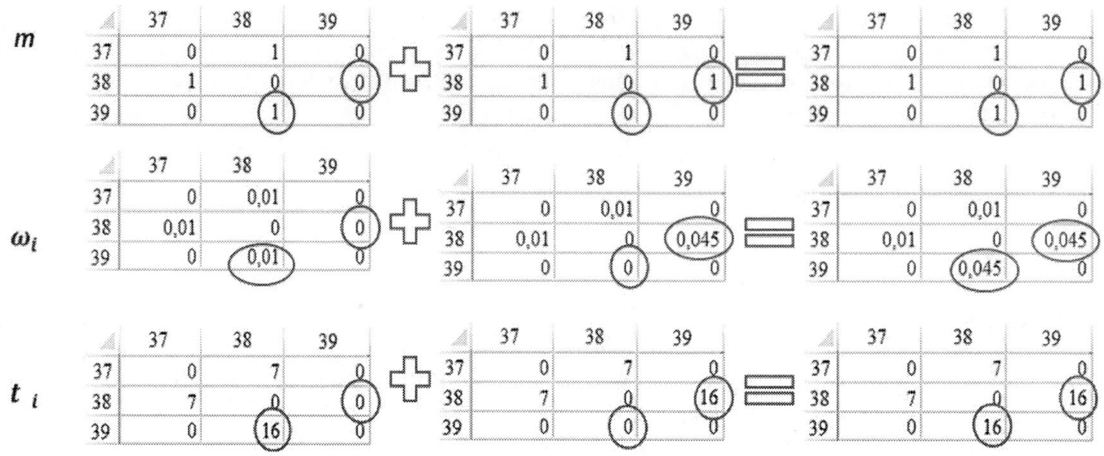

Figure 1. Implementation of the decomposition method for calculating the reliability indicators of nodes.

3. Results of a distribution network reliability analysis

The study was conducted on the example of a 10 kV network section, consisting of two energy areas. Possible operational restrictions in the conditions of redundancy and their influence on the functional reliability of the network are revealed.

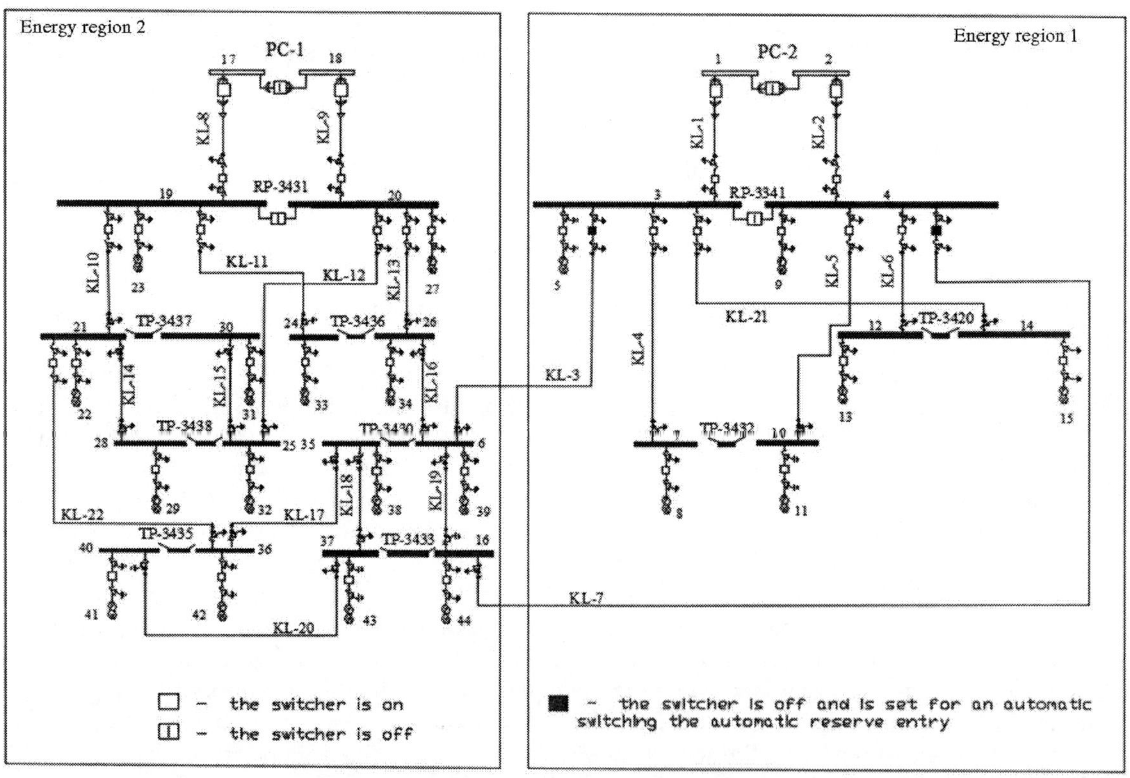

Figure 2. Schematic diagram of the electrical connections of the studied network section.

IOP Conf. Series: Materials Science and Engineering **791** (2020) 012041 doi:10.1088/1757-899X/791/1/012041

Power supply to consumers of the analyzed 10 kV network is carried out from two independent power centers (PC - 1, PC - 2). The schematic diagram of the electrical connections of the distribution network is shown in figure 2.

The distribution network section is represented by ten load nodes with a total number of connected consumers equal to 925 pcs. By load nodes we mean consumer connection points – transformer points (TP), the state of which allows characterizing the reliability of power supply. The total power of the connected load of the first energy region is 7.3 MVA, and the second 3.6 MVA.

Mutual reservation of energy areas in the network is carried out by backup lines KL-3 and KL-7, the inclusion of which is due to the action of automatic switching on the reserve.

Calculations of the electrical modes made it possible to identify the existing regime restrictions on the implementation of the function of mutual reservation between energy regions. In the event of a power failure of the PC-2, the available redundancy in the network is sufficient to maintain power supply to all consumers of both energy regions, while maintaining at least one of the two backup lines in working condition.

However, in case of power failure of the PC-1 and the failure of one of the backup lines, the power supply of the local energy district of the PC-1 will be interrupted due to a mismatch of the operating parameters - exceeding the maximum allowable current on the backup line.

To determine the indicative indicators of the reliability of power supply using the method presented in [8], [9], the nodes are determined, disconnecting the load in which allows for complete mutual redundancy. Therefore, for the selected nodes, equivalent reliability indicators will be determined not only by structural reliability, but also by functional reliability, which will lead to a change in the failure frequency and recovery time of these nodes.

Based on the criterion of minimizing the number of disconnected supply points (which is consistent with the incentive mechanism in Government Decision № 1220) and the need to reduce the reserve line load by 2.05 MW, the following points were selected:

- TP-3431;
- TP-3436;
- TP-3438.

The failure rates and recovery time of power supply in these nodes will be affected by the reliability indicators of the backup lines, depending on the backup path. The above is taken into account when compiling the incident matrix [8], [9], which reflects the topology and possible direction of power flows, determined as a result of the analysis of the circuit and calculation of electrical modes.

Due to the complexity of the topology that determines the flow distribution, the decomposition and superposition of the correlation matrices is used to determine the structural and functional reliability of the load nodes.

Table 1 presents the equivalent reliability indicators of the load nodes, where it is necessary to disconnect the load for inter-reservation in case of power failure from the PC-1.

Table 1. Equivalent reliability indicators of nodes.

Lode node		Excluding the regime factor		With the regime factor	
Nomination	Number	ω_i, 1/year	ti, h	ω_i, 1/year	ti, h
TP-3431	71	$4.259 \cdot 10^{-6}$	19.272	$5.168 \cdot 10^{-5}$	7.999
TP-3436	47	$3.575 \cdot 10^{-6}$	21.684	$5.168 \cdot 10^{-5}$	7.999
TP-3438	56	$3.575 \cdot 10^{-6}$	21.680	$5.168 \cdot 10^{-5}$	7.999

Based on the equivalent reliability indicators of the load nodes, indicative indicators of the reliability of power supply are determined, both without taking into account and in the presence of operational restrictions on the performance of the function of mutual reserving of energy regions, table 2.

Table 2. Indicative indicators of reliability of power supply.

Energy region	Excluding the regime factor		With the regime factor	
	SAIDI, h	SAIFI, r.u.	SAIDI, h	SAIFI, r.u.
The First	0.6658	0.0294	0.6658	0.0294
The Second	0.6847	0.0322	1.9400	0.2119
All over the network	0.6791	0.0313	1.5577	0.1571

4. Discussion

As a result of the calculation of indicators, in the case of the assumption that there are no regime restrictions on the performance of the mutual reservation function, we can state a high level of reliability, since the values of the indicators did not exceed 1, which corresponds to 1 shutdown per year for no more than 1 hour, and an average level reliability of power supply when taking into account the influence of the regime factor, since the values of SAIDI and SAIFI indicators do not exceed 4, which corresponds to 1 shutdown per quarter for no more than 1 hour [14].

However, the influence of the regime factor is great. When considering the regime factor, an increase in both indicators is observed. SAIDI increased 2.83 times, and SAIFI increased 6.59 times for the second energy region. For the network as a whole, SAIDI increased 2.29 times, and SAIFI increased 5.02 times.

The load nodes identified as a result of the study, the disconnection of which allows ensuring the mode when backing up on one cable line - these are actually "bottlenecks" of the network where measures for ensuring reliability are necessary. These nodes can be considered as points of connection of micro generation or installation of energy storage systems [15, 16] which may be the subject of further research.

5. Conclusion

In the work, the possibility of mutual reservation of power supply of two energy region of a 10 kV network was investigated. The presence of operational restrictions in case of interruption of power supply from one power center and the failure of one backup line is revealed.

Due to the complexity of the topology and possible flow distribution, it is proposed to use the decomposition method to calculate network reliability indicators.

Comparison of indicative indicators of reliability, taking into account operating restrictions and without, showed the importance of operating restrictions on the functional reliability of the distribution network, which confirms the hypothesis that it is necessary to take into account operating factors in distribution networks 10-0.4 kV.

The necessity of taking into account indicative reliability indicators at the stage of designing distribution networks and making decisions on measures aimed at improving reliability, which will improve the technical and economic efficiency of the network, is shown.

Acknowledgments

This work was supported by the Novosibirsk State Technical University (project C-19-26).

References

[1] Batueva D E 2017 Issledovanie intellektual`ny`x sistem upravleniya v raspredelitel`ny`x setyax 10 kV na osnove reklouzerov PSS-10 [The study of intelligent control systems in 10 kV distribution networks based on PSS-10 reclosers] *Collection of Articles of the Scientific Forum Technical and Physics Mathematical Sciences* **5** (Moscow: MTsNO) pp 181-96 [In Russia]

[2] Parent J, Meyer T, Volin J, Fahey R and Witharana 2019 An analysis of enhanced tree trimming effectiveness on reducing power outages *J. of Environmental Management* **241** 397-406

[3] Byk F, Kityshin V, Myshkina L and Kaminskaya E 2019 Composite structural elements of overhead lines as a means of increasing technical efficiency *IOP Conf. Ser.: Mater. Sci. Eng.* **552** 012012

[4] Guide for Electric Power Distribution Reliability Indices URL https://ieeexplore.ieee.org/document/6209381

[5] Warren M and Ward D 1992 The effect of reducing momentary outages on distribution reliability indices *IEEE Trans. Power Delivery* **7** 1610-17

[6] Kumar G et al. 2013 Reliability improvement of radial distribution system with incorporating protective devices-case study *Int. J. of Science & Emerging Technologies* **4** 60-74

[7] Sultana B, Mustafa M, Sultana U and Bhatti A 2016 Review on reliability improvement and power loss reductionin distribution system via network reconfiguration *Renewable and Sustainable Energy Reviews* **66** 297-310

[8] Byk F, Myshkina L and Khokhlova K 2017 Power supply reliability indexes *Advances in Engineering Research* **133** 525-30

[9] Byk F and Myshkina L 2018 Texnologicheskaya i e`kspluatacionnaya nadezhnost` sistemy` e`lektrosnabzheniya [Technological and operational reliability of the power supply system] *Nadežnost' i bezopasnost' ènergetiki* [*Safety and Reliability of Power Industry*] **11(3)** 200-7 [In Russia]

[10] Zhang P, Li W and Wang S 2012 Reliability-oriented distribution network reconfiguration considering uncertainties of data by interval analysis *Int. J. of Electrical Power & Energy Systems* **34** 138-44

[11] Ray S, Bhattacharya A and Bhattacharjee S 2016 Optimal placement of switches in a radial distribution network for reliability improvement *Int. J. of Electrical Power & Energy Systems* **76** 53-68

[12] Kumar K and Jayabarathi T 2012 Power system reconfiguration and loss minimization for an distribution systems using bacterial foraging optimization algorithm *Int. J. of Electrical Power & Energy Systems* **36** 13-17

[13] Savier J and Das D 2011 Loss allocation to consumers before and after reconfiguration of radial distribution networks *Int. J. of Electrical Power & Energy Systems* **33** 540-9

[14] Getting Electricity Methodology URL http://www.doingbusiness.org/Methodology/Getting-Electricity

[15] Ehsan A and Yang Q 2018 Optimal integration and planning of renewable distributed generation in the power distribution networks: A review of analytical techniques *Applied Energy* **210** 44-59

[16] Byk F, Frolova Y and Myshkina L 2019 The efficiency of distributed and centralized power system integration *E3S Web of Conf.* **114** 05007

IOP Conf. Series: Materials Science and Engineering **791** (2020) 012042 doi:10.1088/1757-899X/791/1/012042

Analysis of energy-saving measures in industrial steam supply systems

E G Gasho and A I Kiseleva

Department of Industrial Heat Energy Systems, National Research University MPEI, Krasnokazarmennaya St., 14, Moscow, 111250, Russia

E-mail: 290461@bk.ru, sashulka_kiseleva@mail.ru

Abstract. The high level of heat losses in Russia is determined by both the excessive centralization of many heat supply systems and the poor state of heat networks and the low quality of their service. For this reason, the development and application of energy-saving measures to reduce heat loss is a priority and urgent area in the power industry. The article is devoted to the assessment and analysis of energy-saving measures of industrial steam supply systems using the example of an industrial steam pipeline from Smolensk CHPP-1. The study examined the main factors affecting the parameters of the heat transfer agent in the consumer. The main measures to reduce heat losses in steam networks are considered and their calculation is carried out. Of the considered methods of energy saving, it is preferable to increase the thickness of the thermal insulation of steam pipelines. Moreover, the calculations showed that the optimal value of the insulation thickness for the investigated steam pipeline was 0.17 m, which corresponds to the highest quality of the supplied steam and the shortest payback period.

1. Introduction

In most cases, the distribution of heat among consumers, regardless of the equipment used, is carried out with deviations from the standards. This is due to the fact that real heat supply systems differ significantly from the design in terms of the number and characteristics of consumers, characteristics of pumping equipment, pipeline diameters, length of sections, laying methods, routing, etc. All these deviations ultimately lead to a change in the thermal and hydraulic modes of operation of the heating network.

In industrial steam supply systems, the majority of losses are heat losses in steam networks for a number of reasons [1, 2]:

- poor technical condition and maintenance of heating networks;
- violation of thermal insulation;
- disconnecting old or connecting new consumers, as well as changing the heat load required by consumers;
- long service life of pipelines.

These factors lead to a decrease in the reliability of heat supply, a mismatch between the parameters of the heat transfer media and contractual ones, an increase in heat losses, and a decrease in the quality of heat supply as a whole [3]. In this regard, energy-saving measures aimed at reducing heat and coolant losses in steam networks were considered.

2. Materials and Methods

As an object of study, the steam heating system of the city of Smolensk was considered, providing the needs for a few of industrial enterprises [4]. All consumers have a steam technological load. At many enterprises, water vapor is used as a coolant, as well as for heating containers and heat pipes. Most often, water vapor is used as a heat transfer agent. This is due to the fact that during the reverse process of the phase transition of water into steam - condensation, a large amount of energy is released. The obtained energy can be used for heating various media and chemical reactors, for heating rooms, in the food industry, and for heating viscous fluid working in pipelines, which tend to thicken at lower temperatures (fuel oil, various chemicals, oils, etc.). For these reasons, it is especially important that steam is supplied to heat-consuming installations of subscribers with specified contractual values (pressure, temperature). In the winter mode of operation, consumers use steam for technological load, heating and ventilation system, as well as to cover the needs for hot water supply. In summer operation, the heating system does not function while the remaining loads are constant.

The scheme of the steam supply system of the city of Smolensk with the indication of the lengths and diameters of the sections and also the load of consumers is presented in Figure 1.

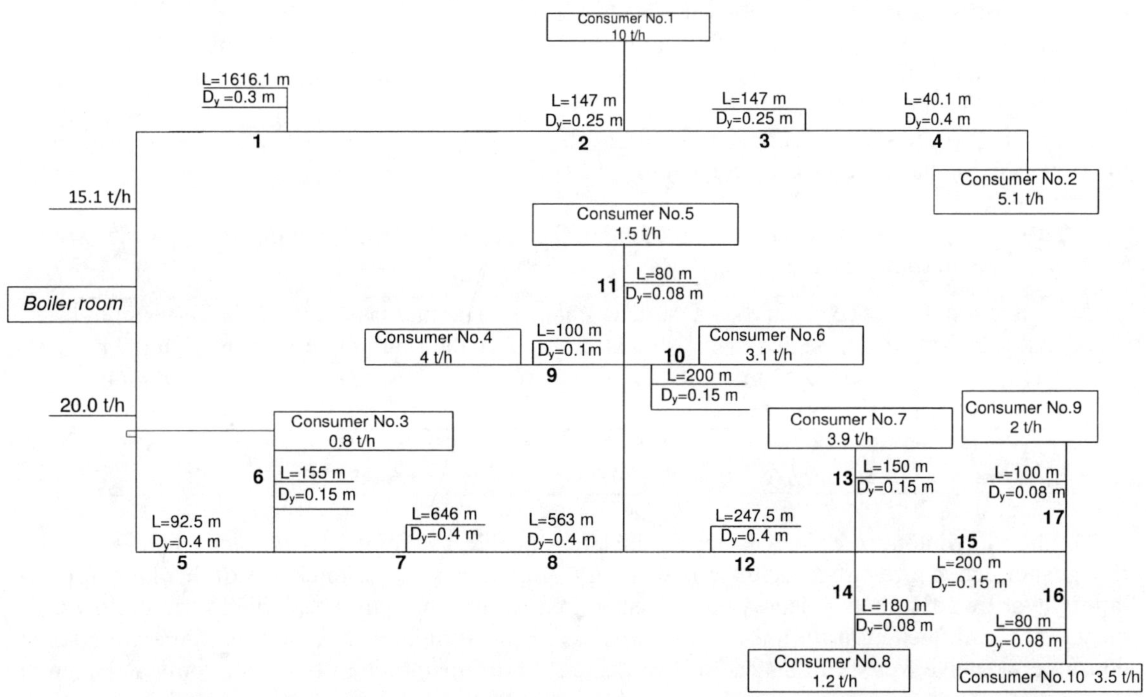

Figure 1. Scheme of connecting heat consumers to Smolensk CHPP-1.

The following were considered as energy-saving measures: changing the diameter of the pipeline and increasing the thickness of the insulation. All calculations in the work are based on thermodynamic dependencies.

The calculation results for the winter period are presented in table 1.

In the winter operating mode, the steam pipeline from the CHPP-1 fully copes with the technological load. Due to the economically feasible speed of the coolant and a large flow speed of steam, all consumers receive a pair of specified parameters, namely superheated. Therefore the implementation of energy-saving measures is possible only with the aim of reducing the total and consequently, specific losses [5]. We choose an arbitrary section of the steam pipeline and compare the efficiency of the applied energy-saving measures.

Table 1. The results of the calculation of heat losses when applying energy-saving measures in winter mode.

Section 9 L = 200 m	Technical design D_y=0.15 m, δ_{ins}=0.12 m .	Diameter change D_y: 0.15 m → 0.1m	The increase in insulation thickness δ_{ins}: 0.12 m → 0.17m	Joint increase in insulation thickness and pipe diameter
Q, W	15851.22	15847.41	13328.19	4161.52
q, W/m	79.26	79.24	66.64	20.81

Thus, the considered energy-saving measures, such as changing the diameter and increasing the thickness of the insulation separately, do not significantly affect the specific losses on the area. However when these parameters change (D_y, δ_{ins}) together, it can be seen that the value of specific losses decreases significantly and amounts to only 26.25% of the losses with the initially specified steam pipeline design. But carrying out these measures requires capital expenditures and may be economically disadvantageous, therefore, it is first necessary to assess whether these measures will be economically justified and appropriate for this project.

In the summer operating mode, despite the high steam parameters, some consumers do not receive the heater of the required parameters [6, 7]. Therefore, the implementation of energy-saving measures in this mode is a necessary.

Similarly, for the summer mode, a section of the steam pipeline was selected and the efficiency of energy-saving measures was calculated (Table 2)

Table 2. The results of the calculation of heat losses when applying energy-saving measures in summer mode.

Section 9 L = 200 m	Technical design D_y=0.15 m, δ_{ins}=0.12 m	Diameter change D_y: 0.15 m → 0.1m	The increase in insulation thickness δ_{ins}: 0.12 m → 0.17m	Joint increase in insulation thickness and pipe diameter
Q, W	15851.22	15847.41	13328.19	4161.52
q, W/m	79.26	79.24	66.64	20.81

Unlike the winter mode increasing the thickness of the insulation at a constant diameter is not the most effective measure to save thermal energy. An increase in the thickness of the thermal insulation layer by 42% gives a decrease in the specific heat loss by only 8%, which shows a low efficiency of the influence of the insulation thickness in the summer mode of operation, while some consumers receive wet steam instead of overheated [8]. In this mode of operation, only a change in the diameter of the pipeline can be performed constructively without changing the thickness of the heat-insulating layer, since the heat losses in these cases are practically equal, and an increase in thermal insulation requires the greatest capital outlay.

Based on the specifics of the object under consideration, of all energy-saving measures, it is only possible to strengthen the thermal insulation of the steam pipe, since the complete rearrangement of the steam network will lead to disruption of the technological process among consumers [9]. Since the network is already existing, it was decided not to change the material of thermal insulation.

For this, the payback period was calculated. The calculation was carried out based on the following conditions:

1) increase in insulation thickness from 0.12 to 0.17 m;

2) the calculation was carried out for the branches of the steam pipeline, in which the steam comes to consumers wet.

The following data were taken as initial data:

The insulation material is mineral wool.

The cost of 1 m^3 of insulation (C_{ins}) is 1470 rubles.

We will calculate for section No. 5, L = 92.5 m, D_y = 0.4 m, δ_w = 0.008 m.

The amount of insulation required is calculated by the following formula:

$$V_{ins} = L \cdot \pi \cdot (D_{o.ins} - D_{i.ins}), m^3 \quad (1)$$

where $D_{o.ins}$ – the outer diameter of the pipeline with a final insulation thickness, m;
δ_w – wall thickness, m;
$D_{i.ins}$ – the inner diameter of the pipeline with the initial insulation thickness, m.

Since there is an increase in the thickness of the pipe wall insulation by 0.05 m, it can be assumed that the value $(D_{o.ins} - D_{i.ins})$ = 0.1 m, and it remains constant in all sections of the steam pipeline.

The cost of this volume of insulating material is calculated as:

$$C = V_{ins} \cdot C_{ins}, \text{rubles} \quad (2)$$

The calculation results are presented in table. 3.

Table 3. Cost calculation increased thermal insulation.

No. section	Length section, m	Insulation volume, m^3	Cost, rubles
5	92.5	29.1	42696
6	155	48.7	71589
7	646	202.8	298181
8	563	176.8	259896
9	100	31.4	46158
10	200	62.8	92316
11	80	25.1	92316
12	274.5	86.2	126714
13	150	47.1	69237
15	200	62.8	92316
17	100	31.4	46158
TOTAL:		**804.2**	**1 182 158**

The final cost of this event will consist of the cost of material C_m and the costs of the work C_w (design, installation, commissioning, etc.). We accept that the cost of such work is about 150% of the cost of the material.

Then, the cost of increasing thermal insulation by 0.05 m will be equal to:

$$C = C_m + C_w, \text{rubles} \quad (3)$$

$$C = 1\ 182\ 158 + 1.5 \cdot 1\ 182\ 158 = 2\ 955\ 395 \text{ rubles.}$$

Determine the heat loss from the steam pipe with increased thermal insulation.

We accept that the duration of the heating period (winter mode) is 210 days and summer - 155: heat loss during insulation 0.12 m

$$Q = (558642.6 \cdot 24 \cdot 210 \cdot 3600 + 269273.8 \cdot 24 \cdot 155 \cdot 3600) / 10^6 = 13\ 742\ 126 \text{ MJ;}$$

heat loss during insulation 0.17 m

IOP Conf. Series: Materials Science and Engineering **791** (2020) 012042 doi:10.1088/1757-899X/791/1/012042

$$Q = (494351.8 \cdot 24 \cdot 210 + 243815 \cdot 24 \cdot 155) / 10^6 = 3398.5 \text{ MW}.$$

To calculate the heat savings, we translate these values in Gcal / h.

$$Q_{0.12} = 3282.1 \text{ Gcal / h}; Q_{0.17} = 2919.0 \text{ Gcal / h}.$$

With an average cost of 1 Gcal = 1485.5 rubles. It was found that heat saving with an increase in the thickness of the pipeline insulation is 363.1 Gcal / h or in the price equivalent of 539385.1 rubles per year. During the work of Smolensk CHPP-1, we get that the payback period for such an energy-saving event is about 2.2 years. Thus, this event can be recommended as energy-saving [10].

3. Results

Despite the fact that heat losses are actually reduced by 13%, final consumers (2 out of 8) still do not receive steam of specified parameters. Below are the calculations for heat saving with an increase in insulation thickness by 0.1 m (Table 4).

Table 4. Comparison of the effectiveness and economic feasibility of increasing the thickness of the insulation.

Insulation layer thickness δ_{ins}, m		Construction insulation 0.12	Insulation thickness 0.17	Insulation thickness 0.22
Heat loss Q, GJ	Summer mode	3 606.11	3 165.00	2 978.50
	Winter mode	10 136.01	8 087.45	7 444.97
TOTAL	**GJ:**	13 742.12	11 252.45	10 423.47
	Gcal:	3 279.74	2 685.55	2 487.71
Reduced heat loss compared to factory insulation, %			18.2	24.1
Heat saving while increasing insulation, Gcal/year			594.2	792.0
Annual savings by reducing heat loss, rubles			882 684.1	1 176 516
Cost of energy-saving measures, rubles			2 955 395	5 910 790
Payback on energy-saving measures, years			3.4	5.0
The number of consumers receiving pairs of specified parameters			6/8	7/8

The calculations presented in the table allow us to conclude that an increase in thermal insulation over 0.17 m is an economically and technically disadvantageous measure. Annual heat loss is reduced by only 5.9%, and the cost of increasing thermal insulation to 0.22 m compared to increasing thickness to 0.17 m is 200%.

Despite the fact that the payback period of each considered energy-saving measure does not exceed 5 years, an increase in the thickness of thermal insulation to 0.17 m should be recognized as optimal.

In order to determine the optimal value of the thermal insulation thickness, a graphical dependence of the heat loss on the increase in the insulation thickness was constructed, which is shown in Fig. 2 [11].

Each step of increasing the selected insulation is 0.05 m, since it is this value that corresponds to the standard value presented by manufacturers on the market.

The obtained dependence suggests that an increase in the thickness of the thermal insulation of the pipeline has its limits. For each pipeline diameter and steam parameters, there is an extremely effective insulation thickness exceeding which will entail only material costs with a constant value of losses. For the considered section, the calculations show that with an insulation thickness of 0.22

IOP Conf. Series: Materials Science and Engineering **791** (2020) 012042 doi:10.1088/1757-899X/791/1/012042

m or more, the magnitude of the heat loss will be constant. The research results show that only with a comprehensive solution is it possible to achieve the maximum energy-saving effect.

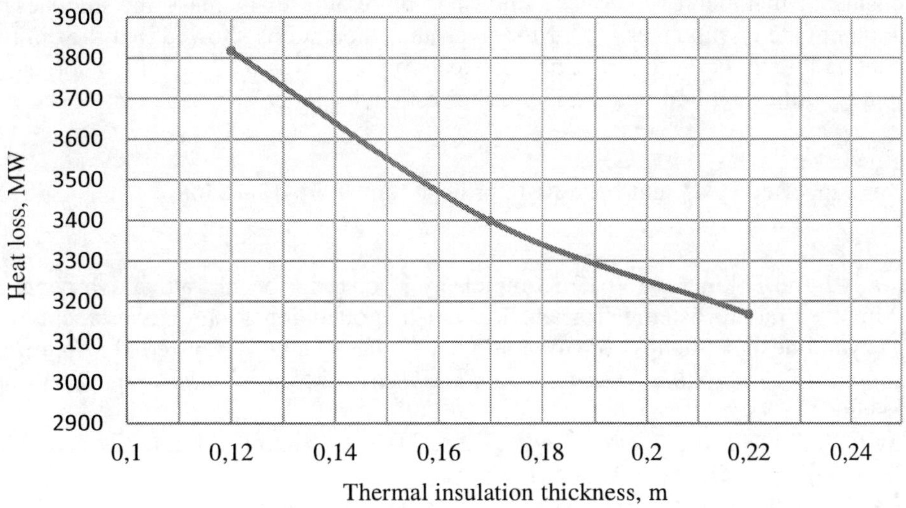

Figure 2. The dependence of heat loss on the thickness of the insulating material.

From the analysis of the calculations, it follows that the increase in the insulation thickness to 0.17 m should be recognized as optimal, since a further increase in the insulation thickness leads to an extension of the payback period of the event with a slight increase in the quality of the steam received by consumers [12].

4. Discussion

The steam parameters at the consumer are determined by the following factors:

1. Steam parameters at the source (pressure, temperature);
2. The operation mode of the steam pipe (design / non-design mode);
3. The state of thermal insulation of the steam pipe (material of thermal insulation, degree of deterioration).

In the course of the study, it was revealed that the final parameters of steam at the consumer in some modes of operation of the steam pipeline cannot correspond to the contractual values even when the steam parameters at the heat source increase [13, 14]. Given all these factors, it is necessary to increase the thermal resistance of thermal insulation relative to the standard value recommended by technical and regulatory bases.

5. Conclusion

Energy saving in the energy system is the application of a number of energy-saving solutions and measures at the levels of receipt, transportation and use of thermal energy. Heating networks are the most important and at the same time the most vulnerable part of the heat supply system. Industrial steam supply systems are an important part of the heat supply scheme of each of the cities of Russia. However, due to the economic and political situation, the load on thermal energy in the form of steam decreased, which led to the emergence of such a mode of operation of steam pipelines as non-project mode. Consumers ceased to receive pairs of specified parameters, which in turn leads to a violation of the process and entails a decrease in the quality of products. That is why industrial steam supply systems require careful study and research. The study examined energy-saving measures when transporting thermal energy in the form of steam

For industrial steam networks 2 energy-saving measures aimed at increasing the parameters of steam at consumers were considered:

1) increase in insulation thickness;

2) ensuring the movement of steam at an economical speed.

Of the considered methods of energy saving, it is preferable to increase the thickness of the thermal insulation of steam pipelines [15]. Moreover, the calculations showed that the optimal value of the insulation thickness for the investigated steam pipeline was 0.17 m, which corresponds to the highest quality of the supplied steam and the shortest payback period.

Acknowledgments
This article was supported by a grant from RSF, project number 16-19-10568

References

[1] Kiseleva A I and Fokin A M 2019 Kompleksny`j podxod k ocenke vliyaniya neproektny`x rezhimov na rabotu sistem parosnabzheniya [Integrated approach to assessment of impact of beyond-design modes on operation of steam supply systems] *Nadezhnost` i bezopasnost` e`nergetiki* [*Safety and Reliability of Power Industry*] **12(1)** 10-7 [In Russian]

[2] Sokolov E Ya 2009 *Teplofikaciya i teplovy`e seti* [Heating and heating networks] (Moscow: Izdatel`skij dom ME`I) [In Russian]

[3] Zare S and Jamalkhoo M H 2018 Experimental study of direct contact condensation of steam jet in water flow in a vertical pipe with square cross section *Int. J. of Multiphase Flow* **104** 74-88

[4] Sy`chev A V 2017 Uzkie mesta v organizacii paroprovodov [Bottlenecks in the organization of steam pipelines] *Spiraskop* **1** 8-13 [In Russian]

[5] Pérez-Uresti S I, Martín M and Jiménez-Gutiérreza A 2019 Estimation of renewable-based steam costs *Applied Energy* **250** 1120-31

[6] Fu W, Liu Z, Wu H and Tang Y 2018 Numerical study for the influences of primary steam nozzle distance and mixing chamber throat diameter on steam ejector performance *Int. J. of Thermal Sciences* **132** 509-16

[7] Li X, Min Z and Zhu P 2018 Storage and Recycling of Interfacial Solar Steam Enthalpy *Joule* **2** 247-8

[8] Cay A 2018 Energy consumption and energy saving potential in clothing industry *Energy* **159** 74-85

[9] Xu Q and Liang C 2017 Interfacial characteristics of steam jet condensation in crossflow of water in a vertical pipe *Applied Thermal Engineering* **113** 1266-76

[10] Lucas D and Liao Y 2016 Poly-disperse simulation of condensing steam-water flow inside a large vertical pipe *Int. J. of Thermal Sciences* **104** 194-207

[11] Ebrahimi M 2017 The environ-thermo-economical potentials of operating gas turbines in industry for combined cooling, heating, power and process (CCHPP) *J. of Cleaner Production* **142** 4258-69

[12] Amini A, Miller J and Jouhara H 2017 An investigation into the use of the heat pipe technology in thermal energy storage heat exchangers *Energy* **136** 163-72

[13] Zhu K, Chen X, Wang Y and 2017 Operation characteristics of a new-type loop heat pipe (LHP) with wick separated from heating surface in the evaporator *Applied Thermal Engineering* **123** 1034-41

[14] Tan R and Zhang Z 2016 Heat pipe structure on heat transfer and energy saving performance of the wall implanted with heat pipes during the heating season *Applied Thermal Engineering* **102** 633-40

[15] Therkelsen P and Mc Kane 2013 A Implementation and rejection of industrial steam system energy efficiency measures *Energy* **57** 318-28

Economic effectiveness of the aspiration systems in fiber by-products recycling technology

T N Il'ina[1] and I A Kuznetsova[2]

[1]Department of Heat and Gas Supply and Ventilation, Belgorod State Technological University named after V.G. Shukhov, Kostyukov St., 46, Belgorod, 308012, Russia
[2]Department of Economics and Production Organization, Belgorod State Technological University named after V.G. Shukhov, Kostyukov St., 46, Belgorod, 308012, Russia

E-mail: ilina50@rambler.ru

Abstract. The paper presents a wide variety of by-products with low bulk density and high dispersibility and ways of their utilization. To obtain compacted products from this type of materials the preliminary micro-granulation stage is necessary, which can be done by the roller compactor. It is suggested to carry out the process of powdered by-products compaction in the vibration-centrifugal granulator, which includes a device for pre-compacting and three drums, where sequentially conducted processes of pellet formation in waterfall-cascade mode, classification by vibration and compacting of the surface layer in the drum with toroid-shape chambers. The manufacturing technology of compacted products made of paper, cardboard and other fiber-type material wastes includes a waste shredding stage in the shredder and rotary crusher that follows by a pneumatic conveying of the gas-dispersion mix into the cyclone unit for dispersed phase settling. Therefore, the mentioned-above method of its compacting is acceptable for the collection in cyclone grinded particles. To increase the settling efficiency of the fiber type by-products it is worthwhile pre-compacting it, which follows with the microgranules obtaining by the pneumomechanical method in the airflow. The paper presents the technology of the granulated materials obtaining by pulp-and-paper waste recycling. It demonstrates the economic effectiveness of the introduction into the aspiration system of technological complex the pneumomechanical device for dispersed mix agglomeration using subquality products and bonding additives.
Keywords: fiber by-product, microgranulation, pneumomechanical device, cyclone unit, baghouse filter, degree of purification.

1. Introduction

Complex processing of raw material resources is an important issue in the modern industry, including recycling of different wastes resulting in commercial product output for different purposes. Industrial plant waste can be liquid or solid in the form of dispersed materials with different physical-and-mechanical and physical-and-chemical properties. There is a special group that consists of fiber waste from the woodworking and pulp-and-paper industries, agriculture, paper, and cardboard waste, etc. Due to their low actual and bulk density, their branched structure and high «volatility» it is difficult to catch them in the aspiration system plants of technological complexes.

The objective of this paper. Reveal economic effectiveness of the technological complex operation for paper and cardboard waste recycling using the pneumomechanical device in the aspiration

IOP Conf. Series: Materials Science and Engineering **791** (2020) 012043 doi:10.1088/1757-899X/791/1/012043

system to microgranulate grinded material for the purpose of the settling degree increase in the cyclone unit. The use of the subquality products in pneumomechanical agglomeration as a center of pellet formation decreases material losses and increases complex operation effectiveness as a whole.

2. The by-product characteristics

One of the promising ways of the powdered by-products recycling is a manufacture of compacted products using granulation, extrusion or compaction methods.

It should be noted that the manufacture of the compacted products is a topical issue in many technological processes, for example, glasswork industry [1-3], metal and mining industry [4], because granules contain the most of useful properties in less possible volume. Compacting of the dispersed mixes and materials is being done both at the intermediate and also at the final stages of the market-ready products manufacture, which allow material dusting reduction, simplify storage conditions, transportation, etc.

However, some by-products have low bulk density (less than 200 kg/m^3), high dispersity. The granulated product manufacture from such batches requires a step-by-step process of formation, which must start from the compacting stage to create centers for granulation [5]. Such materials include powdered waste from the production of expanded perlite, vermiculite, lime, and the other construction industry waste. The obtained granules can be used in the heat-insulating mixes, lime-potash and organomineral fertilizers [6]. The process of the powdered by-products compacting is suggested to perform in the developed by us vibration-centrifugal granulator (RF patent #2412753). The vibration-centrifugal granulator includes a pre-compacting device and three drums, where in a sequential manner conducted processes of pellet formation in waterfall-cascade mode, classification by vibration and compacting of the surface layer in the drum with toroid-shape chambers (Figure 1).

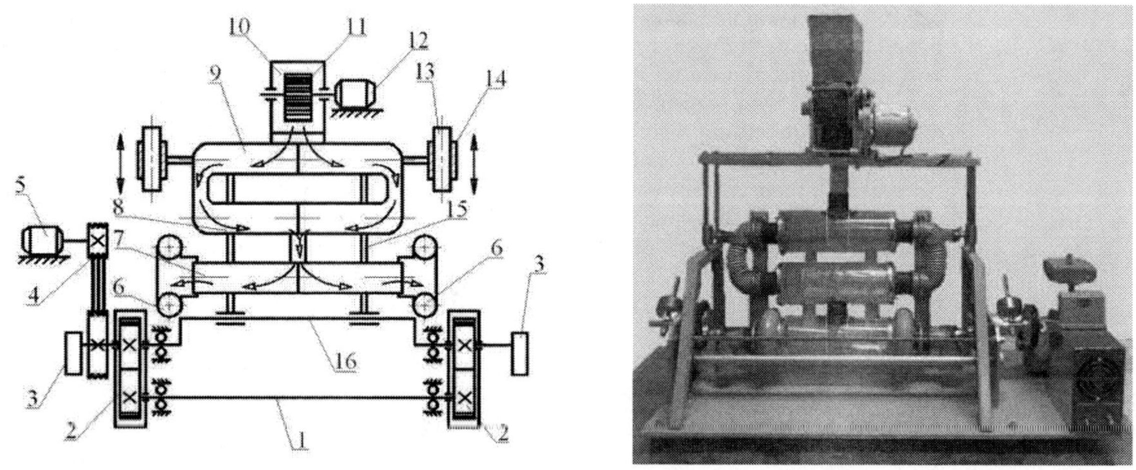

a) b)

Figure 1. Kinematic diagram (a), overall view of the vibration-centrifugal granulator (b): 1 - connecting shaft; 2 - pair of gears; 3 - counterbalances; 4 - V-belt drive; 5 - electric motor of the main drive; 6 - toroid-shape chambers; 7, 8, 9 - drums for material granulation; 10 - vibrating chute; 11 - elastic compacting roll for microgranulation; 12 - electric motor of the device for material pre-compacting and microgranulating; 13 - guide columns; 14 - sliders; 15 - moving frame; 16 - eccentric shaft.

It should be noted that the process of the polydisperse mix formation is very complicated, technology and equipment choice of which depends on the physicomechanical and physicochemical characteristics of the moulded mixes, and their rheological properties. Theoretical and practical research of these processes can be found in many Russian and foreign scientific papers [7-12].

IOP Conf. Series: Materials Science and Engineering **791** (2020) 012043 doi:10.1088/1757-899X/791/1/012043

The fiber by-products represent a special group of wastes. The by-products that have a fiber structure consists of woodworking industry waste (in the form of sawdust, wood shavings), construction industry (manufacture of the heat-insulating materials, asbestos cement products); processing industry (textile production, fiberglass); agriculture industry (plant production, agro-industrial complex) and so on. Most of the fiber by-product materials (FBM) is a pulp-and-paper waste (PPW). Cellulose is the most common polymer on Earth, which plays a very important role in carbon circulation in nature. However, the pulp-and-paper industry waste annual increase and environmental contamination by it require search for new PPW recycling methods.

Figure 2 presents the fiber by-products classification and methods of their recycling.

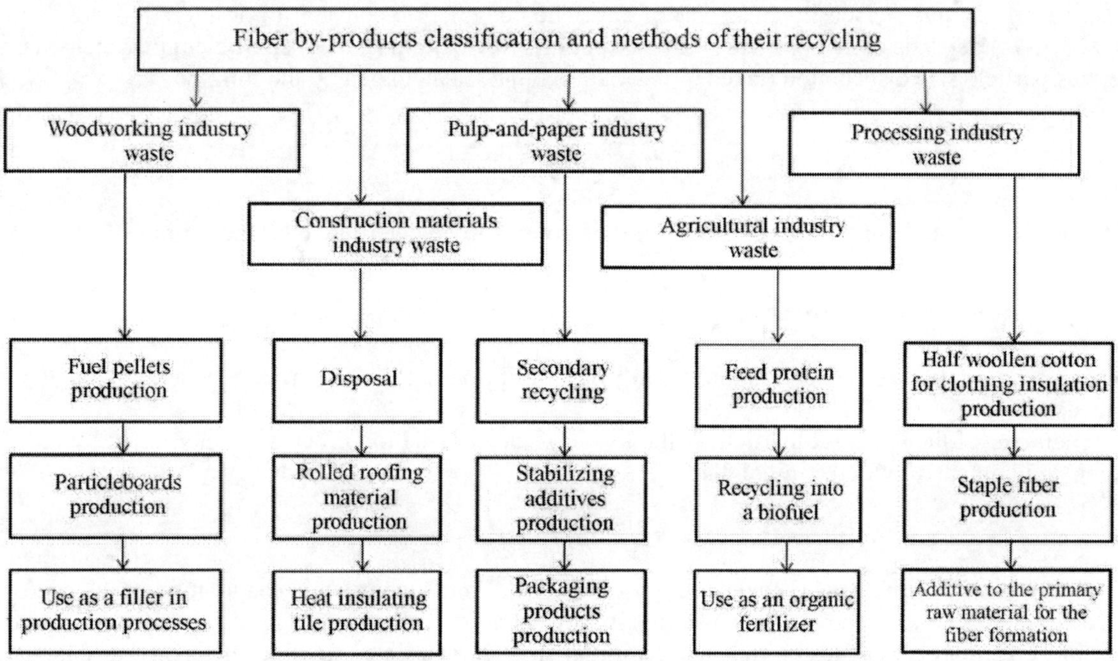

Figure 2. Fiber by-products classification and methods of their recycling.

One of the ways of the pulp-and-paper waste recycling is a granulated products manufacture. At the tech park of the Belgorod State Technological University named after V.G. Shukhov, the resource-saving complex was developed, which is dedicated to the production of the granulated stabilizing additive for stone-mastic asphalt-concrete mix based on pulp-and-paper waste (Patent #2542010 RF, C04B 26/26, C08L 95/00). In this paper analysis of the complex operation has been conducted and the technological complex was discussed, in order to develop methods of its operational effectiveness improvement.

3. Technological complex operation analysis

The technological complex for pulp-and-paper waste (PPW) recycling is a set of stages performed in sequential order: initial PPW storage, batching and transporting it to the shredder, two-stage grinding, first in the shredder and then in the rotary-hammer crusher. The grinded particles, size of 1-2 mm, with the airflow enters the cyclone-discharger, where material is settling and air goes to the baghouse filter where dust is removed. Settled in the cyclone unit material goes into the unit where materials mixing takes place and then into the chamber of plane-matrix granulate extrusion. The exiting granules go through drying and classification by spillage in the spiral-drum dryer unit. After classification, the product enters a feeding box of bucket elevator and then goes to the storage bunker.

IOP Conf. Series: Materials Science and Engineering **791** (2020) 012043 doi:10.1088/1757-899X/791/1/012043

The analysis of material flows in the fiber by-product recycling technology demonstrated that the settling degree of finely-dispersed grinded cardboard particles in the cyclone unit is less than 70% and most of the material is lost in the aspiration system. To minimize material loss, it is worthwhile to add a device for preliminary microgranulation into the technological complex for the purpose of the grinded cardboard particles settling degree increase in the cyclone unit.

It was established that in dust collecting and pneumotrasporting equipment the particle coagulation in the flow is very important, which can be thermo, gradient and turbulent, which is determined by the acceleration mechanism of particles with different size and density.

The conducted analysis of all components influence demonstrated that a dominant impact on the grinded pulp-and-paper particles agglomeration in the moving flow is done by the turbulent coagulation, which is determined by the density difference, kg/m^3, disperse phase (ρ_m) and disperse medium (ρ_g), speed (υ_g, m/s), viscosity (ν_g, m^2/s) of gas. The coagulation constant of acceleration mechanism K_{accel}, (m^3/s) depends on the particle size also (dp, m), flow length (l, m) – coefficient β, which characterizes particle size distribution (taken to be equal 1), and is calculated by the formula:

$$K_{accel} = \pi \frac{\rho_m}{\rho_g} \frac{\upsilon_g^{9/4}}{l^{3/4} \nu_g^{5/4}} \beta d_p^4, \tag{1}$$

The mean particle mass m_p, kg, at the point of time τ can be determined by the formula:

$$m_p = \frac{z}{n} = \frac{z\left(1+0.5K_{accel} n'\tau\right)}{n'} \approx \frac{K_{accel} z\tau}{2}, \tag{2}$$

where z - particle mass concentration, kg/m^3; n' - initial particle concentration, 1/m^3; τ - coagulation time, s.

Diameter of the increased particle as the result of the coagulation process is determined by the formula and compared with the initial diameter:

$$d_p = \sqrt[3]{6m_p / \pi\rho_p}. \tag{3}$$

Based on the formulas (1-3) calculation of the 0.5-2.0 mm size particle coagulation in the cyclone-discharger was conducted, which is installed in the technological complex. As can be seen in Figure 2, particle coagulation depends on time and its initial size. The particle growth coefficient was calculated as a relation of the agglomerate diameter, obtained as the result of their adhesion, to the initial diameter of particles. As can be seen in Figure 3, for the fiber particles with the size of 0.5-2.0 mm the increase of coagulation time, flow speed and particle size create more favorable conditions for agglomeration, which can be achieved in the equipment with toroid-shape chambers.

The most favorable equipment for this process is a vortex plant [13-16]. For this purpose, we developed the device for pneumomechanical granulation of by-products (Patent of invention #2538579 RF). Particularly the fiber structure and enhanced particle volume concentration of grinded materials allow conducting the process of its agglomeration in material-and-air flow [17, 18].

The fiber size in grinded PPW (mostly cardboard) is 0.5-2.0 mm, actual material density 800±50 kg/m^3.

To determine the solid material concentration in the airflow and speed of two-phase flow the measurements of the airflow speed (υ, m/s) in the technological complex at different productivity (G_s, kg/hr.) were conducted.

Table 1 presents the measurement results of two-phase flow speed in the pipe of pneumotransport system (υ_1, m/s) and in the cyclone-discharger (υ_2, m/s), and also mass concentration ($\mu = G_s/G_a$, kg$_s$/kg$_a$).

IOP Conf. Series: Materials Science and Engineering **791** (2020) 012043 doi:10.1088/1757-899X/791/1/012043

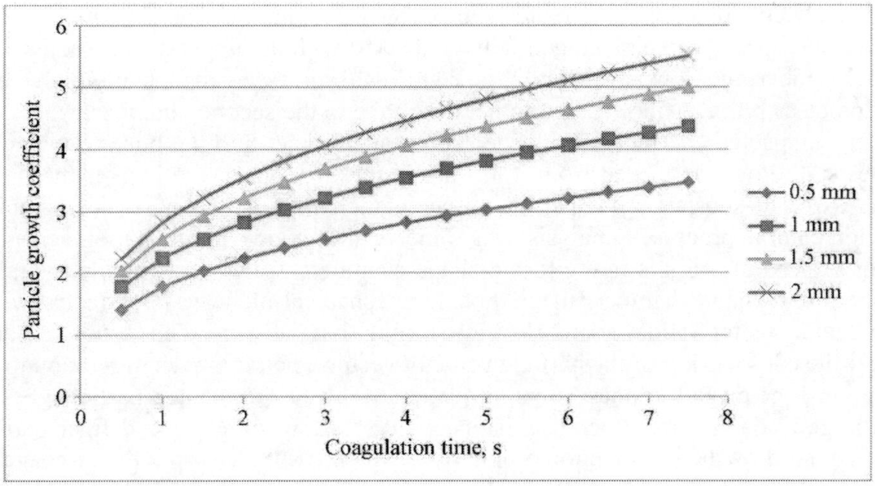

Figure 3. The dependence of particle growth coefficient on coagulation time.

Table 1. Measurement results of two-phase flow parameters of the technological complex

№	F, m^2	υ, m/s	L_a, m^3/s	G_a, kg/hr.	G_s, kg/hr.	μ, kg/kg	G_{2p}, kg/hr.	ρ_{2p}, kg/m^3	υ_1, m/s	υ_2, m/s
1	0.044	5.6	0.246	1132	0	0	1132	1.28	20.20	2.54
2	0.0407	4.0	0.163	748	160	0.21	908	1.54	13.38	1.68
3	0.033	3.7	0.122	561	180	0.32	741	1.68	10.02	1.26
4	0.0296	3.5	0.104	476	200	0.42	676	1.81	8.49	1.07

Table 1 indicates that the mass flow rate (G_{2p}, kg/hr.) and two-phase flow speed decrease when productivity (G_s) increases, and also mass concentration (μ) from 0.2 to 0.5 kg$_s$/kg$_a$. This range is the most promising for further by-product properties study (suspension velocity, particle drag coefficient). For the physical modeling and study of the pulp-and-paper waste granulation process, we developed the semi-industrial experimental plant (Figure 4).

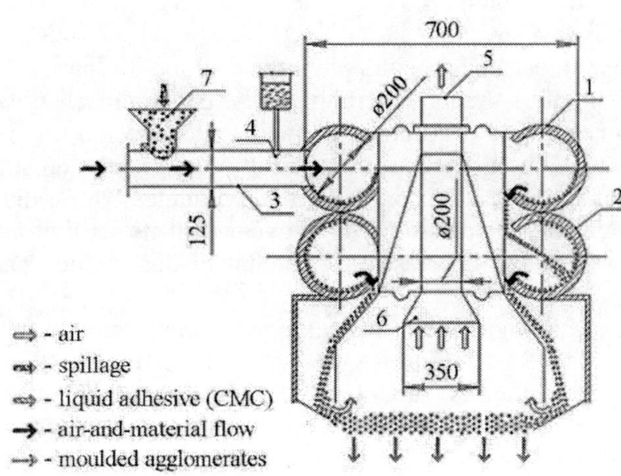

Figure 4. Plant for by-product agglomeration in the airflow.

The plant consists of two toroid-shape chambers, upper (1) and lower (2), which are situated horizontally, one above the other. The air-material mix feeds through the pipe (3) with installed nozzle (4) for liquid adhesive spray. The pipe tangentially connected with the upper toroid-shape chamber. In the toroid-shape chamber, the flow is subject to the centrifugal forces, by means of which material particle agglomeration takes place. After this, agglomerates move to the second chamber, where they are additionally being compacted and increase in size. Then particles get into the bunker, where moulded microgranulates spill down, and freed from it air moves upward, out of the bunker through the pipe (5) and the probe (6). The substandard material (spillage) enters through the auger dosing unit (7).

The agglomeration processes analysis was done in the device for pneumomechanic granulation (Figure 4) at a speed of 4-6 m/s in the toroid-shape chambers. Calculation of the two-phase flow trajectory in the toroid-shape chambers of the pneumomechanical plant was conducted with the help of computer modeling using SolidWorks 2017 Flow Simulation software. The results of numerical simulation and profile construction of the particle velocity vector when it moves in relation to the gas phase demonstrated that the dispersed phase slips in relation to the gas phase, and particle goes in spiral motion with a large lead. The agglomeration happens due to flow spiraling and fiber particle adhesion, that can be enhanced by the introduction of additives which ensure adsorptive-and-coagulative particle interaction.

As the binding additive for grinded PPW was chosen a solution of carboxymethylcellulose (CMC), which demonstrates high adsorption to paper and creates a viscous coagulation structured system. The subquality product additive (spillage) in the dispersed system acts as the centers of granulation and leads to the formation of agglomerates with increased density and size. The research of the agglomeration processes was conducted with the additive CMC (0.8-1.4) % and spillage (5-20) % of the initial material.

The following equation was obtained as a result of the conducted researches:

$$Y = 202.57 + 11.08X_1 + 58.61X_2, \; R_{xy} = 0.997, \tag{4}$$

where Y - density of the obtained agglomerates (output parameter), kg/m^3; X_1 – addition of spillage into the initial material, %; X_2 - addition of CMC into solid, %. The research results were processed by software Statistica.

Figure 5 presents the improved complex, technological line of which has the plant for pneumomechanical agglomeration.

The conducted study revealed that for obtaining of the microgranules with a diameter of 3-5 mm and a density of 400±30 kg/m^3, in the pneumomechnical granulation plant the initial mix should contain 10±5% of spillage (Retour), CMC additive is 1.0±0.2% of the grinded cardboard. Based on the rheological study it was established that CMC should be fed in a form of 5% water solution into the upper chamber in the same direction as moving gaseous dispersed flow. The recommended speed in the toroid-shape chambers is (4-5) m/s.

Positioning of the pneumomechanical plant before the cyclone-discharger (item 9) facilitated a size increase of particles, which enters the cyclone unit. In the pneumomechanical plant under the influence of the centrifugal forces grinded cardboard particle agglomeration takes place. Subquality product (10±5% of spillage) is being used to create centers for microgranulation and component mix density increase. The process of agglomeration produces microgranules with a diameter of 3-5 mm and a density of 400±30 kg/ m3, which increases the effectiveness of its settling in the cyclone unit of the technological complex by 20% and decreases the amount of dust, which enters the baghouse filter [19].

To prevent dusting the technological line was equipped with an aspiration hood on the feeding funnel (Useful model patent #154559 RF) (Figure 5, item 5).

Using the developed by us method the aerodynamic calculation of the pneumomechanical plant and aspiration system of the technological complex was conducted, considering the concentration and properties of the dispersed fiber particles [20]. The calculation results pointed out the necessity of an additional fan installation, determined required parameters and model of the fan (Figure 5, item 12).

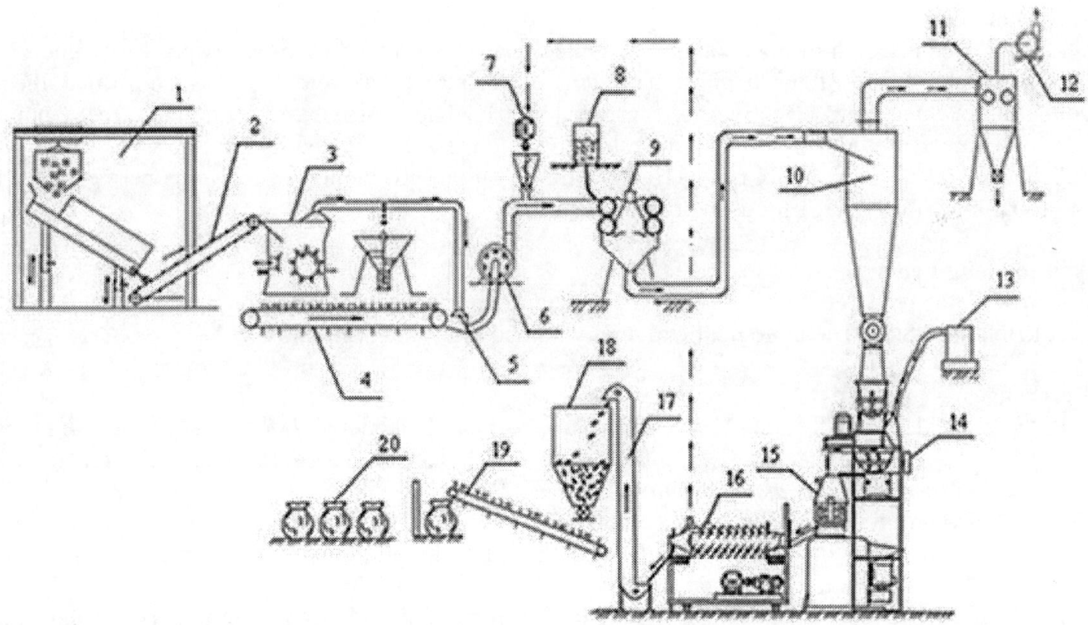

Figure 5. Improved low-tonnage technological complex for granulated products manufacture: 1 - unit for initial material feeding (cardboard, paper), 2 - belt conveyor, 3 - unit for PPW grinding (first stage in shredder), 4 - conveyor, 5 - aspiration hood, 6 - rotary crusher, 7 - auger dosing unit, 8 - tank with adhesive additive, 9 - pneumomechanical unit for microgranulation, 10 - cyclone unit, 11 - baghouse filter, 12 - fan, 13 - tank with additives, 14 - set of mixers, 15 - plane-matrix granulator, 16 - spiral-drum dryer unit with classification function, 17 - bucket elevator, 18 - storage bunker, 19 - drag conveyor, 20 - finished product packaging.

4. Technical and economic calculation

The conducted calculation of the improved complex includes: determination of the additional capital investment funds, which is needed for project development; granulated stabilizing additives output cost calculation; annual economic effect calculation.

The economic effect calculation was conducted in accordance with «Methodological recommendation for investment project effectiveness evaluation (second edition). Official edition. - M.: Economic, 2000 (Approved: Ministry of Economic Development of RF, Ministry of Finance of RF, State Committee of RF on construction, architectural and housing policy. No. VK 477 21.06.1999) ».

Effectiveness of the technological complex use for production of the granulated stabilizing additives (GSA) for stone-mastic asphalt-concrete mix based on pulp-and-paper waste is conditioned by the product output cost decrease as the result of material cost reduction, firstly, due to microgranulation process use in the pneumomechanical plant with introduction of the liquid organic adhesion carboxymethylcellulose (CMC), spillage after the classifier and material losses decrease, respectively.

The results of the conducted calculations, based on the discounted cash flow model, were used to determine the main measures of the suggested investment project effectiveness, which confirmed its feasibility. At the total additional non-recurrent expenses of 500 thousand rub. the economic integration effect, which serves as the final absolute result of the investment project realization, will be 581 thousand rub., the profitability index value, which is the degree of project profitability, - 2.1, i.e. more than 1. The payback period of the project is 1.2 years. The sum of additional profit from product sales at the wholesale price of 35000 rub. for 1 ton of (GSA) will be 481.3 thousand rub, or calculated for 1 ton – 1471.8 rub.

5. Conclusion

The conducted research of the pneumomechanical plant for grinded pulp-and-paper waste microgranulation, which is part of the aspiration system of technological complex, has confirmed the dust-collecting cyclone unit's effectiveness increase by 20%. The approximate economic effect from material losses decreases during aspiration, reloading and also the subquality products recycling will be 1471.8 rub/tons. The complex productivity with the output production of 327 tons per year will bring the economic effect of 581 thousand rubles per year.

Acknowledgments

The article was prepared within development program of the Flagship Regional University on the basis of Belgorod State Technological University named after V.G. Shukhov.

References

[1] Rieschel H 1971 Uber den Verdichtungsvorgang beim Briketieren *Aufbereitungs-Technik* **11** 691-8
[2] Min'ko N I and Laz'ko E A 2009 Doroganov Effect of finely disperse cullet on glass batch briquetting *Glass and Ceramics* **1** 305-9
[3] Mulevanov S V, Min'ko N I, Kemenov S A, Osipov A A and Bykov V N 2009 Vibrational spectroscopy investigation of the structure of multicomponent phosphorus-containing silicate glasses *Glass and Ceramics* **66(3-4)** 117-9
[4] Il'ina T N 2009 Structural and mechanical properties of pelletized fine materials *Chemical and Petroleum Engineering* **45(3-4)** 115-8
[5] Sevost'yanov M V, Il'ina T N, Sevost'yanov V S and Uvarov V A 2013 Mechanical Compaction Process Models of Materials and Their Instrumentation Software *World Applied Sciences J.* **25(12)** 1770-5
[6] Il'ina T N 2019 *Processy` aglomeracii v texnologiyax pererabotki dispersny`x materialov* [Agglomeration processes in technologies of dispersed materials recycling] (Saratov: Aj Pi Ar Media) [In Russian]
[7] Nikol S K and Adamiak Z P 1973 Rolle of bentoniton in wet pelletiring pocosses *Trans. Jnst. Mining and Met.* **82** 26-33
[8] Rice D A and Stone R L 1972 *Trans. Soc. Mining Eng AJME* **252(1)** 1-6
[9] Kortman H and Mai A 1970 *Aufbereitungs - Technik* **5** 251-6
[10] Wada M and Tsuchuja O 1970 The Rolle of Hidrophile colloid in Jron Ore Pelletisation *Procceding IX Int. Mineral, Proccessing Congres* (Praga) pp 22-53
[11] Sevostyanov M V, Il'ina T N, Sevostyanov W S and Emelyanov D A 2014 Methodological Principles of Agglomeration Processes Improvement in Technologies of Disperse Materials Processing *Research J. of Applied Sciences* **9(11)** 738-44
[12] Sevostyanov M V, Il'ina T N, Sevostyanov W S and Emelyanov D A 2015 About the Role of Surface Phenomena in the Processes of Processing of the Dispersed Materials *Research J. of Applied Sciences* **10** 684-9
[13] Sevost'yanov M V, Il'ina T N, Kuzneczova I A, Osokin A V and Martakov I G 2016 Resursosberegayushhij texnologicheskij kompleks dlya proizvodstva granulirovanny'x stabiliziruyushhix dobavok shhebenochno-mastichnogo asfal'tobetona [Resource-saving technological complex for granulated stabilizing additives for stone-mastic asphalt-concrete mix manufacture] *Vestnik TGTU* [*Vestnik Tambovskogo Gosudarstvennogo Texnicheskogo Universiteta*] **22** 272-9 [In Russian]
[14] Daly B I and Harlow F N 1970 Transport Equations of Turbulence *The Physics of Fluids* **13** 2634-49
[15] Il'ina T N, Bojchuk I P and Emel'yanov D A 2016 O vzaimodejstvii chasticz texnogenny`x voloknisty`x materialov v vozdushnom potoke [On fiber by-product particles interaction in

the air flow] *Vestnik BGTU im V. G. Shukhova* [*Bulletin of BSTU named after V G Shukhov*] **6** 116-21 [In Russian]

[16] Artyuxov A E, Fursa A S and Moskalenko K V 2015 Klassifikaciya i separaciya granul v vixrevy`x granulyatorax [Granules classification and separation in the vortex granulator] *Ximicheskoe i Neftega-zovoe Mashinostroenie* **5** 15-8 [In Russian]

[17] Emel'yanov D A and Plotnikov K V 2016 O vliyanii ob``emnoj koncentracii texnogenny`x voloknisty`x materialov na koe`fficient formy` chasticz v material`no-vozdushnom potoke [On the influence of a fiber by-product volume concentration on a particle drag coefficient in the air-material flow] *Vestnik BGTU im. V.G. Shukhova* [*Bulletin of BSTU named after V G Shukhov*] **3** 91-5 [In Russian]

[18] Il'ina T N and Emel'yanov D A 2019 O roli poverxnostny'x vzaimodejstvij chasticz voloknistoj struktury` pri aglomeracii v vozdushnom potoke [On role of fiber particles surface interaction during agglomeration in the air flow] *Vestnik BGTU im. V.G. Shukhova* [*Bulletin of BSTU named after V G Shukhov*] **5** 106-11 [In Russian]

[19] Emel'yanov D A, Il`ina T N and Sevost'yanov M V 2018 Razrabotka komponentnogo sostava smesi dlya aglomeracii texnogenny'x voloknisty'x materialov v pnevmomexanicheskom apparate [Component composition of mixture development for fiber by-product agglomeration in the pneumomechanical plant] *Innovacionny'e Puti Resheniya Aktual'ny'x Problem Prirodopol'zovaniya i Zashhity' Okruzhayushhej Sredy'* (Alushta) vol III pp 29-36 [In Russian]

[20] Minko V A, Emel'yanov D A and Il'ina T N 2016 Metodological bases fo-calculation of the aspiration ventilation systems *Int. J. of Pharmacy Technology* **8(4)** 26644-52

IOP Conf. Series: Materials Science and Engineering **791** (2020) 012044 doi:10.1088/1757-899X/791/1/012044

Review of technologies for suppressing emissions of benzo(a)pyrene at energy enterprises

Yu O Rikker[1] and M V Kobylkin[2]

[1]Department Science, Transbaikal State University, Aleksandro-Zavodskaya St., 30, Chita, 672039, Russia
[2]Department of Energy, Transbaikal State University, Aleksandro-Zavodskaya St., 30, Chita, 672039, Russia

E-mail: yrikker@mail.ru

Abstract. Power engineering enterprises have a great impact on the environment. Flue gas emissions contain solid ash particles, sulfur oxides SO_2, SO_3, nitrogen oxides NOx, carbon monoxide CO (monoxide) CH hydrocarbons; in some cases, there may be a danger of radiation contamination, but the most dangerous substance emitted into the atmosphere is benzo(a)pyrene $C_{20}H_{12}$, which has a negative effect on human health. Polycyclic aromatic hydrocarbons (PAH), including benzo(a)pyrene contained in flue gas emissions from energy enterprises, have a serious impact on human health. Taking measures to suppress other harmful emissions from thermal power plants (nitrogen oxides and sulfur oxides) can increase the content of benzo(a)pyrene, which poses a serious problem for modern researchers. As part of this study, various technologies for suppressing benzo(a)pyrene emissions at energy enterprises were considered. The main disadvantages of the described methods are highlighted. The possible danger of increasing benzo(a)pyrene emissions when trying to suppress other emissions is noted.

1. Introduction

Air quality has a direct impact on human health, especially those living in a large city. First of all, air pollution affects the respiratory system. So, in 2018, an increase in the incidence of cancer was noted in the Trans-Baikal Territory. In the Aginsky district, growth was 1.76 times, in Chita - 1.21, in Petrovsk-Zabaykalsky - 1.24, in Ulyotovsky, Shelopuginsky and Shilkinsky - 1.1. Moreover, according to the analysis of the regional Rospotrebnadzor, in the list of incidence of oncology in the period from 2011 to 2015, the first place was occupied by cancer of the trachea, bronchus and lung. [11]

In 2018, Chita was included in the list of cities with the dirtiest air. The substance that caused the high pollution was benzo(a)pyrene, the vast majority of the emissions of which accounted for by numerous energy enterprises - the Chita TPP-1 and the Chita TPP-2, as well as a number of boiler houses that also operate on coal. [1-18]

Polycyclic aromatic hydrocarbons (PAH), including benzo(a)pyrene, are widespread in all areas of the environment, formed during the combustion of fuel.

Benzo(a)pyrene ($C_{20}H_{12}$) (Fig. 1) - an aromatic compound, a representative of the polycyclic hydrocarbon family, which is formed during the combustion of hydrocarbon fuels - causes a high level of atmospheric pollution (to a lesser extent during gaseous combustion). Benzo(a)pyrene is one of the most toxic combustion products belonging to the group of polycyclic aromatic hydrocarbons. This is a

IOP Conf. Series: Materials Science and Engineering **791** (2020) 012044 doi:10.1088/1757-899X/791/1/012044

substance of the first hazard class, with an extremely high level of hazardous effects on the environment and human health (there is a high carcinogenicity in contact with benzo(a)pyrene). In addition to the special danger, benzo(a)pyrene is chemically and thermally stable, and when it enters the body it can accumulate. At the same time, the mechanism of PAH occurrence during fuel combustion has not been sufficiently studied to date. [12]

2. Materials and Methods

In addition to the particular danger, benzo(a)pyrene is widespread, chemically and thermally stable, but it is fat-soluble, and when it enters the body it can accumulate.

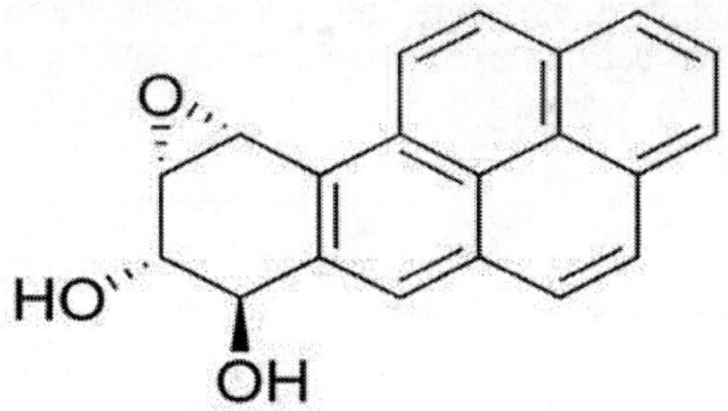

Figure 1. Chemical formula of benzo(a)pyrene.

The presence of benzo(a)pyrene in the emissions of the CHPP poses a rather serious complex problem of minimizing harmful emissions in flue gases, among the ways to solve which it can be indicated:

- Introduction of energy-saving technologies for heat consumers (including control and metering devices);
- Decrease in heat losses in networks;
- Optimization of boiler operation modes [18];
- Improving the collection efficiency of fly ash (including finely divided fractions).

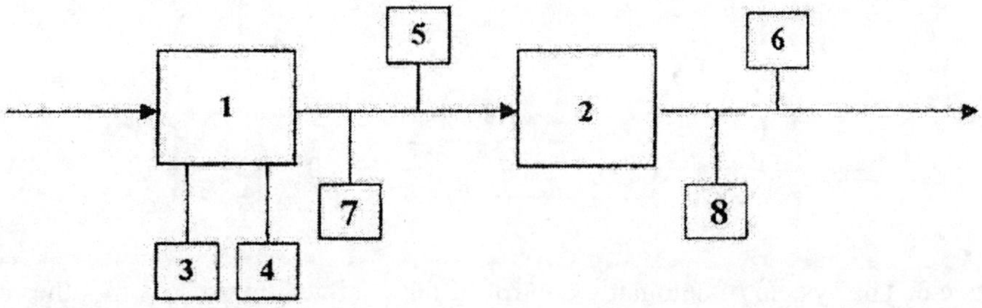

Figure 2. The method of purification of gas emissions from polycyclic aromatic hydrocarbons, including benzo(a)pyrene. 1 - nitrogen and carbomide injection device, 2 - ultraviolet irradiation device, 3 - ultraviolet irradiation device, 4 - pre-exposure chamber, 5-6 - sample selection, 7 - solid waste disposal device.

Among the technologies for suppressing emissions of benzo(a)pyrene at energy enterprises, it is possible to characterize the following:

Technology for purification of exhaust gases from PAHs by irradiating with a stream of accelerated electrons in the presence of mineral acid vapors, taken in a mass ratio to polycyclic aromatic hydro-

IOP Conf. Series: Materials Science and Engineering **791** (2020) 012044 doi:10.1088/1757-899X/791/1/012044

carbons, equal to (1-1.2): 1. [5] This technology is very energy-intensive, since molecular transformations occur due to the collision of all components of the gas emissions with the electron beam introduced into the exhaust gases. At the same time, there is a slight decrease in the content of benzo(a)pyrene in the exhaust gases subjected to irradiation.

Method of reducing the content of PAHs (including benzo(a)pyrene) in the exhaust gases by photooxidation of PAHs when irradiated with electric discharge radiation in the wavelength range of 340-410 nm with an average light energy density of 10-3-310-1 J / cm2 at operating temperatures from -20 ° C to + 80 ° C. [1]

Method quite effective due to low energy consumption. Upon irradiation, PAH molecules "pass into an excited singlet state with their subsequent transition due to collisions into a triplet state and the generation of singlet oxygen, which reacts with PAHs, with which some components of gas emissions also react." [1] Moreover, the rate of destruction of benzo(a)pyrene and the degree of purification of industrial emissions is not high, and reaches no more than 30-35%, and the specific energy consumption for each gram of destroyed benzo(a)pyrene reaches 0.5 kW · h and more. A relatively low reaction rate of PAH destruction is also noted.

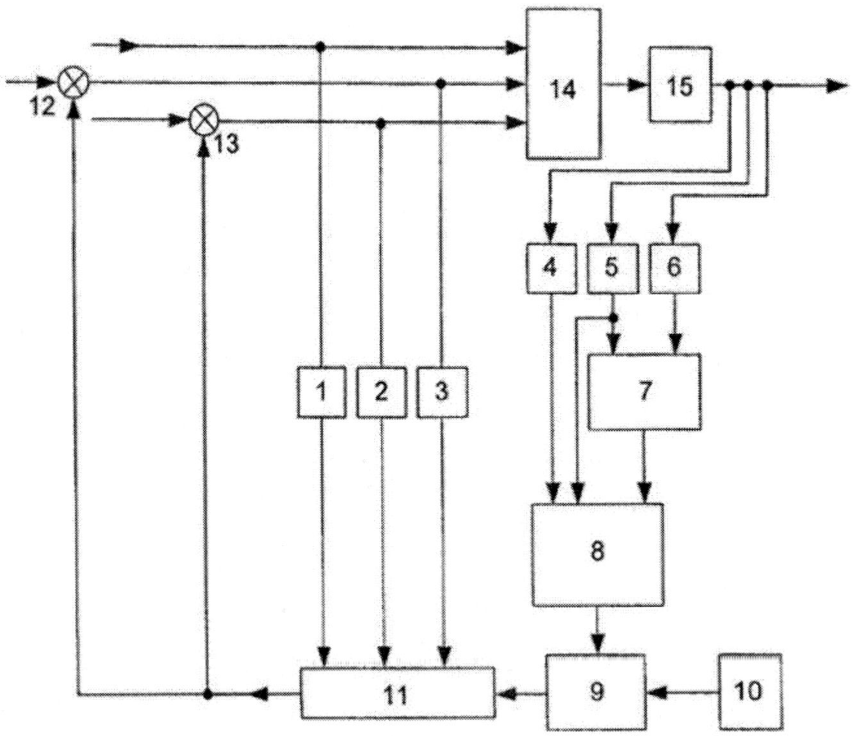

Figure 3. The system of automatic control of the combustion process in a thermal unit. The system contains a fuel consumption sensor 1, an air consumption sensor 2, a water or liquid consumption sensor 3, a nitrogen analyzer 4, a carbon monoxide analyzer 5, an oxygen analyzer 6, a device 7 for calculating the concentration of benzo(a)pyrene, an adder 8 conditional emission toxicity, block 9 for calculating the difference between excesses of the conditional emission toxicity over its permissible standard value, a regulator 10 of the normative component of the emission toxicity, a regulator 11 receiving signals from the sensors 1, 2 and 3 of the fuel, air and water consumption s, secondary regulators 12 and 13, the thermal unit 14 and the exhaust 15.

IOP Conf. Series: Materials Science and Engineering **791** (2020) 012044 doi:10.1088/1757-899X/791/1/012044

Method of irradiating gas emissions with ultraviolet radiation of an electric discharge in the working range of wavelengths with an average light energy density of 10-3-3 · 10-1 J / cm2, while irradiation is seen in the presence of saturated liquid vapor. [2] The working temperature of gas emissions is from 0 ° C to 250 ° C, ozone is added to the water.

This method is also distinguished by low efficiency, since the degree of purification of flue gases from benzo(a)pyrene is small. Also, this method leads to acid corrosion of the internal surfaces of the installation, which leads to additional operational costs for equipment repair.

Widely studied way to reduce emissions of benzo(a)pyrene is to introduce water or water vapor into the active combustion zone, as well as to burn water-fuel emulsions. This method is good for its low cost, however, the combustion of water-fuel emulsions is applicable only when using liquid fuels (fuel oil). [4]

Various studies conducted on pulverized coal boilers confirm an increase in the amount of benzo (a) pyrene after measures to reduce emissions of nitrogen oxides (according to some reports, up to 4-5 times). The use of effective ash traps can to some extent correct the situation, but effective ash collection (especially finely divided fractions of fly ash on which benzo(a)pyrene is actively sorbed) can capture only up to 20% of the total emissions of benzo(a)pyrene. [18]

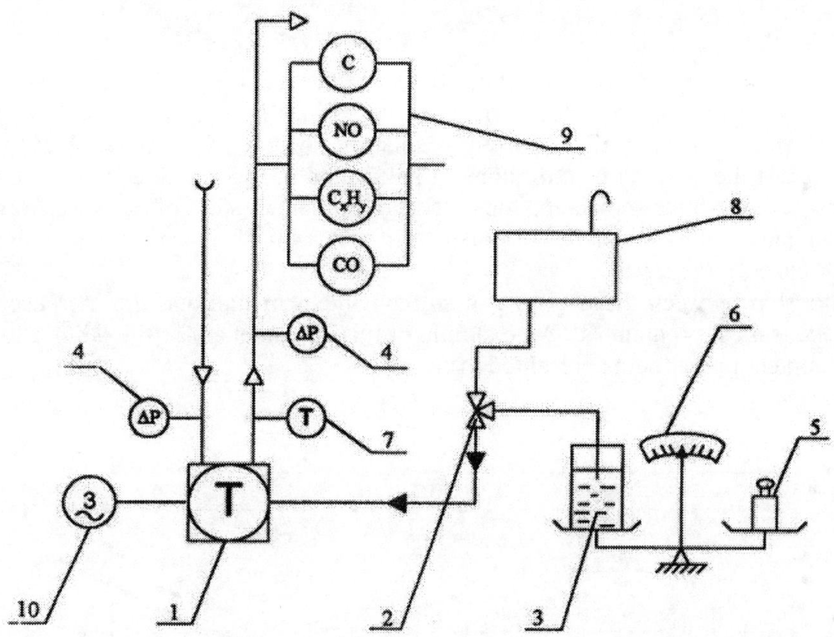

Figure 4. Fuel additive usage scheme: consisting of a diesel engine with standard systems 1, a measuring tank 3 and a fuel supply tank 8 connected to each other and the fuel system of the engine with a three-way valve 2, pressure gauges 4 mounted on the intake and exhaust manifolds of the diesel engine, weights 6 and weighs 5 mounted on the balance, and a thermometer 7 to determine the temperature of the exhaust gases.

There is still a way to automatically control the combustion process in a thermal unit, including measuring the consumption of fuel and air, measuring the concentration of nitrogen and carbon oxides in the flue gases leaving the atmosphere, determining the total allowable conditional emission toxicity, determining the difference in excess of the conditional emission toxicity over its allowable standard value and using this difference as a correction signal. In this method, using an additional sensor, the flow rate of water is measured in the form of liquid or steam, using an additional gas analyzer, the oxygen concentration in the flue gases leaving the atmosphere is measured, the concentration of ben-

IOP Conf. Series: Materials Science and Engineering **791** (2020) 012044 doi:10.1088/1757-899X/791/1/012044

zo(a)pyrene is calculated with the help of an additional device, and the total allowable conditional emission toxicity is determined taking into account the influence of the above parameters. The invention allows to reduce the total toxicity of emissions into the atmosphere of pollutants in the flue gas. [6]

The disadvantage of this method is that this method of regulation should be secondary to the main system for regulating the load of the thermal unit and in the event of a conflict of control commands, the commands of the primary regulation system will be executed first, ignoring the environmental component. [9]

Of particular interest is a separate type of solution for the suppression of harmful substances - the use of special fuel additives. An example of this type of solution is additive adapted for internal combustion engines. [7] According to the author's data, the additive contains tetraethoxysilane and 0.1-1.0% transition metal acetylacetonates in the ratio. Its introduction into the fuel at a concentration of 0.01% leads to a decrease in the content of harmful substances in the fuel combustion products: carbon monoxide by 20-85%; unburned hydrocarbons by 15-80%; nitrogen oxides by 15-55%; 40-90% smoke; benzo(a)pyrene by 15-70%. [3]

The use of such additives at thermal power plants is potentially possible. However, it is necessary to conduct additional research aimed at adapting additives to the characteristics of technological processes for the combustion of various fuels.

3. Results

The main problem in this situation is that the content of benzo(a)pyrene in the gross emissions of pulverized coal and gas-oil boilers in the surface air layer does not exceed 0.05 maximum permissible concentration, which, on the basis of current regulatory documents, is not subject to standardization and accounting. At the same time, emissions of sulfur and nitrogen oxides are not only standardized and taken into account, but also become the objects of implementation of measures to reduce them (for example, changing combustion modes in boiler furnaces), which, in turn, can easily lead to significant emissions of benzo(a)pyrene.

The relationship between the decrease in sulfur content in fuel and the increase in emissions of benzo(a)pyrene is most pronounced. An example of this dependence for the BKZ-220 boiler with different sulfur content in the fuel is presented in figure 5.

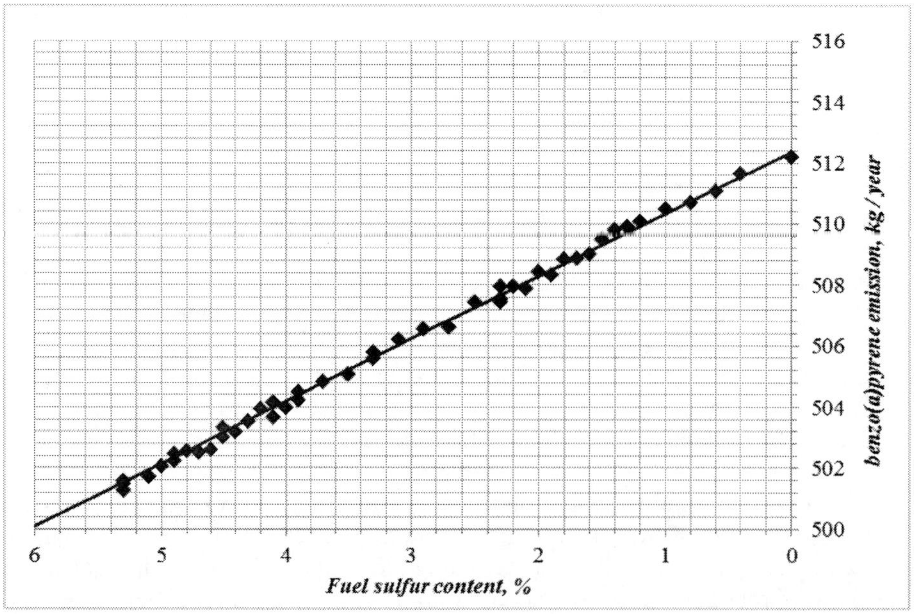

Figure 5. Dependence of benzo(a)pyrene emissions on sulfur content in fuel.

Thus, the relationship between a decrease in the content of sulfur oxides and nitrogen oxides and an increase in the content of benzo(a)pyrene poses a rather serious complex problem of minimizing harmful emissions in flue gases.

4. Conclusion

The insufficient knowledge of the mechanisms of the occurrence and suppression of benzo(a)pyrene, the high risk to human health, the absence of regulatory documents that control its emissions pose a challenge for modern researchers - to find a balance between the suppression of some emissions of thermal power plants, while not increasing others.

The mechanism of occurrence of polycyclic aromatic hydrocarbons during fuel combustion has not been sufficiently studied to date. A number of authors note the existence of a correlation between measures to suppress emissions of sulfur and nitrogen oxides and the concentration of benzo(a)pyrene in the flue gases of thermal power plants, but it is impossible to establish the connection between these processes without extensive experimental studies.

References

[1] Burkat V S 1998 Sposob snizheniya v gazovyh vybrosah koncentracii benzo(a)pirena i drugih po-liciklicheskih aromaticheskih uglevodorodov [The method of reducing the concentration of benzo(a) pyrene and other polycyclic aromatic hydrocarbons in gas emissions] (Patent RU No 2118913) [in Russian]

[2] Ivanickij M S and Griga A D 2013 Sposob ochistki gazovyh vybrosov ot policiklicheskih aromaticheskih uglevodorodov, v tom chisle benz(a)pirena [The method of purification of gas emissions from polycyclic aromatic hydrocarbons, including benzo(a)pyrene] (Patent RU No 2541320) [in Russian]

[3] Kashnikov G N, Krasnen'kov V M and Burkat V S 1998 Sposob snizheniya v gazovyh vybrosah kon-centracii benz(a)pirena i drugih policiklicheskih aromaticheskih uglevodorodov [The method of reducing the concentration of benz (a) pyrene and other polycyclic aromatic hydrocarbons in gas emissions] (Patent RU No 2118913) [in Russian]

[4] Kashnikov G N, Kashnikov E G and ZHuravlev S F 2003 Sposob ochistki gazovyh vybrosov ot poli-ciklicheskih aromaticheskih uglevodorodov, v tom chisle benzo(a)pirena [The method of purification of gas emissions from polycyclic aromatic hydrocarbons, including benzo(a)pyrene] (Patent RU No 2257256) [in Russian]

[5] Rose L, Mary X A, Issac J S and Thomas B 2018 Power plant optimization for energy efficiency enhancement *Int. J. of Mechanical Eng. and Technology* **9** 700-7

[6] Oikawa Katsuo, Yongsiri Chaturong, Takeda Kazuo and Harimoto Takayoshi 2003 Seawater flue gas desulfurization: Its technical implications and performance results *Environmental Progress* **22** 67-73

[7] Caiazzo Giuseppe, Langella Giuseppe, Miccio Francesco and Scala Fabrizio 2013 An experimental investigation on seawater SO2 scrubbing for marine application *Environmental Progress & Sustainable Energy* **32** 1179-86

[8] Bing Li and Chun Yuan Ma 2018 SO2 removal by powder activated carbon in a drop tube furnace experimental system *Key Eng. Materials* **783** 34-40

[9] Li-Bin L, Yan L, Jin-Ming L, Ning T, Kazuichi H and Tsuneaki M 2007 Development of analytical methods for polycyclic aromatic hydrocarbons (PAHs) in airborne particulates: a review *J. of Environmental Science and Health* **19** 1-11

[10] Zhang Zhillong and Tang Mao-Ping 2002 Applications of scrubber desulphurisation of dim gases using the required sea water for six power units with a capacity of 600 MW el. Stancii zhangzhou houshi *Zhongguo Dianli-Elec. Power* **35(7)** 69-71

[11] Pyat` gorodov Rossii s samym gryaznym vozduhom [Five cities of Russia with the dirtiest air] https://news.rambler.ru/ecology/37554801/?utm_content=rnews&utm_medium=read_more &utm_source=copylink

[12] Gray J P 2014 *Benzo(a)pyrene. Encyclopedia of Toxicology* (London: Academic Press)

[13] Uno Shigeyuki and Makishima Makoto 2009 Benzo[a]pyrene Toxicity and Inflammatory Disease *Current Rheumatology Reviews* **5** 266-71

[14] Zaccaria Kimberly and Mcclure Peter 2013 Using Immunotoxicity Information to Improve Cancer Risk Assessment for Polycyclic Aromatic Hydrocarbon Mixtures *Int. J. of Toxicology* **32** 236-50

[15] Zavoruev V and Zavorueva E 2019 Instrumental determination of the location of benzo[a]pyrene emission sources *IOP Conf. Series: Materials Science and Eng.* **537** 062070

[16] Beloglazova Natalia, Lenain Pieterjan, Rycke Esther, Goryacheva Irina, Knopp Dietmar and Saeger Sarah 2018 Capacitive Sensor for Detection of Benzo(a)pyrene in Water *Talanta* **190** 219-25

[17] Tanaka Nobuyuki and Sakata Masahiro 2002 Effect of photooxidation on δ13C of benzo(a)pyrene and benzo(e)pyrene in the atmosphere *Geochemical J.* **36(3)** 235-45

[18] Lee-Ruff Edward, Kazarians-Moghaddam H and Katz M 2011 Controlled oxidations of benzo[a]pyrene *Canadian J. of Chemistry* **64** 1247-53

IOP Conf. Series: Materials Science and Engineering **791** (2020) 012045 doi:10.1088/1757-899X/791/1/012045

Feasibility study of collector-distributor district heating system

D O Romanov, Y V Yavorovsky, E V Zhigulina and A Y Shelginsky

Department of industrial heat engineering systems, National Research University "Moscow Power Engineering Institute", Krasnokazarmennaya str. 14, Moscow, 111250, Russia

E-mail: dimrom09@yandex.ru

Abstract. One of the ways to improve efficiency of district heating systems is to use thermo-hydraulic dispatchers (THD) along with distributed variable speed pumps (DVSP) for reducing electricity consumption. First part of this paper deals with the application of THD at consumers to evaluate its technical and economic advantages. Although the result of the analysis shows that savings are quite small for Moscow conditions, the solution still might be applicable due to low additional capital expenditures. The second part of the paper uses the obtained result and summarizes previous works in this field to introduce collector-distributor district heating system which extensively incorporates THD and DVSP. Feasibility study of the proposed system was performed in comparison with a conventional district heating system. Electricity saving potential of collector-distributor district heating system is estimated to be 42-129 MW*hr/yr, while discounted payback period of additional capital expenditures is less than 2 years.

1. Introduction

District heating systems, in general, have branch or ring structures requiring significant amount of electricity for pumping a heat carrier. Usually required hydraulic head of central circulating pumps is defined by the most distant consumer from the heat supply source. It leads to redundant power of the pumps which is wasted by throttle valves in substations and consumers located closer to the heat supply source. In order to make district heating systems more efficient, new designs and concepts of hydraulic network are being proposed. New hydraulic scheme of district heating system based on distributed variable speed pumps (DVSP) and hydraulic connector was investigated both analytically and experimentally in [1-8]. It was shown in [1] that DVSP system consumes at least 31% less than conventional system. Similar research works [2-4] showed that savings reach almost 50%. Hybrid scheme, in which both DVSP and throttling valves are used, is considered in [5]. In this case, annual electricity consumption is estimated to be 28.5% less than in conventional system.

The hydraulic scheme with DVSP can be also applied to district cooling systems. District cooling system with cooled water storage and DVSP is shown in [6]. The cooled water storage in the system can be considered as large hydraulic connector. The analysis showed that the system allows for 10% of electricity saving in comparison with conventional district cooling system. District heating and cooling system with DVSP was considered in [7] and, in result, annual equivalent costs were reduced by 20.8-27.7%. An interesting idea of coupling district heating system with district cooling system by means of energy bus system was proposed in [8]. Energy bus system consists of heat pumps, ring topology pipe network, DVSP, hydraulic connectors, and heat and cold consumers. The results showed that this system can utilize waste or low-exergy heat during the whole year. Moreover, DVSP along

IOP Conf. Series: Materials Science and Engineering **791** (2020) 012045 doi:10.1088/1757-899X/791/1/012045

with hydraulic connectors allow for electricity saving as well as thermal and hydraulic decoupling of sources and consumers.

Thermohydraulic dispatcher (THD) is a device similar to hydraulic connector, but the difference is that usually THD has several manifolds adjusted to connect several consumers or (and) sources with their own pipe network. Several papers concerning experimental research of THD and its potential application to district heating systems have been recently published [9-13]. Different schemes of heat substation (direct connection, indirect connection, scheme with THD) were analyzed and electricity saving effect was calculated as well as capital costs in [11]. Different types of heat substation connection to heat supply network (conventional scheme, DVSP scheme, THD-based scheme) were analyzed in [13]. So, before performing overall analysis of THD implementation in district heating systems, which is the end goal of this paper, analysis of consumers, namely, space heating (SH) systems connected to THD is needed to be done.

2. Methods

Figure 1 shows two ways of organizing SH systems at consumers' side. Conventional system is depicted on the left side, while the system with THD is on the right side of Figure 1. The latter suggests replacement of automatic balancing valve along with partner valve with variable frequency pump. This allows for avoiding interdependence of SH loops. If a wet rotor pump is applied, then showed in Figure 1 sensor, frequency converter, and controller are not needed, as they are integrated into the pump. The electricity consumption of two schemes differs by the value that is throttled in the balancing valve. So, electric output that is reduced can be estimated by equation (1):

$$\Delta N = \frac{0.0278 \cdot V^3}{\eta \cdot kvs^2} \tag{1}$$

V – volume flow rate, m^3/hr.
η – coefficient of performance (COP) of pump.
kvs – amount of flow through a fully-open valve under pressure difference of 1 bar, $m^3/(hr \cdot bar^{0.5})$.
0.0278 – coefficient for conversion to kW.
The reduction of electricity consumption can be calculated by equation (2):

$$\Delta E = \Delta N \cdot \tau \cdot 24 \tag{2}$$

ΔE is the reduction of electricity consumption, kW·hr.
τ is duration of heating period.

While Figure 1 suits for two-pipe SH system where radiators are connected in parallel, Figure 2 represents one-pipe SH system (connection of radiators in series).The difference between one-pipe and two-pipe SH systems lies in using different balancing valves. However, aforementioned principles are correct for one-pipe SH system as well.

COP of pump = 0.6 and was assumed to be constant value. Although this value might change, it was shown in [14] that pump efficiency change can be neglected if speed change is less than 33%. Density of water = 1000 kg/m^3. Duration of heating period in Moscow is 205 days. Discount rate = 0.08. For the sake of simplicity, calculations were performed under condition that balancing valves are fully open during the whole heating period. So, calculated values of electricity economy can be considered as minimum ones. However, during heating period it is most likely that thermal and hydraulic regimes of consumers change and so does the valve opening. So, the goal of the analysis is to show minimal energy saving potential.

Average prices of automatic balancing valves (for one-pipe and two-pipe systems) and wet rotor pumps were received from manufacturers. These values are represented in Figure 3.

IOP Conf. Series: Materials Science and Engineering **791** (2020) 012045 doi:10.1088/1757-899X/791/1/012045

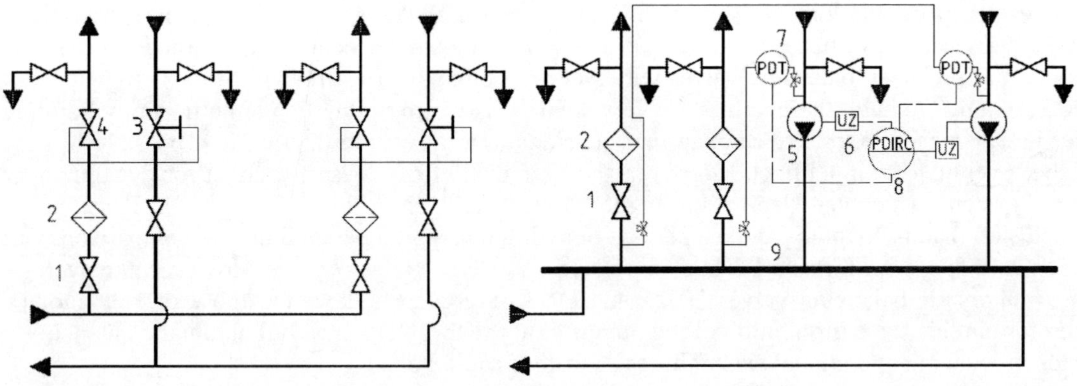

Figure 1. Principal schemes of space heating system (conventional scheme – on the left, scheme with THD – on the right); 1 – shut-off valve, 2 – strainer, 3 - automatic balancing valve (two-pipe version), 4 – partner valve, 5 – pump, 6 - variable speed drive, 7 – pressure difference sensor, 8 - programmable logical controller, 9 – THD.

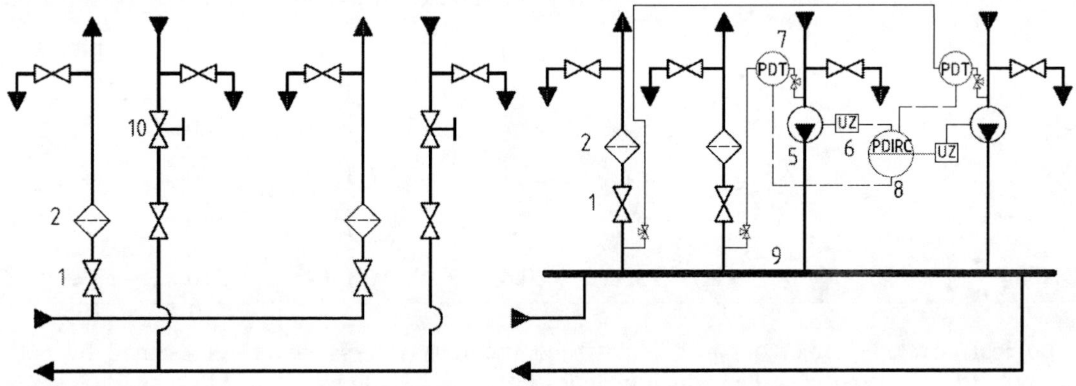

Figure 2. Principal schemes of space heating system (conventional scheme – on the left, scheme with THD – on the right); 10 – automatic balancing valve (one-pipe version).

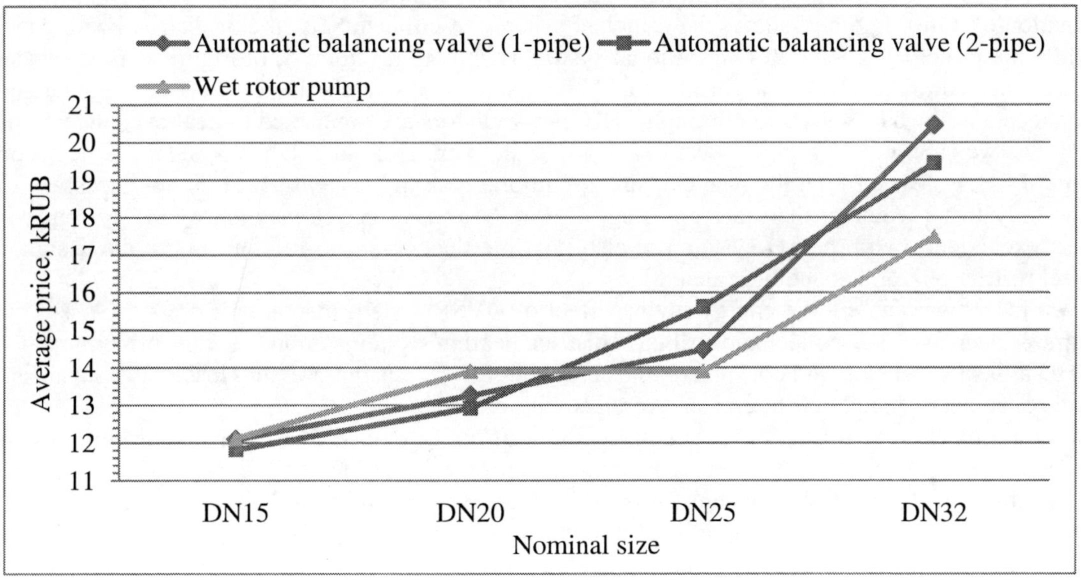

Figure 3. Average prices of automatic balancing valves and wet rotor pumps replacing them.

IOP Conf. Series: Materials Science and Engineering **791** (2020) 012045 doi:10.1088/1757-899X/791/1/012045

3. Results

The results of calculations are shown in Table 1, in which ΔE is electricity economy during a year, ΔOPEX is decrease of operational expenditures due to electricity economy. It turned out that economic effect of replacing balancing valves with wet rotor pumps varies from 130 to 510 RUB/yr for a single SH loop. Although these values are very small, it is worth noting that a multi-family building usually has many SH loops. And taking into consideration the whole district, it is possible to get much higher overall economic effect. Moreover, the estimated effect doesn't account for dynamic operation of SH system, which can also lead to higher values. Analyzing Figure 3 and Table 1 one can draw a conclusion that discounted payback periods of additional capital expenditures in comparison with conventional SH system for small diameters (DN15 and DN20) are 1-7 years. However, the average prices of automatic balancing valves DN25 and DN32 are even higher than those of wet rotor pumps; therefore in this case, installation of the pumps along with THD instead of automatic balancing valves leads to reducing operational expenditures as well as capital ones.

Table 1. Feasibility study of single SH loop of the scheme with THD in comparison with conventional scheme.

	V (m^3/hr)	kvs (m3/hr·bar0.5)	ΔN (W)	ΔE (kW·hr/yr)	Tariff (RUB/kW·hr)	ΔOPEX (RUB/yr)	Discounted payback period (yr)	
							1-pipe SH	2-pipe SH
DN15	0.62	1.6	4.3	21.2	6.1	130	1	3
DN20	0.97	2.5	6.7	33.1	6.1	200	4	7
DN25	1.55	4	10.8	52.9	6.1	320	–	–
DN32	2.44	6.3	16.9	83.4	6.1	510	–	–

Logical development of obtained results and previous works can be represented by collector-distributor district heating system which includes DVSP and THD installed at heat supply source, heat substation and consumers. The system suggests that losses of electric energy are minimal due to replacement of throttling valves with variable frequency pumps. At the same time, such system allows for implementing different control strategy (either supply temperature control or flow control, or combination of both). Figure 4 shows the principal scheme of collector-distributor district heating system, while the principal scheme of conventional system is shown in Figure 5. In Figure 4, heat substations are connected to the network according to DVSP scheme [1], while SH and DHW heat exchangers are connected to the heat substation through THD, and radiators are connected to heat exchangers through THD as well. Shown in Figure 4 features of the system are two-stage DHW system, indirect connection of SH system through the heat exchanger and one-pipe SH system. But it is not necessary to follow only these features, therefore one-stage DHW system, direct connection of SH system without heat exchanger, two-pipe SH system are also possible depending on reference data for design, technical restrictions, and economic calculations.

In [4], it was shown that energy saving effect of DVSP system increases if pressure losses at consumers decrease. So, collector-distributor district heating system embodies this principle by using THD at heat substation and consumers, thus allowing for decreasing pressure losses at heat substations and consumers which leads to greater overall saving effect.

Based on the previous results, feasibility study of collector-distributor district heating systems in comparison with conventional district heating systems was performed. One-pipe SH system, indirect connection of SH and DHW systems through THD, one-stage DHW system were chosen for calculations. Systems with 3 different total heat outputs were considered: 2.4 MW, 7.2 MW, and 14.4 MW. The results of calculations can be seen in Table 2, 3, and 4, in which ΔCAPEX is increase of capital expenditures. Analysing the results, it is clear that the network and source have the greatest electricity saving effect, while consumers have the least one. Overall result demonstrates that collector-distributor district heating system has less operational expenditures and more capital expenditures than conventional system. Electricity economy varies from 42 to 129 MW*hr/yr. However, discounted payback period of additional capital expenditures for all three cases is less than 2 years which is a good indicator for choosing collector-distributor district heating system over the conventional one.

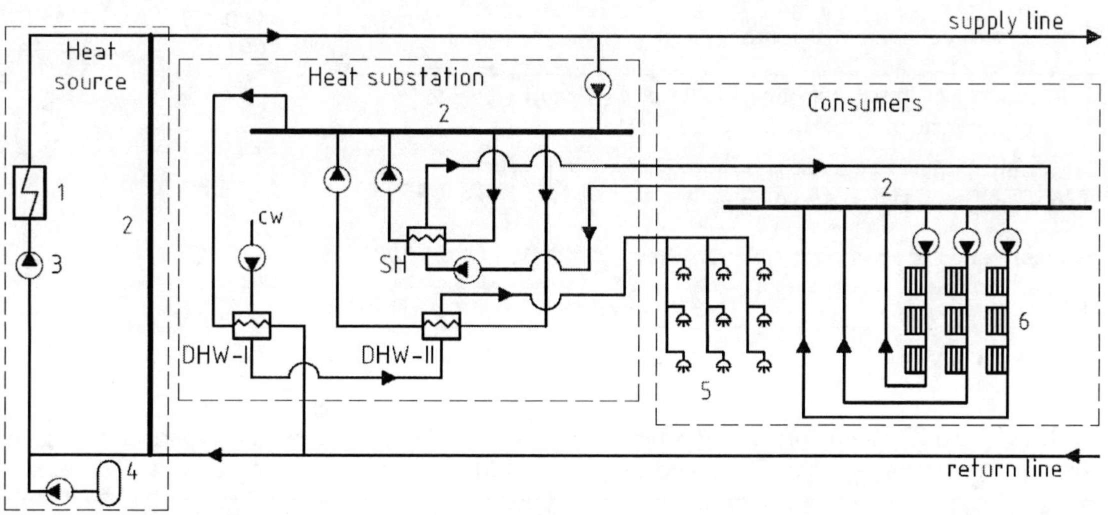

Figure 4. Principal scheme of collector-distributor district heating system. 1 – heat supply source, 2 – THD, 3 – pump, 4 – water tank, 5 – tapping points, 6 – radiators. SH – space heating system heat exchanger. DHW-I – domestic hot water supply heat exchanger (the first stage). DHW-II – the second stage, cw – cold water.

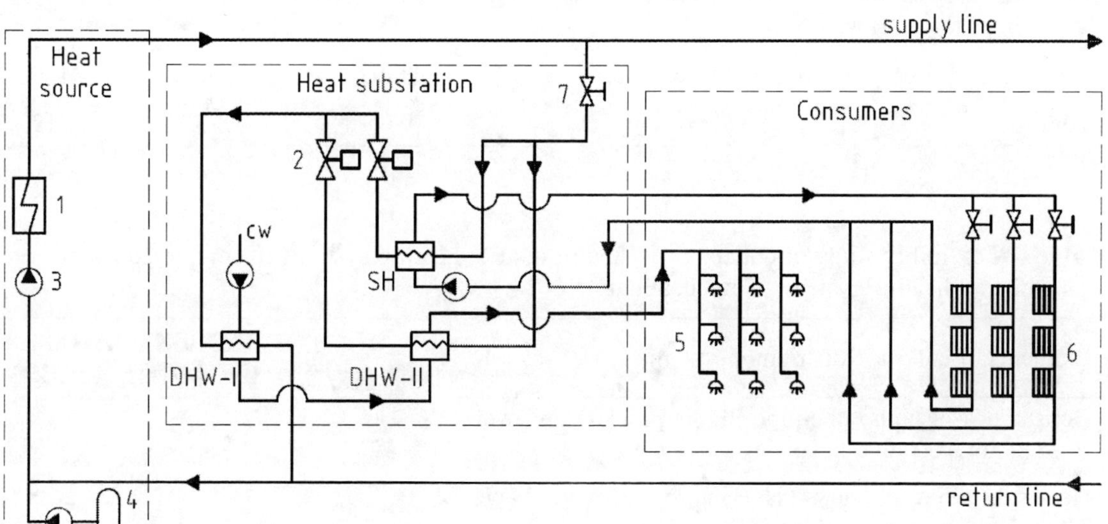

Figure 5. Principal scheme of conventional district heating system. 1 – heat supply source, 2 – control valve, 3 – pump, 4 – water tank, 5 – tapping points, 6 – radiators, 7 – balancing valve. SH – space heating system heat exchanger. DHW-I – domestic hot water supply heat exchanger (the first stage). DHW-II – the second stage, cw – cold water.

IOP Conf. Series: Materials Science and Engineering **791** (2020) 012045 doi:10.1088/1757-899X/791/1/012045

Also capital expenditures of consumers might be less than in conventional system which is shown in the ΔCAPEX column of the tables. It can be seen in Table 2 that capital expenditures on consumer units and heat substation in collector-distributor system is equal to those of the conventional one. So, the system with total heat output of 2.4 MW has smaller payback period than others.

Table 2. Feasibility study of collector-distributor district heating systems in comparison with conventional district heating systems. Total heat output is 2.4 MW.

Elements of collector-distributor system	ΔE (kW·hr/yr)	ΔOPEX (kRUB/yr)	ΔCAPEX (kRUB)
Consumers (21 units; one-pipe SH loop; THD and wet rotor pump instead of balancing valve DN15)	445	3	1
Heat substation (3 units; THD+indirect SH, 0.3 MW; THD+one stage DH, 0.5 MW)	2244	14	0
Network and supply source (DVSP scheme; 2.4 MW)	39011	238	291
Total	41701	254	292

Table 3. Feasibility study of collector-distributor district heating systems in comparison with conventional district heating systems. Total heat output is 7.2 MW.

Elements of collector-distributor system	ΔE (kW·hr/yr)	ΔOPEX (kRUB/yr)	ΔCAPEX (kRUB)
Consumer (39 units; one-pipe SH loop; THD and wet rotor pump instead of balancing valve DN20)	1291	8	25
Heat substation (3 units; THD+indirect SH, 0.9 MW; THD+one stage DHW, 1.5 MW)	9594	59	465
Network and supply source (DVSP scheme; 7.2 MW)	83030	506	429
Total	93915	573	919

Table 4. Feasibility study of collector-distributor district heating systems in comparison with conventional district heating systems. Total heat output is 14.4 MW.

Elements of collector-distributor system	ΔE (kW·hr/yr)	ΔOPEX (kRUB/yr)	ΔCAPEX (kRUB)
Consumer (48 units; one-pipe SH loop; THD and wet rotor pump instead of balancing valve DN25)	2539	15	-28
Heat substation (3 units; THD+indirect SH, 1.8 MW; THD+one stage DHW, 3.0 MW)	29289	179	597
Network and supply source (DVSP scheme; 14.4 MW)	97629	596	527
Total	129457	790	1096

IOP Conf. Series: Materials Science and Engineering **791** (2020) 012045 doi:10.1088/1757-899X/791/1/012045

4. Discussion

There are 3 possible operating regimes of THD:

- mixing regime (when total flow rate in primary circuits is less than that in secondary circuits);
- nominal regime (when total flow rate in primary circuits is equal to that in secondary circuits);
- bypassing regime (when total flow rate in primary circuits is bigger than that in secondary circuits).

It is worth pointing out that THD should work in nominal regime. Otherwise, electricity saving effect may not be reached. Therefore, when changing flow rates in secondary circuits, flow rates in primary circuits should be changed accordingly. This can create some difficulties for control system of collector-distributor district heating system. At the same time, regulation method of a system based on multiple distributed pumps was described in [15] and can be used for collector-distributor district heating system as well. The system is likely to have both flow control and supply temperature control strategies because variable frequency pumps have limited range of regulation. Nevertheless, regulating algorithms of collector-distributor district heating system should be further investigated.

Dynamic operation of the proposed system was neglected in this paper, but it is important for a hydraulic network to cope with changing flow rates and heat loads. So, dynamic models are necessary for more accurate calculations and developing control strategies. Although some models have been developed in [1,16], this subject can also be further investigated and new improved models can be developed.

5. Conclusion

After having analyzed the possibility of connecting SH system through THD and replacing automatic balancing valves with DVSP, it was clear that this solution is feasible for Moscow conditions, especially for diameters of balancing valves DN25 and DN32. For DN15 and DN20, discounted payback period is 1-7 years.

Then collector-distributor district heating system including THD and DVSP was proposed in this paper. Its feasibility study shows that it is possible to save 42 - 129 MW*hr/yr of electric energy in district heating systems of 2.4 - 14.4 MW heat output. The savings can be higher if dynamic operation and flow control strategy are taken into account. At the same time, additional capital investments in comparison with conventional system are relatively low and the discounted payback period is less than 2 years.

The concept of using THD along with DVSP might be also applicable to district cooling systems or to district heating and cooling systems. Another possible area of its implementation is low temperature district heating. These subjects might be of great interest for future research works.

Acknowledgements

The work was performed under financial support of the state order No 13.9613.2017/8.9 and the state order No 13.13471.2019/13.2.

References

[1] Yan A, Zhao J et al 2013 Hydraulic performance of a new district heating systems with distributed variable speed pumps *Applied Energy* **112** 876-85

[2] Sheng X and Duanmu L 2016 Electricity consumption and economic analyses of district heating system with distributed variable speed pumps *Energy and Buildings* **118** 291-300

[3] Sheng X and Duanmu L 2016 Energy saving analyses on the reconstruction project in district heating system with distributed variable speed pumps *Applied Thermal Eng.* **101** 432-45

[4] Sheng X and Duanmu L 2017 Energy saving factors affecting analysis on district heating system with distributed variable frequency speed pumps *Applied Thermal Eng.* **121** 779-90

[5] Gu J, Qi C et al 2019 Analysis of a hybrid control scheme in the district heating system with distributed variable speed pumps *Sustainable Cities and Society* **48** 101591

[6] Zhang Y, Chen C and Xia J 2017 Energy performance analysis of an integrated distributed variable-frequency pump and water storage system for district cooling systems *Applied Sciences* **7** 1139

[7] Zeng J, Han J and Zhang G 2016 Diameter optimization of district heating and cooling piping network based on hourly load *Applied Thermal Eng.* **107** pp 750-7

[8] Wang H, Meng H and Long W 2017 A district energy planning method with mutual interconnection and interchange of thermal grids *Procedia Eng.* **205** pp 1412-9

[9] Yavorovsky Y V, Romanov D O et al. 2017 Application of thermohydraulic dispatcher in low temperature district heating systems for decreasing heat carrier transportation energy cost and increasing reliability of heat supply *IOP Conf. Series: J. of Physics: Conf. Series* **891** 012167

[10] Yavorovsky Y V, Romanov D O et al. 2018 Experimental research of thermo-hydraulic separators and dispatchers in heat supply systems *Solid State Phenomena* **284** 1385-9

[11] Romanov D O, Yavorovsky Y V and Smirnov V V. 2019 Application of thermohydraulic dispatcher for existing and new district heating systems *IOP Conf. Series: Mater. Sci. Eng.* **552** 012025

[12] Romanov D O, Yavorovsky Y V and Smirnov V V. 2019 Experimental'nye issledovaniya rezhimov raboty termogidravlicheskogo raspredelitelya [Experimental research of thermo-hydraulic dispatcher operating regimes] *Nadezhnost' i Bezopasnost' Energetiki* **2** 106-12 [in Russian]

[13] Romanov D O and Yavorovsky Y V 2019 Feasibility study of conventional, DVSP, and THD-based heat supply systems *E3S Web of Conf.* **124** 04007

[14] Sarbu I and Valea E S 2015 Energy savings potential for pumping water in district heating stations *Sustainability* **7** 5705-19

[15] Wang H, Wang H and Zhu T 2017 A new hydraulic regulation method on district heating system with distributed variable speed pumps *Energy Conversion and Management* **147** 174-89

[16] Wang H, Wang H et al 2017 Optimization modeling for smart operation of multi-source district heating with distributed variable-speed pumps *Energy* **138** 1247-62

IOP Conf. Series: Materials Science and Engineering **791** (2020) 012046 doi:10.1088/1757-899X/791/1/012046

Energy efficient multi-circuit cogeneration plant

S F Stepanov[1], V V Kovalenko[1], P V Kovalenko[1], I I Artyukhov[1], E E Mirgorodskaya[2] and N P Mityashin[2]

[1]Department of Power and Electrical Engineering, Yuri Gagarin State Technical University of Saratov, 77 Politechnicheskaya str., Saratov, 410054, Russia
[2]Department of Electronic Devices and Systems Engineering, Yuri Gagarin State Technical University of Saratov, 77 Politechnicheskaya str., Saratov, 410054, Russia

E-mail: stepanov460915@yandex.ru

Abstract. The paper reviews an energy efficient multi-circuit power plant with three independent power generation circuits and a generator control system. The proposed three-circuit autonomous power plant was constructed in accordance with a modular design. The structural scheme, design, operating modes, and simulation model in the Matlab+Simulink environment is presented. The increase in electrical efficiency at this cogeneration plant is due to the use of low-temperature boiling point heat-transfer fluids. This article proposes the method of increasing the efficiency of an existing installation from 60 kW to 100 kW by equipping it with additional equipment. This equipment provides deeper utilization of flue gas heat. If the base power plant is 60 kW, the heat input is about 200 kWh (an efficiency of the station is 30%), then the total electricity generation will be 100 kW. In this case, an additional 40 kW of power is without increasing fuel consumption.
Keywords. Multi-circuit power plant, low-temperature boiling point heat-transfer fluid, organic Rankine cycle, modeling and simulation in MATLAB

1. Introduction

The object of our study was an autonomous electric equipment complex comprising of a three-circuit gas turbine power plant using the technology based on low-temperature boiling point heat-transfer fluids. The complex had an automatic control system distributing electrical loads among the circuits in a given proportion. The proposed three-circuit autonomous power plant was constructed in accordance with a modular design. Each circuit of the power plant could operate as an independent electric generator. The second and the third modules had similar sets of their elements.

To simulate the various modes of operation in a three-circuit power plant, a simulation model was developed in the Matlab Simulink system.

The advantage of the developed autonomous gas turbine power plant with a multi-circuit thermal power plant was the possibility of a fuller utilization of the thermal energy from the gases generated in the gas turbine.

The proposed circuit solution allowed increasing the electrical efficiency of the device up to 50-55%, which affected the effectiveness of a fuel gas usage. An increase of electrical efficiency by 1% results in 3% fuel savings.

It is assumed that the first circuit of the cogeneration plant operates on the basis of existing microturbines (for example, of Capstone brand) and has an electrical power of 60 kW. The second and third circuits operate on the Rankine cycle. They differ from each other in the following details: the turbine on the second circuit operates on the Rankine gas-steam cycle, while purified water is used as the coolant. The third circuit has the lowest temperature, although it operates on the Rankine cycle as well.

The transformation of low-potential thermal energy can be carried out in various thermodynamic cycles. However, attention of the specialists is attracted by organic Rankine cycle. This cycle differs from conventional Rankine vapor-gas cycle in various organic substances (such as freons) rather than water used as a working medium.

The technology using the Rankine organic cycle can work on any heat source with a minimum temperature difference of 65 °C between the heat source and the heat sink.

Calculations conducted by the specialists have shown that radial-axial centripetal turbines can be used as expansion machines and drives of electric generators in the low-power Rankine cycle. This type of turbine provides a fairly high energy conversion efficiency at low cost and relatively high degree of working fluid expansion.

Use of water as a working medium in power plants operating in the Rankine cycle with a single-stage radial-axial turbine at low output power necessitates an increase in the operating frequency of the turbo generator shaft to extremely high values and, as a result, unreasonable complication of the electric parts of the power plants.

The use of the Rankine organic cycle allows for lower operating temperatures and relatively high operating pressures in the cycle.This leads to decrease in geometric dimensions of the turbine.

This results in technologically acceptable dimensions of the impellers, relatively high values of the thermal efficiency of the cycle and isentropic efficiency of the turbine [1].

As an example, table 1 provides the parameters of single-stage radial-axial turbines operating on organic Rankine cycle on R114 freon in the power range from 1 to 60 kW.

Table 1. Physical parameters of organic Rankine cycle.

Condensation temperature (°C)			30		
Boiling point (°C)			120		
Temperature in front of turbine (°C)			130		
Turbine inlet pressure (bar)			20.8		
Turbine outlet pressure (bar)			2.5		
Turbine shaft power (kW)	**1**	**15**	**30**	**45**	**60**
Mass flow through the turbine (kg/s)	0.04	0.59	1.16	1.70	2.22
Impeller outer diameter, mm	16	62	86	105	120
Turbine rotor speed, thousand rpm	200	52	38	31	28
Isentropic efficiency of turbines, %	79	81	82	84	85
Thermal efficiency of the cycle, %	15.7	16.0	16.1	16.5	16.8

Acceptable values of thermodynamic conversion efficiency in low-power power plants are provided at a relatively high frequency of the turbine rotor rotation. This implies the implementation of a gearless design of microturbine cogeneration plant. A high-frequency generator is used as an electromechanical energy converter. The heat energy of the flue gases in the Capstone C65 microturbine is 559 BTU/h [2]. This value corresponds to heat output of 164 kW. Let us take this indicator as the basis for our calculations.

In the second circuit, generation of 30 kW of electrical power is expected from such heat energy of flue gases.In the third circuit, the heat energy of flue gases of the second stage is expected to produce 10 kW of electrical power. Ethanol (ethyl alcohol) is used as a low boiling point working substance.

The choice of ethanol as a working substance is largely determined by its low capability to serve as a solvent for various materials, such as gaskets, sealing rings, glands and various insulating materials.

IOP Conf. Series: Materials Science and Engineering **791** (2020) 012046 doi:10.1088/1757-899X/791/1/012046

The calculation of the Rankine cycle on ethanol for the third generation loop is performed using the information resource [3] based on MathCad software. Such calculation is required to identify the the efficiency of converting thermal energy of flue gases into electricity.

When calculating a microturbine of the second circuit with a temperature drop from 350°C to 100°C using the steam cycle, the theoretical efficiency of cogeneration plant is 40%. To achieve this, it is necessary to expend 75 kW of thermal power to generate electricity.

When calculating a microturbine of the third circuit with a temperature drop from 100° C to 30 °C using an ethanol cycle , cogeneration plant efficiency is 15% [4-6]. Here, it is necessary to expend heat power equal to 67 kW to generate electricity.As a result, total thermal power output from the first circuit for additional power generation should be at least 142 kW.

As mentioned earlier, the heat output from the gas turbine unit of the first circuit is 164 kW. Thus, it can be stated that with such output of thermal power, it is possible to provide deeper heat utilization with an increase in cogeneration plant efficiency up to 50% by adding two additional circuits.

2. Materials and Methods

Based on the analysis of existing microturbine technologies and the developed projects of autonomous power plants, the following concept of increased efficiency multi-circuit gas turbine power plants is proposed [7-13]:

Three independent power generation circuits:

- in each circuit on the same shaft with the turbine a high-speed electric generator with gas-dynamic bearings is installed;
- to obtain electric power of industrial frequency, a frequency converter is installed;
- a rectifier is connected to each electric generator to rectify the generated voltage and current;
- in order to summarize the power of electric generators generated in each circuit, stabilize the voltage and obtain voltage of a given quality, an inverter is installed.

The scheme of the proposed multi-circuit gas turbine power plant is shown in figure 1.

The positions in the drawing indicate the following: 1 - compressor intake; 2 - compressor; 3 - the first turbine; 4 - second gas turbine; 5 - the first electric generator; 6 - second electric generator; 7 - recuperator; 8 - the first evaporator; 9 - smoke exhaust; 10 - the first pump; 11 - the first shell-and-tube heat exchanger-condenser; 12 - supply pipeline fuel; 13 - combustion chamber; 14 - third turbine; 15 - the third electric generator; 16 - the first rectifier; 17 - the second rectifier; 18 - the third rectifier; 19 - second evaporator; 20 - second shell-and-tube heat exchanger-condenser; 21 - the second pump; 22 - inverter; 23 - microprocessor control unit, 24 - current sensor after the first rectifier, 25 - current sensor after the second rectifier, 26 - current sensor after the third rectifier, 27 - voltage sensor after the first generator, 28 - voltage sensor after the second generator, 29 - voltage sensor after the third generator.

It is assumed that the first circuit of the installation is a microturbine of a finished industrial production (for example, Capstone brand) with an output electric power of 60-65 kW.

Gases exhausted in the primary circuit are used as a heating medium, including a circuit with a low-boiling medium, which are used as freons, aqueous ammonia, pentane, isopentane, isobutane, etc.

The second and third circuits operate on organic Rankine cycle, but depending on the choice of the heating medium and the output heat parameters, the second circuit can operate onRankine steam-gas cycle (in which purified water is assumed to be the heat transfer medium) or on organic Rankine cycle, and the third, low temperature, circuit - on organic Rankine cycle with a deeper heat recovery.

Each circuit has its own high-speed electric generator installed on the same shaft as the turbine.

It is assumed that the first circuit of the installation is a microturbine of a finished industrial production (for example, Capstone brand) with an output electric power of 60-65 kW [14].

IOP Conf. Series: Materials Science and Engineering **791** (2020) 012046 doi:10.1088/1757-899X/791/1/012046

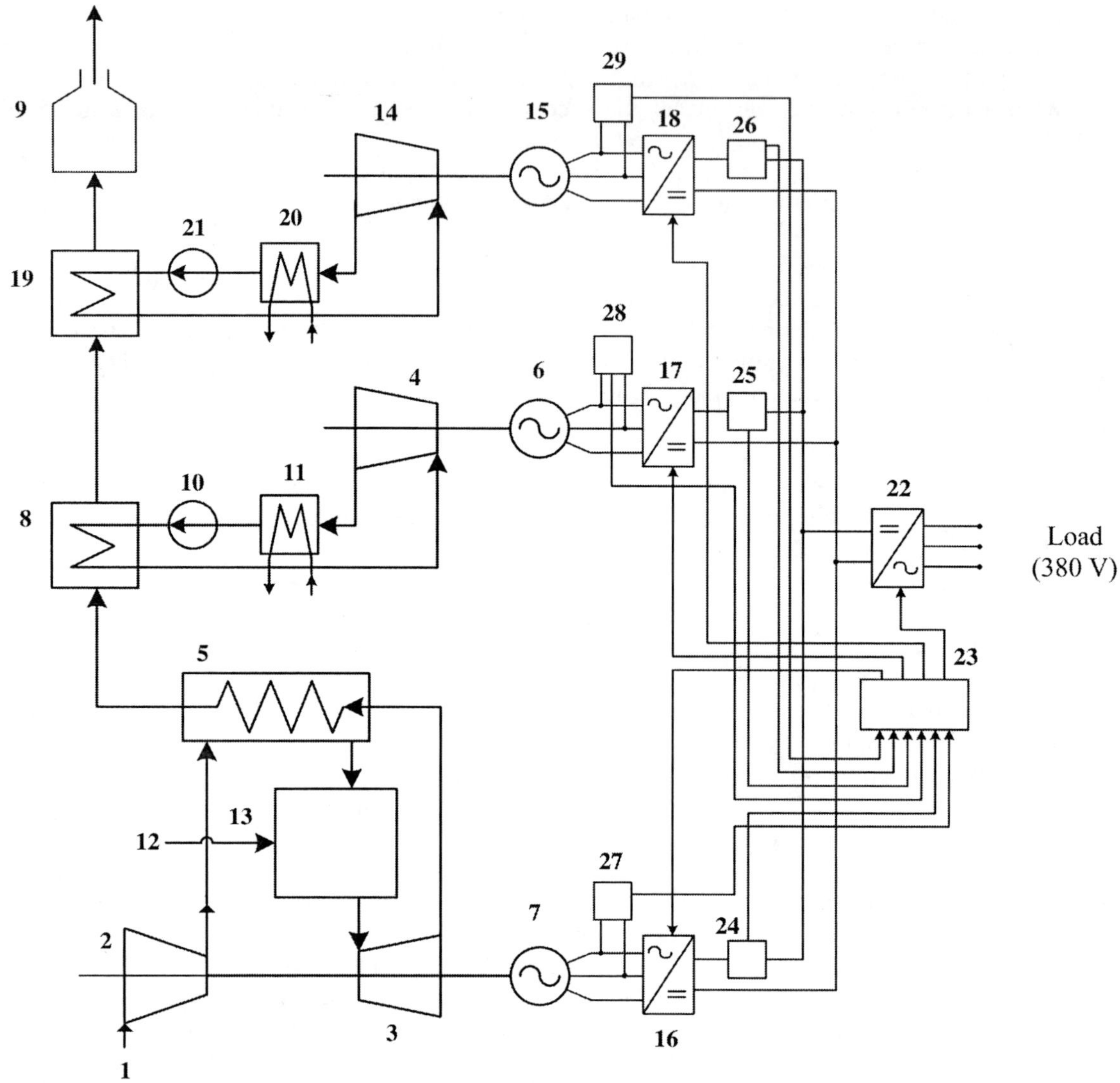

Figure 1. The scheme of the proposed multi-circuit gas turbine power plant.

Gases exhausted in the primary circuit are used as a heating medium, including a circuit with a low-boiling medium, which are used as freons, aqueous ammonia, pentane, isopentane, isobutane, etc.

The second and third circuits operate on the Rankine cycle, but depending on the choice of the heating medium and the output heat parameters, the second circuit can operate on Rankine steam-gas cycle (in which purified water is assumed to be the heat transfer medium) or on the organic Rankine cycle, and the third, low temperature, circuit - on organic Rankine cycle with a deeper heat recovery.

Each circuit has its own high-speed electric generator installed on the same shaft as the turbine.

For the proposed circuit, the following values of the output electric power were set: the first circuit - 60 kW, the second circuit - 30 kW, the third circuit - 10 kW.

IOP Conf. Series: Materials Science and Engineering **791** (2020) 012046 doi:10.1088/1757-899X/791/1/012046

3. Results

As mentioned above, on the first circuit, the electrical efficiency of an existing microturbine is on average 30% (for various manufacturers). In this case, with the given electrical installation parameters, the power of the consumed thermal energy of the primary source is 60 / 0.3 = 200 kW, on the second circuit with the consumed thermal power of 200 kW, the efficiency can be increased to (60 + 30)/200 = 45%, on the third accordingly, up to (60 + 30 + 10)/200 = 50%.

Thus, when constructing a system using a three-circuit scheme with a total power of 100 kW, it is possible to increase the electrical efficiency by 1.5 times compared to a conventional single-circuit microturbine system, without additional emission of harmful substances into the atmosphere.

The thermal output of the Capstone C65 microturbine is 591 000 kJ/h. This indicator corresponds to a thermal power of 164 kW (591000 kJ/h = 591000/3600 kJ/s = 16 kJ/s = 164 kW).

This indicator was taken as the basis of theoretical calculations. When calculating a microturbine of 2 circuits with a temperature difference from 350 °C to 100 °C when using a cycle on water vapor, the theoretical efficiency of the installation was 40%. Here, it is necessary to expend heat power equal to 30 kW/0.40 = 75 kW to generate electricity. When calculating microturbines of 3 circuits at a temperature difference from 100°C to 30°C when using a cycle on ethanol, the plant efficiency was 15%.

Here, it is necessary to expend heat power equal to 10 kW / 0.15 = 67 kW to generate electricity.

As a result, the total heat output at the output from the primary circuit to obtain additional power generation should be at least 142 kW. As mentioned earlier, the output thermal power from the gas turbine installation of the first circuit is 164 kW.

Thus, it can be stated that with such heat output, it is possible to provide a deeper heat recovery with bringing the efficiency of the installation by adding two additional circuits to 50%.

In order to be able to supply power to the final consumer of the proposed power plant, a number of tasks need to be solved, such as providing the required parameters of power quality in terms of voltage and frequency.

Acceptable values of the thermodynamic conversion efficiency in low-power power plants are provided at a sufficiently high frequency of rotation of the turbine rotor. This circumstance involves the use of high-frequency generators as an electromechanical energy converter.

Assessing the advantages and disadvantages of various types of synchronous and asynchronous generators, as well as drawing on the experience gained by domestic and foreign experts in studying the operation of high-speed electric turbomachines with powers not exceeding 200-250 kW, synchronous machines with excitation from permanent magnets are offered as electric generators, which eliminates the problems arising from the parallel operation of generators. In this case, the inability to control the output voltage is compensated by the use of semiconductor power converters.

Thus, the consumer is connected to the power generating installation through a frequency converter with a DC-link (D_1) the generated alternating current is first rectified, and then transmitted through the inverter I1, which provides the required frequency, to the consumer (figure 2). In the concept of a three-circuit power plant, for each generating circuit (generators G_1, G_2, G_3), a rectifier device R_1, R_2, R_3 with an individual control unit (ICU) is used, operating on a common DC bus, by which the power of individual generating circuits is summed.

In this case, it is necessary to ensure:

- the same value of the rectified voltage for each circuit;
- synchronous control of rectifiers in a complex of three parallel-running generators of different power.

Rectifiers R_2 and R_3 are controlled by feedback on the load current determined by the current sensor CS_n.

IOP Conf. Series: Materials Science and Engineering **791** (2020) 012046 doi:10.1088/1757-899X/791/1/012046

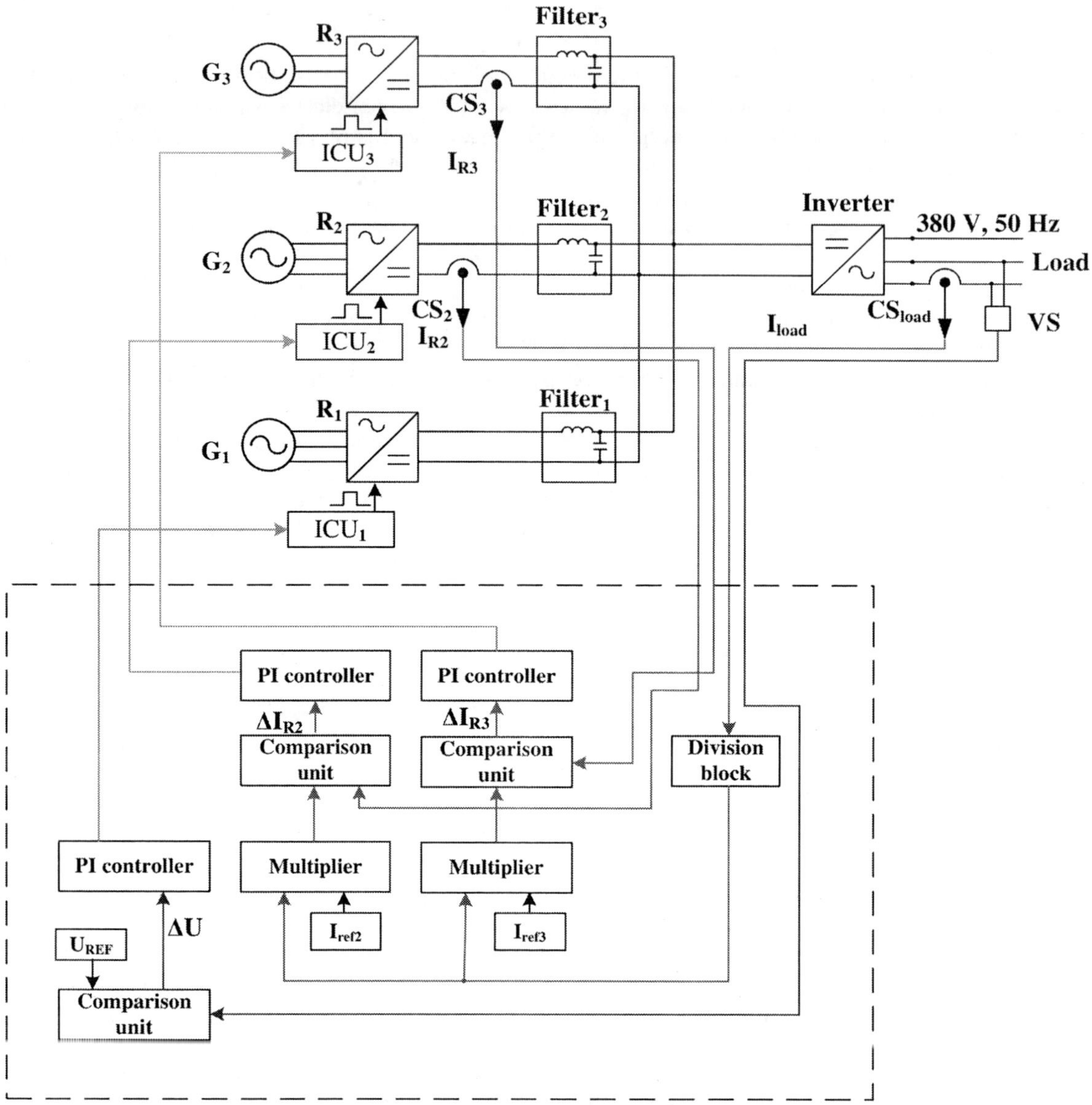

Figure 2. Block diagram of the formation of control rectifiers for work on the total load.

The signal from the CSn is transmitted to the division unit, in which the load current value is compared with the set value of the rated current generated by the electrical installation. To obtain information on the values of the generated current IR_2, IR_3, the current sensors CS_2, CS_3 are connected to the output circuits of the rectifiers. In the comparison blocks, the deviation of the value of the generated current of the generators from the load required by the conditions is determined and a control signal is set, which is transmitted to the controlled thyristor electrode of the rectifier unit and unlocks it. Since the initial feedback signal according to the magnitude of the load current is supplied simultaneously to all rectifier control units RCU and the process of generating the control signal is the same, the percentage distribution of the generated power by each generator is stored at any load value.

Thus, the load of the generators is the same and varies synchronously in proportion to its rated power.

IOP Conf. Series: Materials Science and Engineering **791** (2020) 012046 doi:10.1088/1757-899X/791/1/012046

The rectifier R1 is controlled by feedback on the voltage determined by the voltage sensor VS at the terminals of the inverter. Information about the voltage value from the voltage sensor VS is compared in the comparison unit with the value of the specified reference voltage U_R. The signal from the comparison unit enters through the regulator to the control unit of the rectifier RCU_1, which forms the control signal necessary to maintain the voltage at the required level, regardless of the load. Thus, a voltage stabilization system has been formed, which allows providing the required voltage value at the rectifier terminals, including in the idle mode.

The use of a two-bridge autonomous inverter (current or voltage) with a thyristor-reactor compensation circuit allows you to get the output voltage at the consumer's terminals with the required quality in magnitude and harmonic composition.

4. Discussion

The ideas embodied in the presented power plant can find implementation in the design of autonomous power supply systems for remote consumers, for example, the far north, where the issue of energy supply is quite acute and the efficiency of their use is the most relevant.

5. Conclusion

The paper proves the technical feasibility of constructing multi-module power generation systems on low-boiling coolants in combination with gas turbine power plants, which allows increasing electrical efficiency up to 50%.

References

[1] Van Buijtenen J, Larjola J et al. 2003 Design and validation of a new high expansion ratio radial turbine for ORC application *5th European Conf. on Turbo-machinery* (Praha) pp 1-14

[2] Van Buijtenen J 2009 The tri-o-gen organic Rankine cycle: development and perspectives *Power Engineer* **13(1)** 4-9

[3] van Buijtenen P, Heinimo J, Larjola J, Backman J 2003 Small electricity production with high speed ORC technology *Int. Nordic Bioenergy Conf.* (Jyvaskyla) pp 1-4

[4] Heinimo J, van Buijtenen J, Backman J, Ojaniemi A. and Malinen H. 2004 High-speed ORC technology for distributed electricity production *2nd World Conf. on Biomass for Energy, Industry and Climate Protectio* (Rome: WIP-Munich) pp 1761-4

[5] Harinck J, Pasquale D, Pecnik R, van Buijtenen J, Colonna P 2013 Performance improvement of a radial ORC turbine by means of automated CFD design *Proceedings of the Institution of Mechanical Engineers, Part A: J. of Power and Energy* **227(6)** 637-645

[6] Soleman A, Saifiuriddin P and M. 2006 Dynamic analysis of microturbine/fuel cell for peak power saving *IEEE Power Eng. Society General Meeting* pp 1-7

[7] Vankeirsbilck I, Vanslambrouck B, Gusev S and De Paepe M 2011 Organic Rankine Cycle as efficient alternative to steam cycle for small scale power generation *Proceedings of the HEFAT 2011 Conf.* (Mauritius) pp 785-92

[8] Kang S H 2012 Design and experimental study of ORC (organic Rankine cycle) and radial turbine using R245fa working fluid *Energy* **41** 514-24

[9] Delgado-Torres A M and Garcia-Rodriguez L 2010 Analysis and optimization of the low-temperature solar organic Rankine cycle (ORC) *Energy Conversion and Management* **51** 2846-56

[10] Yu G, Shu G, Tian H, Wei H and Liu L 2013 Simulation and thermodynamic analysis of a bottoming Organic Rankine Cycle (ORC) of diesel engine (DE) *Energy* **51** 281-90

[11] Velez F, Segovia JJ, Martin MC, Antolin G, Chejne F and Quijano A 2012 A technical, economical and market review of organic Rankine cycles for the conversion of low-grade heat for power generation *Renewable and Sustainable Energy Reviews* **16** 4175-89

[12] Quoilin S, Van Den Broekb M, Declayea S, Dewallefa P, Lemorta V 2013 Techno-economic survey of Organic Rankine Cycle (ORC) systems *Renewable and Sustainable Energy*

Reviews **1** 168-86
[13] Rentizelas A, Karellas S, Kakaras E and Tatsiopoulos I 2009 Comparative techno-economic analysis of ORC and gasification for bioenergy applications *Energy Conversion and Management* **50(3)** 674-81
[14] Sprouse C, Depcik C 2013 Review of organic Rankine cycles for internal combustion engine exhaust waste heat recovery *Applied Thermal Eng.* **51** 711-22
[15] Bertani G, Bossi C, Fornari F, Massucco S, Spelta S and Tivegna F 2004 A microturbine generation system for grid connected and islanding operation *IEEE PES Power System Conf. and Composition* pp 360-5

IOP Conf. Series: Materials Science and Engineering **791** (2020) 012047 doi:10.1088/1757-899X/791/1/012047

Analysis of the unsteady temperature condition of a building during non-working hours when heating load is decreased

[1]M V Svirin, [2]S A Bychikhin, [2]P A Trubaev and [1]A S Seminenko

[2]Department of Heat and Gas Supply and Ventilation, Belgorod State Technological University named after V.G. Shukhov, Kostyukov St., 46, Belgorod, 308012, Russia
[1]Department of Energy Engineering of Heat Technologie, Belgorod State Technological University named after V.G. Shukhov, Kostyukov St., 46, Belgorod, 308012, Russia

E-mail: trubaev@gmail.com

Abstract. This paper examines the issue of fuel and energy resources conservation through the non-working hours heating load decrease. The paper`s objective is the determination of savings when non-working hours heating load is decreased, considering the unsteadiness of the heat transfer processes through the enclosing structures, heating modes determination, which ensure the required microclimate parameters in the building during working hours. The analysis of the unsteady temperature condition has been conducted using systems of differential equations, the heating load decrease method has been suggested with further calculation for the existing building and technical-and-economic evaluation of the suggested energy conservation measure. To solve this problem the mathematical model was constructed, which considers the unsteadiness of weather conditions and temperature conditions of the building when the heating load is decreased. The method of heating load decrease was suggested with further savings calculation for the High School №4 in Belgorod city. The modeling helped to establish that to ensure steady microclimate parameters in the building after period of heating load decrease to 12°C ends, it is necessary to conduct short-duration overheating of the building by increasing inside temperature to 21°C for the period from 5:00 till 9:00, to prevent loss of heat by the enclosing structures. When the room temperature during non-working hours is decreased to 12°C the calculated economy is 10%.

1. Introduction

One of the main problems in budget expenses control in the Belgorod region is constantly rising prices for the energy resources, costs for transportation to the consumer and its ineffective end use [1]. At present, most of the budget institutions have heat losses, which are results of the following factors [2]:

- poor condition of the building`s thermal envelope (doorways and window openings);
- insufficient maintenance of the supply and exhaust ventilation;
- out of adjustment heat load, which leads to substantial temperature overshoot in buildings against the normative, and as the result – overconsumption of the thermal energy for heating.

Building`s energy consumption forecast is necessary for planning, control, and conservation of energy. The data-based models provide a practical approach for the energy consumption forecast [3]. Forecast is very important for the control issues of heating system, ventilation, and air conditioning. On a practical level, there is a significant reduction of energy consumption in the building`s heating after the automated domestic heating plant (DHP) is installed, which controls the heating load depending on the weather conditions [4].

Energy consumption for the buildings is around 35% of the world`s total energy production. The educational institutions consume 20-40% of heat energy [5]. Therefore, they contribute significantly to the expenses of the municipal governments. A high level of energy consumption is caused, mostly, by the low building`s characteristics, modes of operation, etc.

One of the alternatives that give energy consumption reduction in buildings is an energy service contracting for energy efficiency improvement [6]. The energy service contract allows to reduce the operating costs. The heating load control automation, which is based on the energy service contract, ensures energy conservation with minimal expenses on the consumer`s side.

Evaluation of the heat energy savings value depends on the efficiency level of the reconstructed heating system and requires detailed information about the heat energy consumption value in relation to the weather parameters [7]. The scientific literature estimates savings value in the range from 3 to 30%.

2. Materials and Methods

In the budget institutions, which are in use mostly during working hours, the heating load can be adjusted, which results in a reduction of heat energy consumption. The calculation should be done under comparable conditions. The objective of the comparable conditions use in calculations is a recalculation of the values for the possibility of energy consumption of different buildings and conditions comparison.

The comparable conditions – combination of technical, climatic and mode factors, which have numeric values, that influence fuel-and-energy resources consumption in physical terms, but don`t influence the savings after implementation of measures for energy conservation and energy resources conservation, which are for recalculation of the energy consumption during the reporting period for value that could have been in conditions of base (standard, specified) period factors value.

Below presented the factors (comparable conditions) which influence the volume of TER consumption [8]:

- weather conditions change (average outside air temperature over a period, average duration of the daylight over a period, duration of the heating period);
- an institution`s working hours change (rate of working days and/or working hours in a period, number of employees and/or customers);
- institution`s room (rooms) purpose change (average temperature inside the room);
- change of the set-up, number and heated rooms (heated rooms area or heated volume);
- set-up, number or capacity of the energy-consuming equipment.

The unsteady heat transfer through the walls of a building can be calculated in the following way [9]:

$$\rho c\frac{\partial T}{\partial \tau}-\frac{\partial}{\partial x}\left(\lambda\frac{\partial T}{\partial x}\right), \tag{1}$$

where ρ – density kg/m^3; c – specific heat capacity, kJ/(kg·°C); τ – time, s.; T – temperature, °C, x – grid step, m; λ – heat transfer coefficient, kW/(m·°C).

A solution of the differential heat transfer equation lies in its discretization, i.e. substitution of the value, which is given by the differential function, for several values in certain points – nodes of a spatial grid [10]. Determination of the temperature values in the nodes of a grid is done by sequential calculation of temperature in every node using temperature value in the preceding node. To increase calculation precision, it is possible to use different discretization methods, i.e. substitution of the differential for increment (finite difference method, balance method, etc.). The disadvantage of the mentioned above methods is in low accuracy, therefore, mainly the two-point boundary problem is considered in the heat transfer modeling when the heat transfer conditions are given in right and left boundaries [11].

To solve the differential heat transfer equation of the enclosing structure the spatial grid should be divided into 10 points (0-9) (see Figure 1). When a building is insulated 6 points are used (0-5) to de-

IOP Conf. Series: Materials Science and Engineering **791** (2020) 012047 doi:10.1088/1757-899X/791/1/012047

termine the temperature field through the wall and 3 points (7-9) to determine the temperature field through insulation.

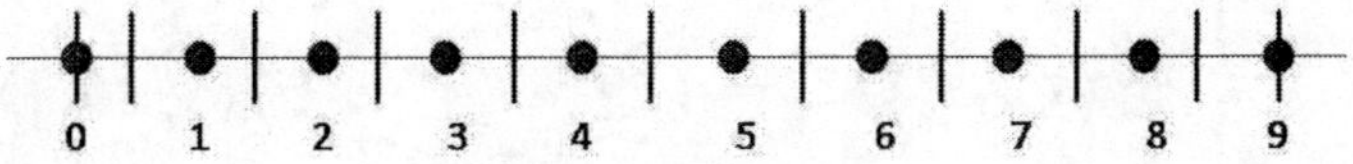

Figure 1. Spatial grid

For the endpoints of the spatial grid the boundary conditions of the third kind were given [12]:
for the left boundary:

$$q = \alpha_{in}\left(T_0^1 - T_{in}^1\right), \tag{2}$$

where α_{in} – heat transfer coefficient of the enclosing structure's inner surface, W/(m^2·K), T_0^1 – temperature in a corresponding point at the current time, °C; T_{in}^1 – temperature inside the building in a corresponding point at the time of preceding iteration, °C.
for the right boundary:

$$q = \alpha_{out}\left(T_{out} - T_n^1\right), \tag{3}$$

where α_{out} – heat transfer coefficient of the enclosing structure's outer surface, W/(m^2·K); T_{out} – outside temperature, °C; T_n^1 – temperature in a corresponding point at the current time, °C.

As a result of the equation (1) discretization on the grid, presented in Figure 1, the equations are obtained for zero point, *i-th* point and *n-th* point.

Heat transfer equation for zero point, T_0:

$$\frac{\rho c}{\Delta\tau}\cdot\left(T_0^1 - T_0^0\right) = \frac{\lambda}{\Delta x}\cdot\left(T_1^1 - T_0^1\right) - \alpha_{in}\cdot\left(T_0^1 - T_{in}^1\right), \tag{4}$$

where ρ – material density, kg/m^3; c – specific heat capacity of a material, kJ/(kg·K); Δτ - time step, s; λ – heat transfer coefficient of a material, W/(m·K); Δx – grid step, m; α_{in} – heat transfer coefficient of the enclosing structure's inner surface, W/(m^2·K); T^1 – temperature in a corresponding point at the current time, °C; T^0 – temperature in a corresponding point at the time of preceding iteration, °C.

Heat transfer equation for *i-th* (*i*=1-8) point, T_i:

$$\frac{\rho c}{\Delta\tau}\cdot\left(T_i^1 - T_i^0\right) = \frac{\lambda}{\Delta x}\cdot\left(T_{i+1}^1 - T_i^1\right) - \frac{\lambda}{\Delta x}\cdot\left(T_i^1 - T_{i-1}^1\right). \tag{5}$$

Heat transfer equation for *n-th* point, T_n:

$$\frac{\rho c}{\Delta\tau}\cdot\left(T_n^1 - T_n^0\right) = \alpha_{out}\cdot\left(T_{out} - T_n^1\right) - \frac{\lambda}{\Delta x}\cdot\left(T_n^1 - T_{n-1}^1\right), \tag{6}$$

where ρ – material density, kg/m^3; c – specific heat capacity of a material, kJ/(kg·K); Δτ - time step, s; λ – heat transfer coefficient of a material, W/(m·K); Δx – grid step, m; α_{out} – heat transfer coefficient of the enclosing structure's outer surface, W/(m^2·K).

As a result, the system is obtained, which contains $(n+1)$ equation:

$$
\begin{cases}
\frac{\rho c}{\Delta\tau}\cdot\left(T_0^1 - T_0^0\right) = \frac{\lambda}{\Delta x}\cdot\left(T_1^1 - T_0^1\right) - \alpha_{in}\cdot\left(T_0^1 - T_{in}^1\right) \\
\dots \quad \dots \dots \quad \dots \dots \\
\frac{\rho c}{\Delta\tau}\cdot\left(T_i^1 - T_i^0\right) = \frac{\lambda}{\Delta x}\cdot\left(T_{i+1}^1 - T_i^1\right) - \frac{\lambda}{\Delta x}\cdot\left(T_i^1 - T_{i-1}^1\right) \\
\dots \quad \dots \dots \quad \dots \dots \\
\frac{\rho c}{\Delta\tau}\cdot\left(T_n^1 - T_n^0\right) = \alpha_{out}\cdot\left(T_{out} - T_n^1\right) - \frac{\lambda}{\Delta x}\cdot\left(T_n^1 - T_{n-1}^1\right)
\end{cases} \tag{7}
$$

To find the solution using the sweep method [13] the system is presented in the standard way:

$$
\begin{cases}
a_0\cdot T_1^1 + b_0\cdot T_0^1 + d_0 = 0 \\
\dots \quad \dots \quad \dots \quad \dots \quad \dots \\
a_i\cdot T_2^1 + b_i\cdot T_1^1 + c_i\cdot T_0^1 + d_i = 0 \\
\dots \quad \dots \quad \dots \quad \dots \quad \dots \\
b_n\cdot T_n^1 + c_n\cdot T_{n-1}^1 + d_n = 0
\end{cases} \tag{8}
$$

where $a_0 = \frac{\lambda}{\Delta x}$; $b_0 = \frac{\rho c}{\Delta\tau} + \alpha_{in} + \frac{\lambda}{\Delta x}$; $d_0 = \left(-\frac{\rho c}{\Delta\tau}\cdot T_0^0 + \alpha_{in}\cdot T_{in}^1\right)$;

$a_i = \left(-\frac{\lambda}{\Delta x}\right)$; $b_i = \left(\frac{\rho c}{\Delta\tau} + 2\cdot\frac{\lambda}{\Delta x}\right)$; $c_i = \left(-\frac{\lambda}{\Delta x}\right)$; $d_i = \left(-\frac{\rho c}{\Delta\tau}\cdot T_1^0\right)$;

$b_n = \left(\frac{\rho c}{\Delta\tau} + \alpha_{out} + \frac{\lambda}{\Delta x}\right)$; $c_n = \left(-\frac{\lambda}{\Delta x}\right)$; $d_n = \left(-\frac{\rho c}{\Delta\tau}\cdot T_n^0 - \alpha_{out}\cdot T_{out}\right)$

The temperature field through the insulation layer is calculated by a similar method.
Heat transfer equation for zero point, T_0:

$$\frac{\rho_1 c_1}{\Delta\tau}\cdot\left(T_0^1 - T_0^0\right) = \frac{\lambda_1}{\Delta x_1}\cdot\left(T_1^1 - T_0^1\right) - \alpha_{\text{in}}\cdot\left(T_0^1 - T_{\text{in}}^1\right) \tag{9}$$

Heat transfer equation for *i-th* (i=7-9) point, T_i:

$$\frac{\rho_1 c_1}{\Delta\tau}\cdot\left(T_i^1 - T_i^0\right) = \frac{\lambda_1}{\Delta x_1}\cdot\left(T_{i+1}^1 - T_i^1\right) - \frac{\lambda_1}{\Delta x_1}\cdot\left(T_i^1 - T_{i\ 1}^1\right) \tag{10}$$

Heat transfer equation for *n-th* point, T_n:

$$\frac{\rho_1 c_1}{\Delta\tau}\cdot\left(T_n^1 - T_n^0\right) = \alpha_{out}\cdot\left(T_{out} - T_n^1\right) - \frac{\lambda_1}{\Delta x_1}\cdot\left(T_n^1 - T_{n-1}^1\right) \tag{11}$$

As the result the system of equations was obtained:

$$\begin{cases} \frac{\rho_1 c_1}{\Delta\tau}\cdot\left(T_0^1 - T_0^0\right) = \frac{\lambda_1}{\Delta x_1}\cdot\left(T_1^1 - T_0^1\right) - \alpha_{in}\cdot\left(T_0^1 - T_{in}^1\right) \\ \dots \qquad \dots\dots \qquad \dots\dots \\ \frac{\rho_1 c_1}{\Delta\tau}\cdot\left(T_i^1 - T_i^0\right) = \frac{\lambda_1}{\Delta x_1}\cdot\left(T_{i+1}^1 - T_i^1\right) - \frac{\lambda_1}{\Delta x_1}\cdot\left(T_i^1 - T_{i-1}^1\right) \\ \dots \qquad \dots\dots \qquad \dots\dots \\ \frac{\rho_1 c_1}{\Delta\tau}\cdot\left(T_n^1 - T_n^0\right) = \alpha_{out}\cdot\left(T_{out} - T_n^1\right) - \frac{\lambda_1}{\Delta x_1}\cdot\left(T_n^1 - T_{n-1}^1\right) \end{cases} \tag{12}$$

or

$$\begin{cases} a_0\cdot T_1^1 + b_0\cdot T_0^1 + d_0 = 0 \\ \dots \quad \dots \quad \dots \quad \dots \quad \dots \\ a_i\cdot T_2^1 + b_i\cdot T_1^1 + c_i\cdot T_0^1 + d_i = 0 \\ \dots \quad \dots \quad \dots \quad \dots \quad \dots \\ b_n\cdot T_n^1 + c_n\cdot T_{n-1}^1 + d_n = 0 \end{cases} \tag{13}$$

where $a_0 = \frac{\lambda_1}{\Delta x_1}$; $b_0 = \frac{\rho_1 c_1}{\Delta\tau} + \alpha_{in} + \frac{\lambda_1}{\Delta x_1}$; $d_0 = \left(-\frac{\rho_1 c_1}{\Delta\tau}\cdot T_0^0 + \alpha_{in}\cdot T_{in}^1\right)$;
$a = \left(-\frac{\lambda_1}{\Delta x_1}\right)$; $b = \left(\frac{\rho_1 c_1}{\Delta\tau} + 2\cdot\frac{\lambda_1}{\Delta x_1}\right)$; $c = \left(-\frac{\lambda_1}{\Delta x_1}\right)$; $d = \left(-\frac{\rho_1 c_1}{\Delta\tau}\cdot T_1^0\right)$;
$b_n = \left(\frac{\rho_1 c_1}{\Delta\tau} + \alpha_{out} + \frac{\lambda_1}{\Delta x_1}\right)$; $c_n = \left(-\frac{\lambda_1}{\Delta x_1}\right)$; $d_n = \left(-\frac{\rho_1 c_1}{\Delta\tau}\cdot T_n^0 - \alpha_{out}\cdot T_{out}\right)$

The discrete analog of the heat transfer equation [16] on the border of layers (point 6) can be described as follows:

$$a_i\cdot T_{i+1}^1 + b_i\cdot T_i^1 + c_i\cdot T_{i-1}^1 + d_i = 0, \tag{14}$$

where $a = \left(-\frac{\lambda_1}{\Delta x_1}\right)$; $b = \left(0{,}5\cdot\frac{\rho c\Delta x}{2\Delta\tau} + 0{,}5\cdot\frac{\rho_1 c_1\Delta x_1}{2\Delta\tau} + 2\cdot(0{,}5\cdot\frac{\lambda}{\Delta x} + 0{,}5\cdot\frac{\lambda_1}{\Delta x_1})\right)$;
$c_i = \left(-\frac{\lambda}{\Delta x}\right)$; $d = \left(-(0{,}5\cdot\frac{\lambda}{\Delta x} + 0{,}5\cdot\frac{\rho_1 c_1\Delta x_1}{\Delta\tau})\cdot T_i^0\right)$

2.1. Method of savings determination

The inside room temperature calculation uses dependences between air heat capacity, heated space, time interval under consideration, dynamically changing incoming and outgoing heat energy quantities [14]. Dynamics of the air temperature change inside the building under consideration can be examined with the help of the following formula:

$$t_{i+1} = \frac{c_{air}\cdot V_{heat}\cdot t_i - \Delta\tau\cdot(Q_{vent} + Q_{attic} + Q_{basem} + Q_{wind} + Q_{wall} - Q_{inc})}{c_{air}\cdot V_{heat}}, \tag{15}$$

where t_{i+1} – inside room temperature in the next point of time, °C; c_{air} – air specific heat capacity, kJ/(m^3·K); V_{heat} – heated space in a building, m^3; t_i – inside room temperature in the preceding point of time, °C; $\Delta\tau$ – time step, s.; Q_{vent} – heat losses through the ventilation, kW; Q_{attic} – heat losses through the attic floor, kW; Q_{basem} – heat losses through the basement, kW; Q_{wind} – heat losses through the windows, kW; Q_{wall} – heat losses through the walls, kW; Q_{inc} – incoming heat from the building`s heat supply system, kW.

It is also necessary to monitor the inner surface of the wall when the heating load is reduced because excessive temperature reduction can lead to condensation on the wall`s surface [15]. The inner wall`s surface temperature is calculated as follows:

$$t_{in.wall} = \frac{t_{in}-(t_{in}-t_{out})}{\alpha_{in} \cdot R_{wall}}, \tag{16}$$

where $t_{in.wall}$ – inner temperature of a wall, °C; t_{in} – inside room temperature, °C; t_{out} – outside temperature of the atmosphere, °C; α_{in} – heat transfer coefficient of the enclosing structure`s inner surface, W/(m^2·°C); R_{wall} – thermal resistance of a wall, (m^2·°C)/W.

Outer surface temperature of the wall is calculated as follows:

$$t_{out.wall.} = \frac{t_{out}-(t_{in}-t_{out})}{\alpha_{out} \cdot R_{wall}}, \tag{17}$$

The idea behind heat energy conservation is in heating load reduction. It is realized in the calculations by the heating load coefficient:

$$K = \frac{Q_{heat}}{Q_{heat\ init}}, \tag{18}$$

where K – heating load coefficient; Q_{heat} – current consumed heat energy amount for heating of the rooms, kW; $Q_{\text{heat init}}$ – initial consumed amount of heat for building heating, kW.

The mathematical model allows observing the building`s parameters change as the result of a reduction or increase of the heating load and search of the optimal choice of the heating mode when savings are maximum [16].

3. Results

The heating load adjustment during 1 week has been studied, which allows conducting a more accurate analysis of temperatures. The studies were carried out at several actual buildings. The calculation results presented here are from the High School № 4 in Belgorod city.

The obtained room air temperature changes and average wall`s temperature is presented in Figure 2.

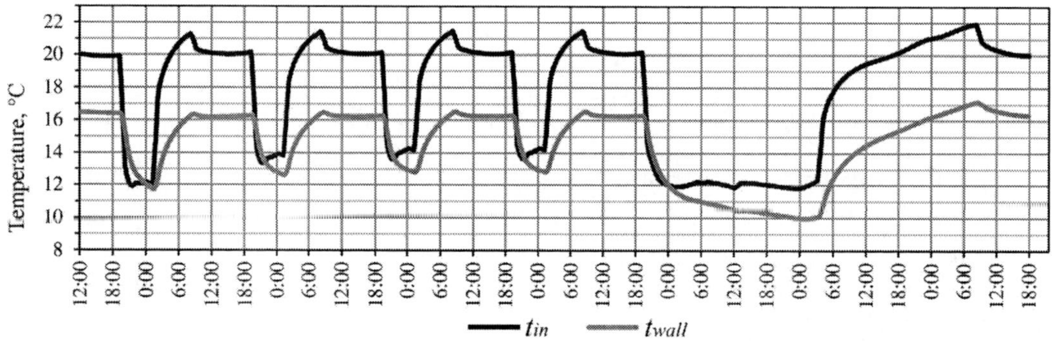

Figure 2. Dependence diagram of the air temperature in the High School №4 with temperature reduction to 12°C and average wall`s temperature within one week on the heating load adjustment.

The heating load reduction starts at 20:00 and then gradually increases until 2:00. From 2:00 the heating load increases until the heating load coefficient reach 1.2, so the initial air temperature parameters in the building are back at the beginning of the working day at 9:00. Therefore, the diagram shows short-duration overheating of the building until 9:00, however, at the beginning of the working day the temperature will be the same as the day before. The temperature changes are presented in Figure 3.

IOP Conf. Series: Materials Science and Engineering **791** (2020) 012047 doi:10.1088/1757-899X/791/1/012047

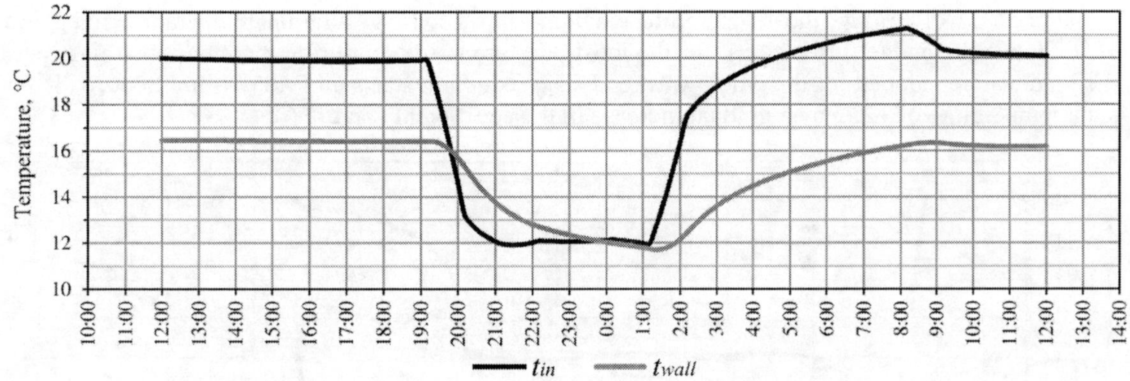

Figure 3. Dependence diagram of the air temperature in the High School №4 with temperature reduction to 12°C and average wall`s temperature during the working day on heating load adjustment.

Short-duration overheating of the building before the beginning of the next working day is necessary to prevent enclosing structure cooling, which afterward can lead to condensation forming on the inside surface of the walls. The average temperature of the outer enclosing structures change during the week without short-duration overheating is presented in Figure 4.

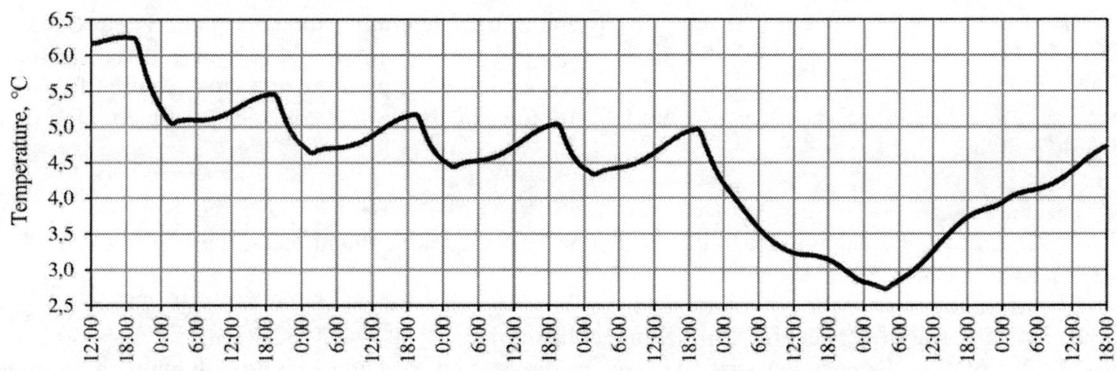

Figure 4. Change of the average outer wall temperature in the High School №4 without short-duration overheating.

This problem can be solved by the slight temperature increase from 20 to 21°C for the period from 5:00 to 9:00. The result of the adjustment is presented in Figure 5.

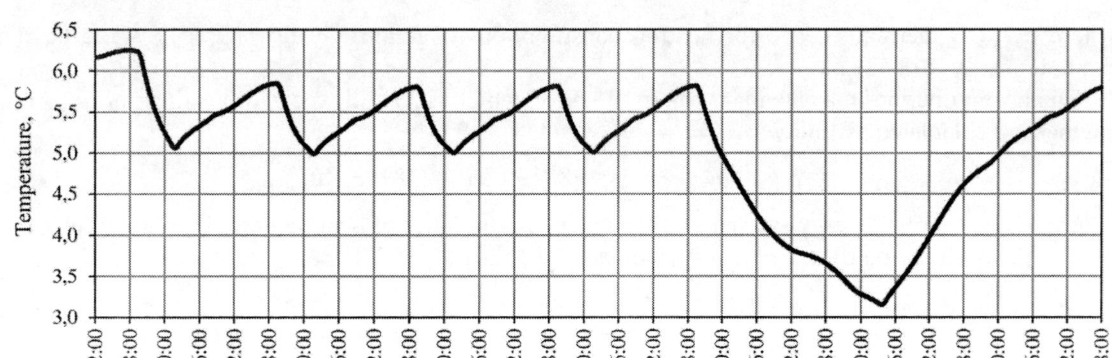

Figure 5. Change of the average outer wall temperature in the High School №4 with short-duration

overheating.

At the end of Friday, the temperature diagram shows a long-duration temperature reduction to 12°C. Since the budget institutions, for the most part, are not open during the mentioned time period, the load can be reduced for the whole weekend. The reduction starts on Friday from 20:00 with a constant temperature of 12°C by a gradual increase in the coefficient (Figure 6).

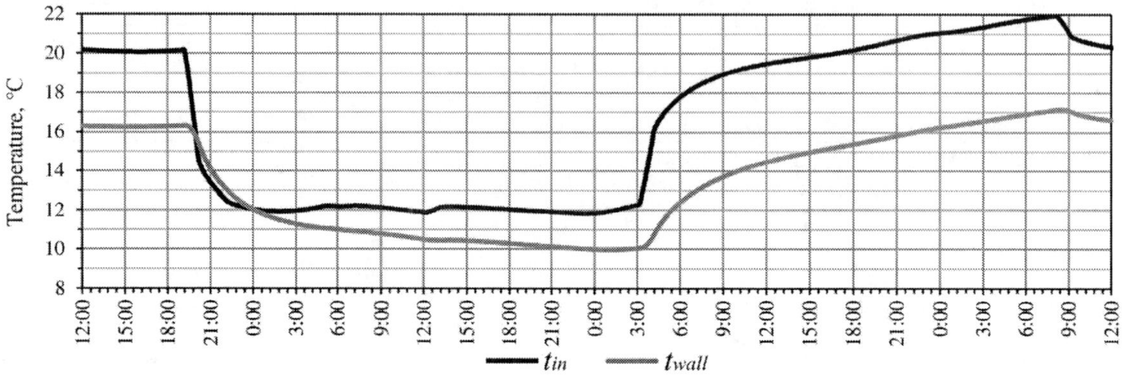

Figure 6. Dependence diagram of the air temperature in the High School №4 with temperature reduction to 12°C and average wall temperature during weekend day on heating load adjustment.

The calculation has shown that returning to the normal operating mode requires heating load increase starting on Sunday at 4:00 and the heating load coefficient should stay at 1.18 until Monday 9:00 of the next week. During this period the inside building temperature and temperature of the outer wall return to the initial values, which are comfortable for the employees and full functioning of the whole building.

4. Discussion

Let`s determine the economic feasibility of the heating load adjustment using the High School №4 in Belgorod city as an example.

The average price for the automated DHP in Belgorod city is around 400.000 Rub. The average annual cost of 1GCal of heat energy for the budget institutions in 2019 is 1882 Rub.

The calculations have shown that one week of the heating load adjustment with temperature reduction to 12°C gives savings in heat energy consumption on:

$$dQ_{cons} = 19620940 \text{ J/week} = 19.621 \text{ GJ/week} = 4.686 \text{ GCal/week}$$

Financially the weekly savings can be calculated as follows:

$$E_{week} = dQ_{cons} \cdot P_{heat} = 4.686 \cdot 1882 = 8819 \text{ Rub/week},$$

where dQ_{cons} – thermal energy for heating consumption reduction, GCal/week; P_{heat} – the tariff for thermal energy, Rub/GCal.

The heating period in a calendar year is 191 days, which is 27 full weeks of heating period. Annual savings are calculated as follows:

$$E_{year} = E_{week} \cdot n_{week} = 8819 \cdot 27 = 238113 \text{ Rub},$$

where E_{week} – thermal energy savings, Rub/week, n_{week} – number of heating period weeks, week.

The payback period of equipment is calculated as follows:

$$PP = C / E_{year} = 400000 / 238113 = 1.68 \text{ year},$$

where C – expenses for measures implementation, Rub; E_{week} – savings from measures implementation, Rub/year.

The obtained payback period indicates the satisfactory feasibility of the suggested system of adjustment implementation.

5. Conclusion

In the process of work the mathematical model was constructed with analysis purpose of the unsteady temperature condition of the building and savings determination from heating load reduction during non-working hours considering unsteady heat transfer processes through the enclosing structures, heating modes determination, which allow establishing required microclimate parameters during working hours. The model allows modeling of the unsteady processes of heating and cooling of the building during a change of weather conditions and heat energy consumption, that is supplied to the building, which allows determining the heat saving values when heating load is reduced during non-working hours.

The study has shown that the biggest savings can be achieved by air temperature reduction to 12°C. The modeling helped to establish that to provide stable microclimate parameters in the building after the end of heating load reduction period, it is necessary to carry out short-duration overheating of the building with inside temperature increase to 21°C for the period from 5:00 till 9:00, and after long-duration reduction (weekends or holidays) to 22°C for the period from 18:00 till 9:00.

The method of heating load reduction also has been suggested, with the following savings calculation for the High School № 4 in Belgorod city. When the room temperature during non-working hours is reduced to 12°C the calculated saving is 10%.

Acknowledgments

The article was prepared within development program of the Flagship Regional University on the basis of Belgorod State Technological University named after V G Shukhov.

References

[1] Bychikhin S A, Svirin M V and Trubaev P A 2019 Evaluation of energy-saving effect during the installing process of automated heating stations in educational institutions *IOP Conf. Ser.: Mater. Sci. Eng.* **552** 012019

[2] Zomorodian Z S, Tahsildoost M and Hafezi M 2016 Thermal comfort in educational buildings: A review article *Renewable and Sustainable Energy Reviews* **59** 895-906

[3] Amasyali K and El-Gohary N 2018 A review of data-driven building energy consumption prediction studies *Renewable and Sustainable Energy Reviews* **81** 1192-205

[4] Chen C, Neng Z, Kun Y and Fan Y 2018 Data and analytics for heating energy consumption of residential buildings: The case of a severe cold climate region of China *Energy and Buildings* **172** 104-15

[5] Tumanova K and Borodinecs A 2016 The Extensive Analysis of Circumstances Between Heat Consumption of Multi-apartment Buildings and Information Campaigns *Energy Procedia* **96** 945-52

[6] Sorrell S 2007 The economics of energy service contracts *Energy Policy* **35** 507-21

[7] Zhou X, Yan J, Zhu J and Cai P 2013 Survey of energy consumption and energy conservation measures for colleges and universities in Guangdong province *Energy and Buildings* **66** 1128

[8] Juodis E, Jaraminiene E and Dudkiewicz E 2009 Inherent variability of heat consumption in residential buildings *Energy and Buildings* **41** 1188-94

[9] Kuznetsov V A and Trubaev P A 2018 Resources and Problems of the Mathematical Simulating Thermo-Technological Processes *J. Phys.: Conf. Ser* **1066** 012024

[10] Peng X, Fan Y ang Ye Z 2019 Quantitative study on adjacent room heat transfer: Heating load and influencing factors *Sustainable Cities and Society* **51** 101720

[11] Lindberg K B, Bakker S J and Sartori I 2019 Modelling electric and heat load profiles of non-

residential buildings for use in long-term aggregate load forecasts *Utilities Policy* **58** 63-88
[12] Puning X, Yi J and Zhigang Z 2019 Multi-step ahead forecasting of heat load in district heating systems using machine learning algorithms *Energy* **188** 116085
[13] Nigitz T and Golles M 2019 A generally applicable, simple and adaptive forecasting method for the short-term heat load of consumers *Applied Energy* **241** 73-81
[14] Mehdi E, Khattabi E and Mharzi M 2017 Effect of locations and thicknesses for the different material constituting a building wall *Energy Procedia* **139** 328-33
[15] Anisimova E 2017 Building Thermal Regime Modeling *Procedia Engineering* **206** 795-9
[16] Seminenko A S, Sheremet E O, Gushchin S V, Elistratova J V and Kireev V M 2018 Hydraulic stability of heat networks for connection of new consumers *IOP Conf. Series: Materials Science and Eng.* **327(4)** 042105

IOP Conf. Series: Materials Science and Engineering **791** (2020) 012048 doi:10.1088/1757-899X/791/1/012048

The system of indicators of street lighting systems energy efficiency

K R Valiullin

Department of electrical and thermal power, Orenburg State University, Pobedy Av., 13, Orenburg, 460018, Russia

E-mail: ValiullinKamil91@gmail.com

Abstract. One of the important tasks that arise when choosing a project for the modernization of any electric power system is estimation of its energy efficiency. Criterion of energy efficiency is one of the most important for development plans of electrical systems, as well as other criterions. Evaluation of energy efficiency of street lighting systems is complex task, because it should take in account effectivity of street lighting that cannot be directly measured. This paper propose a system of indicators of street lighting systems, based on process of energy transformation and requirements of regulatory documents. Suggested indicators rate every elements of street lighting systems from electrical grid to luminaires, and can be used for estimation of energy efficiency of existing street lighting systems or selection of the most energy efficient option of electrical street lighting system modernization. Proposed indicators can be also used for determination of the most effective geometrical position of luminaire and can be used for energy efficiency optimization of street lighting systems on the design stage.

1. Introduction

At present, the government of the Russian Federation has marked a trend to increase the energy efficiency of electrical systems, to reduce their energy consumption with remained or increased useful effect. So, the criterion of energy efficiency is one of the most important for development plans of electrical systems, as well as reliability, economic and ecological criterions. Assessment of energy efficiency allows you to choose the most optimal option of energy system development. At the macro economy level good results show methods of energy saving projects evaluation, based on Data Environment Analysis [1-3]. For electrical systems and appliances, the evaluation of energy efficiency is a complex task, because this indicator takes into account not only the amount of energy consumed to create a useful effect, but also the useful effect itself, which in many cases cannot be directly quantified. There are researches dedicated to estimation of energy efficiency on the different levels from large plants [4] and pumping stations [5] to particular induction motors [6].

Street lighting systems (SLS) are the one of the most important parts of city energy system [7]. Nowadays in many cities SLS are being upgraded: usually modernization projects include replacement of old mercury or sodium vapor lamps to LED or implementation of modern lighting control systems. As a rule there are several modernization projects and choice of one of them is performed by economic criterions [8]. But cheapest project usually is not the best choice, so there is a necessity of development of additional criterions for choice of the best SLS modernization option. One of this criterions could be the energy efficiency criterion.

IOP Conf. Series: Materials Science and Engineering **791** (2020) 012048 doi:10.1088/1757-899X/791/1/012048

Energy efficiency of electrical system is usually estimated as ratio of useful effect to energy consumption of this system. Energy consumption of SLS can be easily calculated, but useful effect of SLS is safety and comfort of citizens, that cannot be directly measured. For this reason, development of energy efficiency indicators for street lighting systems is actual problem.

GOST 54993-2012 that declares energy efficiency indicators and classes can be applied only for household light sources with luminous flux no more than 6500 lm. Because the most of street lighting installations have luminous flux much more than 6500 lm, this standard cannot be used for estimation of their energy efficiency. In accordance with Code of rules SP 52.13330.2016 "Daylighting and artificial lighting", energy efficiency of street lighting (SLI) is evaluated by the indicator of the SLI relative power density, which is insufficient to estimate the energy efficiency of the entire SLS, because it consists of several components: power source, lighting networks, street lighting installations.

There are several studies of foreign scientists dedicated to problems of energy efficiency estimation of street lighting systems [9-15], that suggest indicators, based on energy consumption, economical values and street illuminance, but proposed indicators estimate energy efficiency of street lighting installations but not of entire SLS. And it is also worth noting that there is no researches made in accordance with Russian Federation street lighting standards.

2. Proposed methods of street lighting systems energy efficiency evaluation

For development a system of indicators that estimates the energy efficiency of each SLS component, consider the process of energy transfer between SLS elements, Figure 1.

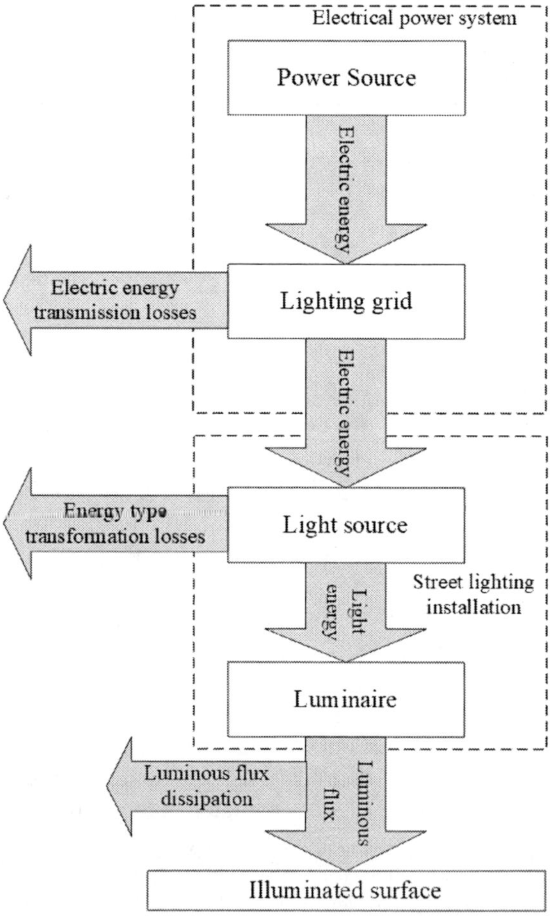

Figure 1. Energy transformation in street lighting system.

IOP Conf. Series: Materials Science and Engineering **791** (2020) 012048 doi:10.1088/1757-899X/791/1/012048

2.1. Evaluation of energy efficiency of electrical power systems
Transmission of electrical energy from the power source to the SLI in the power lines is accompanied by energy losses that depend on the flowing current, resistance of the conductors and length of the supply lines. Energy losses in the power supply system can be estimated through the value of its conductor's resistance and the voltage drop from the nominal value in the most distant point:

$$\Delta W = \frac{(U - U_0)\cdot Z}{Z^2} = \frac{\Delta U^2}{Z}, \tag{1}$$

where U – voltage at the most remote from energy source lighting installation, V;
U_0 – energy source voltage, V;
Z – full resistance of power transmission line, Ω.

Resistance of the power lines cannot be used as a reliable indicator of energy efficiency, because in such case short line will be always more energy efficient that is incorrect. So, it is suggested to evaluate the efficiency of electric power transmission from the power source to the SLI through the amount of the voltage drop at the most remote from energy source lighting installation:

$$\Delta U = \frac{U - U_{rated}}{U_{rated}} \cdot 100\%, \tag{2}$$

where U – voltage at the most distant lighting installation, V;
U_{rated} – rated voltage of the network, V.

The higher value will correspond to the greater value of the loss and lesser value of energy efficiency. It should be noted that the demands to voltage drop are declared in regulatory documents. According to them, the voltage loss in these networks should not exceed 5% for the most remote SLI.

2.2. Luminous efficacy indicator
Conversion of electric energy into light energy in SLI, can be characterized by well-known indicator of luminous efficacy:

$$\eta = \frac{\Phi}{P}, \tag{3}$$

where Φ – light flux emitted by the light source, lm;
P – power consumption of the SLI, W.

Luminous efficacy indicator characterizes the efficiency of conversion of electrical energy into light energy in a SLI. A higher value of the indicator corresponds to a higher energy efficiency. For fluorescent lamps the value of this indicator lies in the range of 78-115 lm/W, for LED - 90-150 lm/W [10].

2.3. Street lighting installation relative power density indicator
In the Code of rules SP 52.13330.2016 proposed the value of SLI relative power density which is an integral value linking electrical and lighting characteristics of SLI. The indicator of SLI relative power density is measured in $mW/m^2{\cdot}lx$ determined by the formula:

$$D_p = \frac{P}{\sum_{i=1}^{n} E_i \cdot S_i}, \tag{4}$$

where P – total rated power of SLI in the area, W;
E_i – illuminance of a single section of the area, W;
S_i – the area of a single section of the considerated area, m^2.
n – number of sections in the area.

IOP Conf. Series: Materials Science and Engineering **791** (2020) 012048 doi:10.1088/1757-899X/791/1/012048

For evaluation of this indicator, illuminated area should be divided into sections with area S_i with calculated mean value of illuminance E_i (Figure 2).

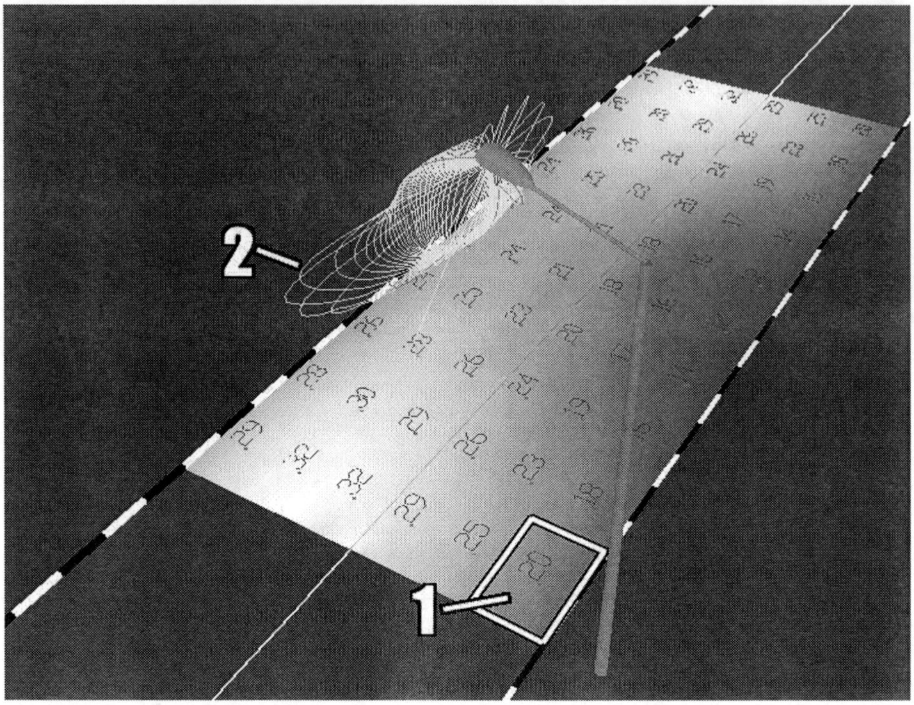

Figure 2. Illustration for relative power density indicator evaluation. 1 – single section and its mean illuminance; 2 – luminous flux distribution.

Bigger values of indicator correspond to lower energy efficiency. The maximum allowable values of the indicator are presented in Code of rules SP 52.13330.2016 and depend on the class of the illuminated street.

2.4. Indicator of efficiency of transmission of a light flux from street lighting installation to an illuminated surface (Luminous flux transmission coefficient)

The illuminance of the surface is measured in lux. One lux is illuminance (E) of the surface, with an area (S) of 1 square meter, when a light flux (Φ) of 1 lumen falls on it. From this definition, it is possible to find the light flux that falls on the considerated area:

$$\Phi = E \cdot S, \tag{5}$$

Part of the light flux generated by the SLI dissipates in the environment, so it is proposed to introduce an indicator that takes into account the efficiency of the transmission of the light flux from the SLI to the illuminated surface. The value of the indicator can be determined by the formula:

$$\eta_{tr} = \frac{\sum_{i=1}^{n} E_i \cdot S_i}{\Phi_0}, \tag{6}$$

where Φ_0 is the luminous flux produced by the SLI, lm.

Luminous flux transmission coefficient is a relative dimensionless value equal to the ratio of the light flux that reached the surface to the light flux created by the SLI. For an ideal SLI, η_{tr}=1, that means, that all luminous flux, created by light source reach illuminated surface.

IOP Conf. Series: Materials Science and Engineering **791** (2020) 012048 doi:10.1088/1757-899X/791/1/012048

2.5. The indicator of the efficiency of the distribution of the light flux over the illuminated area (coefficient of regulatory compliance of the light flux)

It is difficult to obtain a qualitative estimate of the distribution of the light flux over the area, so in the work suggested evaluation approach based on the quantitative indicators given by the illumination standards. For an ideal SLS, the illumination at all points of the illuminated area will be equal to given in normative documents. Up upwards deviations indicate a possible decrease in the luminous flux, and hence the power of the lighting installations, and the downwards deviations indicate the inefficiency of the SLS. For evaluation of the efficiency of the distribution of the luminous flux proposed the value of the coefficient of regulatory compliance of the light flux:

$$\eta_{norm} = \frac{\sum_{i=1}^{n} E_i \cdot S_i}{E_{norm} \cdot S}, \tag{7}$$

where E_{norm} – illumination, according to the standarts and regulatory documents, lx;
S – illuminated area, m^2.

Coefficient of regulatory compliance of the light flux is a strictly positive dimensionless value. The most energy efficient SLS corresponds to the value of the indicator closest to 1. Values less than 1 indicates mismatch with the standards of street lighting. The values of the coefficient greater than 1 correspond to the situation in which the considerate surface illumination is higher than the standards. That situation indicates an overestimated light flux and the power of the SLI, which leads to lower energy efficiency.

The proposed system of indicators is presented in Table 1.

Table 1. Proposed indicators of street lighting energy efficiency.

Energy efficiency indicator	Indicator meaning	Influencing SLS parameters
Lighting network voltage loss	Efficiency of electric power transmission from the power source to the SLI	- SLI power; - lighting network length; - lighting network wires cross section.
Luminous efficacy	Efficiency of conversion of electrical energy into light energy in a SLI	- energy conversion efficiency of a light source and its ballast/driver.
Luminous flux transmission coefficient	Efficiency of the transmission of the light flux from the SLI to the illuminated surface	- light intensity distribution curve; - parameters of luminaire; - geometric position of the SLI.
Coefficient of regulatory compliance of the light flux	Efficiency of the distribution of the luminous flux	- illumination standarts; - light intensity distribution curve; - parameters of luminaire; - geometric position of the SLI.
Street lighting installation relative power density	Efficiency of use of electric energy for creating illuminance on the surface	- SLI power; - light intensity distribution curve; - parameters of luminaire; - geometric position of the SLI.

3. Results

To approbate the proposed system of indicators, five street lighting projects with different types of luminaires have been calculated with use of the Light-in-Night Road software. In all projects was con-

IOP Conf. Series: Materials Science and Engineering **791** (2020) 012048 doi:10.1088/1757-899X/791/1/012048

sidered a section of a two-lane road, 150 meters long, 7.5 meters wide, belonging to class B2 (highways and streets of district importance in the city center), the average horizontal illumination should be at least 15 lx. The arrangement of the luminaires is one-side, the height of the support is 8 meters, the step between the luminaires is 20 meters. Parameters of luminaires are presented in Table 2, values of energy efficiency indicators for each of the projects are presented in Table 3.

Table 2. Parameters of luminaires for different projects.

Parameter	Number of project				
	1 (mercury lamp)	2 (sodium vapor lamp)	3 (LED)	4 (LED)	5 (LED)
Light flux, lm	13000	15000	11180	11000	9500
Rated power, W	250	150	98	120	100
Light intensity distribution curve	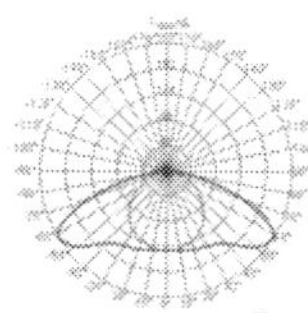	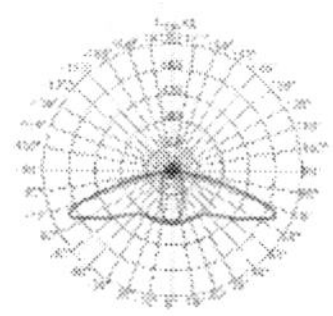	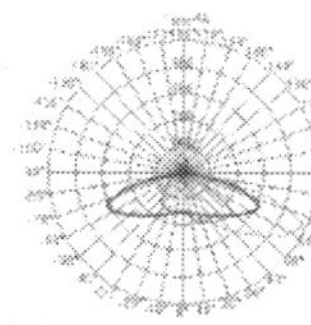	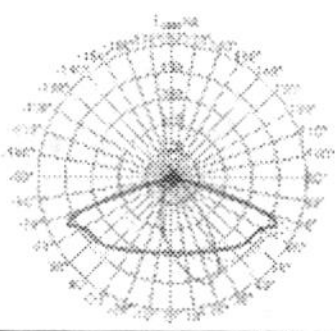	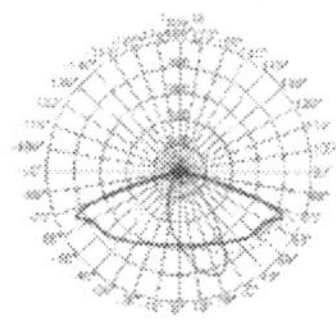

Table 3. Comparision of street lighting projects with different luminaires

Indicator	Number of project				
	1 (mercury lamp)	2 (sodium vapor lamp)	3 (LED)	4 (LED)	5 (LED)
Lighting network voltage loss, %	0.89	0.53	**0.35**	0.43	0.36
Luminous efficacy, lm/W	52	100	**114**	91.67	95
Luminous flux transmission coefficient	0.17	0.29	0.207	**0.327**	**0.327**
Coefficient of regulatory compliance of the light flux	0.982	1.934	**1.031**	1.596	1.379
Street lighting installation relative power density, mW/m^2·lx	113.1	34.46	42.25	33.4	**32.22**

Proposed system of indicators makes it possible to determine which of the possible variants of the luminaires will be the most energy efficient under given conditions.

4. Discussion

Also, the proposed energy efficiency parameters were calculated for a given luminaires (luminous flux 13,000 lm, wide light intensity curve) and various variants of the geometric position of the SLI. Variation of values of energy efficiency indicators for different versions of the geometric position of the luminaries is shown in Figures 3-5.

According to Figure 3, as the distance between the SLI changes, the value of the luminous flux transmission coefficient increases, because the total illuminated area increases. However, at a certain moment the maximum of the illuminated area is reached and the characteristic increase stops. From the point of view of this parameter, energy efficiency of different distances between luminaires exceeding 10 meters is equivalent.

A similar situation is observed when considering the change in the SLI relative power density - variants with a distance between SLI exceeding 10 meters are equivalent in their energy efficiency. In this regard, to determine the optimal distance, it is proposed to use the coefficient of regulatory compliance of the light flux - the highest energy efficiency is achieved with the value of this parameter closest to 1, which corresponds to the distance between the SLI at 19 meters.

According to Figure 4, with increase of luminaire installation height luminous flux transmission coefficient decreases, because of increasing luminous flux dissipation. Similarly decreases the coefficient of regulatory compliance of the light flux and relative power density.

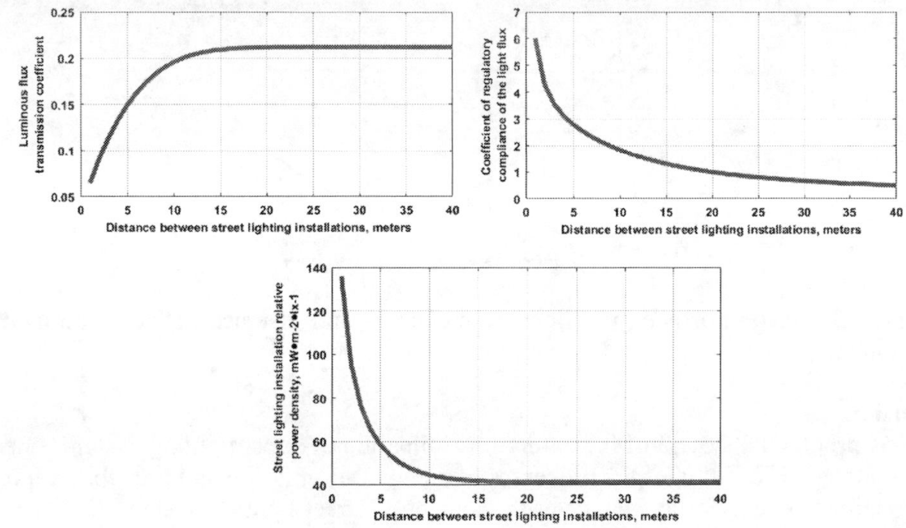

Figure 3. Relation of energy efficiency indicators and distance between street lighting installations.

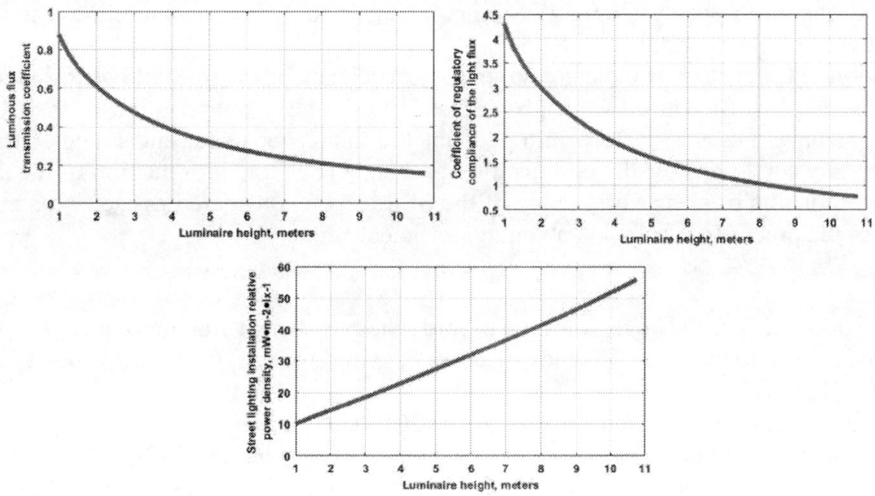

Figure 4. Relation of energy efficiency indicators and height of luminaire installation height.

Figure 5 shows relations of energy efficiency indicators and distance between luminaire and pavement. According to them can be found optimal value of 4,5 meters between luminaire and pavement.

Thus, the proposed system of indicators allows us to consider the task of increasing the energy efficiency of the SLS as a multidimensional optimization task, taking into account the boundary conditions imposed by technical and economic requirements.

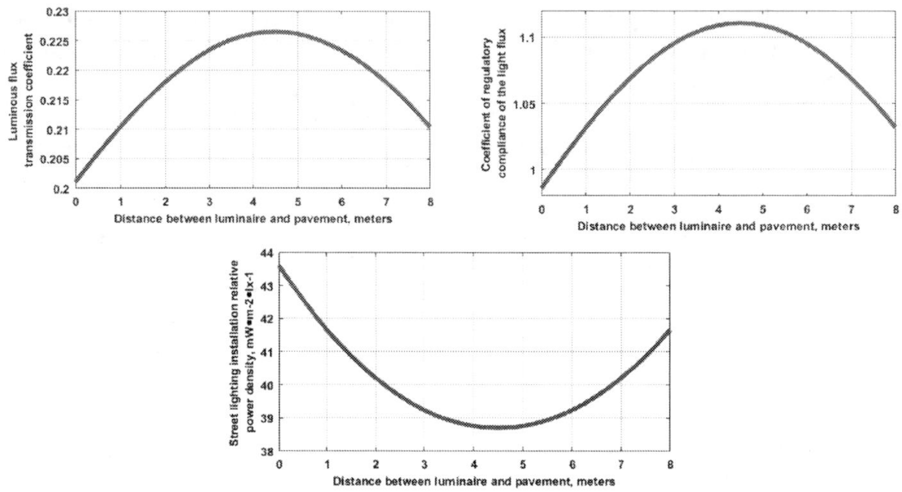

Figure 5. Relation of energy efficiency indicators and distance between luminaire and pavement.

5. Conclusion

In paper was proposed a system of indicators for estimation of street lighting systems energy efficiency. Based on the consideration of the processes of energy transmission and transformation, a system of indicators estimate energy efficiency of each element of street lighting system. Voltage losses in the lighting network rate energy efficiency of electrical power system; the light output of the light sources can be used as measure of light source energy efficiency; the transmission coefficient of the light flux, the coefficient of compliance of the light flux to the standard and the relative specific power installations of street illumination estimate energy efficiency of luminaire and its geometric position. Experiments proved, that proposed system of indicators is consistent and allows to choose the most energy efficient option for the development of SLS.

Introduction of the system of indicators of energy efficiency of the SLS allows to reduce the task of increasing its energy efficiency to the problem of multicriteria optimization with known solution methods. It becomes possible to take into account economic, technical and other constraints in the form of boundary conditions for the optimization problem. Thus, the introduction of the system of criteria for the estimation of energy efficiency of the SLS can be considered as the first step towards the automation of the process of designing an energy efficient SLS.

References

[1] Derevyanov M Yu, Pleshivtseva Yu E, Kordyukova L S and Rapoport A V 2019 Using DEA models to measure the efficiency of energy saving projects *IOP Conf. Ser.: Mater. Sci. Eng.* **552** 012020

[2] Shi, Guang-Ming, Jun Bi, and Jin-Nan Wang 2010 Chinese regional industrial energy efficiency evaluation based on a DEA model of fixing non-energy inputs *Energy Policy* **38(10)** 6172-9

[3] Vlontzos G Niavis S and Manos B 2014 A DEA approach for estimating the agricultural energy and environmental efficiency of EU countries *Renewable and Sustainable Energy Reviews* **40** 91-6

IOP Conf. Series: Materials Science and Engineering **791** (2020) 012048 doi:10.1088/1757-899X/791/1/012048

[4] Zhu, L and Chen, J 2019 A dynamic approach to energy efficiency estimation in the large-scale chemical plant *J. of Cleaner Production* **212** 1072-85

[5] Grigoriu M, Gheorghiu H, Crai A, Dinu D, and Visan D 2008 Energy efficiency evaluation method *Annals of DAAAM & Proceedings* 569-71

[6] Sertsöz M, Fidan M and Kurban M 2019 Three autoregressive-neural network hybrid models for energy efficiency estimation of induction motors *COMPEL - The Int. J. For Computation and Mathematics in Electrical and Electronic Eng.* **38** 431-51

[7] Semenova N G, Valiullin K R 2015 Intellektual'naya sistema energoeffektivnogo upravleniya ulichnym osveshcheniem na osnove nejrosetevyh tekhnologij [Intellectual system of energy efficient street lighting control based on neural network technologies] *Vestnik Orenburgskogo Gosudarstvennogo Universiteta* **4** 183-8 [In Russian].

[8] Vyazigin V L, Legostaev A V 2016 Ekonomicheskie kriterii vybora istochnikov sveta dlya ulicy [Economic criteria for choosing street light sources] *Sovremennyj Vzglyad na Budushchee Nauki* 66-9 [In Russian].

[9] Ylinen A M, Tähkämö L, Puolakka M and Halonen L 2011 Road lighting quality, energy efficiency, and mesopic design–led street lighting case study *Leukos* **8** 9-24

[10] Kostic M and Djokic L 2009 Recommendations for energy efficient and visually acceptable street lighting *Energy* **34** 1565-72

[11] Burgos-Payán M, Correa-Moreno F J, and Riquelme-Santos J M 2012 Improving the energy efficiency of street lighting. A case in the South of Spain *9th Int. Conf. on the European Energy Market IEEE* pp 1-8

[12] Lobão J A, Devezas T and Catalão J P S 2015 Energy efficiency of lighting installations: Software application and experimental validation *Energy Reports* **1** 110-5

[13] Pracki P 2011 A proposal to classify road lighting energy efficiency *Lighting Research & Technology* **43** 271-80

[14] Delaney D T, O'Hare G M and Ruzzelli A G 2009 Evaluation of energy-efficiency in lighting systems using sensor networks *Proceedings of the First ACM Workshop on Embedded Sensing Systems for Energy-Efficiency in Buildings* 61-6

[15] Covitti A. Delvecchio G. Neri F, Ripoli A and Labini M S 2005 Road lighting installation design to optimize energy use by genetic algorithms *EUROCON 2005-The Int. Conf. on" Computer as a Tool"* **2** 1541-4

IOP Conf. Series: Materials Science and Engineering **791** (2020) 012049 doi:10.1088/1757-899X/791/1/012049

The study of ecological impact of fly ash-based geopolymer binders on soil and aquatic biota

M I Vasilenko[1], M S Lebedev[2], E N Goncharova[1], N I Kozhukhova[3] and M I Kozhukhova[4]

[1]Department of Industrial Ecology, Belgorod State Technological University of V.G. Shukhov, Kostyukov St., 46, Belgorod, 308012, Russia
[2]Department of Construction and Municipal Economy, Belgorod State Technological University of V.G. Shukhov, Kostyukov St., 46, Belgorod, 308012, Russia
[3]Department of Materials Science and Materials Technology, Belgorod State Technological University of V.G. Shukhov, Kostyukov St., 46, Belgorod, 308012, Russia
[4]Department of Civil Engineering and Mechanics, Belgorod State Technological University of V.G. Shukhov, Kostyukov St., 46, Belgorod, 308012, Russia

E-mail: kozhuhovanata@yandex.ru

Abstract. Coal power plants represent the large source of environmental pollution in soils, ground and surface water by leaching out toxic compounds from open landfilled and impounded coal combustion residuals (CCR) such as fly ash. The use of fly ash as a main component in geopolymer binder (GB) systems is one of the effective methods of CCR beneficial utilization. The toxicity assessment as well as leachability of harden geopolymers were tested in aqueous media where oat seeds were used as affected test objects. Bio-resistance of the developed geopolymers was evaluated by the resistance to mold fungi growth. The developed geopolymers demonstrated a very low toxicity and leachability to the biota based on the low effect to germinating ability and growth retardation of the exposed oat seeds. Therefore, the studied fly ash products were proved to be environmentally safe when used in GB systems. The absence of fungi growth observed on the surface of geopolymer samples demonstrates high bio-resistance.

1. Introduction

Enormous amount of unused industrial secondary and by-products pose serious global environmental issues. In this regard, currently a big effort from research institutes is led towards reduction of ecological risk, recycling and beneficial use of the industrial by-products [1-9].

An efficient use of industrial by-products focuses to solve the following problems:

- reduction the amount of industrial wastes released to the environment;
- cleaning up large areas used for landfills and impoundments, and therefore elimination the sources of soil and water contamination;
- beneficial use of industrial hazardous by-products and waste in different applications. Coal combustion residuals considered to be environmentally hazardous contaminants due to elevated amounts of toxic, radioactive components and heavy metals [10, 11].

Based on the research data [12–18], production and application of fly ash-based geopolymers is a very efficient approach to solve ecological issues mentioned above. In addition, due to its structure, geopolymers are enabled to encapsulate certain hazardous elements and oxides and, so, preventing their leaching into the environment.

Early study [19] have investigated the influence of low-calcium fly ash products on life activities of flora and fauna. In some cases, [20] it was observed a negative effect to biota due to release of the ions such as Na^+ and Cl^- as well as SO_3 and CaO oxides.

Nevertheless, the use of fly ash as a major component in geopolymer binders assumes an encapsulation and, so, prevention the release of toxic oxides and chemical elements into the environments.

At the same time, the research focused on fungicidal properties of some fly ash products [2, 21–23] experimentally demonstrated no obvious negative effect on mold fungi and, so declaring a safe co-existence of the developed geopolymers with the ecosystem. However, fungi growth can cause a bio-corrosion and durability issues of the developed binders.

This study was focusing on the toxicity of fly ash-based geopolymers by studying the response of vegetation on the example of oat seeds and bio-resistance to mold fungi growth.

2. Materials and Methods

2.1. Materials

Four low-calcium fly ash products from different coal power plants were used in this study to produce geopolymers. Portland cement CEM I 42,5N was used as a reference binder material. Chemical composition of the cementitious components is shown in Table 1.

Aerated distilled water as a reference aqueous media and NaOH with the purity of 98 % as an alkaline activating agent were used for the preparation of geopolymer binder systems.

2.2. Methods

Evaluation of the growth retardation of higher plants was performed using a method of phytotoxicity determination according to Russian Standard GOST 33777-2016 2016 «Surface-active agents. Method for determination of phytotoxicity for higher plant seeds». Bio-resistance of the developed geopolymer binders was evaluated by evaluating resistance to mold fungi growth according to Russian Standard GOST 9.048-89 «Unified system of corrosion and aging protection. Technical items. Methods of laboratory tests for mold resistance».

Table 1. Chemical composition of fly ash products and Portland cement.

Material ID	SiO_2	Al_2O_3	TiO_2	Fe_2O_3	MnO	MgO	CaO	Na_2O	K_2O	P_2O_5	SO_3	TiO_2	ZrO_2	LOI
FA1	59.0	28.3	1.0	4.6	0.1	1.0	3.7	0,6	0.7	0.4	0	0	0	6.1
FA2	60.2	30.9	.	3.4	0.1	0.6	1.3	0,5	0.8	0.5	0	0	0	1.9
FA3	47.3	28.7	1.1	16.2	0	1.0	3.4	0,9	1.7	0.5	0	0	0	1.9
PC	19.1	5.2	0.3	3.6	0	1.3	65.4	0	0.6	0	3.5	0.3	0	0.2

3. Results

3.1. Phytotoxicity

For phytotoxicity evaluation of the fly ashes and the developed binder systems, water extracts of the fly ashes and crashed geopolymer samples at the age of 28 days were taken and then used to prepare water solutions in ratios of 1/10, 1/100 and 1/1000. The prepared solutions were placed to the separate Petri dish covered with filter paper. On top of each Petri dish with filter paper were placed a certain number of oat seeds and stored in a thermostat for 7 days.

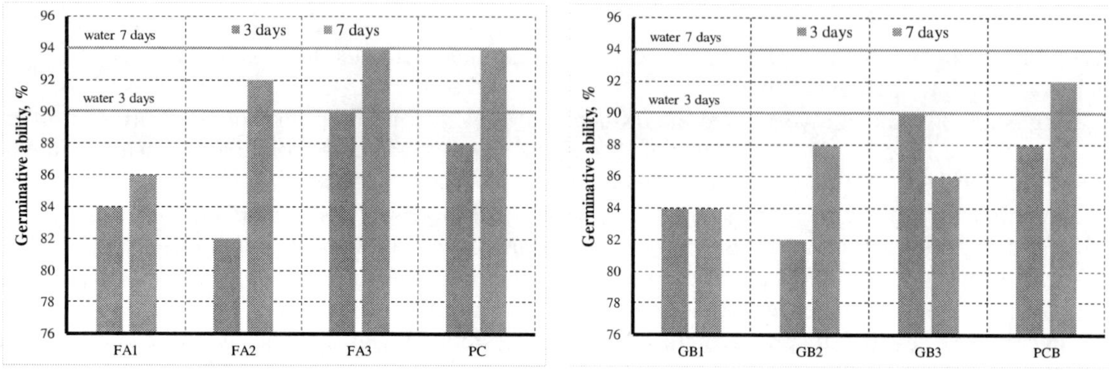

a b

Figure 1. Phytotoxicity level of a) cementitious components and b) fly ash-based GBs and PCB based on value of germinative ability.

Visual inspection of the oat roots growth was done after 3 and 7 days of exposure to the aqueous extract of fly ash and fly ash-based geopolymers. The results of the experiment showed that the oat roots exposed to 1/10 extract had about a 35 % decrease in growth rate as shown in Figures 1 and 2. For phytotoxicity evaluation of fly ash products and geopolymer binders average germinative ability (Figure 1), average coat root length (Figure 2) and retardation effect (Figure 3) were used as controlling parameters.

Retardation effect E_r in this study was calculated as a difference between root lengths for oats exposed in reference (L_o) and experimental (L_1) aqueous media using Equation (1):

$$E_r = 100\,(L_o - L_1)\,/\,L_o \qquad (1)$$

where E_r is a retardation effect, %;
L_1 is average oat root length in experimental aqueous medium, mm;
L_o is average oat root length in reference aqueous medium, mm.

According to the Guidelines MR 2.1.7.2297-07 «Hazard assessment for production and use of industrial wastes and by-products by the level of phytotoxicity» phytotoxic effect takes place if the E_r is higher than 20 %.

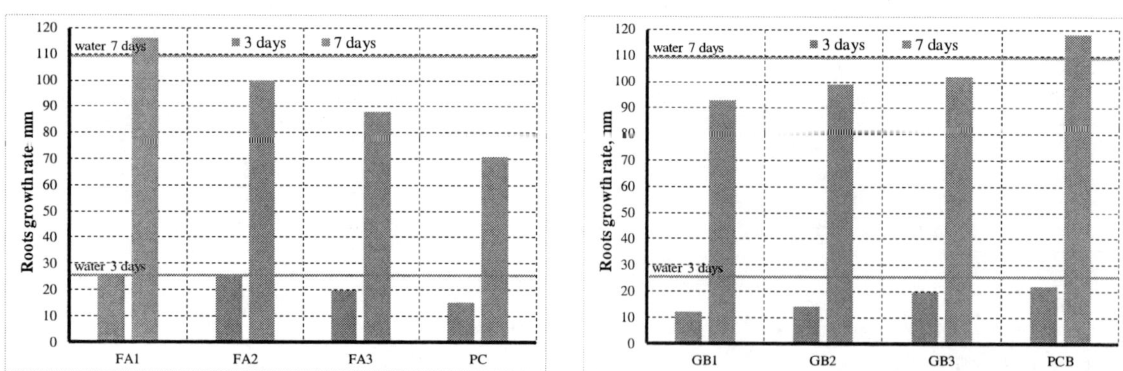

a b

Figure 2. Phytotoxicity level of a) cementitious components and b) fly ash-based GBs and PCB based on value of roots growth rate.

The oat seeds exposed to more dilute solutions have not displayed apparent acceleration or retardation in roots growth. Based on the phytotoxicity study the fly ash-based binders showed a more apparent retardation effect on oat seeds growth rate than pure fly ash products (Figure 3). The highest degree of retardation was observed for the FA3 and GB3 samples, which is 19.3 % and 6.4 %, respectively (Figure 3, a).

Regarding fly ash products, the lowest or zero retardation effect was observed for the FA1 and FA2 samples. At the same time, PC demonstrated the highest up to 34.9 % of growth retardation. However, the opposite effect was observed for the GB samples (Figure 3, b), although the growth retardation is for the geopolymers considerably lower in comparison with that for the PC sample. Thus, the highest E_r value 14.7 % was detected for GB1 and the lowest E_r of 6.4 % was observed for GB3. In this study, the retardation effect for PCB had a negative value of - 8.2 %.

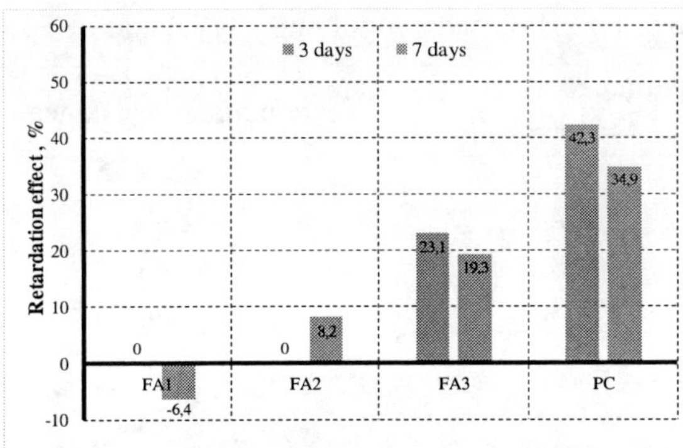

a

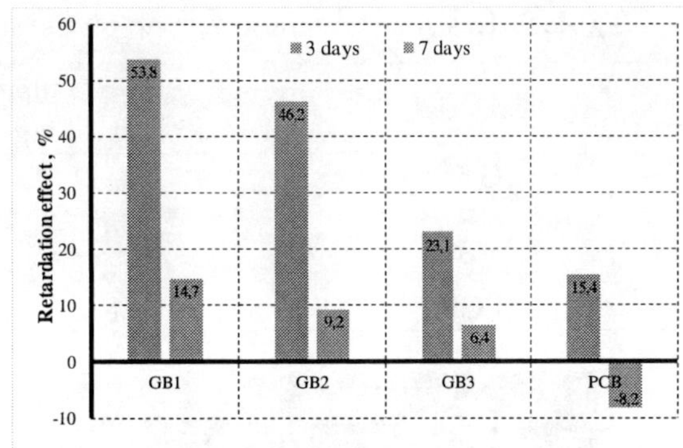

b

Figure 3. The retardation effect of fly ash products and PC and the binders on their basis on oat seeds growth rate.

3.2. Fungicide resistance

Bio-resistance of the studied cementitious components and developed binders was evaluated through resistance to mold growth according to the Russian Standard GOST 9.048-89 (Table 2). Using standard procedure, the fungicide resistance was evaluated based on the growth index that ranges from 0 to 5. In this experiment, the GB and Portland cement binder (PCB) samples were first saturated with water suspension that contained fungi spores *Aspergillus niger* and then placed in a chamber with 98 % humidity for 28 days.

Table 2. Resistance to mold growth on the surface of low-calcium fly ash products.

Material	Growth index		Fungicide resistance
	Before washing	After washing	
Cementitious components			
FA1	5	5	Low
FA2	5	4	Low
FA3	5	5	Low
PC	5	0	High (after washing)
Binder systems			
GB1	5	0	High (after washing)
GB2	0	0	High
GB3	0	1	High
PCB	0	0	High

The samples were then removed from the chamber and evaluated for fungicide resistance using optical microscope Axio Scope Al (Carl Zeiss Jena). A sample was considered highly resistant to fungi

IOP Conf. Series: Materials Science and Engineering **791** (2020) 012049 doi:10.1088/1757-899X/791/1/012049

growth if the growth index or the population of the mold on the sample surface less than 1. The results of the experiment are presented in Table 2.

The intensive mold growth on the surface of the studied fly ash products (Figure 1 a) indicates the metals and other compounds leaching from CCRs [20] do not have a harmful effect on mold population, and the mineral compounds are suitable for metabolisms of microorganisms.

For fly ash-based geopolymers and PCB samples there is no evidence of mold growth (Figure 4 b), which indicates a high fungicide resistance of the samples. In the Petri dish, there was observed a clearly defined fungicidal zone around binder sample with no sign of mold population. The presence of fungicidal zone (Table 3) in the study is related to high pH leaching from the sample that creates an unfavorable condition for mold growth.

Table 3. Resistance to mold growth on the surface of fly ash-based geopolymers (GB) and PCB.

Material	Fungicidal zone, %		Fungicide resistance
	Before washing	After washing	
GB1	–	7.0	High
GB3	13.5	8.0	High
GB2	11.3	8.0	High
PCB	9.8	8.0	High

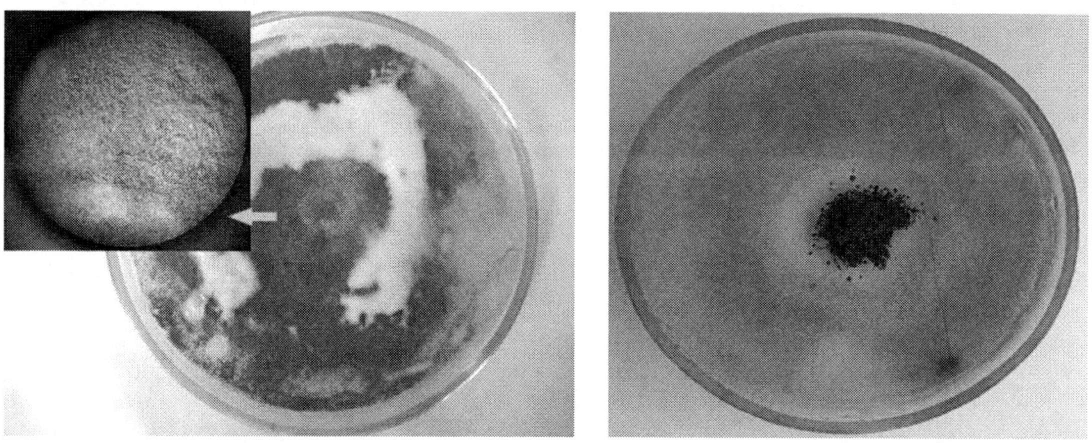

a b

Figure 4. Mold growth on the surface of (a) FA3 and (b) not washed GB3

In order to eliminate the influence of high pH and, at the same time, focus on modeling the exploitation conditions when construction material has a long-term exposure to rain and ice water, the cubic samples of GB and PC binders washed in water for 72 hours with continuous stirring. After, the samples were soaked in the suspension of mold spores. After 28 days, all investigated samples were visually inspected and did not show any sign of mold growth on the surface observed with a naked eye (Figure 5).

It should be noted that with the reduction of pH level of crashed samples by washing with water, the fungicide zone around the samples was gone, however, the surface was still free of mold fungi. The fungicide zone for all investigated samples is compiled in Table 2. Interestingly, the mold fungi use solid samples just as mechanical support to further growth upward (Figure 5 a) as a more favorable environment for the microorganisms.

a b c

Figure 5. Different views of mold growth on the surface of water washed binder samples a) GB2; b) GB3; c) PCB

Visual inspection of the GB samples after exposure to a mold fungi media showed no apparent signs of bio-corrosion or any visual mold growth on the surface of the GB and PCB (Figure 5 b, c). Only by means of optical microscopy, insignificant mold growth was observed on the surface of one of the samples (GB2, GB3). Although, this phenomenon is yet to be investigated.

4. Discussion

The developed geopolymers demonstrated very low toxicity and leachability to the biota based on the low effect on germinating ability and growth retardation of the exposed oat seeds.

To a greater degree, this is observed for the fly ash product FA3, where the retardation effect achieves 19.3 %. The lowest or zero retardation effect among the studied fly ash products was observed for the FA1 and FA2.

The developed fly ash-based GB systems have demonstrated lower retardation effect on oat seeds growth, so, the E_r value for those ranges within 6.4 – 14.7 %. The highest value 14.7 % of Er in this study was observed for geopolymer composition GB1 and the lowest 6.4 % *Er* is for GB3.

Fungicide resistance test showed an intensive mold growth on the surface of the studied fly ash products which indicates a low bio-resistance and, therefore low toxicity with respect to microorganisms.

In the case of the fly ashes-based GB systems, the apparent absence of the *Aspergillus niger* mold fungi growth on the surface and a clearly defined fungicidal zone around the binder samples indicated their high fungicide resistance in terms of direct contact with microorganisms due to high pH of the binder systems. At the same time, the intensive population of the mold fungi around the studied samples confirms a low leachability of toxic components comprising fly ash products.

5. Conclusion

Overall results of the leachability and bio-resistance study showed that the developed binders using coal combustion products such as low-calcium fly ash products are eco-friendly materials towards biota, and, at the same time, appear a high fungicide resistance even for long-term exploitation. This outcome arises a particular interest in large-scale applicability of CCPs as cementitious binders, because up to 80 % of produced fly ash products have demonstrated to be safe for flora and fauna, and, therefore, can be readily utilized in the production of geopolymer binders.

Acknowledgments

This work was realized under the support of the President Scholarship; in the framework of the Program of flagship university development on the base of the Belgorod State Technological University named after V G Shukhov, using equipment of High Technology Center at BSTU named after V G Shukhov.

References

[1] Shekhovtsova J, Kearsley E P and Kovtun M 2016 Using fly ash in cement: cement & concrete *IMIESA* **41(2)** 67-8

[2] Strigáč J, Števulová N, Mikušinec J and Sobolev K 2018 The fungistatic properties and potential application of by-product fly ash from fluidized bed combustion *Constr. and Build. Mater.* **159** 351-60

[3] Alex T C, Kalinkin A M, Nath S K, Gurevich B I, Kalinkina E V, Tyukavkina V V and Kumar S 2013 Utilization of zinc slag through geopolymerization: Influence of milling atmosphere *Int. J. of Mineral Processing* **123** 102-7

[4] Ageeva M S, Shapovalov S M, Botsman L N and Ishchenko A V 2016 K voprosu ispolzovania promyishlennyih othodov v proizvodstve vyazhushchih veshchestv [Regards the use of industrial wastes in the production of binders] *Vestnik BGTU im V. G. Shukhova* [*Bulletin of BSTU named after V G Shukhov*] **9** 58-62 [In Russian]

[5] Mahdikhani M and Reza Khanban M 2019 Utilization of industrial waste residue containing heavy metals as a substitute for fine aggregates *Constr. and Build. Mater.* **221** 772-86

[6] Bulusu S, Aydilek A H, Petzrick P and Guynn R 2005 Remediation of abandoned mines using coal combustion by-products *J. Geotech. Geoenviron. Eng. ASCE* **131** 958-69

[7] Sapronova Z A, Sverguzova S V, Fomina E V and Fokina E V 2018 Carbonate-containing precipitate of sugar production from sugar beet as filler for polymer concrete *IOP Conf. Series: Mater. Sci. and Eng.* **365** 032042.

[8] Starostina I, Simonov M, Volodchenko A, Starostina Yu, Fomin A and E Fokina 2018 The usage of iron-containing sludge wastes in ceramic bricks production *IOP Conf. Series: Mater. Sci. and Eng.* **365** 032066

[9] Kozhukhova N., Kadyshev N, Cherevatova A, Voitovich E and Lushin K 2018 Reasonability of Application of Slags from Metallurgy Industry in Road Construction *Advances in Intelligent Systems and Computing* **692** 776-82

[10] Lesovik R V, Leshchev S I, Ageeva M S, Karatsupa S V and Alfimova N I 2015 The Use Of Zeolite-Containing Rottenstone Powder For The Composite Binding Production *Int. J. of Appl. Eng. Research (IJAER)* **10(24)** 44889-95

[11] Pedro Bolívar J, García-Tenorio R, Luis Mas J and Vaca F 2002 Radioactive impact in sediments from an estuarine system affected by industrial wastes releases *Environment Int.* **27(8)** 639-45

[12] Dawczyński S, Górskim M and Krzywon R 2016 Geopolymer as an alternative ecological material for buildings *14th Int. Conf. on New Trends in Statics and Dynamics of Buildings* (Slovakia, Bratislava)

[13] Duxson P, Provis J, Lukey G C and Van Deventer J S J 2007 The Role of Inorganic Polymer Technology in the Development of 'Green Concrete *Cem. and Conc. Res.* **37(12)** 1590-7

[14] Wei-Hao Lee, Ta-Wui Cheng, Yung-Chin Ding, Kae-Long Lin and Chun-Ping Huang 2019 Geopol-ymer technology for the solidification of simulated ion exchange resins with radionuclides *J. of Environmental Management* **235** 19-27

[15] Strigáč J and Martauz P 2016 Fungistatic properties of granulated blast furnace slag and related slag-containing cements *Ceramics-Silikáty* **60** 19-26

[16] Soebpong Tuntachon, Khanita Kamwilaisak, Theerasak Somdee, Wiyada Mongkoltanaruk, Vanchai Sata, Kornkanok Boonserm, Ampol Wongsa and Prinya Chindaprasirt 2019 Resistance to algae and fungi formation of high calcium fly ash geopolymer paste containing TiO_2 *J. of Build. Eng.* **25** 100817

[17] Habert G, d'Espinose De Lacaillerie J B and Roussel N 2011 An environmental evaluation of geopolymer based concrete production: reviewing current research trends *J. of Clean. Prod.* **19** 1229-38

[18] Rozineide A Antunes Boca Santa, Cíntia Soares and Humberto Gracher Riella 2016 Geopolymers with a high percentage of bottom ash for solidification/immobilization of different toxic metals *J. of Hazard. Mater.* **318** 145-53

[19] Kozhukhova N I, Zhernovsky I V and Strokova V V 2015 Evaluation Of Geopolymer Binders Biopositivity Based On Low-Calcium Fly Ash *Int. J.of Appl. Eng. Res.* **10(15)** 35527-9

[20] Kozhukhova N I, Lebedev M S, Vasilenko M I and Goncharova E N 2019 Toxic effect of fly ash on biological environment *IOP Conf. Series: Earth and Environmental Science (EES)* **272** 022065

[21] Kozhukhova N I, Lebedev M S, Vasilenko, M I and Goncharova E N 2018 Fly Ash Impact from Thermal Power Stations on the Environment *J. of Phys: Conf. Series* **1066(1)** 012010

[22] Sanjeet K, Verma Kripal Singh, Anand K Gupta, Vimal Chandra Pandey, Pragya Trivedi Rajesh and KumarVerma D D Patra 2014 Aromatic grasses for phytomanagement of coal fly ash hazards *Ecol. Eng.* **73** 425-8

[23] Strigácˇ J and Martauz P 2016 Fungistatic properties of granulated blast furnace slag and related slag-containing cements *Ceramics-Silikáty* **60** 19-26

IOP Conf. Series: Materials Science and Engineering **791** (2020) 012050 doi:10.1088/1757-899X/791/1/012050

Problems of energy resources allocation at decision-making stages in architecture and urban planning activities

N A Vasilenko

Department of Architectural Constructions, Belgorod State Technological University of V.G. Shukhov, Kostyukov St., 46, Belgorod, 308012, Russia

E-mail: vasilenko.na@bstu.ru

Abstract. All the levels of architecture and urban planning activities (management, designing, execution of construction works) bring up various questions of the optimal distribution of energy resources. This paper shows that the energy resources allocation is connected with different classes of tasks in optimal decisions making and belongs to mathematical programming problems. Optimization tasks at the stages of scientific research, substantiation and version design should be performed on the basis of dynamic system models of the type «population ↔ environment». The article justifies the need for deductive approach and system methods and evaluation criteria of energy resources distribution at the stages of design and quality evaluation of architectural objects. It is demonstrated that in architectural theory and practice, for the purpose of quantitative measurement of the quality of architectural and construction activity products, the scientific research of the new area – architectural metrology – is needed. The energy interpretation of the general optimality criterion of integral architectural objects as systems, functioning in accordance with general system principles, is discussed.

1. Introduction

The global energy demand at the beginning of the 21st century has exceeded $5 \cdot 10^{17}$ kJ per year, or over 12 bln tons of oil equivalent [1 p 6]. Satisfying the growing energy demands of the modern society together with the rational distribution of energy and preserving the environment is one of the most crucial tasks nowadays. A wide range of various research and design works in the sphere of architecture and urban planning activities can be considered as transforming resources into the result.

The problems of energy resources distribution in urban development and national-economic activity are considered in a number of fundamental works [1–5]. Research in the sphere of urban and regional planning methodology is considered in various works [6–8]. Optimization of structure and functioning of architectural activity objects is dealt with in a number of works [9–15]. The aspects of parameter optimization of an architectural design object are presented in the works [16–18]. The issues of adapting the system methods of research and design of architectural objects are considered in the papers of the authors [8, 12, 17–19] etc. The issues of optimizing the parameters of technological processes and the sequence of technological operations in the sphere of design and construction are covered in a wide range of scientific and practical research works [20–25].

The analysis of the above-mentioned scientific publications and the experience of project design have allowed us making a conclusion that at present there is no unified approach to formulating tasks of energy resources allocation in architectural activity. This is a result of discrepancies in terms of research methods and

IOP Conf. Series: Materials Science and Engineering **791** (2020) 012050 doi:10.1088/1757-899X/791/1/012050

architecture and urban planning activities, including those at the levels of settlement systems and general plans of cities.

2. Materials and Methods

The purpose of this research is determining and substantiating the tasks of optimizing energy resources allocation in the process of formation of architectural and urban planning objects as integral systems.

Research objectives:

- determination and substantiation of design stages as tasks of optimizing energy resources allocation in the process of architectural and urban planning systems formation;
- substantiation of the system approach and system methods of studying evaluation criteria of energy resources allocation at the stages of design and quality assessment of architectural objects;
- developing a structure model of an integral architectural system with specifying the place and role of the energy component of its subsystems;
- developing a thermodynamical interpretation of development processes of integral architectural systems of the type «population ↔ environment» (demoecosystems).

The object of research is design stages as the levels of optimization tasks in architectural and urban planning systems formation. The subject of research is optimization tasks of energy resources distribution in architectural systems at different levels of their organization. The methodology of research is based on methodological provisions of the general system theory with regard to architectural objects of various hierarchy levels.

2.1. Optimization tasks of energy resources allocation at various design levels

Design is a process of creating a model of a future architectural object, as an eventual result of conceptual, functional and physical modeling. The combination of engineering, natural and social sciences in studying architectural systems has resulted in developing a new methodology of their research. There also appeared the synthetic, «hybrid» branches of learning, including complex systems modeling [5 p 186].

Figure 1 presents a scheme, which depicts stepwise decision making at designing architectural systems. The scheme shows a possible variant of errors in decision-making process. Field of the III level falls outside the limits of the field in the II level, which indicates miscalculation of limitations at the III level, or an error at the II level of decision making.

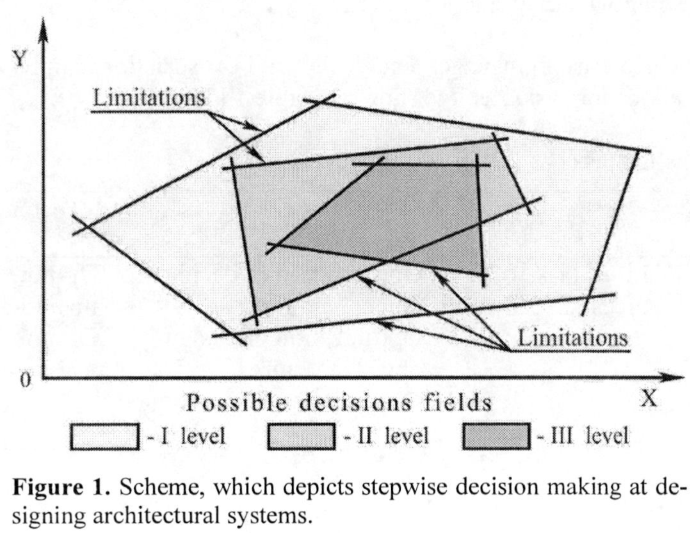

Figure 1. Scheme, which depicts stepwise decision making at designing architectural systems.

IOP Conf. Series: Materials Science and Engineering **791** (2020) 012050 doi:10.1088/1757-899X/791/1/012050

Figure 2 presents a graph of comparative assessment of possible errors range at various stages of decision making (energy resources allocation). The optimal capital investments and optimal energy resources allocation at each stage of decision making with achieving the optimal quality are taken as 100%. At the stages of scientific research and substantiation, as well as at version design, the possibility to alter the quality of an architectural object is within the range from 100 to 25%. After a project goes to the construction stage the alteration of its quality is possible only within the range of ±10% [12 p 10]. So, design is one of the critical stages, determining the achievement of proper architectural objects' quality and the rational energy resources distribution. At the level of urban development and regional planning the requirements for systemacity and scientific validation are basic and reasonably necessary.

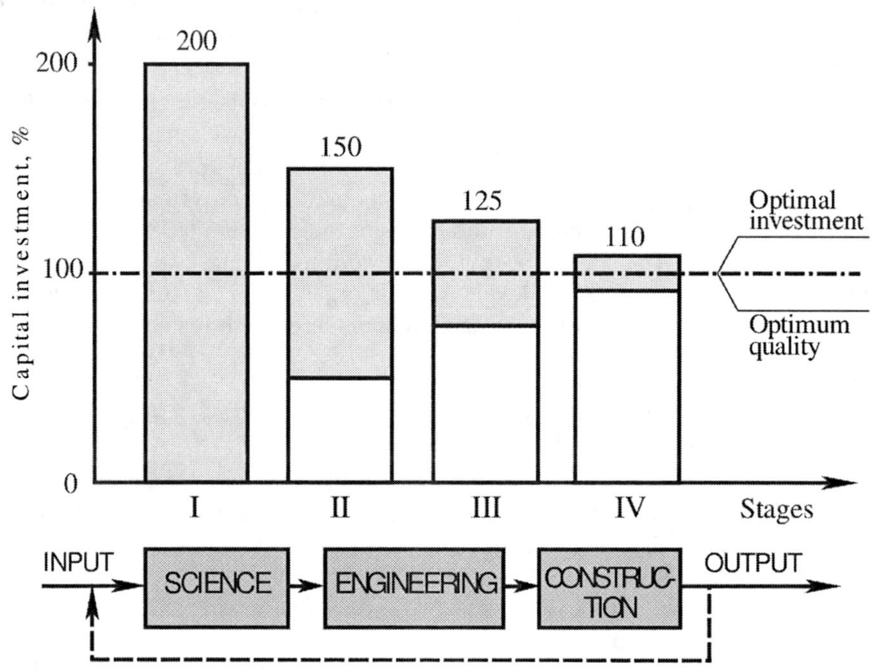

Figure 2. Graph of the comparative assessment of possible errors range at various stages of decision making: I – basic; II – functional; III – design-and-engineering; IV – engineering and manufacturing.

The main tasks, which arise in management, design and construction, can be included into the class of energy resources allocation tasks, as it is shown in table 1 [3 p 26].

Table 1. Optimization tasks at various design levels.

Application area	Management	Design	Technological processes development
Main tasks	Various tasks of resources allocation	Optimization of design object parameters Optimization of design object structure Optimization of functioning	Optimization of product manufacturing route Optimization of technological processes parameters

IOP Conf. Series: Materials Science and Engineering **791** (2020) 012050 doi:10.1088/1757-899X/791/1/012050

2.2. Methods and criteria of evaluating the quality of energy resources allocation at the stages of design and quality evaluation of architectural objects

To methods of solving the tasks of urban development and national-economic activities the special requirements of objectivity are specified. The present-day methods of architectural objects quality evaluation, and, consequently, of the resources allocation, are presented with four main approaches: complex approach, system approach, expert-based (intuitive) approach and technical-and-economic approach. In table 2 the methods and criteria of architectural objects quality evaluation are given. Considering these four existing approaches (methods), we can make a conclusion that the most efficient is the system approach, which meets the requirements of promptness, universality, application of computer technologies and possibilities of successful implementation into the Building Information Modeling (BIM).

Table 2. Methods and criteria of architectural objects quality evaluation.

Characteristic of methods	Methods			
	System	Complex	Expert evaluation	Technical and economic
Method of decision taking (conclusion)	Mostly deductive	Mostly inductive	Depends on an expert's experience	Mostly inductive
Number of considered parameters (factors)	Determining (main) factors	Maximum possible number of factors	Maximum possible number of factors	Maximum possible number of factors
Nature of the criterion structure	Nonadditivity	Additivity	Mostly additivity	Additivity
Approach to the degree of achieving goal	Single-criterial	Multi-criterial (for a complex in general)	Mostly multi-criterial (for a complex in general)	Multi-criterial (for a complex in general)
Approach to decision making	Hierarchical (from a higher, more general level, to a lower, more particular)	Mostly single-level	Depends on an expert's experience	Mostly single-level

In Figure 3 the stages of system models development from determining the system components to the experimental testing of models are presented.

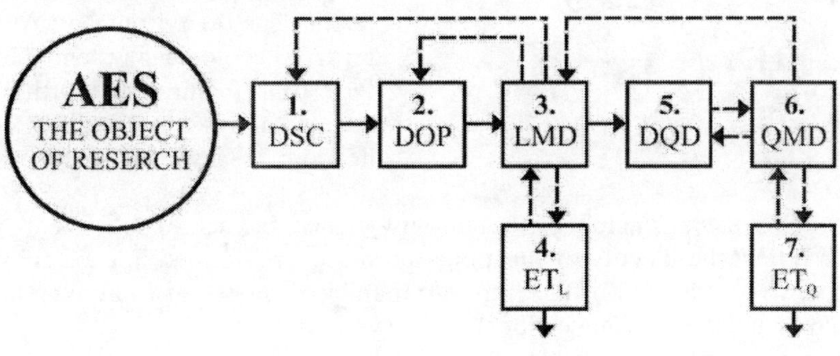

Figure 3. System models development stages: AES – demoecosystem: DSC – determining of system components; DOP – determining of operation principles of a system; LMD – logical models development; DQD – determining quantitative data and criteria; QMD – quantitative models development; ET_L, ET_Q – experimental testing of models (logical and quantitative, respectively).

IOP Conf. Series: Materials Science and Engineering **791** (2020) 012050 doi:10.1088/1757-899X/791/1/012050

At each stage of problem solving the taking of an optimal decision is required. The optimal energy resources allocation and the successful forecasting in the sphere of architecture and urban planning is possible only on the basis of developing and applying mathematical models, which describe objects from the perspective of system approach and general system principles. For this purpose, first of all, it is necessary to understand and implement in mathematical models the principles of architectural and urban-planning objects behavior (functioning and development).

2.3. The structural model of the energy component of architectural systems

An important aspect of an architectural activity object is its definition as an ecological system of the type «population ↔ environment», including three functional-spatial subsystems: artificial (architectural and urban-planning) environment, natural environment and the population. In figure 4 the principal structural model of an ecological system of the type «population ↔ environment» is presented. This scheme singles out three main subsystems:

- P – population (society), social resources of the population involved and not involved in the production;
- AE – artificial (architectural) environment, the second nature of the human society, including scientific and technical opportunities at the given period, objects of productive and nonmanufacturing activities of the population, solutions of communications between the architectural system's elements;
- NE – natural environment, including the elements of biosphere and geosphere, climatological, topological and geological peculiarities of the landscapes' territories, natural resources for the spheres of production and consumption.

Unlike the biological ecosystems, which adapt to the changing conditions, in architectural ecological system humans actively influence their natural surroundings, creating an artificial environment, a «second nature» [5, 8, 12, 17, 18]. In figure 5 a functional model of energy subsystems and their inter-influence and interdependency principle is shown. This scheme presents a slightly generalized picture of the interaction mechanism of the main subsystems at the energy level.

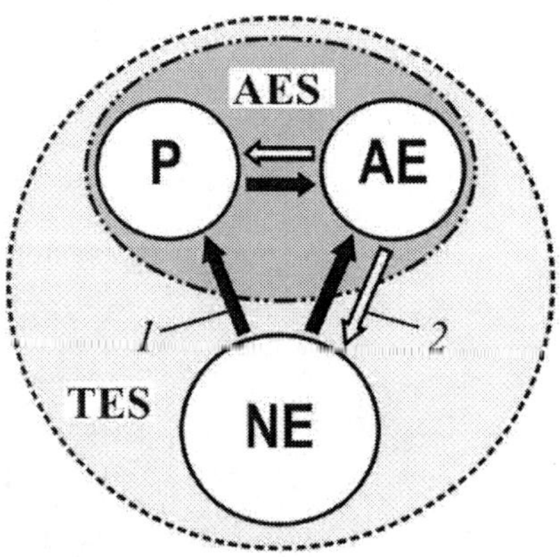

Figure 4. The principal structural model of an ecological system of the type «population ↔ environment»: P – population; NE – natural environment; AE – artificial (architectural) environment; AES – artificial ecological system; TES – terrestrial (natural and artificial) ecological system; 1 – direct connection; 2 – reverse connection

According to the system principle of structure invariance, in each of subsystems we can single out elements, which reflect the identity of functional processes, performed in each subsystem. These are industrial processes («production of the I kind»), municipal processes («production of the II kind»), recreation and communications (connections) of the system (figure 6).

Figure 7 presents a structural scheme, depicting the relations between types of activity in each of subsystems, however different they may be in qualitative or quantitative determination of elements or connections.

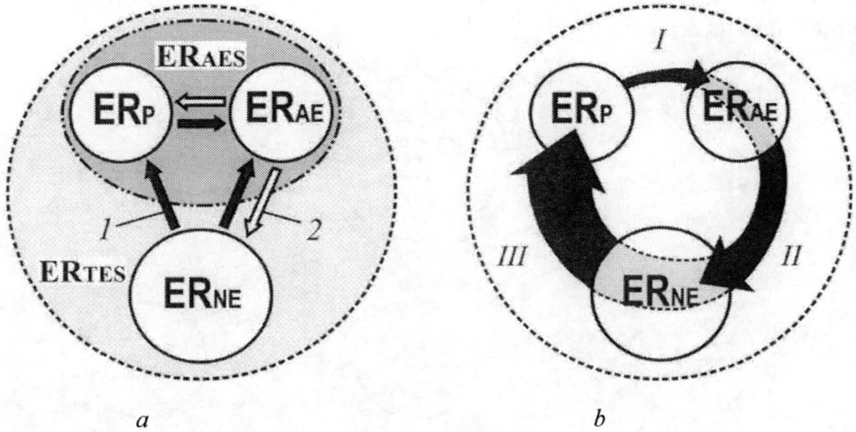

Figure 5. The structural model of an energy component of the «population ↔ environment» systems: *a* – functional model of energy subsystems; *b* – inter-influence and interdependency principle of elements (subsystems). ER_P – energy resources of the «population» subsystem; ER_{NE} – energy resources of the «natural environment» subsystem; ER_{AE} – energy resources of the «artificial (architectural) environment» subsystem; ER_{AES} – energy resources of the artificial ecological system; ER_{TES} – energy resources of the terrestrial (natural and artificial) ecological system; 1 – direct connection; 2 – reverse connection; *I* – error at the stage of research and/or design; *II* – an error, materializing and operating in time, adversely affects the natural environment resources; *III* – systematic adverse effect of energy resources of the damaged natural landscapes on the human health.

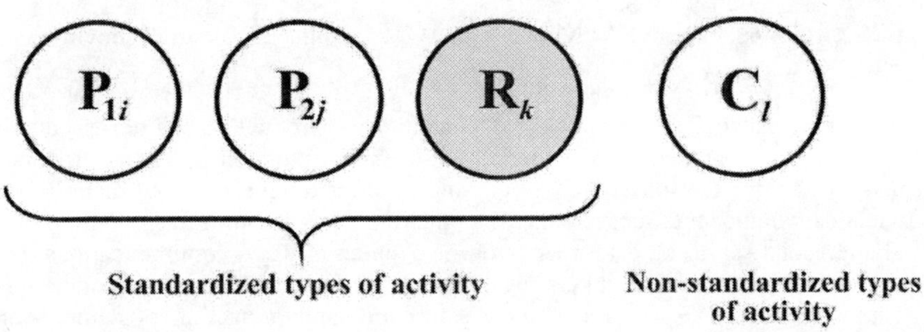

Figure 6. System-forming components of integral architectural systems: P_1 – production («production of the first kind»), i – 1, 2, … m; P_2 – municipal processes, social infrastructure («production of the second kind»), j – 1, 2, … n; R – recreation, k – 1, 2, … m; C – communications (connections) of an architectural system, l – 1, 2, … s.

3. Discussion

Exhaustibility of natural resources and the resulting influence of population on the natural environment's mechanisms make the problem of studying the regularities in reactions of human and natural environment to the impacts of artificial (urbanized) environment rather acute. Figure 5 illustrates the inter-influence and interdependency of the architectural environment subsystems at the energy level. Knowledge of these regularities is necessary in various aspects: ecological, social-economic, physiological and moral.

The energy and its potential for living (biological) and non-living (technical) subsystems should be evaluated in different values. It is hardly possible to evaluate in the same values the energy resources of the subsystems «Population», «Artificial environment», «Natural environment». Quality evaluation

IOP Conf. Series: Materials Science and Engineering **791** (2020) 012050 doi:10.1088/1757-899X/791/1/012050

of the taken design solutions and architectural objects is possible in case of creating a new area in architectural theory and practice – architectural metrology [12 p 11]. One of the aspects of this branch is a science about the quantitative measurement of the architectural and construction activity product quality – architectural qualimetry.

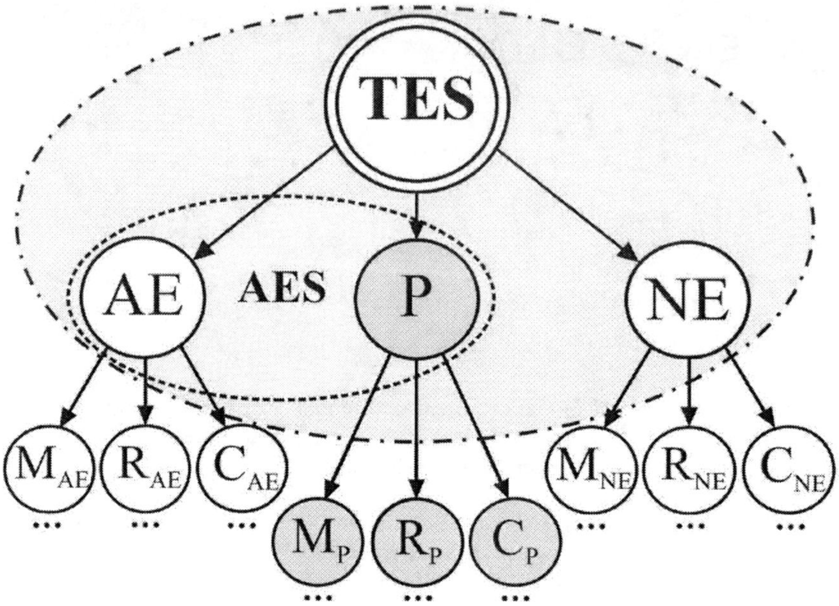

Figure 7. The structure of types of activity as elements of the terrestrial ecological system (TES): AES – artificial ecological system. Subsystems of artificial ecological system: AE – artificial (architectural) environment $AE = M_{AE} \cup R_{AE} \cup C_{AE}$; P – population, $P = M_P \cup R_P \cup C_P$; NE – natural environment, $NE = M_{NE} \cup R_{NE} \cup C_{NE}$. Subsystems of artificial environment: M_{AE} – production of material-and-technical component of the productive and non-productive spheres; R_{AE} – recreation of material-and-technical component of the productive and non-productive spheres; C_{AE} – communications (connections) of material-and-technical component. Subsystems of population: M_P – social infrastructure (cultural and social services); R_P – recreation of population; C_P – communications (connections) of population. Subsystems of natural environment: M_{NE} – reproduction of natural resources; R_{NE} – recreation (restoration) and protection of natural landscapes (geosphere and biosphere); C_{NE} – intra-subsystem connections (communications) of the natural environment.

Trying to evaluate the energy costs and the quality of an architectural object by means of technical-and-economic indices, natural indices or gross figures can't be a substantial basis for measuring the decision making efficiency on mathematical models of architectural systems. So, a number of specialists in the sphere of systems analysis [3, 8, 12] consider it promising to use energy measuring units in evaluating the efficiency of energy resources allocation and of the national economy. P.G. Kuznetsov [3] suggested the idea of the national economy planning and management on the basis of determining the so-called «energy balance of the society».

Integral architectural systems are the systems, which function in accordance with general system principles and the third law of thermodynamics (Prigozhin principle). So, the optimality criterion of architectural systems can be interpreted in terms of energy. A system, varying with time, changes its energy potential. The behavior and development of a system is its passing from one state to another. All the existing systems, which have a higher energy potential, than the surrounding systems, give them a certain amount of energy. So, all the things in existence can be classified by the types of energy output. According to the third law of thermodynamics, biological and social systems, accumulating information, increase the level of their orderliness and at the same time decrease the entropy level.

According to the «maximum work principle», suggested by E.S. Bauer, and the «free energy growth law», suggested by G.I. Lavrik [8, 12], development of a system is a result of increasing the external work of system as related to an energy unit, received from the environment. So, in the course of time the entropy of a system is reduced. The energy analysis of development processes in systems «population ↔ environment» (integral architectural systems – demoecosystems) was suggested by G.I. Lavrik [12 p 86], according to which F – energy received by the system from the environment and A – external work of the system are related as follows:

$$\frac{A}{F} \to \max \tag{1}$$

$$\Delta S = \frac{\partial Q}{\partial t} = \frac{\partial\left(\frac{A}{F}\right)}{\partial t} > 0, \tag{2}$$

where ΔS – entropy of system; ∂Q – system state function; ∂t – unit of time; F – energy received by the system from the environment, and A – external work of the system.

F – the energy, received by the system from outside, – can be presented as a sum:

$$F = F_1 + F_2, \tag{3}$$

where F_1 – part of energy, spent directly for performing the work «A» ; F_2 – part of energy, absorbed by the system. So,

$$A(F) = A(F_1 + F_2) = A(F_1),\ A(F_2) = 0. \tag{4}$$

Taking into account the «maximum work principle» and indentifying the values F_1 and A, we obtain the following relations:

$$\frac{A}{F} \to \max,\quad \frac{F_1}{F_1 + F_2} \to \max,\quad \frac{F_2}{F_1} \to \min. \tag{5}$$

Inserting into the relations (5) the determinations of subsystems' structure as elements of a terrestrial ecological system of population, we obtain:

$$\frac{A(P \cup AE \cup NE)}{F(P \cup AE \cup NE)} \to \max,\quad \frac{F_1(P \cup AE \cup NE)}{F_1(P \cup AE \cup NE) + F_2(P \cup AE \cup NE)} \to \max,$$
$$\frac{F_2(P \cup AE \cup NE)}{F_1(P \cup AE \cup NE)} \to \min, \tag{6}$$

where P – population; AE – artificial environment; NE – natural environment. Identifying these relations with types of functional processes, performed in subsystems, according to the general system principle of structure invariance, we obtain:

$$\frac{A(M \cup R \cup C)}{F(M \cup R \cup C)} \to \max,\quad \frac{F_1(M \cup R \cup C)}{F_1(M \cup R \cup C) + F_2(M \cup R \cup C)} \to \max,$$
$$\frac{F_2(M \cup R \cup C)}{F_1(M \cup R \cup C)} \to \min, \tag{7}$$

where M – manufacturing processes («production of the first kind» and «production of the second kind»); R – recreational processes; C – communicational processes (connections).

So, by the «free energy growth law» the ratio of the amount of energy, absorbed by the system at performing the work *A*, to the energy, spent directly for performing the work *A*, should decrease. This correlation reflects the «general» optimality criterion of architectural systems of the type «population ↔ environment». The «free energy growth law» and the law of capacity maintenance are invariant, uniting the natural, social (industrial) and mental processes into a unified system.

4. Conclusion

According to the findings of the carried-out research we can arrive at the following conclusions.

1. The tasks of resources allocation in architecture and urban planning activities have a wide application area – from the level of technological processes development and parameters optimization of a small-scale design object to the level of managing the development and functioning of large urban-planning objects. The energy resources allocation tasks, which are essentially optimization tasks at the stage of decision making in architecture and urban planning activities, in mathematical terms belong to mathematical programming problems.

2. Among the determining ones in optimization tasks of resources allocation are the stages of scientific research, substantiation and version design. These stages are aimed at achieving the optimal quality of architectural environment. At each stage the optimization should be performed on the basis of the existing limitations, taking into account its functional-dynamic character, i.e. on the basis of dynamic models of systems of the type «population ↔ environment».

3. The structural model of an energy component of the systems «population ↔ environment» includes the resources of the following basic subsystems: «Population», «Artificial environment», and «Natural environment». The energy resources allocation tasks at the stage of decision making in architecture and urban planning activities require using deductive approach and system methods and criteria of architectural object quality evaluation.

4. The system-forming components of integral architectural systems are the following types of functional processes: production («production of the first kind»), municipal processes («production of the second kind»), recreation and communications (connections) of an architectural system.

5. The problems of quantitative measurement of the architectural and construction activity product quality and the optimal energy resources allocation are the tasks of a new scientific area – architectural qualimetry, as one of the aspects of architectural metrology.

Acknowledgments

The work is realized in the framework of the Program of flagship university development on the base of the Belgorod State Technological University named after V.G. Shoukhov.

References

[1] Vasilenko A B and Tetelmin V V 2018 *Sovremennaya e`nergetika i e`nergetika budushhego: Texnologii proizvodstva. Netradicionny`e istochniki. E`kologicheskaya bezopasnost`* [*The modern and future power industry: Production technologies. Unconventional sources. Ecological safety*] (Moscow: URSS) [In Russian]

[2] Kuritsky B Ya 2018 *Optimizaciya vokrug nas. Kak matematika pomogaet prinimat` resheniya* [*Optimization around us: How mathematics helps in decision making*] (Moscow: URSS) [In Russian]

[3] Kuznetsov P G 1968 Vozmozhnosti e`nergeticheskogo analiza osnov organizacii obshhestvennogo proizvodstva [Opportunities of energy analysis of the social production organization basics] *E`ffektivnost` Nauchno-Texnicheskogo Tvorchestva* [Efficiency of scientific and technical creativity] (Moscow: Nauka) pp 38-59 [In Russian]

[4] Ettinger J v and Sittig J 1968 *More ... through quality* (Rotterdam: International Quality Centre)

[5] Reimers N F 1994 *E`kologiya (teorii, zakony`, pravila, principy` i gipotezy`)* [Ecology (theories, laws, rules, principles and hypotheses)] (Moscow: Rossiya Molodaya) [In Rissian]

[6] Gutnov A E 1986 System approach in studyin a city: basics and outlines of urban development

theory *System research. Methodological problems* (Moscow: Nauka) pp 211-32

[7] Elkin T, McLaren D and Hillman M 1991 *Reviving the city:* Towards Sustainable Urban Development (London: Friends of the Earth, Policy Studies Institute) p 280

[8] Lavrik G I 2007 *Metodologicheskie osnovy` rajonnoj planirovki. Vvedenie v demoe`kologiyu* [The Methodological Basis of Settlement Planning. Introduction to Demoecology] (Belgorod: BGTU im. V.G.Shukhova)

[9] Ilyichev V, Emelianov S, Kolchunov V and Bakaeva N 2018 Principles of Urban Areas Reconstruction Ensuring Safety and Comfortable Living Conditions *IOP Conf. Ser.: Mater. Sci. Eng.* **463** 032011

[10] Perkova M V, Baklazhenko E V and Vaytens A G 2018 Urban Conflicts of the Belgorod Regional Settlement System and its Elements *IOP Conf. Ser.: Mater. Sci. Eng.* **463** 032014

[11] Yastrebova N A and Stetsenko S E 2018 Principles of Ecological Settlement in the Zone of Influence of a Unique Natural Complex on the Example of the Volgograd Agglomeration *IOP Conf. Ser.: Mater. Sci. Eng.* **463** 032021

[12] Lavrik G I 2007 *Metody` ocenki kachestva zhilishha. Issledovanie, proektirovanie, e`kspertiza* [*Methods for Assessing the Quality of Housing. Research, Design, Expertise*] (Belgorod: BGTU im. V.G.Shukhova)

[13] Attia S, De Herde A, Gratia E and Hensen J L M 2013 *Achieving informed decision-making fornet zero energy buildings design using building performance simulation tools* (Springer International Publishing AG: Building Simulation) vol 6(1) pp 3–21

[14] Clarke J A 2001 *Energy Simulation in Building Design, 2nd edition* (Butterworth-Heinemann, Oxford)

[15] Hossain F 2018 *Sustainable Design and Build* (Butterworth-Heinemann Publisher)

[16] Bhooshan Sh 2017 Parametric design thinking: A case-study of practice-embedded architectural research *Design Studies* **52** 115–43

[17] Vasilenko N A 2018 The System Principle of Compactness as a Defining One in the Formation of Functional and Planning Structure and in the Forming of Power-Saving, Power-Active, Environmentally-Friendly and Hi-Tech Buildings *IOP Conf. Series: Materials Science and Engineering* **463** 032086

[18] Vasilenko N A 2019 Parametric optimization of an architectural object's form as a method to improve its energy efficiency *IOP Conf. Series: Materials Sci. and Eng.* **552** 012027

[19] Gharajedaghi J and Ackoff R L 1985 Toward Systemic Education of System Scientists *Systems Research* **2(1)** 21-7

[20] Bychikhin S A, Svirin M V and Trubaev P A 2019 Evaluation of energy-saving effect during the installing process of automated heating stations in educational institutions *IOP Conf. Ser.: Mater. Sci. Eng.* **552** 012019

[21] Garg V, Mathur J, Tetali S and Bhatia A 2017 *Building Energy Simulation: A Workbook Using Design Builder*™ (Boca Ration: CRC Press. Taylor and Francis Group)

[22] Harmati N, Jakšić Z and Vatin N 2015 Energy consumption modelling via heat balance method for energy performance of a building *Procedia Eng* **117(1)** 791-9

[23] Hooke J 2004 *Energy Management Information Systems: Achieving Improved Energy Efficiency: A Handbook for Managers, Engineers and Operational Staff* (Ottawa: Office of Energy Efficiency of Natural Resources Canada)

[24] Liu J and Wu X 2019 Controller and architecture co-design of wireless cyber-physical systems *J. of Systems Architecture* **94** 42-59

[25] Trubaev P A, Bulanin A V, Shirrime K J and Koshlich Yu A 2018 Particular Qualities of Execution of the Program Component for Energy Resources Management of the Belgorod Region *J. Phys.: Conf. Ser.* **1066** 012020

IOP Conf. Series: Materials Science and Engineering **791** (2020) 012051 doi:10.1088/1757-899X/791/1/012051

The usage of organic wastes of malt and lysine-sulphate production for reclamation of soils, contaminated with petroleum products

T A Vasilenko and O V Efremova

Department of Industrial Ecology, Belgorod State Technological University named after V.G. Shukhov, Kostyukov St., 46, Belgorod, 308012, Russia

E-mail: Land-VNA78@list.ru

Abstract. This paper presents the research findings of using two organic wastes – from feed additives production and malt production, which can optimize the process of soil remediation from petroleum products. The experiment was performed *in situ* with applying in the soil, contaminated with petroleum products (industrial oil), the excess sludge of malt production waste biological purification (1.8 or 3.6 kg/m^2 was applied). The organic matter content in the sludge was 31.79 %. With account of the high initial concentration of petroleum products (3100–8600 mg/kg) and their degradation during a short-term experiment (21 days), a conclusion was made, that the excess sludge of wastewater biological treatment had the potential for remediating soils from petroleum products. The percent reduction made up from 16.12 to 85.64% (at applying 0.44 kg/m^2 of petroleum products) and 20.37; 33.72 and 89.03 % (at applying 0.66 kg/m^2 of petroleum products). From the second waste product, which contained 83.0 % of organic matter, – corn-steep extract purification sludge, – a water extract was obtained. It has been determined that the cell colonies, which appeared from this extract (8-fold dilution) in the agarized growth medium in the presence of used petroleum products, are not inhibited, and petroleum products become a source of biogenic carbon for them.

1. Introduction

One of the most important areas of resource-saving activity is the efficient utilization of production waste, which makes possible the more rational usage of material resources, reduction of production waste amounts and the considerable decrease of raw materials consumption. Waste products can be used in building and chemical industry, and in land reclamation.

In the world practice the considerable experience of using organic wastes for soil reclamation has been accumulated. The conventional organic wastes, used for this purpose, include the decomposed agricultural waste: decomposed cattle and pig manure, chicken and duck manure. But with the development of biochemical and food industries, waste products may include valuable components, which find no practical application and are dumped at disposal sites. This is due to the fact that providing recommendations for the usage of waste requires time for its research. Organic waste can be used in polluted areas treatment technologies both *in situ* (the polluted site is treated on the spot) or *ex situ* (the polluted soil is treated in special equipment).

Traditionally soils are purified from petroleum products in suspended-media bioreactors. An example of such method is described in the publication, where the bioremediation of oil-contaminated soil (initial concentration up to 200 g/kg) was performed by means of such reactor. To the reactor with contaminated soil the bio-surfactants of actinobacteria of the *Rhodococcus* genus were added. Biotreatment of the contaminated soil resulted in the decrease of petroleum concentration by 88% after two months. Further on, to improve the efficiency of biological purification from petroleum products, the ploughing, woodchips, wetting and biofertilizers were used. The pollution level was reduced from 1.0 to 1.5 g/kg after 5–7 weeks. In conclusion, the phytoremediation by means of three species of perennial grasses was performed [1].

To remove petroleum hydrocarbons from soil it is necessary to know the amount of bacteria in it, to regulate the moisture content and pH of the soil, to apply fertilizers and carriers [2]. The authors of the survey paper indicate the three strategies of treating contaminated soils: biological, which implies the application of bacterial strains (bioaugmentation); adding special admixtures and nutrients (biostimulation); physical-chemical methods (solvent extraction; surfactant addition; thermal pretreatment and oxidative pretreatment) [3]. At oil pollutions of soil the concentrations of polycyclic aromatic hydrocarbon (PAH) and heavy metals in it are increased. Such soil can be treated by such technology as landfarming [4].

As a rule, the combined remediation of soils, contaminated with petroleum hydrocarbons, is performed with the usage of bioaugmentation and biostimulation. As a bacterial strain the yeast *Candida-tropicalis SK21* was used, which can operate in the pH range 3–9 in the liquid medium [5]. To purify soils from petroleum products a bacterial consortium, sampled from oil-contaminated soils, (consisting of *Pseudomonas* sp.) and rice husk can be recommended. The experiment has demonstrated that the purification efficiency from petroleum hydrocarbons amounted to 95 % [6]. The similar research was carried out with using rice husk, sawdust and woodchips for treating soils, contaminated with diesel oil. The results of the research have shown that the combination of the above-mentioned additives has influence on the bioremediation of contaminated soil [7]. For the efficient removal of total petroleum hydrocarbons (TPHs) the composted chicken manure was applied to the soil. The content of lubricating oil in soils was 5; 10 and 20 % by weight. After 42 days the content of TPHs in the soil decreased by more than 60 % in case of their original content 5 % by weight [8]. In the oil-contaminated soil (Kuwaiti) two biosurfactant-producing *Pseudomonas aeruginosa* strains (KISRC1 and KISRB1) were experimentally singled out. The strains differ in carbon source, nitrogen concentration, and pH of the medium [9]. It has been determined that washing soil with surface-active agents results in the loss of fine particles in it, which influences its mechanical properties [10]. It has been proved that a large amount of microbial communities, which can reduce the petroleum products concentration in soil, is observed at petroleum oil content 3–8%. During a ~12 months of natural accumulation, the quantity of oil-degrading microbes increased from 10^5 to 10^8 cells/g of soil [11].

There are examples of other physical-chemical methods. To remove petroleum products from soil, hydrogen peroxide can be also used (the Fenton reaction). In the process of hydrogen peroxide decomposition (H_2O_2) hydroxyl radicals are formed, which are strong oxidizers. The soils under study contained a large amount of sorbed petroleum products (10009; 4850 and 2850 mg/kg). The stepwise oxidation by the Fenton reaction increased the removal efficiency of sorbed petroleum products [12]. The similar research was carried out for the efficient purification of soils from petroleum hydrocarbons (TPHs). The optimal conditions were as follows: 4.27 wt% of catalyst content and 2.17 mol/L of hydrogen peroxide, 57% and 67% of TPH removal were achieved by magnetite (naturally mineral) and zero-valent iron (ZVI). Magnetite and ZVI can catalyze heterogeneous Fenton-like reaction [13]. To remove crude oil from the soil it was saturated with soluble free Fe (II) and humic acid (HA). This allowed catalyzing the hydrogen peroxide in the soil. It was found out that the formation of oxygen and the decomposition rate of H_2O_2 is reduced in case of presoaking with Fe (II) + HA as comparison with only pre-impregnated Fe(II) and modified Fenton (MF) [14]. The properties of subcritical water were used for purifying soil, contaminated with organic pollutants. In this state of the water there is no difference between its gaseous and liquid phase. It has been experimentally deduced that at purifying soils from crude oil by extraction method with the use of subcritical water at 300°C within 1 hour with the flow rate 1.5 ml /min, the removal efficiency is 95% [15]. The magnetic petroleum sorbent on the basis of waste kieselguhr sludge and the carbonaceous sorbents on the basis of vegetable oil extraction

IOP Conf. Series: Materials Science and Engineering **791** (2020) 012051 doi:10.1088/1757-899X/791/1/012051

industry waste sludge were also suggested for purifying wastewaters from petroleum products [16, 17].

2. Materials and Methods

2.1. Physical-chemical and microbiological properties of lysine-sulphate production waste

The work studies a waste product – sludge, formed at decanting corn-steep extract (lysine-sulphate production waste). In the original extract, which is a raw stuff for microbiological lysine-sulphate production, suspended solids content is 19 %; its purification provides their removal up to the final content 1% in the decanter. The properties of corn-steep extract sludge are given in Tab. 1. The energy-dispersive research of the incinerated waste is presented in Fig. 1.The chemical analysis of ash was performed with a scanning election microscope X-MaxOxford Instrument with the integrated energy-dispersive X-ray microanalyzer.

Table 1. Physical-chemical and microbiological properties of corn-steep extract purification sludge.

Parameter	Measuring unit	Content
pH	pH unit	4.11–4.30
Moisture	%	44.8
Ash	%	17.0
Organic matter	%	83.0
Total nitrogen	%	4.71
Total phosphorus(P_2O_5)	%	4.14
Total potassium(K_2O)	%	2.59
Lead	mg/kg	0.32
Cadmium	mg/kg	0.009
Copper	mg/kg	16.5
Zinc	mg/kg	269.7
Nickel	mg/kg	3.30
Chrome	mg/kg	0.66
Mercury	mg/kg	0.008
Arsenic	mg/kg	0.25
Salmonella	Cells per gram	not found
Total coliform bacteria	Cells per gram	below 10

2.1.1. Microbiological study of the extract of corn-steep extract purification sludge in the presence of used engine oil

For cultivating microorganisms' colonies the Sabouraud's growing medium was prepared (pH = 5.6). To inhibit the bacteria growth an antibiotic (chlorfenicol) was added to the medium in amount of 0.03 g for 600 ml of distilled water (dehydrated medium weight – 43.2 g). The Petri dishes were sterilized; 1 ml was placed into each dish, in the center of some of the dishes 1 ml of a used petroleum product (used engine oil) was placed. The experiment was performed in triplicate.

IOP Conf. Series: Materials Science and Engineering **791** (2020) 012051 doi:10.1088/1757-899X/791/1/012051

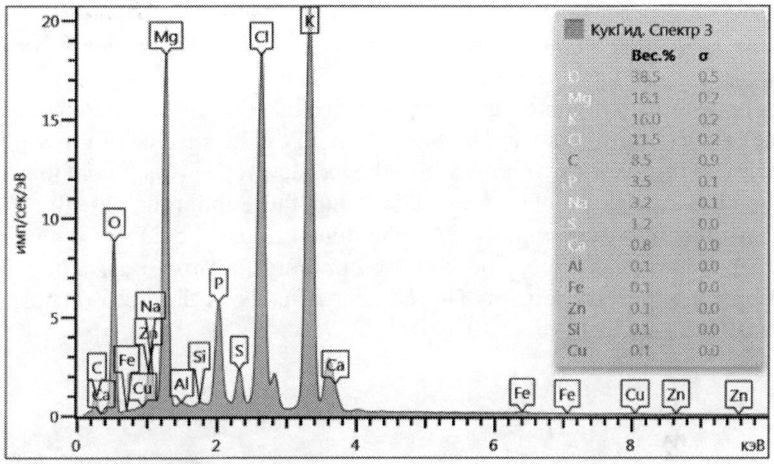

Fig. 1. Energy-dispersive spectrum of the incinerated waste.

2.1.2. Preparing of extract dilutions

To prepare dilutions the sterile distilled water was poured in amount of 9 ml into dry sterile test tubes. Then 1 ml of the extract under research was transferred with a sterile pipette into a test tube with 9 ml of sterile water – this is the first dilution (10−1). The following dilutions were prepared in the same way. At first the extract of corn-steep extract purification sludge was obtained, which had pH = 4.20, with the subsequent 8-fold dilution. Before the surface inoculation a melted agarized growth medium was poured into a number of sterile Petri dishes, from 15 to 20 ml into each. Microorganisms in the colonies, depending on their growth rate, were counted after 7 days of incubation.

2.2. Physical-chemical properties of malt production waste

In the experiment the waste product «Dewatered excess sludge of malt production wastewater biological purification» was used. Excess sludge is formed only in industrial process wastewater purification, performed at biological treatment plants. This waste finds no practical application and has the following composition (Tab. 2).

Table 2. Physical-chemical and microbiological properties of excess sludge.

Parameter	Measuring unit	Content
Moisture	%	66.24±4,64
Ash	%	< 5.00
Organic matter	%	31.79
Calcium	%	0.46±0,12
Magnesium	%	0.050±0,017
Phosphates	%	0.16±0,02
Chlorides	%	0.010±0,001
Nitrates	%	0.034±0,009
Nitrites	%	0.011±0,004
Iron	%	0.380±0,14
Aluminium	%	< 0.05
Sulphates	%	0.37±0,06
Potassium	%	0.28
Nickel	%	< 0.005
Lead	%	< 0.001

3. Results

3.1. Count of cells, cultivated on the growth medium with adding the extract of corn-steep extract purification sludge and the used engine oil

In Figure 2 the photos of cell colonies are presented. In figures 3 and 4 in Petri dishes we can see aggregations of bacterial colonies of a similar shape, formed on the surface of growth medium. The color of the colonies varies from yellow to brown. So, the extract of the waste contains a certain amount of natural oil-oxidizing bacteria, which, being placed into the Sabouraud's medium in the presence of petroleum products, begin to grow rapidly. Morphological traits of colonies in the samples are the following: round shape, convex surface, smooth edge contour, uniform consistency. The count was performed without opening the Petri dishes. The number of cells in the Petri dishes (№ 3 and 4 Fig. 2) amounts to $12.3 \cdot 10^9$ cells, in № 2 – $0.42 \cdot 10^7$ cells.

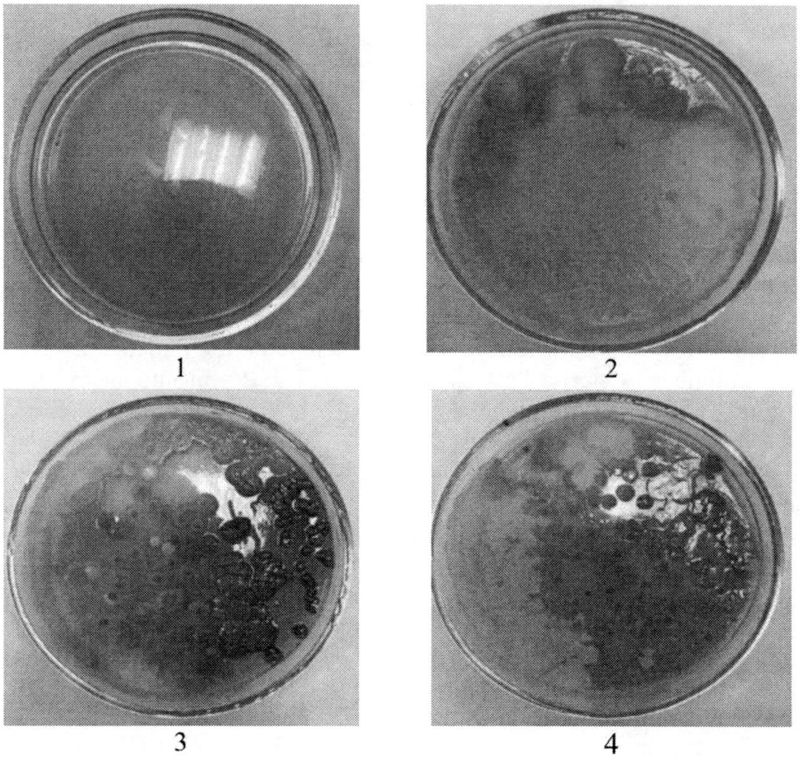

Figure 2. Photos of Petri dishes with microorganisms' colonies after 15 days of the experiment: *1* – reference sample, *2* – extract of the waste (8-fold dilution); *3, 4* – extract of the waste (8-fold dilution) + 1 ml of used petroleum products.

3.2. Experiment on purifying soil, contaminated by petroleum products, in the presence of wastewater excess sludge

In the university's land lot 10 testing sites 1 × 1 m were made. In Tab. 3 characterization of soil is presented. The type of soil – leached chernozem – was determined by its parameters. To the depth of 15 cm petroleum products (industrial oil) were applied in amount of 0.44 or 0.66 kg/m^2. Petroleum products were distributed evenly throughout the whole area and dug over with the waste – the excess sludge. The amount of sludge was also different: 1.8 or 3.6 kg/m^2 were applied, which corresponded to 18 and 36 t/ha. The reference sites were those, where there were petroleum products.

In Table 4 different variants of soil pollution with petroleum products are presented, with information about the amount of the applied dewatered excess sludge and the petroleum products content. After the application of petroleum products and excess sludge, the sites were watered with tap water. For the uniform distribution of the waste and the petroleum oil in the contaminated soil, and for aeration, the soil was hoed to the depth no less than 25–30 cm. The petroleum products content was studied in an accredited laboratory with fluorimetric method. This method is based on extracting from a sample with chloroform, purification of the extract by column chromatography method and the subsequent measuring of the fluorescence intensity of the extract with an analyzer.

Table 3. Characterization of soils.

Parameter	Measuring unit	Content
pH_{KCl}	pH unit	6.46
pH_{H2O}	pH unit	7.69
Dry residue	%	1.152
Mineralization	%	0.420
Moisture	%	11
Nitrates (NO_3^-)	mg/kg	40.3
Nitrites (NO_2^-)	mg/kg	1.64
Ammonium ions(NH_4^+)	mg/kg	8.4
Mobile nitrogen	mg/kg	16.49
Phosphates (P_2O_5)	mg/100 g	32.9
Humus	%	5.29
Exchangeable acidity	mg-eq/100 g	1.166
Cation exchange capacity	mg-eq/100 g	46.62

Table 4. Petroleum products concentration on the 1st and the 21st day of the experiment in the soil with the added excess sludge.

Biological treatment sludge content, kg/m^2	Petroleum products content, kg/m^2	Petroleum products concentration, mg/kg		Decrease of petroleum products concentration, %
		1st day of experiment	21st day of experiment	
0	0.44	5900	3800	35.59
0	0.66	7700	5400	29.87
1,8	0.44	5000	2905	41.90
1,8	0.44	3100	2600	16.12
1,8	0.66	8600	5700	33.72
1,8	0.66	6900	757	89.03
3,6	0.66	5400	4300	20.37
3,6	0.44	5900	847	85.64

3.3. Determination of biological activity of soils, contaminated with petroleum products, by the Mishustin method

In the experiment a natural linen canvas was used. In Fig. 3–6 the findings of the research for some soils, contaminated with petroleum products, are presented, with account of petroleum products removal efficiency (E_{re}, %). The variant, presented in Fig. 3, contained no excess sludge. During the experiment a linen canvas (10×10cm of size) was placed into the soil to the depth 15 cm, using a sterile spade. The linen is applied to a smooth section wall laterally. After the 20-day exposure the glass was excavated with the linen, and the soil particles were removed from it. Before the experiment and after finishing it the initial weight of the linen was determined and the decomposition ratio was calculated (Dr, %).

4. Discussion

The findings of the research, concerning the influence of used petroleum products on microorganisms' cells colonies after applying the extract of corn-steep extract purification sludge (8-fold dilution) on the agarized growth medium, indicate that the used petroleum products not only don't inhibit the colonies' growth, but provide a source of biogenic carbon. The average number of colonies, grown after inoculation of this cultivation in the presence of petroleum products, amounted to 123 (№ 3 and 4 Fig. 2); without petroleum products – 28 (№ 2 Figure 2).

As we can see in Tab. 4, at the initially high concentration of petroleum products in the soil, on the 21st day of the experiment their concentration decreases more intensively at the site, where 0.44 kg/m^2 of petroleum products were applied and the created initial concentrations were within the range from 3.0 to 6.0 thousand mg/kg. The reduction of petroleum products content is directly related to the weight of the applied sludge. The percent reduction made up from 16.12 to 85.64%, respectively (at adding 0.44 kg/m^2 of petroleum products). In case of applying 0.66 kg/m^2 of petroleum products the percent reduction was 20.37; 33.72 and 89.03 %. This indicates that there are microorganisms in the biological treatment sludge, which can decompose petroleum products. The biological activity of soils increases with the increase of the applied sludge amount.

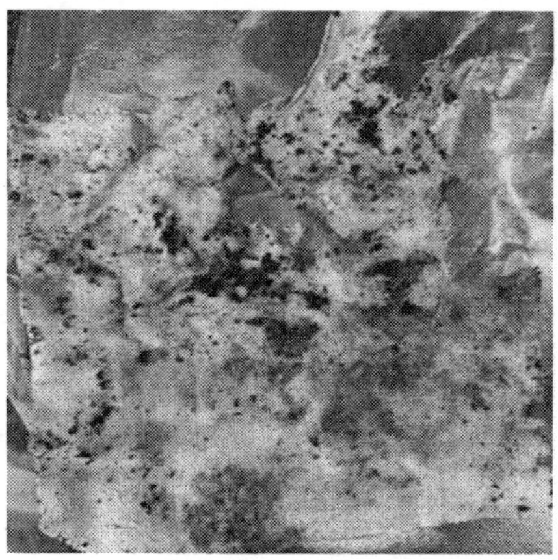

Figure 3. Dr = 38.4 %. E_{re} of petroleum products = 29.87 %

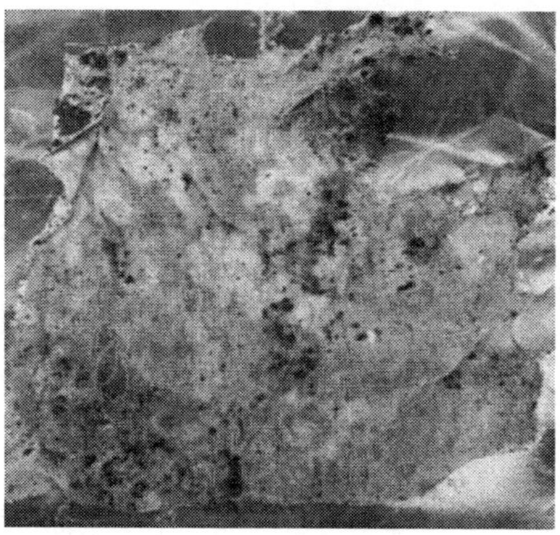

Figure 4. Dr= 34.2 %. E_{re} of petroleum products = 16.12 %

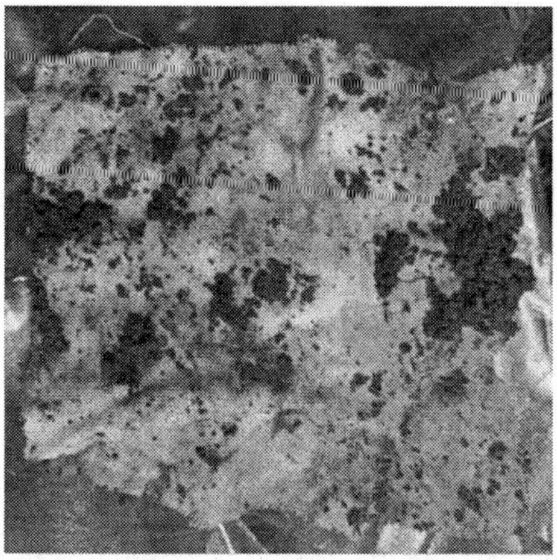

Figure 5. Dr = 20.37 %. E_{re} of petroleum products = 20.37 %

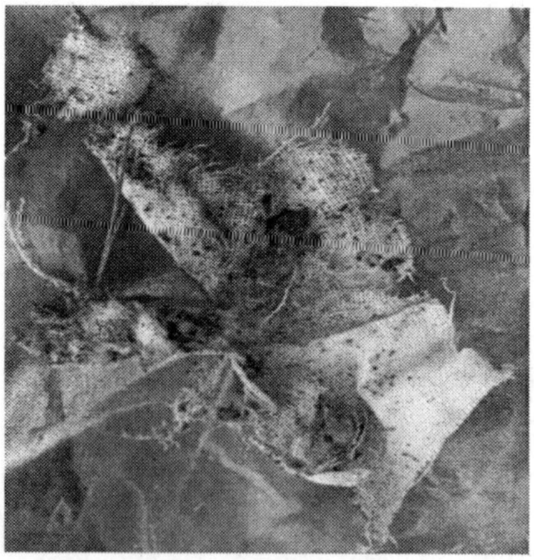

Figure 6. Dr = 92.60 %. E_{re} of petroleum products = 85.64 %

5. Conclusion

On the basis of the carried-out experiments we can make a conclusion that the application of waste products – excess biological treatment sludge and corn-steep extract purification sludge – is promising for the purpose of reducing petroleum products concentration in soil. Taking into account the high initial dose of petroleum products and their degradation during a short-term experiment, a conclusion is made, that the excess sludge of wastewater biological treatment has the potential for remediating soils, contaminated with petroleum products.

Acknowledgments

The work is realized in the framework of the Program of flagship university development on the base of the Belgorod State Technological University named after V.G. Shoukhov, using equipment of High Technology Center at BSTU named after V.G. Shoukhov.

References

[1] Kuyukina M, Ivshina I, Ritchkova M, Philp J, Cunningham C and Christofi N 2003 Bioremediation of Crude Oil-Contaminated Soil Using Slurry-Phase Biological Treatment and Land Farming Techniques *Soil and Sediment Cont.: An Int. J.* **12(1)** 85-99

[2] Michael H 1994 Guidelines for land-treating petroleum hydrocarbon-contaminated soils *J. of Soil Contamination* **3(3)** 299-318

[3] Tomei C and Daugulis A 2013 Ex Situ Bioremediation of Contaminated Soils: An Overview of Conventional and Innovative Technologies *Critical Reviews in Environmental Science and Technology* **43(20)** 2107-39

[4] Al-Awadhi N, Al-Daher R, EINawawy A and Salba M 1996 Bioremediation of oil-contaminated soil in Kuwait. I. landfarming to remediate oil-contaminated soil *J. of Soil Contamination* **5(3)** 243-60

[5] Fan MY, Xie RJ and Qin G 2014 Bioremediation of petroleum-contaminated soil by a combined system of biostimulation–bioaugmentation with yeast *Environmental Technology* **35(4)** 391-9

[6] Xu G L, Liu H, Li M J, Li Z M, Peng Z H, Zuo L M, He X, Liu W W and Cai L G 2016 In situ bioremediation of crude oil contaminated site: A case study in Jianghan oil field, China *Petroleum Science and Technology* **34(1)** 63-70

[7] Onwosi C O, Odibo F J, Enebechi C K, Nwankwegu A S, Ikele A I and Okeh O C 2017 Bioremediation of Diesel-contaminated Soil by Composting with Locally Generated Bulking Agents *Soil and Sediment Contamination: An Int. J.* **26(4)** 438-56

[8] Naowasarn S and Leungprasert S 2016 Bioremediation of Oil-contaminated Soil Using Chicken Manure *Soil and Sediment Contamination An Int. J.* **25(7)** 739-56

[9] Yateem A, Balba M, Al-Shayji Y and Al-Awadhi N 2002 Isolation and Characterization of Biosurfactant-Producing Bacteria from Oil-Contaminated Soil *Soil and Sediment Cont.: An Int. J.* **11(1)** 41-55

[10] Vázquez B, Bandala E, Reyes R and Torres L 2010 Variation of Mechanical and Hydraulic Properties of Oil-Contaminated Soil Due to a Surfactant-Enhanced Washing Process *Soil and Sediment Cont.: An Int. J.* **19(5)** 531-46

[11] Jia J, Zong S, Hu L, Shi S, Zhai X, Wang B, Li G and Zhang D 2017 The Dynamic Change of Microbial Communities in Crude Oil-Contaminated Soils from Oil Fields in China *Soil and Sediment Cont.: An Int. J.* **26(2)** 171-83

[12] Xu J, Yang C, Li L, Huang T and Huang R 2018 Enhancing the Removal of Sorbed Crude Oil from Soil Through Multiple Oxidation Steps in Stepwise Fenton Processes *Soil and Sediment Cont.: An Int. J.* **27(5)** 369-82

[13] Jamialahmadi N, Gitipour S, Jamialahmadi O and Baghdadi M 2015 Remediation of a Diesel-Contaminated Soil Using a Fenton-Like Advanced Oxidation Process: Optimization by Response Surface Methodology *Soil and Sediment Cont.: An Int. J.* **24(6)** 609-23

[14] Xu J, Li X and Huang T 2017Abatement of Sorbed Crude Oil by Heterogeneous Fenton Process Using A Contaminated Soil Pre-Impregnated with Dissolved Fe(II) and Humic Acid *Soil and Sediment Cont.: An Int. J.* **26(2)** 195-209

[15] Nazrul Islam M, Taki G, Jung Y-J, Jung S-K and Park J-H 2018 Remediation of Gulf War Oil Spill Contaminated Soil by a Subcritical Water Extraction Process: Oil Removal, Recovery, and Degradation, *Soil and Sediment Cont.: An Int. J.* **27(2)** 120-30

[16] Starostina I V, Nikitina A E and Porozhnyuk E V 2018 Magnetic petroleum sorbent based on waste kieselghur sludge from oil extraction industry for removing petroleum products from water surface *Solid State Phenomena* **284** 754-60

[17] Starostina I Nikitina A, Kosukhin M, Starostina Yu 2018 Efficient carbonaceous sorbent based on the waste sludge of oil extraction industry for purifying wastewater from petroleum products *Int. J. of Eng. and Technology* **7(2)** 266-9

Productivity of Enhanced Solar Still Under Various Environmental Conditions in Yekaterinburg city / Russia

N T Alwan[1, 2], S E Shcheklein[1] and O M Ali [3]

[1]Ural Federal University named after the first President of Russia B. N. Yeltsin 19 Mira St., Yekaterinburg 620002, Russia
[2]Kirkuk Technical College, Northern Technical University, 36001 Kirkuk, Iraq
[3]Renewable Energy Research Unit, Northern Technical University, 36001 Kirkuk, Iraq

E-mail: nassir.towfeek79@gmail.com

Abstract. Water is the most important source for human being and other living creatures. The naturul sources of water started to decline especially in the recent years. Researchers have focused on the productivity of drinking water as one of the main live requirements. Conversion of salt water into drinking water using solar distillation technology can be considered as the most viable methods in the dry climate regions. The productivity of solar stills effected by various parameters such as design, operational and environmental parameters. The current study aims to study the effect of different environmental conditions on improvement rate of productivity for suggested solar still. The tests were conducted at the Ural Federal University / Russia, during four months from June to September 2019 under different environmental conditions. The study results showed that there is a noticeable impact on the improvement rate in the productivity of the enhanced solar still, which depends mainly on solar radiation and ambient temperature. The rate of improvement fluctuated during the four months; the highest rete of enhancement obtained by 200% in September.

1. Introduction

Solar distillation is the best way to provide clean water to rural and remote areas one of the most important issues for human survival. Therefore, the demand for fresh water is increasing day by day due to the population growth and continuous urbanization. At present there are several different techniques for converting polluted and salty water into fresh water such as solar distillation technology [1]. The productivity of solar stills effected by different parameters includes; design, operational and environmental parameters [2]. The control of these environmental parameters is not easy, therefore, the researchers tried to increase the productivity of solar stills by the controlling operational and design parameters [3]. The design parameters are important factors that have strong impact on the productivity of solar stills which can be easily controlled and developed. The design parameters include thermal storage materials, solar tracking, reflectors, water depth, evaporation area, transparent cover angle, thermal insulation and other parameters that have an impact on productivity [4,5]. The thermal storage materials have important role in solar stills, increase the thermal capacity led to increases the productivity of solar still [6]. Solar tracking system was used to follow the sun, in order to improvement the productivity of solar still [7]. By used internal and external refractors to focus intensity of solar radiation on the basin solar still, the solar still efficiency enhanced by 48% [8]. The water depth parameter has also important effect on the productivity of solar still [9]. Increased evaporation area increases solar still productivity, meaning that the productivity of the solar distillers is proportional to the surface area of evaporation [10]. The solar still cover is usually made of glass or plastic and must be has a highly transmittance in order to absorb salt water in basin solar still radiation that passes through the transparent cover. The solar still cover should also have a good thermal insulation to reduce thermal losses to the surrounding. Several attempts have been made to increase the cover transmittance and therefore increase the solar still productivity [11]. The thermal insulation of the solar still is a necessary parameter to be considered, the proper insulation reduces wasted thermal losses from the sides and bottom of the solar still basin [12]. The current study aims to show the effect of environmental parameters on the productivity of an enhanced solar still with a hollow rotating drum.

2. Methodology

Description of the enhanced and conventional solar stills

Single-basin single-slope solar stills are the cheapest and most widespread of all currently available solar distillation systems, its main problem is low productivity. Therefore, in the current research a simple and inexpensive solar distillation system has been designed. Figures 1 shows a photograph view of the experimental station, which includes two types of solar stills. The first type is enhanced single slope solar still with a rotating hollow drum (SSS with RHD) and the second type is a conventional single slope solar still without a rotating hollow drum (SSS without RHD). The two stills have the same dimensions 100*50*60*24.8 cm, with a wooden frame of dimensions 100*50*10 cm with thickness of 0.18cm, with a fixed cover of 0.3cm plexiglass at single slope of 35 degree, from back side. The plexglass was mounted on MDF wooden panel with dimensions 60 * 100 * 0.18 cm and fixed on the wooden frame by using an aluminum channel of 0.1*0.1*0.1 cm to collect condensed water droplets through the plexiglass. The condensed water passed through the aluminum channel to a gradient plastic cylinder installed at the bottom of solar still. The basin was made of galvanized iron with a dimension of 94cm*46cm*10cm, coated with black colour to absorb a large amount of solar enrgy. Silicone glue has been used to instilled all parts of solar still and to prevent air leakage. Within solar still mounted a hollow drum (32 cm diameter, 90 cm length), which was made of Galvanized iron sheet (100 cm × 100 cm × 0.06 cm). Hollow drum mounted on the low carbon steel shaft (diameter 0.8 cm x length 95 cm) by using 0.8 cm ball bearings on the two ends. The hollow cylinder is rotated by using a small 12 v, DC motor, as which used to move glass of car with low power 0.1 A, by a small rubber belt, the rotational motion is transferred from the motor to the hollow cylinder shown in figure 2. The DC motor was fed by a small 30-watt photovoltaic panel during the sunshine days, and connected with a storage battery to power the DC motor in times of shadow and sunsetas.

A cylinder tank 50 cm*100 cm used to feed the solar stills with a brin water, a ball valve has been used to controll on the flow water from a tank and at entrance the solar still, also instilled a ball valve at the bottom of the water basin to clean the basin liner, the water level inside the basin maintained at 5 cm by using a mechanical float. The other technical specifications of SSS with RHD and SSS without RHD are listed in Table 1. 0.3 mm thermocouples K-type used to measure the temperatures at basin water , basin liner, air and vapour temperature within solar still and inside and outside plexiglass cover by using SD data logger 4 channel K thermocouple device (model 88598), all thermocouples have been calibrated between (100-0) °C. In order to measure out surface hollow drum temperature has been used digital laser infrared thermometer temperature model (TEGMART TE-TEM-LS-PRB). also was used humidity and temperature meter (GM1362) to measure the ambient temperature and relative humidity and It has been placed at a height of 1 m from the surface level in the shade to protect it from the sun radiation. a solar power meter device (TENMARS TM-207) in W/m^2 units used to measure the solar radiation intensity throughout at the sunrise times . Anemometer device (ut363) has been used to measure the wind speed, which was surrounded the stills as shown in table 2.

Test conditions and All tests were carried out during 4 months period from June to Desember at Ural Federal University of Yekaterinburg city/Russia, and at different environmental conditions. Current study consisted of two stages, first stage is the primer testing on the system, in order to

IOP Conf. Series: Materials Science and Engineering **791** (2020) 012052 doi:10.1088/1757-899X/791/1/012052

evaluate performance and the possibility of the amendment in the designs in order to reach the best case. Second stage is an actual test of the system, which included the calculation of the average daily productivity of solar stills every 30 minutes. The results has been compared at different environmental conditions, all tests were started at 8:00 am continued until 20:00 pm (12 hours), the total number of tests was 40.

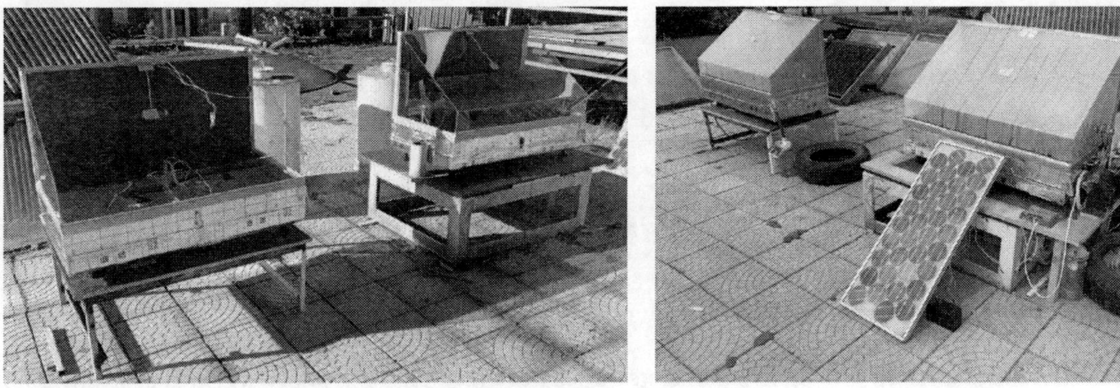

Figure 1. Photographic view of the experimental station (SSS with RHD and SSS without RHD).

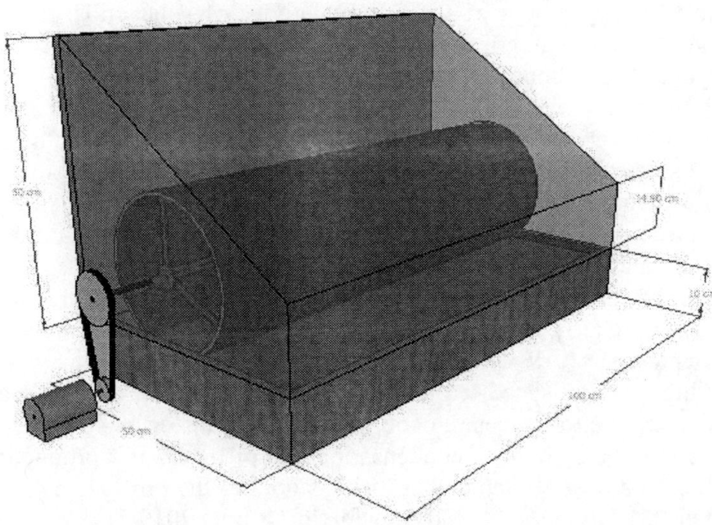

Figure 2. Schematic diagram of SSS with RHD.

Table1. Technical specifications of SSS with RHD and SSS without RHD.

Specifications	Dimensions
Basin area, m^2	0.5
Plexiglass area, m^2	0.825
Hight of basin, m	0.6
Height solar still from ground, m	0.5

Table2. The accuracy of various experimental devices.

No	Device
1	SD data logger 4 channel K thermocouple device (model 88598) ± 0.3% rdg+1°C
2	Digital laser infrared thermometer temperature model (TEGMART TE-TEM-LS-PRB). ± 1.5%
3	Humidity and temperature meter (GM1362). Humidity 3%
4	Solar power meter device (TENMARS TM-207). ±10 $W\ /\ m^2$
5	Anemometer device (ut363). ±5% rdg+0.5°C

3. Results and discussion

The current study used an analytical program (Origin pro 2018) to illustrate effects. Experimental weather data has been studied in detail, which included solar radiation density, relative humidity, ambient temperature and wind speed. Figure 3 shows the experimental results taken for a two days at 12 Jun 2019 and 15 September 2019, which represent the a perfect days of improvement in the rate of productivity between SSS with RHD and SSSwithout RHD at different environmental conditions. It was obvious that the environment parameters varies depended on the weather which is always variable and unstable in the Yekaterinburg city/Russia, rarely the weather was clear and without clouds during the four months of testing.

Figure 3, shows that the amount of water productivity of solar stills greatly influenced by the environmental parameters, the figure illustrated the comparison in the amount of productivity between the SSS with RHD and SSS without RHD during a two perfect days (12.06-15.09.2019). It can be seen that the amount of fresh water productivity during 12 June 2019 is greater than day 15 September 2019. The amount of productivity of fresh water during the 12 hours from 8 am to 20 pm, in 12 June 2019 from enhanced solar still found to be 2350 ml and from conventional solar still 900 ml, while in 15 September 2019 amount of fresh water produced from enhanced solar still is 1500 ml and from conventional solar still was 500 ml. The reason for these differences is that the productivity of solar stills is affected significantly by the intensity of solar radiation, increased the rate solar radiation intensity which reached to the solar still surfaces leads to an increased fresh water production and similar to those found in the literature [13]. Figure 4 shows that the intensity of solar radiation in 12 June 2019 is greater than in 15 September 2019. The other reason was wind speed, which effects on the Plexiglass cover temperature, higher wind speed led to increase the rate of heat transferred from the Plexiglass cover to the atmosphere, due to increase in coefficient heat transfer by convection. This in turn leads to increased evaporation, the amount of condensation and of freshwater production [14]. On the other hand, the improvement ratio in 15 September 2019 is greater than in 12 July 2019, the improvement ratio in 15 September 2019 found to be 200%, while in 12 June 2019 161%, that explanation is due to that the ambient air temperature after 12 noon in 15 September 2019 is greater than in 12 June 2019. The ambient air temperature represents one of the most influential factors on the performance and productivity of solar stills, high ambient air temperature leads to enhanced output [15].

The ambient temperature effects on the parameters within the solar stills, such as glass temperature, vapor temperature within the solar still, basin liner temperature and salt water temperature. Figure 5 illustrates that the temperature of the four parameters in 15 September 2019 almost afternoon was greater than in 12 June 2019. This leads to higher rate of solar radiation intensity which increase the surface temperature of the rotating cylinder and increase the evaporation rate and Plexiglass temperature, and Plexiglass temperature. This leads to reduced the temperature difference between the water surface within solar still and the inside Plexiglass temperature and reduced the condensation rate. However, when the relative humidity is relatively low and intensity of the solar radiation was appropriate and the ambient air temperature relatively high. This leads to increase the rate of fresh water production from SSS with RHD in compression with SSS without RHD.

IOP Conf. Series: Materials Science and Engineering **791** (2020) 012052 doi:10.1088/1757-899X/791/1/012052

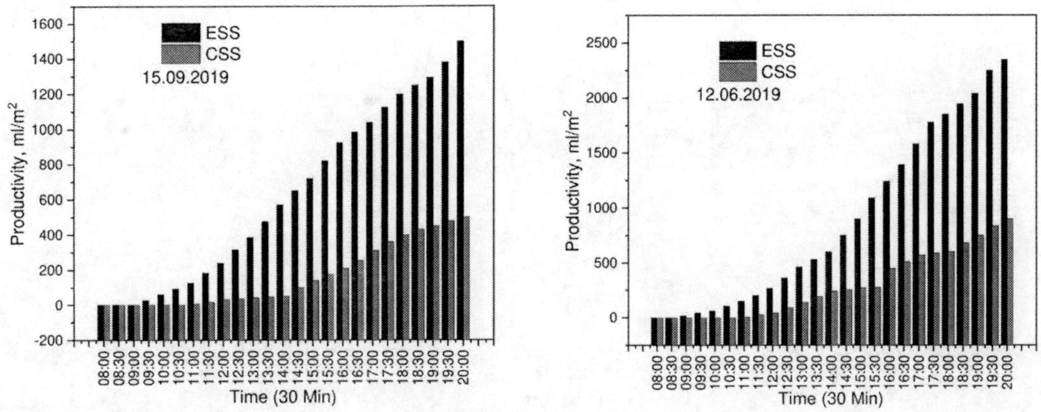

Figure 3. Relation between the time for each 30 minutes and fresh water production for both types of solar stills for two a perfect day 12.06-15.09.2019.

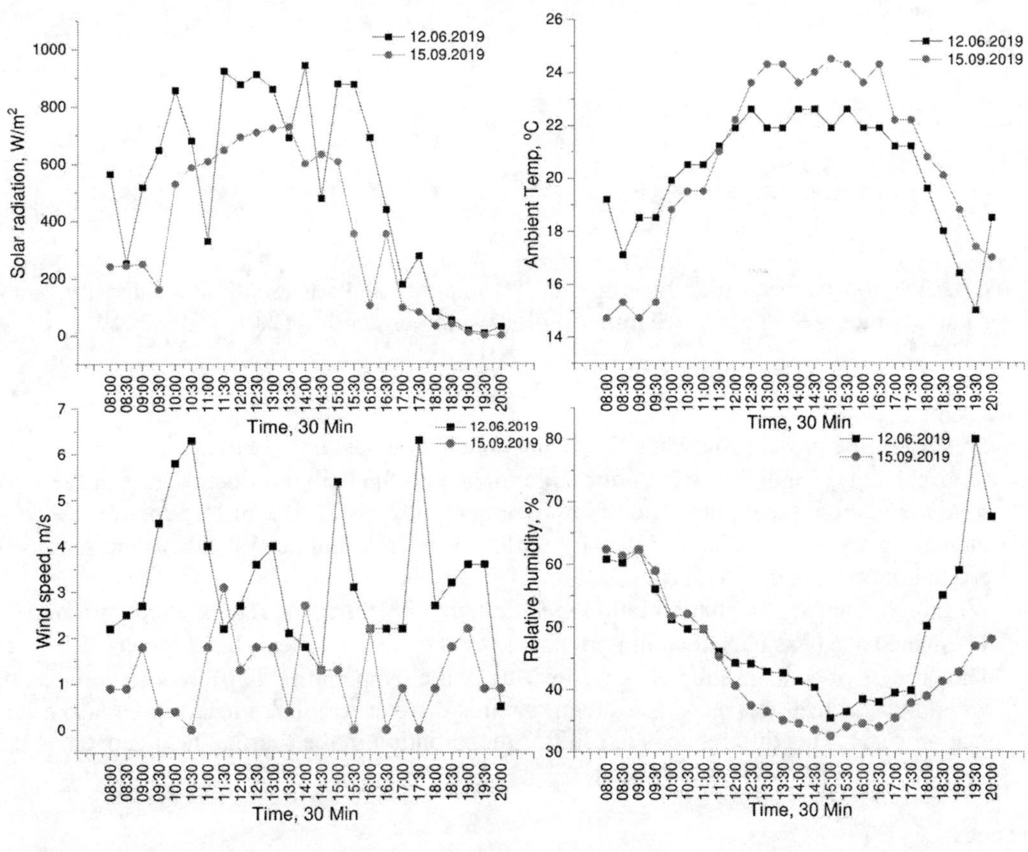

Figure 4. Relation between the time for each 30 minutes and solar radiation intensity, ambient temperature, wind speed and relative humidity for two perfect days 12.06-15.09.2019.

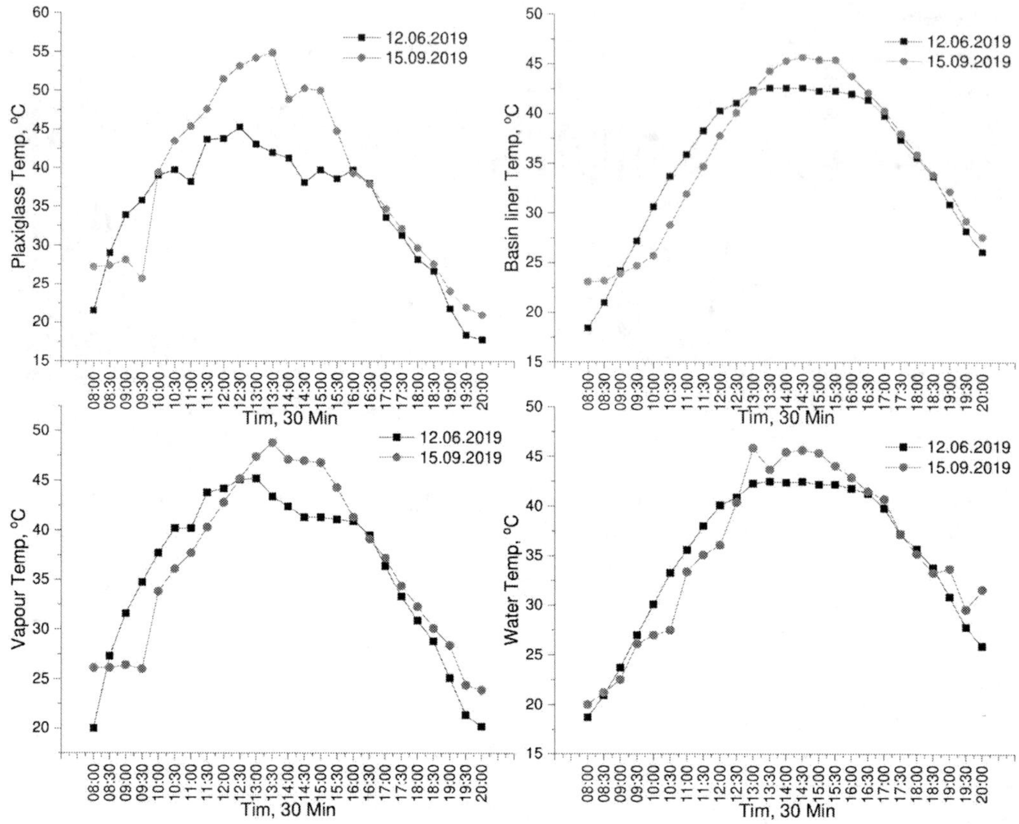

Figure 5. Relation between the time for each 30 minutes and solar radiation intensity, ambient temperature, wind speed and relative humidity for two a perfect day 12.06-15.09.2019.

4. Conclusion

From the results of the present study, the following conclusion has been addressed:

- Environmental conditions have noticeable impact on the daily production rate of the improved solar distillation; the highest rate of improvement ratio was 200% in 15 September 2019.
- Increasing the solar radiation intensity with low relative humidity leads to an increase in the productivity of solar still.
- With high intensity of solar radiation greater than 750 W/m^2, the rate of evaporation increased.
- The wind speed has played an important role to decrease the Plexiglass temperature.
- The process of condensation is occurring due to the temperature's differences between the water and glass cover (an increase of temperatures differences, that's lead to increase of condensate process). This difference is a result of the variation in the thermal heat capacities between both materials (Plexiglass and water).

References

[1] Zarzoum K, Zhani K, Ben Bacha H and Koschikowski J 2019 Experimental parametric study of membrane distillation unit using solar energy *Solar Energy* **188** 1274-82

[2] Al-Hinai H, Al-Nassri M S and Jubran B A 2002 Effect of climatic, design and operational parameters on the yield of a simple solar still *Energy Convers. Manag.* **43** 1639-50

[3] Panchal H N and Patel S 2017 An extensive review on different design and climatic parameters to increase distillate output of solar still *Renew. Sustain. Energy Rev.* **69** 750-8

[4] Abujazar M S S, Fatihah S, Rakmi A R and Shahrom M Z 2016 The effects of design parameters on productivity performance of a solar still for seawater desalination: A review *Desalination* **385** 178-93

[5] Khalifa A J N and Hamood A M 2009 On the verification of the effect of water depth on the performance of basin type solar stills *Solar Energy* **83** 1312-21

[6] Asbik M, Ansari O, Bah A, Zari N, Mimet A and El-Ghetany H 2016 Exergy analysis of solar desalination still combined with heat storage system using phase change material (PCM) *Desalination* **381** 26-37

[7] Abdallah S and Badran O O 2008 Sun tracking system for productivity enhancement of solar still *Desalination* **220** 669-76

[8] Tanaka H and Nakatake Y 2006 Theoretical analysis of a basin type solar still with internal and external reflectors *Desalination* **197** 205-16

[9] El-Sebaii A A, Ramadan M R I, Aboul-Enein S and El-Naggar M 2015 Effect of fin configuration parameters on single basin solar still performance *Desalination* **365** 15-24

[10] Prakash P and Velmurugan V 2015 Parameters influencing the productivity of solar stills - A review *Renew. Sustain. Energy Rev.* **49** 585-609

[11] Selvaraj K and Natarajan A 2018 Factors influencing the performance and productivity of solar stills - A review *Desalination* **435** 181-7

[12] Sampathkumar K, Arjunan T V, Pitchandi P and Senthilkumar P 2010 Active solar distillation-A detailed review *Renew. Sustain. Energy Rev.* **14** 1503-26

[13] Almuhanna E A 2013 Evaluation of Single Slop Solar Still Integrated With Evaporative Cooling System for Brackish Water Desalination *J. Agric. Sci.* **6** 48-58

[14] El-Sebaii A A 2011 On effect of wind speed on passive solar still performance based on inner/outer surface temperatures of the glass cover *Energy* **36** 4943-9

[15] Sharshir S W et al. 2014 Factors affecting basin type solar still productivity: A detailed review *Renew Sustain. Energy Rev.* **38** 430-47

IOP Conf. Series: Materials Science and Engineering **791** (2020) 012053 doi:10.1088/1757-899X/791/1/012053

Algorithm and program for calculating the facility's power supply system based on photovoltaic modules

E A Bekirov, M M Asanov and D V Polyanovsky

Electrical Energy and Electrical Engineering Department, V.I. Vernadsky Crimean Federal University, Academician Vernadsky Ave., 4, Simferopol, 295007, Russia

E-mail: asanov.m.m@cfuv.ru

Abstract. The popularity of renewable sources in design of the facility's energy supply system is mainly due to their environmental safety. They are the sources of the so-called "clean energy", the production of which is not accompanied by harmful emissions into the atmosphere. Renewable energy plants are complex systems with a large number of interconnected elements. Their design requires the solution of a number of problems, which include the selection of the functional, technical and technological structure of the system. The power plants components parameters depend on many factors, including the system's installation location characteristics, the consumers operating mode features, the system components compatibility with each other. Design automation of power plants based on renewable sources will help reduce the workload on the designer during the system development, thereby will promote wider use of such sources in facility's energy supply systems. In this work, the authors developed an algorithm and a program based on it for designing a facility's power supply system with photovoltaic modules as an energy source. The program allows user to select the necessary system components based on databases; evaluate the system effectiveness depending on the facility location; calculate the selected power supply system cost. When simulating the system operation a check for the input data correctness is made to exclude possible errors.

1. Introduction

The share of energy generated by renewable sources in the energy systems of a number of countries is constantly growing. According to statistics, for 5 years from 2013 to 2017 the growth in renewable sources capacity in the world amounted to about 610 000 MW, and energy production over the same period increased by 1 149 633 GWh [1]. Therefore, at the end of 2017, energy production by renewable sources in the world amounted to 6 190 948 GWh, which is 25% within the global power generation mix. Norway is the world leader in this area, where at the end of 2017 the share of energy produced by renewable sources was 97.8%. In Russia, this indicator for the same period amounted to 17.5% [2].

The power plants based on renewable energy sources popularity growth is connected not least with the desire to reduce the negative impact of humanity on the environment. UN Secretary General Antonio Guterres called climate change "the main issue of our time" [3]. According to the results of the UN Climate Action Summit, which took place on September 23, 2019 in New York, 65 countries and major sub-national economies committed to reduce greenhouse gas emissions to zero by 2050, and another 70 states announced that they would either expand their national action plans by 2020, or have already begun this process [4]. A number of countries and private companies have announced a significant increase in financing activities to promote the renewable energy sources use.

Power plants based on renewable energy sources are complex technical systems, and their design requires solving a number of problems. The difficulty lies in choosing the functional, technical and technological structure of systems with renewable sources. It is necessary to form a general structural and functional model of a power plant at the initial design stage, which is determined by a number of factors, including resource capabilities of a given region, consumer characteristics, and system components compatibility. In addition, the project feasibility should be evaluated from a technical and economic point of view [5-7].

The above mentioned problems, the growing popularity of power plants based on renewable sources caused the need for their design automation, as well as evaluating effectiveness of such systems at the initial development stage. For this purpose, a number of programs have been developed and are successfully used [8-10].

The first of the considered applications for modeling a renewable energy system is the Solar Advisor Model (SAM). It was developed by the National Renewable Energy Laboratory in collaboration with Sandia National Laboratories in 2005 for U.S. Department of Energy's Solar Energy Technologies Program. It is designed to predict energy production and estimate its cost for renewable energy systems connected to a common network, depending on the selected configuration and installation location. As the energy sources can be selected solar cells, including those with solar energy concentrators, wind turbines, solar systems, geothermal systems, biomass combustion systems, etc. The program allows user to interact with external models made in Excel and TRNSYS. In addition, SAM has its own scripting language, which makes it possible to write your own program to control the simulation, change the values of the input signals and write information to text files [8, 9, 11, 12].

HOMER (Hybrid Optimization of Multiple Energy Resources) software was developed by U.S. National Renewable Energy Laboratory for modeling the operation of low-power energy systems connected to the common network and isolated. Together with such systems functioning analysis, HOMER allows user to evaluate its installation and operation costs. The simulated system can provide energy to both electric and thermal loads, powered by photovoltaic modules, wind power, small hydropower, biomass power, reciprocating engine generators, microturbines, fuel cells, batteries, and hydrogen storage. The main tasks of HOMER are simulation, optimization and sensitivity analysis. In the simulation process, the program analyzes the operation of a given system to assess its development and cost. In the optimization process, the program analyzes a number of different systems in order to find the only one that fully meets the requirements set by the user, as well as that has lowest price. In the sensitivity analysis process, the program evaluates how sensitive the output is to the changes in input. At the same time, the user himself sets those input variables, which values will change. According to the available data, the program searches for the most optimal system configuration taking into account the accepted assumptions [9, 13, 14].

The Renewable Energy Technologies Screen was developed by the Government of Canada and is designed to evaluate the effectiveness, the ability of several sources to work together, and to analyze the energy production of the selected sources. The program allows user to work with a number of databases, including climate database from NASA ground stations and satellites. It is possible to integrate cost, benchmark, hydrology, and other databases. RETScreen makes it possible to analyze the entire working cycle of the selected project in order to evaluate the effectiveness of its work anywhere in the world under any environmental conditions [9, 15].

There are number of software that simulates the systems with renewable energy sources operation. Among them, the following programs used to analyze the energy systems based on photovoltaic cells can be noted:

- Photovoltaic systems (PVSyst) is used to analyze, simulate operation and evaluate the size of the system [9, 16];
- PV*SOL is used for dynamic analysis of the system, which includes energy storage, with the 3D visualization possibility and detailed analysis of the panels shading [17];
- Solarius PV allows user to analyze the work and evaluate the economic component of systems with a large equipment database and the ability to 3D visualization [18];

IOP Conf. Series: Materials Science and Engineering **791** (2020) 012053 doi:10.1088/1757-899X/791/1/012053

- Solar Pro simulates the systems operation, including ones with a dual axis Sun tracking system, with the assessment of shading and 3D visualization [19].

2. Materials and Methods

The aim of this work is to create an algorithm and a program based on it, allowing simulating the operation of a photovoltaic system with the ability to evaluate its technical and economic characteristics.

When developing the program, the choice was made in favor of VisualBasic.NET, which now is one of the most popular programming languages. Despite the existing shortcomings and the fact that it is very inferior in popularity to C ++, Java, C #, VisualBasic.NET functionality does not differ much from the programming languages listed above.

The main feature of VB.NET is its object orientation. VisualBasic.NET itself is a full-fledged object-oriented language that supports polymorphism, static typing, inheritance, and operator overloading. The object-oriented approach gives user the opportunity to solve the problem of developing the scalable, flexible and extensible applications [20].

Another difference between VisualBasic.NET and the standard Visual Basic is the use of the modern .NET platform.

Let us single out the main, in our opinion, VisualBasic.NET properties:

- Various languages support. This environment supports several languages C#, F#, C++, VB.NET, as well as languages that are tied to .NET, for example: Delphi.NET. The developer can choose the language that is suitable to him. It became possible thanks to the Common Language Runtime (CLR), which is the basis of the .NET platform. All code in any of the listed languages is compiled into an assembly in the CIL (Common Intermediate Language).
- Cross-platform. The .NET system is a portable platform. For example, the latest version of the platform at the moment is NETFramework 4.8 is supported on most modern Windows OS (Windows / Vista / 7/8 / 8.1). In addition, thanks to the Mono project, user can create applications that will continue to work on other OSs of the Linux family, including Android and IOS mobile systems.
- Large class library. Another feature of .NET is a single class library for all supported languages. During any application development process using VB.NET, in any event, interaction with the .NET class library takes place.
- Lots of auxiliary technology. .NET has a whole stack of technologies that can be applied when writing various applications. For example, ADO.NET technologies and the Entity Framework platform are designed to work with databases; WPF technology is used for the development of graphic programs with a rich interface; Windows Forms is used for the simplest applications; ASP.NET is used to create websites or web services.

As described above, VisualBasic.NET code compiles into applications or assemblies with exe or dll extensions in the Common Intermediate Language. Then, after the application starts, JIT compilation (Just - In - Time) is done into machine code, which is then directly performed. At the same time, since the launched application can be large and have many instructions, only part of the application that is being accessed will be compiled. When accessing other parts of the code, the system will also be compiled from CIL to machine code. The already compiled part of the application is saved until the program terminates. It is necessary in order to increase program performance, increase speed and perform instant computing functions.

3. Results

Authors developed an algorithm and program that allow calculating any autonomous power supply systems based on solar panels for any facility. The block diagram of the algorithm is shown in figure 1.

The program development began with the libraries analysis, preparation of windows, forms, and buttons. A power supply system components database was made: solar panels, batteries, inverters, diesel generators, as well as a database of cities– system installation location data. The first version of the program had a small selection of cities and minimal interaction with databases.

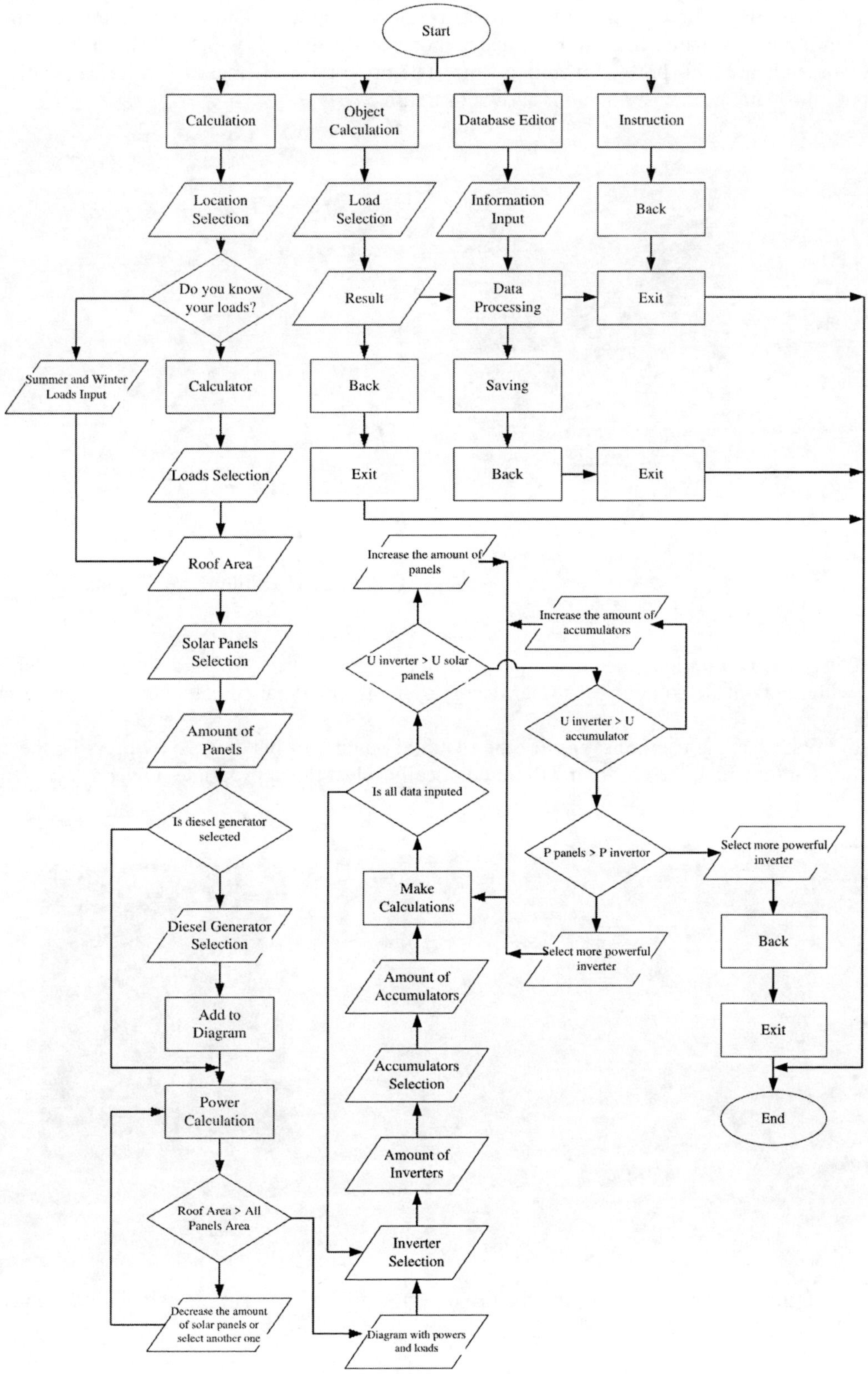

Figure 1. The developed algorithm of the described program.

IOP Conf. Series: Materials Science and Engineering **791** (2020) 012053 doi:10.1088/1757-899X/791/1/012053

Subsequent versions provided more information, the input information was checked for errors, the access to additional sources of information was opened to the user, and a slight simplification for information input was made. In addition, a calculator was developed, which allows user to select their own loads, indicate their power, operating time. The program itself makes further calculations based on input information. The user menu is shown in figure 2.

Figure 2. User interface of a start menu.

The user has two options for interacting with the program - performing a simple calculation of the house with ready-made solutions for solar energy systems or a detailed calculation of the facility power supply.

In the first case, the user must specify the required number of his devices by pressing the appropriate buttons, after which the program will automatically select the appropriate power system (figure 3).

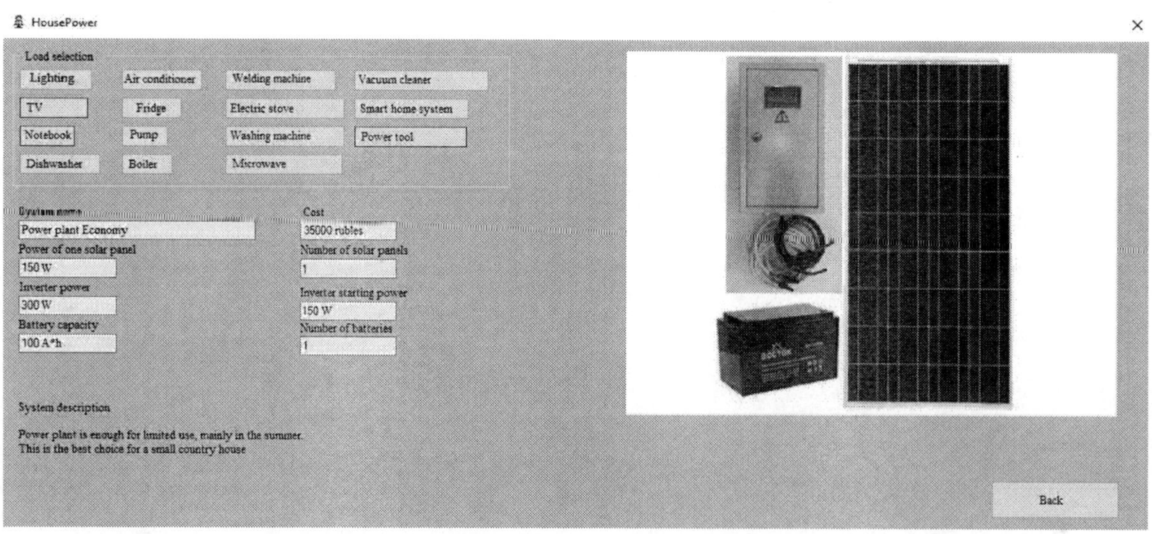

Figure 3. User interface of the facility power supply system simple calculation menu.

IOP Conf. Series: Materials Science and Engineering **791** (2020) 012053 doi:10.1088/1757-899X/791/1/012053

If the user needs to perform a more detailed calculation, he needs to choose the location, in which the system will be installed, indicate his daily loads in winter and summer. If daily loads are unknown, there is a special calculator, which helps user to specify this information. The program saves entered data for further plotting.

The next step is the selection of power system components (solar panels, inverters, batteries and diesel generators) from a database created in Microsoft Access. Firstly, the necessary solar panels and their number are selected, after which the data is saved. When user clicks the "Calculate power" button, a graph will be built showing loads in winter and summer. Seasonal loads on the graph are marked with the corresponding color; an annotation is indicated on the right of them.

User can choose a diesel generator, if necessary. For this, user must click on the "Add diesel generator" checkbox, after which a list of diesel generators available in the database with their parameters appears. To select a suitable generator, must click on the "Add Generator" button.

The next step is the inverter selection. The inverter power, as known, depends on the solar panels power, and must exceed it in order to withstand the load that goes through it. The selection is made from the table, while the number of inverters is also indicated. After the data is recorded, the user receives information about the inverters power; and user proceeds to the batteries choice.

After selecting the batteries from the corresponding table and their number, the system itself determines the capacity and displays the result on the screen (figure 4).

When all the data is selected, user must click on the "Calculate" button, and a confirmation message will appear. Pressing the "Yes" button will start checking the entered data. If all the criteria and conditions are correct, the system will display a message "Calculation completed!", otherwise an error message will be displayed. In this case, it is necessary to change the data according to the specified information (figure 4).

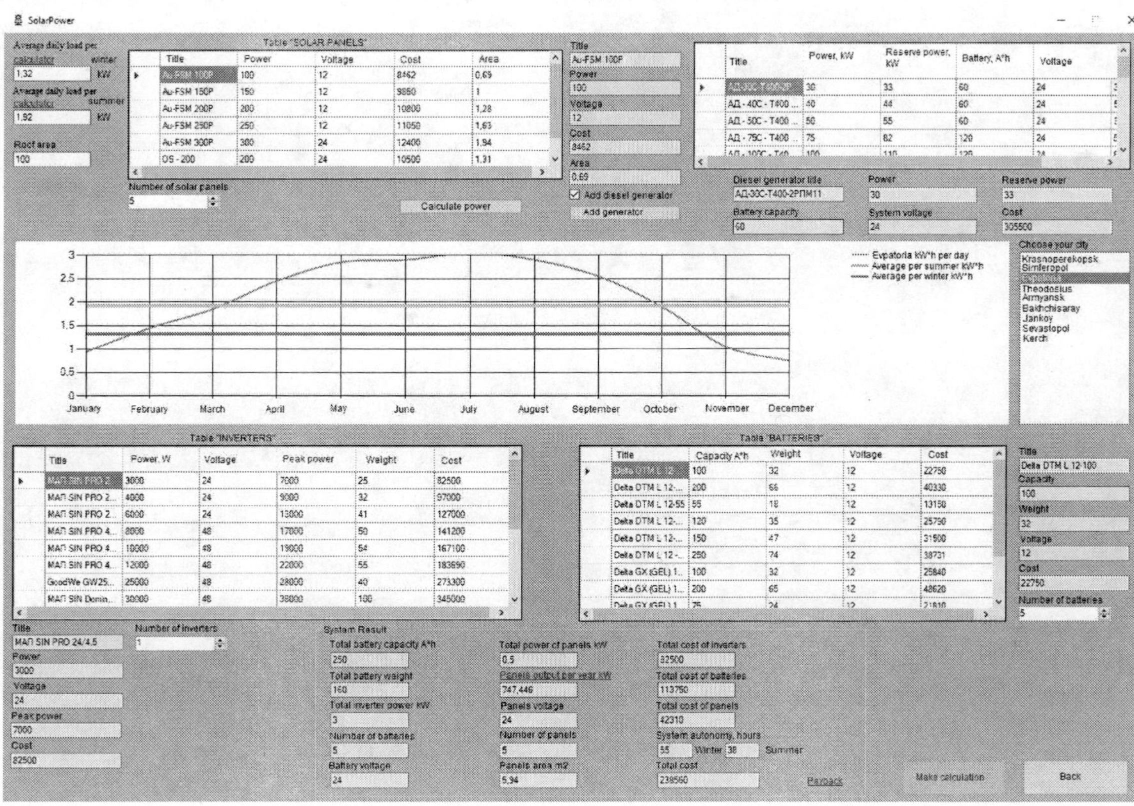

Figure 4. User interface of the facility power supply system detailed calculation menu.

IOP Conf. Series: Materials Science and Engineering **791** (2020) 012053 doi:10.1088/1757-899X/791/1/012053

Based on the calculated data, the "System Result" window will appear, in which the following data will be listed:

- total battery capacity;
- total mass of batteries;
- number of batteries;
- battery voltage in accordance with the connection method;
- total power of inverters;
- total power of the panels;
- voltage of panels;
- number of panels;
- panel production for the year;
- area occupied by panels;
- total cost of inverters;
- total cost of batteries;
- total cost of panels;
- total amount;
- autonomy of the system for winter and summer.

The program has a database editor, in case there is a need to remove something from the database or add a new element (figure 5). If necessary, user can call the instructions for working with the program.

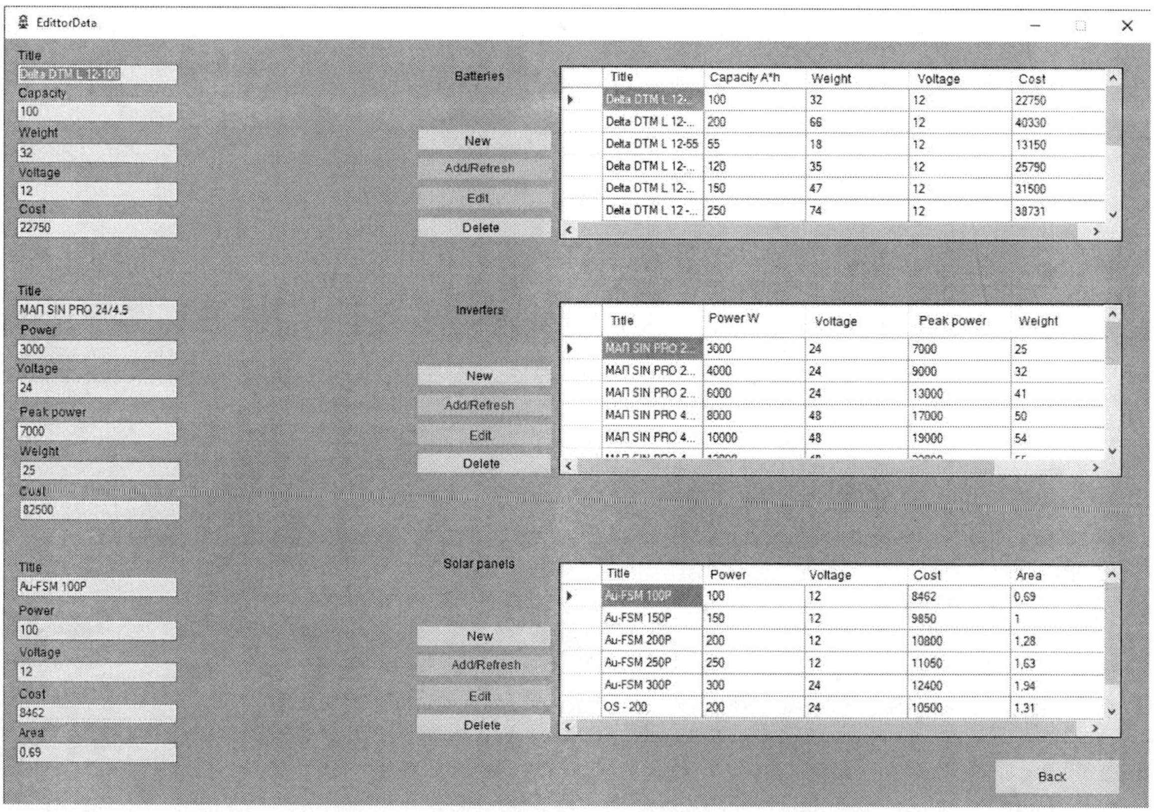

Figure 5. User interface of a database editor.

4. Conclusion

As a result of the study, the authors developed an algorithm and a program that implements it, used to automate the design process of a facility's power supply system based on photovoltaic modules. VisualBasic.NET was chosen as the program language, which, thanks to a number of advantages, makes it possible to successfully realize all of the set goals with low human and machine resources. The developed software solves a number of problems faced by the power supply system's designer. In particular, it helps to calculate the facility load, facilitates the selection of system components and their parameters, inputs the selected facility location characteristics, checks the entered results correctness to eliminate errors when modeling the system. As a result of calculations based on the entered data, among other, the program gives information about the selected system components and their quantity, their cost and the whole system cost, as well as the power production of the system in summer and winter. The software also includes detailed operating instructions.

References

[1] Whiteman A, Esparrago J, Elsayed S, Rueda S, Arkhipova I and Escamilla G 2019 *Renewable Energy Statistics* (Abu Dhabi: IRENA) URL https://www.irena.org/-/media/Files/IRENA/Agency/Publication/2019/Jul/IRENA_Renewable_energy_statistics_2019.pdf
[2] Global Energy Statistical Yearbook 2018 URL https://yearbook.enerdata.net/2017/renewables/renewable-in-electricity-production-share.html
[3] Climate Action Summit 2019 URL https://www.un.org/en/climatechange/
[4] Climate Action Summit. Closing Release URL https://www.un.org/en/climatechange/assets/pdf/CAS_closing_release.pdf
[5] Trubaev K P 2018 Study of Power Dependence of Wind Power from Wind Speed *J. Phys.: Conf. Ser* **1066** 012026
[6] Zubova N V and Rudykh V D Optimization of power output for a wind turbine using methods of artificial intelligence *IOP Conf. Ser.: Mater. Sci. Eng.* **552** 012034
[7] Zaliskyi M, Petrova Y, Asanov M and Bekirov E 2019 Statistical data processing during wind generators operation *Int. J. of Electrical and Electronic Eng. & Telecommunications* **8** 33-8
[8] Gurupira T and Rix A 2017 PV simulation software comparisons: PVSYST, NREL SAM and PVLIB *SAUPEC 2017* (Stellenbosch: University of Stellenbosch)
[9] Umar N, Bora B, Banerjee Ch and Panwar B S 2018 Comparison of different PV power simulation softwares: case study on performance analysis of 1 MW grid-connected PV solar power plant *Int. J. of Eng. Science Invention* **7** 11-24
[10] Ringkjob H-K, Haugan P M and Solbrekke I M 2018 A review of modelling tools for energy and electricity systems with large shares of variable renewable *Renewable and Sustainable Energy Reviews* **96** 440-59
[11] Blair N, Dobos A P, Freeman J, Neises T and Wagner M 2014 *System Advisor Model, SAM 2014.1.14: General Description* (Denver: NREL)
[12] NREL System Advisor Model (SAM) URL https://sam.nrel.gov/
[13] Farret F A, Simoes M G 2006 *Integration of Alternative Sources of Energy* (New Jersey: John Wiley & Sons)
[14] HOMER Energy URL https://www.homerenergy.com/
[15] Natural Resources Canada URL https://www.nrcan.gc.ca/energy/retscreen/7465
[16] PVSYST Photovoltaic Software URL https://www.pvsyst.com/
[17] Valentin Software URL https://www.valentinsoftware.com/en/products/photovoltaics/ 57/pvsol-premium
[18] Solar Design Software URL https://www.accasoftware.com/en/solar-design-software
[19] Laplace Systems URL https://www.lapsys.co.jp/english/products/pro.html
[20] Grundgeiger D 2002 *Programming Visual Basic NET* (Sebastopol: O'Reilly)

Efficiency research of free electricity (power) transfer zones integration by distributed generation

F L Byk, L S Myshkina, N S Terlyga and Y A Frolova

Department of Automated Electrical Power Systems, Novosibirsk State Technical Universality, Prospekt K. Marksa, 20, Novosibirsk, 630073, Russia

E-mail: terlyga.nikita@gmail.com

Abstract. The possibility of combining free electricity (power) transfer zones by distributed generation is considered in this paper. The relevance of a subject is caused by the existing condition of network infrastructure and a high part of a network component in the final electricity cost, change of a fuel and energy landscape in territorial subjects of the federation, increase in ecological requirements and need of increase in power production efficiency. The object of the study is the second price zone of the wholesale market located on the territory of Siberia, where the task of increasing the availability of electricity is combined with the problem of reliable heat supply due to climatic conditions. It causes the presence of a large number of heating boiler houses. It is shown that the medium and large boiler houses transfer to the cogeneration mode of operation in the cities of the Altay Reginon and the Republic of Altai will allow to integrate the free electricity (power) transfer zones Altai and Siberia among other significant effects. Distributed generation was seen as an effective alternative to strengthening intersystem connections. It is obvious that the new network construction leads to an increase in the electricity transmission tariff and the transition to distributed generation leads to an increase in competition in retail regional markets, which reduces the cost of electricity.
Keywords. Free electricity (power) transfer zone, gasification, boiler house, distributed generation, gas-piston engine, cogeneration

1. Introduction

The concept of free electricity (power) transfer zone (FETZ) appeared in 2010, when the Federal Antitrust Service defined 29 FETZes, where competitive power selection is carried out using the power price limit.

The FETZ allocation is justified by the presence of technical restrictions on the transmission of power systems through intersystem connections. In each FETZ the generating power can be distributed without any restrictions. It's possible to conclude that there is internal competition with the absence of external competition due to restrictions on external connections load. Such zones may be both redundant and scarce.

Currently there are 21 FETZes. In the second price zone there are 5 zones, where the largest is Siberia (Figure 1). The reduction in the number of FETZes by integrating them increases competition among electricity suppliers [1]. Those suppliers are members of the wholesale electricity and power market (WEPM). Usually the integration is achieved by the construction and reconstruction of networks to remove network restrictions. This method requires large investments and consequently leads to an increase in the electricity transmission tariff imposed by the executive bodies of the federal entity state power.

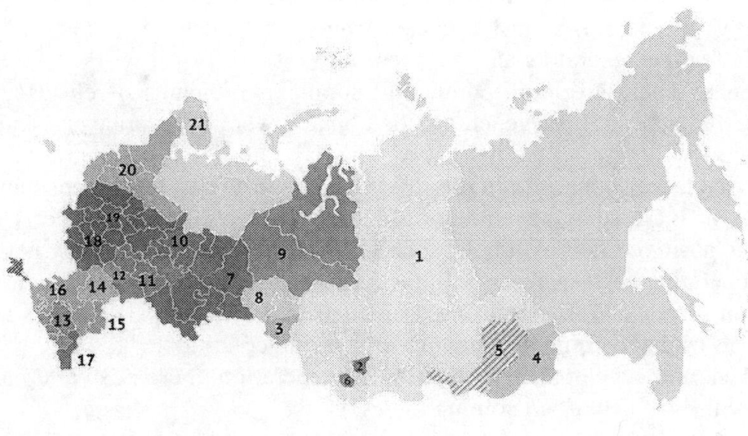

Figure 1. Current map of the free electricity (power) transfer zones.

It is obvious that for scarce FETZes, network restrictions can be removed by planting new generating capacities. Construction of large power plants in recent years was carried out by power supply contracts, which were concluded by a system operator with unified and territorial generating companies [2]. However, this mechanism also resulted in an increase in the electricity cost for end-users, who are charged with the 12.5% return of investments through the power consumption payment mechanism [3].

The rise in the final electricity cost for consumers, compounded for industry by the existence of cross-subsidies in favour of the population, required them to seek an alternative to a centralized electricity system. As a solution to the problem, large and energy-intensive consumers have chosen the option of creating their own energy supply systems based on distributed generation. Much of them were made possible by scientific and technological advances that made distributed generation competitive and technically effective.

Efficiency of distributed generation allows setting and solving the problem of power engineering transition. It will lead to increase of availability, reliability and quality of power supply. The considerable length of the Russian territory, the heterogeneity of the population and the distributed placement of resource and production capacities require new methods on solving the problems of power engineering development. All the prerequisites and conditions for the restructuring of Russian power system on the basis of a combination of centralized and local power supply systems are appeared. It is obvious that distributed generation technology, as well as non-traditional and renewable sources of energy, will be widely used in local power systems.

The aim of the work is to explore the possibility of using distributed generation in integration of FETZes. To date, the literature hasn't met such theme, which makes it possible to say about the novelty of the obtained results for solving the tasks considered in the work.

2. Materials and Methods

2.1. Free electricity (power) transfer zones' boundaries and their integration effects

According to [2], the free electricity (power) transfer zone is a part of the Unified Power System of Russia. Electricity and power produced or planned for supply can be redistributed between the power plants located inside this zone in accordance with their technical characteristics, providing the flow of electricity and power restrictions between such zones.

The Regulation for free electricity (power) transfer zones parameters determination and updating of Unified Power System (UPS) states that the list of FETZes is determined in accordance with the "Pro-

IOP Conf. Series: Materials Science and Engineering **791** (2020) 012054 doi:10.1088/1757-899X/791/1/012054

cedure of free electricity (power) transfer zone determination". It's stated there that it is necessary to reduce the load of intersystem connections by 1/3 from the peak load to integrate FETZes.

Several positive effects can be distinguished from FETZ integration. The integration will allow to reduce the load of some generators and put them into reserve [4]. It will increase reliability of power supply in the zone. There are more competing supply participants of electricity and power to the WEPM in integrated zones. Those generators that have lower power price and energy production costs compared to others will pass on the market for the day ahead and competitive power selection. Production prices, which are about 50% in the final cost of electricity for the consumers, can be expected to decline due to the increase in the number of bidders for electricity and power.

However, this positive effect will be reduced if the FETZ is combined by network construction. It is obvious that operating costs increase with the increase in investment in the network complex. Therefore it leads to an increase in the electricity transmission tariff. The size of this tariff is largely determined by the costs of maintaining the network in working condition.

With the advent and development of distributed generation, there is an alternative to network construction to solve the problem of combining the FETZes.

2.2. Local power systems and effects of their integration with Unified Power System

First of all, it is necessary to point out not only the high technical efficiency of distribution generation [5], but also the existence of imperfect market relations in retail markets, which they relay to as the main prerequisites for the emergence of local power supply systems.

For small and medium-sized businesses covered by the cross-subsidization mechanism, centralized energy supply was largely becoming a critical factor, often forcing them to leave the law field and seek illegal routes for power supply [6]. With the advent of distributed generation (DG), they have the opportunity to become subjects of the local power supply system, where DG is being the main source of power. The cost of electricity in such systems is usually lower, since being in the immediate proximity of the supplier and user of electricity significantly reduces the network component [7].

The development of local power systems is facilitated by achievements in the implementation of the program of gasification of the country. To date, the level of gasification in Russia has reached 68.6% [8]. It indicates a radical change in the fuel and energy landscape on the territory of the European part of the country and partially in Siberia. But in the near future, with commissioning of the pipeline gas supply system "Power of Siberia" and "Power of Siberia 2" the situation will change [9]. This program promotes the conversion of boiler houses to an ecological fuel and creates the prerequisites for their conversion in cogeneration operating mode. It is stimulated by the Government to improve their energy efficiency, energy saving and non-volatile regional power supply systems.

These prerequisites and other factors create investment attractiveness for private capital and the expediency of creating local power systems that have a positive impact on the social and economic development of the federation entities. It should be noted that distributed generation leads to a reduction of network construction volume for the development of the regional power supply system. It also has a positive impact on the size and growth rate of the network component of centralized power supply [10].

It is obvious that if the part of DG in the regional power and energy balances included in the scarce FETZ increases, it can lead to intersystem connections unloading. And therefore it will lead to fulfilment of the conditions necessary for their integration. But to do this, there is a need to find the answer to the question: How many and where is it needed to install distributed generation objects to ensure the integration of FETZes?

2.3. Research object

The second price zone of the wholesale market was chosen as the research object. This zone is located in Siberia so there is a need for affordable heat and electricity. Therefore, a large number of boiler houses located on the territory of Siberia are operating on various types of fuel. Often they are subsidized from the federation entity budget. It's caused due to very high cost of released heat. The only

IOP Conf. Series: Materials Science and Engineering **791** (2020) 012054 doi:10.1088/1757-899X/791/1/012054

possible way to reduce the cost is to put the boiler house into cogeneration operating mode [11], which provides combined generation of heat and electricity, i.e. leads to the emergence of mini thermal power plant.

Cogeneration can be carried out by any kind of fuel, but the most effective is the usage of gas and the technology based on gas-piston and gas turbine engines [12]. The gasification program of regions contributes to this transfer and also the growth of environmental requirements gives acceleration to this process. It should be noted that gas-piston engine is more effective than gas turbine engine by the following indicators:

- Higher electrical efficiency and its little change while load changing (40-47% against 17-36%);
- Advantageous ratio of heat generation to electrical (1/1.5 versus 1/2.5, suggesting more efficient fuel use);
- Price per 1 kW is two times lower (400-600$ versus 1000-1400$);
- Motor resource of such engines is higher (40-100 thousand hours of work until failure for gas-piston and 30-60 thousand for gas turbine engine);
- Gas-piston engine can use close to domestic low pressure gas, while gas turbine engines can only operate at high pressure (near 16-20 bars).

"Altai" is the best suited research zone among all FETZes from the second price zone. It's the part of power system located on the territory of Altai Region and the Republic of Altai. The part of thermal power plants reaches 99% of all generation capacities in that zone (it's 1163 MW of installed power). Also there are 5 solar power plants with power of 40 MW which don't participate in power balance. It causes the coverage of load schedule peaks due to the flows from FETZ "Siberia". The presence of intersystem connections allows to ensure reliability of power and energy balances.

The research object is dominated by non-productive consumers, such as tourist facilities. It should be noted that Altai is a sanatorium and resort area with a large number of nature reserves and the presence of agricultural industry, which determinates the importance of ecology for the region. Due to distributed consumption, networks have a large length and it leads to high losses. The prospect of gas pipeline "Power of Siberia-2" construction allows to count on changes in the fuel landscape, where natural gas will become the main resource.

In addition, the interest of the executive power in increasing the availability of electricity is obvious, as in FETZ "Altai: even the value of electricity cost for the population [13] is higher than in the nearest FETZ "Siberia" for 1.5-2 times [14]

Nowadays there is not enough installed power of power plants in FETZ "Altai" for independent load coverage, so the part of this power comes from other system. In table 1 the power balance for FETZ "Altai" is shown.

Table 1. Power balance for FETZ "Altai"

Generation, MW	Load, MW	Losses, MW	Summary consumption, MW	Flow from the over systems, MW
1163	1560	36	1596	433

It's necessary to reduce the flow by 1/3 (145 MW) according to the "Procedure of free electricity (power) transfer zone determination" to integrate FETZ "Altai" with "Siberia". It will remove network restrictions that limit the social and economic development of the regions included in the FETZ "Altai".

2.4. Data for calculations

2.4.1. Determination of cogeneration plants locations. In order to determine the installed power that can be obtained during the installation of cogenerating plants, it is necessary to identify boiler houses which are possible to transfer to mini thermal power plants. The composition of the houses is specified in the heat supply schemes of cities, where it is possible to choose suitable one according to such criteria as:

- Installed capacity (not less than 5 Gcal/h and not more than 50 Gcal/h, which characterizes boiler rooms of average capacity);
- Performance of heating functions.

2.4.2. Choice of the cogeneration units. Cogeneration units of the company GazEcos with 1 MW and 2 MW capacity were chosen as the main units. According to the manufacturer, these units have the highest electrical efficiency - 44%, and the total reaches 82%. These installations are produced in Russia in Ivanteyevka. Plants are characterized by high maneuverability (70-80% of load drop/surge in 2 seconds), possibility to operate at low gas pressure (600 mbar) and their warranty service life reaches 20 years [15]. Chosen units are shown in the table 2.

Table 2. Characteristics of chosen units

Unit name	Installed active power, MW	Rotation speed, rpm	Electrical efficiency, %	Total efficiency, %
GazEcos 16GDG49	2	1000	44	82
GazEcos 8GDG49	1			

2.4.3. Operation modes calculation. For mode parameters analysis the section connecting FETZ "Altai" and "Siberia" was determined, and for more accurate assessment it was divided into 4 separate sections: Novosibirsk – Altai (1), Kuzbass – Altai (2), Rubtsovsk – Altai (3), Omsk – Altai (4). These sections are shown on figure 2.

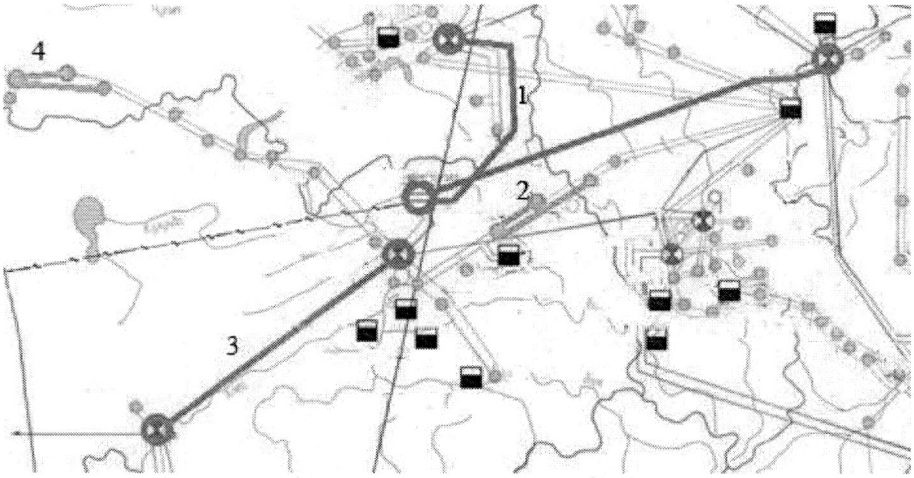

Figure 2. Highlighted controlled sections.

During calculations in calculation program complex RastrWin it is necessary to determine the flow rate for each of the sections and the limit for static stability. The maximum allowable flow rate is estimated to provide a static aperiodic stability margin factor of 20% for active power in normal mode and for each section was determined by the following formula:

$$P_M = 0{,}8 \times P_L - \Delta P_{IF} \qquad (1)$$

here P_P – limit flow value when static stability is maintained,

IOP Conf. Series: Materials Science and Engineering **791** (2020) 012054 doi:10.1088/1757-899X/791/1/012054

ΔP_{IF} – Value of amplitude of irregular fluctuations in the controlled section, calculated as:

$$\Delta P_{IF} = K \times \sqrt{\frac{P_1 \times P_2}{P_1 + P_2}} \tag{2}$$

here K – coefficient characterizing the method of active power flow control in the controlled section (1.5 for operational dispatching control),
P_1, P_2 – active power consumption from the power system at each side of controlled section.

For estimation of cogeneration efficiency, the calculations were made in the mode of winter peak load and the limits on distributed generation power in dedicated circuit nodes were taken into account:

$$\begin{cases} \sum P_{DG} \leq P_{Tr} \\ \sum P_{DG} \leq 0{,}5 P_{Load} \\ P_{DG} \leq 25\ MW \end{cases} \tag{3}$$

here P_{DG} – installed power of cogeneration plants of the forming local power systems,
P_{Tr} – active transformers power with the condition of n-1 (n – quantity of transformers),
P_{Load} – load power in the node where local power systems are formed.

3. Results

The heat supply schemes analysis of such cities as Barnaul, Biisk, Kamen-na-Obi, Gorno-Altaysk, Slavgorod and Belokuricha showed that there are 42 suitable boiler houses in the region with a total installed power of 790 MW. Taking into account the specifics of the region, where the ratio of heat consumption to electricity consumption is 2:1, which, taking into account the emergency reserve of 20%, causes 280 MW of available capacity.

Accordingly, the maximum allowable flow was calculated and the load of each section was determined for the cogeneration efficiency estimation. These calculations are shown below in the table 3.

Table 3. Highlighted sections and their load

Section	Flow, MW	Maximum allowable flow, MW	Load, %
Novosibirsk – Altai	-63.5	233	41
Kuzbass – Altai	310.6	380	82
Rubtsovsk – Altai	18	92	19
West (Omsk) – Altai	141	566	25

According to calculations the most loaded section Kuzbass – Altai was selected as the main one. Therefore, the calculations for cogenerations efficiency estimation were made according to calculation methods and there are results below. Calculations for efficiency estimation are placed in table 4.

Table 4. Calculations results

Parameter	The value before cogeneration usage, MW	The value after cogeneration usage, MW	Decrease, %
Flow from the other systems	432.7	224.8	48
Kuzbass – Altai section load	310.6	174	36
Losses	36	27	24.9

IOP Conf. Series: Materials Science and Engineering **791** (2020) 012054 doi:10.1088/1757-899X/791/1/012054

4. Discussion

According to results, there are more boiler houses suitable for reconstruction in the region than necessary for combining FETZes. But it is obvious that there will be known problems associated with their connection to high-voltage networks. These problems can be solved if local power systems with 10 kV network will be built on their basis. According to [16], such systems should operate on the principle of self-balance, but integrated into regional power supply systems, which is effective from the point of view in increasing the fuel efficiency, the utilization factor of installed power, reliability, quality, reliability of power supply and performing the multi-voltage regulation [17].

Also, the calculations show that cogeneration usage allows to reduce various parameters and generally the flow from the other systems more than by 1/3. According to "Procedure of free electricity (power) transfer zone determination" FETZ "Altai" and "Siberia" can be integrated without network construction.

5. Conclusion

This research has indicated the effectiveness of using distributed generation for integration of free electricity (power) transfer zones. In addition to integration of the free electricity (power) transfer zones, the use of distributed generation makes it possible to create local power systems, which will improve the reliability and quality of power supply and reduce the cost of regional distribution network building. Also the use of distributed generation leads to increased persistence, energy efficiency and independent power supply to consumers in the investigated region. It's necessary to note that the occurrence of distributed generation does not increase the electricity transmission tariff. In general, it was obtained that the price determined for the long-term capacity market for 1 MW can be halved (up to 100 thousand rubbles). As a result, it is possible to expect a decrease in the cost of electricity and power for the end-users in the Altai free electricity (power) transfer zone.

Acknowledgments

This research was supported by the Novosibirsk State Technical University (project C-19-26).

References

[1] Oficial website of PJSC "Federal Network Company of Unified Power System" URL https://www.fsk-ees.ru
[2] Oficial website of OJSC "System Operator of Unified Power System URL https://so-ups.ru
[3] Letova K, Yao R, Davidson M and Afanasyeva E 2018 A review of electricity markets and reforms in Russia *Utilities Policy* **53** 84-93
[4] Karpov Y 2014 Razvitie setey kak vazhneyshiy faktor rosta konkurentsii na optovom rynke elektroenergii i moschnosti [Networks development as the most important growth factor growth of competition in the wholesale market of electricity and power] *Vestnik GUU* **21** 77-9 [In Russian]
[5] Razavi S-E, Rahimi E, Javadi M S, Nezhad A E, Lotfi M, M Shafie-khah M and Catalão J P S 2019 Impact of distributed generation on protection and voltage regulation of distribution systems: A review *Renewable and Sustainable Energy Reviews* **105** 157-67
[6] Erdogdu E 2011 The impact of power market reforms on electricity price-cost margins and cross-subsidy levels: A cross country panel data analysis *Energy Policy* **39(3)** 1080-92
[7] Hinz F, Schmidt M and Möst D 2018 Regional distribution effects of different electricity network tariff designs with a distributed generation structure: The case of Germany *Energy Policy* **113** 97-111
[8] Oficial website of PJSC "Gazprom" URL https://www.gazprom.ru
[9] Sidortsov R, Ivanova A and Stammler F 2016 Localizing governance of systemic risks: A case study of the Power of Siberia pipeline in Russia *Energy Research & Social Science* **16** 54-68
[10] Blázquez de Paz M 2018 Electricity auctions in the presence of transmission constraints and transmission costs *Energy Economics* **74** 605-27

[11] Wu Y, Fu L, Zhang S and Tang D 2019 Study on a novel co-operated heat and power system for improving energy efficiency and flexibility of cogeneration plants *Applied Thermal Engineering* **163** 114429
[12] Gambini M and Vellini M 2015 High Efficiency Cogeneration: Electricity from Cogeneration in CHP Plants *Energy Procedia* **81** 430-9
[13] Official website of JSC "Altaienergosbyt" URL https://altaiensb.com
[14] Electricity tariffs for the population in Russia in 2019 URL https://energybase.ru/tariff/electricity/2019
[15] Oficial website of LLC "GazEcos" URL https://www.gazecos.ru
[16] Byk F, Frolova Y and Myshkina L 2019 The efficiency of distributed and centralized power system integration *E3S Web of Conf.* **114** 05007
[17] Bedawy A and Yorino N 2018 Reactive Power Control of DGs for Distribution Network Voltage Regulation Using Multi-Agent System *IFAC-PapersOnLine* **51(28)** 528-33

IOP Conf. Series: Materials Science and Engineering **791** (2020) 012055 doi:10.1088/1757-899X/791/1/012055

Evaluation of the hazardous emissions from different types of RDF combustion in low power boilers

N V Kornilova[1] and P A Trubaev[2]

[1]Department of Electric Power Engineering and Automation, Belgorod State Technological University of V.G. Shukhov, Kostyukov St., 46, Belgorod, 308012, Russia
[2]Department of Energy Engineering of Heat Technologie, Belgorod State Technological University of V.G. Shukhov, Kostyukov St., 46, Belgorod, 308012, Russia

E-mail: trubaev@gmail.com

Abstract. Content evaluation of the hazardous substance emission from different types of waste combustion in two solid-fuel water heating boilers was carried out. For 100 kW pyrolysis boiler the content in exhaust gases of the following compounds was determined: NO_2, NO, CO, CH_4, phenol, benzol, SO_2, formaldehyde, HCl. For 200 kW stoker grate boiler the content in the exhaust gases of the following compounds was determined: CO, NO_2, NH_3, mercaptans, H_2S, phenol. For the targeted factors, the calculated hazardous substance ground-level concentrations don`t exceed the maximum allowable concentration limit. The hazardous substance emissions in low power boilers are comparable with high power analogs but have lower combustible efficiency. Waste combustion in the pyrolysis boilers creates 2-10 times less hazardous substances than in grate firing. It was established that combustion mode influences only the carbon monoxide content. The content of other hazardous compounds in the exhaust gases doesn`t depend on the combustion parameters. The additive criterion of total emissions was suggested, which presents the sum of the ratio of the calculated ground-level hazardous substance concentrations to maximum allowable concentrations. Unlike the existing dimensionless concentration, the criterion considers hazard class and compound fraction in gases. The highest criterion value of the total emissions was noted during wood waste combustion (windows, doors, furniture) and RDF fuel, which contains a large amount of plastics

1. Introduction

At this time, there are two main technologies of waste thermal recycling that exist, which most of the waste incinerating plants are using – combustion, mostly in the grate stokers, and gasification [1].

As recently as at the 20th-century beginning, people were complaining about smell from the waste incinerating plants, and doctors were saying that smoke from the waste combustion is not more harmful than a waste landfilling [2]. The problems of emissions control at the waste incinerating plants grown worse in 1960s, after discovery of strong organic poisons and cancerogens – dioxins (PCDD) and furans (PCDF), and large-scale researches in the US, England, and Netherlands, which detected an increased number of diseases in the areas where waste incinerating plants were located.

Concern about the harmful effect on human health from the dioxin and furan emissions generated a large number of studies on this kind of emissions control, in order to keep it under the allowable limits [3]. At the same time, on a practical level, this situation is not that simple - popular and ecological magazines claim that waste incinerating plant emissions have a very harmful effect on the environment, while scientific papers, based on the experiments, claim that emissions are much lower than the allowable limits.

In 1987, the United States Environmental Protection Agency (US EPA) published a report, where it noted very high risks from the waste incinerating plant emissions for human`s health. But the papers [4, 5] showed that the expected cancerogen and non-cancerogen risks from the emissions are by several orders of magnitude less than the values presented in the US EPA report, and by several orders of magnitude less than concentrations that can be harmful to human`s health. Substantial differences are caused by the fact that the US EPA report used unrealistic and overestimated conditions.

The main harmful substances, which combustion products contain, and their limit values are presented in Table 1. But unlike coal burning, where the main contaminating substances are SO_2, NO_x, HCl, SO_3, HF, and macroparticles, in MSW and RDF combustion the main contaminating substances are particles of heavy metals and organic emissions [6].

Table 1. Limit values of emissions, mg/m^3.

Emissions	Content in the uncleaned combustion products [7]	Limit values [7]			Threshold values of emissions into the air for waste incinerating plants (Directive 2010/75/EU), mg/m^3	
		USA	China	Japan	Average-daily	Average half-hourly
Dust	1000-5000	24	80	10-50	10	30
TOC[a]	1-10	–	–	–	10	20
HCl	500-2000	25	75	15-50	10	60
HF	1-10				1	4
SO_2	150-400	30	260	10-30	50	200
NO_x[b]	200-500	150	400	30-125	200 400	400
CO	<10-30	100	150	50	50	100
Hg	0.1-0.5	0.08	0.2	0.03-0.05	0.05	
Cd (and Tl)	0.1-0.5	0.02	0.1		0.05	
Other[c]	–	–	–	–	0.5	
PCDD/F	$1\text{-}10\cdot10^{-6}$	$0.3\cdot10^{-6}$	$0.1\cdot10^{-6}$	$0.1\cdot10^{-6}$	$0.1\cdot10^{-6}$	

[a] Organic substances expressed as a total organic carbon content
[b] Calculated in terms of NO_2
[c] Sb, As, Pb, Cr, Co, Cu, Mn, Ni, V (total)

As stated in the paper [1], the environmental impact from modern waste incinerating plants is comparable with the industrial emissions and electrical plants, which use mineral fuel. The paper [8], published in 1998, presents a comparison of the sorted municipal solid waste (MSW) and fuel, made of recycled waste (RDF) combustion in the same low-tonnage furnace with inclined grate and with capacity 0.5-0.8 MW. The authors concluded that the combustion of RDF, made by 25-100 mm fractionation of MSW, is characterized by much higher energy efficiency and less amount of emissions, but the content of harmful emissions during MSW and RDF combustion is substantially lower than the allowable values [8]. The paper [9] examined the emissions during combustion in the furnace with 11 kW capacity of four fuel mixtures, which consisted of wood construction waste, MSW and RDF. The results showed that the presence in the fuel of the wood construction waste has a dominant influence on the content of harmful emissions, their presence increased emission amount more than three times. Food waste didn`t have much influence on the emission amount. The opposing data was presented in the paper [10], which studied the combustion of waste collected in two Sweden regions – with a separate collection of waste and without separation. This paper established that an increase in food waste content leads to increased dioxin and furan emissions.

Analysis of the emissions shouldn`t be concentrated on its content in the exiting gases, but instead, consider emission dispersion in the atmosphere [7]. Based on the actual measurements from the literature data, the paper [11] presented emissions comparison for seven thermal recycling technologies of MSW and RDF. Distinct advantages on emissions have technology of RDF pyrolysis, where emissions three times lower than at direct MSW combustion. RDF combustion is characterized by a slight emission reduction in comparison with MSW combustion. The authors determined the ratio of the emission concentration in the environmental air and at the ground level to limit values, which can be found in the Directive 2008/50/EC. All technologies had emissions that are much lower than the values that can create health risks (assumed as not more than 5% of the maximum allowable concentrations for residential areas).

The difficulty of monitoring dioxin and furan emissions is in high prices of the equipment for their detection [7]. Therefore, several papers have carried out the determination of factors that influence their formation. The paper [3], based on the actual data from 14 factories in the US, established that the lowest dioxin and furan content (concentration in the uncleaned gases less than 1 ng/m^3) can be achieved when the combustion product temperature is 800°C and CO content in the exiting gases is 10 ppm (which corresponds to combustion efficiency 99.9%). In some factories, the temperature increase over 900°C lead to dioxin and furan content increase [3].

Besides CO and organic compounds, emissions of which are determined by the combustion mode, the rest emissions are determined individually by the specific technology and fuel [12]. The paper [3, 13] concluded that the achievement of the allowable emission values is possible by creating the required combustion mode and monitoring it. At the same time, the reduction of energy and exergy losses leads to a reduction of the emission's impact on the environment [14]. The paper [9] suggests for the MSW combustion efficiency increase and emissions reduction use not only the steady required combustion mode but also perform afterburning using an additional fuel, which allows the reduction of NOx and organic compounds emissions.

The paper [15] claimed that as of now, the dioxin emissions are being reduced not by limiting chlorine content in the incinerated waste but by the combustion process control. This requires not only furnace operating mode control but also getting full information using mathematical modeling.

Therefore, it can be concluded that:

a) most of the measured emissions from the waste combustion are within limits and don`t pose risks for human health;

b) none of the technologies give an obvious ecological advantage during waste combustion;

c) obtaining the allowable emission values is possible by keeping the required combustion mode (combustion with excess air, when oxygen content is 6-12% in the combustion products for MSW incineration and 3-9% for RDF incineration; ensuring complete mixing of fuel and combustion products with air; keeping gas temperature at 800-900°C; CO content in the exiting gases less than 50 ppm).

2. Measured substances and the method of calculations

In this paper, evaluation of the individual emissions during different types of wastes combustion in two low power water heating boilers was carried out. In the process of the industrial experiment the measurements of harmful substances in the chimney stuck after boiler were conducted, the calculated ground-level substances concentration was also determined, which was then compared with the norms of maximum allowable concentration (MAC) of emissions in the atmosphere of the populated place.

The tests were conducted using two boilers, which TC «Ecotrans» (Belgorod city) operates:

- 100 kW pyrolysis heating boiler «KO-100» made by LLC «Borkotlomash» (Voronezh region);
- 200 kW solid-fuel heating boiler with the stoker grate and forced-flow fan made by TC «Ecotrans».

The values of emissions were compared with the following allowable values:

- MAC s.t. – maximum allowable short-term concentration of a harmful substance in the atmospheric air (concentration that throughout whole life doesn`t have direct or indirect harmful effect on the present or future generation, doesn't degrade human performance, doesn't worsen their health conditions and sanitary-and-household living conditions);

- MAC a.d. – the maximum allowable average-daily concentration of a chemical substance in the air of populated places (this concentration shouldn't have a direct or indirect harmful effect on human when inhaled for an indefinitely long period);
- MAC w.s. – maximum allowable concentration of harmful substances in the workspace air (concentrations that affect humans daily, except weekends, during 8 hour work period (or different duration, but not more than 41 hrs. per week) within whole working life and doesn't induce disease or health problems, which can be detected by the modern health examination methods, in workers during professional activities and after retirement, and also in the next generations).

MAC values are given in accordance with the Russian normative:

- Hygienic normative GN 2.1.6.3492-17 «Maximum allowable concentration (MAC) of pollutants in the atmospheric air of urban and rural settlements», approved by a decision of the Chief State Medical Officer of the Russian Federation on 22.12.2017 № 165;
- State standard GOST 12.1.005-88. Occupational safety standards system. General sanitary requirements for working zone air;
- Hygienic normative GN 2.2.5.3532-18 «Maximum allowable concentration (MAC) of harmful substances in the working zone air», approved by a decision of the Chief State Medical Officer of the Russian Federation on 13.02.2018 № 25.

To calculate the surface concentration of the harmful substance, which was estimated relative to the MPC, the paper "Methods for calculating the dispersion of harmful (polluting) substances in air" (2017) approved by the Ministry of Natural Resources and Ecology of the Russian Federation was used.

Maximum value of the ground-level harmful substance concentration c_m, mg/m^3, during air-gas mixture discharge from the single point source with a round opening, was calculated by the formula:

$$c_{\mathrm{m}} = \frac{AMFmn\eta}{H^2 \sqrt[3]{V_1 \Delta T}}, \qquad (1)$$

where A – coefficient that depends on the thermal stratification of the atmosphere (for Belgorod city A = 180); M – mass of the harmful substance, which is discharged into the atmosphere per unit time, is calculated by using the measured concentration in the smokestack g/s; F – non-dimensional coefficient, which considers settling velocity of harmful substances in the atmospheric air (for gaseous harmful substances F = 1); m and n - coefficients which consider conditions of the gas-air mixture discharge from the opening of emission source; H – above ground height of the emission source, m; η – non-dimensional coefficient, which considers surface topography influence, in case of flat or moderately rugged terrain with elevation difference not more than 50 m per 1 km, h = 1; ΔT – difference between the discharged air-gas mixture temperature T_{gas} and temperature of the outside atmospheric air T_{air}; V_1 – volumetric flow rate of the air-gas mixture, m^3/s,

$$V_1 = \frac{\mathrm{p}D^2}{4}\omega_0, \qquad (2)$$

where D – diameter of the emission source opening, m; ω_0 – average exit speed of the air-gas mixture at the emission source opening, m/s.

In case of concurrent combined presence in the atmospheric air of n substances, which possess summation of the harmful effect, the paper "Methods for calculating the dispersion of harmful (polluting) substances in air" suggest to calculate the non-dimensional summarized concentration q or reduce conditionally the concentration values to the concentration value of one of them.

The non-dimensional concentration q in the paper is determined by the formula

$$q = \sum \frac{C_i}{\mathrm{MAC}_{\mathrm{s.t.}i}}, \tag{3}$$

where Ci – calculated concentrations of harmful substances in the air; $\mathrm{MAC}_{s.t.\,i}$ – corresponding maximum allowable short-term concentration of harmful substances in the atmospheric air.

3. Emission value evaluation during different type of wastes combustion in the pyrolysis boiler

In the process of tests, that lasted for 3 days, content analysis of the following compounds in the gases exiting a boiler was carried out: carbon oxide, methane, phenol, benzol, sulfur dioxide, formaldehyde, chlorine hydride. Measurements were conducted using the multicomponent portable gas analyzer GANC-4 (Gas Analyzer Cassette based).

Based on the measurement results according to the method paper "Methods for calculating the dispersion of harmful (polluting) substances in air" the maximum value of the ground-level harmful substances concentration was calculated for conditions of the air-gas mixture emission from the single point source with a round opening. The following data was used for the calculation: smokestack opening is 10 m above the ground; smokestack opening diameter 0.25 m; gas temperature in the opening is 200°C; outside air temperature 0°C; average speed of the air-gas mixture exiting from the emission source opening 1 m/s.

The obtained values were compared with the maximum allowable concentration (MAC) of harmful substances which can be found in the norms [14]. For all considered measures the calculated ground-level concentration of harmful substances doesn`t exceed MAC. In most of the cases, it is much lower than the allowable values. This confirms mentioned in some papers statement that the advantage of pyrolysis, in comparison to direct waste combustion, is in substantial emission reduction [15-17].

Table 2. Ratio of the calculated ground-level concentration of harmful substances to maximum short-term MAC in the atmospheric air of the populated places.

№	Item name	NO_2	NO	CO	CH_4	Phenol	Benzol	SO_2	Formaldehyde	HCl
2-a	Railway sleepers	0.7%	0.3%	0.3%	–	0.0%	3.7%	0.1%	0.4%	0.1%
2-b	Wood (pine tree)	0.0%	0.0%	0.5%	–	0.9%	0.1%	0.0%	0.0%	0.4%
2-c	Wood waste (window, door, furniture)	0.5%	1.5%	0.6%	–	13.5%	6.2%	0.1%	3.3%	0.1%
2-d	Wood pellets	0.7%	0.4%	0.6%	–	0.9%	0.1%	0.0%	0.2%	0.4%
3	RDF fuel	0.1%	0.2%	0.2%	–	0.1%	1.1%	0.0%	1.8%	0.0%
5	Mixture of wood pellets and RDF fuel	0.5%	1.6%	0.6%	–	0.0%	2.5%	0.0%	2.1%	0.2%

Table 3. Ratio of the calculated ground-level concentration of harmful substances to average-daily MAC in the atmospheric air of the populated places.

№	Item name	NO_2	NO	CO	CH_4	Phenol	Benzol	SO_2	Formaldehyde	HCl
2-a	Railway sleepers	3.4%	2.1%	0.5%	–	0.0%	11.1%	1.5%	2.2%	0.4%
2-b	Wood (pine tree)	0.1%	0.1%	0.8%	–	1.5%	0.3%	0.01%	0.0%	1.3%
2-c	Wood waste (window, door, furniture)	2.5%	10.0%	1.0%	–	22.5%	18.4%	1.1%	16.4%	0.2%
2-d	Wood pellets	3.3%	2.6%	1.0%	–	1.5%	0.3%	0.01%	1.0%	1.2%
3	RDF fuel	0.3%	1.2%	0.4%	–	0.1%	3.3%	0.01%	8.9%	0.1%
5	Mixture of wood pellets and RDF fuel	2.4%	10.7%	1.0%	–	0.0%	7.3%	0.01%	10.2%	0.7%

IOP Conf. Series: Materials Science and Engineering **791** (2020) 012055 doi:10.1088/1757-899X/791/1/012055

Table 4. Ratio of the harmful substance concentrations in flue gases to MAC in the workspace air.

№	Item name	NO_2	NO	CO	CH_4	Phenol	Benzol	SO_2	Formaldehyde	HCl
2-a	Railway sleepers	6.1	2.3	6.2	–	0.0	6.6	–	3.9	0.1
2-b	Wood (pine tree)	0.3	0.1	10.2	–	2.6	0.2	–	0.0	0.3
2-c	Wood waste (window, door, furniture)	4.5	10.7	13.4	–	40.0	11.1	–	29.2	0.1
2-d	Wood pellets	5.8	2.7	13.4	–	2.6	0.2	–	1.8	0.3
3	RDF fuel	0.5	1.3	4.7	–	0.2	2.0	–	15.8	0.0
5	Mixture of wood pellets and RDF fuel	4.3	11.5	13.4	–	0.0	4.4	–	18.2	0.2

The relationship between harmful emissions content is presented in Table 5.

Table 5. Correlation between content of different harmful substances.

Compound	NO_2	NO	CO	CH_4	Phenol	Benzol	SO_2	Formaldehyde	HCl
NO_2	–	0.40	0.38	0.01	0.16	0.43	0.54	0.03	-0.05
NO		–	0.64	0.59	0.55	0.67	0.16	0.80	-0.31
CO			–	0.27	0.43	0.14	-0.19	0.24	0.47
CH_4				–	0.70	0.50	0.00	0.88	-0.55
Phenol					–	0.77	0.44	0.71	-0.32
Benzol						–	0.79	0.74	-0.67
SO_2							–	0.21	-0.51
Formaldehyde								–	-0.69
HCl									–

Based on the analysis of the measured compounds, it may be concluded that waste combustion doesn`t have a noticeable impact on the environment.

4. Results of measurements in the boiler with grate firing

To determine the possibility of using RDF briquettes for combustion the industrial experiment was conducted for the purpose of gas content analysis, which is generated when briquettes incinerated in the 200-kW heat output water-heating boiler with grate firing, which is located at the LLC TC «Eco-trans» (Belgorod city) [16].

To determine emissions characteristics a series of measurements were carried out during different boiler`s operational modes. The following fuel was used:

RDF: wood 70% and plastic 30% (fuel № 6);

RDF: wood 50% and plastic 50% (fuel № 7).

The fuel was prepared in the form of briquettes with 50 mm diameter and random length (up to 400 mm). The measurements were done during combustion at the boiler`s exit using the measuring system GANC-4 and lasted 3 days. In the process of tests, the analysis of the boiler`s operational efficiency was also carried out.

The temperature of the fuel`s layer, measured on the grate by pyrometer, was 670-760°C.

The obtained values were compared with the maximum allowable concentration (MAC) of harmful substances; which can be found in the norms [14]. For all considered measures the calculated ground-

IOP Conf. Series: Materials Science and Engineering **791** (2020) 012055 doi:10.1088/1757-899X/791/1/012055

level concentration of harmful substances doesn`t exceed MAC. In most cases, it is much lower than the allowable values.

Table 6. Ratio of the calculated ground-level concentration of harmful substances to maximum short-term MAC in the atmospheric air of the populated places.

Fuel №	Item name	CO	NO_2	NH_3	Mercaptan R-SH	H_2S	Phenol
6	Ratio of maximum concentration to MAC	2.5%	2.8%	3.0%	0.05%	2.7%	0.1%
6	Ratio of average concentration to MAC	1.2%	1.7%	2.3%	0.2%	1.9%	0.0%
7	Ratio of maximum concentration to MAC	3.5%	0.8%	8.2%	3.0%	2.4%	2.1%
7	Ratio of average concentration to MAC	1.8%	0.6%	4.7%	1.9%	1.4%	1.1%

Table 7. Ratio of the calculated ground-level concentration of harmful substances to average-daily MAC in the atmospheric air of the populated places.

Fuel №	Item name	CO	NO_3	NH_3	Mercaptan R-SH	H_2S	Phenol
6	Ratio of maximum concentration to MAC	4.1%	14.2%	14.9%	3.5%	13.6%	0.1%
6	Ratio of average concentration to MAC	2.0%	8.6%	11.6%	1.2%	9.5%	0.0%
7	Ratio of maximum concentration to MAC	5.9%	3.8%	40.9%	20.7%	12.0%	3.5%
7	Ratio of average concentration to MAC	2.9%	3.0%	23.4%	13.3%	6.8%	1.9%

Table 8. Ratio of the harmful substance concentrations in flue gases to MAC in the workspace air.

Fuel №	Item name	CO	NO_2	NH_3	Mercaptan R-SH	H_2S	Phenol
6	Ratio of maximum concentration to MAC	62	29	3	3	0.2	0.2
6	Ratio of average concentration to MAC	32	18	3	1	0.2	0.1
7	Ratio of maximum concentration to MAC	72	8	10	19	0.2	8.1
7	Ratio of average concentration to MAC	42	6	5	13	0.1	4.0

To analyze the amount of emissions dependence on the fuel combustion mode, the data from series of tests were used, where the values of harmful substances content are the results of one-time measurements, and the temperatures and excess air coefficients were averaged for the time interval when the gas analysis was performed [17].

Figures 1 and 2 present the emissions content at different gas temperatures after the boiler and different excess air coefficient.

IOP Conf. Series: Materials Science and Engineering **791** (2020) 012055 doi:10.1088/1757-899X/791/1/012055

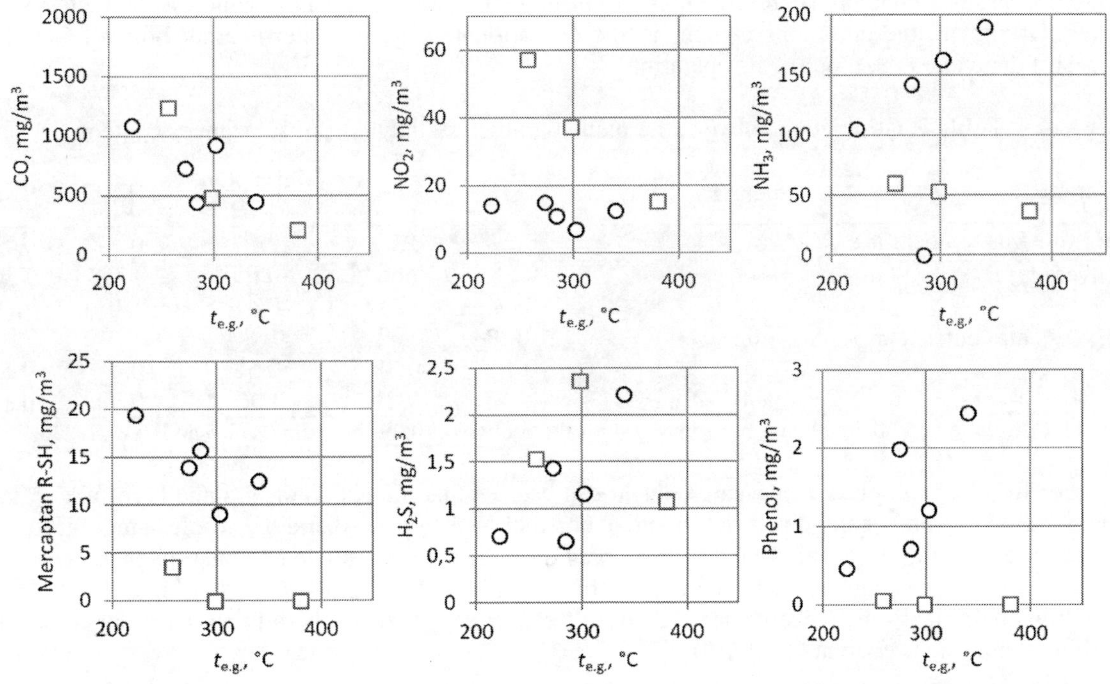

Figure 1. Dependence of the emissions amount on the exiting gas temperatures.

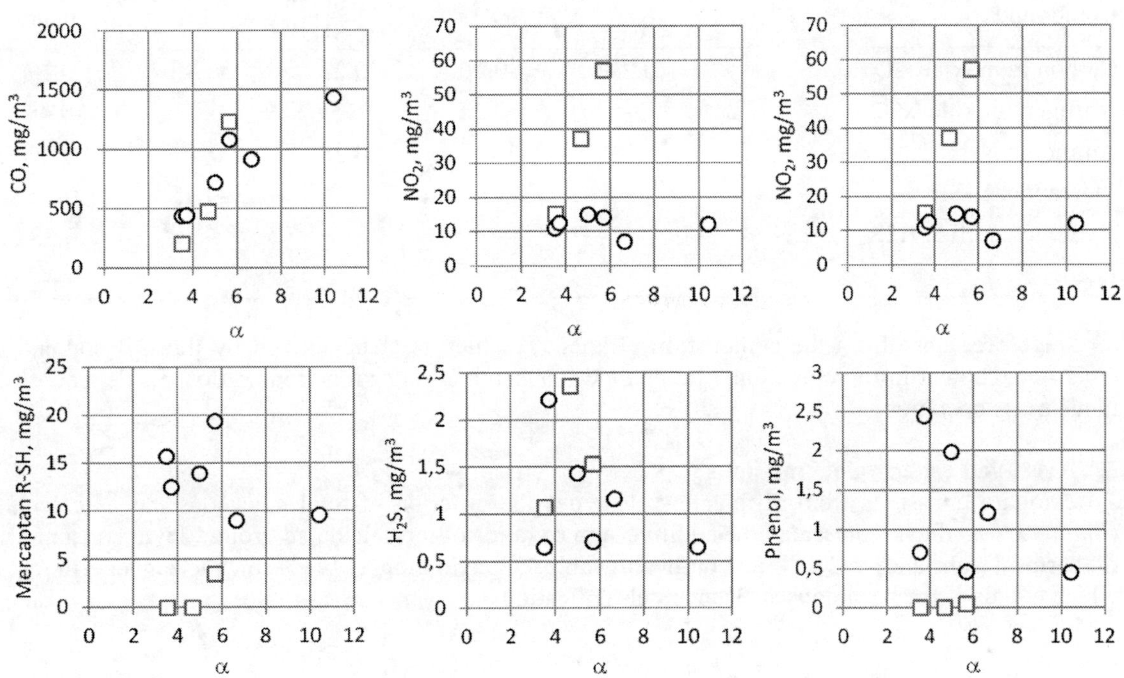

Figure 2. Dependence of the emissions amount on the excess air coefficient.

Dependence of the content of the harmful substances on the mode parameters of the boiler was detected only between CO content and the exiting gas temperature (correlation coefficient – 0.9) and excess air coefficient (0.9). Between the rest of emissions content and $t_{e.g.}$ (α) there is no dependence

IOP Conf. Series: Materials Science and Engineering **791** (2020) 012055 doi:10.1088/1757-899X/791/1/012055

(module of the correlation coefficient doesn`t exceed 0.39). Also, there is no dependence between different emissions content. Therefore, apart from the combustion efficiency, which is characterized by the CO content, the emissions content in the operational range of water-heating boiler functioning doesn`t depend on the mode of its operation.

Table 9. Correlation between the mode parameters and harmful substances content.

Parameter	Fuel №	CO	NO_2	NH_3	Mercaptan R-SH	H_2S	Phenol
Exiting gas temperature (average) $t_{e.g.}$	6	-0.90	-0.99	-1.00	-0.76	-0.52	-0.76
	7	-0.88	-0.28	0.68	-0.05	0.74	0.76
	6 and 7	-0.88	-0.06	0.31	-0.41	0.46	0.18
Excess air coefficient α	6	0.95	1.00	0.99	0.84	0.39	0.84
	7	0.94	-0.15	-0.38	-0.52	-0.45	-0.51
	6 and 7	0.85	-0.07	-0.21	0.06	-0.38	-0.16

Note: Correlation coefficient between $t_{e.g.}$ and α – for the fuel № 6; -0.99; №7: -0.80; № 6 and 7: -0.8

For fuel № 6 (3 points of measurement) it can be seen that mode parameters and harmful substances content have practically direct dependence. For fuel № 7 (6 points) and the whole sample (9 points) dependence on the combustion mode is observed only for the incomplete combustion characteristics – CO. There is no dependence between the rest of the emissions content and $t_{e.g.}$ (α) (module of the correlation coefficient for the whole sample doesn`t exceed 0.31). Also, there is no dependence between different emissions content (Table 10).

Table 10. Correlation between the content of different harmful substances.

Compound	CO	NO_2	NH_3	Mercaptan R-SH	H_2S	Phenol
Carbon monoxide CO	–	0.21	-0.10	0.23	-0.38	-0.17
Nitrogen dioxide NO2		–	-0.22	-0.55	0.43	-0.48
Ammonia NH3			–	0.28	0.46	0.77
Mercaptan R-SH				–	-0.45	0.52
Hydrogen sulfide H2S					–	0.28
Phenol						–

Therefore, apart from the combustion efficiency, which is characterized by the CO content, the emissions content in the operational range of water-heating boiler functioning doesn`t depend on the mode of its operation.

5. Averaged criterion of emissions

To compare emissions from different waste types the criterion of total emissions q was calculated (Figure 3, 4), which represented the additive sum of ratios of the calculated ground-level concentration of harmful substances C_i to maximum allowable concentrations MAC_i considering coefficient k_i, which equals to the hazardous substance's classification:

$$q = \frac{\sum k_i C_i \left(\frac{C_i}{\mathrm{MAC}_i}\right)}{\sum k_i C_i} 100\% = \frac{\sum \frac{k_i C_i^2}{\mathrm{MAC}_i}}{\sum k_i C_i} 100\%.$$

The biggest emissions were observed during the fuel № 2-c combustion (wood waste – windows, doors, furniture), which can be explained by resins and paint presence in the waste. The fuels № 6 and 7 also showed large emissions, because of the large amount of plastic.

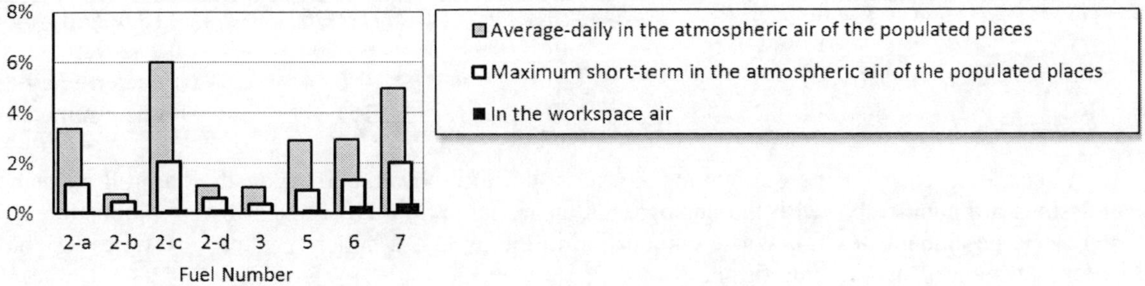

Figure 3. Criterion of the total emissions for the calculated ground-level concentration of harmful substances, which is calculated on values of MAC in the atmospheric air of the populated places and workspace.

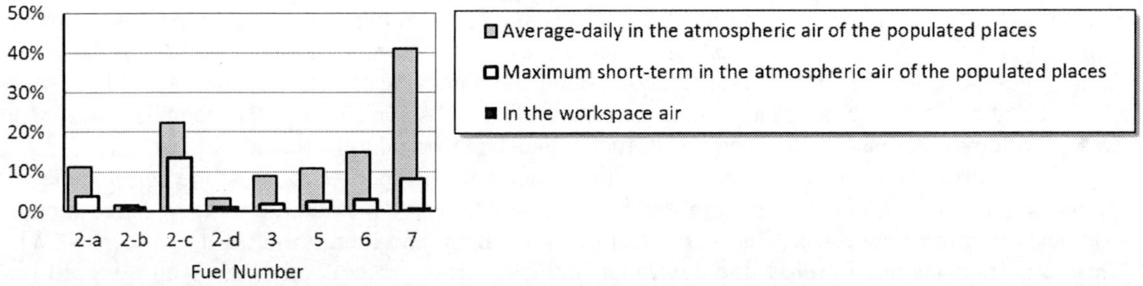

Figure 4. Maximum value of the ratio of the calculated ground-level harmful substances concentration content to MAC in the atmospheric air of the populated places and workspace.

The obtained results cannot fully characterize the above-mentioned fuels, because besides the fuel`s content the mode of combustion (temperature and excess air) also has an impact on the emissions content, but they allow a qualitative comparison of the ecological characteristics of fuels.

Conclusion. The results of the gas content measurements after the boiler, which incinerates a mixture of wood and municipal wastes, showed that on all targeted measures the calculated ground-level concentrations are within MAC limits. Therefore, the combustion of wood waste and MSW doesn`t have a harmful impact on the environment and humans. But it should be noted that the content of dioxins and furans wasn`t measured during the experiments.

6. Conclusion

1. The results of the gas content measurements after the boilers, which incinerate different types of wastes, showed that on all targeted measures the calculated ground-level concentrations are within MAC limits.

Table 11 presents harmful substances concentration during MSW combustion both at the foreign and domestic waste incineration plants and also obtained by the authors during the experiments at low power water-heating boilers, located at the TC «Ecotrans» (Belgorod city) when different types of waste were incinerated.

IOP Conf. Series: Materials Science and Engineering **791** (2020) 012055 doi:10.1088/1757-899X/791/1/012055

Table 11. Harmful substances concentration in the uncleaned exiting gases during MSW incineration, mg/m^3.

Substance	Typical concentrations at the foreign plants [9]	Waste incinerating plant № 2 [9]	Low power water-heating boiler at the TC «Ecotrans»	
			Pyrolysis	With a grate stocker
NO_x	75-600	320	1.2-96	18.6-27.4
CO	<30	30	94-267	582-920
SO_2	50-400	120	0-6.44	Not determined
HCl	50-1000	120	0.151-1.73	Not determined

As can be seen from the measurement results, the low power boilers on the harmful substance emissions are comparable with the analogs, but at the same time have much lower combustion efficiency (large amount of CO). Waste combustion in the pyrolysis boilers creates 2-10 times less hazardous substances than in grate firing.

The emissions content in the exiting gases, apart from the carbon monoxide that characterizes the quality of combustion, doesn`t depend on the boiler`s mode parameters.

But it should be noted that the dioxin content wasn`t measured during the experiments.

2. The additive criterion of total emissions was suggested in the paper, which presents the sum of the ratios of the calculated ground-level hazardous substance concentrations to maximum allowable concentrations, and, unlike the existing dimensionless concentration defined in the normative GN 2.1.6.3492-17, considers hazard class and compound fraction in gases.

The highest value of the total emissions criterion was observed during the combustion of wood waste (windows, doors, furniture) and RDF fuel, which contains a large amount of plastic.

3. Therefore, it may be concluded that the thermotechnical properties of waste are comparable with the traditional types of fuel. The main requirement for the waste incineration is the desired mode of combustion enforcement, meaning thermotecnical organizing and monitoring of the processes. When these requirements are fulfilled, the environmental risks from waste combustion won`t exceed risks from its disposal at landfills. Therefore, wood waste and MSW incineration in the low power boilers doesn`t have a harmful impact on the environment and humans.

Considerable reduction of emissions allows using the recycling method by waste combustion in the pyrolysis boiler for residential heating, for example in the automation systems of heat supply for buildings and residential communities.

Acknowledgments

The article was prepared within development program of the Flagship Regional University on the basis of Belgorod State Technological University named after V.G. Shukhov.

References

[1] Arena U and Gregorio F D 2013 Element partitioning in combustion- and gasification-based waste-to-energy units *Waste Management* **33(5)** 1142-50

[2] de Fodor E. 1911 *Elektrizität aus Kehricht* (Budapest: K.u. K. Hofbuchhandlung von Julius Benkö) [in German]

[3] Hasselriis F 1987 Optimization of combustion conditions to minimize dioxin emissions *Waste Management & Research* **(5)3** 311-26

[4] Roffman A and Roffman H K 1991 Air emissions from municipal waste combustion and their environmental effects *Science of The Total Environment* **104(1–2)** 87-96

[5] Kilgroe J D 1989 Combustion control of trace organic air pollutants from municipal waste combustors *Environmental Impact Assessment Review* **9(3)** 199-222

[6] Ruth L A 1998 Energy from municipal solid waste: A comparison with coal combustion technology *Progress in Energy and Combustion Science* **24(6)** 545-64

[7] Vehlow J 2015 Air pollution control systems in WtE units: An overview *Waste Management* **37** 58-74

[8] Chang Y-H, Chen W C and Chang N-B 1998 Comparative evaluation of RDF and MSW incineration *J. of Hazardous Materials* **58(1–3)** 33-45

[9] Edo M, Ortuño N, Persson P-E, Conesa J A and Jansson S 2018 Emissions of toxic pollutants from co-combustion of demolition and construction wood and household waste fuel blends *Chemosphere* **203** 506-13

[10] Myrina E S, Persson P-E and S Jansson S 2014 The influence of food waste on dioxin formation during incineration of refuse-derived fuels *Fuel* **132** 165-9

[11] Ragazzi M and Rada E C 2012 Multi-step approach for comparing the local air pollution contributions of conventional and innovative MSW thermo-chemical treatments *Chemosphere* **89(6)** 694-701

[12] Werther J 2007 Gaseous emissions from waste combustion *J. of Hazardous Materials* **144(3)** 604-13

[13] Lombardi L and Carnevale E A 2018 Evaluation of the environmental sustainability of different waste-to-energy plant configurations *Waste Management* **73** 232-46

[14] Rosen M A 2004 Exergy Analysis of Waste *Emissions Encyclopedia of Energy* ed C J Cleveland (Elsevier) pp 623-32

[15] Costa M, M Dell'Isola M and Massarottic N 2012 Temperature and residence time of the combustion products in a waste-to-energy plant *Fuel* **102** 92-105

[16] Kornilova N V and Trubaev P A 2019 Thermotechnical tests of the low power solid fuel heating boilers while incinerating briquetted RDF *IOP Conf. Ser.: Mater. Sci. Eng.* **552** 012030

[17] Suslov D Yu, Ramazanov R S and Shvydkaya M A 2018 To the Question of Biogas Combustion in Vortex Burners *J. Phys.: Conf. Ser.* 1066 012028

Development of calculation and design of rotor-blade engine N.N. Tverskoy with a built-in electric generator

A S Korobets and S F Stepanov

Department of Power and Electrical Engineering, Yuri Gagarin State Technical University of Saratov, Politechnicheskaya St., 77, Saratov, 410054, Russia

E-mail: Andrey-wow.64@mail.ru

Abstract. Current requirements are considered for an autonomous source of power supply for powering a consumer located in areas remote from a centralized power system in comparison with the method of connecting to a centralized power system, using a petrol, diesel and power plant using a Stirling engine with external heat supply in this article. The main aspects are identified power plants with external heat supply, analysed materials that support combustion, which can be used to operate an autonomous generator. The history is described and the purpose of the rotary engine of the Russian engineer N.N. Tverskoy. An economic assessment and engine performance have been considered, in which superheated steam is not required to be used, but steam with droplets of liquid for use is permissible. The design proposed rotary vane engine N.N. Tverskoy with a built-in electric generator, in which the expansion part (turbine) and the generator represent a single joint part. The design methodology is considered, in which the main sections on the calculations and design of the rotary machine with the main empirical formulas are highlighted. Further ways of using the engine are described in the low-grade heat recovery system.

1. Introduction

All spheres of human activities associated with electricity today. One day without electricity, like a difficult challenge for the modern man and for the company, which operates large amounts of data, in real conditions of a disaster. Serious demands on the reliability and continuity of power supply are applied to the centralized power supply on the basis of the relationship between various kinds of human activities from electricity at the moment. The question arises: people in the areas as to be remote from centralized energy system and those who impact natural disasters create long breaks in power supply.

Electricity by attaching to the centralized energy system of the object line 6-10 kV is very limited in power and length of paving of the line, the more power consumption of the consumer (up to 250 kW), the shorter may be the power lines. The building is inexpedient, the power lines 35 kV and above, the line will work in the regime close to the idling speed at small transmission capacities, the economic efficiency of capital investments is reduced to zero [1].

Scientific-technical progress develops the production of electricity using power plants based on renewable sources of electricity to use their autonomy from the centralized energy system. Renewable energy sources are developing, and this is due to the efforts and concern about greenhouse gases that influence global climate. Reduction of dependence on imports of energy and raw materials to power plants is also a very important aspect of renewable energy development. Events, analyzing the latest time you may notice a reduction in the consumption amount of raw material needed for energy production by the countries of Europe. It can be concluded on the basis that renewable energy sources play an important role in the energy policy of the European Union. But given the fact that renewable energy can create some difficulties in offline and distribution network due to its oscillatory nature, use plants for organic fuels for redundancy [2].

2. Materials and Methods

Diesel and petrol are the most common small (up to 10 kW) autonomous power plants now that use internal combustion engines, and we also highlight the not so common one - the Stirling engine, which can be used in a cycle with external combustion. The engines presented have a number of advantages and disadvantages, which we highlight in Table 1 and Table 2.

Table 1. Advantages of a power plant with a petrol, diesel or Stirling engine.

Diesel	Petrol	Stirling with external heat input
Lack of ignition system	Affordable cost	Closed loop work
Work resource	Service	Low noise
Economical consumption	Noise level	Ability to work on various types of fuel
Quick spread	Quick spread	Partial load characteristics
Power		Efficiency at low ambient temperatures

Table 2. Disadvantages of a power plant with a petrol, diesel or Stirling engine.

Diesel	Petrol	Stirling with external heat input
Noise level	Fine tuning the intake / start system	System warm up required
Whimsical to fuel quality	Whimsical to fuel quality	Efficiency 15-25%
Service cost	Fuel consumption	Slow application progress
Crank mechanism	Crank mechanism	Crank mechanism
		Communication with the generator through the shaft
		Overall dimensions

The use of a small autonomous power station (petrol, diesel fuel) creates storage restrictions, and requires constant monitoring of the amount of fuel in conditions of remoteness from gas stations. The use of a power plant with a Stirling engine with an external heat supply allows us to rely on materials that support combustion, the specific heat of combustion in atmospheric air of which is presented in Table 3. Taking into account the shortcomings of the Stirling engine shown in Table 2, it can be concluded that the operation of this system and its resource of work, but one should not miss the advantages that are decisive in the face of a limited supply of petrol (diesel fuel).

The advantages of the system on an external heat supply, taking into account the design flaws of the Stirling engine, should be taken into account that the engine is necessary for efficient operation, in which there is no: a crank mechanism and a shaft between the engine and the generator through which torque is transmitted. An engine which does not have the above items has a longer service life due to the reduction in the number of components of the motor structure and the loss on the shaft will be excluded when transmitting torque. It remains only to determine the choice of working fluid for the engine to work with an external supply of heat.

IOP Conf. Series: Materials Science and Engineering **791** (2020) 012056 doi:10.1088/1757-899X/791/1/012056

Table 3. Characteristics of materials that support combustion.

Substance	Calorific value, MJ/kg
Propane	47.54
Petrol	44
Diesel fuel	42.7
Household gas	31.8
Firewood	14.4-17.4
Waste *	≈17.0

*the calorific value when burning waste of fabric, nylon, nylon, rubberized fabric, impregnated PVC fabric, polystyrene foam, household waste, paper, plastic film, rags, PVC, wood waste is at least 17 MJ/kg.

3. Results

The idea arose of the design of a rotary machine for a submarine by the engineer-mechanic of the fleet of the Russian Empire Nikolai Nikolayevich Tverskoy in 1876 (Figure 1) [3]. A true rotary machine differs from engines with a crank mechanism in its symmetry, small number of rotating parts, simplicity of design and rotation by a working fluid (freon), which is heated outside the closed loop of the system according to the ORC (Organic Rankine Cycle) duty cycle.

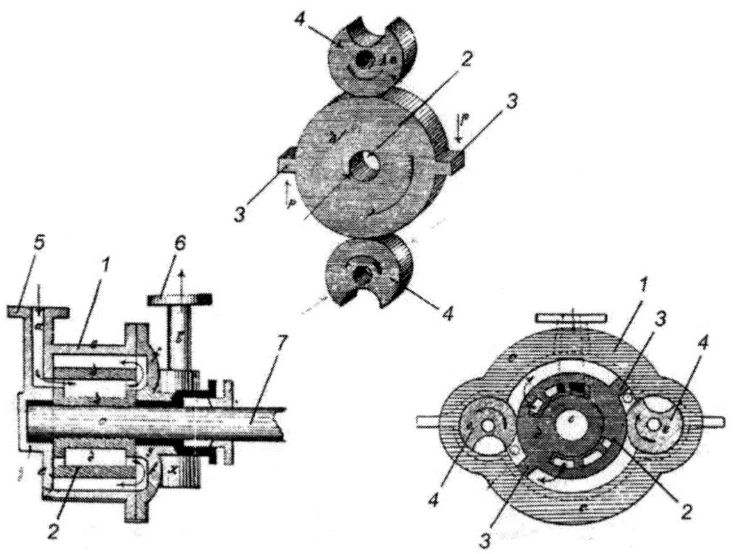

Figure 1. Diagram of a rotary machine designed by N.N. Tverskoy (1 - housing, 2 - rotor, 3 - pistons, 4 - sealing rollers, 5,6 - steam supply and exhaust pipes, 7 - shaft).

Rotary vane engine N.N. Tverskoy with an integrated electric generator combines the advantages of a rotary machine, and also has a longer service life due to the reduction in the number of components of the motor structure, increased efficiency due to the use of heat losses from the generator, which additionally heats the working fluid and eliminates shaft losses during transmission of the rotary moment (Figure 2) [4]. The unified design of the engine - generator is an integrated module.

The spread of devices with a blade structure is slow at present, but I would like to note the invaluable potential of these machines. Unlike turbomachines, the production of this design is relatively inexpensive, and as a result, the price will be attractive for the final product using a vane engine. The mode of operation is weakly dependent on dynamic effects, in contrast to the transmission of torque in a turbine associated with the movement of a liquid, steam or gas, which means that overheating of the working fluid is not required. If we don't need overheating of the working fluid, then the area of the calculated heat exchanger will be less, and this increases the economic viability of the system, calculated according to formula 1. Thus, this apparatus is suitable for efficient operation under any working fluid, as well as for working with mixtures. On the contrary, working with a working fluid in which there will be liquid seals the gaps between the rotor and the housing and the rotor, and the sealing rollers improving lubrication. Capital costs for the production of such an apparatus are reduced due to improved lubrication. In addition, adequate requirements should be noted when designing the pistons (blades) of the engine N.N. Tverskoy and the possibility of combining with a generator without gear links [5].

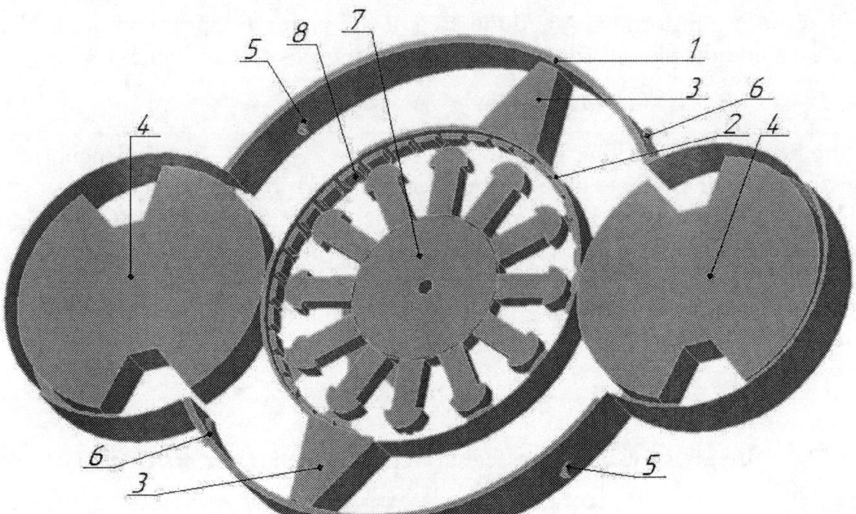

Figure 2. Diagram of a rotary vane engine N.N. Tverskoy with a built-in electric generator (1 - housing, 2 - rotor, 3 - pistons, 4 - sealing rollers, 5,6 - supply and exhaust pipes of the working fluid, 7 - stator, 8 - permanent magnets).

$$E = \frac{Q}{N}, \quad (1)$$

$$E = \frac{G \cdot c \cdot \Delta t_{\max}}{V \cdot \Delta p},$$

where E is the total heat capacity of the apparatus, Q is the heat transferred through the heat exchange surface, N is the work spent to overcome the hydrodynamic resistance, G is the mass flow rate of the coolant, kg / s, s is the heat capacity of the coolant, kJ/kg * K, Δt_{max} - the maximum temperature head or the difference between the initial temperatures of the coolants, K, V is the volumetric flow rate of the coolant, m^3 / s, Δp is the hydrodynamic resistance, Pa [6].

The movement is converted from the pressure of the working fluid, which drives the rotor with pistons (blades) on which the permanent magnets are located on the inside. When the engine is running, the pressure of the working fluid on all the parts, the magneto dynamic coercive force, the inertia and friction forces, which depend on the various operating modes of the apparatus, act. Therefore, the following design sequence is possible in the calculation method of the rotor-vane engine N.N. Tverskoy with a built-in electric generator [7, 8].

Input of initial data for designing the apparatus: power N, angular velocity ω and torque on the rotor M. Power N is determined by the load schedule N (t) for a given time interval by the formula 2. Given the number of revolutions per unit time, ν we determine the torque M by the formula 3.

$$N_{и} = \frac{1}{\eta \cdot \tau} \int_{0}^{\tau} N(t) dt, \tag{2}$$

$$M = \frac{N}{2 \cdot \pi \cdot v}, \tag{3}$$

Calculation of the parameters of the expansion part of the rotary vane engine: the flow rate of the working fluid according to formula 4 and the area of the piston (blade) working surface [9]. The gap between the casing and the pistons (blades) is set to 50 microns (0.002 inches).

$$M \cdot c_{v} \cdot dT + p \cdot dV + dQ_{w} = 0, \tag{4}$$

where M is the mass of the working fluid in the chamber; c_v is the specific isochoric heat capacity of the working fluid; T is the temperature of the working fluid in the chamber; p is the pressure of the working fluid in the chamber; V is the volume of the working chamber; dQ_w is the heat given (received) by the working fluid as a result of heat exchange with the walls of the chamber.

The choice of the working fluid, the calculation and the selection of the heat exchanger for the working fluid for the parameters of the amount of heat input from the type of material (fuel) that supports combustion according to Table 3 according to the formula (5).

$$Q = K \cdot F \cdot \Delta t_{cp} \cdot \tau, \tag{5}$$

where Q is the heat flux, F is the heat transfer surface, K is the kinetic coefficient of heat transfer characterizing the rate of heat transfer, Δtav is the average temperature difference over the heat transfer surface, and τ is the time.

The pressure of the working fluid at the inlet to the apparatus is calculated, the thermodynamic parameters necessary for calculating the geometry of the structure are determined.

Design and construction of rotor-blade engine parts.

Performing power, strength and dynamic calculations for the engine design with the aim of determining inertial parameters, centering the rotor stroke and other strength characteristics.

Modeling engine operating modes in the software package.

Electric power is calculated by the obtained thermodynamic and inertial parameters to determine the geometric data and the stator winding of a synchronous permanent magnet generator.

The inductive parameters of the synchronous generator are calculated, and the inductances are calculated using the formula 6. Mathematical modeling of the engine - generator system is performed in the Matlab Simulink software package to obtain the machine operating mode.

$$L = \frac{\Psi}{I}, \tag{6}$$

$$\Psi = \sqrt{\frac{1}{n} \sum_{i=1}^{n} (\psi_i - \psi)^2},$$

$$I = \sqrt{\frac{1}{n} \sum_{i=1}^{n} i_i^2},$$

where L is the inductance, is the rms value of the flux linkage of the phases, I are the phase currents.

Design of the design of the stator and external rotor with slots for permanent magnets, as well as a bearing operating at high temperatures.

IOP Conf. Series: Materials Science and Engineering **791** (2020) 012056 doi:10.1088/1757-899X/791/1/012056

The thermal calculation of the synchronous generator to determine the necessary characteristics of the materials used in the generator, taking into account the temperature in the expansion part of the rotor-blade engine N.N. Tverskoy.

Design and assembly of the structure to determine the integral weight and size characteristics of the rotary vane engine N.N. Tverskoy with a built-in electric generator.

Design and calculation of the engine-generator control system according to the scheme depicted in Figure 3 [10].

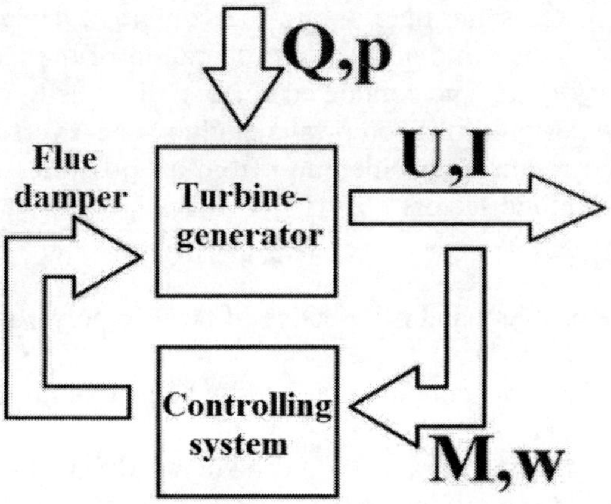

Figure 3. The control system of the rotor-vane engine N.N. Tverskoy with a built-in electric generator indicating the main parameters.

where Q is the heat flux, p is the pressure of the working fluid in the chamber, U is the voltage, I is the current, M is the torque, ω is the angular velocity.

4. Discussion

Modern policies of many countries aimed at the production of electricity in a way that will not harm the ecological background in this area and terrain. The issue of "clean energy" remains open and active for discussions at the world conferences and forums on the background of global warming and world climate change. I would like to note that the policy is widely distributed in the field of ecology, particularly in the industrialized States, but differs from the actual use virtually much. Significant result are possible to obtain when planning ecological development involving several sectors of the state fuel and energy complex, economy and Finance, natural farming, urban planning, agriculture and transport. It is possible to achieve maximum efficiency from the operations only at the intersection of interaction of state sectors [11-13].

Rational use of natural resources distributed generation when using recycled materials from human and animal production of biogas is rapid direction in the development of environmental protection now. This is a Prime example of the synergy of the sector development of the fuel and energy sector and agriculture. The aim of the projects is the creation, development and application of technologies, equipment and systems for the efficient production, collection and processing of various biological and manmade materials. Example of a research program in this direction: the increasing use of alternative energy source – biogas. The biogas produced from processing agricultural waste: animal and vegetable origin. Further development of the technology of using biogas as an alternative to considering natural gas, should take into account the efficiency of the first 30-50% less in comparison with the second [14-16].

The development of alternative sources of energy will have a positive impact on the environment, but also of note is the energy that is produced by thermal, nuclear, power plants, and then just disappears into the atmosphere in the form of heated air. Energy are not fully utilized and a great loss not only lowers the efficiency of the equipment, as well as adversely affect global climate change because of the huge number of low-grade thermal energy, emitted into the atmosphere. Application of technologies for utilization of low potential energy production is the solution, thereby increasing the overall efficiency of the station, and also positively affects the development of the municipal districts.

Technology for utilization of low-potential heat based on ORC. In the ORC is used as the working fluid with a low boiling working body (freon), which have a low boiling point at atmospheric pressure that allows to reach the required system parameters when using waste heat from power plants. The application of technologies for utilization of low-grade heat hides additional economic and ecological effect from the introduction of this heat recovery system the exhaust [17-19].

5. Conclusion

Design of a rotary engine should be considered at the same time due to the combined design of the thermal and electrical parts, as well as the ability to take into account the operation of the expansion apparatus for a couple in which there are drops of liquid. The conducted studies allow us to conclude that the potential and practical applicability of the design of the rotor-vane engine N.N. Tverskoy. The topic of the environmental problem is topical and requires consideration from all possible sides and involving modern technologies, new solutions and public sectors.

References

[1] Surzhikova O A and Nikulina I E 2006 The problem and perspective of electric power supply of isolated consumers *European J. of Natural History* **2** 151-3

[2] Yasmine A, Sharon J and Rosemary N 2012 Low grade thermal energy sources and uses from the process industry in the UK *Applied Energy* **89** 3-20

[3] Rassol I R 2011 Podvodny`e lodki inzhenera-mexanika N.N. Tverskogo i dvigateli dlya nix [Submarines of machine engineer N.N. Tverskoy and their engines] *Sydostroenie* **4** 67-71 [in Russia]

[4] Korobets A S and Stepanov S F 2018 Efficient reclamation of low-grade heat with the help of a small multifuel autonomous power-plant *J. Phys.* **1111** 012042

[5] Leibowitz H, Smith I K and Stosic N 2006 Cost effective small scale orc systems for power recovery from low grade heat sources *ASME Int. Mechanical Eng. Congress and Exposition* (Chicago: ASME) pp. 521-527

[6] Caputo A C, Pelagagge P M and Salini P 2008 Heat exchanger design based on economic optimization *Applied Thermal Engineering* **28** 1151-9

[7] Qing-ming C, Fang-wen H, Deng-hai T, Fang-lin H and Lin-zhang L 2012 Prediction of loading distribution and hydrodynamic measurements for propeller blades ln a rim driven thruster *J. of Hydrodynamics* **24** 50-7

[8] Andreev M, Zhuravlev Y, Lukyanov Y and Perminov L 2013 Autonomous power station based on rotary-vane engine with an external supply of heat *Proc. of the 9th Int. Scientific and Practical Conf.* (Rēzekne: Rezekne Academy of Technologies) **2** pp 97-100

[9] Zhuravlev Y N, Semenov S N and Ivanov A N 2014 Raschyot temperatur i davlenij v rotorno-lopastnom dvigatele s vneshnim podvodom teploty` [Calculation of temperatures and pressures in a rotary vane engine with external heat input] *Vestnik PskovGU* **5** 170-5 [in Russia]

[10] Chinchilla M, Arnaltes S and Burgos J C 2006 Control of permanent-magnet generators applied to variable-speed wind-energy systems connected to the grid *IEEE Transactions on Energy Conversion* **21(1)** 130-5

[11] Jordan A and Lenschow A 2010 Policy paper environmental policy integration: a state of the art Review *Environmental Politics* **20** 147-58

[12] Duit A, Feindt P H and Meadowcroft J 2016 Greening Leviathan: the rise of the environmental state *Environmental Politics* **25** 1-23

[13] Burns C, Eckersley P and Tobin P 2019 EU environmental policy in times of crisis *J. of European Public Policy* **27(1)** 1-19

[14] Shein N T, Sevostyanov V S, Obolonsky V V, Sevostyanov M V, Goryagin P Yu and Babukov

IOP Conf. Series: Materials Science and Engineering **791** (2020) 012056 doi:10.1088/1757-899X/791/1/012056

V A Resource and energy-saving technologies of complex processing and utilization of technogenic materials *IOP Conf. Ser.: Mater. Sci. Eng.* **552** 012042

[15] Suslov D Y and Ramazanov R S The study of energy performance of biogas from agricultural waste *IOP Conf. Ser.: Mater. Sci. Eng.* **552** 012032

[16] Suslov D Y, Ramazanov R S, Temnikov D O and Lobanov I V Development and research of low pressure injection burner for biogas combustion *IOP Conf. Ser.: Mater. Sci. Eng.* **552** 012031

[17] Dai Y, Wang J and Gao L 2009 Parametric optimization and comparative study of organic Rankine cycle (ORC) for low grade waste heat recovery *Energy Conversion and Management* **50** 576-82

[18] Ebrahimi K, Jones G F and Fleischer A S 2014 A review of data center cooling technology, operating conditions and the corresponding low-grade waste heat recovery opportunities *Renewable and Sustainable Energy Reviews* **31** 622-38

[19] Srinivasan K K, Mago P J and Krishnan S R 2010 Analysis of exhaust waste heat recovery from a dual fuel low temperature combustion engine using an Organic Rankine Cycle *Energy* **35** 2387-99

IOP Conf. Series: Materials Science and Engineering **791** (2020) 012057 doi:10.1088/1757-899X/791/1/012057

Energy generation by reverse electrodialysis

S A Loza, A N Korzhov, N V Loza and N A Romanyuk

Department of Chemistry and High Technology, Kuban State University, Stavropolskaya St., 149, Krasnodar, 350049, Russia

E-mail: s_loza@mail.ru

Abstract. In this work, the process of generating electricity from the water salinity gradient (RED) using the Russian produced ion exchange membranes MK-40 and MA-41 was investigated. The experiments were carried out with a saturated brine of sodium chloride in a chamber with salt water and with a solution of sodium chloride with a concentration range 0.017-3.39 g/l in a chamber with fresh water. The maximum value of the specific power obtained was 0.3 W/m^2. The dependence of the specific power of the salt content in the chamber with fresh water was investigated and its optimal value (0.5 – 1.5 g/l) was found. The dependence of the received power on the flow velocity in chamber with fresh water was studied. The RED process with three different RedOx systems was studied and their effect on the received power was determined. The highest power was obtained for the RedOx system of potassium ferro-ferricyanide. In case of not using the ferro-ferricyanide system leads to a significant decrease in specific power and the release of gaseous chlorine in the anode chamber.

1. Introduction

Recently significant interest in technology of the direct conversion of the salt concentration gradient between salty and fresh water into electricity using Reverse Electrodialysis (RED) is not diminishing, due to the increased demand for alternative and autonomous energy sources. The gradient of salt concentration in salty and fresh water, separated by an ion selective membrane, leads to the formation of a potential difference on membrane assembly.

In the installation, shown in figure 1, consisting of alternating cation and anion exchange membranes, between which salty and fresh water flows, the potential difference appearing on the membranes is summed up and can be used as an energy source. To convert the ion current into the electricity a RedOx reaction must take place in the electrode chambers [2-4]. Thermodynamic calculations show that when 1 m^3 of river water (1 g/l NaCl) is mixed with 1 m^3 of sea water (30 g/l NaCl), 1.4 MJ of energy can be obtained (equivalent to 0.4 kWh). The obtained energy can be increases when using feed water more saline than seawater. In case of mixing 1 m^3 saturated NaCl brine (350 g/l NaCl) and 1 m^3 river water, approximately 17 MJ (equivalent to 4.7 kWh) can be obtained. Theoretically, 2400 GW of electric power can be obtained by mixing the global runoff of river water with sea water, 1000 GW from which can be technically used [5-7]. The main problem that prevents the mass use of the reverse electrodialysis process is the large internal electrical resistance of the apparatus, which limits the power received [8-10].

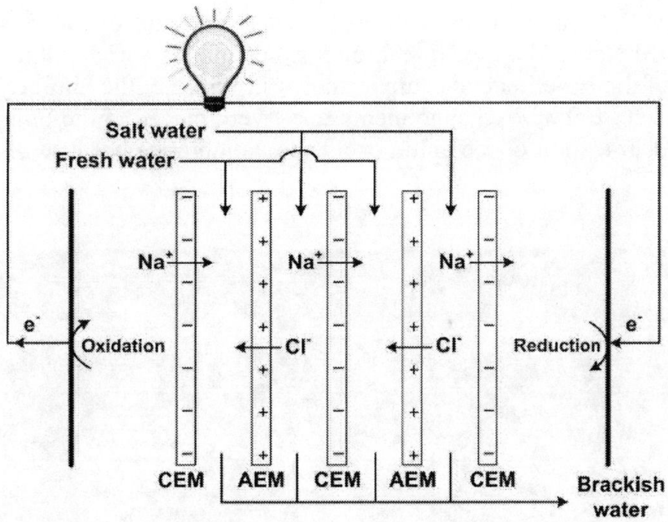

Figure 1. Principle of reverse electrodialysis (RED) [1].

Of great importance is the selectivity of ion-exchange membranes, as well as the design of the separator, which provides turbulent flow and fixation of the intermembrane distance. As the thickness of the fresh water chamber increases, ohmic losses increase significantly, and with a small intermembrane distance partially filled with an inert separator, the cost of pumping the solution can exceed the useful work of the process [11, 12].

It is known that the efficiency of electricity production by reverse electrodialysis can be improved using profiled membranes [7]. This allows to abandon the inert spacer and reduce the cost of pumping the solution [7, 13]. Also, the increased surface area positively affects the efficiency of the process. At the same time, it is possible to use highly efficient and inexpensive membrane materials for RED. As such materials, Russian produced heterogeneous ion-exchange membrane materials are promising, which are relatively inexpensive and at the same time have all the necessary properties, which will increase the specific power of the installation and reduce the cost of electricity.

Thus, the production of renewable energy using sea and river water by the method of reverse electrodialysis is very promising, but not well studied.

The aim of this work is to study the process of generating electrical energy by use of a salt concentration gradient using ion exchange membranes MK-40 and MA-40.

2. Materials and Methods

The objects of study in this work were Russian produced heterogeneous ion exchange membranes MK-40 – a polymer composition of cation exchange resin KU-2 (polystyrene, cross-linked by divinylbenzene), polyethylene and reinforcing fabric. Functional group $–SO_3^-$, and MA-41 – a polymer composition of anion exchange resin AB-17 (polystyrene, cross-linked by divinylbenzene), polyethylene and reinforcing fabric. Functional group $–N^+(CH_3)_3$.

Heterogeneous membranes MK-40, MA-41 passed standard conditioning [14]: the surface of the membrane was wiped with carbon tetrachloride, then it was treated for 24 hours each time with ethyl alcohol, saturated NaCl solution, NaCl solution with a concentration of 100 g/l, NaCl solution with a concentration of 30 g/l. Then the membrane was washed with distilled water until its constant resistance after a 24 h of contact with the membrane.

The experimental setup consisted of an electrochemical cell, 3 LS-301 peristaltic pumps, 2 Agilent U1252B multimeters, a resistance bridge (0.1 Ω –100 kΩ) and solution containers. Working area of membranes was 5x20 cm^2. Electrodes were made of titanium covered with platinum. As a model solution of salt water saturated NaCl was used (approximately 350 g/l), model solution of fresh water:

NaCl with a concentration of from 0.017 to 3.39 g/l. The reverse electrodialysis process was carried out with a MK-40/MA-41 membranes with an electrode solution of 0.025 M $K_4Fe(CN)_6$, 0.025 M $K_3Fe(CN)_6$ and 0.25 M NaCl. The flow rate through all chambers varied within 72-384 ml/min.

The appearance of the experimental setup is shown in figure 2, the number of pair chambers in the cell varied from 5 to 20. Cation exchange membranes were placed at the both sides of the membrane package to prevent degradation of the anion exchange membranes because of contact with potassium hexanoferrate.

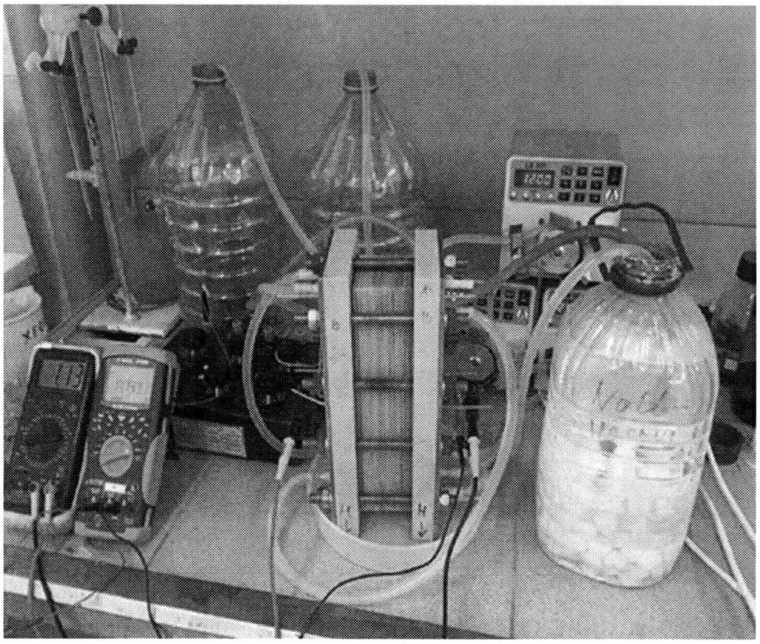

Figure 2. The appearance of the experimental setup.

To determine the power characteristics, a different load was connected to the cell under study and the current strength and voltage were measured. The load range was from 0 (Short circuit) to 500 Ω. In this way, power versus current density curves were obtained.

With increasing concentration of salt in fresh water, an increase in power occurs due to a decrease in the ohmic resistance of the cell. If the threshold concentration is exceeded, the power characteristics decrease due to a decrease in the concentration gradient.

3. Results and Discussion

The process of generating electrical energy by the reverse electrodialysis method using MK–40 and MA–41 membranes was studied. The power characteristics of an electrochemical generator with a different salinity of fresh water were determined.

The numbers on the right side of figure 3 shows the sodium chloride concentration in the chamber with fresh water. With an increase of the salt concentration in the chamber with fresh water, an increase in the obtained specific power occurs due to an increase of the electric conductivity of the solution. A further increase of the concentration of sodium chloride leads to a decrease of the received power due to a decrease in the concentration gradient in chambers with fresh and salt water.

IOP Conf. Series: Materials Science and Engineering **791** (2020) 012057 doi:10.1088/1757-899X/791/1/012057

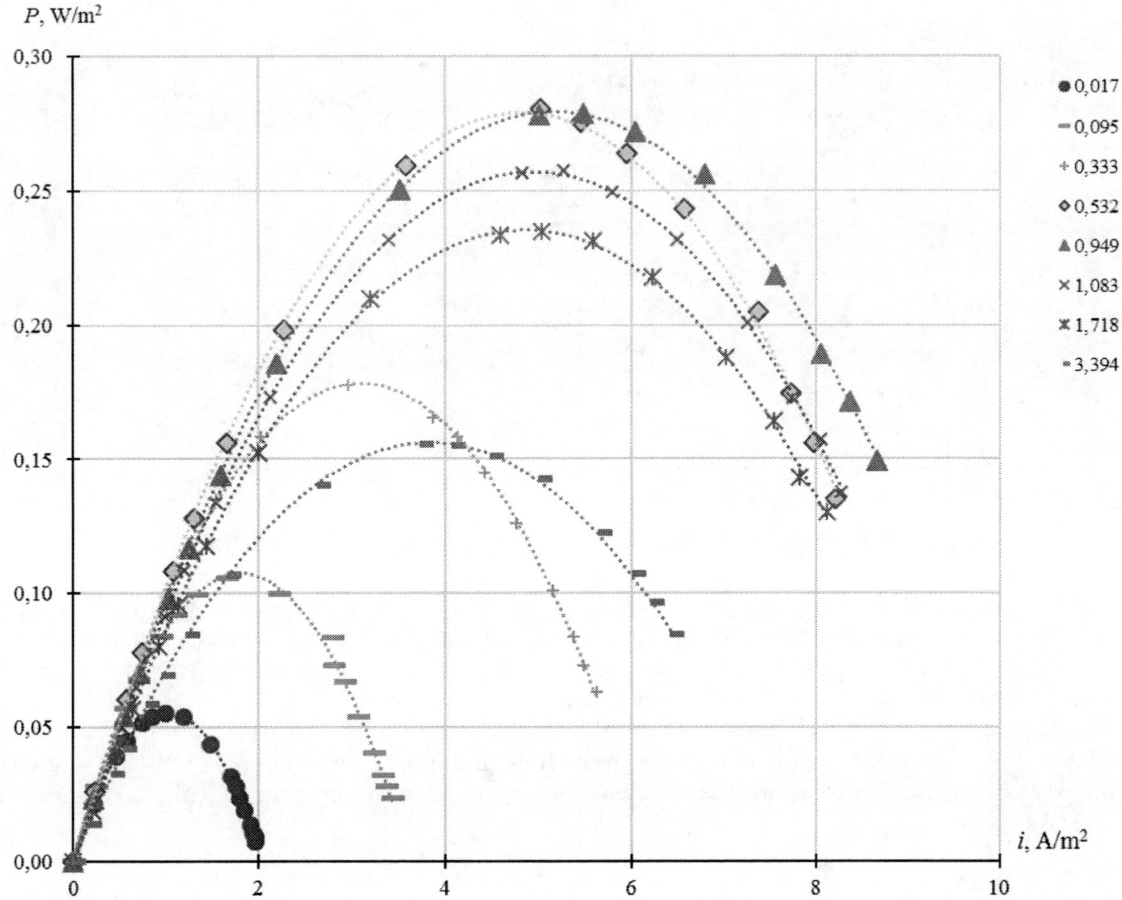

Figure 3. Specific power density versus load current for different salt concentration in "fresh" water.

The maximum specific power obtained was 0.28 W/m^2 with concentration of sodium chloride in fresh water in range 0.5-1 g/l.

Open circuit voltage (OCV) (with zero load) of RED cell with 5 pair chambers were measured (figure 3). Theoretical OCV can be calculated by (1).

$$E_{OCV} = \alpha \cdot N_m \cdot \frac{R \cdot T}{z \cdot F} \cdot \ln\left(\frac{\gamma_s \cdot c_s}{\gamma_f \cdot c_f}\right) \tag{1}$$

In which R is the universal gas constant (8.314 J/(mol·K)), T is the absolute temperature (K), F is the Faraday constant (96485 C/mol), z is the valence of the ion species (-), γ is the activity coefficient of the salt species (-) and c is the concentration of the salt (mol/liter). The subscripts (s) and (f) indicate salty water or fresh water, respectively.

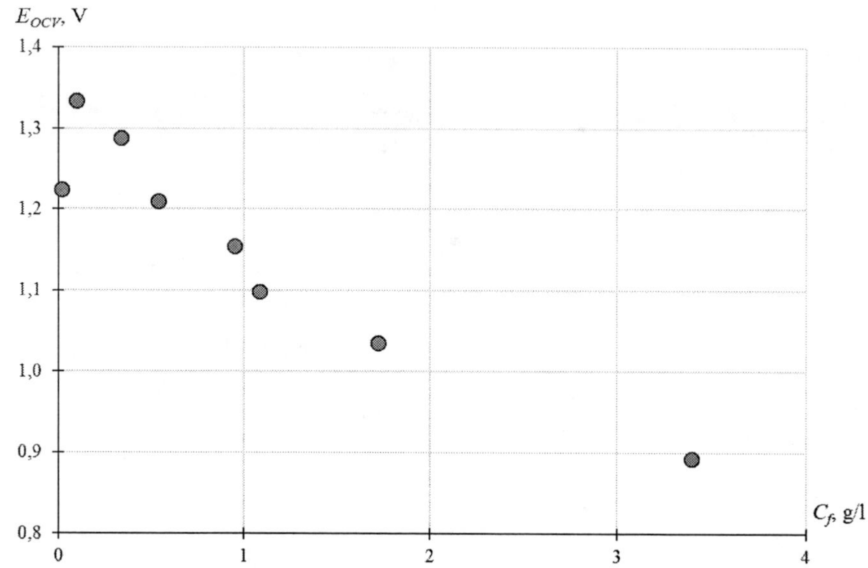

Figure 4. Open circuit voltage dependence of salt concentration in fresh water.

The value of OCV decreases with increasing salt concentration in the chamber with fresh water due to a decrease in the salt concentration gradient. The maximum value was 1.34 V (0.11 V per membrane).

3.1. Specific power dependence of flow rate

The dependence of the specific power of the flow velocity in the chamber with fresh water was investigated. It was shown that with an increase in the flow rate from 72 ml/min to 384 ml/min, the resulting power increases by 15%, however, the energy consumption of pumping the solution significantly increases.

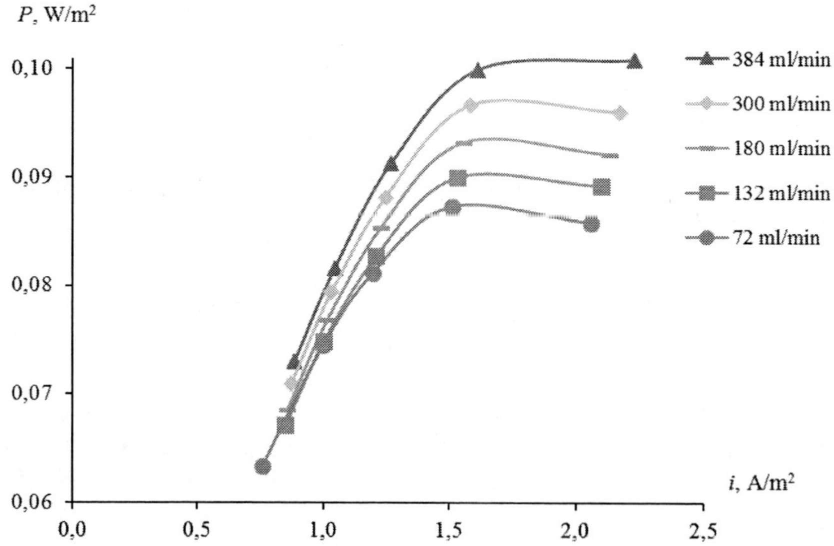

Figure 5. Specific power density versus load current for different flow rates of "fresh" water.

IOP Conf. Series: Materials Science and Engineering **791** (2020) 012057 doi:10.1088/1757-899X/791/1/012057

3.2. The effect of the RedOx system on RED process

In the present work, three RedOx systems were used: two reversible (equimolar mixture of $K_3[Fe(CN)_6]$ and $K_4[Fe(CN)_6]$ (2)) and Quinone/Hydroquinone (3). In the other case, in the absence of a special RedOx system, water and salt electrolysis occurred in the electrode chambers (4).

$$Fe(CN_6)^{3+} + e \leftrightarrow Fe(CN_6)^{2+} \quad (2)$$

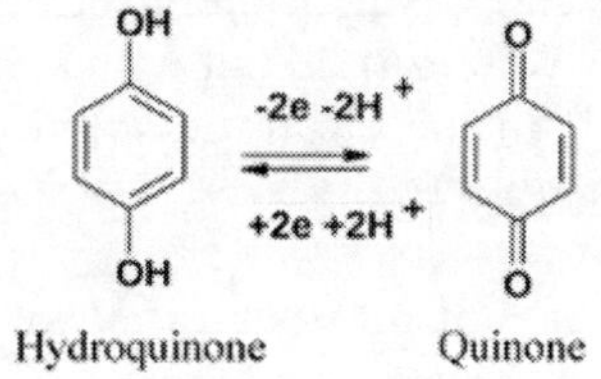

(3)

$$\text{cathode } (-): 2H^+ + 2e \rightarrow H_2,$$
$$\text{anode } (+): 2Cl^- - 2e \rightarrow Cl_2;\quad 2H_2O - 4e \rightarrow 4H^+ + O_2 \quad (4)$$

From literature data it is known [15] that potassium hexacyanoferrate is most often used, but it's quite toxic and can affect the electrical conductivity of anion exchange membranes.

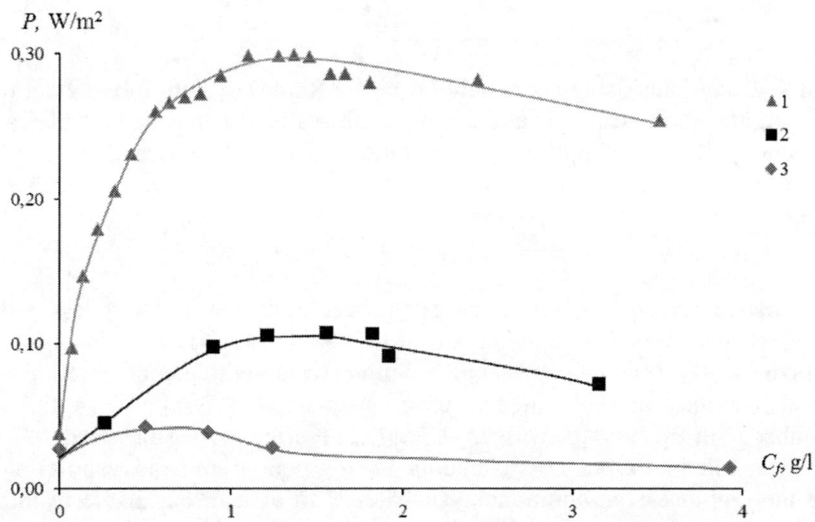

Figure 6. Specific power dependence of salt concentration in fresh water for different RedOx system: 1 - $K_3[Fe(CN)_6]/K_4[Fe(CN)_6]$ 0.025+0.025 mol/l + NaCl 0.25 mol/l; 2 - NaCl 0.25 mol/l; 3 - Hydroquinone/Quinone (saturated + NaCl 0.25 mol/l).

RED process was carried out using various RedOx systems (Figure 6). For systems (1) and (3), a cell with 5 pair chambers was used. Since the potential drop on the cell is lower than the water decomposition voltage, the cell with 20 pair chambers was used for the RedOx system (2). Studies have shown that when using the Quinone/Hydroquinone system, the resulting specific power is reduced by

IOP Conf. Series: Materials Science and Engineering **791** (2020) 012057 doi:10.1088/1757-899X/791/1/012057

7 times, which makes this system unsuitable for RED process. It was also noted that when using RedOx systems (1) and (3) anion-exchange membranes significantly change their color after the experimets. When RED process carried out with water electrolysis in the electrode chambers, the power decreases by 3 times in comparison with the use of hexacyanoferrate, however, additional toxic components are not used. Also, the exception of a special RedOx system should increase the lifetime of the anion exchange membranes. However, in this case, it is also possible the evolution of chlorine gas due to electrolysis.

3.3. The effect of the hexacyanoferrate concentration on RED process

The effect of the concentration of the RedOx pair in the electrode solution on the obtained specific power was investigated. The experiments were carried out with a total concentration of the RedOx pair $K_3[Fe(CN)_6]$ and equimolar $K_4[Fe(CN)_6]$ in the electrode solution in the range of 0-0.1 mol/l with 0.25 M NaCl, the results are presented in table 1.

Table 1. Dependence of the specific power on the concentration of RedOx pair $K_3[Fe(CN)_6] + K_4[Fe(CN)_6]$ in the electrode solution (concentration of sodium chloride in fresh water 1,94 g/l)

Concentration of redox pair, mol/l	Obtained power density
0.000	0.107
0.010	0.281
0.025	0.289
0.050	0.292
0.100	0.310

It was found that with increasing concentration of the RedOx system in the electrode solution, the resulting power slightly increases. However, to a certain value, then goes to a plateau. We can conclude that it's necessary to find an optimal concentration of hexacyanoferrate, and it will be possible to reduce the toxicity of the electrode solution by several times, while the power will change slightly.

4. Conclusion

In the present work the process of electric energy production by the RED process with using Russian produced heterogeneous ion-exchange membranes MK-40 and MA-41 with saturated solution of sodium chloride (approx. 350 g/l) and a solution of sodium chloride with a concentration of 0.017-3.39 g/l were investigated. The maximum obtained specific power was 0.3 W/m^2. The optimal salt concentration in the chamber with fresh water was 0.5-1.5 g/l. The effect of various RedOx systems on RED process was studied. Maximum power was obtained using potassium hexacyanoferrate. It was shown that the use of the Quinone-Hydroquinone system leads to a decrease in the resulting power by 7 times. In case of refusal to use a special RedOx system when using a sufficiently large number of membranes in a package, the electric current flows due to the electrolysis of water. The resulting power is about 3 times lower than when using potassium hexacyanoferrate, but there is a decrease in the ecological load on the environment.

Acknowledgments

The reported study was funded by RFBR and Administration of Krasnodar region, according to the research project number 19-48-230047.

IOP Conf. Series: Materials Science and Engineering **791** (2020) 012057 doi:10.1088/1757-899X/791/1/012057

References

[1] Vermaas D A, Kunteng D, Saakes M and Nijmeijer K 2013 Fouling in reverse electrodialysis under natural conditions *Water Research* **3** 1289-98

[2] Han J, Hwang K, Jeong H, Byeon S, Nam J, Kim C,·Kim H,·Yang S,·Choi J·Y and Jeong N 2019 Electrode system for large-scale reverse electrodialysis: water electrolysis, bubble resistance, and inorganic scaling *J. of Applied Electrochemistry* **49** 517-28

[3] Daniilidis A, Vermaas D A, Herber R and Nijmeijer K 2014 Experimentally obtainable energy from mixing river water, seawater or brines with reverse electrodialysis *Renew Energy* **64** 123-31

[4] Di S J, Cosenza A, Tamburini A, Micale G and Cipollina A 2018 Long-run operation of a reverse electrodialysis system fed with wastewaters *J. Environ Manag* **217** 871-87

[5] Post J W, Hamelers H V M and Buisman C J N Energy 2008 Recovery from controlled mixing salt and fresh water with a reverse electrodialysis system *Environmental Science & Technology* **42(15)** 5785-90

[6] Yip N Y and Elimelech M 2012 Thermodynamic and energy efficiency analysis of power generation from natural salinity gradients by pressure retarded osmosis *Environmental Science & Technology* **46(9)** 5230-9

[7] Vermaas D A, Saakes M and Nijmeijer K 2011 Power generation using profiled membranes in reverse electrodialysis *J. Membr. Sci.* **385-386** 234-42

[8] Zhu X, He W and Logan B E 2015 Reducing pumping energy by using different flow rates of high and low concentration solutions in reverse electrodialysis cells *J. Membr. Sci.* **486** 215-21

[9] Pawlowski S, Crespo J G and Velizarov S 2014 Pressure drop in reverse electrodialysis: experimental and modeling studies for stacks with variable number of cell pairs *J. Membr. Sci.* **462** 96-111

[10] Vermaas D A, Saakes M and Nijmeijer K 2014 Early detection of preferential channeling in reverse electrodialysis *Electrochim Acta* **117** 9-17

[11] Post J, Goeting C, Valk J, Goinga S and Veerman J 2010 Towards implementation of reverse electrodialysis for power generation from salinity gradients *Desalination and Water Treatment* **16** 82-93

[12] Ghaffour N, Lattemann S, Missimer T, Sinha S and Amy G 2014 Renewable energy – driven innovative energy – efficient desalination technologies *Appl. Energy* **1136** 1155-65

[13] Zabolotskii V I, Loza S A and Sharafan M V 2005 Physicochemical properties of profiled heterogeneous ion-exchange membranes *Russian J. of Electrochemistry* **41(10)** 1053-60

[14] Berezina N P, Kononenko N A, Dyomina O A and Gnusin N P 2008 Characterization of ion-exchange membrane materials: properties vs structure *Advances in Colloid and Interface Science* **139** 3-28

[15] Veerman J, Jong R, Saakes M and Metz S J 2009 Reverse electrodialysis: Comparison of six commercial membrane pairs on the thermodynamic efficiency and power density *J. Membr. Sci.* **343** 7-15

IOP Conf. Series: Materials Science and Engineering **791** (2020) 012058 doi:10.1088/1757-899X/791/1/012058

Hydrogen permeability of surface-modified Pd-Ag membranes at low temperatures

I S Petriev[1,2], I S Lutsenko[1], K A Voronin[1], P D Pushankina[1], M G Baryshev[1,2]

[1]Department of Physics, Kuban State University, Stavropolskaya St., 149, Krasnodar, 350040, Russia
[2]Laboratory of problems of stable isotope spreading in living systems, Southern Scientific Centre of the RAS, Rostov-on-Don, 344000, Russia

E-mail: petriev_iliya@mail.ru

Abstract. Hydrogen-permeable membranes of the Pd77% Ag33% alloy were obtained. Membranes were modified with a catalytically active coating, which consist of a crystalline nanostructured palladium black. Membranes were obtained by magnetron sputtering using mosaic targets. The surface modification with a catalytically active coating was performed by electrochemical deposition from an aqueous solution of H_2PdCl_4. The low-temperature kinetic characteristics of the hydrogen permeability of metallic Pd77% Ag23% membranes with a modifying coating of crystallized palladium black were analysed. During the analysis of the obtained experimental data about the kinetic characteristics of the obtained membranes, it was revealed that the surface effects are the limiting stage of hydrogen transport through palladium membranes at low temperatures. It was experimentally confirmed that the high specific surface of the synthesized catalysts causes the acceleration of the kinetic characteristics of hydrogen transport. The possibility of intensifying the processes of hydrogen transport through metallic membranes by creating active centers of chemisorption and dissociation on their surface is shown.

1. Introduction

In recent years the use of hydrogen as energy carrier is a very promising area in the energy sector. High-purity hydrogen (>99,999%) is the most demanded in production of optical fibers, semiconductors, medicaments, and also as an energy source for fuel cells [1]. It is assumed that significant part of hydrogen will be produced due to methane conversion, and also due to its separation other hydrocarbon gases, which are not involved in energy production process nowadays (petroleum production associated gases, petrochemical production discharge gases and other) [2, 3]. For this purpose, palladium-based and palladium-alloy membranes are perfectly fits. Those membranes have almost total hydrogen selectivity, high ability to cleave C-C-bonds, high thermal stability and mechanical strength [4]. Those membranes can be used in membrane reactor, that allows to combine chemical reaction for obtaining hydrogen with purification stage and in hydrogen isotope separation plants for the production of water with a modified isotopic composition [5-9]. For mass integration of metallic membranes into industrial hydrogen separation processes, it is very important to estimate their productivity and stability in working conditions. Since its considered, that those types of membranes are working at relatively high temperatures (300–600 °C) and high pressure (2–50 bar or even higher), and also in transient conditions, in which they can undergo rapid, it is important to understand the nature of changes of the material properties and under the conditions they were made.

IOP Conf. Series: Materials Science and Engineering **791** (2020) 012058 doi:10.1088/1757-899X/791/1/012058

Currently the main effort are directed at lowering working temperatures, membranes cost, increasing mechanical strength, durability and catalytic activity. Palladium is rare and expensive material, and its cost is expected to rise because of growing demand for its use in large scales. Minimizing of palladium quantity contributes to lower membrane.

Producing of ultra-thin palladium layers allows to minimize discharge intensity of precious metals However this entails two problems: limitation of membrane mechanical strength and the difficulty of obtaining films without defects. Therefore, during the deposition of a selective layer, three main tasks must be performed: lowering of the selective layer thickness (the aim is to increase permeability and lower membrane cost while ensuring of complete absence of defects); decreasing the number of defective membranes (the aim is to control the total cost of the production process); producing of Pd-based alloys (the aim is to improve permeability in working conditions, i.e. in thermal cycles or in the sulfur presence). According to this, we further use the electrolytic deposition of finely-divided palladium from H2PdCl4 water solution on impulse low-voltage potentiostat-galvanostat P–250I («Elins», Russia).

Hydrogen permeation of membranes is heavily decreasing at the temperatures below 200 °C because equilibrium is achieved very slowly [10]. This condition can be partially overcome by modifying the working surface with palladium-containing film coatings [11]. Usually, pure palladium suffers from hydrogen embrittlement because of the lattice expansion caused by the $\alpha - \beta$ phase transition which occurs when metal is exposed to hydrogen atmosphere at the temperature 298 °C and the pressure 2 MPa [12, 13]. This phase transition generates tensile stress, especially in the case of a tubular geometric shape, which often leads to cracking of the Pd layer and, consequently, to subsequent loss of membrane hydrogen selectivity. This disadvantage can be avoided by working under the conditions of the above critical point, when the membrane is exposed to hydrogen, or by changing the phase diagram of Pd [14]. The last option can be implemented by alloying pure Pd with other metals: silver, copper, ruthenium, gold. Pd-based alloys with specific concentrations of these metals change the metal hydride phase diagram, avoiding the mentioned embrittlement problem [15].

Hydrogen permeation in palladium-silver alloys varies according to the extreme law and reaches a maximum at a 15-25 % content of silver [16]. Phase diagram of the Pd-Ag system presented in Figure 1. A large amount of silver in alloys reduces the purity of the diffused hydrogen, doesn`t allow to lower a dew point temperature below - 70 °C and poisons the membranes. Relatively frequent in the operation of capillaries and membranes from a complexly doped alloy are most likely associated with unequal distribution of doping elements [16]. We have chosen an alloy of the composition Pd77% Ag23%, the choice of which caused by the most optimal mechanical and catalytic properties [17].

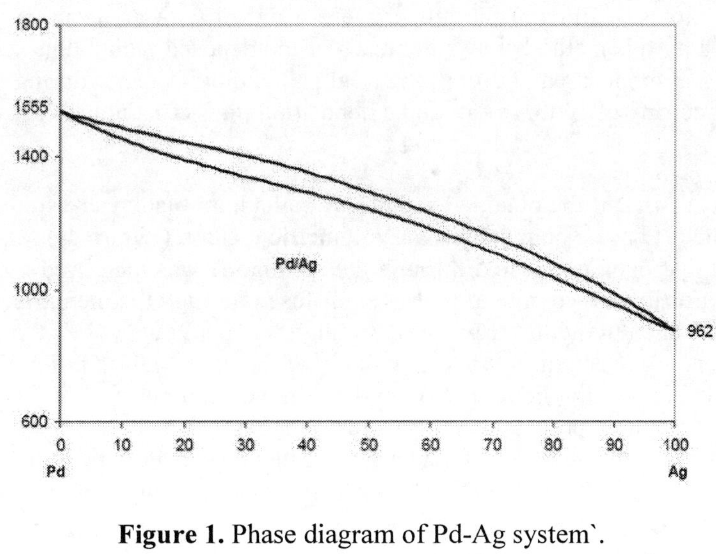

Figure 1. Phase diagram of Pd-Ag system`.

IOP Conf. Series: Materials Science and Engineering **791** (2020) 012058 doi:10.1088/1757-899X/791/1/012058

Thus, the aim of this research was to study the efficiency of the modification of the Pd-Ag alloy membrane to increase the kinetic characteristics of hydrogen transport in the low-temperature operating range (25-90°C) for the use of membranes in ultrapure hydrogen and low-temperature electrochemical current sources.

2. Materials and Methods

2.1. Membrane

A magnetron sputtering method for obtaining experimental Pd77% Ag23% alloy membranes was developed based on mosaic targets. The sputtering itself was performed on a Q150T ES sputtering vacuum system (Quorum Technologies, UK). To create a composite target for a magnetron sputtering system, chemically pure refined palladium and silver ingots were rolled on DRM – 130 electromechanical roll mills (Durston, UK) to a thickness of 25 μm with intermediate annealing in a muffle furnace at the temperature 250 °C. Silver had been previously rough-forged with a wedge-shaped anvil on the anvil along the length of the ingot to a thickness of 3–3.5 mm, followed by smoothing with a flat pane to a thickness of 3 mm and intermediate annealing at 300 °C for 20 seconds. Then, the metal plates were connected into a target packet between the contact pressure plate and the ring with the ratio of the surface areas, which are free for the magnetron sputtering: S (Ag) / S (Pd) = 20.8 / 79.2. The difference in the areas ratio of the elements to their ratio in the film composition is justified by the difference in the atomization energies of palladium and silver (ΔH (Pd) = 353 kJ/mol, ΔH (Ag) = 251,5 kJ/mol).

2.2. The synthesis of catalyst

To modify the surfaces of the Pd77% Ag23% alloy films, pretreatment in a 6 M concentrated NaOH solution was performed for 30 minutes, followed by etching in a 60% HNO3 solution for 30 seconds. Then, the films were transferred to an electrolytic cell connected for subsequent electrodeposition through a silver wire (99.99% silver) to a potentiostat / galvanostat P250I («Elins», Russia) Anodic polarization was carried out in 0.1 M HCl at a current density of 10–20 mA / cm^2. Then, cathodic polarization was carried out in 0.05 M H_2SO_4 at a current density of 10–20 mA / cm^2. A series of palladium black samples were electrolyticaly deposed in a 2% H_2PdCl_4 solution at current densities of 2–14 mA / cm^2 for 0.1 to 1.5 hours, and after the deposition films were washed with bidistilled water and cathodically polarized in 0.05 M H_2SO_4. As a result of surface analysis of the experimental series of s surface modifiers synthesis, two types of coating were obtained (with a deposition time of 0.5 hours and 1 hour), which are the most interesting from a scientific and applied point of view. Coatings from these two series are structures that potentially have high catalytic activity values.

The amount of the dispersed phase of palladium was controlled by direct gravimetry using an analytical balance. Samples of films in the holder were weighed with an accuracy of 10-6 g. The mass difference obtained by this method shows the mass of the deposed palladium, i.e., its discharge intensity. Thus, with a membrane area of 2 cm^2, the total palladium discharge during electrodeposition at a deposition current density of 10 mA / cm^2 and a deposition time of 1 1 hour is about 5.2 mg.

2.3. Hydrogen Permeation

The kinetic characteristics of the obtained samples of palladium black were studied on the installation we developed, which is basing on the microgasvolumetric method (Figure 2).

The amount of hydrogen permeated through the membrane was measured by the microvolumetric method, i.e. by measuring the volume of hydrogen diffused through the membrane at constant pressure The entering of hydrogen to the lower gas loop through the membrane when the purge valve is closed causes an equivalent enlargement in the total volume of the gas mixture in the gas outlet loop and the meniscus level of the measuring liquid in the capillary moves in time.

The hydrogen permeation process consists of three main stages [18]:

1. dissociation of hydrogen on the input surface of membrane, flowing with velocity v_i,
2. dissociation of atomic hydrogen through the membrane, flowing with velocity v_D,

IOP Conf. Series: Materials Science and Engineering **791** (2020) 012058 doi:10.1088/1757-899X/791/1/012058

3. atomic recombination of hydrogen into molecules on the output membrane surface, flowing with velocity v_o.

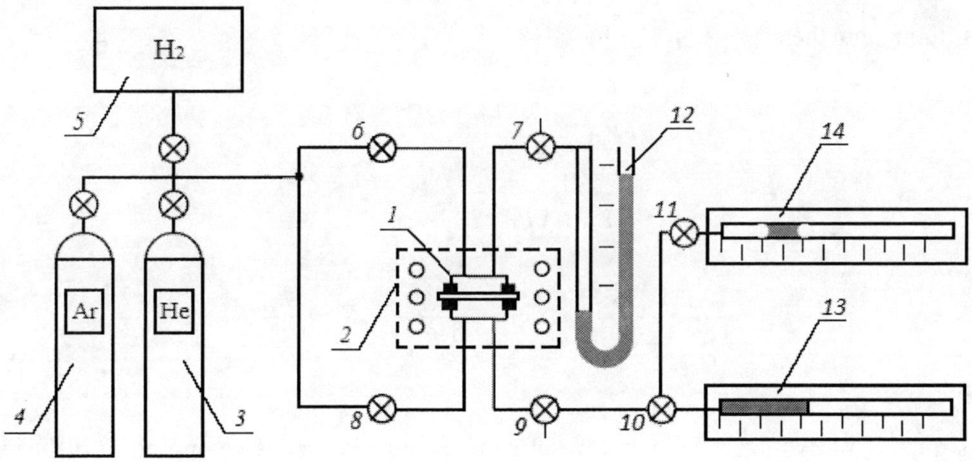

Figure 2. Installation for hydrogen permeation measurement: 1 – permeability cell; 2 – liquid thermostat; 3 – helium cylinder; 4 – argon cylinder; 5 – hydrogen generator «Spektr»; 6-11 – gas valves; 12 – mercurial U–shaped manometer; 13 – gas analyzer; 14 – measuring capillary.

Under the condition of equilibrium between hydrogen in the gas phase and atomic hydrogen in the adsorbed layer:

$$v_i = k_D p_1 - k_P p_{1A}^2; \tag{1}$$

$$v_o = k_P p_{2A}^2 - k_D p_2; \tag{2}$$

$$v_D = D\frac{k_S\left(p_{1A} - p_{2A}\right)}{h} = D\frac{C_1 - C_2}{h}. \tag{3}$$

k_D and k_P – dissociation rate constant and recombination constant, D – diffusion coefficient of hydrogen in metal, k_S – solubility coefficient of atomic hydrogen in metal, p_1 and p_2 – molecular hydrogen pressure on the input and output membrane surfaces, p_{1A} and p_{2A} – atomic hydrogen pressure on the input and output membrane surfaces, h – membrane thickness,
$C_1 = k_S \cdot p_{1A}$ and $C_2 = k_S \cdot p_{2A}$ – concentration of dissolved atomic hydrogen to near-surface layer of metal on the input and output membrane surfaces.

Under stationary conditions, these velocities are equal and corresponds to the hydrogen permeation rate:

$$v = v_i = v_o = v_D.$$

From the equations (1) and (2) we can express:

$$p_{1A} - p_{2A} = \frac{k_D\left(p_1 - p_2\right) - 2v}{k_P\left(p_{1A} + p_{2A}\right)},$$

$$C_1 - C_2 = \frac{k_S^2}{2k_P C_{cp}} \left(k_D \left(p_1 - p_2\right) - 2v\right),$$

where $C_{cp} = (C_1+C_2)/2$.

Substituting into the expression (3):

$$v = D\frac{k_S^2}{2hk_P C_{cp}} \left(k_D \left(p_1 - p_2\right) - 2v\right),$$

$$v = k\left(p_1 - p_2\right),$$

$$k = \frac{1}{1/\alpha_{diff} + 2/k_D}, \quad \alpha_{diff} = \frac{Dk_S^2 K_P}{2hC_{cp}}, \quad K_p = \frac{k_D}{k_p}.$$

Mass transfer of hydrogen through a metal membrane resistance is composed of diffusion (determined by α_{diff}) and “chemical” resistances. Diffusion resistance depends on the membrane thickness δ, diffusion and solubility coefficients, which depend on temperature, and on the concentration of dissolved hydrogen, which depends on pressure. If the limiting stage of the process is diffusion, then k_D is much larger than α_{diff}, and:

$$k = \alpha_{diff} = \frac{Dk_S^2 K_P}{2hC_{cp}}.$$

An equilibrium is established on the surface: $v_i = v_o = 0$, accordingly:

$$p_{1A} = \sqrt{K_p p_1},$$

$$p_{2A} = \sqrt{K_p p_2},$$

$$C_{cp} = \frac{k_S\sqrt{K_P}}{2}\left(\sqrt{p_1} + \sqrt{p_2}\right),$$

$$k = \frac{Dk_S\sqrt{K_P}}{h\left(\sqrt{p_1} + \sqrt{p_2}\right)},$$

$$v = \frac{Dk_S\sqrt{K_P}}{h}\left(\sqrt{p_1} - \sqrt{p_2}\right).$$

Thus, the rate of the permeation process is proportional to the square root of the pressure of gaseous hydrogen, which generally corresponds to the Richardson equation [19]:

$$v = k\left(\sqrt{p_1} - \sqrt{p_2}\right)e^{-\frac{E_A}{RT}}.$$

IOP Conf. Series: Materials Science and Engineering **791** (2020) 012058 doi:10.1088/1757-899X/791/1/012058

If the limiting stage of the process is dissociation, then k_D is much less than α_{diff}, and accordingly:

$$k = \frac{k_D}{2},$$

$$v = \frac{k_D}{2}(p_1 - p_2).$$

Thus, it this case hydrogen permeation rate is to pressure of gaseous hydrogen.

2.4. SEM Image

The surface of the modified membranes was examined using a scanning electron microscope with a field emission cathode JSM–7500F («JEOL», Japan). The surfaces of both coatings were investigated from the permeate side of the membrane and from the retentate side. The coating homogeneity of the applied coatings was confirmed by a series of microphotographs (a sample of at least 10 microphotographs of the surface per cm^2).

3. Results

3.1. Hydrogen permeation

Figure 3 and Table 1 show the hydrogen permeation data for a Pd77% Ag23% alloy with formed palladium black structures and undercrystallized palladium black structures:

- flux density J, mmol / $c \cdot m^2$ – flow through a unit area of a sample;
- hydrogen permeation coefficient P, μmol / $(c \cdot m^2) \cdot cm$ / $Pa^{0,5}$- value numerically equal to the flux density through the sample of unit thickness at Δp.

Table 1. Hydrogen permeation data.

Δp, MPa	J, mmol/$c \cdot m^2$	P, μmol/$(c \cdot m^2) \cdot cm/Pa^{0,5}$
(b) Undercrystallized palladium black structures		
0.1	0.0037	0.00052
0.2	0.0107	0.00107
0.3	0.0173	0.00142
0.4	0.0243	0.00173
0.5	0.0324	0.00204
0.6	0.0393	0.00228
(a) Formed palladium black structures		
0.1	0.006	0.00148
0.2	0.016	0.00253
0.3	0.026	0.00322
0.4	0.037	0.00383
0.5	0.048	0.00438
0.6	0.059	0.00486

IOP Conf. Series: Materials Science and Engineering **791** (2020) 012058 doi:10.1088/1757-899X/791/1/012058

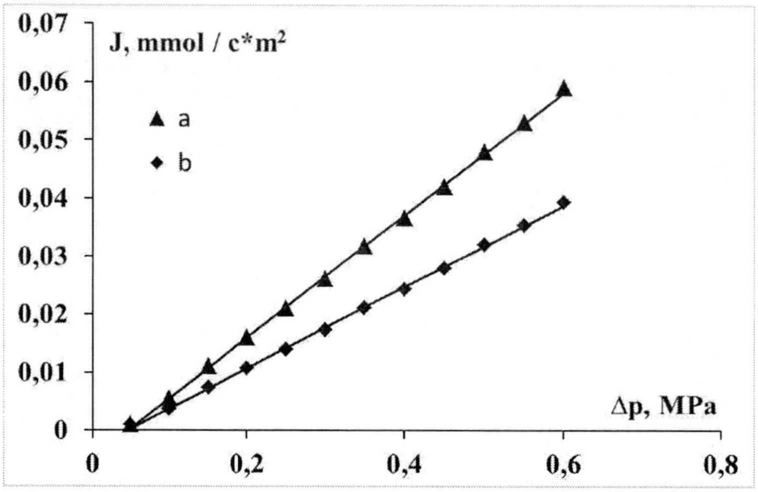

Figure 3. Hydrogen permeation graph for Pd77% Ag23% alloy membranes with two modifying coatings 0.5 h (a) and 1 h (b).

As can be seen from the data presented in Figure 3 and Table 1, the hydrogen flux density in the formed structures of palladium black is 1.5 times higher than hydrogen flux density of undercrystallized at an gauge pressure of 0.6 MPa.

Figure 4 shows that the dependence of the flux density for a membrane with a modified surface is well approximated by a first-order line, which, according to [20] indicates that the hydrogen penetration rate is limited by dissociation of hydrogen on the surface.

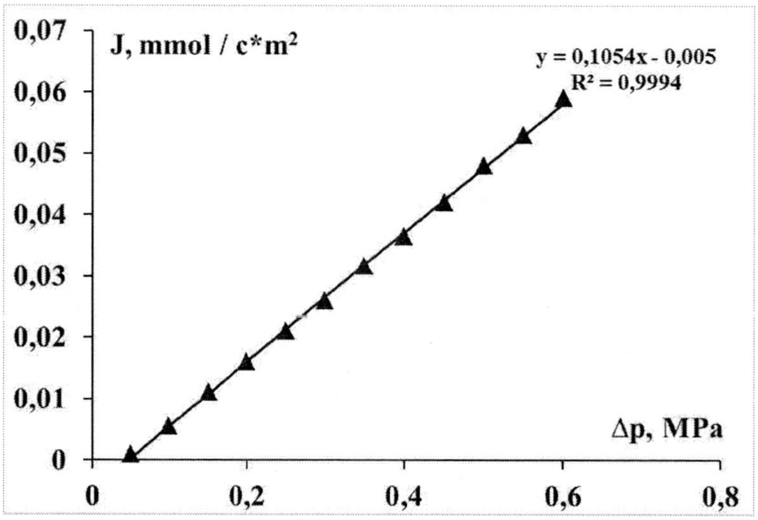

Figure 4. Dependence of flux density on gauge pressure on the input membrane surface for membrane with modified surface.

Figure 5 shows the data on the measurement of the hydrogen flux density through a palladium-silver membrane (Pd 77% Ag 23%) with electrochemically modified surfaces at different temperatures.

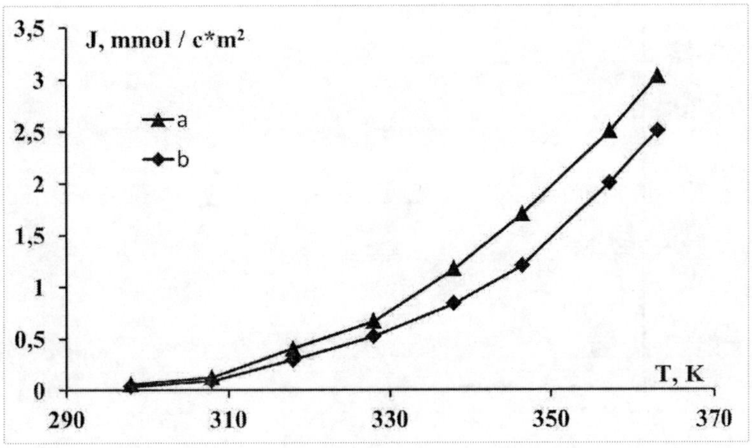

Figure 5. Dependence of hydrogen flux density on the temperature for Pd77Ag23 alloy membranes with two types of modifying coatings 0.5 h (a) and 1 h (b).

As can be seen from the data presented in Figure 5, the flux density of the palladium black structures increases by 1.2–2 times for every 10 K. In the formed palladium black structures, the hydrogen flux density at a temperature of 363 K is 1.2 times higher than of hydrogen flux density undercrystallized structures at an excess pressure of 0.6 MPa.

Figure 6 shows the data on measuring the dependence of hydrogen permeability on pressure for membranes with formed palladium black structures and undercrystallized palladium black structures.

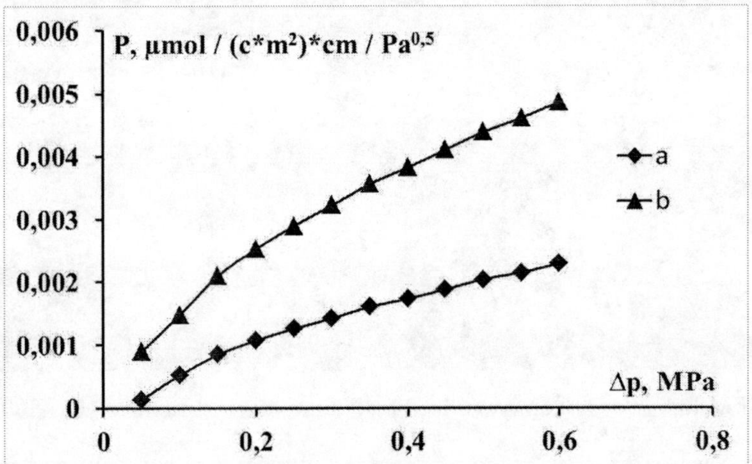

Figure 6. Dependence of hydrogen permeation on pressure for Pd77Ag23 alloy membranes with two types of modifying coatings 0.5 h (a) and 1 h (b).

As can be seen from the data presented in Fig. 6, the hydrogen permeation of the formed structures of palladium black is 2.1 times higher than hydrogen permeation of undercrystallized structures at an excess pressure of 0.6 MPa.

Figure 7 presents the data of measuring the hydrogen permeability of the formed palladium black structures and undercrystallized palladium black structures with temperature change.

IOP Conf. Series: Materials Science and Engineering **791** (2020) 012058 doi:10.1088/1757-899X/791/1/012058

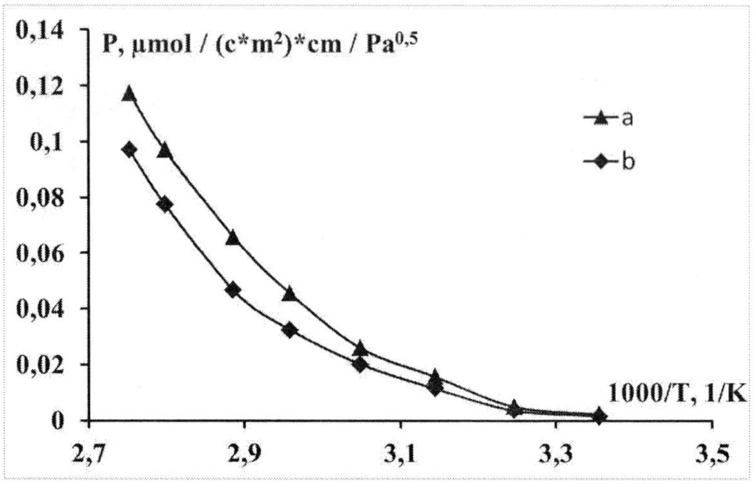

Figure 7. Dependence of hydrogen permeation on temperaturer pressure for Pd77Ag23 alloy membranes with two types of modifying coatings 0.5 h (a) and 1 h (b).

3.2. SEM Image

Micrographs of coatings on Pd77% Ag23% alloy membranes were obtained by the method, described in paragraph 2.4 SEM-Images of the formed palladium black structures and undercrystallized palladium black structures are shown in Figure 8 and Figure 9 at magnifications of x5000 and x50000.

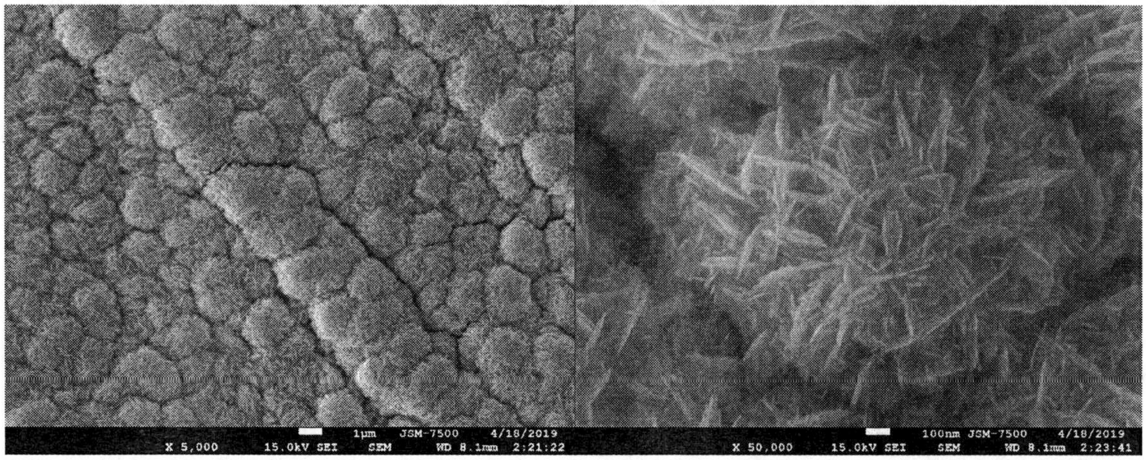

Figure 8. SEM Image of undercrystallized palladium black structures (a - x5000, b - x50000).

Undercrystallized palladium black structures are nanoparticles with a size of 30-120 nm in length and 15-20 nm in width. The average sizes of spherical ensembles of palladium black particles in undercrystallized structures are 1-3 μm^2. Completely formed palladium black structures have the form of crystallites with a size of 200-800 nm. The average size of the ensembles of the formed palladium black structures is 4-6 μm^2.

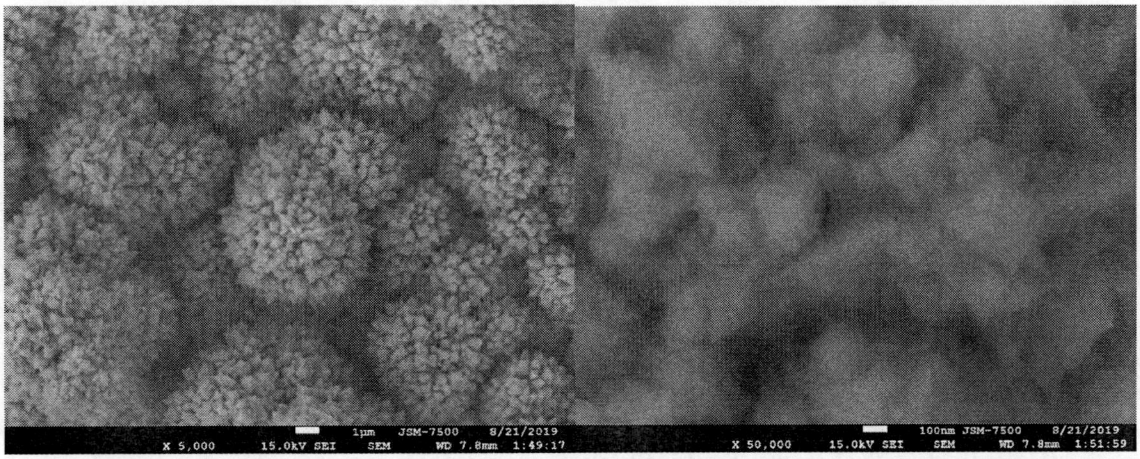

Figure 9. SEM Image of formed palladium black structures (a - x5000, b - x50000).

4. Discussion

The investigation of the correlation between the structure of the modifying coating and the deposition parameters provides an understanding of the evolution of microstructures during the growth and re-crystallization of surface modifiers. Further, this makes it possible to synthesize surface modifiers for specific technological conditions, for example, under transient conditions of working temperatures and pressure, membranes selectivity for various compositions of hydrogen-containing mixtures of gases and toxic impurities.

In this research, the low-temperature kinetic characteristics of the hydrogen permeability of metallic Pd77% Ag23% membranes with a modifying coating of crystallized palladium black were analyzed. During the electrolytical deposition of the modifying coating, a different spectrum of microstructures was obtained. The structures differed not only in the grain size of palladium black, but also in surface morphology, expressed by crystallographic orientation and structural defects. In order to develop low-temperature, high-performance membranes and to understand the correlation between the kinetic characteristics of the hydrogen transport process and the microstructure of surface modifiers, further researches will be carried out.

5. Conclusion

The Pd77% Ag23% alloy was obtained on the Quorum Q150TS / E / ES sputtering system by the developed method of magnetron sputtering using mosaic targets. Catalysts in the form of finely dispersed palladium were synthesized on the surface of prepared substrates by electrolytical deposition — two series with different deposition times. A comparative investigation of the kinetic characteristics of hydrogen transfer in modified membranes in the low temperature range was carried out. Samples, obtained at a 1 hour deposition time, have a hydrogen flux density and hydrogen permeability coefficient of almost two times higher than a hydrogen flux density and hydrogen permeability coefficient of undercrystallized palladium black structures, obtained at a 0.5 hours deposition time at the same current density. The dependence of the hydrogen flux density on the gauge pressure is approximated by a first-order line, which indicates the limitation of the hydrogen transport process through modified membranes by surface phenomena. The analysis of microphotographs of the obtained films surface using a JEOL JSM-7500F scanning electron microscope allowed us to conclude, that the acceleration of chemisorption processes caused by the finely dispersed coating. The active centers of the surface processes are the angles and faces of palladium black crystallites, which provide high activity and selectivity of the membrane surface in relation to hydrogen. Has been proposed a method for intensifying hydrogen transport through Pd77% Ag23% alloy membranes by developing the surface by creating chemisorption centers.

Acknowledgments
The reported study was funded by RFBR and Krasnodar region to the research project № 19-42-233004 and by the scholarship of the President of the Russian Federation to young scientists and post-graduate students SP-1237.2018.1 and by the state assignment of the SSC RAS (Project AAAA-A19-119040390083-6).

References
[1] Lu G Q, Diniz da Costa J C, Duke M, Giessler S, Socolow R, Williams R H and Kreutz T 2007 *J. Colloid Interface Sci.* **314** 589–603
[2] Alimov V N, Bobylev I V, Busnyuk A O, Kolgatin S N, Kuzenov S R, Peredistov E Yu and Livshits A I 2018 *Int. J. of Hydrogen Energy* **43(29)** 13318-27
[3] Alimov V N, Bobylev I V, Busnyuk A O, Notkin M E, Peredistov E Yu and Livshits A I 2018 *J. of Membrane Science* **549** 428-37
[4] Liu J, Sun B, Hu J, Pei Y, Li H and Qiao M 2010 *J. Catal.* **274** 287-95
[5] Lytkina A A, Orekhova N V, Ermilova M M, Petriev I S, Baryshev M G and Yaroslavtsev A B 2017 *Petroleum Chemistry* **57(13)** 1219-27
[6] Lytkina A A, Orekhova N V, Ermilova M M, Petriev I S, Baryshev M G and Yaroslavtsev A B 2019 *Int. J. Hydrogen Energy* **44** 13310-22
[7] Dzhimak S S, Basov A A, Volchenko N N, Samkov A A, Fedulova L V and Baryshev M G 2017 *Doklady Biochemistry and Biophysics* **476** 323-5
[8] Basov A A, Elkina A A, Samkov A A, Volchenko N N, Moiseev A V, Fedulova L V, Baryshev M G and Dzhimak S S 2019 *Iran. Biomed. J.* **23** 129-41
[9] Fedulova L V, Basov A A, Vasilevskaya E R and Dzhimak S S 2019 *Curr. Pharm. Biotechnol.* **20** 245-53
[10] Petriev I S, Frolov V Y, Bolotin S N, Baryshev M G and Kopytov G F 2015 *Russian Physics J.* **58(8)** 1044-8
[11] Petriev I S, Bolotin S N, Frolov V Y, Barishev M G and Sozayev V A 2018 *Bulletin of the Russian Academy of Sciences* **82(7)** 807-10
[12] Petriev I S, Bolotin S N, Frolov V Y and Baryshev M G 2019 *Doklady Physics* **64(5)** 210-3
[13] Petriev I S, Frolov V Y, Bolotin S N, Baryshev M G, Kopytov G F and Isaev V A 2019 *Russian Physics J.* **61(10)** 1894-8
[14] Shu J, Grandjean B P A and van Neste A 1991 *The Canadian J. of Chemical Eng.* **69(5)** 1036-60
[15] Rahimpour M R, Samimi F, Babapoor A, Tohidian T and Mohebi S 2017 *Chemical Engineering and Processing: Process Intensification* **121** 24-49
[16] Burkhanov G S, Gorina N B, Kolchugina N B, Roshan N R 2006 Splavy` palladiya dlya vodorodnoj e`nergetiki [Palladium alloys for hydrogen energetics] *Rossijskij Ximicheskij Zhurnal* **L(4)** 36-40 [In Russian]
[17] Petriev I S, Bolotin S N, Frolov V Y, Barishev M G, Isaev V A and Kopytov G F 2016 *Bulletin of the Russian Academy of Sciences* **80(6)** 624-6
[18] Baichtok Iu K, Sokolinskii Iu A and Aizenbud M B 1976 O limitiruyushhej stadii proniczaemosti vodoroda cherez membrany` iz palladievy`x splavov [On the limiting stage of hydrogen permeability through membranes from palladium alloys] *Zhurnal Fizicheskoj Ximii* **50(6)** 1543-6 [In Russian]
[19] Richardson O W, Nicol J and Parnell T 1904 *Philosophical Magazine Series 6* **8(43)** 1-29
[20] Baichtok Iu K, Aizenbud M B, Kurkovskii V A 1972 Vodoproniczaemost` palladievy`x splavov pri vy`sokix davleniyax [Water tightness of palladium alloys at high pressures] *Ximicheskaya Promy`shlennost`* **9** 679-82 [In Russian]

IOP Conf. Series: Materials Science and Engineering **791** (2020) 012059 doi:10.1088/1757-899X/791/1/012059

Investigation of the effect of mixing and biomass temperature in the anaerobic digester on the volume of biomethane

D Y Suslov

Department of Heat Supply and Ventilation, Belgorod State Technological University of V.G. Shukhov, Kostyukov St., 46, Belgorod, 308012, Russia

E-mail: Suslov1687@mail.ru

Abstract. The article is devoted to the study of the process of biomethane production during anaerobic fermentation in a biomethane installation. The most effective methods for increasing the efficiency of biomethane plants are to increase the intensity of mixing and the temperature of the biomass. However, currently there are various data on the optimal temperature of the fermentation process and the frequency of mixing of biomass per day. A program for conducting experimental studies and an experimental setup for producing biomethane with a bubble-type mixing system has been developed. For research, the central composite rotatable plan of the full factorial experiment was used. Experimental studies of the process of biomethane production during anaerobic fermentation of pig manure in mesophilic temperature conditions were carried out. In the process of implementing the plan, a regression equation was obtained that characterizes the effect of the bubbling mixing intensity and biomass temperature on the biomethane output. Plots of the dependence of the specific yield of biomethane on varied factors are obtained. A rational mode of operation of the biomethane plant has been established, which is achieved with a mixing intensity of 5...6 days^{-1} and a biomass temperature of 39.5...40.5 °C with a maximum specific yield of biomethane of 393.58 l/kg. This mode can be recommended for the operation of biomethane plants in the processing of organic waste from agricultural enterprises.

1. Introduction

One of the promising areas for the development of alternative energy, which is widely used in the European Union (EU) countries, is biogas [1-3]. The difference of biogas from the natural one consists in low methane (40-70%) content and the presence of impurities such as: carbon dioxide (30-40%), hydrogen sulfide (0-2%) and other gases (0-5%). After purification, biogas is similar in chemical composition to natural gas and is called biomethane [4-6]. Biomethane is produced by biomethane units in the process of anaerobic fermentation of organic substances. Energy plants and agricultural wastes are used as an initial substrate for biomethane production [7-9].

To increase the intensity of the fermentation process, the following methods are used:

- the addition of various substances in the anaerobic digester, promoting the growth of bacteria;
- the retention of bacteria in the anaerobic digester using anaerobic biofilter;
- intensification of heat and mass transfer using mixing systems;
- increase in process temperature.

The most effective methods to increase the efficiency of the anaerobic digester are to increase the intensity of mixing and increase the temperature.

IOP Conf. Series: Materials Science and Engineering **791** (2020) 012059 doi:10.1088/1757-899X/791/1/012059

Modern biomethane installations operate under two temperature conditions: mesophilic (30-45 °C) and thermophilic (50-60 °C). Most existing biomethane plants operate in the mesophilic mode [10-12]. This allows you to get a large amount of biomethane at low energy costs for heating the anaerobic digester. However, there are various data on the optimal temperature of the anaerobic fermentation process in the mesophilic mode. In the earliest works [13-17], it is indicated that the greatest amount of biogas is released at temperatures of 32–35 °C. The authors of [18, 19] write that the optimum temperature is 36-37 °C. According to modern data, the optimal temperature range is 38–42 °C [20-22].

For uniform distribution of the biomass temperature and the concentration of bacteria throughout the methane tank, 3 types of mixing systems are used [23-25]: mechanical, hydraulic and bubble (pneumatic). The most widely used are mechanical systems with paddle and propeller mixers. Hydraulic mixing is carried out by recirculation of biomass from one part of the digester to another using a pump. Bubbling mixing is carried out by supplying biogas to the lower part of biomethane. Bubbles of gas rising upward cause the circulation of biomass and its intense mixing.

There are also conflicting data on the mixing frequency. In [7, 17-20] it is indicated that the ideal mode of mixing of the substrate is several times a day (4-12 $days^{-1}$). In large-volume anaerobic digesters, mechanical stirrers of continuous or quasi-continuous operation are mainly used [22-25].

The aim of this work is to study the effect of bubbling mixing intensity and biomass temperature in the anaerobic digester on the efficiency of the fermentation process.

2. Materials and Methods

To conduct experimental research, an experimental setup was developed (Figure 1) [26]. The installation consists of a anaerobic digester, in which fermentation takes place, a gasholder for collecting biomethane, devices for heating and mixing biomass and instrumentation.

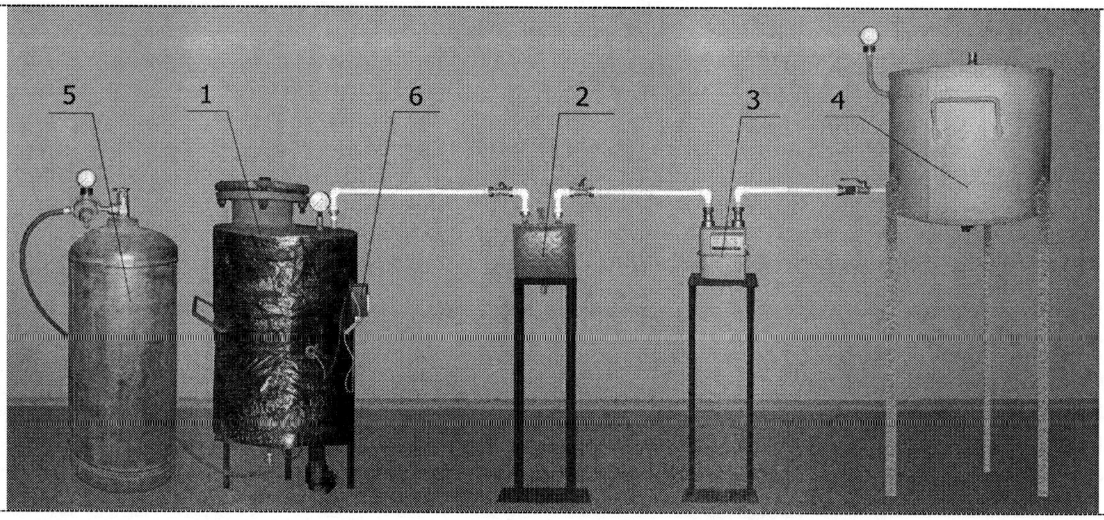

Figure 1. General view of the experimental setup for biomethane production: 1 – anaerobic digester, 2 – hydraulic shutter, 3 – gas meter, 4 – gas tank, 5 – gas bottle.

In the experimental setup, a bubbler type anaerobic digester is used, the distinguishing feature of which is the design of the mixing system (Figure 2). The bubbler mixing device is a uniformly perforated pipe in the form of a vertical spiral, the turns of which form a cone. The anaerobic digester also contains a pipe for supplying the initial substrate, a pipe for draining biomass, a pipe for removing the resulting biomethane.

IOP Conf. Series: Materials Science and Engineering **791** (2020) 012059 doi:10.1088/1757-899X/791/1/012059

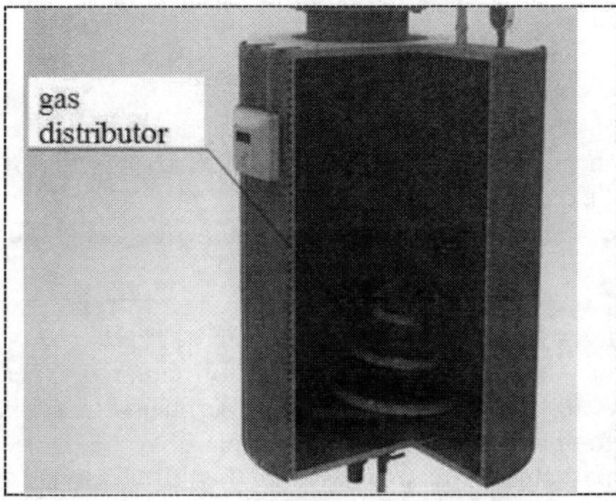

Figure 2. The anaerobic digester with system of hashing of bubbling type.

The heating device of the anaerobic digester is a thermocouple in the form of a cord, evenly located outside the anaerobic digester shell. This position of the thermocouple allows you to make the heating uniform and ensure the temperature regime within the specified limits. The automation system of the experimental setup maintains a predetermined temperature in the anaerobic digester constant.

The principle of operation of the installation is as follows. The prepared initial substrate is fed into the anaerobic digester 1 through the inlet pipe, after filling the anaerobic digester is hermetically closed. Using a heating device 6, the biomass is heated to the desired temperature and maintained at a given level. In the fermentation process, biomethane is released, which is collected in the upper part of the anaerobic digester and through the connecting pipelines, passing through the hydraulic shutter 2 and gas meter 3, enters the gas tank 4 for further consumption. From the gas bottle 5, the gas of the required pressure is supplied to the bubbler mixing device installed in the lower part of the anaerobic digester.

Pig manure with a volumetric humidity of 90% was used as the initial substrate. The temperature regime is mesophilic. The cycle time of the biomethane production process was 30 days.

When conducting experimental studies, the following parameters were subject to direct measurement:

- volumetric flow rate of biomethane;
- temperature in the anaerobic digester;
- pressure in a anaerobic digester, gas tank and gas bottle.

The main indicator of the experimental study is the specific yield of biomethane V (l/kg), determined in liters per kilogram of dry matter:

$$V = V_B/m, \tag{1}$$

where V_B – the volume of biomethane produced, l; m – the mass of absolutely dry biomass matter in the anaerobic digester, kg.

The volumetric flow rate of biomethane was determined using a VK – G1.6T gas volumetric meter. Since the volume of gas used for mixing also passes through this counter, the volume of biogas produced is determined as the difference between the counter readings on the gas holder (V2) and the counter installed on the cylinder (V1), which supplies gas to bubble mixing:

$$V_B = V_2 - V_1 \tag{2}$$

The substrate temperature in the anaerobic digester was measured with a thermometer of the technical grade TTZh-M. To control the pressure in the gas tank, anaerobic digester, and bubbler mixing system, pressure gauges DM 02-100-2-M are used.

The experimental research program provides for the determination of the effect of bubbling mixing and substrate temperature in the anaerobic digester on the fermentation process and the release of bi-

IOP Conf. Series: Materials Science and Engineering **791** (2020) 012059 doi:10.1088/1757-899X/791/1/012059

omethane. The specific yield of biomethane is taken as the optimization parameter, then the response function has the form:

$$V = f\,(t,\ \eta),\tag{3}$$

t – biomass temperature in anaerobic digester, °C; η – mixing frequency, day^{-1}.

For conducting experimental research, we will use the methods of matrix design of experiments, which allows us to reduce the number of experiments.

Taking into account the coding of the studied factors, the response function has the form:

$$V = f\,(X_1,\ X_2),\tag{4}$$

X_1 – biomass temperature in anaerobic digester, °C; X_2 – mixing frequency, day^{-1}.

For research, we use the central composite rotatable plan for the full factorial experiment (CCRP FFE). Application of such a plan allows to reduce the number of experiments to a minimum and to obtain the regression equation in the form of a second-order polynomial.

The matrix of the central compositional rotatable plan is based on the full factorial experiment – 2^n, where n – the number of variable variables (in our case n = 2), with the addition of 2n stellar points at the ±α level and 5 zero points (N_0). The total number of experiments according to plan for n = 2 will be:

$$N=2^n+2n+N_0=13\tag{5}$$

The values of the levels of factors on a conditional scale are set as follows: maximum +1, average 0 minimum –1 and "star" points +1.41, –1.41.

Based on the foregoing, a second-order rotatable plan matrix has been compiled for CCRP FFE 2^2 for two independent variables – the frequency of mixing of the biomass and the temperature of the biomass (Table 1).

Table 1. Planning matrix CCRP FFE 2^2

Experience number	Experience number		Target function
	X_1	X_2	V (l/kg)
1	–1	–1	223
2	+1	–1	265
3	+1	+1	279
4	–1	+1	232
5	+1.41	0	412
6	–1.41	0	65
7	0	+1.41	260
8	0	1.41	124
9	0	0	199
10	0	0	205
11	0	0	189
12	0	0	192
13	0	0	207

The values of all factors required by the central compositional rotatable plan of the full factorial experiment according to the selected scheme are given in Table 2.

Table 2. Factor Variation Levels

Factors	Variation interval	Factor Variation Levels				
		-1.41	-1	0	+1	+1.41
X_1 – biomass temperature in anaerobic digester (°C)	2.5	33.5	34.5	37	39.5	40.5
X_2 – mixing frequency (day^{-1})	2	0	1	3	5	6

3. Results and Discussion

The results of a multivariate experiment of the function $V = f(\eta,\ t,)$ are presented in the form of a regression equation in physical quantities:

$$V = 5802.62 - 333.03t + 4.88t^2 + 2.13\eta + 0.02\eta^2 + 0.25t\eta. \quad (6)$$

Analysis of the regression equation shows that the greatest influence on the yield of biomethane has temperature, to a lesser extent - the frequency of mixing.

Based on the results of experimental studies, based on the regression equation (6), graphical dependencies are constructed. The graphs make it possible to more fully determine the influence of technological factors on the operation of the anaerobic digester with a bubble-type mixing system.

Figure 3 shows the effect of the mixing frequency of the biomass in the anaerobic digester on the volumetric yield of biomethane at different biomass temperatures.

With an increase in the stirring frequency from 0 to 1 day^{-1}, the yield of biomethane increases by 12.27 l/kg. With an increase in the mixing frequency from 1 to 3 day^{-1}, the yield of biomethane increases by 24.67 l/kg. With an increase in the mixing frequency from 3 to 5 day^{-1}, the yield of biomethane increases by 24.83 l/kg. With an increase in the mixing frequency from 5 to 6 day^{-1}, the yield of biomethane increases by 12.48 l/kg. A change in the mixing frequency from 0 to 6 day^{-1} entails an increase in the volumetric yield of biomethane by 74.25 l/kg or by 18.9%.

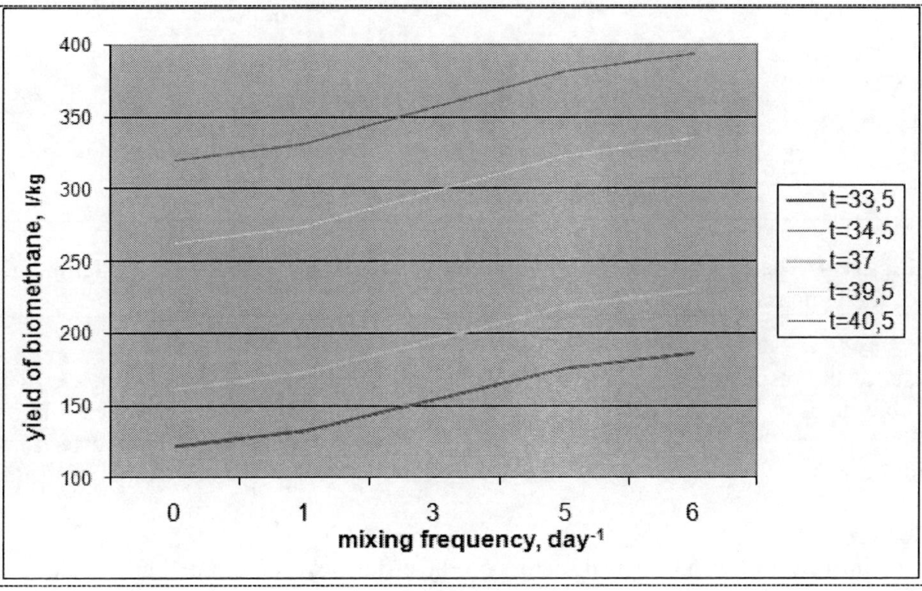

Figure 3. The dependence of the volumetric yield of biomethane on the mixing frequency.

The graph shows that an increase in the mixing frequency promotes a uniform temperature distribution over the volume of the anaerobic digester, which allows to increase the volumetric yield of biomethane. However, with more intensive mixing ($\eta > 5$), the volumetric yield of biomethane increases slightly.

Figure 4 shows a graph of the dependence of the volumetric yield of biomethane on the biomass temperature in the anaerobic digester for various values of the mixing frequency.

With an increase in the biomass temperature from 33.5 to 34.5 °C, the yield of biomethane increases by 0.31 l/kg. With an increase in the biomass temperature from 34.5 to 37 °C, the yield of biomethane increases by 43.47 l/kg. With an increase in the biomass temperature from 37 to 39.5 °C, the yield of biomethane increases by 104.48 l / kg. With an increase in the biomass temperature from 39.5 to 40.5 °C, the yield of biomethane still increases by 58.87 l/kg. A change in the biomass temperature from 34.5 to 40.5 °C entails an increase in the volumetric yield of biomethane by 207.13 l/kg or by 52.63%.

This relationship is complex. With an increase in temperature from 33.5 °C to 34.5 °C, the volume of biomethane produced remains virtually unchanged. Further, there is a significant increase in biomethane yield over the entire temperature range from 34.5 to 40.5 °C. A further increase in the yield of biomethane is caused by the fact that an increase in temperature leads to an intensive growth of bacteria.

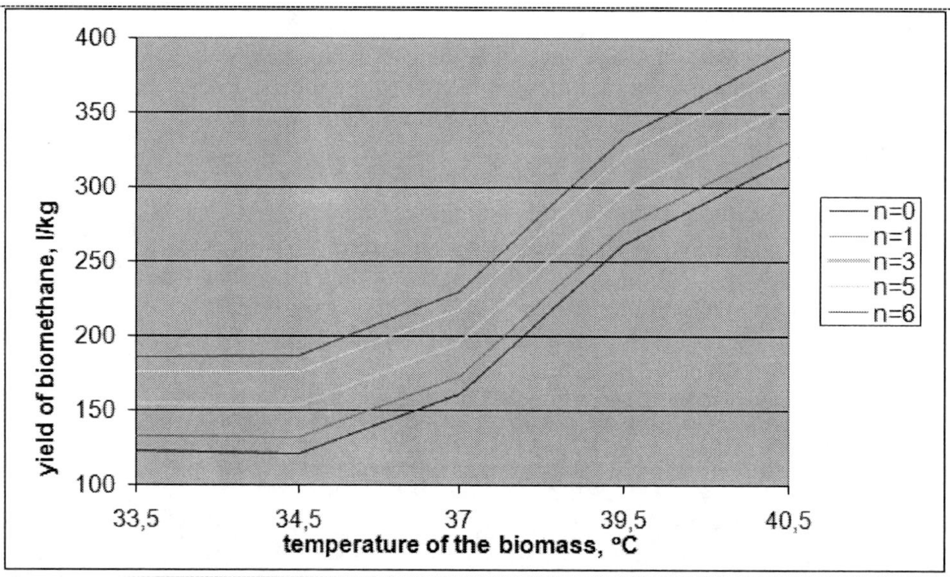

Figure 4. The dependence of the volumetric yield of biomethane on the temperature of the biomass.

Analysis of the results of experimental studies shows that the most optimal mode of operation of the anaerobic digester is achieved with $X_1 = +1.41$; $X_2 = +1.41$, which provides a high volumetric yield of biomethane. This mode can be recommended when using a anaerobic digester for the disposal of agricultural waste.

4. Conclusion

To study the influence of technological factors on the efficiency of a biomethane plant, a research program has been developed and an experimental plant has been manufactured. The research program provides for the determination of the effect of temperature and the degree of bubbling of the substrate in the anaerobic digester on the yield of biomethane.

As a result of the implementation of the research program, a regression equation was obtained that characterizes the influence of the studied factors on the biomethane yield during anaerobic fermentation in a anaerobic digester equipped with a bubble-type mixing system.

Graphic dependences of the specific yield of biomethane on variable factors were obtained, conclusions were drawn on the effect of each variable factor on the optimization parameter. Rational ranges of values of variable factors for a anaerobic digester with bubbling mixing are established. The most rational mode of operation is achieved with stirring 5-6 days^{-1}, the biomass temperature of 39.5-40.5°C.

Acknowledgments
The study was carried out in the framework of the scholarship of the President of the Russian Federation SP-180.2019.1.

References

[1] Wagner L 2015 Trends from the use of biogas technology in Germany *VIV Asia Biogas Conf. on March 12th 2015* (Bangkok) p 50
[2] Linke B 2015 *Country Report Germany IEA Bioenergy* T 37 (Berlin, Germany)
[3] Scarlat N, Dallemand J-F , Fahl F 2018 Biogas: Developments and perspectives in Europe *Renewable Energy* **129(A)** 457-72
[4] Beil M and Beyrich W 2013 Biogas upgrading to biomethane *The Biogas Handbook* (Cambridge: Woodhead Publishing) pp 342-77
[5] Geletukha G, Kucheruk P, Matveev Yu 2014 Prospects of production and use of biomethane in Ukraine *Analyticalnote BAU* vol 11 URL https://saf.org.ua/wp-content/uploads/2014/07/Position-paper-UABIO-11-EN.pdf
[6] Suslov D 2018 Application of Biomethane for Gas Supply Within the Settlements *J. Phys.: Conf. Ser.* **1066** 012004
[7] Vedeneev A and Vedeneva T 2011 *Rukovodstvo po biogazovym texnologiyam [Biogas Technology Guide*] (Bishkek: DEMI) [In Russian]
[8] Kovalev A 2001 Effektivnost` proizvodstva biogaza na zhivotnovodcheskix fermax [Efficiency of biogas production on livestock farms] *Texnika v sel`skom xozyajstve* **3** 30-3 [In Russian]
[9] Kondaurov P and Marinenko E 2006 O celesoobraznosti proizvodstva i ispol`zovaniya biogaza v zamknutom cikle obrabotki otxodov sel`xozpredpriyatij s utilizaciej pobochnogo produkta anae`robnoj fermentacii [Biogas production and use in a closed cycle of agricultural waste processing using anaerobic fermentation by-product] *Izv. vuzov. Sev.-Kavk. region. Texn.* **10** 105-10 [In Russian]
[10] Da costa Gomez C 2013 Biogas as an energy option: an overview *The biogas handbook* (Cambridge: Woodhead Publishing) pp 1-16
[11] Zupančič G and Grilc V 2012 Anaerobic Treatment and Biogas Production from Organic Waste *Management of Organic Waste* Ed S Kumar and A Bharti (London: IntechOpen) pp 3-28
[12] Wu B, Zhang X, Di Bao, Xu Y, Zhang S, Deng L 2016 Biomethane production system: Energetic analysis of various scenarios *Bioresour Technol* **206** 155-63
[13] Baader W, Dohne E and Brenndörfer M 1987 *Biogas in Theorie und Praxis - Behandlung organischer Reststoffe aus der Landwirtschaft durch Methangärung* [*Biogas in theory and practice - treatment of organic residues from agriculture by methane fermentation*] (Münster – Hiltrup: KTBL) [In German]
[14] Mata-Alvarez J 2002 Fundamentals of the anaerobic digestion process *Bimethanization of the organic fraction of municipal solid wastes* Ed J Mata-Alvarez (Amsterdam: IWA) pp 1-20
[15] Kuschev L A and Suslov D Yu 2015 Teoreticheskoe opisanie processa anaerobnoj fermentacii v biogazovyh ustanovkah [Theoretical description of the process of anaerobic fermentation in biogas plants] *Vestnik BGTU im. V.G. Shukhova* [*Bulletin of BSTU named after V G Shukhov*] **6** 227-30 [In Russian]

IOP Conf. Series: Materials Science and Engineering **791** (2020) 012059 doi:10.1088/1757-899X/791/1/012059

[16] Hashimoto A G 1983 Thermophilic and mesophilic anaerobic fermentation of swine manure *Agricultural Wastes* **6** 175–91
[17] Gyunter L and Gol'dfrab L 1991 *Metantenki* [Digesters] (Moscow: Strojizdat) [In Russian]
[18] Vedeneev A and Vedeneva T 2006 *Biogazovye tekhnologii v Kyrgyzskoj Respublike* [*Biogas technologies in the Kyrgyz Republic*] (Bishkek: Evro) [In Russian]
[19] Vasilov R 2007 Perspektivy razvitiya proizvodstva biotopliva v Rossii. Soobshchenie 3: biogaz [Prospects for the development of biofuel production in Russia. Message 3: biogas] *Vestnik Biotekhnologii i Fiziko-Himicheskoj Biologii im. YU.A. Ovchinnikova* [*Yu.A. Ovchinnikov Bulletin of Biotechnology and Physical and Chemical Biology*] **3(3)** 54-61 [In Russian]
[20] Eder B, Schulz H 2007 *Biogas-Praxis: Grundlagen, Planung, Anlagenbau, Wirtschaftlichkeit, Beispiele* [*Biogas practice: basics, planning, plant construction, economy, examples*] (Staufen im Breisgau: Ökobuch) [In German]
[21] Gemmeke B, Rieger C and Weiland P 2009 *Biogas-Messprogramm II, 61 Biogasanlagen im Vergleich* (Gülzow: FNR)
[22] Paterson M and Kuhn W 2010 *Guide to Biogas - From production to use* (Gülzow: FNR)
[23] Appels L, Baeyens J, Degrève J and Dewil R 2008 Principles and potential of the anaerobic digestion of waste activated sludge *Progress in Energy and Combustion Science* **34** 755-81
[24] Grazia Leonzio 2018 Study of mixing systems and geometric configurations for anaerobic digesters using CFD analysis *Renewable Energy* **123** 578-89
[25] Karim K, Klasson K T, Hoffmann R, Drescher S R, DePaoli D W and Al-Dahhan M H 2005 Anaerobic digestion of animal waste: Effect of mixing *Bioresource Technology* **96(14)** 1607-12
[26] Kushchev L and Suslov D 2011 High-speed technology for processing organic wastes in bubble type bioreactors *Chemical and Petroleum Engineering* **47(1-2)** 70-3

IOP Conf. Series: Materials Science and Engineering **791** (2020) 012060 doi:10.1088/1757-899X/791/1/012060

The Development of Solar-Powered Gas Metering Units Equipped with Radio or 3G Based Telemetry Systems

M A Velichko, Yu P Gladkikh and I B Kostina

Department of Computer Science, Natural Sciences and Teaching Methods, Belgorod State National Research University, Pobeda St., Belgorod, 85, 308015, Russia

E-mail: velichko@bsu.edu.ru

Abstract. In this article we have described the development of solar-powered gas metering units including modern ultrasonic and vortex flowmeters for volume and mass measurement of natural gas, associated petroleum gases, inert gases, oxygen, hydrogen, air, nitrogen, carbon dioxide and any other single and multicomponent gases that are not aggressive to the flowmeter materials. We have shown implementation of such units in the existing gas producing stations in the desert of Kara-Kum. Confirmed was the ability of our gas metering units to operate more than a year without network power supply and staff intervention. Also pictured was creation of telemetry system for collecting and transmitting metrological information, including gas flow rate, temperature, pressure, humidity to the central terminal. Different physical layers have been used to transfer data from the metering units, including radio channels at 433 MHz frequency between units and central station. This telemetry system implemented self-maintenance and self-monitoring features, facilitating remote control of metering unit, so important in rigid conditions of solar-rich deserts and semi-deserts.
Key words: solar energy, telemetry, gas flow rate, gas metering unit, Web Socket

1. Introduction

Natural gas of Central Asia and Transcaucasia has long played an important role in meeting the needs of the domestic market of Russia, countries of the Commonwealth of Independent States and also far abroad. Russia plans to continue gas purchases in Central Asia and Azerbaijan, strengthening its position in this region [1].

Kazakhstan, Turkmenistan, Uzbekistan and Azerbaijan are also objects of increased interest from China, whose energy consumption has enlarged significantly due to the growth of industrial production. China also plans to increase its influence in this market, catching up profit that was lost in the Soviet period [2].

In Central Asian countries, gas pipelines are often laid through such areas as deserts or semi-deserts, that are difficult to access for electrical networks. Gas metering units and gas-distributing stations are spatially spaced and in most cases it is expensive to maintain and monitor them. Even if metering units are located within the reach of electrical lines, sandstorms and other weather and technogenic conditions can often lead to power supply failures.

The Central Asian region is also rich with oil, where production meets the same power and maintenance problems. In some countries infrastructure has not been seriously modernized since the Soviet Union. We propose the combined approach that can significantly relieve metering units monitoring and dramatically reduce power consumption so that constructing or upgrading existing power infrastructure in many cases will not be necessary.

IOP Conf. Series: Materials Science and Engineering **791** (2020) 012060 doi:10.1088/1757-899X/791/1/012060

The main origin of such approach is that Central Asian region has no shortage of solar energy. The annual duration of sunlight here is 2200-3000 hours per year, and the estimated power is 1300-1800 kW per 1 square meter per year. This is enough for the construction of efficient solar power stations. And, of course, it is more than enough to maintain the smooth operation of gas metering units. Mobile networks are deployed along gas and oil transportation routes and face minimum technological difficulties (besides they are rather modern), so remote metering units monitoring can almost substitute necessity of specialists' presence in the fields. In cases when metering units are by some reasons out of mobile network coverage, radio channel transmission can be organized.

So the aim of our work was to develop a solar powered gas metering unit equipped with the telemetry system, that in its turn, could transmit metrological data to the remote terminals and could also provide self-maintenance features. This unit should be able to measure and control gas flow rate under different conditions. The whole system should be energy effective and robust in the rigid conditions of desert and semi-desert. These units should include flowmeters able to measure flows of natural gas, associated petroleum gases, inert gases, oxygen, hydrogen, air, nitrogen, carbon dioxide and any other single and multicomponent gases that are not aggressive to the flowmeter materials. That is, it should be able to cover the entire spectrum of gases produced in Central Asia, as well as in most other solar-rich regions. It could also be used in oil production process, e.g. measuring associated petroleum gas in gas-lift methods.

Figure 1. The model of solar-based gas metering unit using IRGA-Helio flowmeter.

2. Materials and Methods

IRGA-TS telemetry system was developed and used to monitor and automate gas and power consumption at the fields of Turkmenneft group. Ultrasonic and vortex IRGA gas meters with software were installed at the wells in "Goturdepenebit" oil and gas production unit. Gas was used in the gas-lift method of well operation, which is a logical continuation of the fountain method of oil production. The missing amount of associated petroleum gas for oil recovery was pumped into the well from the surface. It was necessary to constantly monitor the temperature, pressure and volume flow of gas, as

IOP Conf. Series: Materials Science and Engineering **791** (2020) 012060 doi:10.1088/1757-899X/791/1/012060

well as remotely control solenoid valves and gates for effective management of oil production in wells.

For operation of gas lift valves in the conditions of the field, compressed gas was supplied from the compressor station with a maximum (peak) pressure of 8.0 MPa. In order to comply with safety conditions, a safety margin of K = 1.25 was accepted. So the flowmeter sensor was designed for pressure P = 8.0 x 1.25 = 10.0 MPa. This requirement was fulfilled by the sensor of the PDTVH-1Eh-02-10,0 MPa mark No. 12-03369 installed in 2 wells of the field.

Solar-powered version of IRGA flowmeters was called IRGA-Helio (Figure 1). It implemented the solar panel as a power source, that converted light energy into electricity, and a battery that stored that energy. One not very sunny day of battery charging at the latitude of the device homeland – Belgorod – was enough for continuous operation of the metering unit for two days. The solar panel had a small size and was placed in a strong aluminum frame with structured tempered glass.

The design of the gas metering unit allowed us to expose the panel at an optimal angle to the solar flow for fast battery charging. The helium batteries installed in the system had high temperature stability and high power output, as well as good recovery after deep discharge. The controller maintained the energy system. It didn't not allow its overload or reverse current at night.

All elements of the system had a very high service life and had been tested for a long time in various climatic conditions. IRGA "Helio" could operate without human intervention for months, sending the necessary information about the parameters of the gas environment with the help of a telemetry system.

Telemetry is an automated communications process by which measurements and other data are collected at remote or inaccessible points and transmitted to receiving equipment for monitoring [1]. Systems that need external instructions and data to operate require the counterpart of telemetry, telecommand [2].

Although the term commonly refers to wireless data transfer mechanisms (e.g., using radio, ultrasonic, or infrared systems), it also encompasses data transferred over other media such as a telephone or computer network, optical link or other wired communications like power line carriers. Many modern telemetry systems take advantage of the low cost and ubiquity of GSM networks by using SMS to receive and transmit telemetry data.

A telemeter is a device used to remotely measure any quantity. It consists of a sensor, a transmission path, and a display, recording, or control device. Telemeters are the physical devices used in telemetry. Electronic devices are widely used in telemetry and can be wireless or hard-wired, analog or digital. Other technologies are also possible, such as mechanical, hydraulic and optical.

Telemetry may be commutated to allow the transmission of multiple data streams in a fixed frame.

Different telemetry systems used in metrology are described in [1-6].

To measure the volume flow rate of associated petroleum gas, full-pass vortex flow meters with nominal flow diameters from 50 to 200 mm were selected. The vortex motion in such flow meters is provided by a flow body located in the path of the moving flow, which in the cross section has a shape close to the trapezoid. The system of vortices arising behind this obstacle is called the Karman vortex path, and the frequency of the collapsing vortices in the first approximation is proportional to the flow velocity [8]. The flow rate, in turn, allows to calculate the gas flow rate under operating conditions with an accuracy of 1%. The conversion of the flow rate into operating conditions into a unified flow rate under standard conditions was carried out in the computing device, which received the necessary metrological data from the pressure and temperature sensors. These sensors were located in the pipeline together with the flow meters.

Given the instability of electricity metering units, as well as the abundance of solar energy in the desert, instrumentation cabinets were equipped with autonomous power supply from solar panels (panel controllers and batteries were also installed in instrumentation cabinets).

For effective management of the gas lift process, monitoring of metering stations and control of associated gas flows had to be carried out continuously and without interruptions. Also, for the full automation of the process in the future, a thorough analysis of the accumulated statistical data was required, for which a detailed archiving of the transmitted metrological information was necessary [9].

3. Results

The data was transmitted to the server via the Internet. HTTP and Web Socket protocols were used as the application layer protocols. Users with the appropriate access rights, using modern Web-browsers sent GET-requests to the server. The server stored various temporary archives that contained all the information necessary for monitoring and decision-making (environment temperature, air, gas temperature, atmospheric pressure in the area of the well, pressure drop at different points of the pipeline, the presence or absence of electricity, battery charge, occurrence and description of emergency situations, etc.). These archives were sent to the server by POST requests from the computing devices located in the accounting nodes. POST requests were formed automatically in accordance with the established time schedule [10]. Thus, the server had two types of clients: web-browsers of telemetry system users and special software of the client device developed by our group.

The module developed by our group was used as a transmitting POST-requests device, which was installed in the control point at a distance of no more than 10 km from each of the wells. This IRGA-LT device served as an intermediate between the accounting nodes and the server. It included a 3G module controlled by a special microcontroller with the help of automated AT-commands. It was decided not to install similar 3G modules in each metering station due to 3G communication failures in the vicinity of the fields and the need to pay for and maintain SIM cards, as well as to simplify the transmission system.

Each metering point contained another self-developed device IRGA-NT connected to the computing device using UART protocol. Transfer of data from IRGA-NT to IRGA-LT was carried out via 433 MHz radio frequency [11]. Code and time modulation was used. Each radio module placed in the unit IRGA-NT, was assigned its own unique number, and the data was passed in turn from each node after the receipt of the request from the IRGA-LT radio module. In case of transmission errors signals of emergency situations were generated.

Both IRGA-LT and IRGA-NT radio-boards implemented E32-TTL-100 (Figure 2) boards with power of 100 mW, that were able to transmit and receive data over more than 10 km in the desert without additional amplifiers. It was possible to use such range of frequencies at such power according to special administrative permissions in Turkmenistan. Despite its high-distance transmission E32-TTL-100 had ultra-low power consumption of about 30 μA. It also offered broadcast transmission that was useful when data had to be transferred from IRGA-LT to all IRGA-NTs.

Figure 2. E32-TTL-100 radio board as electronic part of IRGA-LT and IRGA-NT modules.

When creating models of IRGA-NT and IRGA-LT devices, AVR microcontrollers installed in Arduino boards (UNO for IRGA-LT and NANO for IRGA-NT) were used to control the operation of radio modules, 3G modules and other electronic components. Arduino board software was developed in the Arduino IDE in the C++ programming language [12].

The server contained a PHP file to handle requests from the client IRGA-LT, as well as the HTML page passed as the response to the Web browsers. These HTML pages used JavaScript language, especially for a periodic AJAX poll to the server to inquire if the new data from IRGA-LT had arrived.

We have recently developed the new industrial time-of-flight ultrasonic gas flow meter IRGA-RU that has already been entered into the State register of measuring instruments [13].

Its operation principle is based on the time-to-pulse method of gas flow measuring. It consists in measuring the time of ultrasonic pulses passage in the direction of the gas flow in the pipeline and against it. Excitation and reception of pulses is carried out by piezoelectric transducers, which are installed in an all-metal case of the flowmeter at an angle (from 30° to 45°, depending on the version) to the flow direction. The speed of ultrasound in the medium depends on the physical and chemical properties of this medium: temperature, pressure, etc. At the same time, it is much greater than the speed of the medium, so that the actual speed of ultrasound in a moving medium is not much different from its speed in a stationary medium. The difference in transit times Δt even at flow rates of about 10 m/s is a fraction of a microsecond, while the measurement error should not exceed a few nanoseconds [14].

These circumstances necessitate the use of complex electronic circuits in combination with microprocessor technology, providing compensation for the influence of these factors.

Structurally, the flow meter consists of three blocks:

- primary flow converter, which is a housing with built-in ultrasonic transceivers;
- electronic unit, which carries out the reception and transmission of signals through ultrasonic transceivers, their conversion, processing and calculation of the volume flow of gas under operating conditions, followed by the formation of the output signal;
- power supply unit with built-in spark protection barrier to provide explosion protection circuits if necessary.

The electronic unit controls ultrasonic transceivers, performs receiving, processing, converting and transmitting signals to the computing device. These signals contain, in particular, information about the propagation time of the ultrasonic pulses, that is required to calculate the volumetric gas flow rate in operating conditions [15].

4. Discussion

The design of flow meter eliminates the possibility of unauthorized influence on the flow meter software and measurement information. We have developed and successfully tested our own control circuits for piezoelectric transceivers, and also created new original algorithms for processing received signals [16].

It allowed us to

- significantly expand the dynamic range of gas flow measurement to values of the order of 1:2000; increase the accuracy of measurements to 1% in most tasks;
- use our flowmeter on pipelines of almost any diameter;
- measure the flow rate of various physical and chemical properties of gas media;
- carry out measurements at high speed, in a wide range of temperatures and pressures;
- not to introduce additional pressure losses into the pipeline.

At the same time, our flow meter has a fairly simple design and does not require significant lengths of straight sections before and after its installation.

We have also mastered and successfully tested the technology of applying anti-adhesive coating on the inner walls of the flowmeter to avoid sticking of the solid fraction of the measured medium. This helped us to reduce the probability of non-reproducibility of measurement results by 0.2% [17]. In future it will also lead to calibration interval decrease.

In our flowmeter design we have used modern time-to-digital microcontroller convertors, that apply energy saving technologies. We have also developed energy efficient algorithms of transmission, receiving and processing of ultrasonic pulses. Power consumption has decreased to several μW. It helped us to implement autonomous power supply from rechargeable batteries.

Our colleagues from Turkmenistan gave us a possibility to test our device powered by solar panels in desert conditions with almost cloudless weather and periodic changes in energy supply. Experiment has shown that our flowmeter is able to operate without direct maintenance for months [17].

Information from IRGA-RU was transferred to the computing device and then transmitted via above described telemetry system. The same system was also used to unite our flowmeters into the network.

5. Conclusion

We have developed and effectively tested the solar-powered metering unit with telemetry system in a set of applications and conditions. No extra power except solar power was needed to maintain the system.

According to Turkmenneft report the use of flowmeters was effective for accounting for the flow of the working agent (gas) and calculation of associated gasin order to increase oil production from a certain category of wells currently operated by the gas lift method.

A significant reserve for improving the efficiency of the use of metering of the working agent (gas) and calculation of associated gas was to improve the selection of wells, taking into account the readings of the IRGA gas meters with services and to expand the range of adjustment and optimization works at specific wells.

The telemetry system united gas flow monitoring and control nodes at oil fields, giving us the opportunity to manage oil production process from another city. It also combined several high dynamic range ultrasonic flowmeters in the network, providing gas consumption map of the region. At the same time, we managed to test our new flowmeters in real operating conditions. This telemetry system also helped us to remotely verify domestic gas flowmeters by our new ultrasonic mobile stand.

References

[1] Velichko M, Satler O, Krasovskaya L, Bronnikova M, Belyaeva I, Kostina I and Gladkikh Yu 2018 Calibration of Ultrasonic Flow Meter on Wi-Fi Network Using a Web Browser *J. Advanced Research in Dynamical and Control Systems*, **08-Special Issue** 1593-6

[2] Velichko M, Satler O, Krasovskaya L, Erina T, Belyaeva I and Gladkikh Yu 2017 Using Drone as WiFi Access Point During Infrared Thermography for Subsidiary Data Acquisition *J. Fundam. Appl. Sci.* **9(1S)** 1279-88

[3] Velichko M, Chernyavskikh S, Kostina I, Gladkikh Y, Krasovskaya L 2018 Improving educational process quality in the lessons of natural and mathematical cycle by means of STEM-training *Cypriot J. of Educational Sciences* **13(4)** 501-10

[4] Hamidullin V, Malakhanov R and Khamidoullina E 2001 Statics and Dynamics of Ultrasonic Flowmeters as Sensing Elements for Power Control Systems. *IEEE Int. Conf. on Control Applications, CCA '01* (Mexico: IEEE) pp 680-5

[5] Frana D, Ono Y and Jen C 2011 Contrapropagating Ultrasonic Flowmeter Using Clad Buffer Rods for High Temperature Measurements *J. Dynamic Systems, Measurement and Control, Transactions* **1** 011007

[6] Hu L, Qin L, Mao K, Chen W and Fu X 2016 Optimization of Neural Network by Genetic Algorithm for Flowrate Determination in Multipath Ultrasonic Gas Flowmeter *J. IEEE Sensors* **5** 1158-67

[7] Luca A, Marchiano R and Chassaing C 2016 Numerical Simulation of Transit-time Ultrasonic Flowmeters by a Direct Approach *J. IEEE Transactions on Ultrasonics, Ferroelectrics, and Frequency Control* **6(63)** 886-97

[8] Gryshanova I 2007 The Effect of Duct Cross-sectional Shapes on Metrological Performance of Ultrasonic Flowmeters *Proceedings of the 5th Joint ASME/JSME Fluids Eng. Summer Conf., FEDSM2007* (San Diego) pp 27-31

[9] Hamidullin V 1995 Dynamics of Ultrasonic Flowmeters *Proceedings of the IEEE Ultrasonics Symp.*(Washington) pp 1109-13

[10] Gerasimov S, Glushnev V and Serov N 2018 Analysis of Calibration and Verification Indirect Methods of Ultrasonic Flowmeters *Digest: 13th Int. Conf. on Advanced Technologies, Systems and Services in Telecommunications, TELSIKS2017* (Niš: University of Niš, MTT) pp 451-4

[11] Bates R, et al. 2013 Development of a Custom On-line Ultrasonic Vapour Analyzer/Flowmeter for the Atlas Inner Detector, with Application to Gaseous Tracking and Cherenkov Detectors *J. of Instrumentation* **1(8)** C01002

[12] Gershman E M, Kruglo D S, Fafurin V A, Borisov A M and Sabirzyanov A N 2017 Flow measurement of two-phase flow by ultrasonic flow meters. *J. Transactions Of Academenergo* **1** 30-41

[13] Kuranov K N and Sergeev S S 1996 Avtomatizaciya rascheta parametrov ul`trazvukovy`x rasxodomerov [Automated calculation of parameters of ultrasonic flowmeters] *Sozdanie Resursosberegayushhix Mashin i Texnologij* [*Creation of Resource-Saving Machines and Technologies*] (Mogilev: Belorussko-Rossijskij universitet) pp 99-112 [In Russian]

[14] Strelnikov E A and Eremin I Yu 2016 K voprosu o povy`shenii tochnosti ul`trazvukovy`x rasxodomerov [The issue of improving the accuracy of ultrasonic flowmeters] *Zakonodatel`naya i Prikladnaya Metrologiya* [*Legislative and applied Metrology*] **1** 140-5 [In Russian]

[15] Abzalilova E R and Galimullina E 2017 Umen`shenie vliyaniya vibracii na tochnost` izmereniya ul`trazvukovy`x rasxodomerov [Reducing the influence of vibration on the measurement accuracy of ultrasonic flowmeters] *Vestnik Nauki i Obrazovaniy* **1(25)** 33-5 [In Russian]

[16] Tyryshkin R A, Sabirzyanov A N and Fafurin V A 2016 Chislennoe modelirovanie korrektiruyushhego koe`fficienta ul`trazvukovy`x rasxodomerov [Numerical simulation of the correction coefficient of ultrasonic flowmeters] *Matematicheskie Metody` v Texnike i Texnologiyax, MMTT* [*Mathematical Methods in Technics and Technologies, MMTT*] vol 9(91) (Saratov: SGTU im Yu A Gagarina) pp 17-23 [In Russian]

[17] Biryukov E A 2018 Preobrazovateli ul`trazvukovy`x rasxodomerov [The transducers of ultrasonic flowmeters] *Tradicionnaya i Innovacionnaya Nauka: Istoriya, Sovremennoe Sostoyanie, Perspektivy`* (Ufa: Omega Sains) pp 10-2 [In Russian]

IOP Conf. Series: Materials Science and Engineering **791** (2020) 012061 doi:10.1088/1757-899X/791/1/012061

Electricity in retail markets as a commodity in smart energy systems

Yu N Zatsarinnaya[1], A G Logacheva[2] and M O Grigoreva[3]

[1]Department of Power stations named after V.K. Shibanov, Kazan State Power Engineering University, Krasnoselskaya str., 51, Kazan, 420066, Russia
[2]Department of Power supply systems of Industrial Enterprises, Kazan State Power Engineering University, Krasnoselskaya str., 51, Kazan, 420066, Russia
[3]Department of Power stations named after V.K. Shibanov, Kazan State Power Engineering University, Krasnoselskaya str., 51, Kazan, 420066, Russia

E-mail: waysubbota@gmail.com

Abstract. Currently, the worldwide trend has been the transition of many countries to low-carbon energy sources. These changes are dictated both by external factors, such as the rapid growth of information and communication technologies, as well as the general policy of world powers advocating a clean environment and ensuring energy independence for the future. In Russia, the introduction of small generating capacities on renewable energy sources has great prospects due to the huge number of regions that are difficult to access for electricity supply. But the integration of generating capacities based on renewable energy sources requires not only the availability of special technical and legislative means, but also the competent selection of a tariff for a prosumer. This article presents the concept of an intelligent system that allows individual households to be involved in a progressive type of energy consumption and generation and creates a new type of energy system and market participant - prosumer. Two systems are offered for households equipped with solar panels. It is proposed to extend the practice of settlements at unregulated prices to individuals and households with the possibility of choosing the most favorable price category for the considered prosumer.

1. Introduction

One of the challenges of the energy security doctrine, approved by the Presidential decree 216 of May 13, 2019, is an increase in the share of renewable energy sources (RES) in the global fuel and energy balance. The challenge identified in the doctrine is a consequence of the commitment Russia made to reduce CO emissions by 2020. As an adequate response to the challenge, Russian manufacturers of equipment for renewable sources need to provide support and assistance in its implementation. To create a convenient platform and make appropriate amendments to the legislation for making settlements between the consumer and the producer, in one person, who will be called a prosumer and the already existing system in which payments for electricity are carried out now. Such a shift will be possible due to the development of intelligent technologies in the field of energy and the steady increase in the amount of renewable energy, which are the objectives of the above-mentioned doctrine.

Moreover, besides it is necessary not only to adopt the climate agreement and national legislative acts, but also to conduct an active policy in our state aimed at supporting the development of renewable sources by the population. All of the above requires the support of investment and venture funds, a flexible tax policy of the state, as well as the support of prosumers in the consumption and production

IOP Conf. Series: Materials Science and Engineering **791** (2020) 012061 doi:10.1088/1757-899X/791/1/012061

of clean energy [1]. Of great importance is the readiness of scientific organizations to develop competitive domestic equipment and local factories to begin production of equipment intended for the use of renewable energy sources.

The current trend facing the energy policy of any state is the search for optimal relation between the availability of electricity technically and from economic point of view. In most markets, both wholesale and retail, electricity demand has a low price sensitivity, since fluctuations in electricity prices that occur in the market hourly are not communicated to the retail consumer. Retail prices in Russia are regulated, and sales companies do not seek to use prices in real time and inform consumers of the household sector about the opportunity to save. Now, in the context of price regulation in retail markets, consumers of utility services, according to the annex to the Federal Tariff Service N 20-e/2, adopted in 2004, are charged at a fixed price, unlike commercial consumers with multi-rate tariffs. The government has repeatedly attempted to introduce different tariffs for the population and piloted them in several regions of the Russian Federation, but for the most part, payment for the tariff was made depending on the amount of energy consumed, and not on price fluctuations in the wholesale electricity market. Since in the short term, energy consumption is almost unchanged, and demand is variable, there is a shortage of it during peak periods and surplus at night. If the consumer is aware of real-time prices, then the price in the retail market will accurately follow the load level at the same time [2]. As a result of public awareness of tariff increases during the peak period, the need for generating capacity would be reduced by 10% with the help of only price factors. As a result, there would be an equalization of the demand curve for electricity, due to a decrease in its peak and an increase in the base parts of the daily schedule. If the consumer is informed of large fluctuations in electricity prices, this may lead to flattening of the consumption curve peak. To do this, it is necessary to amend the existing legislation and make calculations of the population at unregulated tariffs [3].

According to the International Energy Agency [4] the share of renewable energy sources is expected to grow by 20% and reach 12.4% of the total world consumption in 2023. Electricity generation at power plants using renewable energy sources has been showing steady growth in 170 countries for the past ten years, so in 2018 commissioning of capacities using renewable energy sources amounted to 181 GW, in 2017 - 8% less than in 2018, and a year earlier, only 85 GW [5]. Currently, hydropower industry remains the largest renewable source based type of generation in the world, with a share of 16% of global electricity generation, wind power is in the second place accounting about 6% and solar photovoltaic energy is in third place (4%).

Nevertheless, the last three years, the leaders in commissioned capacities on the basis of renewable energy sources have been accounted for the solar stations, which is explained by the desire of investors to invest in an acceptable payback period. This is due to scientific and technological progress in the development of materials for photo panels and, as a result, to a decrease in their cost, which has continued over the past decade.

In many developed countries, as well as places where centralized power supply is unavailable, for example, islands or, as is typical for Russia, a huge number of regions that have difficulties with access to power supply, in which RES will be competitive, it is necessary to pay attention to the development of distributed power supply to consumers [6]. To meet this energy need, 77% of global demand refers to buying a home solar system. Therefore, the use of distributed renewable energy, intellectualization of infrastructure and the transition of consumers to active, prosumer behavior patterns is obvious [7, 8]. At the same time, effective tariff setting for the prosumer will allow to get additional profit and adjust its consumption.

2. Materials and methods

To solve the problem of effective tariff setting for the prosumer in Russia the current system of retail tariffs was researched. According to clause 5 of Decree of the Government of the Russian Federation dated 04.05.2012 No. 442, there are six price categories of electricity in the retail electricity markets.

Based on this document, the Federal Tariff Service, the Ministry of Economic Development of the Russian Federation, the Ministry of Energy of the Russian Federation and the Federal Antimonopoly

Service developed rules for the application and determination of unregulated electricity prices (capacity) by guaranteeing suppliers. These rules are approved in the Decree of the Government of the Russian Federation dated December 29, 2019 No. 1179.

In accordance with these documents, the choice of the category according to which the calculation will be made between the legal entity and the sales organization is the duty and right of the consumer of electric energy (power) itself. Let us to consider the pricing principles and calculation methods which are applicable for each of the categories, according to the above regulatory legal acts.

The first price category includes small consumers with a maximum power of less than 670 kW. This article discusses the object of energy consumption, which can be classified in this category, due to the fact that the amount of electricity consumed by it is relatively small. Payment for electricity in this category is carried out at the same price for the whole amount of electricity consumed per month. The price already includes costs for the purchase of electricity in the wholesale market and for its transmission through networks of the corresponding voltage. In the invoice for payment for electricity, there is only one item at a single-rate tariff, and this category is one of the simplest, since hourly planning of energy consumption is not carried out, but it is often the most expensive for legal entities.

It is worth noting that in relation to legal entities (industries) that did not notify the guaranteeing supplier of the choice of a category on time, the calculation is made according to the first category. But it is possible to automatically transfer the price category used in the previous year to the current year.

The only difference of the second price category from the first one is that when calculating within this category, the division into zones is carried out to account for electricity consumption. The control zone is day and night consumption - two-zone consumption, as well as periods of peak, base and half-peak load, which are related to the three-zone consumption metering system. The object considered in the article has its own energy source - the solar panel, so the calculation of the two zones will be relevant for this article, since the solar panel operates in the mode of power output only in the daytime. As well as for the first price category, the cost of power and transmission costs are inherent in the price and consumer power does not exceed 670 kW.

When using the third price category, the electricity charge is hourly, therefore, appropriate metering of electricity is necessary. The transfer tariff is considered single-rate, as in the first and second price categories. Starting with the third, the range of application of price categories is expanding: they can be applied to legal entities that consume both less and more than 670 kW. Based on these factors, the third price category as well as the first and second ones may be applied to the object in question.

Common to the fourth, fifth and sixth price categories is that the electricity consumed by a legal entity is charged every hour at a different price. The main distinguishing feature between them is the need for consumers to plan their hourly consumption a day ahead (for the fifth and sixth). In case of deviations, the legal entity is obliged to pay for them (they are included in the price of electricity). This is not always feasible, therefore we will not calculate the fifth and sixth categories for calculating the consumption of electric energy by the object under consideration.

As for the tariff for the transmission of electricity, the fourth and sixth price categories can be combined into one group. They are different from other categories. Only in the calculations for these price categories the two-rate transmission tariffing is applicable. Calculation of payment by price categories between legal entities and guaranteeing suppliers is established by the law - the use of other mechanisms is not legal now in Russia. The real price of electric energy and capacity for each guaranteeing supplier is published before the 14th day of the month following the estimated one on the website of the «Administrator of the Trading System of the Wholesale Electricity Market», Joint Stock Company. Tariffs for transmission services and marketing premiums for guaranteeing suppliers are set annually by the regional energy commissions. Thus, having information on the data of actual and planned consumption, it becomes possible to verify the cost of electricity production, which is set by the guaranteeing supplier for the month, with the highest degree of probability [9].

What to do if, when checking the expediency of using a category, it proved to be unprofitable? The answer is switch to another category. The transition in the framework of the first, second, third and

IOP Conf. Series: Materials Science and Engineering **791** (2020) 012061 doi:10.1088/1757-899X/791/1/012061

fifth categories is possible once a month for one year. Similarly, between the fourth and sixth. For the transition, it is necessary to notify the guaranteeing supplier at least 10 working days before the start of the settlement period. The company can only switch from the first, second, third and fifth to the fourth or sixth categories once a year. For the transition, it is necessary to inform the guaranteeing supplier and replace the tariff for the transfer of electric energy within one month after the publication of tariffs for a new period. Usually, updated information is available at the end of December.

Thus, a legal entity can choose between a single-rate or two-rate tariff for the transfer of electric energy only once a year [10]. As for the choice within this division, the situation is much better. You can change the price category on a monthly basis if it meets the requirements of economic feasibility.

For our object - a residential building in the city of Kazan - we assumed the choice of one of several price categories: first, second, third and fourth, and also performed a calculation for each of them. Based on the data received, recommendations were made for the electricity consumer. The choice fell on these categories of data not accidentally. The price categories for which independent planning of consumption by the enterprise is provided (fifth and sixth) are not beneficial to the vast majority of consumers, because it is impossible with 100% probability to plan hourly consumption for the day ahead, resulting in deviations from the declared volume, which are necessary pay to a legal entity [11]. Thus, for consumers with a maximum power of less than 670 kW, the determination of the price category is reduced to the first, second, third or fourth, which differ in the tariff option for electric power transmission services, as well as taking into account electric energy. Unlike the second, the first is characterized by a more "ragged" load profile. According to this characteristic, in accordance with the schedule of electrical load, a residential building can be attributed to the first price category.

3. Results

In view of the foregoing, a technical solution should be developed for the integration of renewable energy sources into the energy system, which takes into account market signals and is accessible not only to large industrial facilities, but also to single households or housing societies. Then each person can be involved in a system of progressive energy consumption. The main components of this system are smart metering devices with built-in data transmission devices, communication lines, a hardware-software complex for data processing and computing, and user applications. The structural diagram of the system is shown in Figure 1.

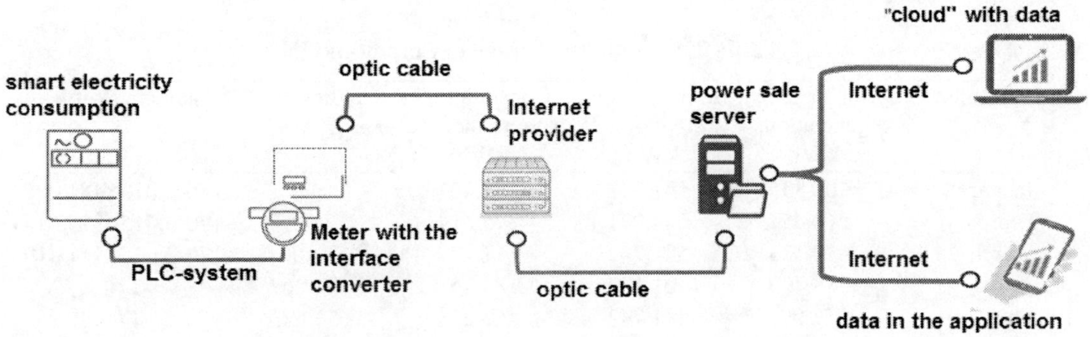

Figure 1. Diagram of a smart billing system.

Information about the consumption or generation of electricity on various devices of the facility is collected in a single smart settlement system located on the server of the energy supplying organization or guaranteeing supplier. This server carries out financial calculations for electricity. One of the requirements for the system is to support the cloud storage function to provide users with access to information about their consumption and the state of the electricity bill. The user will be able to view

IOP Conf. Series: Materials Science and Engineering **791** (2020) 012061 doi:10.1088/1757-899X/791/1/012061

data in a mobile application that supports communication with cloud storage at any time of the day. In Figure 1, the implementation of communication channels is shown by PLC technology and optical communication lines, but it is possible to use twisted pair or wireless data transmission methods.

The financial calculations carried out by the system are carried out according to two models. The first model is the energy credit model. When choosing this model, the prosumer gets the opportunity to take into account the electricity generated and delivered to the network on his account in the form of marketable products. In periods when own consumption exceeds generation, it can be used to offset the consumed electricity from the network. The second model is called energy billing. In energy billing, the electricity generated and consumed from the network has its own set prices. The consumer sells electricity generated in excess of his own consumption and takes it into account in monetary terms.

An energy credit model requires a bi-directional meter, which, when the electricity is consumed from the grid, will add its volume to the value recorded in the meter, and when the electricity is generated, it will subtract its amount. The prosumer pays in monetary terms only for the purchase of electricity from the grid at a retail price.

Energy billing requires two meters that separately measure the energy consumed and delivered to the grid. In this model, a prosumer buys electricity from the grid from a utility at a retail price and sells its own produced energy at a wholesale price. At the same time, in order to track real-time pricing, the meter must take into account the current market price so that the prosumer has up-to-date information for choosing a retail tariff [12, 13].

In order to assess the economic feasibility of the proposed calculation models, we carried out calculations of the estimated generation obtained from the solar panels installed on the roof of a residential building in the city of Kazan, and the income from their use. Realization of the generated energy was carried out according to the two models described above.

According to the model of energy credit, energy was consumed by a residential building, both from the network and from its own sources of generation. In periods when the generation of electricity by solar panels exceeded the needs of a residential building, the surplus was given to the network at the expense of future consumption. According to the energy billing model, surplus energy generated by solar panels was sold to the network for money.

The rightmost column in tables 1 and 2 displays the monthly status of the prosumer payment. The amount in the payment is determined based on the average consumption rate of electricity by the household (the base cost for the tariff in Russia is 170 rubles), and the amount of electricity sold or additionally consumed. Electricity consumption is paid at a price of 170 rubles/kWh, sale at a price of 3.015 rubles/kWh.

Table 1. Calculation for energy credit model.

Month	Solar panels generation, kWh	Building consumption, kWh	Consumption from/to the grid, kWh	Credit of energy, kWh	Payment, Rubles
January	100.335	581.742	-481.407	0	-170-3.015×481.407
February	268.457	644.071	-375.614	0	-170-3.015×375.614
March	470.323	581.742	-111.419	0	-170-3.015×111.419
April	788.400	601.133	187.267	187.267	-170
May	1051.000	581.742	469.258	656.525	-170
June	1283.000	601.133	681.867	1338.392	-170
July	1106.000	581.742	524.258	1862.65	-170
August	919.742	581.742	338.000	2200.65	-170
September	511.920	601.133	-89.213	0	-170
October	242.477	581.742	-339.265	0	-170
November	108.000	601.133	-493.133	0	-170
December	52.258	581.742	-529.484	0	-170
Total	**6901.912**	**7120.797**	**-218.885**	**6245.484**	**-2700**

IOP Conf. Series: Materials Science and Engineering **791** (2020) 012061 doi:10.1088/1757-899X/791/1/012061

Table 2. Calculation for energy billing model.

Month	Solar panels generation, kWh	Building consumption, kWh	Consumption from/to the grid, kWh	Payment, Rubles
January	100.335	581.742	-481.407	-170-3.015×481.407
February	268.457	644.071	-375.614	-170-3.015×375.614
March	470.323	581.742	-111.419	-170-3.015×111.419
April	788.400	601.133	187.267	-170+3.015×187.267
May	1051.000	581.742	469.258	-170+3.015×469.258
June	1283.000	601.133	681.867	-170+3.015×681.867
July	1106.000	581.742	524.258	-170+3.015×524.258
August	919.742	581.742	338.000	-170+3.015×338
September	511.920	601.133	-89.213	-170-3.015×89.213
October	242.477	581.742	-339.265	-170-3.015×339.265
November	108.000	601.133	-493.133	-170-3.015×493.133
December	52.258	581.742	-529.484	-170-3.015×529.484
Total	**6901.912**	**7120.797**	**-218.885**	**-103.0949**

In this case, the total consumption is 7120.797 kWh, the total electricity generation is 6901.912 kWh. At the same time, according to the results of the month, the user's debt to the supplying organization under the credit system amounted to 2700 rubles, and under the billing system 7213 rubles. Similar calculations for households with different ratios of generation and consumption showed that an energy credit is beneficial for those who generate less than they consume. And energy billing is appropriate for prosumers with generating capacities, the generation of which often exceeds the amount of household needs.

In the above calculations, fluctuations in the price of electricity in the retail market were not taken into account. However, as mentioned above, for the full inclusion of all consumers in a progressive consumption model, the financial settlement system includes the function of translating market price signals. Thus we consider and evaluate several models of calculations using price categories that are currently valid in Russia for legal entities, and the essence of which is proposed to be transferred to the financial calculations of prosumers.

Electricity generation from renewable energy sources is inconsistent, so the consumer has to combine power from the network and renewable energy sources [14]. If the apartment building is not equipped with batteries or other sources of alternative energy, except for the sun, that is, consumption from the network will be made every day at night, the comparison chart of the selected categories will look like shown in Figure 2.

For the case when the calculation is carried out by category, but the consumer does not give "surplus" to the network, the comparison chart of the selected price categories will look like shown in Figure 3. Figure 4 shows a comparison of the third and fourth price categories.

4. Discussion

It can be seen from Figure 2 that the use of a two-rate transmission tariff compared to others is more profitable by 1,310 rubles. Thus, under these conditions, it is more profitable to use the second price category, when electricity is calculated in two zones of the day.

In the period from April to August, the amount of electricity consumption from the network is zero. The diagram in Figure 3 shows that when comparing these categories, the first turned out to be the most profitable. The price of electricity for the first price category is calculated according to the residual principle. That is, all the expenses of the guaranteeing supplier after subtracting of the bills of all consumers in categories from second through sixth, the remaining money is distributed evenly among consumers using the first price category.

IOP Conf. Series: Materials Science and Engineering **791** (2020) 012061 doi:10.1088/1757-899X/791/1/012061

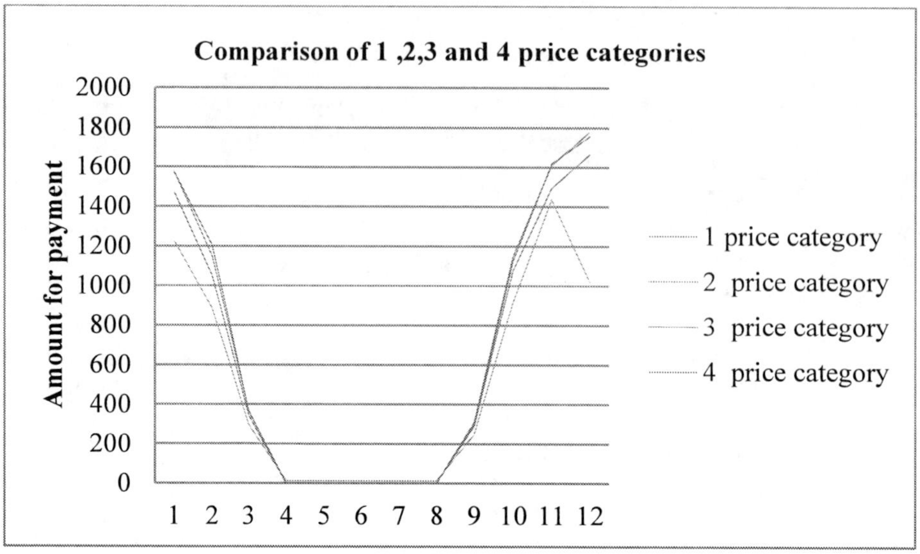

Figure 2. Comparison of annual payment amounts for settlements for the first, second and third price categories for the energy credit model.

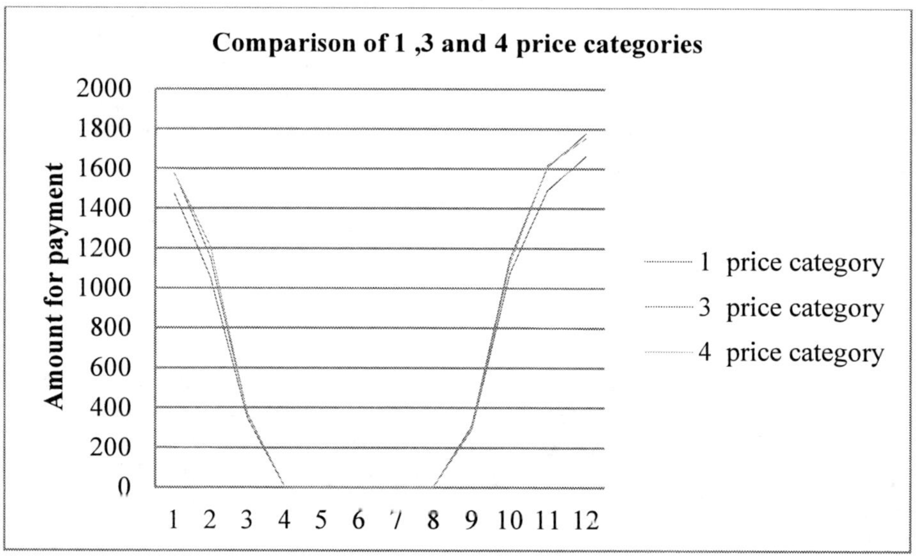

Figure 3. Comparison of annual payment amounts for settlements in the first, third and forth price categories for the energy billing model.

The first price category should be more expensive than the others, should encourage consumers to install hourly metering devices and choose other price categories with more accurate methods for determining actual consumption. However, due to the residual mechanism for calculating the price of the first price category, in some regions there are "distortions" at which it is the cheapest category for consumers. When using the first category, the consumer benefits for the year compared to the third and fourth, respectively, 545 and 544 rubles, respectively.

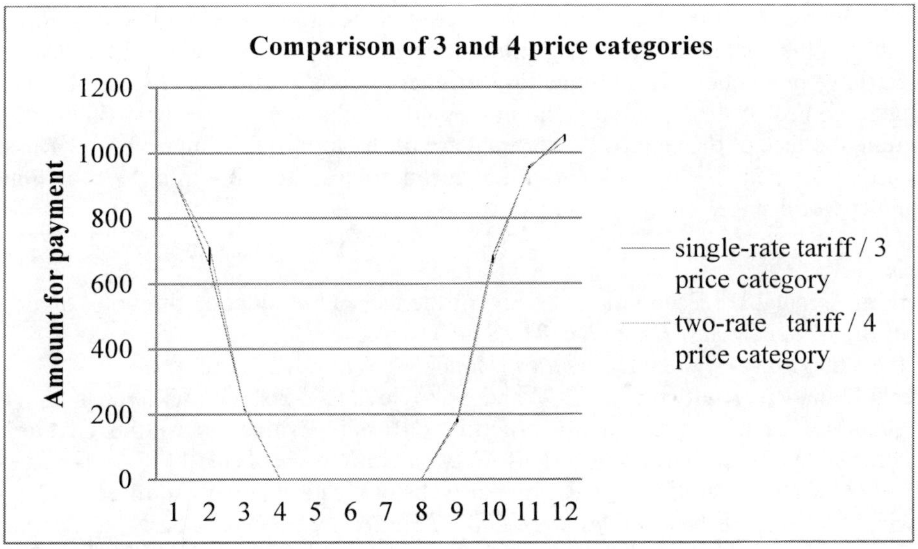

Figure 4. Comparison of annual payment amounts for settlements for the third and fourth price categories.

The diagram in Figure 4 shows that the two-rate tariff turned out to be, although not by much, but more profitable than the single-rate one. Consumer savings in this case amounted to 117 rubles.

If the facility has an accumulator of unlimited capacity, with the help of which accumulation is carried out, it is possible to assume the possibility of selling excessively generated electricity to the network during peak hours in a balancing market.

However, it should be noted that the profitability of any price category is variable. It cannot be said unequivocally that any one specific category will be beneficial for all regions. Moreover, for each prosumers within the region various price categories are beneficial. Therefore, in order not to fall for an increase in price, it is necessary to periodically analyze the conditions of energy supply of the consumer and, if necessary, make changes. It is worth considering that if the period of maximum electricity consumption occurs at night, it is more profitable to use the second price category. If from an economic point of view it is more profitable to use two-rate tariffing for energy transfer, then a fourth or sixth price category should be used. But we must not forget about the need for consumption planning in the calculation of the fifth and sixth categories.

5. Conclusion

As a first step towards the implementation of the proposed concept of the energy trade system, amendments should be made to the tax legislation, which will be aimed at stimulating the development of renewable energy sources. Such measures, in relation to prosumers, may include the introduction of tax benefits on value added for the transportation of solar panels, their purchase and installation, property tax, tax holidays for the first years of use of equipment, and introduce state subsidies for interest rates on loans in in relation to renewable energy facilities [15].

The next step is to agree on the procedure for technological connection to the network object. Here, the technological process of integrating renewable energy generation should be supported by network equipment and the capabilities of the network operator. After this, the principles of contractual relations with the sales organization are established. In general, this means choosing a billing model for the user. To implement this step, it is necessary to extend the practice of settlements at unregulated prices to individuals and households with the possibility of choosing the most favorable price category for the considered prosumer and make changes to paragraph 5 of the Resolution 442 issued by the Government of Russian Federation and equate the population with consumers who pay for electricity at unregulated prices.

Today, the degree of RES integration in the energy system is an indicator of the technological and industrial development of the state. Renewable energy is the driver of economic development, science and education. In the context of the transformation of global energy systems with a deep decentralization of electricity production, the introduction of energy conservation practices, digital technologies and intelligent control systems, cloud technologies, block chain and other new financial technologies will determine the face of the electric power industry of the future. It is important for Russia to stay on the wave of development and transition to a new technological structure in order to maintain competitiveness on the world stage.

References

[1] Ferreira A et al 2018 Economic overview of the use and production of photovoltaic solar energy in Brazil *Renew Sust Energ Rev* **81** 181–91
[2] Stoft S 2002 *Power System Economics* (Hoboken: Wiley-IEEE Press)
[3] Grechukhina I A, Kudryavtseva O V and Yakovleva E Y 2016 E`ffektivnost` razvitiya ry`nkov vozobnovlyaemy`x istochnikov e`nergii v Rossii [Evaluation of the development of the renewable energy markets in russia] *E`konomika Regiona* **12(4)** 1167-77 [In Russian]
[4] *Renewables 2018* (Paris: IEA) URL https://webstore.iea.org/download/direct/2322
[5] Sawin J L et al 2018 *Renewables 2018 Global Status Report* (Paris: REN21 Secretariat)
[6] Eroshenko S A, Samoylenko V O and Pazderin A V 2016 Renewable energy sources for perspective industrial clusters development *2nd Int. Conf. on Industrial Eng., Applications and Manufacturing, ICIEAM-2018* vol 1 (Chelyabinsk) pp 821-824
[7] Zhiznin S Z and Timokhov V M 2017 Energy impact on sustainable development *World Economics and International Relations* **61(11)** 34-42
[8] Eroshenko S A and Khalyasmaa A I 2018 Intelligent model of decision support system of distributed generation integration *Proc. of the IEEE 8th Int. Conf. on Software Eng. and Service Science, ICSESS 2017* (Beijing: IEEE Computer Society) pp 79-82
[9] Dmitriev A V, Dmitrieva O S and Zinurov V E 2018 Influence of elements thickness of separation devices on the finely dispersed particles collection efficiency *Int. Conf. on Modern Trends in Manufacturing Technologies and Equipment, ICMTMTE 2018* (Sevastopol) vol 224 pp 02073
[10] Dmitriev A V, Dmitrieva O S and Zinurov V E 2018 Intensification of gas flow purification from finely dispersed by means of rectangular separator *Int. Conf. on Construction, Architecture and Technosphere Safety, ICCATS 2018* (Chelyabinsk) vol 451 p 012211
[11] Akhmetov A A, Akhmetova R V 2016 Development of water saving technology for water supply system of industrial enterprises *Int. Scientific-Technical Conf. on Innovative Eng. Technologies, Equipment and Materials 2015, ISTC-IETEM-2015* (Kazan) vol 134 pp 012001
[12] Bulatov Y N, Kryukov A V and Suslov K V 2017 Multi-agent technologies for control of distributed generation plants in the isolated power systems *Far East J. of Electronics and Communications* **17(5)** 1197
[13] Lombardi P et al 2012 Optimal storage capacity within an autonomous micro grid with a high penetration of Renewable Energy Sources *3rd IEEE PES Innovative Smart Grid Technologies Europe, ISGT Europe* (Berlin) p 1
[14] Staroverova N A, Shustrova M L and Satdarov M R 2020 Development of a Cyber-Physical System for the Specialized On-Track Machine Operators Training *Cyber-Physical Systems: Industry 4.0 Challenges* ed A G Kravets, A A Bolshakov and M V Shcherbakov (Cham: Springer) 315-25
[15] Abdureshitova D V 2018 Nalogovy`e metody` regulirovaniya i stimulirovaniya razvitiya vozobnovlyaemoj e`nergetiki Rossijskoj Federacii [Tax methods of regulation and stimulation of renewable energy development Russian Federation] *Nauchny`j Vestnik: Finansy`, Banki, Investicii* [*Scientific Herald: Finance, Banks, Investments*] **1** 57-64 [In Russian]

IOP Conf. Series: Materials Science and Engineering **791** (2020) 012062 doi:10.1088/1757-899X/791/1/012062

Energy efficiency tools of buildings design solutions in information and analytical management systems for construction and overhaul process

I P Avilova and M O Krutilova

Department of Assessment and Management in Real Estate, Belgorod State Technological University named after V.G. Shukhov, Kostyukova St., 46, Belgorod, 308012, Russia

E-mail: kafeun@mail.ru

Abstract. The energy efficiency of construction and overhaul of buildings, including the optimization of energy consumption in production of building materials, construction and technological support of construction, has a significant role in energy efficiency assessment of construction as a whole, being the most important component of the implementation of green building principles. At the same time, the organizational-economic balance of interests of construction investors and authorities in the field of environmental rationalization in structural design as well as the motivation of developers to use green building materials and technologies, as assessed within economic tools, has still not developed in the Russian construction community. The paper proposes to consider the basic principles of the author's methodology for assessment of energy efficiency of construction and optimizing the structural design, which allows one to implement the tool into the information-analytical systems for managing the construction and overhaul process of buildings an instrument of promotion of green building, operating with quantitative energy efficiency assessments and cost indicators of design solutions.

1. Introduction

Construction industry is one of the most energy-intensive industrial sectors, including energy consumption at the phase of construction, mechanization, logistic and maintenance of buildings. Lifecycle construction project duration highlights a significant research potential of technical and technological decisions for resource optimization taken at the design stage of construction projects.

Energy audit of completed buildings and energy efficiency analysis of building functioning, determining its energy efficiency class and rate of compliance with national and international green standards are a common modern practice of assessing energy efficiency in completed construction projects and its subsequent facility and infrastructure improvements. At the same time economic evaluation (cost effectiveness or better cost utility analysis) of using green building materials and technologies often is taken into account empirically or declaratively. At the national level, using of green constructive solutions incenting by economic mechanisms for developers is insufficient, especially for BTS-developers who are not involved with the process of further maintenance and do not share the owner's potential risks of insufficient energy efficiency [1]. An essential obstacle to creation of effective national system of environmental and energy audit of construction materials and construction projects is the absence in the normative field of domestic construction of legislative directives. It is important to

IOP Conf. Series: Materials Science and Engineering **791** (2020) 012062 doi:10.1088/1757-899X/791/1/012062

establish specific financial obligations incurred with respect to manufacturers of construction materials and investors that comply or not with green building standards [2].

The purpose of the paper is to propose the key principles and offer for consideration and implementation the tool for reasonable economic stimulation of using green building technologies in information and analytical management systems for construction and overhaul process.

2. Materials and Methods

A distinctive feature of environmental problems is their global nature. The Russian Federation is one of the countries with the worst environmental situation [3]. In recent years, environmental pollution has deteriorated on a number of indicators, despite ongoing activities and programs for environmental monitoring and protection. Environmental protection activities in construction industry of Russia is local in nature, and is implemented at the level of normative legal acts of municipalities, i.e. do not contain universality focused on preservation of overall eco-balance on the scale of territories, countries, continents. Economic assessment of environmental impact is defined as practical and permissible resource and monetary losses due to anthropogenic exposure on qualitative and quantitative parameters of environment affecting in general and its individual eco-resources (hydrosphere, lithosphere, atmosphere, biosphere) [4]. However, the regulatory framework of pricing in construction in Russia and in most developing countries does not provide adequate quantitative assessment of civil construction environmental impact [5]. According to Russian legislation, environmental expertise is mandatory element of the Comprehensive State Expert Examination, and economic project assessment is declared from perspective of sustainable development and resource conservation at the initial stages of any project. Nevertheless, methods for quantifying the cost of environmental potential harm during construction phase still remain the subject of scientific discussions and research.

There are several methodological approaches for assessing environmental harm caused by industrial enterprises, including construction sector. There are two main ones: indirect (enlarged) and recipient (based on direct account) [6]. Indirect approach is characterized by simplicity and methodological completeness but provides low validity of the result and subjectivity of expert peer reviews. Exposure assessment on the basis of recipient approach is a time-consuming and long-term process requiring significant expenditures of time and resources, inconvenient for widespread use in economic calculations. This approach is used as a tool to create an information base in indirect methods development.

At the same time, three main methods are used to determine the components of pollution damage: focus districts (based on comparison of indicators of conventionally polluted and clean areas), analytical dependencies (based on obtaining mathematical dependencies, e.g. by means of multivariate analysis between indicators of the state of corresponding economic system and level of environmental pollution), and combined [5]. The methods listed are solving different problems and differ in their functionality.

Energy efficiency management of buildings should begin with the optimization of technical and technological constructive solutions at the start of invest construction projects - at the design phase. At this stage, information modeling technologies allow contractors to provide more environmentally friendly technical solutions without inefficient labor costs and to achieve efficiency on a number of criteria for sustainable development: reduction of resource intensity, minimizing energy inputs for building materials production, machinery and mechanisms; increasing of using recycled materials, united by a common efficiency assessment [7]. A key feature of such assessment should be it focusing on environment and economic – evaluation and quantitative accounting of negative environmental factors from extraction of raw materials till production, building-and-assembly works, etc. The method based on full information accounting of all construction process resource components is necessarily accompanied by obtaining an economic assessment of negative environmental impact of the construction project on through standard objective and cross-checked audit algorithms [8].

We will provide key aspects of the proposed eco-oriented assessment of constructive project solutions by comparing alternative design solutions under construction with traditional and green technologies.

IOP Conf. Series: Materials Science and Engineering **791** (2020) 012062 doi:10.1088/1757-899X/791/1/012062

Primary comparative cost estimation of construction is determined by standard methods of estimated pricing based on data of information modeled bill of quantities and classifiers of current resources market value.

Table 1. Volumes of the main resource- and energy-intensive building materials

Resources	Regular constructive technologies	Green constructive technologies
	Building materials	
Concrete, m^3	3672	3024
Steel reinforcements, t	250	241
…		
	Machines	
Tower Cranes, hrs	4856	4475
Mobile Cranes, hrs	356	41
…		
Total cost, thousand rub.	65600	72000

Comparison of alternative design solutions at the pre-project stage is most effective method from the standpoint of minimum cost of appropriate changes to the project. At this stage, compared projects already have preliminary space-planning solutions, however, there is an acute lack of information about details of design solutions that are subject to quantitative comparison. In order to improve this process the method of quantitative assessment of key building energy-harmful resources was proposed upon the established regression between resources and indicators of space-planning solutions (area, number of stories, etc.). For approximation in techno-economical processes what are always consist of non-linear (irregular, stochastic) and linear (regular, systematic) parts, K-polynomials proposed by author are useful, relevant and confident [9].

K-polynomial of nth degree means the symmetrical mathematical expression of normal and inverse powered variables as follows:

$$Y = a_{(-n)} x^n + a_{(-(n-1))} x^{n-1} + \ldots + a_0 x^0 + \ldots + a_{n-1} x^{-(n-1)} + a_n x^{-n}, \quad (1)$$

where Y – resources estimated costs,
x – quantitative indicator of the building space-planning decision,
a_i – constants,
x^0 – dummy term (always equal to 1), used for structure's clearness.

Left part of K-polynomials ($a_{(-n)} x^n + a_{(-(n-1))} x^{n-1}$) used for approximation of non-linear parts in approximated processes, right one ($a_0 x^0 + \ldots + a_{n-} x^{-(n-1)} + a_n x^{-n}$) — for approximation of linear parts in approximated processes.

Proposed K-polynomial could be easily converted to linear function ($i = 1$; $a_{(-1)} \ldots a_0 = \text{const}$, $a_1 = 0$); polynomial of nth degree ($a_{(-n)} \ldots a_0 = \text{const}$; $a_1 \ldots a_n = 0$); exponential of nth degree ($a_n \ldots a_0 = \text{const}$; $a_{(-n)} \ldots a_{(-1)} = 0$) or used to combine advantages of all above types.

In the study, the authors confirmed universality and expediency of use of K-polynomials for regression between key energy-intensive building materials and space-planning indicators and high correlation of used regression exceeding on R2 default trend lines of standard spreadsheet applications as well. The results are given in the authors' figures (figure 1).

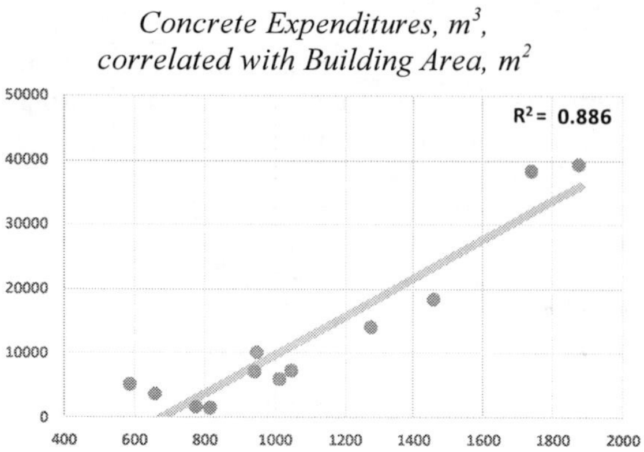

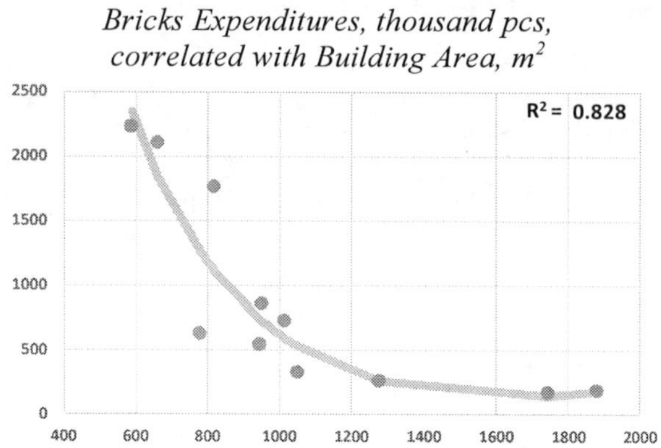

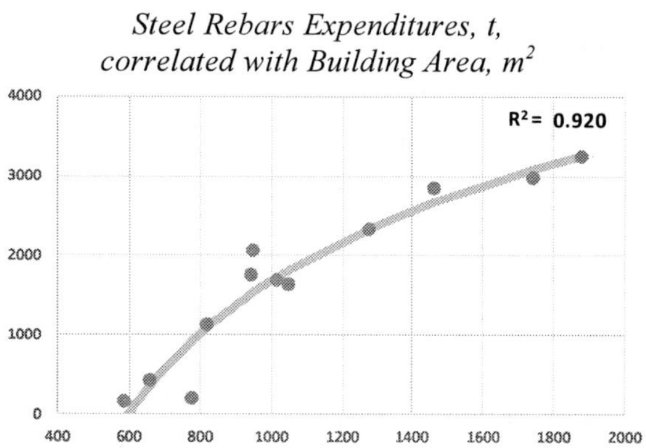

Figure 1. Specimens of regressions for expenditures of k key energy-intensive building materials and building area (author's dataset for multistory residential buildings with full concrete framework, K-polynomials of 1st degree used).

Table 2. Embedded Energy Calculator, expected emissions of aggregated GHG converted to CO_2.

Energy source	Regular constructive solution	Green constructive solution
	Materials	
	Concrete	
Expected emission of concrete producing, g/kg	100	
Concrete density, kg/m^3	2100	
Concrete volume, m^3	3671	3024
Expected embedded CO_2, t	771	635
…		
	Machines and appliances	
	Tower crane (67 kWt, loading capacity 8 t)	
Estimated working time, hr	4856	4475
Estimated GHG emission, g/kWt*hour	790	
Estimated GHG emission total, t	257	237
…		
Total GHG emission caused by materials, t	**1643**	**1507**
Total GHG emission caused by machines, t	**1901**	**1747**

Based on obtained resources' consumptions and correlation of greenhouse gases (GHG) converted to CO_2 produced by technical and technological sources of energy at the construction site, which are currently being developed by the authors, including information on the energy intensity of materials extraction and production from the standpoint of life cycle assessment analysis (LCA) [10]. The information on materials and energy required during mining to material production, transportation of material to construction site, construction activities and operational energy during use stage has been considered to conduct LCA analysis. It is proposed to include in analysis the calculator of embedded energy (shown above). By means of it one evaluates the expected weight of aggregated GHG emitted, caused and embedded within alternative design solutions, materials and technology of construction what are being compared (Table 2).

Presented estimate of reduced GHG emission in physical terms directly characterizes the degree of energy efficiency of the compared design decisions and needs to be transformed into a comparative cost estimate added to the estimated earlier construction cost. An algorithm for such transformation is the tool of conventional environmental taxation (EcoT) of construction projects (Table 3), which are alternatively implemented according to the compared design decisions [11].

Table 3. Tools of conventional environmental taxation (EcoT) of construction projects.

Group of EcoT	Reasons for EcoT
Construction material (for 1 measurement unit of consumed material)	Concrete, m3 Steel for rebars, t Steel for framework, t
Transportation (for 1000 t/km of average distance to construction plants)	up to 20 km 21-49 km more than 50 km
Installation (for 100 machine-hours of machine/vehicle in installation processes)	up to 20 machines/vehicles 21-49 machines/vehicles more than 50 machines/vehicles
In-situ energy consumptions for facilities and workers accommodation (for 10 worker-day at construction site)	up to 50 workers 51-199 workers more than 200 workers
In-situ energy consumptions for tools and engines (for 1 day of 10 tool/engine at construction site)	up to 20 tools/engines 21-49 tools/engines more than 50 tools/engines

EcoT, being an instrument that objectively assesses the environmental safety and energy efficiency of construction, is also an effective state regulatory mechanism that allows for the implementation of an additional project assessment that is obvious to all project participants and system customers in the information and analytical system of building management, taking into account penalties for the development of the project proposed by the regional market of green materials and solutions and the regional specifics of the construction industry [9]. For effective and transparent algorithm for the formation of such an assessment the marginal assessment is proposed, obtained from the available boundary data on the estimated construction costs (ECC_{max} and ECC_{min}) and reduced energy efficiencies (CO_{2max} and CO_{2min}) of construction projects implemented or being implemented in the region similar real estate. Local authorities are given the opportunity to adjust the amount of tax on a quarterly published multiplier *M*, which allows one to selectively assess and regulate the degree of economic interest of developers in green construction by industry and type of construction products being created.

$$EcoT = \frac{ECC_{max} - ECC_{min}}{CO_{2max} - CO_{2min}} \cdot M \qquad (2)$$

IOP Conf. Series: Materials Science and Engineering **791** (2020) 012062 doi:10.1088/1757-899X/791/1/012062

3. Results

The presented tool for assessing environmental and economic effectiveness of alternative design decisions of construction, placed in the information-analytical system of construction management, includes an automated assessment of the information on the materials and technologies used in the compared design decisions, generates individual and general project estimates environmental potential harm and conventional energy efficiency of projects, transforms them into a universal amount of conventional environmental tax on public authorities that control the regional construction market and the environmental situation on construction sites, quantitatively highlights the design decisions that are insufficient from the standpoint of sustainable development and green building. Table 4 shows authors dataset of compared building frames referring to probable alternative green constructive solution and EcoT calculated on the basis of contractors' equipment, construction site amenities (Table 3), embedded energy calculations (Table 2) and *M* multiplier recommended for local authorities. Calculated efficiency of using an alternative green technology, expressed in ratio of eco taxes and extra green cost commonly is quite significant (up to 51%, 22% in average). Obviously, this creates an understandable and acceptable by market participants and universally applicable by the regulator flexible tool for the economic motivation of regional building communities to active and purposeful implementation of green construction and support the production and use of green building materials [12].

Table 4. Comparative assessment of construction costs in alternative design solutions taking into account conventional environmental taxation

Building frame #	Cost before EcoT, mln. rub.	Cost of alternative green frame, mln. rub.	EcoT, %	Cost after EcoT, mln. rub.	Green efficiency
1	81.3	101.4	31.8%	107.2	7%
2	113.7	122.9	35.2%	153.7	27%
3	114	136.5	11.0%	126.5	-9%
4	62.4	74.4	56.4%	97.6	37%
5	89.1	109.8	57.0%	139.9	34%
6	57.8	68.1	20.6%	69.7	3%
7	54.9	67.8	11.2%	61.0	-12%
8	72.6	77.0	46.6%	106.4	41%
9	113.5	119.5	37.6%	156.2	32%
10	54.7	65.8	39.4%	76.3	19%
11	63.2	68.8	35.4%	85.6	27%
12	78	82.0	56.4%	122.0	51%
13	96.5	115.7	47.8%	142.6	28%
14	76.7	86.3	10.0%	84.4	3%
15	58.2	66.1	26.8%	73.8	13%
16	110.9	122.0	56.6%	173.7	47%
17	107.9	119.9	29.4%	139.6	18%
18	66.8	69.5	14.4%	76.4	10%
19	91.4	114.1	40.8%	128.7	16%
20	115.3	128.8	52.2%	175.5	41%
21	97	98.4	19.2%	115.6	18%
Average	**84.6**	**95.9**		**114.9**	**22%**

4. Discussion

Analysis of existing methods and approaches to environmental potential harm assessment allows us to conclude that there is no universally approved system of indicators of environmental harm, which allows to consolidate data on possible harm caused by enterprises of different production technologies, products and operating modes, etc. [13]. It does not allow to bring ecological damage to uniform universal equivalent (conventional unit of environmental harm) and to construct on this basis formalized algorithms of practically eco-oriented pricing and estimated rationing in construction. In Russian pricing and estimated system in construction the factor of "environmental pollution" associated with the production of construction works and the use of building materials, products and structures is taken into account only declaratively. According to [14] in design and budget documentation section of environmental harm assessment caused by construction project is not included. The concept and definition of environmental harm, mentioned in [15] do not have a full comprehensive justification and are not enshrined in law. The state examination of project documentation in accordance with [16] does not provide for a full and comprehensive analysis of possible environmental harm. Appropriate methodological approach to this issue is the proposed universal sets of generalized natural indicators (on the main pollutants, the main harmful effects, etc.) and the method of converting these figures into monetary equivalent ("eco-cost") with subsequent implementation of the resulting algorithms in software for compiling building estimation . Thus, in practice, it can be implemented relevant and promising eco-oriented pricing, based on the methodology of multifactorial and step-by-step predictive environmental harm assessment as part of the existing system of construction estimates.

5. Conclusion

The tools considered in the work of energy-efficient optimization of design decisions of buildings and structures in information-analytical systems for managing construction and overhaul process allow to optimize structural, space-planning decisions and technological solutions of construction, repair, reconstruction of buildings and structures from the perspective of minimizing future production and materialized in building materials expended energy expressed in aggregated GHG emissions. This is a key factor threatening irreversibility of global negative environmental changes. An effective environmental audit in investment and construction activities should be based on popularization of green standards, supported by a transparent and effective methodology for valuation of environmental harm caused by construction industry activities. Proposed methodology will allow quantitatively substantiating rational approaches for design in construction projects, forming a motivating system for the effective management and regulation of regional construction markets with a standpoints of green processing. It makes environment more sustainable, will be an effective supplement to the practice of green design and real estate certification established in the country and the world.

Acknowledgments

The work is realized in the framework of the implementation of a comprehensive project to create high-tech production "Development of new methods and tools for management of property in the budget sector and their implementation in the software package of the information-analytical system for centralized management of property owned by the constituent entities of the Russian Federation, municipalities, as well as state property companies" (agreement No. 074-11-2018-026 of 07/11/2018), with the financial support of Ministry of Science and Higher Education of the Russian Federation.

References

[1] Avilova I P and Krutilova M O 2018 Methodology of ecooriented assessment of constructive schemes of cast in-situ RC framework in civil engineering *IOP Conf. Series: Earth and Environmental Science* **3** 012127
[2] Avilova I, Krutilova M and Peresypkina E 2017 Economic incentives of green standards in civil and municipal engineering *Int. Multidisciplinary Scientific Geo Conf. SGEM* vol 53 pp 551-7
[3] Wen-Hsien Tsai, Chih-Hao Yang, Cheng-Tsu Huang and Yen-Ying Wu 2016 The impact of the carbon tax policy on green building strategy *J. of Environmental Planning and Management* **60(8)** 1-27
[4] Avilova I, Naumov A and Krutilova M 2017 Methodology of cost-effective eco-directed

IOP Conf. Series: Materials Science and Engineering **791** (2020) 012062 doi:10.1088/1757-899X/791/1/012062

structural design *Int. Multidisciplinary Scientific Geo Conf. SGEM* vol 53 pp 255-61

[5] Naumov A E, Shchenyatskaya M A and Tovstiy V P 2016 Kachestvennye pokazateli ob"ekta nedvizhimosti kak faktor ekonomicheskoy optimizatsii organizatsionno-tekhnologicheskikh resheniy investitsionno-stroitel'nogo proekta [Qualitative indicators of the property as a factor in the economic optimization of organizational and technological decisions of the investment and construction project] *Vestnik BGTU im. V G Shukhova* [*Bulletin of BSTU named after V G Shukhov*] **5** 202-6 [In Russian]

[6] Kara K A, Dolzhenko A V and Zharikov I S 2018 Influence of processing factors over concrete strength *IOP Conf. Series: Materials Science and Engineering* **327** 032027

[7] de Beer J, Harnisch J and Kerssemeeckers M 2000 Gas Emissions From Major Industrial Sources - III Iron and steel production (RK Utrecht: ECOFYS)

[8] 2016 *Conference of the Parties Report of the Conference of the Parties on its twenty-first session, held in Paris from 30 November to 13 December 2015. Part two: Action taken by the Conference of the Parties at its twenty-first session. Decisions adopted by the Conference of the Parties* URL http://unfccc.int/resource/docs/2015/cop21/eng/10a01.pdf

[9] Krass D, Nedorezov T and Ovchinnikov A 2013 Environmental Taxes and the Choice of Green Technology *Production and Operations Management* **2 (5)** 1035-55

[10] Grabovyy P G and Manukhina L A 2014 Natsional'naya strategiya vnedreniya energoresursov i ekologicheski bezopasnykh (zelenykh) tekhnologiy i proizvodstv v stroitel'stvo i ZhKKh [The national strategy for the implementation of energy resources and environmentally friendly (green) technologies and industries in construction and utilities] *Nedvizhimost': ekonomika, upravlenie* **1-2** 6-8 [In Russian]

[11] Keane M, Siert A, Stone S and Chen B T 2016 Profiling stainless steel welding processes to reduce fume emissions, hexavalent chromium emissions and operating costs in the workplace *J. of Occupational and Environmental Hygiene* **13** 1-8

[12] Metz B, Davidson O R, Bosch P R, Dave R and Meyer L A 2007 *IPCC, 2007: Climate Change 2007: Mitigation. Contribution of Working Group III to the Fourth Assessment Report of the Intergovernmental Panel on Climate Change* (New York: Cambridge University Press)

[13] Krutilova M O 2018 Napravleniya sovershenstvovaniya ekonomicheskikh mekhanizmov minimizatsii vybrosov parnikovykh gazov v techenie zhiznennogo tsikla zdaniya [Directions for improving economic mechanisms to minimize greenhouse gas emissions during the life cycle of a building] *Ekonomika stroitel'stva i prirodopol'zovaniya* **1(66)** 63–71 [In Russian]

[14] Avilova I P, Naumov A E and Krutilova M O 2019 Methodology of GHG emissions assessment caused at the construction of energy facilities. Case study: Hydropower *IOP Conf. Ser.: Mater. Sci. Eng.* **552** 012018

[15] Abakumov R G and Naumov A E 2018 Building information model: advantages, tools and adoption efficiency *IOP Conf. Series: Materials Science and Eng.* **11** 022001.

[16] Telichenko V I and Benuzh A A 2016 Sostoyanie i razvitie sistemy tekhnicheskogo regulirovaniya v oblasti zelenykh tekhnologiy [Status and development of the technical regulation system in the field of green technologies] *Academia. Arkhitektura i stroitel'stvo* **1** 118-21 [In Russian]

IOP Conf. Series: Materials Science and Engineering **791** (2020) 012063 doi:10.1088/1757-899X/791/1/012063

Resource-saving optimal control of energy-intensive process of work part thermochemical treatment

M Yu Derevyanov[1], M Yu Livshits[1] and E A Yakubovich[2]

[1]Department of Heat-and-Power Engineering, Samara State Technical University, Molodogvardeyskaya St., 244, Samara, 443100, Russia
[2]Department of Metal science, powder metallurgy, nanomaterials, Samara State Technical University, Molodogvardeyskaya St., 244, Samara, 443100, Russia

E-mail: eyakubovich@mail.ru

Abstract. Paper discusses approach to the resource-saving optimal control algorithms development, which includes problem oriented mathematical modeling of the process, determination of the tolerance limits for controlling action changes and condition of thermochemical treatment process as a subject under control, optimality criteria formalization and determination. Specified and solved critical problems for resource-saving optimal control of energy-intensive thermochemical process of work parts treatment on the maximum accuracy and speed of operation criteria. Alternance optimization method was used for optimal control problems solution, which allows to determine not only interval of constancy size but also maximum achievable control accuracy in every subset of intervals. Analytical mathematical model of vacuum carburizing in an acetylene atmosphere with variable coefficients of mass transfer was developed, which takes into consideration different mechanisms of carbon mass transfer from the atmosphere into the work part surfaces at the saturation and diffusion stages. Dependences of the carburization accuracy on the optimal control parameters were constructed. Influence of the technological constraints on the control quality was analyzed. Suggested approach demonstrated high energy efficiency.
Key words: optimal control, mathematical model, carburizing, parametric identification, carbon distribution, surface hardening.

1. Introduction

Modern industry in order to improve wear resistance of metal parts uses hardening technologies, which lead to physical and mechanical properties of the part's surface change [1-3]. Carburizing is the most common among this kind of technologies, during which part surface saturates with carbon. Process can be accelerated significantly as the result of high temperature carburizing implementation (with high carbon potential at the first stage) under pressure lower than atmospheric (vacuum carburizing) [4-6].

Vacuum carburizing process control on the optimal algorithms can provide not only high equipment productivity but also improve quality of the part surface under hardening [7-9].

Objective of this paper is development of the optimal control algorithms, that consist of problem oriented mathematical process modeling, tolerance limits determination for controlling action changes and condition of carburizing process as a subject under control, optimality criteria formalization and determination.

IOP Conf. Series: Materials Science and Engineering **791** (2020) 012063 doi:10.1088/1757-899X/791/1/012063

2. Mathematical model of the vacuum carburizing process

Model of the carbon diffusion from the atmosphere through the part surface into the depth at a constant temperature, based on Fick's second law, can be described in the form of the differential equation of mass transfer [10-11]:

$$\frac{\partial C(x,\tau)}{\partial \tau}=\frac{\partial}{\partial x}\left(D(C,T)\cdot\frac{\partial C(x,\tau)}{\partial x}\right),$$
$$\forall \tau \in (0,t_k], \forall x \in (0,\infty), \tag{1}$$

where $C(x,\tau)$ - carbon concentration in the work part; T - carburizing temperature; $D(C,T)$ - coefficient of carbon diffusion into steel [10].

Assuming that coefficient of diffusion dependence on concentration is insignificant, the equation of diffusion within narrow range of carburizing process temperatures can be linearized:

$$D(C,T)=D=const; \tag{2}$$

$$\frac{\partial C(x,\tau)}{\partial \tau}=D\cdot\frac{\partial^2 C(x,\tau)}{\partial x^2}, \forall \tau \in (0,t_k], \forall x \in (0,\infty), \tag{3}$$

As a boundary conditions, which most accurately represent physics of carbon transfer from a gaseous phase to the surface, apparently, can be taken boundary conditions of the III kind [10]:

$$-D\cdot\left.\frac{\partial C(x,\tau)}{\partial x}\right|_{x=0}=\beta(\tau)\cdot\left(\varphi(\tau)-C(x,\tau)\big|_{x=0}\right). \tag{4}$$

Here $\beta(\tau)$ - coefficient of mass transfer; $\varphi(\tau)$ - carbon potential of an atmosphere, φ_{max} - maximum permissible level of the carbon potential of an atmosphere on process capability of furnace and soot formation:

$$\varphi_{max} \geq \varphi(\tau) \geq 0. \tag{5}$$

Taking into the consideration insignificant depth $x=h_d$ of the diffusion layer in comparison with work part dimensions and relatively short time of carburizing $t=t_\kappa$, much shorter than full saturation time, one can consider work part as a semi-infinite solid with adiabatic conditions:

$$\left.\frac{\partial C(x,\tau)}{\partial x}\right|_{x\to\infty}=0. \tag{6}$$

Initial distribution of carbon concentration in the layer can be considered constant:

$$C(x,0)=C_0=const. \tag{7}$$

It is necessary to prevent intense carbide forming, that reduces surface wear resistance, in the process of carburizing.

$$C^{\kappa}{}_{max} \geq C(\tau). \tag{8}$$

Here $C^{\kappa}{}_{max}$ - maximum permissible, on carbide content, level of carbon concentration in the work part, which is defined by the carburized layer embrittlement.

Boundary condition of the third kind (4) determines diffusing element flow through the separation surface «metal – saturating atmosphere» and considers it proportional to the difference of concentration on the surface $C(0,\tau)$ and equilibrium with an environment $\varphi(\tau)$.

3. Optimal control problem statement

During diffusion saturation there is no necessity and technical capabilities to provide the given profile $C^*(x)$ accurate realization at the process end, because some uncontrolled disturbances always present in the work environment: initial carbon content variations in steel C_0, atmosphere's gas composition instability, atmosphere flow irregularity and so on [10,12,13]. In addition, there is a possibility that given profile $C^*(x)$ is not a part of the boundary value problem solutions (3), (4), (6), (7), which, in general, suggests that it is inaccessible in principle.

Therefore, in real conditions permissible range of parameters change and model status (3), (4), (6), (7), required resulting state of the carburizing process transforms from the given concentration distribution $C^*(x)$ into some region Ω_c - «tube» of permissible deviations $C^*(x) \pm \varepsilon$, that is characterized by Chebyshev measure [14-16]:

$$\rho_c = \max_{x \in [0,\infty)} \left| C^*(x) - \varepsilon \right|. \tag{9}$$

Naturally, in order to obtain a maximal wear resistance deviation value of ε must be minimal. Therefore, for controlled object (3), (4), (6), (7), under constraint conditions (5) – (8), problems are being formulated:

Speed of an operation problem:

$$J_\tau^{opt} = \min_{\varphi(\tau)} t_k \Big|_{C(x,t_k) \in \Omega_c} ; \tag{10}$$

Maximum accuracy problem:

$$J_\varepsilon^{opt} = \min_{\varphi(\tau)} \max_{x \in [0,h_d]} \left| C(x,t_k) - C^*(x) \right|. \tag{11}$$

Stated problems are problems of optimal control with right variable end point of trajectory in the infinite dimensional non-smooth region of permissible resulting conditions.

$$\Omega_c = \left\{ C(x,\tau) : \max_{x \in [0,h_d]} \left| C(x,t_k) - C^*(x) \right| \le \varepsilon \right\}$$

4. Optimal control problems solving method

Solution to the problems (10), (11) by the alternance optimization method (AOM) [15-17], determine both the optimal control $\upsilon^0(\tau) = \varphi(\tau) \equiv \varphi\left(\Delta_n^{(i)}\right) \subset R^n$, $i = 1,2,...,n$, in the form of piecewise constant function with i intervals of constancy Δ_i, and maximal attainable in every i - th subset accuracy

$$\varepsilon = \varepsilon_{min}^{(i)} = \max_{\varphi\left(\Delta_n^{(i)}\right)} \left| C(x,t_k,\Delta_i) - C^*(x) \right| \tag{12}$$

of uniform approximation to the profile which is given as satisfying conditions of operational requirements $C^*(x)$.

AOM procedure assumes the determination of unknown optimal parameters Δ_i, $\varepsilon_{min}^{(i)}$, t_k, on the basis of solving auxiliary definiens system of transcendental equations:

$$\begin{cases} \theta(z,t_k)\big|_{z=z_{kj}} = \pm\varepsilon_m^{(i)} \\ \left.\dfrac{\partial\theta(z,t_k)}{\partial z}\right|_{z=z_{kj}} = 0, \end{cases} \tag{13}$$

where $\theta(z,t_k)=\left(C(x,t_k)-C*(x)\right)/\left(\varphi_{max}-C_0\right)$; $\varepsilon_m=\varepsilon/\left(\varphi_{max}-C_0\right)$; $z=x/h$, $\varphi_{max}=\max_{\tau}\varphi(\tau)$.

Here z_{ej} - points of extremum of the function $\theta(z,t_k)$, $z_b=1,h_d$ - boundary points of the function $\theta(z,t_k)$, while $i=1,2,...S$; $S=i, at\ \varepsilon_{m_{min}}^{(i-1)}\geq\varepsilon_m^*\geq\varepsilon_{m_{min}}^{(i)}$; $S=i+1,\ at\ \varepsilon_m=\varepsilon_{m_{min}}^{(i)}$; $\underset{j=1,2...i}{z_{kj}}=z_{ej}\cup z_b$; $t_\kappa=\sum_{i=1}^{S}\Delta_i$.

Therefore, in order to solve the definiens system (13) it is necessary to obtain direct solution of the boundary problem (3), (4), (6), (7), for control $\varphi(\tau)$ [18]:

$$\varphi(\tau)=\begin{cases}\varphi_{max},\tau\in(t_i,t_{i+1}),\ at\ i=0,2,4...\\ 0,\tau\in(t_i,t_{i+1}),\ at\ i=1,3,5...\end{cases} \tag{14}$$

In order to take into consideration an initial carbon distribution before each following interval, it is reasonable to use Green function method [10].

In this manner, solution of the linear non-homogeneous boundary problem (3), (4), (6), (7), at any point of time $\tau>t_\gamma$, can be written as follows:

$$C(x,\tau)=\int_0^\infty f(\xi)\cdot G(x,\xi,\tau)d\xi-D\cdot\int_0^\tau g(\tilde{\tau})\cdot G(x,0,\tau-\tilde{\tau})d\tilde{\tau}, \tag{15}$$

where $f(\xi)=C(\xi,\tau)\big|_{\tau=t_\gamma}$ - carbon concentration distribution at a switching moment $\tau=t_\gamma$, where $\gamma=0,1,2...$ - interval number, while $t_0=0$ - starting moment of time; $\xi,\tilde{\tau},\eta$ - intermediate values, $G(x,\xi,\tau)$ - Green function for the problem (3), (4), (6), (7) [10, 18]:

$$G(x,\xi,\tau)=\frac{1}{2\cdot\sqrt{\pi\cdot D\cdot\tau}}\cdot\left\{e^{\left[-\frac{(x-\xi)^2}{4\cdot D\cdot\tau}\right]}+e^{\left[-\frac{(x+\xi)^2}{4\cdot D\cdot\tau}\right]}-2\cdot\beta(\tau)\cdot\int_0^\infty e^{\left[-\frac{(x+\xi+\eta)^2}{4\cdot D\cdot\tau}-\frac{\beta(\tau)}{D}\right]}d\eta\right\}, \tag{16}$$

$$g(\tau)=-\frac{\beta(\tau)}{D}\cdot\varphi(\tau). \tag{17}$$

Then, to obtain the distribution function $C(x,\tau)$ from n - th intervals it is necessary to convolute the solution for $n-1$th interval with Green function (16), moreover, as can be seen from (15), on the odd intervals it will be needed an introduction of the complimentary summand in the form of convolution flow $g(\tau)$ (17) and Green function $G(x,0,\tau-\tilde{\tau})$ with intermediate variable $\xi=0$, which is the equivalent of zero depth of carbon penetration:

$$C_n(x,\tau)=\int_0^\infty C_{n-1}(\xi,\tau)\cdot G(x,\xi,\tau)\big|_{\tau=t_n}d\xi-D\cdot\int_0^\tau g(\tilde{\tau})\cdot G(x,0,\tau-\tilde{\tau})d\tilde{\tau}. \tag{18}$$

Formula (18) describes carbon distribution in the carburized layer at an arbitrary point of time with n - th number of controlling actions.

IOP Conf. Series: Materials Science and Engineering **791** (2020) 012063 doi:10.1088/1757-899X/791/1/012063

Due to the fact that integrand in the first summand (18) is a convolution of a complex function $C_{n-1}(\xi,\tau)$ with a Green function $G(x,\xi,\tau)$, the calculation of such integrals becomes a sophisticated problem. To solve this problem, it is worthwhile to use integrand approximation $C_{n-1}(\xi,\tau)$.

In particular case, for example, for a single interval control ($i=1$), according to (7):

$$f(\xi)=C(\xi,\tau)\big|_{\tau=t_0}=C_0. \tag{19}$$

In this manner, from the equation (15) with consideration of (19) solution for the single interval control is obtained:

$$C_I(x,\tau)=\int_0^\infty C_0\cdot G(x,\xi,\tau)d\xi-D\cdot\int_0^\tau g(\tilde{\tau})\cdot G(x,0,\tau-\tilde{\tau})d\tilde{\tau}. \tag{20}$$

For the double interval control ($i=2$) initial distribution at the switching moment $\tau=t_1$:

$$f(\mu)=C_I(\mu,\tau)\big|_{\tau=t_1}, \tag{21}$$

and $\tau=t_2$ should be introduced into (16), here μ - intermediate variable.

Carbon distribution for the double interval control is described by the general equation (15) with consideration that for the second interval controlling action is $\Phi(\tau)\big|_{\tau=t_2}=0$, and then in (17) $g(\tau)=0$. For the double interval control we obtain:

$$C_{II}(x,\tau)=\int_0^\infty C_I(\mu,\tau)\big|_{\tau=t_1}\cdot G(x,\mu,\tau)\big|_{\tau=t_2}d\mu. \tag{22}$$

In (22) instead of $C_I(\mu,\tau)\big|_{\tau=t_1}$ approximation can be used and that will simplify further calculations significantly.

For the triple interval control ($i=3$) solution can be written as follows:

$$C_{III}(x,\tau)=\int_0^\infty C_{II}(\varpi,\tau)\cdot G(x,\varpi,\tau)\big|_{\tau=t_3}d\varpi-D\cdot\int_0^\tau g(\theta)\cdot G(x,0,\tau-\theta)d\theta. \tag{23}$$

Likewise, analytical problem solution for the quadruple interval control ($i=4$) is as follows:

$$C_{IV}(x,\tau)=\int_0^\infty C_{III}(\psi,\tau)\cdot G(x,\psi,\tau)\big|_{\tau=t_4}d\psi. \tag{24}$$

5. Results and Discussion

In order to solve formulated extremum problems (10), (11) by solving the auxiliary system (13) it is necessary to identify parameters (coefficients of diffusion D and mass transfer $\beta(\tau)$) of the mathematical model (3), (4), (6), (7) [13].

Identification procedure is based on the least squares method, which determined values $\beta(\tau)$ and D. Identification results with the constant coefficient of mass transfer $\beta(\tau)=\beta_1=const$ and with carbon potential $\varphi_{max}=4.1\ \%C$, determined by a foil test method, presented in Figure 1.

Comparison of the estimated and experimental carbon profiles for steel 14HN3MA (14CrH3Mo) suggests that the difference between experimental and estimated profile with mass transfer coefficient $\beta(\tau)=\beta_1=const$ is around 40%, which is intolerable.

Cause for this discrepancy can be explained by vacuum carburizing specifics as opposed to, for example, gaseous, firstly, because of the mass transfer mechanism of carbon from an atmosphere to the work part surfaces. Mechanism of the carbon mass transfer process, during gas carburizing, remains constant at all process stages. Only carbon saturation intensity changes. During vacuum carburizing processes of the mass transfer at saturation stage $i = 1,3,5,...$ and diffusion stage $i = 0,2,4,...$ vary significantly [13].

At the diffusion stage acetylene molecules interact with steel surface:

$$C_2H_{2(gas)} \rightarrow C_2H_{2(ads)} \xrightarrow{catalysis} \underset{\uparrow Fe}{CH} + \underset{\uparrow Fe}{CH} \xrightarrow{catalysis} \underset{\uparrow Fe}{C} + \underset{\uparrow Fe}{C} + H + H \rightarrow H_2 \qquad (25)$$

Carburizing atmosphere at a diffusion stage of a vacuum carburizing is not present, and carbon sublimates from the work part surface into vacuum. Physics of a carbon removal from a metal surface into atmosphere during vacuum and gas carburizing is different. Therefore, in technological mode calculations of vacuum carburizing should be used different coefficient values $\beta(\tau) = \beta_1|_{i=1,3,5...}$ and $\beta(\tau) = \beta_2|_{i=2,4,6...}$ for even $i = 0,2,4...$ and for odd $i = 1,3,5...$ intervals due to different conditions of mass transfer on the surface, at the saturation and diffusion stages:

$$\beta(\tau) = \begin{cases} \beta_1, \tau \in (t_i, t_{i+1}), at\ i = 0,2,4... \\ \beta_2, \tau \in (t_i, t_{i+1}),\ at\ i = 1,3,5... \end{cases} \qquad (26)$$

Identification of the model (3)-(5) by the least squares method in conditions (25) with different mass transfer coefficients for steel 14HN3MA (14CrH3Mo) (Figure 1), ensures accuracy at least 5% according to (9).

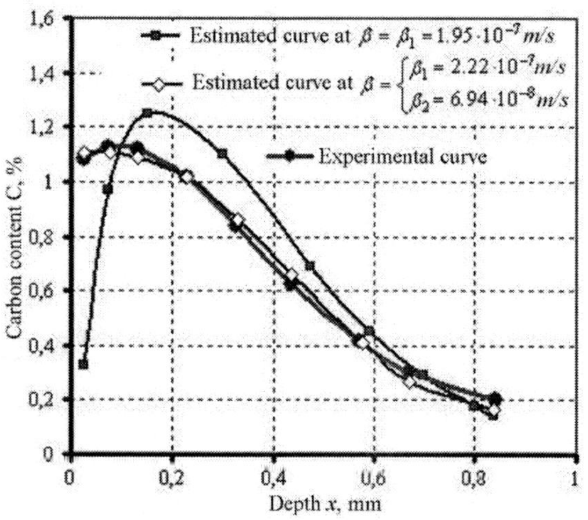

Figure 1. Comparison of experimental and estimated carbon concentration profiles at $D = 6.44 \cdot 10^{-11} m^2 \cdot s^{-1}$, $\varphi_{max} = 4.1\ \%C$.

Examined model of vacuum carburizing (18) under controlling action (14) doesn't take into consideration influence of soot formation, in other words, free carbon precipitation on the surfaces of work parts and structural parts of a furnace, which obstructs uniform access of carbon atoms to a steel surface. Therefore, control $\varphi(\tau)$, in combination with (5), should be under constraints (Figure 2):

$$\varphi_{min} \le \varphi(\tau) \le \varphi_{max}, \qquad (27)$$

$$\varphi_{min} \le \varphi(\tau) \le \varphi_{con}(\tau), \qquad (28)$$

$$\varphi_{con}(\tau) = a\tau^2 + b\tau + c, \quad \tau \in \left(t_i + \Delta^0, t_{i+1}\right), i = 0,2,4... \tag{29}$$

$$\varphi_{min} = const \tag{30}$$

Here φ_{min} - the maximum value of carbon potential, determined by furnace atmosphere irregularity; coefficients a,b,c designated in accordance with a furnace type, work parts characteristics and saturated atmosphere; Δ^0 - duration of constraint (14).

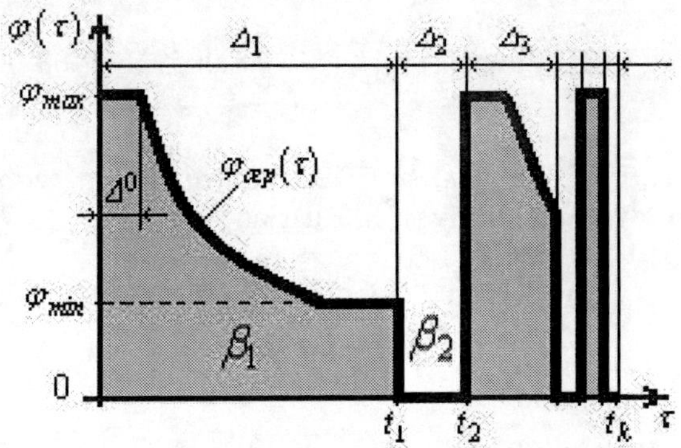

Figure 2. Controlling action $\varphi(\tau)$ for $i = 6$ with a soot formation constraint.

For the real industrial technology of the vacuum carburizing formulated optimal problems (10), (11) have been solved for the model (3), (4), (6), (7) in variable coefficients condition $\beta(\tau)$ under controlling action $\varphi(\tau)$ (5) without soot formation influence. Below presented dependencies of the interval durations Δ_i of optimal vacuum carburizing process on the carburization accuracy ε without soot formation constrains consideration (figure 3) and with technological limitations consideration (5), (27) – (30) on the controlling action $\varphi(\tau)$ of soot formation (figure 4).

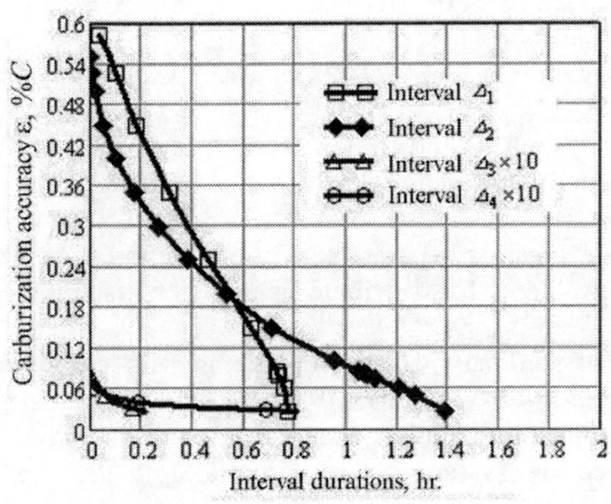

Figure 3. Dependencies of the interval of constancy durations Δ_i on the carburization accuracy ε without soot formation constrains.

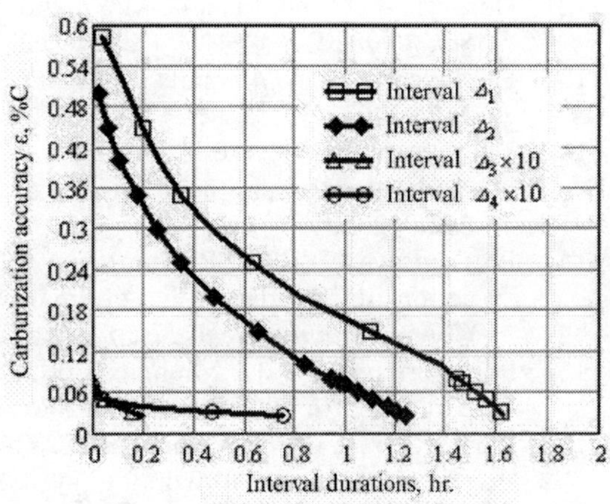

Figure 4. Dependencies of the interval of constancy durations Δ_i on the carburization accuracy ε with soot formation constrains.

Coefficient values, used in calculations, β_1=2.22·10^{-7}m·s^{-1}, β_1=6.94·10^{-8}m·s^{-1}, D=6.19·10^{-11}m^2·s^{-1}, φ_{max} = 4.1 %C.

6. Conclusion

Obtained dependencies analysis (Figure 3 and Figure 4) allows to make a conclusion that consideration of the soot formation influence increases process duration slightly and therefore expenses on fuel. However, long and systematic use of carburizing technology, taking into consideration possible soot formation, allows decrease significantly labor intensity and cost for soot formation removal from the furnace surfaces.

Process optimization allows to synthetize flexible resource saving customized TCT technology, that ensures specific operation requirements with maximum possible accuracy in a minimum time and minimum resource usage [19]. Efficiency of such algorithms is verified by their introduction into the industry [11, 12].

Acknowledgments
Research is conducted with the assistance of RFBR grants (project 17- 08-00593, 19-08-00232)

References

[1] Lakhtin M Yu and Arzamasov B N 1985 *Ximicheskaya termoobrabotka metallov* [Chemical Heat Treatment of Metals] (Moscow: Metallurgiya) [In Russian]
[2] Edenhofer B 1999 *Heat Treatment of Metals* **26(1)** 1-5
[3] Semenov M Yu, Smirnov A E and Ryzhova M Yu 2013 *Metal Science and Heat Treatment of Metals* **1** 38-42
[4] Kula P, Pietrasik R and Dybowski K 2005 *J. of Materials Processing Technology* **164-165** 876-81
[5] Kula P, Kaczmarek Ł, Dybowski K, Pietrasik R and Krasowski M 2013 *Vacuum* **87** 26-9
[6] Lee-Der Liu1 and Fan-Shiong Chen 2003 *Materials Chemistry and Physics* **82(2)** 288-94
[7] Lee-Der Liu and Fan-Shiong Chen 2004 *Surface and Coatings Technology* **183(2-3)** 233-8
[8] Bataev I A, Golkovskii M G, Losinskaya A A, Bataev A A, Popelyukh A I, Hassel T and Golovin D D 2014 *Applied Surface Science* **322** 6-14
[9] Kazuyuki Yada and Osamu Watanabe 2013 *Computers & Fluids* **79** 65-76
[10] Derevyanov M Yu, Lipkind V Ya, Lifshitc M Yu 2010 *Bulletin of Samara State Technical University. Ser. Eng.* **26** 187-95 [In Russian]
[11] Livshitc M Yu and Yakubovich E A 2016 *Materials Science Forum* **870** 647-53
[12] Derevyanov M Yu, Livshitc M Yu and Yakubovich E A 2017 *IOP Conf. Ser.: Earth and Envir. Sci.* **87** 082012
[13] Derevyanov M Yu, Livshits M Yu and Yakubovich E A 2018 *IOP Conf. Ser.: Mater. Sci. Eng.* **327** 022022
[14] Derevyanov M Yu, Pleshivtseva Yu E, Kordyukova L S and Rapoport A V 2019 Using DEA models to measure the efficiency of energy saving projects *IOP Conf. Ser.: Mater. Sci. Eng.* **552** 012020
[15] Pleshivtseva Yu E and Rapoport E Ya 2009 The successive parameterization method of control actions in boundary value optimal control problems for distributed parameter systems *J. of Computer and Systems Sciences Int.* **48 (3)** 351-62
[16] Rapoport E Ya 2013 Minimax optimization of stationary states in systems with distributed parameters *J. of Computer and Systems Sciences Int.* **52 (2)** 165-79
[17] Pleshivtseva Yu E and Rapoport E Ya 2013 Joint optimization of interrelated thermophysical processes in metal working systems based on system quality criteria *Optoelectronics, Instrumentation and Data Processing* **49 (6)** 527-35
[18] Pleshivtseva Yu E and Rapoport E Ya 2018 Parametric Optimization of Systems with Distributed Parameters in Problems with Mixed Constraints on the Final States of the Object of Control *J. of Computer and Systems Sciences Int.* **57(5)** 723-37
[19] ZhaoY et al. 2015 *Alloys, Metallurgical and Materials Transactions* **46** 2218-23

IOP Conf. Series: Materials Science and Engineering **791** (2020) 012064 doi:10.1088/1757-899X/791/1/012064

Approaches to uranium dump processing by bioleaching technology

V M Gavrish[1], T V Chayka[1], N M Derbasova[2] and G A Baranov[1]

[1]Scientific and educational center "Advanced technologies and materials", Sevastopol State University, Universitetskaya St., 33, 299053, Russia
[2]Department of Radioecology and Environmental Safety, Sevastopol State University, Universitetskaya St., 33, Sevastopol, 299053, Russia

E-mail: vmgavrish@sevsu.ru

Abstract. This paper discusses a method of extracting uranium from uranium dumps by bacterial leaching using association of thion bacteria .Bioleaching mechanism is based on interaction of biological, chemical and electrochemical processes. The microstructure and composition of uranium-containing wastes were also investigated. The work presents the results of the study (changes in volumetric activity) before and after leaching. Results of the research proved that at microbiological processing of uranium production waste by association of thionic bacteria the degree of uranium leaching out of dumps at pachuca leaching was 59.3%. Nevertheless the residual radioactivity allows classifying the obtained material to be low-level radioactive waste one. At percolation leaching, having the aim to determine the effect of particle size of feed stock on the degree of uranium extraction, two samples were under study: breakstone in initial condition and ground breakstone. According to results of the study the degree of uranium extraction (a grinded one 0.3-0.5mm fraction) from dumps is 35.18%. Using breakstone in initial condition (10-25 mm fraction) degree of uranium leaching from dumps at percolation leach-ing measures up to 100%. Thus, residual radioactivity corresponds to all national sanitary standards for classifying a solid phase to a nonradioactive one. In addition, using microbial leaching there’s no need in further uranium ore grinding and that reduces costs to process ura-nium dumps. The work proposes a scheme for processing uranium dumps in the technological process. This technology will contribute to transforming the places for uranium waste storage into anthropogenic deposits.

1. Introduction

An essential condition for safe development of nuclear power engineering is creating and adoption of new effective methods of radioactive waste disposal at all stages of a nuclear fuel cycle including uranium ore extraction. Uranium extraction results in waste rock dumps containing natural uranium isotopes (uranium–238, uranium–234, and uranium–235) and decay products (lead–206, thorium–234, thorium–230, radium–222, etc.) [1]. Meanwhile residual radioactivity of developed rocks compared with other enterprises is really low. But in case of long–term storage in dumps and tailing dump, harmful radiation doses occur [2, 3]. That’s why it requires special areas of insulation. The use of standard manufacturing technology is not applicable due to extremely low economic efficiency. In this regard, the problem of uranium dumps disposal is relevant for all countries where uranium is mined. There is a need for environmentally–friendly and cost–effective technology providing extraction of residual uranium and concomitant radioactive elements out of dumps. Meanwhile, waste material from

IOP Conf. Series: Materials Science and Engineering **791** (2020) 012064 doi:10.1088/1757-899X/791/1/012064

processing must comply with sanitary regulations on the level of residual radioactivity. That will allow reusing it in nuclear power engineering and other industries. At the moment, the most promising method for this purpose is microbiological leaching which was considered to be an attractive alternative to traditional physical and chemical methods of ore dressing by reducing energy consumption, transport costs and less harmful effects on the environment [4, 5].

The role of microorganisms in bioleaching is in oxidation of Fe^{2+} to Fe^{3+} (oxidizer of metal sulfides), oxidation of elementary sulphur and reconstructed sulphur compounds, sulfuric acid obtaining, keeping medium acid reaction [6].Today, bioleaching procedure supposes two coexisting mechanisms of bacteria–sulfide minerals interaction: direct (contact) and indirect (non–contact). Direct bacterial leaching occurs through physical contact of the bacterial cells with the mineral surface catalyzed by enzymes [7, 8]. It's supposed that bacteria do not contact the whole mineral surface but choose specific places of defects in crystal lattice [9].

In direct bacterial mechanism, pyrite usually containing in uranium ore is an energy source for bacteria thanks to its oxidation [10].

$$4FeS_2 + 15O_2 + 2H_2 + \textit{bacteria} \rightarrow 2Fe_2(SO_4)_3 + 2H_2SO_4$$

In case of indirect bioleaching, bacteria oxidize ions of bivalent Fe up to trivalent, which oxidizes chemically sulfide mineral [11].

$$FeS_2 + Fe_2(SO_4)_3 \rightarrow \textit{chemical}\ 3FeSO_4 + 2S_0$$

$$4FeSO_4 + O_2 + 4H_2SO_4 + \textit{bacteria} \rightarrow 2Fe_2(SO_4)_3 + H_2O \text{ [12].}$$

Sulfur released in the process is oxidized by bacteria to sulfuric acid [13]

$$2S + 3O_2 + 2H_2O + \textit{bacteria} \rightarrow 2H_2SO_4.$$

At bioleaching of uranium minerals, insoluble tetravalent uranium transforms from insoluble uranium oxides in acid medium to soluble sulphates due to the reaction of trivalent iron and sulfuric acid resulted by microorganisms [14].

$$UO_2 + Fe_2(SO_4)_3 \rightarrow \textit{chemical}\ UO_2SO_4 + 2FeSO_4 \text{ [15, 16]}$$

$$UO_2 + Fe_2(SO_4)_3 + 2H_2SO_4 \rightarrow \textit{chemical}\ H_4[UO_2(SO_4)_3] + 2FeSO_4$$

To maintain a sufficient amount of iron in the solution, chemical oxidation of metal sulfides must be done in acid environment ($pH < 5.0$)[17, 18]. At pH 2.0 – 3.0 bacterial oxidation of Fe^{2+} is about 10^5–10^6 times faster than chemical oxidation[19].

The purpose of the work is to improve the method of microbiological processing of uranium production waste by association of thionic bacteria with high content of Acidithiobacillus ferrooxidans (A. ferrooxidans).

2. Materials and Methods

As the virgin dump samples the ones containing uranium waste (figure 1) of Pridneprovsk Chemical Plant (Ukraine) were used.

Figure1. Photo of uranium-cutting waste samples.

The experiment procedure equipment includes laboratory units for percolation (a) and pachuca (b) uranium leaching (figure2).

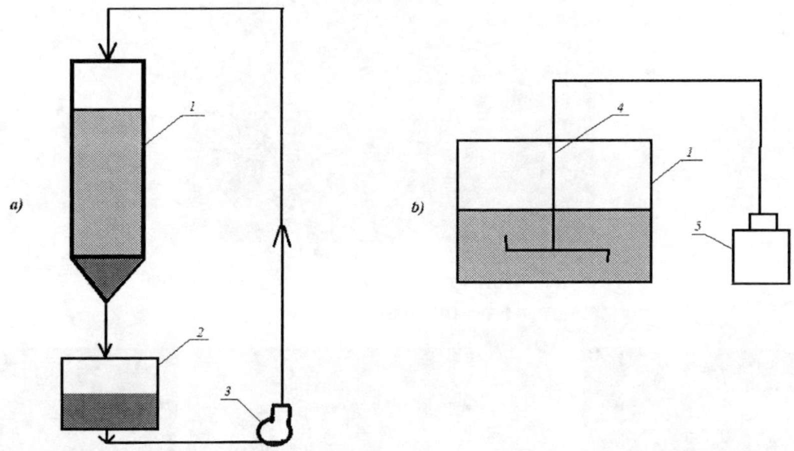

Figure 2. Functional schemes of laboratory units for percolation (a) and pachuca (b) uranium leaching out of base ore and tailing dump rock: 1–reactor; 2–collecting tank; 3– centrifugal pump; 4–mixer; 5–engine.

To study how the size of feed stock particles influences on degree of ore extraction by percolation method we've considered two sample forms – breakstone in initial condition (fraction 10–25 mm) and ground breakstone (fraction 0,3–0,5 mm).

At pachuca leaching we study only feed stock as the solution couldn't pass through ground material and there was a blockage in the result.

The temperature during the process was maintained at 25–30°C, solution pH, content of Fe^{3+}, Fe^{2+} in the solution were monitored.

Uranium wastes were put into the reactor where it was oxidized by association of thionic bacteria with high content of *Acidithiobacillus ferrooxidans* and chemical oxidation by ions of Fe^{3+} (released from microorganisms' activity) with associated reactions of electrochemical metal dissolution. At percolation leaching, bacterial solution of sulfate iron is fed self–flowing into the reactor, filtered through an ore bed and entered the filtrate collecting tank. Later this solution goes back into the sys-

IOP Conf. Series: Materials Science and Engineering **791** (2020) 012064 doi:10.1088/1757-899X/791/1/012064

tem with the help of a centrifugal pump if necessary (solution feed rate is 100 ml/day) (figure2). Pachuca leaching is realized in the reactor by continuous stirring.

The uranium content of the samples was determined by X-ray fluorescence analysis. The microstructure of uranium-containing waste was examined on a PHENOM proX SEM/EDS (Scanning Electron Microscopy with Energy Dispersive Spectroscopy) system by Phenom-World B.V. (Netherlands). Product specifications: the maximum increase of 150,000, a resolution of 10 nm, an accelerating voltage of 5, 10, 15 kV. Elemental analysis of the test samples in the set points was performed using Phenom's «Element Identification» program. The activity of the solid and liquid phase of the sample was measured on a UMF–2000 radiometer

3. Results and discussion

The scanning electron microscopy of the uranium dump samples was performed to determine the microstructure and features of uranium distribution. Figures 3 and 4 show samples before and after leaching.

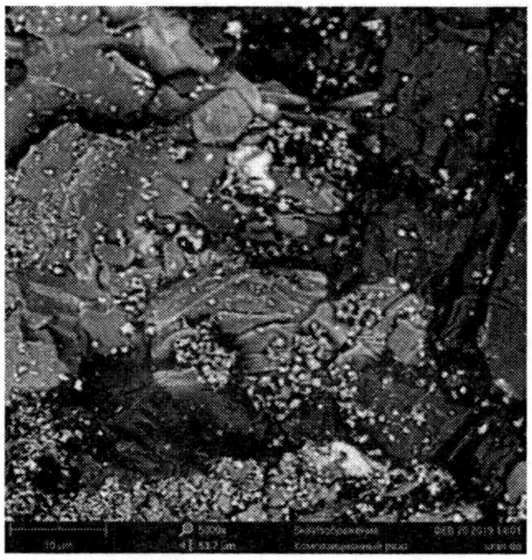

Figure 3. Micrograph of uranium-containing waste sample before leaching.

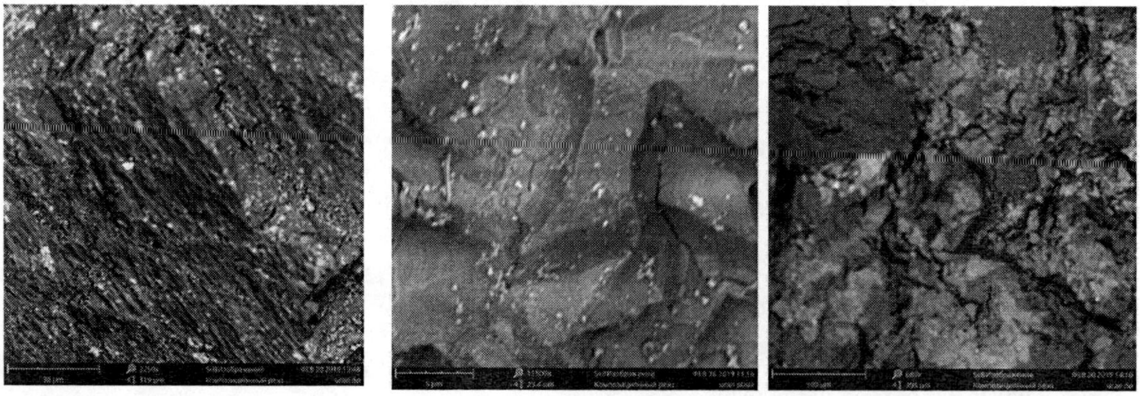

a) b) c)

Figure 4. Micrographs of uranium-containing waste after leaching: a – pachuka method; b – percolation method (fraction 0.3 – 0.5 mm); c – percolation method (fraction 10 – 25 mm).

It can be seen that in figure 4 (after processing) there are considerably fewer areas of light color than in figure 3 (before processing).

Elemental point analysis of the uranium-containing sample prior to leaching is shown in Table 1. The element spectra are shown in figure 5 and figure 6.

Table 1. Results of elemental point analysis of the sample before leaching.

The light color phase			The dark color phase		
Element Symbol	Atomic Conc.	Weight Conc.%	Element Symbol	Atomic Conc.	Weight Conc.%
Fe	23.91	49.81	O	76.87	67.84
O	67.42	40.25	Si	8.62	13.36
Si	4.79	5.02	Na	9.40	11.92
S	3.85	4.60	Al	4.24	6.31
U	0.04	0.32	C	0.86	0.57

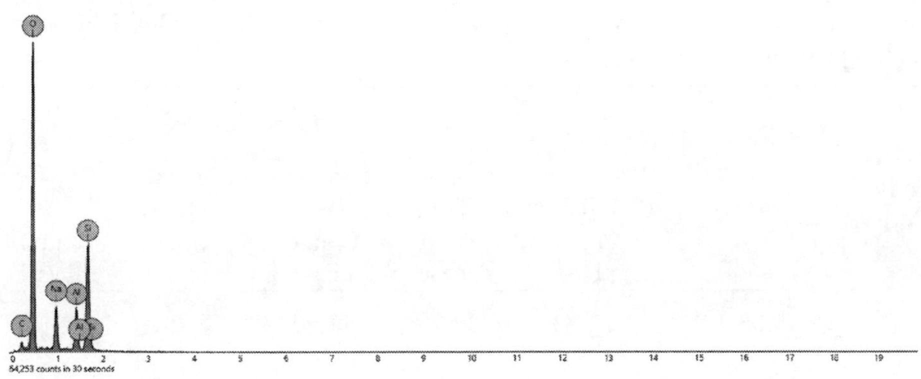

Figure 5. X-ray spectrum lines showing the presence of elements (dark phase).

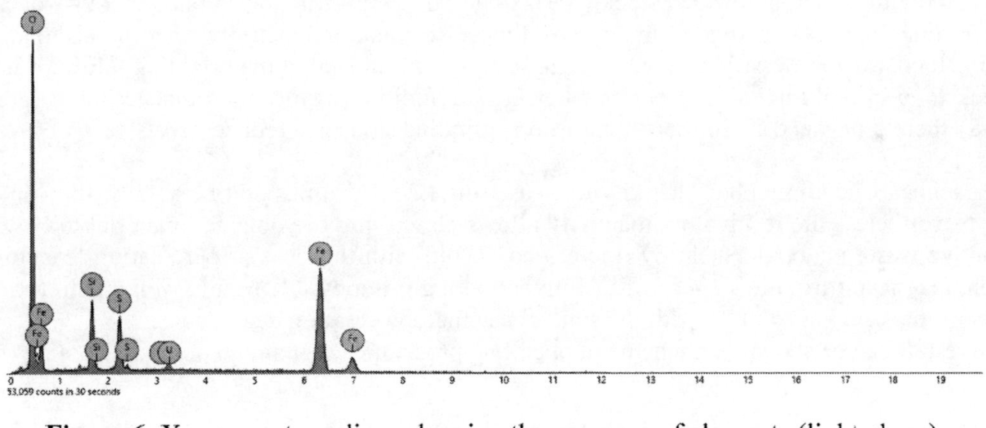

Figure 6. X-ray spectrum lines showing the presence of elements (light phase).

The results of the study of the effectiveness of uranium leaching methods are presented in table 2.

Table 2. Resulting percentage of uranium content in the samples before and after bioleaching.

Leaching method	Uranium content (%) before experiment	after experiment: solid phase	after experiment: liquid phase
Pachuca leaching	0.027	0.0110	0.0230
Percolation leaching(10–25mm fraction)	0.027	0	0.0324
Percolation leaching (0,3–0,5mm fraction)	0.027	0.0175	0.0211

As a result of bacterial leaching (table 2), we have obtained that the degree of uranium extraction from the dumps by the pachuka) leaching method is 59.3%, percolatory up to 100% (fraction 10–25 mm) and 35.18% (fraction 0.3–0.5 mm).

This fact can be explained by agglomeration (adhesion) of ground particles. That's why the leaching solution does not contact thoroughly with the whole ground ore dump.

The results showing change in activity concentration in the samples of solid and liquid phases at bacterial leaching by percolation and pachuca methods are presented in the table 3.

Table 3. Activity concentration of the samples before and after bacterial leaching by percolation and pachuca methods.

Activity concentration Bq/l	Virgin sample	Percolation leaching (0,3–0,5mm fraction) Liquid phase	Percolation leaching (0,3–0,5mm fraction) Solid phase	Percolation leaching (10–25mm fraction) Liquid phase	Percolation leaching (10–25mm fraction) Solid phase	Pachuca leaching Liquid phase	Pachuca leaching Solid phase
A_α	$3.85\cdot10^5$	18.83	3.576	13.27	0	2.1	11.9
A_β	$16.12\cdot10^5$	161.4	134.1	126.8	78	115.8	111

At percolation leaching of feed stock (breakstone 10–25mm fraction) alpha – activity decreased in $3.85\cdot10^5$ times and beta –activity – in $2.06\cdot10^4$ times. Residual radioactivity corresponds to all national sanitary standards for classifying the solid phase to a nonradioactive material that allows it to be used as breakstone in building, as a concrete filler etc. In addition, using microbial leaching (percolation method) there's no need in further uranium ore grinding and that reduces costs to process uranium dumps.

At pachuca leaching alpha – activity decreased in $3.23\cdot10^4$ times and beta –activity – in $1.45\cdot10^4$ times. Nevertheless the residual radioactivity allows classifying the obtained material to be low–level radioactive waste material. Such low indices compared with the ones at percolation leaching, in our opinion, may result from low speed of the mixer making it impossible to mix well a pulp for complete contact of the leaching solution with the solid phase thereby creating dead zones.

The results can be shown in a scheme of operating procedures in uranium dump processing (figure 7).

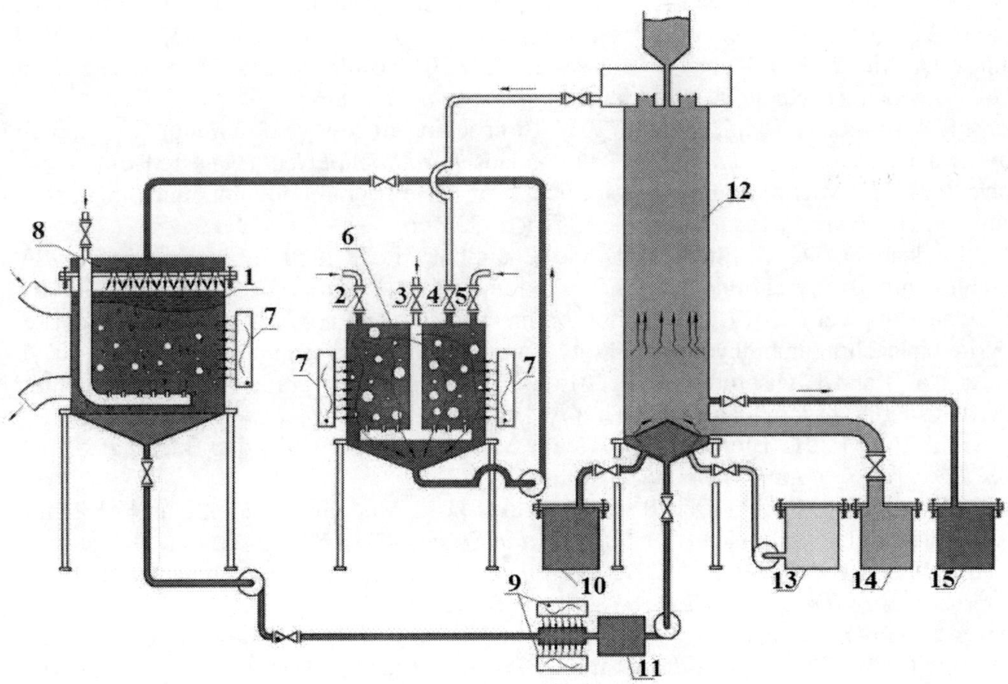

Figure 7. Scheme of microbiological processing of uranium production waste: 1 – reactor; 2 – sulfuric acid supply valve; 3 – air supply valve; 4 – return filtrate valve after the sorption column; 5 – microorganisms supply valve; 6 – unit for preparing bio–solution; 7– infrared radiator; 8 – spray line collector; 9 – ultraviolet radiators; 10 – tank for productive solution; 11 – cavitator; 12 – sorption–stripping column lock valves; 13 – tank for stripping solution; 14 – tank for sorbent; 15– tank for commercial strippant.

Using the proposed scheme of microbiological processing of uranium production waste could solve the problem of places for uranium dumps storage and transform them into anthropogenic deposits to produce uranium and rock being used for building material if needed.

4. Conclusion

There's proposed the method for microbial processing of uranium production waste by association of thionic bacteria with high content of *A.ferrooxidans*. This technology is characterized by a high rate of uranium extraction (up to 100%) and decrease in residual radioactivity for classifying the solid phase to a nonradioactive material.

The result of this work is the technological scheme of microbiological processing of uranium production waste. This technology will contribute to transforming the places for uranium waste storage into anthropogenic deposits.

References

[1] Chareyron B, Živčič L, Tkalec T and Conde M 2014 Uranium mining. Unveiling the impacts of the nuclear industry *EJOLT Report* **15** 116

[2] Brandula O, Lazăr M and Faur F 2015 Ecological reconstruction of radioactive mining waste dumps – case study of natra waste dump *Research J. of Agricultural Science* **47(4)** 19-29

[3] Mishra B, Gubel N R and Bhola R 2013 Uranium Processing *Uranium Processing and*

IOP Conf. Series: Materials Science and Engineering **791** (2020) 012064 doi:10.1088/1757-899X/791/1/012064

Properties 123-72
[4] Szolucha M M and Chmielewski A G 2017 AComparison of uranium recovery from low-grade ore by bioleaching and acid leaching *Physicochem. Probl. Miner. Process* **53(1)** 136-49
[5] Abhilash Dr, Mehta K D, Kumar V and Pandey BD 2011 Bioleaching - An Alternate Uranium Ore Processing Technology for India *Energy Procedia* **7** 158-62
[6] Umanskii A B and Klyushnikov A M 2013 Bioleaching of low grade uranium ore containing pyrite using A. ferrooxidans and A. Thiooxidans *J. Radioanal Nucl Chem* 151-6
[7] Francis A J 2015 Microbes affect the speciation of various uranium compounds in wastes and soils *In Uranium - Past and Future Challenges* 333-46
[8] Pal S, Pradhan D, Das T, Sukla B L and Chaudhury R G 2010 Bioleaching of low-grade uranium ore using Acidithiobacillus ferrooxidns *Indian J. Microbiol* **50** 70-5
[9] Tao H and Dongwei L 2014 Presentation on mechanisms and applications of chalcopyrite and pyrite bioleaching in biohydrometallurgy – a presentation *Biotechnology Reports* **4** 107-19
[10] Umanskii A B and Klyushnikov A M 2013 Bioleaching of low grade uranium ore containing pyrite using A. ferrooxidansandA. Thiooxidans *J. Radioanal Nucl Chem.* **295** 151-6
[11] Roy S and Roy M 2015 Bioleaching of Heavy Metals by Sulfur Oxidizing Bacteria: A Review. *Int. Research J. of Environment Sciences* **4(9)** 75-9
[12] Rashidi A, Safdari J, Roosta-azad R and Zokaei S 2012 Modeling of uranium bi-oleaching by Acidithiobacillus ferrooxidans *Annales Nuclar Energy* **43** 13-8
[13] Ahmed A M and El-Midany A A 2012 Gold recovery from sulphide minerals: a bioprocessing approach *AFINIDAD LXVIIIEnero – Marzo* **55** 762-68
[14] Jahani S, Fatemi F, Firoz–e–zare M A and Zolfaghari M R 2015 Isolation and Characterization of Acidithiobacillus ferrooxidans Strain FJS from Ramsar, Iran *Electronic J. of Biology* **11(4)** 138-46
[15] Baranska J and Sadowski Z 2013 Bioleaching of uranium minerals and biosynthesis of UO_2 nanoparticles *Physicochem. Probl. Miner. Process* **49(1)** 71-9
[16] Umanskii A B and Klyushnikov A M 2013 Bioleaching of low grade uranium ore containing pyrite using A. ferrooxidans and A. thiooxidans *J. of Radioanalytical and Nuclear Chemistry* **295** 151-6
[17] Seifelnassr A A S and Abouzeid A-Z M 2013 Exploitation of Bacterial Activities in Mineral Industry and Environmental Preservation: An Overview *J. of Mining* **2013** 507168
[18] Peńkin M V, Boulyga S, Dabbs B, Fischer D, Humphrey M, Kochetkov A, Koepf A and Sturm M 2018 Isotopic composition of commercially available uranium chemicals and elemental analysis standards *J. of Radioanalytical and Nuclear Chemistry* **316** 791-8
[19] Wang X, Li P, Liu Y, Sun Z, Chai L, Min X, Guo Y, Zheng Z, Ke Y and Liang Y 2018 Uranium bioleaching from low-grade carbonaceous-siliceous-argillaceous type uranium ore using an indigenous *Acidithiobacillus Ferrooxidans J. of Radioanalytical and Nuclear Chemistry* **317** 1033-40

IOP Conf. Series: Materials Science and Engineering **791** (2020) 012065 doi:10.1088/1757-899X/791/1/012065

Method development of the integral specific failure rate determination considering time in service and diameter of gas pipelines

Yu A Golovach[1] and N V Kornilova[2]

[1]Department of Thermal Engineering, Heat and Gas Supply and Ventilation, «Donbas National Academy of Civil Engineering and Architecture», 2 Derzhavina St., Makeevka, 286123, Ukraine

[2]Department of Electric Power Engineering and Automation, Belgorod State Technological University of V.G. Shukhov, Kostyukov St., 46, Belgorod, 308012, Russia

E-mail: july_ttgv@mail.ru

Abstract. Gas consumption increase and also a large scale of gas-distributing networks create new and ever more challenging problems for the engineers, associated with the development and modernization, safety and reliability improvement of the systems. These problems can be solved only by using new technologies and methods of calculation. Analysis of the statistical data on the underground distributing gas pipeline damages in the Gorlovka city of Donetsk region revealed that most of the failures are from the corrosion damage. Based on the statistical data on all types of the underground gas pipeline damages, which is the source of the gas leak, the determination method of integral specific failure rate was developed considering time in service and diameter of the gas pipelines. This method is based on the determination of the optimal step of the gas pipeline time in service, during which the failure rate is taken as constant, and on the integral specific failure rate calculation. It was discovered that the integral specific failure rate depends on the gas pipeline diameter and its time in service. Due to the fact that the coefficient SDR jumps at the diameter 219x7 mm, the dependence equation of the integral specific failure rate on time in service and diameter of the gas pipeline is derived for two diameter ranges 57-159 mm and 219-530 mm. The diameter decrease and the time in service increase lead to the integral specific failure rate increase.

1. Introduction

The modern distributing gas pipeline network is a complex branched system, where the underground gas pipeline reliability is one of the main objectives. It is the most vulnerable element in the system and requires considerable material and human resources for repair and maintenance works.

The reliability of the system can be achieved either through its elements reliability improvement or backups, meaning that the backup should be justified in a way that the level of reliability is in tune with the current requirements.

Rational implementation of measures for network reliability improvement also requires new algorithms and methods development [1].

The probability of failure-free system operation is defined mainly by the failure rate [2], the numerical values of which is impossible to determine without experimental and statistical research, and also its forecast in a long run. This kind of research was done in the papers [3 - 5] and revealed the main causes of failures and influence of different factors on it. It was established that the gas pipeline diam-

IOP Conf. Series: Materials Science and Engineering **791** (2020) 012065 doi:10.1088/1757-899X/791/1/012065

eter and wall thickness of tubing are the major influence factors on the gas network reliability [6, 7]. Dependence of the failure rate on gas pipeline diameter was also confirmed in the research presented in the paper [8].

Having in mind that gas-distributing pipeline systems are systems for long-lasting use, an increase of the time in service increases the probability of failure occurrence [9 - 12]. This fact should be considered when choosing the gas network technical maintenance strategy and risk assessment in its different points [13].

The calculated time, which is used for the reliability justification, should be such that no serious system modernization is planned for this period, that is for its reliability improvement [2]. It should be noted that during the determination of the failure rate dependence on time in service, the experimental data approximation accuracy is influenced by the choice of the step of time in service, during which the failure rate is taken as constant.

The objective of this paper is a development of the determination method of integral specific failure rate considering gas pipeline diameter and time in service.

It is necessary to resolve the following problems to achieve the desired objective:

- choose the optimal step of time in service for reliability measures determination;
- derive the dependence of the integral specific failure rate on the time in service and diameter of the gas pipeline.

2. Materials and Methods

To obtain data of dependences all damage incidents of gas networks with all types of pressure in Gorlovka city over a period 1966-2011 were analyzed. The length of the underground steel distributing gas pipelines is 204.01 km, of which the high pressure first category – 21.4 km (10%), high pressure second category - 51.75 km (25%), medium pressure – 51.99 km (25%), and low pressure – 78.87 km (40%).

The gas-distributing networks are presented by the diameters from 32 to 500 mm.

The analysis of the distributing gas pipelines damages in Gorlovka city during 1966-2011 have shown that 64% of all leaks in the underground gas pipelines are due to the corrosion damages (92 leaks), 31% - due to undermining (45 leaks), 3% - mechanical damages (4 leaks), and 2% - due to weld connection rupture (3 leaks) (see Figure 1).

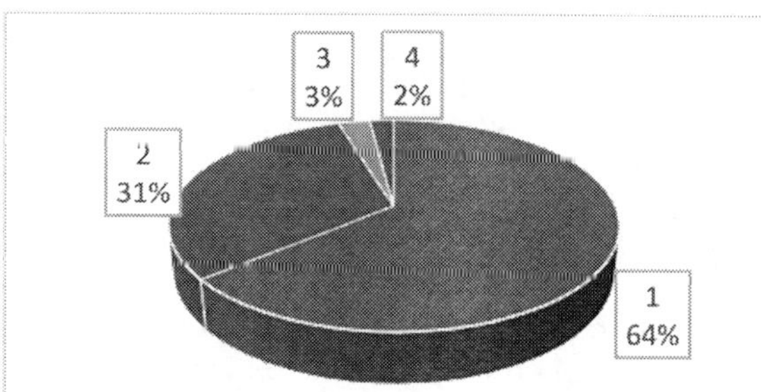

Figure 1. Causes of gas leaks from the underground distributing gas pipelines: 1 – corrosion, 2 – influence of undermining, 3 – mechanical damages, 4 – weld connection rupture.

Table 1 presents the example of initial data on distributing gas pipeline damages in Gorlovka city.

The gas pipelines have the following diameters (outer diameter x wall thickness of tubing, mm): 57x3, 76x4, 89x4, 108x4, 114x4, 133x4, 159x4,5, 219x7, 273x7, 325x8, 426x9 and 530x10 mm.

The coefficient SDR, which presents relation of the outer diameter to wall thickness of tubing, for this range of diameters is as follows, respectively: 19; 19; 22.3; 27; 28.5; 33.3; 35.3; 31.3; 39; 40.6; 47.3; 53.

To increase the approximation accuracy of the experimental data during the study of its damages at gas pipeline failure rate determination, it is necessary to optimize the step of gas pipeline time in service, during which the failure rate is taken as a constant.

Table 1. Data on the distributing gas pipelines damages in Gorlovka city.

№	Year of leak detection, year	Number of leaks n, pcs.	Outer diameter of the gas pipeline D, mm	Year of laying, year	Total gas pipeline length at the laying year, L_k, m	Gas pipeline length at the laying year with diameter L_{ki}, m	Cause
1	1966	1	530	1954	3688	161	Influence of undermining
2	1978	1	133	1962	13052	2199	Influence of undermining
3	…						
143	2011	1	108	1966	5869	2291	corrosion
144	2011	1	108	1963	11269	1272	corrosion

For this purpose, the selection method of the step of gas pipeline time in service was developed.
Let`s introduce the following notations:
i – gas pipeline outer diameter; i = 57, 76, 89, 108, 114, 133, 159, 219, 273, 325, 426, 530 mm;
j – step of the gas pipeline time in service, $j = \overline{1,\ 10}$;
k – calendar year of gas pipeline laying;
c – calendar year of gas pipeline leak detection;
L_k – length of all diameters gas pipelines of k-th year of laying, m;
L_{ki} – gas pipelines length of i-th diameter of k-th year of laying, m;
T_k – gas pipeline time in service of k-th year of laying, at which damage was detected, $T_k = c - k$.

Specific failure rate is calculated by the formula:

$$w_k = \begin{cases} \dfrac{n_k}{T_k \cdot L_k}, & n' = 1, \\ \dfrac{\sum w_k \cdot L_k}{\sum L_k}, & n' \neq 1, \end{cases} \tag{1}$$

where n_k – number of damages in the gas pipeline of k-th year of laying, at which the damage was detected, pcs., during the calculated time period T_k, which is taken as 1 year;
L_k – gas pipeline length of k-th year of laying, m,
n' – number of data with the same values of gas pipeline time in service T_k, calculated by the formula

$$T_k = c - k. \tag{2}$$

Considering that the depreciation period of the steel gas pipelines according to RLALP 0.00-1.20-98 "Safety regulations of the Ukraine`s gas supply systems" is 30 years, one can calculate the failure rate during time in service from p = 1, which corresponds to value $T_{k\ min}$=15 years, to p = 30, which corresponds to value T_k = 44 years, with step $j = (\overline{1, 10})$ by the formula

$$w_{jp-m;j} = \frac{1}{j}\sum_{m=0}^{j-1} w_{jp-m}, \tag{3}$$

where $m = \overline{0, j-1},\ p = \overline{1, 30}$.
Table 2 summarizes the results of the calculation.
The arithmetic mean value of the failure rate, calculated with step j, is determined by the formula

IOP Conf. Series: Materials Science and Engineering **791** (2020) 012065 doi:10.1088/1757-899X/791/1/012065

$$E_{w_p} = \frac{\sum\limits_{j} w_{jp-m;j}}{\alpha}, \tag{4}$$

where α – variant numbers of the chosen step, α = 10.

Mean squared error of the results is calculated

$$\Delta S_w = \sqrt{\frac{p_{\max}\sum\limits_{p}(w_{jp-m;j} - E_{w_p})^2}{p_{\max} - 1}}. \tag{5}$$

Choice of the step is conditioned by $\Delta S_w \longrightarrow \min$.

Table 3 summarizes the results of the calculation.

Table 2. Specific failure rate in relation to the step of gas pipeline time in service.

<table>
<tr><th rowspan="2">p</th><th rowspan="2">T_k, year</th><th colspan="10">w_{ip-m,j}·10⁻⁵, 1/(m·year)</th></tr>
<tr><th>j=1 year</th><th>j=2 year</th><th>j=3 year</th><th>j=4 year</th><th>j=5 year</th><th>j=6 year</th><th>j=7 year</th><th>j=8 year</th><th>j=9 year</th><th>j=10 year</th></tr>
<tr><td>1</td><td>15</td><td>7.7543</td><td rowspan="2">8,0418</td><td rowspan="3">5.3612</td><td rowspan="4">5.1341</td><td rowspan="5">6.5519</td><td rowspan="6">8.2997</td><td rowspan="7">9.5826</td><td rowspan="8">9.8621</td><td rowspan="9">9.7523</td><td rowspan="10">9.3368</td></tr>
<tr><td>2</td><td>16</td><td>8.3292</td></tr>
<tr><td>3</td><td>17</td><td>0.0000</td><td rowspan="2">2.2264</td></tr>
<tr><td>4</td><td>18</td><td>4.4528</td><td rowspan="3">11.2383</td></tr>
<tr><td>5</td><td>19</td><td>12.2234</td><td rowspan="2">14.6311</td><td rowspan="4">14.5901</td></tr>
<tr><td>6</td><td>20</td><td>17.0387</td><td rowspan="5">12.1217</td></tr>
<tr><td>7</td><td>21</td><td>17.2801</td><td rowspan="2">14.5491</td><td rowspan="3">12.6574</td><td rowspan="6">12.4401</td></tr>
<tr><td>8</td><td>22</td><td>11.8181</td><td rowspan="7">14.4440</td></tr>
<tr><td>9</td><td>23</td><td>8.8739</td><td rowspan="2">7.2359</td><td rowspan="4">11.3856</td><td rowspan="8">13.8630</td></tr>
<tr><td>10</td><td>24</td><td>5.5979</td><td rowspan="3">12.2228</td><td rowspan="9">13.6343</td></tr>
<tr><td>11</td><td>25</td><td>8.2291</td><td rowspan="2">15.5353</td><td rowspan="5">17.4116</td><td rowspan="10">13.4240</td></tr>
<tr><td>12</td><td>26</td><td>22.8415</td></tr>
<tr><td>13</td><td>27</td><td>17.8069</td><td rowspan="2">21.8736</td><td rowspan="3">18.6625</td><td rowspan="4">16.3404</td><td rowspan="6">14.3400</td></tr>
<tr><td>14</td><td>28</td><td>25.9403</td></tr>
<tr><td>15</td><td>29</td><td>12.2404</td><td rowspan="2">10.8071</td><td rowspan="7">11.5053</td></tr>
<tr><td>16</td><td>30</td><td>9.3738</td><td rowspan="3">10.0174</td><td rowspan="5">9.4363</td></tr>
<tr><td>17</td><td>31</td><td>12.4906</td><td rowspan="2">10.3393</td><td rowspan="4">9.4520</td><td rowspan="8">10.4303</td></tr>
<tr><td>18</td><td>32</td><td>8.1879</td></tr>
<tr><td>19</td><td>33</td><td>12.6611</td><td rowspan="2">8.5647</td><td rowspan="3">12.7481</td><td rowspan="6">10.4607</td><td rowspan="9">10.6128</td></tr>
<tr><td>20</td><td>34</td><td>4.4683</td></tr>
<tr><td>21</td><td>35</td><td>21.1149</td><td rowspan="2">16.4407</td><td rowspan="4">11.4087</td><td rowspan="5">11.5900</td><td rowspan="10">12.3123</td></tr>
<tr><td>22</td><td>36</td><td>11.7666</td><td rowspan="3">8.1733</td><td rowspan="7">10.3292</td></tr>
<tr><td>23</td><td>37</td><td>12.7532</td><td rowspan="2">6.3766</td></tr>
<tr><td>24</td><td>38</td><td>0.0000</td></tr>
<tr><td>25</td><td>39</td><td>12.3155</td><td rowspan="2">10.2141</td><td rowspan="3">10.9170</td><td rowspan="4">11.9461</td><td rowspan="6">12.9147</td><td rowspan="6">12.9147</td></tr>
<tr><td>26</td><td>40</td><td>8.1127</td><td rowspan="5">13.0346</td></tr>
<tr><td>27</td><td>41</td><td>12.3229</td><td rowspan="2">13.6781</td></tr>
<tr><td>28</td><td>42</td><td>15.0333</td><td rowspan="3">14.9125</td><td rowspan="3">14.9125</td></tr>
<tr><td>29</td><td>43</td><td>24.7097</td><td rowspan="2">14.8520</td><td rowspan="2">14.8520</td><td rowspan="2">14.8520</td></tr>
<tr><td>30</td><td>44</td><td>4.9944</td></tr>
</table>

Table 3. Mean squared error of the results in relation to the step of gas pipeline time in service.

j, year	1	2	3	4	5	6	7	8	9	10
ΔS_w	27.10	16.85	10.76	11.14	9.88	9.12	10.11	9.63	9.92	11.11

Mean squared error of the results takes the minimum value 9.12 when step $j = 6$ years, therefore, for the reliability values calculation the step of time in service is taken as 6 years.

To obtain dependences of the integral specific failure rate on time in service and gas pipeline diameter only corrosion damages are taken into consideration since the insulating wrapping wears out and the risk of corrosion propagation increases as time goes by.

Specific failure rate is calculated by the formula

$$w_{ki} = \begin{cases} \dfrac{n_{ki}}{T_{ki} \cdot L_{ki}}, & n_i' = 1, \\ \dfrac{\sum w_{ki} \cdot L_{ki}}{\sum L_{ki}}, & n_i' \neq 1, \end{cases} \tag{6}$$

where n_{ki} – number of damages in the gas pipeline of k-th year of laying with i-th diameter, pcs., during the calculated time period T_k, which is taken as 1 year;

L_{ki} – length of the gas pipeline of k-th year of laying with i-th diameter, m;

n_i' – number of data with the same values of gas pipeline time in service with i-th diameter $T_{k\,i}$.

Dependence of the specific failure rate on gas pipeline time in service for all diameters of tubing where leaks were detected is calculated.

The determination method of the specific failure rate dependence on time in service of the gas pipelines with an outer diameter 57 mm is presented below.

The failure rate calculation is conducted at the integral distribution within the range from $p = 1$, which corresponds to value $T_{k\,min}$=15 years to $p = 30$, which corresponds to value $T_k = 44$ years, with step 6 years. To form the initial data table the interval $p = \overline{1,30}$ is divided into spacing with data quantity that equals the optimal step of time in service ($r = 6$ years). In this way, five intervals are obtained, where the integral specific failure rate is determined for each of them.

The mean value of time in service is determined for each interval.

For the first interval ($p = \overline{1,6}$) $T_{k\,ave} = (2T_{k\,min} + r\text{-}1)/2$,

for the second ($p = \overline{7,12}$) $T_{k\,ave} = (2T_{k\,min} + 3r\text{-}1)/2$,

for the third ($p = \overline{13,18}$) $T_{k\,ave} = (2\,T_{k\,min} + 5r\text{-}1)/2$,

for the fourth ($p = \overline{19,24}$) $T_{k\,ave} = (2\,T_{k\,min} + 7r\text{-}1)/2$,

for the fifth ($p = \overline{25,30}$) $T_{k\,ave} = (2\,T_{k\,min} + 9r\text{-}1)/2$.

The failure rate in each interval is calculated by the formula

$$w'_{k_i} = \frac{1}{tr}\sum_{s=1}^{tr} w_{ks,} \quad t = \overline{1,5}. \tag{7}$$

Table 4 summarizes the results of the calculation.

CurveExpert Professional software helped to find out that dependence data of the integral specific failure rate on time in service can be described by the formula [14, 15]

$$w_{ki}' = a - be^{-cT_{k\,ave}^{\;d}} \tag{8}$$

where $a = 0.00076171$, $b = 0.000519813$, $c = 9.23027 \cdot 10^{-13}$, $d = 7.409882874$.

Table 4. The integral specific failure rate values in relation to the time in service.

$T_{k\,ave}$, year	w_{ki}', 10^{-5}, 1/(m·year)
17.5	25.3678
23.5	25.3678
29.5	25.3678
35.5	38.0518
41.5	54.9512

Figure 2 shows how this dependence approximates the initial data, which presented in Table 4.

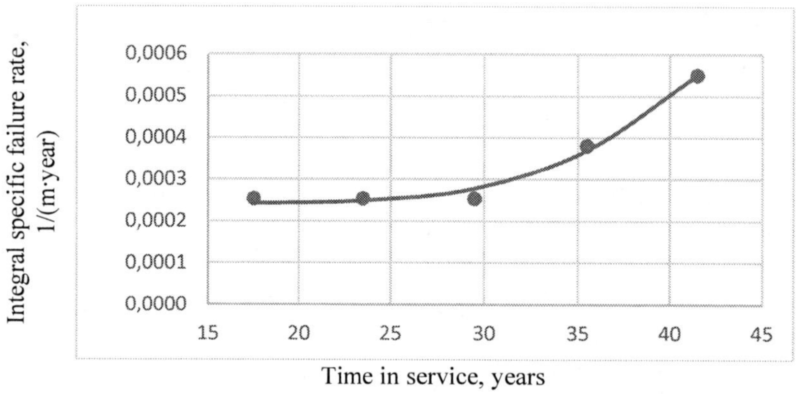

Figure 2. Dependence of the integral specific failure rate on the time in service for diameter 57 mm.

Table 5. Regression equation coefficients, correlation coefficients and Fisher`s number for different gas pipeline diameters.

Gas pipeline diameter, mm	*a*	*b*	*c*	*d*	*r*	*F*
57	0.00076171	0.000519813	9.23027E-13	7.409882874	0.9938	60.4241
76	0.00150811	0.001323264	4.62404E-14	7.942866158	0.9971	128.8695
89	0.000712857	0.000565115	6.60878E-13	7.218912082	0.9991	438.3271
108	0.003900034	0.003840688	2.94848E-12	6.320063587	0.9988	302.0516
114	0.000332768	0.000271129	1.3809E-13	7.607780763	0.9969	121.6554
133	0.003962853	0.003929531	1.11444E-12	6.40375686	0.9723	13.7460
159	0.00011605	9.75718E-05	1.3438E-11	6.881897005	0.9994	610.4012
219	5.54411E-05	2.21226E-05	2.42253E-11	6.722438028	0.9803	19.1943
273	7.66961E-05	5.67662E-05	2.18826E-12	6.835020334	0.9356	6.0153
325	2.82324E-05	2.38965E-05	2.04977E-13	7.534467805	0.9962	98.4803
426	2.58746E-05	1.9152E-05	2.14933E-12	6.838531966	0.9356	6.0200
530	2.62731E-05	7.30762E-06	3.40327E-07	3.400748446	0.9624	10.1640

Verification of the regression equation significance was conducted by using Fisher`s number, calculated by [16], $F = 60.4241$.

The tabular values of the Fisher`s number $F^{m}_{(n-1;n-2;p)}$ are $F^{m}_{(4;3;10\%)} = 5.3427$, $F^{m}_{(4;3;5\%)} = 9.1172$, $F^{m}_{(4;3;2,5\%)} = 15.101$.

$F > F^{m}_{(n-1;n-2;p)}$, 60.4241>15.101, therefore, the equation (8) gives the statistically significant description of the experimental results.

Similar equations were obtained for the other diameters. Coefficients *a, b, c, d* of the obtained regression equations, correlation coefficients *r* and values of the Fisher`s number *F* are presented in Table 5.

Figures 3, 4, 5 shows how the obtained dependences approximate the initial data for different gas pipeline diameters.

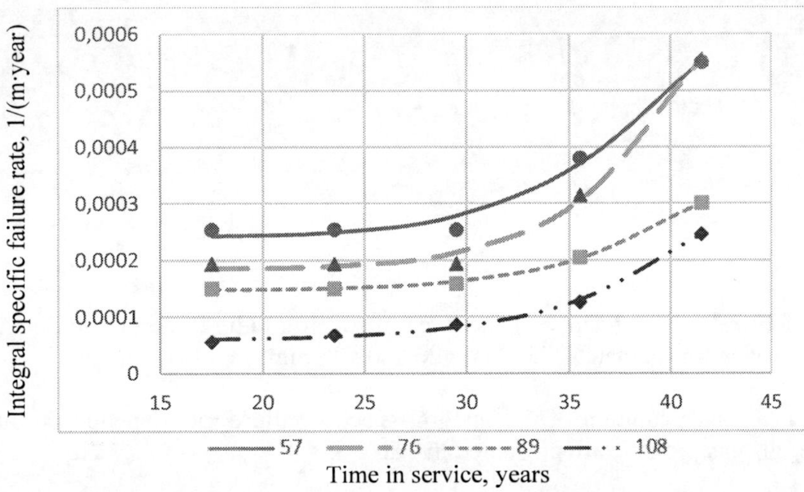

Figure 3. Dependence of the integral specific failure rate on the time in service for diameter 57, 76, 89 and 108 mm.

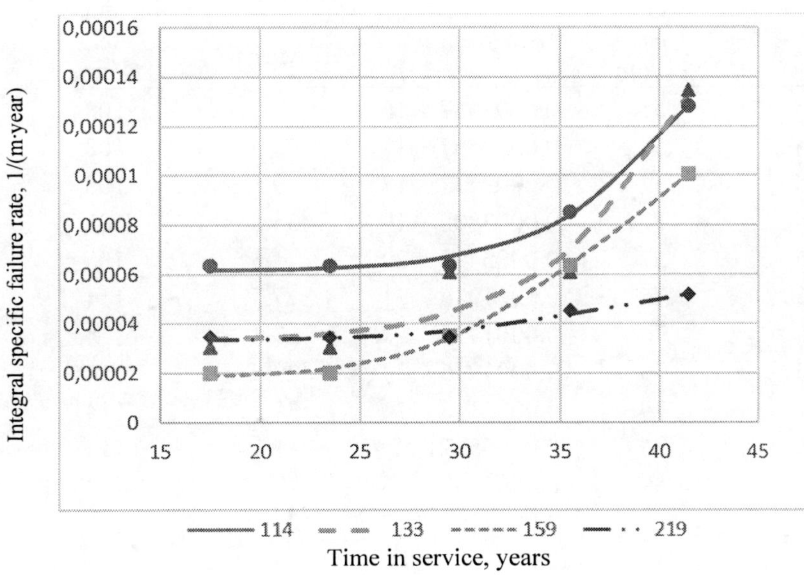

Figure 4. Dependence of the integral specific failure rate on the time in service for diameter 114, 133, 159 and 219 mm.

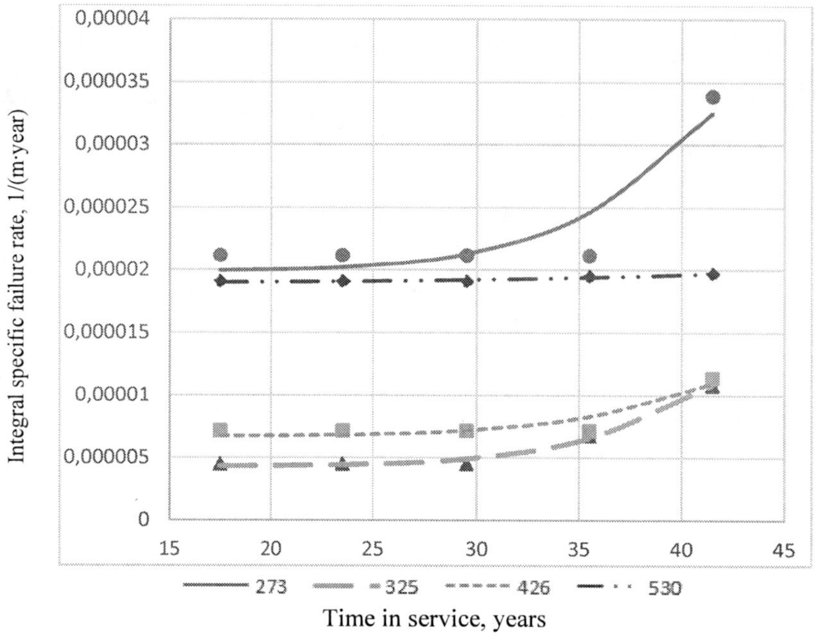

Figure 5. Dependence of the integral specific failure rate on the time in service for diameter 273, 325, 426 and 530 mm.

The initial data for calculation of the integral specific failure rate dependence on time in service and diameter of the gas pipeline are presented in Table 6.

Table 6. Integral specific failure rate at given outer diameters D and gas pipeline time in service $T_{k\text{ ave}}$.

№	Time in service $T_{k\text{ ave}}$, year	Gas pipeline diameter D, mm	w_{ki}', 1/(m·year)	№	Time in service $T_{k\text{ ave}}$, year	Gas pipeline diameter D, mm	w_{ki}', 1/(m·year)
1	17.5	57	0.0002537	16	17.5	108	5.582E-05
2	23.5	57	0.0002537	17	23.5	108	6.716E-05
3	29.5	57	0.0002537	18	29.5	108	8.591E-05
4	35.5	57	0.0003805	19	35.5	108	0.0001263
5	41.5	57	0.0005495	20	41.5	108	0.0002463
6	17.5	76	0.0001936	21	17.5	114	6.353E-05
7	23.5	76	0.0001936	22	23.5	114	6.353E-05
8	29.5	76	0.0001936	23	29.5	114	6.353E-05
9	35.5	76	0.0003146	24	35.5	114	8.518E-05
10	41.5	76	0.0005547	25	41.5	114	0.0001282
11	17.5	89	0.0001502	26	17.5	133	3.047E-05
12	23.5	89	0.0001502	27	23.5	133	3.047E-05
13	29.5	89	0.0001584	28	29.5	133	6.094E-05
14	35.5	89	0.000205	29	35.5	133	6.094E-05
15	41.5	89	0.0003007	30	41.5	133	0.0001347

Continuation of Table 6.

№	Time in service $T_{k\,ave}$, year	Gas pipeline diameter D, mm	w_{ki}', 1/(m·year)	№	Time in service $T_{k\,ave}$, year	Gas pipeline diameter D, mm	w_{ki}', 1/ (m·year)
31	17.5	159	2.011E-05	46	17.5	325	4.514E-06
32	23.5	159	2.011E-05	47	23.5	325	4.514E-06
33	29.5	159	2.011E-05	48	29.5	325	4.514E-06
34	35.5	159	2.011E-05	49	35.5	325	6.771E-06
35	41.5	159	2.011E-05	50	41.5	325	1.083E-05
36	17.5	219	3.457E-05	51	17.5	426	7.148E-06
37	23.5	219	3.457E-05	52	23.5	426	7.148E-06
38	29.5	219	3.457E-05	53	29.5	426	7.148E-06
39	35.5	219	4.534E-05	54	35.5	426	7.148E-06
40	41.5	219	5.181E-05	55	41.5	426	1.144E-05
41	17.5	273	2.119E-05	56	17.5	530	1.908E-05
42	23.5	273	2.119E-05	57	23.5	530	1.908E-05
43	29.5	273	2.119E-05	58	29.5	530	1.908E-05
44	35.5	273	2.119E-05	59	35.5	530	1.948E-05
45	41.5	273	3.391E-05	60	41.5	530	1.972E-05

3. Results

Due to the fact that the coefficient SDR jumps at the diameter 219x6 mm, the behavior of curve is changing, the dependence equation of the integral specific failure rate on time in service and diameter of gas pipeline is derived for two diameter ranges 57-159 mm and 219-530 mm.

The dependence equation of the specific failure rate on time in service and diameter of gas pipeline can be presented as follows

$$w_{ki}' = a + b\cdot T_{k\,ave} + c\cdot D + d\cdot T_{k\,ave}^2 + e\cdot D^2 + f\cdot T_{k\,ave}^3 + g\cdot D^3 + h\cdot T_{k\,ave}\cdot D + i\cdot T_{k\,ave}^2\cdot D + j\cdot T_{k\,ave}\cdot D^2, \quad (9)$$

in the range 57-159 mm

$$w_{ki}' = 0.000614484 - 2.56741\cdot 10^{-5}\cdot T_{k\,ave} - 4.18555\cdot 10^{-7}\cdot D - 3.20662\cdot 10^{-8}\cdot T_{k\,ave}^2 - 9.23756\cdot 10^{-8}\cdot D^2 + 1.86872\cdot 10^{-8}\cdot T_{k\,ave}^3 + 3.32416\cdot 10^{-10}\cdot D^3 + 4.31828\cdot 10^{-7}\cdot T_{k\,ave}\cdot D - 1.0681\cdot 10^{-8}\cdot T_{k\,ave}^2\cdot D + 3.21759\cdot 10^{-10}\cdot T_{k\,ave}\cdot D^2, \quad (10)$$

in the range 219-530 mm

$$w_{ki}' = 0.000235718 - 1.49368\cdot 10^{-7}\cdot T_{k\,ave} - 1.47445\cdot 10^{-6}\cdot D - 1.56316\cdot 10^{-8}\cdot T_{k\,ave}^2 + 2.96547\cdot 10^{-9}\cdot D^2 + 1.10756\cdot 10^{-9}\cdot T_{k\,ave}^3 - 1.97219\cdot 10^{-12}\cdot D^3 + 8.86527\cdot 10^{-11}\cdot T_{k\,ave}\cdot D - 1.55135\cdot 10^{-10}\cdot T_{k\,ave}^2\cdot D + 9.26483\cdot 10^{-12}\cdot T_{k\,ave}\cdot D^2. \quad (11)$$

Figure 6 and 7 shows how the obtained dependences (10) and (11) approximate the initial data, which presented in Table 6.

IOP Conf. Series: Materials Science and Engineering **791** (2020) 012065 doi:10.1088/1757-899X/791/1/012065

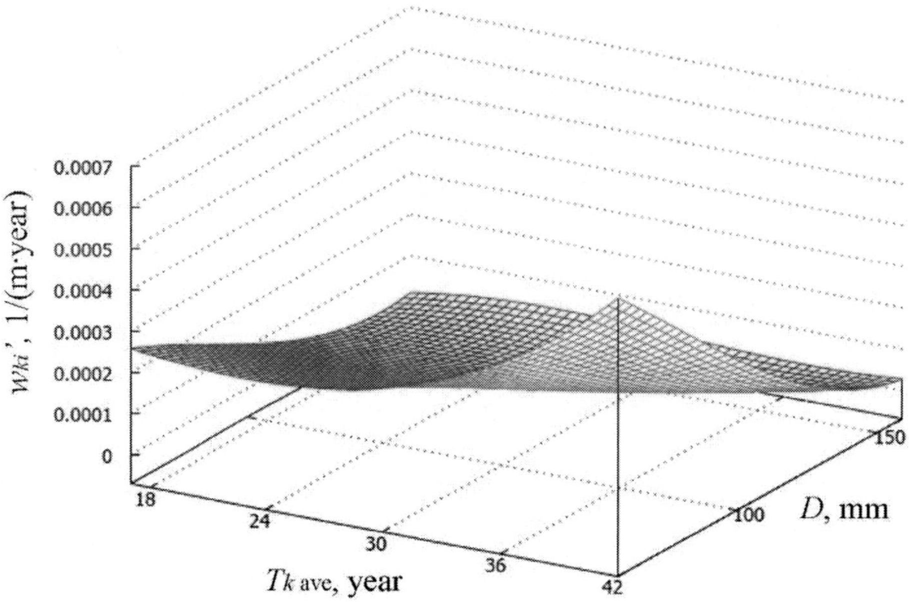

Figure 6. Dependence of the integral specific failure rate on time in service and diameter of gas pipeline in diameter range 57-159 mm.

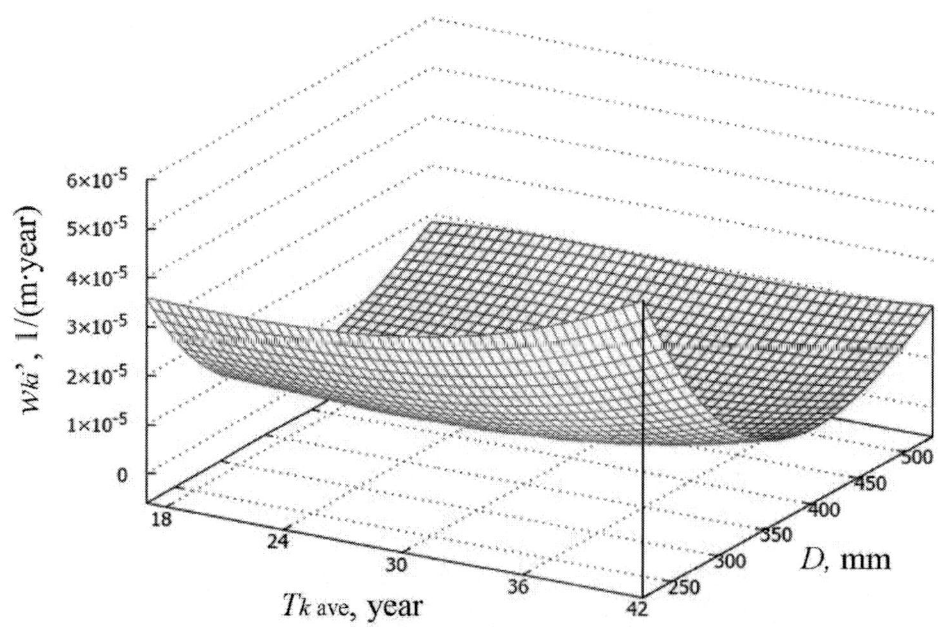

Figure 7. Dependence of the integral specific failure rate on time in service and diameter of gas pipeline in diameter range 219-530 mm.

4. Discussion

The study of the dependence of integral specific failure rate on time in service and diameter of gas pipeline allows to make a conclusion that diameter decrease and time in service increase leads to the failure rate increase. The obtained dependencies allow to calculate values of the integral specific failure rate for diameters that have no data based on the results of operation.

Further study will continue in the direction of the numeric values of reliability measures justification, which are recommended as normative, and also forecast gas pipeline accidents.

5. Conclusion

In modern conditions gas supply system reliability improvement can be achieved by using reliability control tools based on data about the actual technical condition of the gas networks. Statistical analysis of the gas pipeline damages in Gorlovka city has established that the main causes of damage are corrosion and influence of the undermining, while weld connection rupture and mechanical damage give a small number among all accidents. It should be noted that during the calculation of the reliability the experimental data approximation accuracy is influenced by the choice of the step of time in service, during which the failure rate is taken as constant. The developed selection method of the calculated step of gas pipeline time in service, which is 6 years, and calculation of the integral specific failure rate allowed to obtain dependences of the integral specific failure rate on time in service and diameter of the gas pipeline for two diameter range 57-159 mm and 219-530 mm.

References

[1] Rausand M and Hoyland A 2004 System Reliability Theory: Models Statistical Methods and Applications (Hoboken: John Wiley & Sons, Inc.)

[2] Ionin A A, Alibekov K S, Zhila V A and Zatikyan S S 1980 *Nadezhnost` gorodskix sistem gazosnabzheniya* [Reliability of the municipal gas supply systems] (Moscow: Strojizdat) [In Russian]

[3] Cunha S B 2012 Comparison and Analysis of Pipeline Failure Statistics *IPC2012-90186* (Canada) pp 521-30

[4] 2011 Gas pipeline incidents. 8th Report of the European Gas Pipeline Incident Data Group URL https://www.egig.eu/startpagina/$61/$149

[5] 2019 PHMSA Pipeline Failure Investigation Reports URL https://www.phmsa.dot.gov/safety-reports/pipeline-failure-investigation-reports

[6] Lihang Wang 2016 Reliability Analysis of Gas Pipeline with Corrosion Defects *J. of Materials Science and Eng.* **6(3-4)** 101-4

[7] Zvirko O I, Savula S F, Tsependa V M, Gabetta G and Nykyforchyn H M 2016 Stress corrosion cracking of gas pipeline steels of different strength *Procedia Structural Integrity* **2** 509-16

[8] Zaxarov V I and Golovach Yu A 2017 Analiz vliyaniya diametrov i tolshhin stenok truboprovodov na veroyatnost` bezotkaznoj raboty` gazovy`x setej [Analysis of the pipeline diameter and wall thickness of tubing influence on the probability of failure-free gas network operation] *Sbornik trudov II Mezhdunarodnoj Nauchno-Texnicheskoj Konferencii* (Belgorod: BGTU im V G Shukhova) 28-34 [In Russian]

[9] Iesmantas T and Alzbutas R 2011 Age-dependent probabilistic analysis of failures in gas pipeline networks *Proceedings of the of 8th Annual Conf. of Young Scientists on Energy Issues CYSENI 2011* (Lithuania) pp 1-9

[10] Rausand M and Qien K 1996 The basic concepts of failure analysis *Reliability Engineering and System Safety* **53** 73-83

[11] Temnikov D O and Suslov D Yu 2018 Algorithms of the Minimum Spanning Tree in Gas Supply Systems J. Phys.: Conf. Ser. 1066 012001

[12] Nahal M and Khelif R 2019 Numerical model for estimating time-dependent reliability of a corroding pipeline over its lifetime *Int. J. of Advanced Structural Eng.* **11** 1-7

[13] Alzbutas R, Iesmantas T, Povilaitis M and Vitkutė J 2014 Risk and uncertainty analysis of gas

IOP Conf. Series: Materials Science and Engineering **791** (2020) 012065 doi:10.1088/1757-899X/791/1/012065

pipeline failure and gas combustion consequence *Stochastic Environmental Research and Risk Assessment* **28** 1431-46

[14] Meeker W Q and Escobar L A 1998 Statistical Methods for Reliability Data (New York: John Wiley and Sons.)

[15] Johnson N L, Kotz S and Balakrishnan N 1994 Continuous univariate distributions Vol 1 2nd ed. (New York: John Wiley & Sons.)

[16] L`vovskij E N 1982 Statisticheskie metody` postroeniya e`mpiricheskix formul [Statistical methods for the empirical formula construction] *Uchebnoe Posobie* (Moscow: Vy`sshaya shkola) [In Russian]

Trade-off working fluid selection for heat pumps

V V Karnaukh[1], V A Mazur[2], A B Biryukov[3] and K A Rzhesik [1]

[1]Department of Refrigerating and Trade Equipment, Donetsk National University of Economics and Trade named after Mykhaylo Tugan-Baranovsky, Schorsa St.,31, Donetsk, 283050, Ukraine
[2]Department of Thermodynamics, Odessa National Academy of Food Technologies, Dvoryanskaya St. 1/3, Odessa, 65082 Ukraine
[3]Department of Engineering Thermophysics, Donetsk National Technical University, Donetsk, Artema St.,58, 283001,Ukraine

E-mail: karnaukh.vita0629@gmail.com

Abstract. Nowadays heat pump systems are used to recover the heat from waste water from different sources and produce hot water for household and communal purposes. In this research comparative thermodynamic analysis of performance for a water-to-water single-stage vapor compression heat pump using several HCs, HCFCs and HFCs which belong to refrigerants of different generations is presented. A multi-criteria approach for optimal selection of refrigerants was used as scientific tool. Results of theoretical experiments of vapor compression heat pump system with the evaporator temperature range 0…25°C and condenser temperature 60°C and 83°C showed that the COP of R1234ze was higher than that of R600a, R152b by about 6.7…17.3%, 8.25…20.5%, and 1.7…14.4%, respectively, at condenser temperature 60 °C; however results are different if at condenser temperature is 83°C: the average COP of R600a was higher than that of R1234ze and R152b by about 7.7…14.7%, 9.3…15.8%, and 7.5…15.6%, respectively. COP displays a positive correlation with the evaporator temperature as well as a negative correlation with the condenser temperature. The complete analysis of all factors, including environmental safety, indicates that R1234ze and R600a refrigerants should be preferred as the working fluids for using in the water-to-water single-stage vapor compression heat pump for the hot water supply and heating purposes in industrial applications.

1. Introduction

Refrigeration and climate technology in developed countries consume up to 15% or more of produced electricity per year. Today capacity refrigerating and climatic equipment market in the European Union is about 30 billion Euros, and only in Germany - 10 billion Euros. The growth rate of this market in individual countries is 20-30% per year. Russia has a steady growth rate of consumption (25% per year) [1-2].

Refrigerants, coolants and oils - have been and remain the objects of constant thermophysical research throughout the world. Despite the fact that the principle of obtaining artificial cooling based on changing phase state has not changed, the cooling and air-conditioning industry is invariably evolved and expanded, which directly affects the applicability of a particular refrigerant in a certain period of time. Refrigerants have changed over the years, mainly in response to safety and environmental issues (figure 1).

IOP Conf. Series: Materials Science and Engineering **791** (2020) 012066 doi:10.1088/1757-899X/791/1/012066

The standards and codes that guide the application of refrigerants in cooling and air-conditioning industryare getting more strict as well [3].

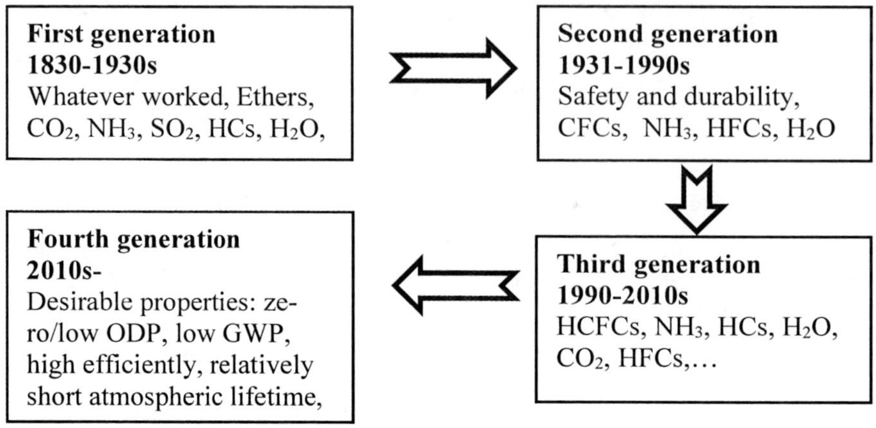

Figure 1.Generation of refrigerants.

Since January 1, 2020 the European Union has introduced a ban on the using of refrigerants with a global warming potential (GWP), greater than 2500, and since 01.01.2025 - a ban on refrigerants with GWP<150. D.Sc. Tsvetkov O.B. et al [1-2] have determined the following requirements for any refrigerants in modern conditions, they are:

- Ozone-Depleting Potential (ODP);
- Global Warming Potential (GWP);
- toxicity;
- fire and explosion hazard;
- ease in detecting leaks;
- critical parameters and thermodynamic properties;
- transfer properties;
- heat and mass transfer characteristics;
- freezing and thermal decomposition temperatures;
- solubility in lubricating oils;
- compatibility with the materials used in engineering and water;
- cost.

Nowadays scientists try to understand if climate change causes from direct consequences or from indirect consequences such as energy-related emissions. There are thousand works showing advantages and disadvantages of the certain refrigerant [4-7].

Clearly, future discovery of ideal refrigerant that combines all the desirable properties is highly unlikely. However, algorithms for searching of a refrigerant with desirable combinations of properties have been described and founded application in ejector cooling cycle [8].

The aim of this study is to compare the thermodynamic performances of the high-temperature water-to-water vapor compression heat pump system using several HCs, HCFCs and HFCs which belong to refrigerants of different generations and to recommend more suitable substance based on theoretical calculation and analysis.

IOP Conf. Series: Materials Science and Engineering **791** (2020) 012066 doi:10.1088/1757-899X/791/1/012066

2. Materials and Methods

In the work we consider the operation of refrigeration system which is simulated by the reverse Rankine cycle. The main processes in the single-stage vapor compression cycle include isentropic compression, isobaric cooling (condensation, subcooling), throttling, and isobaric cooling (evaporation, superheating). The following design specifications are chosen: evaporator and condenser temperatures, t_{ev}= 20°C, t_{cond} = 60°C and t_{cond} = 83°C; cold carrier temperature at the heat pump inlet/ outlet (temperature of the low potentional heat source), t_{low1}/ t_{low2} = 35/30 °C; hot carrier temperature at the heat pump inlet/ outlet, t_{high1}/ t_{high2} = 47/57 °C and 70/80°C; condenser/ evaporator pressure ratio - P_r<9.The temperature range of cold space is the most typical of the industrial cooling water system outlet.

The refrigerants used in this analysis are: natural refrigerants, such as R290A, R600a; third generation refrigerants R142b and R152a, and finally, fourth generation refrigerants, such as R1234yf, R1234ze (table 1). The results obtained from the theoretical calculation are presented and commented in this work from the energetic point of view.

Table 1. Main Characteristics of Refrigerants [9-12].

Refrigerant/ parameter	Natural refrigerants		Ethane Series fluids/ third generation refrigerants		Hydrofluoroolefins fluids/ fourth generation refrigerants	
	R290	R600a	R142b	R152a	R1234yf	R1234ze
Chemical formula	C_3H_8	C_4H_{10}	CH_3CClF_2*	CHF_2-CH_3	CF_3CF=CH_2	CF_3-CH=CHF
Molecular mass, kg/kmol	44.1	58.1	100.5	66.05	114.04	114.04
Normal Boiling Point at p=1atm, K	230.91	261.2	283.15	249.05	243.66	254.17
Critical Temperature, K	370	407.98	410.26	386.41	94.7	109.4
Critical Pressure (absolute), p_{cp}, bar	42.7	36.846	41.4	45.16	33.82	36.4
Critical Density, kg/m³	220.48	225.5	446.0	368	475.55	470.0
Ozon Depletion Potential,ODP	0.0	0.0	0.065	0	0	0
Global Warming Potential, GWP	3	0.01	2000	140	4	7
Safety class[a]	A3	A3	A2	A2	A2L	A2L

[a]Classification: A2 – non-toxit, midly-flammable with higt velocity of flammability; A2L – non-toxit, midly-flammable with low velocity of flammability; A3 – non-toxit, flammable.

The temperature range of hot space is set basing on the most typical values for municipal and industrial high-temperature heating and hot water supply system inlet.

The heat pump thermodynamic cycles diagrams were prepared in the log(p)-h for each of the selected refrigerants. All calculations and diagrams were carried out with the help of freeware CoolPack ("CoolPack– IPU," n.d.) ver. 1.46, freeware REFPROF ver.8 (NIST Standard Reference Database 23) [11-12] and Microsoft Excel 2010.

IOP Conf. Series: Materials Science and Engineering **791** (2020) 012066 doi:10.1088/1757-899X/791/1/012066

The thermodynamic cycle for R600a is shown in the Log(p)-h diagramat t_{ev}= 20°C, t_{cond} = 60°C (figure 2) and at t_{ev}= 20°C, t_{cond} = 83°C (figure 3) as an example.The thermodynamic cycles for the other selected refrigerants have the same order except values of parameter for each state point.

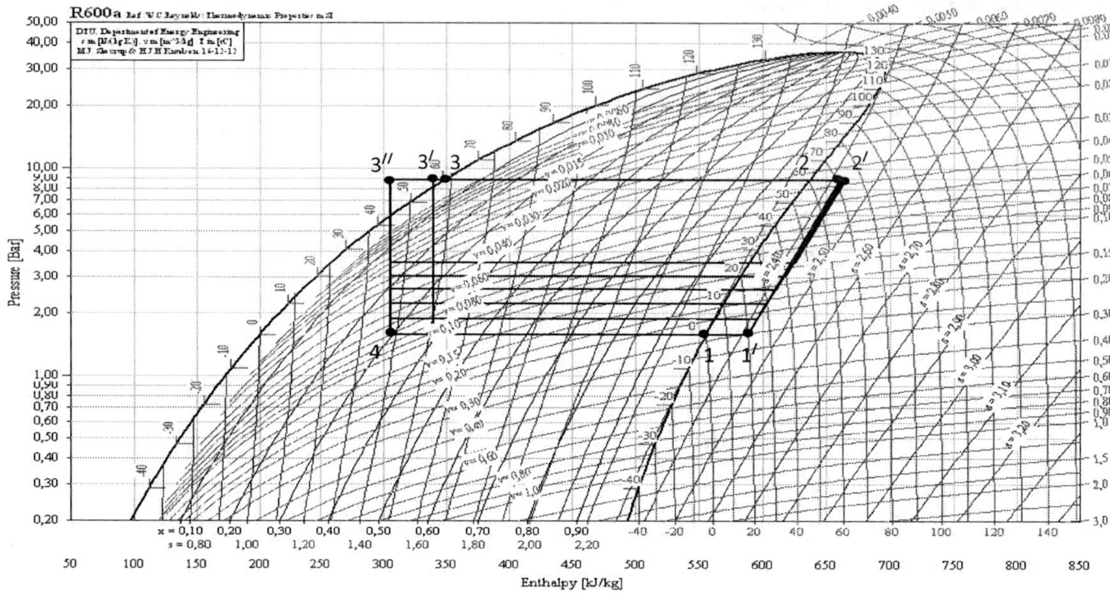

Figure 2.The log(p)-h diagram for R600a at variable evaporation temperature, including t_{ev}= 20°C, t_{cond} = 60°C.

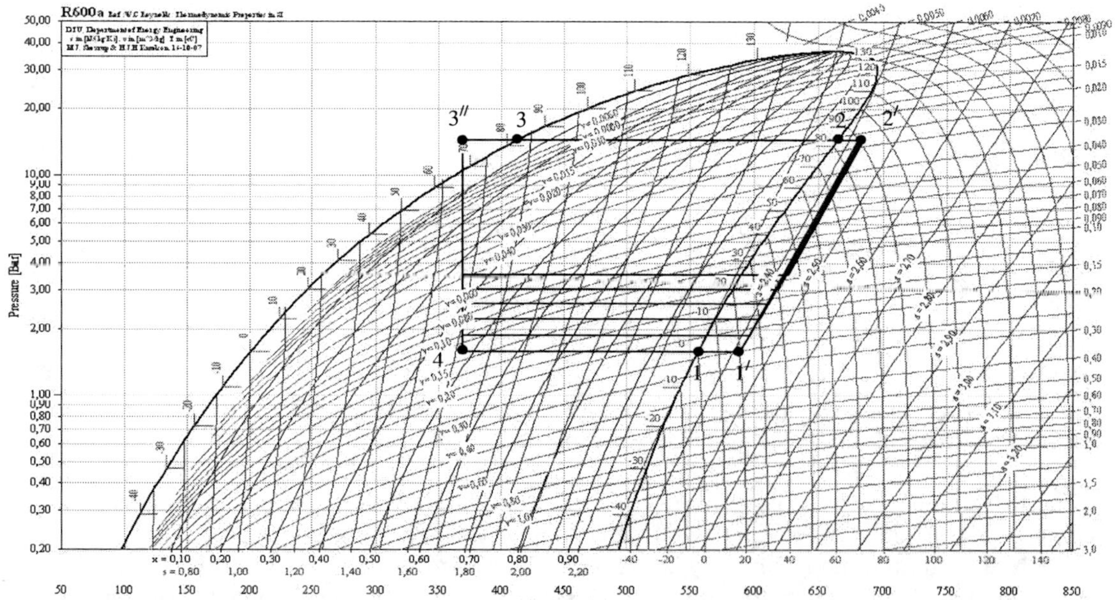

Figure 3.The log(p)-h diagram for R600aat variable evaporation temperature, including t_{ev}= 20°C, t_{cond} = 83°C.

The performance and efficiency of the water-to-water single-stage vapor compression heat pump were estimated as described in [13].

Basing on the parameters from diagrams all the necessary data were calculated for all refrigerants and presented in table 2.

In this work we have tried to adapt a multi-criteria optimization for making comparative analysis in order to choose "desirable" refrigerant for our heat pump system.For the multi-criteria problems the local criteria usually have a different physical meaning, and consequently, they have incomparable dimensions. It complicates the solution of a multi-criteria problem and makes it necessary to introduce the procedure of normalizing criteria or making these criteria dimensionless. There is no unique method for the criteria normalizing and choice of method depends on statement of problem having a subjective character [8].

Table 2. Calculated parameters for using refrigerants at t_{ev}= 20°C, t_{cond} = 60°C.

Refrigerant/parameter	R290	R600a	R142b	R152a	R1234yf	R1234ze
Vaporizer specific heat load, q_{vap}, kJ/kg	269.5	279.0	176.0	251.29	100.0	120.0
Condenser specific heat load, q_{cond}, kJ/kg	316.1	315.0	200.0	283.27	116.0	136.0
Subcooler specific heat load, $q_{s\ cool}$, kJ/kg	8	8	8	8	8	8
Total heat pump specific heat load, q_{hp}, kJ/kg	324.1	323.0	208	291.27	124.4	144.0
Compressor specific work, l_{comp}, kJ/kg	48.2	44.0	32.0	45.83	20.4	18.0
Exergy efficiency ratio, η_e	0.45	0.48	0.4	0.43	0.42	0.52
Coefficient of perfomans COP, μ	6.72	7.34	6.5	6.36	6.1	8.0
Heat equivalent of the electric power, consumed by compressor, q_{cpower}, kJ/kg	53.41	48.75	44.44	50.78	22.6	19.94

Multi-criteria comparative analysis algorithm is realized in the following way.

- Thermodynamic properties and design characteristics of vapor compression cycle are calculated for specified external conditions.
- The "desirable" value of criterion D_i=D_{max} and D_i=D_{min} have been chosen for each of the criterion. The value of D_i depended on thermodynamic mean of each parameter, e.g. for "total heat pump specific heat load" D_{max}=q_{hp}=324.1kJ/kg and D_{min}=q_{hp} =124.41kJ/kg were chosen as data.

If the absolute value of parameter should strive for a minimum value, than $D^0{}_{absREF_i}$ can be obtained as (1)

$$D^0{}_{abs\,REF_i} = \frac{D_i - D_{min}}{D_{max} - D_{min}}. \tag{1}$$

If the absolute value of parameter should strive for a maximum value, than $D^0{}_{absREF_i}$ can be obtained as (2)

$$D^0{}_{absREF_i} = \frac{D_{max} - D_i}{D_{max} - D_{min}}. \tag{2}$$

- The best value of design characteristics D_{absREF} is chosen for each criterion among all concurrent refrigerants.
- Composite criterion D_{absREF} is defined by

IOP Conf. Series: Materials Science and Engineering **791** (2020) 012066 doi:10.1088/1757-899X/791/1/012066

$$D_{absREF} = \sum_{i=1}^{n} D^{0}{}_{absREF_i} . \tag{3}$$

- Minimum value of D_{absREF} – criterion corresponds to best refrigerant among concurrent working fluids.

Trade-off search of working fluid control variables X is formulated as a fuzzy nonlinear programming problem with n non-compatible criteria (economic, environmental, and thermodynamic), m decision variables, and k nonlinear constraints:

$$O_{ptimize}\ K\ [K_{th}(X), K_{ec}(X), K_{en}(X)] \tag{4}$$

subject to

$$G_{Li}\ £\ G_i\ (X)\ £\ G_{Ui},\ I = 1,2,\ldots,\ k \tag{5}$$

$$X_{Li}\ £\ X_i\ £\ X_{Ui},\ i = 1,2,\ldots,\ m \tag{6}$$

where $K_{th}(X)$, $K_{ec}(X)$, $K_{en}(X)$ represent the fuzzy local criteria of thermodynamic, economic, and environmental efficiency; $X\ (X_{1,}\ X_2,\ \ldots,\ X_m)$ is a vector of control variables; G_{Li}, G_{Ui}, are respectively the lower and upper limits for the constraints $G_i(X)$ and X_{Li}, X_{Ui} are respectively the lower and upper bounds for the control variables.

There are several methods of finding "good" solutions to the above problem in thermoeconomic analysis based on scalar optimization. However, as example, the attempts to resolve the CGAM problem [14, 15] via single objective paradigm illustrate a conflict among different approaches and lack of compromise decision [8]. Multicriteria approach is based on synergetic combination of formal and informal making-decision procedures to select a trade-off solution of problem. There are no entirely formal mathematical tools to resolve a multicriteria problem and additional exogenous information is needed.

In the present study, a next sequence of decision-making steps in fuzzy thermoeconomic analysis of refrigeration system is applied [8, 16].

- Determination of the Pareto optimal (or compromise, or trade-off) set X_P as the formal solution of multicriteria problem to minimize a conflict source of uncertainty;
- Fuzzification of goals as well as constraints to represent an ill-structured situation;
- Informal selection of convolution scheme to switch over a vector criterion $K[K_{th}(X)\ K_{ec}(X), K_{en}(X);]$ into scalar combination of the $K_{th}(X)$, $K_{ec}(X)$, and $K_{en}(X)$;

Evaluation of the final decision vector $X_{opt}\hat{I}\ X_P$ to minimize a vagueness source of uncertainty.

3. Results

The complete set of design criteria is considered in table 3 where calculation results for R290A, R600a; R142b, R152a, R1234yf, R1234ze are given.

Results of comparison are shown in figure 4(column a) where thermodynamic advantages of such refrigerants as R600a, R152a and R1234ze are demonstrated obviously. R142b has the worst value of the Global Warming Potential; it was taken into account during the calculation procedure.

The similar calculation stages were made for the same groups of the refrigerants at another condition, videlicet t_{ev}= 20 °C, t_{cond} = 83 °C. Diagram with multi-criteria selection of refrigerants for water-to-water single-stage vapor compression heat pump is shown in figure 4 (column b). In facttheoreticalresults indicate the preservation of the trend in refrigeration choice. Both natural refrigerant R600a and new fourth generation's refrigerant R1234ze have obvious advantages.

In order to obtain the results in the theoretical assessment of a water-to-water single-stage vapor compression heat pump using three chosen refrigerants, two different condenser temperatures were set as 60 °C and 83 °C versus a wide range of evaporator temperatures (range between 0 °C and 25 °C).

Table 3. Assessment of design criteria for using refrigerants.

Refrigerant/ parameter	K	R290A	R600a	R142b	R152a	R1234yf	R1234ze	Dmax	Dmin	ΔD
Total heat pump specific heat load, q_{hp}, kJ/kg	K1	324.1	323.0	208	291.27	124.4	144.0	324.1	124.4	199.7
Exergy efficiency ratio, η_e	K2	0.45	0.48	0.4	0.43	0.42	0.52	0.52	0.4	0.12
Coefficient of perfomans COP, μ	K3	6.72	7.34	6.5	6.36	6.1	8.0	8.0	6.1	1.9
Ozon Depletion Potential, ODP	K4	0.0	0.0	0.065	0	0	0	0.065	0.0	0.065
Global Warming Potential, GWP	K5	3	0.01	2000	140	4	7	2000	0.01	1999.99
Flammability index[a]	K6	1.0	1.0	0.5	0.5	0.0	0.0	1.0	0.0	1.0

[a]Flammability index based on the refrigerant's safety class

It was obtained that the average COP of R 1234ze was higher than that of R600a, R152b by about 6.7-17.3%, 8.25-20.5%, and 1.7-14.4%, respectively, at condenser temperature 60 °C (figure 5a). However results are different if at condenser temperature is 83 °C: the average COP of R600a was higher than that of R1234ze and R152b by about 7.7-14.7%, 9.3-15.8%, and 7.5-15.6%, respectively (figure 5b). COP displays a positive correlation with the evaporator temperature as well as a negative correlation with the condenser temperature.

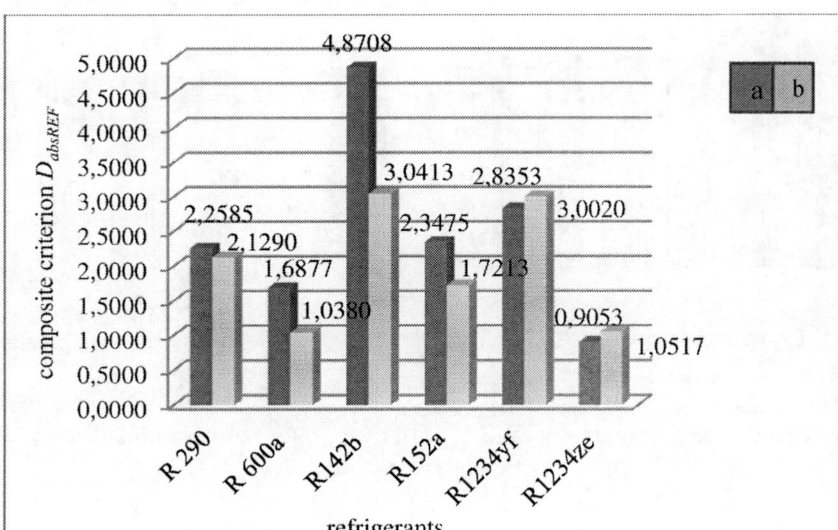

Figure 4. Multi-criteria selection of refrigerants for water-to-water single-stage vapor compression heat pump: a) at t_{ev}= 20°C, t_{cond} =60°C; b) t_{ev}= 20°C, t_{cond} = 83°C.

When the evaporating temperature and the condensing temperature are 20°C and 60°C, COP_h of R600a and R1234ze are 7.37 and 8.0 respectively, so this theoretical cycle is becoming more closer to ideal Karnot cycle. When the evaporating temperature and the condensing temperature are 20°C and 83°C, COP_h of R600a and R1234ze are 4.3 and 3.89 respectively.

The change in exergy efficiency ratio η_e as a function of evaporator temperature for R600a, R152a and R1234ze is demonstrated in figure 6.

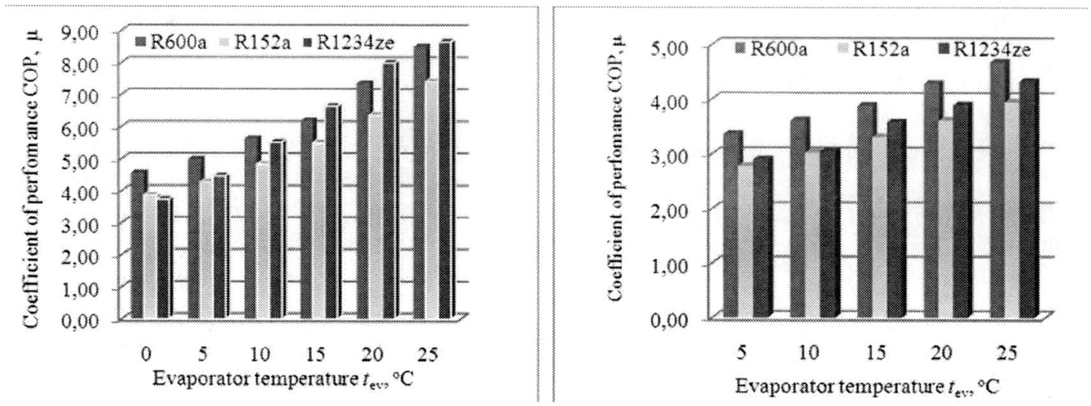

a) b)

Figure 5. Variation of coefficient of performance with evaporator temperature: a) for t_{cond} = 60°C b) for t_{cond} = 83°C.

It was obtained that exergy efficiency ratio η_e doesn't change a lot for each temperature value and η_e difference between R600a, R152a and R1234ze is 7.7-13.5% . η_e of R600a and R1234ze were higher than that of R152a for a range of condenser and evaporator temperatures. It also slightly decreases with the evaporator temperature increase (2-5%).

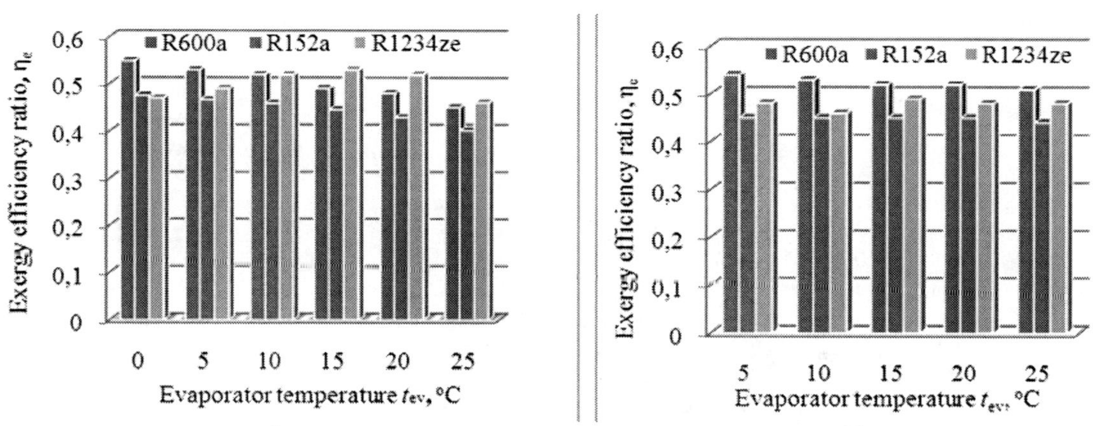

a) b)

Figure 6. Variation of exergy efficiency ratio η_e with evaporator temperature: a) for t_{cond} = 60°C b) for t_{cond} = 83°C.

4. Conclusion

To find balance between high performance and environmental safety of working fluids for HPS is an extremely important goal for heating and cooling industries. In fact, a refrigerant that combines all the desirable properties and has no undesirable properties does not exist.

IOP Conf. Series: Materials Science and Engineering **791** (2020) 012066 doi:10.1088/1757-899X/791/1/012066

Thermodynamic analysis based on a multi-criteria approach to optimum selection of refrigerants is useful tool for engineering calculations.

Theoretical results indicate that both the natural working fluid R600a and new fourth generation’s refrigerant R1234ze have good cycle performances in a water-to-water single-stage vapor compression heat pump cycle. Therefore the complete analysis of all factors, including environmental safety, allows us to draw a conclusion that the R-1234ze and R600a refrigerants should be preferred as the working fluids for using in the water-to-water single-stage vapor compression heat pump for the hot water supply and heating purposes in industrial applications.

Nomenclature

HFC: Hydrofluorocarbon;
HFO: Hydrofluoroolefin;
ODP: Ozone Depletion Potential;
GWP: Global Warming Potential,
COP: Coefficient of Performance;
p: Pressure (kPa);
t: Temperature (°C);

References

[1] Tsvetkov O B, Laptev YU A, Galahova N A and Timofeev B D 2016 Parizhskie ideologemy` i e`nergoe`ffektivny`e rabochie veshhestva texniki nizkix temperatur [Parisian ideologems and energy-efficient working substances of low temperature technology] *Nauchny`j zhurnal NIU ITMO. Seriya "Xolodil`naya texnika i kondicionirovanie"* **2** 6-11 [In Rissian]

[2] Tsvetkov O B, Baranenko A V, Laptev YU A, Sapozhnikov S Z, Khovalyg D M and Pjatakov G L 2014 Ozonobezopasny`e xladagenty` [Ozone layer-safe refrigerants] *Nauchny`j zhurnal NIU ITMO. Seriya "Xolodil`naya texnika i kondicionirovanie"* **3** 98-111 [In Rissian]

[3] 2015 New refrigerants Impact Standards and Code *Carrier Engineering Newletter* vol 3 is 2 URL https://dms.hvacpartners.com/docs/1001/Public/0E/ENG_NEWS_3_2.pdf

[4] Karnaukh V V, Biryukov A B, Mazur V A and Rzhesik K A 2017 Comparative analysis of different refrigerants used in a high-temperature vapor-compression heat pump *J. of Energy for a Clean Environment* **18(2)** 161-74

[5] Meng Z, Zhang H, Qiu J and Lei M 2016 Theoretical analysis of R1234ze(E), R152a, and R1234ze(E)/R152a mixtures as replacements of R134a in vapor compression system *Int. J.of Advances in Mechanical Eng.* **8 (11)** 1-10

[6] Ansari N A, Yadav B and Kumar J 2013 Theoretical exergy analysis of HFO-1234yf and HFO-1234ze as an alternative replacement of HFC-134a in simple vapour compression refrigeration system *Int. J. of Scientific & Eng. Research* **4(8)** 137-44

[7] Gao Y, Zhao H, Peng Y and Roskilly T 2013 Analysis of Thermodynamic Characteristic Changes in Direct Expansion Ground Source Heat Pump Using Hydrofluoroolefins (HFOs) Substituting for HFC-134a *Int. J. Energy and Power Eng.* **5** 11-7

[8] Mazur V A 2003 Optimum Refrigerant Selection for Low Temperature Engineering *Low Temperature and Cryogenic Refrigeration* (Dordrecht: Kluwer Academic Publishers) 101-18

[9] 2019 Designation and Safety Classification of Refrigerants. ASHRAE Standard, ANSI/ASHRAE Standard 34-2010 URL https://www.ashrae.org/technical-resources/standards-and-guidelines/ashrae-refrigerant-designations

[10] Calm J M 1999 *Arti refrigerant database. Primary and recently-added citations* (Arlington: Air-Conditioning and Refrigeration Technology Institute)

[11] Lemmon E W, Huber M L and McLinden M O 2009 *NIST Standard Reference Database 23: Reference Fluid Thermodynamic and Transport Properties* (Gaithersburg: National Institute

of Standards and Technology)

[12] Bell I H, Wronski J, Quoilin S, Lemort V 2014 Pure and Psedo-pure Fluid Thermophysical Property Evalution and the Open-Source Thermophysical Propety Library Cool Prop *Industrial & Engineering Chemistry Research* **53(6)** 2498-508

[13] Trubaev P A and Grishko B M 2010 *Teplovy`e nasosy`* [Heat Pumps] (Belgorog: BGTU im V G Shukhova)

[14] McLinden M O, Domanski P A, Kazakov A, Heo J and Brown J S 2012 Possibilities, limits, and tradeoffs for refrigerants in the vapor compression cycle *ASHRAE/NIST Refrigerants Conference* (Gaithersburg)

[15] Kazakov A, McLinden M O and Frenkel M 2012 Computational Design of New Refrigerant Fluids Based on Environmental, Safety, and Thermodynamic Characteristics *J. Ind. Eng. Chem. Res*. **51(38)** 12537-48

[16] Kazakov A, Muzny C D, Diky V, Chirico R D and Frenkel M 2010 Predictive correlations based on large experimental datasets: Critical constants for pure compounds *J.Fluid Phase Equilib* **298** 131-42

IOP Conf. Series: Materials Science and Engineering **791** (2020) 012067 doi:10.1088/1757-899X/791/1/012067

Experimental study of the silica gel drying in a fluidized bed apparatus

A A Nadeev, A V Barakov, V Yu Dubanin, A Yu Andreev and A V Muravev

Department of Theoretical and Industrial Heat Power Engineering, Voronezh State Technical University, Moscow Ave., 14, Voronezh, 394026, Russia

E-mail: alekn85@mail.ru

Abstract. In this paper, the design and principle of operation of the dryer with a centrifugal fluidized bed is considered. It is a batch apparatus with a vertical drying agent supply and is intended for drying dispersed materials with a particle diameter from 1 to 5 mm, having a high initial moisture content and a strong moisture and solid dry frame bond. Silica gel with a particle diameter from 2.2 to 2.5 mm, used for drying air and industrial gases, was used as a drying object. The scheme of the test bench is given and described. The results of an experimental study of the process of low-temperature drying of silica gel (the temperature of the drying agent ranged from 35 to 65 °C) are presented. The obtained experimental data is presented as a dependence of the thermophysical parameters of the material and the drying agent on the drying time. The comparison of drying efficiency at different modes is given. On the basis of the obtained data, a positive conclusion about the possibility of using this apparatus for low-temperature drying of capillary-porous bodies is made.

1. Introduction

Among all heat and mass transfer units used in industry for dispersed materials drying, a special place is occupied by a fluidized bed apparatus. This is primarily due to the high interphase heat and mass transfer intensity in such devices and their low hydraulic resistance [1-3]. Also their advantages include a relatively simple design, small dimensions, the ability to ensure continuous processes. This led to the presence of a wide range of dryers that implement this technological principle in modern industry [3-5]. Nevertheless, the development and research of new more energy-efficient structures is an urgent task of practical importance.

The results of theoretical and experimental studies of regenerative heat exchanger and absorber with centrifugal fluidized bed were summarized in [6-8]. Such a layer is formed in the annular channel of the apparatus having a cylindrical cross section when air is supplied to the layer at a predetermined angle to the horizontal plane. According to the results of these studies, the use of a centrifugal layer can significantly intensify the heat and mass transfer process and simplify the apparatus design by eliminating a special tool for the dispersed material moving. This paper presents new results for the application of such a layer for capillary-porous dispersed material (silica gel, which is widespread in industry and convenient for experimental research) drying.

2. Materials and Methods

A centrifugal fluidized bed dryer is a batch apparatus with a vertical drying agent supply. Figure 1 shows its layout (a) and appearance (b). It is intended for drying dispersed materials with a particle

IOP Conf. Series: Materials Science and Engineering **791** (2020) 012067 doi:10.1088/1757-899X/791/1/012067

diameter from 1 to 5 mm, having a high initial moisture content and a strong moisture and solid dry frame bond.

The main dryer element is the working chamber 1 with a diameter of 0.4 m and a height of 0.4. The annular channel through which the material moves is formed by a cylindrical case of the apparatus and an inner conical shell with a base diameter of 0.2 m. The transparent cylindrical case of the pilot apparatus, shown in figure 1 (b), allows visual control of the fluidization process during operation. There are eight filling holes in the inner shell which are designed to supply fresh dispersed material to the working chamber from the branch pipe 2. The conical shape of the inner shell reduces the speed of the drying agent in the upper part of the working chamber 1. Unloading of the dried material is carried out through two discharge branch pipes 3 equipped with partitions. The drying agent enters the working chamber through the diffuser 4, and the waste one is discharged into the atmosphere through the chamber 5.

The louver grille 6 is used as a gas distribution device in the drying apparatus. Its blades are inclined to the horizontal plane by 24 degrees and serve to create a directed flow of the drying agent at the entrance to the material layer. A metal gauze is installed above the grille. Its purpose is to prevent particles drop in the diffuser 4 and to ensure a more uniform dried material fluidization.

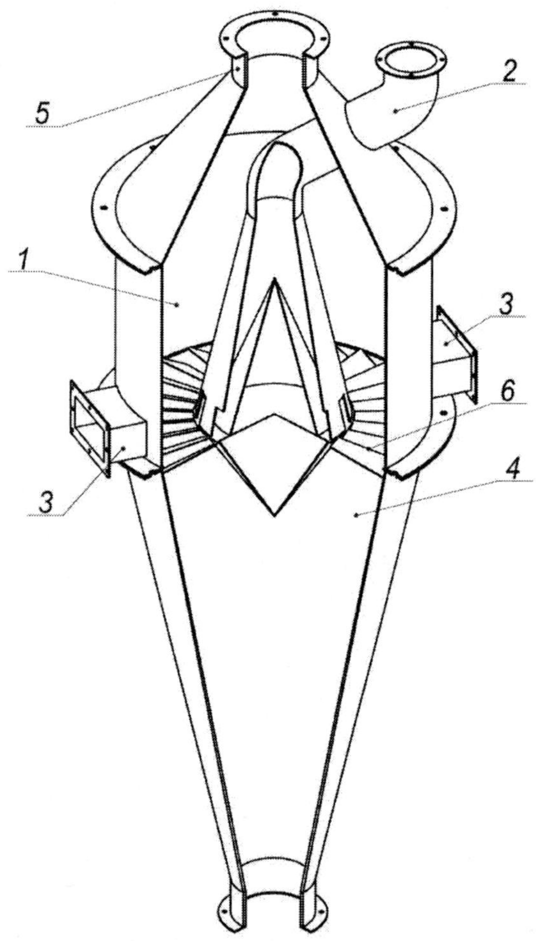

(a)

(b)

Figure 1. The layout (a) and the appearance (b) of the centrifugal bed dryer.

IOP Conf. Series: Materials Science and Engineering **791** (2020) 012067 doi:10.1088/1757-899X/791/1/012067

This dryer due to the annular channel has a very simple design, small dimensions and no stagnant areas. The hydraulic resistance of devices with a centrifugal fluidized bed is insignificant and has a value from 200 to 650 Pa [5, 7]. The heat and mass transfer intensification in such a layer occurs due to the action of centrifugal forces on the material particles [5, 6]. The height of the layer should not exceed 0.1-0.15 m [5-7].

A test bench was constructed to conduct an experimental study of the heat and mass transfer process during dispersed materials drying in this apparatus. Its layout is shown in figure 2.

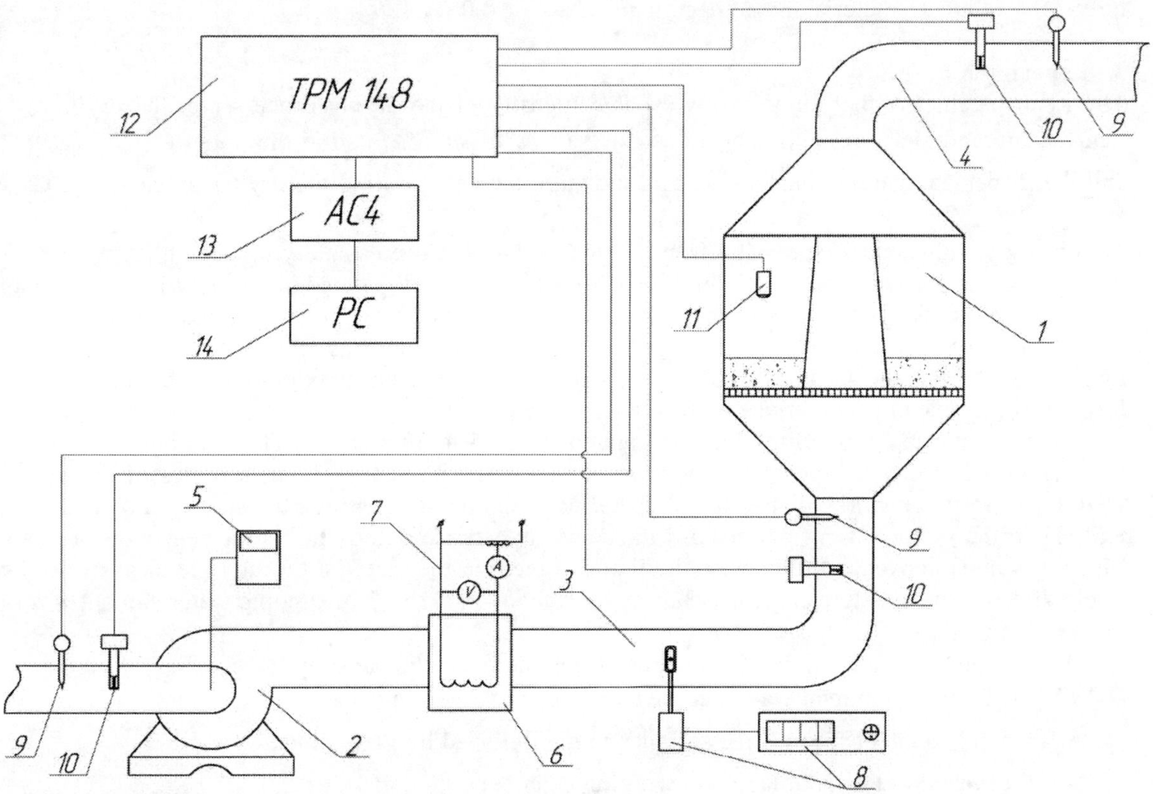

Figure 2. The test bench layout: 1 – dryer; 2 – fan C10-28 No. 4; 3, 4 – air ducts; 5 – frequency converter Delta VFD150E43A; 6 – electric air heater CFOC-25/0,5-I1; 7 – autotransformer PHO-250-5; 8 – thermo-anemometer TTM-2/4-06 and converter TTM-2-04; 9 – thermoelectric converters TP-2088; 10 – moisture sensors Galltec+Mela FRC 3/5; 11 – infrared temperature sensor MoreSunsDIY; 12 – PID controller OWEN TPM 148; 13 – interface converter AC4; 14 – PC.

The drying agent (atmospheric air) is supplied to the apparatus by a centrifugal fan 2 through the duct 3. The fan performance is regulated by a frequency converter 5, which also provides a convenient way to estimate the energy cost of creating a drying agent flow. The exhaust air is discharged into the atmosphere through the duct 4. The cold atmospheric air is heated in an electric air heater 6 having a rated power of 24 kW. Its power is regulated by an autotransformer 7. The drying agent speed is measured by a hot-wire anemometer 8, the primary transducer of which is placed in the duct. Knowing the geometric characteristics of the ducts, it is easy to determine the mass flow rate of the drying agent and its speed at the entrance to the working chamber.

The thermocouples 9 are used to measure the drying agent temperature; duct moisture sensors are used to measure relative moisture 10. Each pair of sensors is installed in front of the fan, in duct 3 in front of the dryer, and in duct 4 after the dryer. The dispersed material temperature is measured using an industrial infrared temperature sensor 11 mounted above the layer in the working chamber of the dryer. The use of this sensor allows continuous recording of data without interfering with the process.

A PID controller 12 is used as a secondary device. An interface converter 13 (RS485-USB) is designed to provide communication between the secondary device and the personal computer 14. The experimental data logging is carried out in the Owen Process Manager SCADA system.

The moisture content of the dispersed material is determined using a moisture analyzer EVLAS-2M.

The limits of the total measuring temperature and moisture error using the appropriate sensors and a secondary device are: air temperature - $\Delta t = \pm 2.7$ °C; air moisture - $\Delta\varphi = \pm 2.3\%$; material temperature - $\Delta t = \pm 2.0$ °C; material moisture content - $\Delta w = \pm 0.05\%$.

3. Experimental results

The main parameters that allow to review the intensity of the drying process are: the drying agent temperature and moisture at the apparatus inlet (t''_{air},°C; φ'', %) and at the outlet (t''_{air},°C; φ'', %), and the dispersed material temperature and moisture changing during the drying process (t_m,°C; w, %).

Silica gel with a particle diameter from 2.2 to 2.5 mm was used as a drying object. Its regeneration was carried out at a temperature of 180-200 °C [9]; however, it was noted in [10] that in devices with a high heat and mass transfer rate it is advisable to regenerate silica gel at a temperature of 40 °C. Thus, the minimum temperature of the drying agent at the apparatus inlet was 35 °C, the maximum temperature was 65 °C, and the accuracy of its maintenance was ± 2 °C. The maximum temperature value was limited by the operating conditions of the air moisture sensors.

The experiments duration in all modes ranged from 25 to 30 minutes. The readings of all sensors were recorded in Owen Process Manager with 5 seconds frequency. Small samples of the material were taken from the working chamber at specified intervals to determine the silica gel moisture content. Then the samples were placed in a moisture analyzer to determine the current moisture value. Since it is more convenient to express the liquid content in the material through the mass of its absolutely dry part, the moisture content w% was being converted into hygroscopic water content u kg/kg db at the same time [2].

The parameters characterizing the various drying modes in the paper are:

- the drying agent temperature at the entrance to the working chamber t'_{air}, °C;
- the drying agent speed at the entrance to the fluidized bed υ_{air0}, m/s;
- the weight of the material in the working chamber M_m, kg;
- the initial moisture content of silica gel u_0, kg/kg db.

In the experimental study of the low-temperature silica gel drying process, these parameters had the following values: t'_{air} = 35; 45; 55 °C; υ_{air0} = 2.0; 2.3 m/s; M_m = 1.0, 1.5 kg; u_0 = 0.32; 0.53; 0.82 kg/kg db. The experimental data processing was carried out according to the recommendations presented in [11].

Figure 3 shows the readings of all installed sensors at t'_{air} = 55 °C; M_m = 1 kg; u_0 = 0.82 kg/kg db (w_0 = 44.92 %); υ_{air0} = 2.0 m/s. It also shows temperature and air moisture in the laboratory (t_{air0},°C; φ_0, %).

It is known that the drying process can be divided into three periods – the material heating, the period of constant drying rate (first) and the period of falling drying rate (second). The type of drying curve 1 depends on the material class [2, 12].

The temperature curve 4 shows that the material temperature gradually decreases to the wet thermometer temperature. This time is the period of the material heating, its duration for this mode is 30-40 seconds from the beginning of drying. Then, up to 2-2.5 minutes from the beginning of drying, the material temperature is approximately constant. It is the period of constant drying rate. Its ending coincides with the local maximum humidity of the exhaust drying agent. After this, the material is heated to the drying agent temperature. It is the period of falling drying rate.

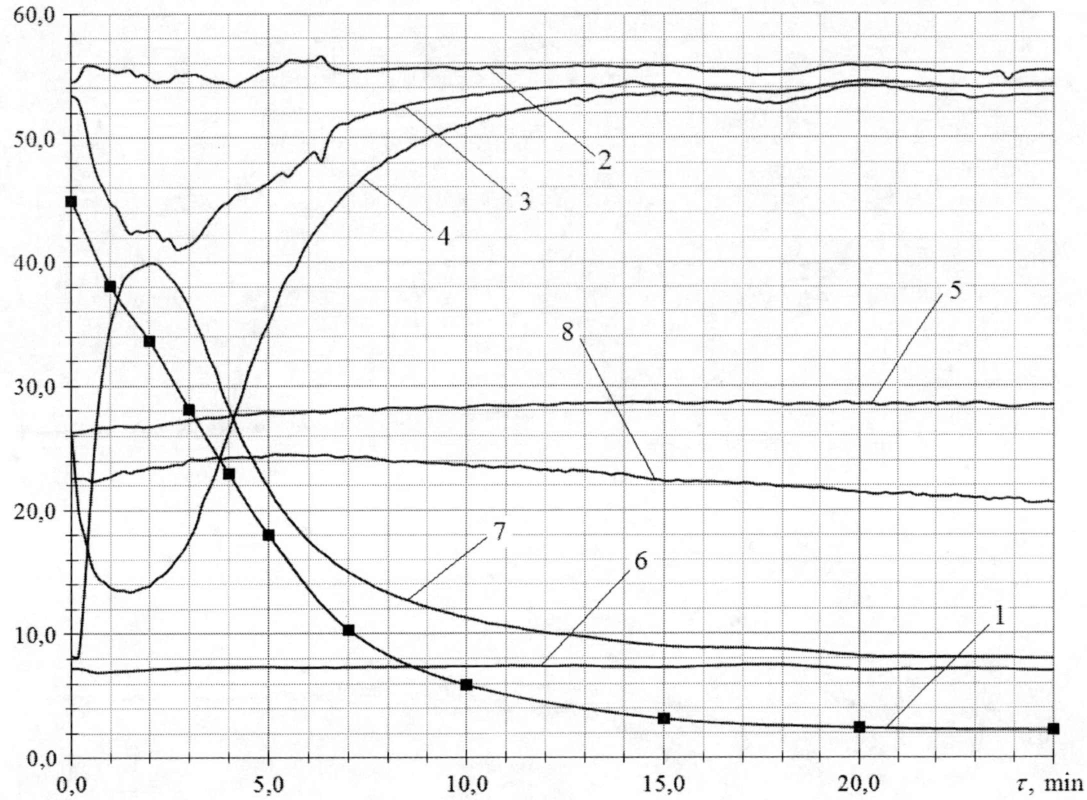

Figure 3. Silica gel drying process: 1 – w, %; 2 – t'_{air}, °C; 3 – t''_{air}, °C; 4 – t_m, °C; 5 – t_{air0}, °C; 6 – φ', %; 7 – φ'', %; 8 – φ_0, %.

Two moisture values can be distinguished on the drying curve, firstly, the critical one u_{cr} (transition from the first period to the second one in the 2nd-2.5 minutes of the process) and, secondly, the equilibrium one u_{eq} (balance with the drying agent moisture). Thus, the moisture removal process for this mode had been completed by the 16th minute.

It should be noted that the drying curve 1 and the material temperature curve 4 are typical of capillary-porous solids to which silica gel belongs. Curve 4 also shows that the material temperature is significantly lower than the ambient temperature. This is due to the high initial material moisture and the intensive moisture removal from it, i.e. evaporative cooling of silica gel occurs.

Three to five experiments were carried out to check the data reproducibility for each temperature mode. Figures 4 - 6 show a comparison of drying and temperature curves obtained from three experiments for one drying silica gel mode (t'_{air} = 55 °C; M_m = 1 kg; u_0 = 0.82 kg/kg db (w_0 = 44.92 %); υ_{air0} = 2.0 m/s).

IOP Conf. Series: Materials Science and Engineering **791** (2020) 012067 doi:10.1088/1757-899X/791/1/012067

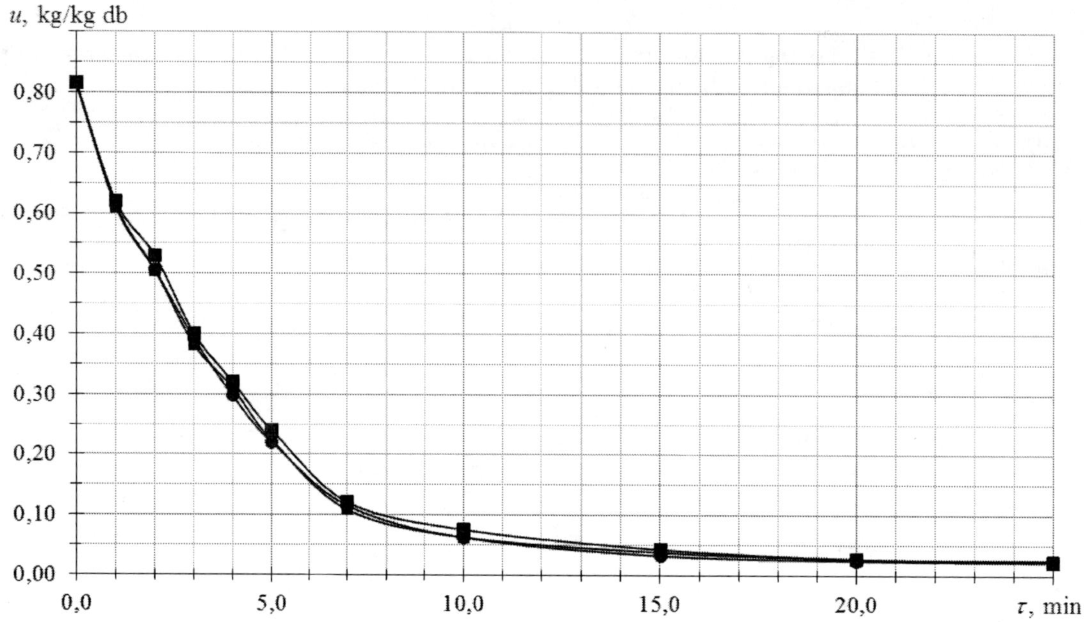

Figure 4. Silica gel drying curves.

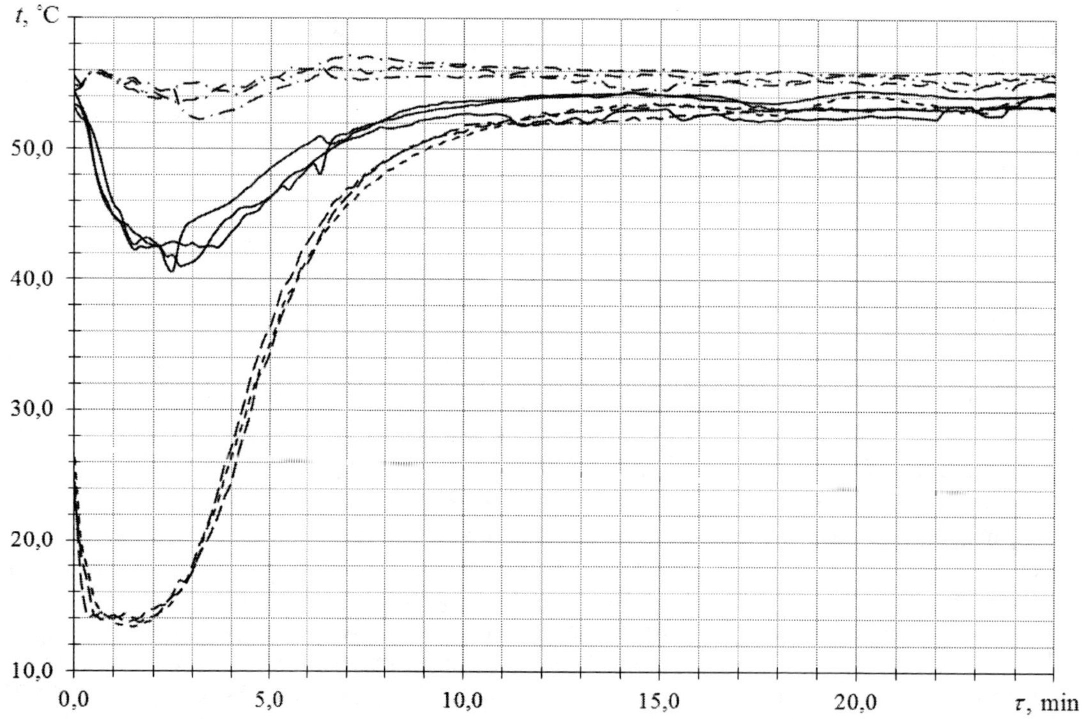

Figure 5. Temperature curves: —·—·— t'_{air} ; ——— t''_{air} ; — — — t_m .

IOP Conf. Series: Materials Science and Engineering **791** (2020) 012067 doi:10.1088/1757-899X/791/1/012067

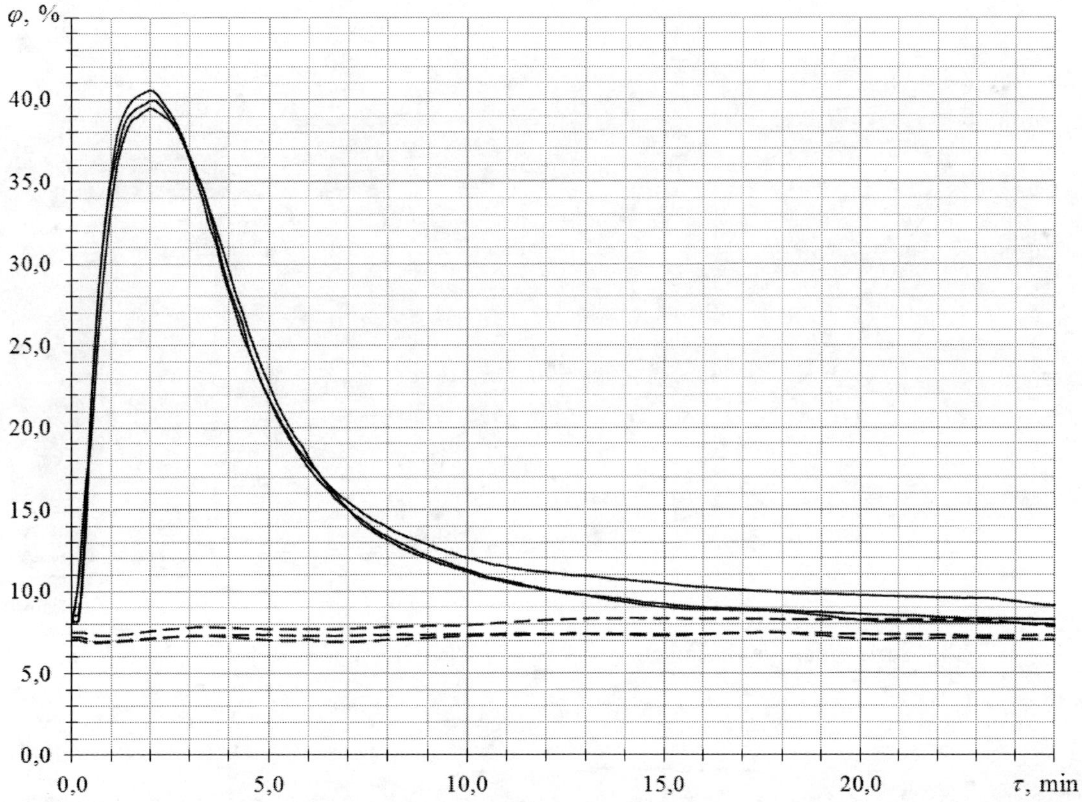

Figure 6. Air humidity curves: – – – – φ' ; ———— – φ'' .

The results of statistical experimental data processing [13] led to the conclusion that the adopted research methodology is adequate and allows to obtain sufficiently reliable data on the material and the drying agent parameters. During the repeated experiments it was observed that moisture and temperature in the laboratory influenced on the measurement results.

Figures 7 and 8 show a comparison of two drying modes that differ in the air temperature at the working chamber inlet t'_{air} – 45 °C and 55 °C. The other mode parameters are: M_m = 1 kg; u_0 = 0.82 kg/kg db; υ_{air0} = 2.3 m/s.

These relations prove that with increasing drying agent temperature, the rate of moisture removal from the material increases and the duration of the first period and the duration of the drying cycle decrease. At t'_{air} = 55 °C the drying process ends by the 14th minute; u_{cr} = 0.48 kg/kg db; u_{eq} = 0.044 kg/kg db. At t'_z = 45 °C u_{cr} the drying process ends by the 16th minute = 0.52 kg/kg db, u_{eq} = 0.041 kg/kg db.

It should also be noted that each temperature mode of silica gel drying is characterized by a coincidence of the final air humidity curves φ'' at the initial stage of the process (the first 1-1.5 minutes).

IOP Conf. Series: Materials Science and Engineering **791** (2020) 012067 doi:10.1088/1757-899X/791/1/012067

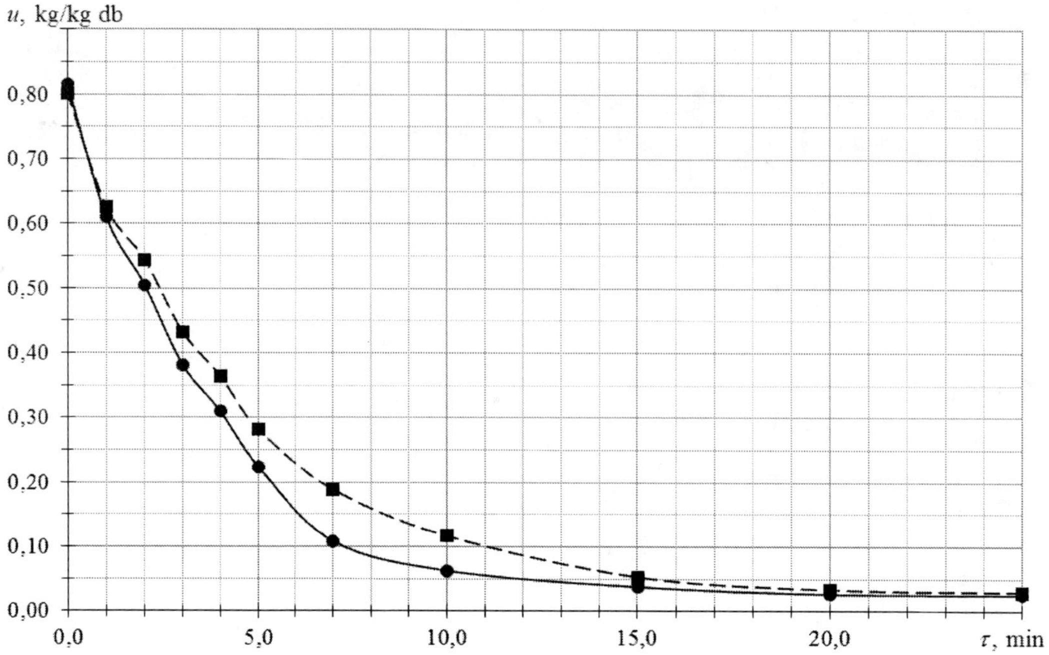

Figure 7. Silica gel drying curves: – – – – t'_{air} = 45 °C; ——— – t'_{air} = 55 °C.

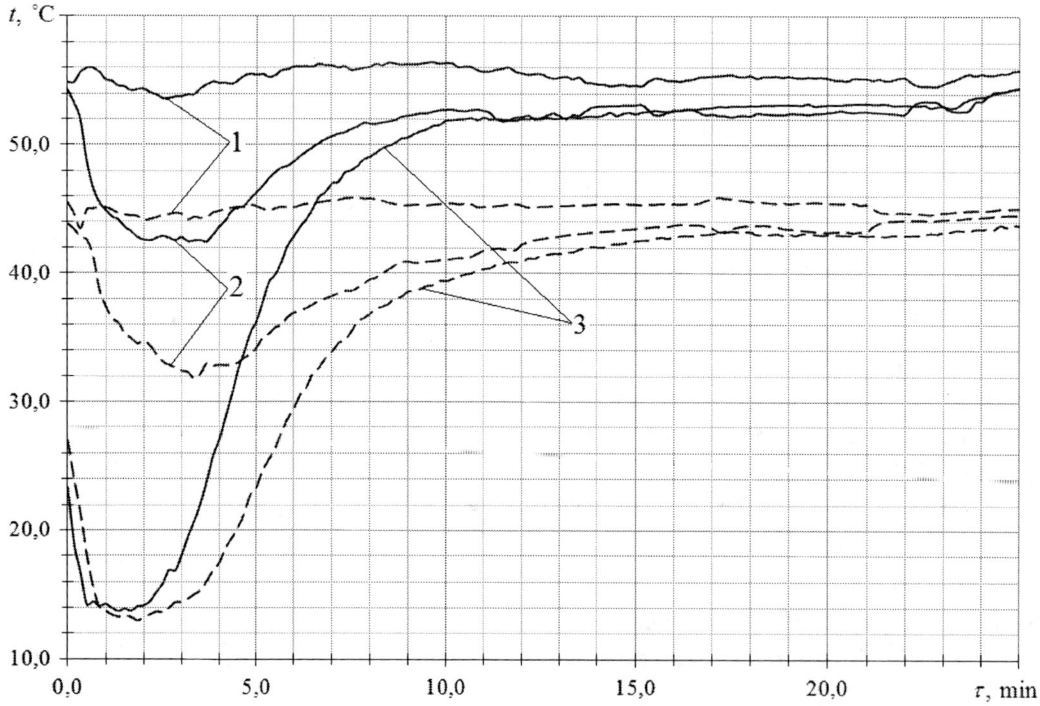

Figure 8. Temperature curves: – – – – t'_{air} = 45 °C; ——— – t'_{air} = 55 °C; 1 – t'_{air} ; 2 – t''_{air} ; 3 – t_m .

IOP Conf. Series: Materials Science and Engineering **791** (2020) 012067 doi:10.1088/1757-899X/791/1/012067

4. Conclusion

Based on the data obtained, we can confirm the conclusion that the heat and mass transfer intensity in a centrifugal fluidized bed apparatus is quite high. This allows to implement the low-temperature drying process, which is especially important for drying of thermolabile materials.

It should be noted that the heat transfer coefficient during moisture evaporation from the material at the beginning of the drying process can be 1.5-2 times higher than at the end of drying [14]. A certain average value of this parameter is used in mathematical models describing the drying process, which impairs the calculations accuracy. The obtained experimental data will also allow to obtain an empirical relation for the heat transfer coefficient, taking into account the material moisture content.

Also a semi-empirical mathematical model that describes the heat and mass transfer process in an apparatus with a moving fluidized bed was proposed in our works [15, 16]. As a result of the analytical model solution we obtained the time relations of the temperature and the material moisture content to the drying agent, containing an empirical coefficient - the drying intensity. The experimental study results will determine this parameter for each drying mode.

References

[1] Gibilaro L G 2001 *Fluidization Dynamics* (Oxford: Butterworth-Heinemann)

[2] Mujumdar A S 2006 *Handbook of Industrial Drying* (Boca Raton: CRC Press)

[3] Yang W-C 2003 *Handbook of Fluidization and Fluid-Particle Systems* (New York: Marcel Dekker, Inc.)

[4] Srivastava S and Mishra G 2010 Fluid Bed Technology: Overview and Parameters for Process Selection *Int. J. Pharm. Sci. Drug. Res.* **2(4)** 236-46

[5] Ambrosio-Ugri M C B and Taranto O P 2007 Drying in the rotating-pulsed fluidized bed *Braz. J. Chem. Eng.* **24(1)** 95-100

[6] Agapov Yu N, Barakov A V, Zhuchkov A V and Sannikov A V 1986 Ocenka gidravlicheskogo soprotivleniya i mezhfaznogo teploobmena v centrobezhnom psevdoozhizhennom sloe [Assessment of hydraulic resistance and interfacial heat transfer in a centrifugal fluidized bed] *Himicheskaya Promyshlennost'* **63(4)** 61-3 [In Russian]

[7] Faleev V V and Barakov A V 2003 Issledovanie mezhfaznogo teploobmena v regenerativnom teploobmennike s dispersnoj nasadkoj [Study of interphase heat transfer in a regenerative heat exchanger with a dispersed material checker] *Prom. Energetika* **6** 35-7 [In Russian]

[8] Barakov A V, Dubanin V Yu and Prutskikh D A 2009 Issledovanie gidrodinamiki regeneratora s dispersnoj nasadkoj [The study of the hydrodynamics of a regenerator with a dispersed material checker] *Energosberezhenie i Vodopodgotovka* **1** 47-8 [In Russian]

[9] Weintraub S 2002 Demystifying Silica Gel *Proceedings of the Objects Specialty Group Session (Miami, 10 June 2002)* vol 9 ed V Greene, P Griffin (Washington, DC: American Institute for Conservation) pp 169-94

[10] Pramuang S and Exell R H B 2007 The regeneration of silica gel desiccant by air from a solar heater with a compound parabolic concentrator *Renew. Energ.* **32(1)** 173-82

[11] Kemp I C, Fyhr B C, Laurent S, Roques M A, Groenewold C E, Tsotsas E, Sereno A A, Bonazzi C B, Bimbenet J-J and Kind M 2001 Methods for processing experimental drying kinetics data *Dry. Tech.* **19(1)** 15-34

[12] Williams-Gardner A 1971 *Industrial Drying* (London: George Godwin)

[13] Mandel J 1984 *The Statistical Analysis of Experimental Data* (New York: Dover Publications)

[14] Watano S, Yeh N and Miyanami K 1999 Heat Transfer and the Mechanism of Drying in Agitation Fluidized Bed *Chem. Pharm. Bull.* **47(6)** 843-6

[15] Barakov A V, Byrdin A P, Dubanin V Y and Nadeev A A 2017 The asymptotic solution of model equations for heat capacities of fluidized bed phases *J. Phys. Conf. Ser.* **891** 012325

[16] Barakov A V, Byrdin A P and Nadeev A A 2016 Study of heat and mass transfer in the dryer with a pseudo liquefied layer of a disperse material *Sci. Herald Vor. Stat. Univ. Architec. Civ. Engineer. Constr. Architec.* **31(3)** 18-26

IOP Conf. Series: Materials Science and Engineering **791** (2020) 012068 doi:10.1088/1757-899X/791/1/012068

Energy intensity decrease in chlorella growth technology by electro ozonation

D A Normov[1], G N Samarin[2], A N Vasilyev[2], A A Shevchenko[1] and R B Goldman[1]

[1]Physics department, FSBEI HE Kuban State Agrarian University, 350012, Krasnodar, 13, Azovskaya street, Flat 40, Russia
[2]Office of Automated Electric-and-Heat Technology, Federal Scientific Agro-Engineering Center VIM, 1st Institutsky passage, House 5, 109428, Moscow, Russia

E-mail: danormov@mail.ru

Abstract. This paper examines process efficiency problems in chlorella microphytic alga growth for quality feed of livestock and poultry. Ozone-air mixture for microphytic alga growth simulation and development is being suggested for use. Paper describes experimental research methods realization and some experimental equipment. Experimental plant to carry out research was built, which produces ozone-air mixture, with required ozone concentration, and delivers it to the tank that contains chlorella microphytic alga suspension. Moreover, equipment was selected in a way so it doesn`t interfere with the main technological process of microphytic alga growth, special attention was given to an energy saving effect of the equipment and technology in general. Experimental research results discussion demonstrates some statistical data and dependence curves, constructed using these results, and also derived on this basis regression equation. Research results were processed by the "Statistika 6.0" software, obtained results indicate sufficiently close correlation relationship between examined data. Some experimental work results were summarized in the conclusions, and substantial efficiency of the suggested technology was verified, which allows to reduce energy consumption in forage production.

1. Introduction

Microphytic alga biomass addition as a vitamin feed supplement is one of many resource-saving technologies for animal feeding.

One of them is chlorella – a microscopic plant, that contains a large amount of protein, complete set of essential amino acids, carbohydrates, fats, vitamins and biological stimulants. Chlorella in agriculture is being used in a suspension form, which is prepared from the strain Chlorella vulgaris IGF S-111. However, in an agriculture in Russia, at present day, due to high energy intensity of a manufacturing process, usage of chlorella suspension is very limited. High energy intensity is explained by the use of a mercury arc lamps, to ensure photosynthesis. In order to produce 40-60 liters of suspension, this type of lamps consume from 12 to 24 kWh of electricity.

Therefore, chlorella manufacturers face necessity to increase productivity at a minimum production costs and eco-friendliness of a process. [1, 7, 10]

IOP Conf. Series: Materials Science and Engineering **791** (2020) 012068 doi:10.1088/1757-899X/791/1/012068

2. Materials and Methods

New experimental pant was built to carry out an experimental research. This equipment allows to produce ozone-air mixture with required ozone concentration and deliver it to the tank that contains chlorella microphytic alga suspension, which is being treated (Figure 1).

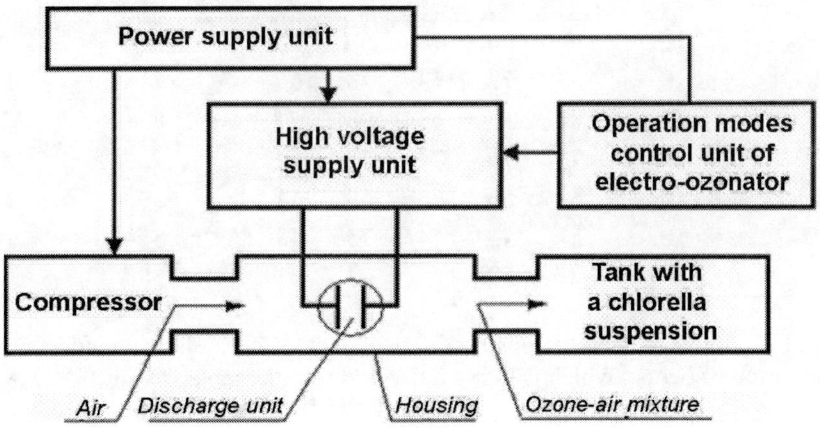

Figure 1. Functional diagram of the experimental plant.

Experimental plant consists of hermetically sealed housing of electro-ozonator, discharge unit, high voltage supply unit (high voltage transformer), laboratory autotransformer (LATR 1M), tank with chlorella suspension, flexible tubings, that connect all parts of the plant, and spray-diffusers.

Experiment was carried out at the FSBI HPE KubGAU laboratory, with the objective to detect influence of ozone-air mixture, at different values of concentration and exposure, on chlorella cells growth and development. First, searching experiment to determine location of ozone effective concentration area was conducted [2, 4, 12, 13]. During this experiment three tanks with chlorella suspension, 1000 ml each, were treated by ozone-air mixture, whose cells went through active fission and became «rejuvenated», with the following ozone concentration values 9.34 mg/m^3; 17.79 mg/ m^3; 40.04 mg/ m^3. Each tank was treated for 10 minutes. Tank number 4 was left untreated for control purpose. Experimental results tabulated in Table 1.

Table 1. Results of the experiment on ozone effective concentration area search.

Ozone concentration $[O_3]$, mg/m^3	Treatment time τ, min	Number of cells under the camera's square N, pcs					Average number of cells under the camera's square Navg, pcs	Concentration of chlorella cells C_{chl}, c/ml
		1	2	3	4	5		
9.34	10	55	49	53	58	61	55	$13.75 \cdot 10^6$
17.79	10	43	55	34	39	57	46	$11.5 \cdot 10^6$
40.04	10	47	40	37	34	52	42	$10.5 \cdot 10^6$
-	-	59	48	63	36	56	52	$13.0 \cdot 10^6$

Analyzed test results suggest that in order to find an optimal area of ozone positive influence on chlorella cells it is necessary to examine carefully influence of the ozone-air mixture with ozone concentration with limits up to 12-13 mg/m^3, at a constant treatment time. [1, 8, 11]

To continue search for an optimal area of ozone positive influence, the following levels of ozone concentration were chosen for treatment: 3.2 mg/m^3; 5.2 mg/m^3; 7.2 mg/m^3; 9.2 mg/m^3; 11.2 mg/m^3; 13.2 mg/m^3. Experimental results presented in Table 2.

IOP Conf. Series: Materials Science and Engineering **791** (2020) 012068 doi:10.1088/1757-899X/791/1/012068

Table 2. Results of the experiment on ozone effective concentration area search.

Ozone concentration $[O_3]$, mg/m^3	Treatment time τ, min	Number of cells under the camera's square N, pcs					Average number of cells under the camera's square Navg, pcs	Concentration of chlorella cells C_{chl}, c/ml
		1	2	3	4	5		
3.2	10	67	58	78	69	62	67	$16.75 \cdot 10^6$
5.2	10	82	72	64	73	65	71	$17.75 \cdot 10^6$
7.2	10	87	85	68	98	74	82	$20.5 \cdot 10^6$
9.2	10	77	76	74	72	62	72	$18.0 \cdot 10^6$
11.2	10	52	53	66	68	54	59	$14.75 \cdot 10^6$
13.2	10	51	55	54	75	48	57	$14.25 \cdot 10^6$
-	-	71	52	63	44	61	58	$14.5 \cdot 10^6$

Experimental results were used to construct dependence of chlorella cells concentration in suspension on concentration of ozone, which is being delivered into the tank. Conclusion was developed after the results analysis, that optimal area of ozone positive influence is within limits from 3.2 mg/m^3 to 9.2 mg/m^3 (Figure 2).

Determination of the effective treatment time by ozone on chlorella cells at concentration 7.2 mg/m^3 required the experiment with the following time levels: 2 min, 4 min, 6 min, 8 min, 10 min, 12 min, 14 min (Figure 3).

Using results of the experiment, also was constructed dependence of chlorella cells concentration in suspension on time of treatment by ozone-air mixture and presented pictures from Goryaev chamber of control and treated samples, made by digital camera Microscope Digital Camera Levenhuk C – Series for microscope, which were processed by Taup View software (Figure 4, 5).

As can be seen from the presented dependence, area of ozone effective influence on chlorella cells falls within limits of 4-10 minutes.

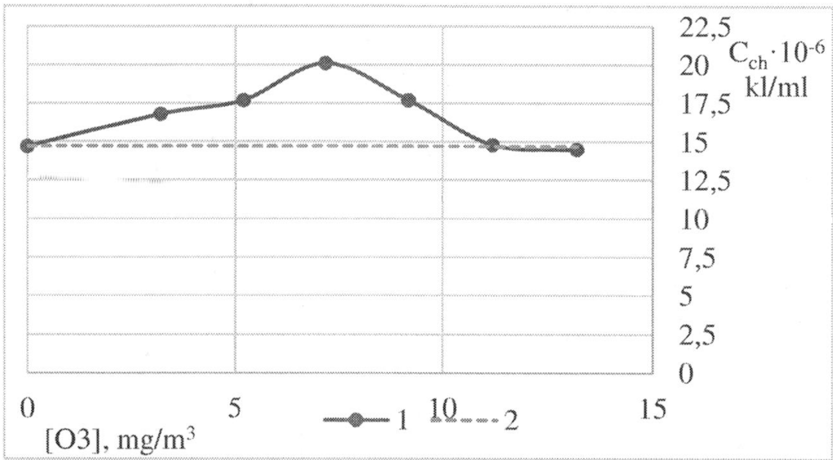

Figure 2. Data of the experiment on ozone effective concentration search, where 1 – cells treated by ozone-air mixture, 2 – untreated cells.

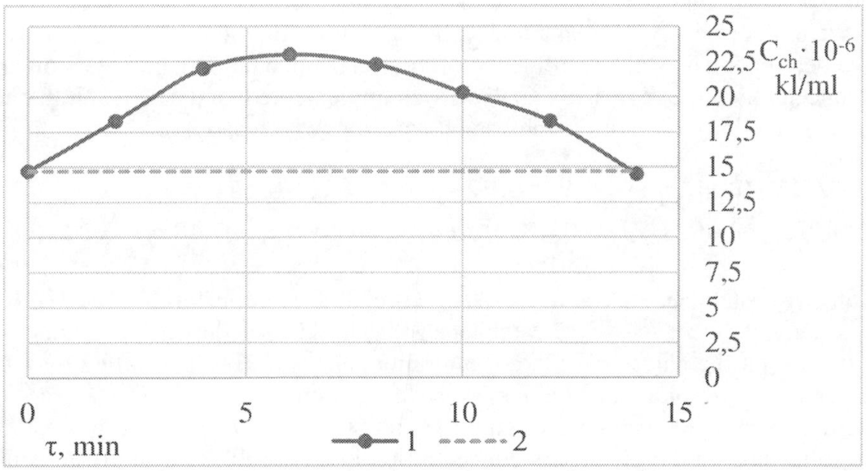

Figure 3. Dependence of chlorella cells concentration in suspension on time of treatment by ozone-air mixture, where 1 – cells treated by ozone-air mixture, 2 – untreated cells.

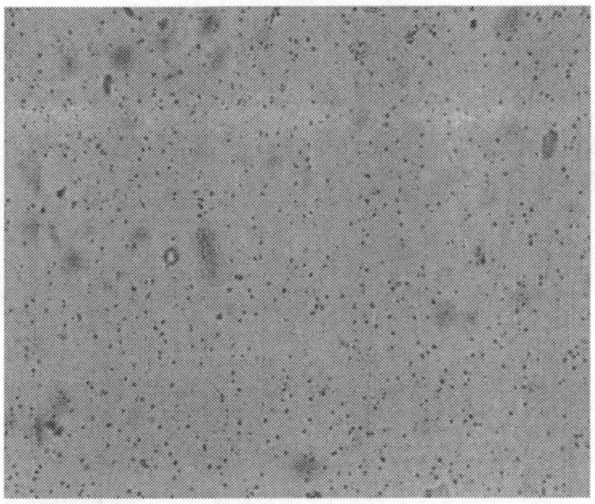

Figure 4. Control sample of chlorella suspension.

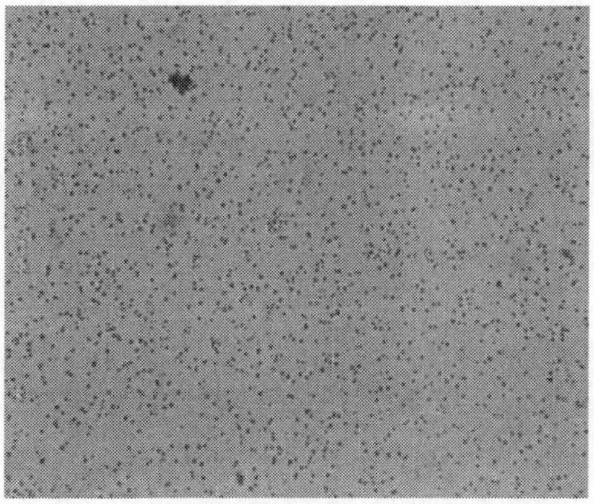

Figure5. Chlorella suspension after electro-ozone treatment with ozone concentration 7.2 mg/m^3 for 6 minutes.

Full factorial experiment was conducted, in order to determine ozone influence effect on chlorella microphytic alga cells more accurately [3, 5, 9, 14, 15]. This experiment was carried out with the help of mathematical statistic methods, the following is taken as an independent variable:

$x_1 - \tau$ – treatment time of chlorella suspension by ozone-air mixture (4 levels: 4; 6; 8; 10 minutes). Level values were taken in the extremum region, obtained from the single-factor searching experiment.

$x_2 - [O_3]$ – ozone concentration in ozone-air mixture, which is being delivered into the tank with chlorella suspension (4 levels: 3.2; 5.2; 7.2; 9.2 mg/m^3). Level values were taken in the extremum region, obtained from the single-factor searching experiment.

$x_3 - T$ – periodicity of chlorella cells suspension treatment by ozone-air mixture (4 levels: 1; 2; 3; 4 times in four days). Level values were taken with consideration of feed and fission cell bio cycles, and also specifics of chlorella microphytic alga cultivation.

During laboratory experiment as a dependent variable was taken y_1 – Cchl – chlorella cells concentration in one milliliter of undiluted suspension, c/ml.

Experiment planning matrix was compiled to carry out the experiment. Obtained from the experiment results were processed by mathematical statistic methods in «Statistika 6.0» software.

In order to determine an extent of influence of the electro-ozonation parameters on chlorella microphytic alga concentration the regression analysis was implemented, and as a result mathematical model was obtained in the form of multiple regression equation, which can be written as follows:

$$y_1 = 17,46 - 0,042x_1 + 0,48x_2 - 0,22x_3 + 0,38x_1x_2 - 0,55x_1x_3 + 0,3x_2x_3 - 0,086x_1x_2x_3 - 0,042x_1^2 + 0,48x_2^2 - 0,22x_3^2 - 0,042x_1^3 + 0,48x_2^3 - 0,22x_3^3 \quad (1)$$

As a result, was established close (R = 0.92) correlation relationship between cells concentration and factors being examined. Also was established that in 85% of the cases, factors, included in the equation, influence on the chlorella cells concentration, and in 15% - controlled by the other factors, which weren`t considered for a mathematical model construction.

Shift determines predicted value y_1, when all variables x equal 0 (in our case a = 17.46), and interpreted in the following way: typical cells concentration in one milliliter of untreated chlorella suspension solution is 17.46 million c/ml.

High significance level for the measure of cells concentration has the regression coefficient at x_2-variable (x_2 = 0.48, p = 0.0001). It shows that, all other things being equal, cells concentration is up by 0.48% due to treatment with certain ozone concentration. Also, substantial influence (0.3% and 0.38 respectively) on the cell concentration increase is done by the combination of factors, such as: x_2 x_3 (x_2x_3 = 0.3, p = 0.015) and x_1 x_2 (x_2 x_3 = 0.38, p = 0.07).

3. Results

Based on the results of the experiments, diagrams were constructed (Figure 6), which reflect influence of the factors being examined (treatment time x_1, ozone concentration x_2, number of treatments in four days x_3) on the independent variable y_1 (chlorella cells concentration in suspension).

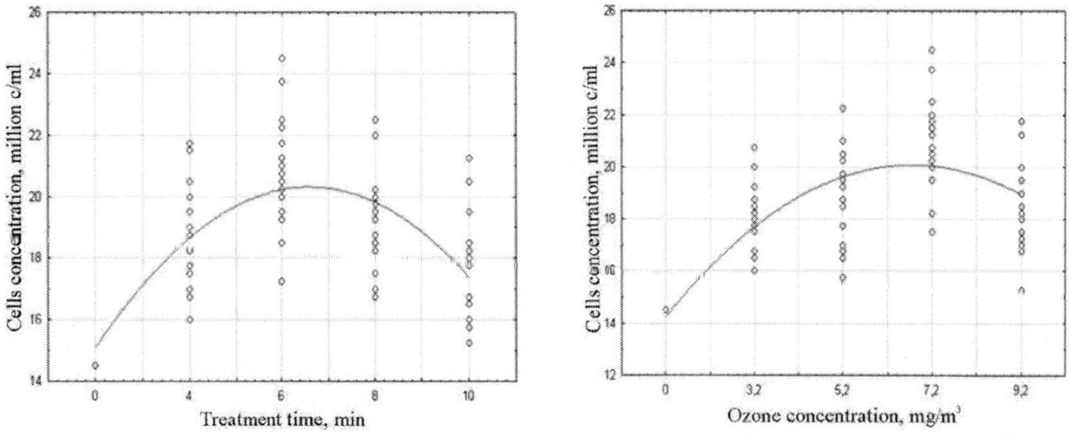

Figure 6. Diagram of influence of number of treatments in four days on the chlorella cells concentration in suspension

Analysis of Figure 6 leads to the conclusion that during treatment of chlorella cells suspension by ozone-air mixture, increase of chlorella cells concentration is observed up to ozone concentration of 7.2 mg/m³ and treatment time 6 minutes. Further increase in ozone concentration and treatment time results in a smooth decrease in chlorella cells concentration, which is, probably, due to cells destruction from strong ozone oxidizing properties.

IOP Conf. Series: Materials Science and Engineering **791** (2020) 012068 doi:10.1088/1757-899X/791/1/012068

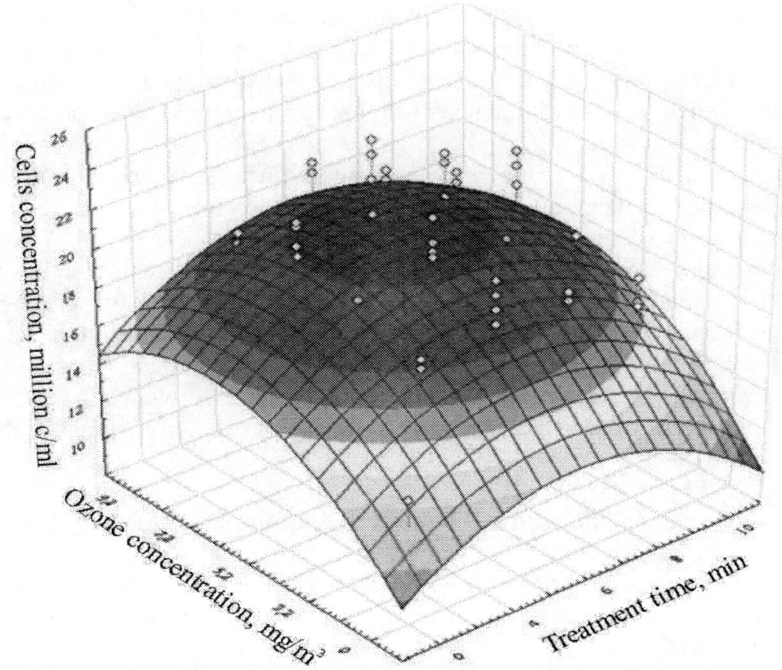

Figure 7. Diagram of the treatment time x_1 and ozone concentration x_2 influence on the chlorella cells concentration in suspension y_1.

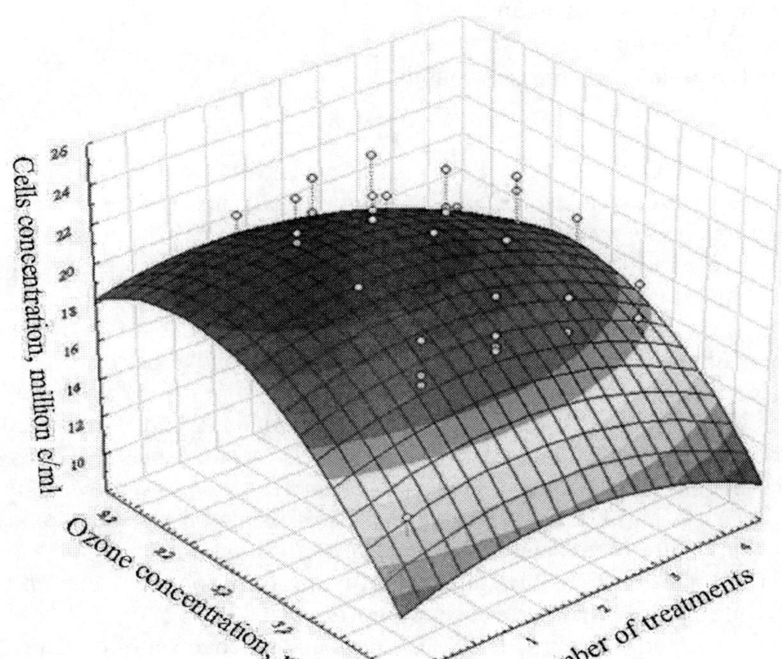

Figure 8. Diagram of the dependence of chlorella cells concentration in suspension y_1 on the ozone concentration x_2 and number of treatments x_3 in four days.

Therefore, the most favorable treatment mode is the mode with the following parameters: ozone concentration – 7.2 mg/m^3 and treatment time 6 minutes. Specified parameters of treatment correspond to chlorella cells concentration in suspension of 24.5 million c/ml, which is 10 million c/ml more, in comparison to the control sample. Using the results of a statistical analysis, multiple regression equation was obtained, which indicates influence of the factors being examined on the independent variable – chlorella cells concentration y_1.

It should be noted that during treatment of the suspension with ozone concentration of 7.2 mg/m^3 , results for chlorella cells concentration were the highest, therefore, in order to find effective ozone influence it is necessary to determine periodicity of such treatment.

For this purpose, diagrams of the dependence of chlorella cells concentration in suspension y_1 on the ozone concentration x_2 and number of treatments x_3 in four days were constructed (Figure 7 and 8).

Analysis of the obtained surfaces indicates, that with ozone concentration of 7.2 mg/m^3 increase of the chlorella cells concentration can be observed if number of treatments goes up from 1 to 2 in four days. Increase in a number of treatments gives decrease of the chlorella cells concentration in suspension. Therefore, it is worthwhile to perform a treatment with periodicity 2 times in four days.

4. Conclusion

Obtained regression equation allows to determine degree of influence of every factor on the process of chlorella microphytic alga stimulation.

Conducted regression analysis of the experimental data indicated that being examined parameters (concentration, treatment time, number of treatments) in 85% of the cases influence on the increase in chlorella cells concentration.

Based on the results of the experimental research, treatment modes of a chlorella strain suspension Chlorella vulgaris IGF S – 111 were determined, which allows to double the increase in productivity of a propagator:

- ozone concentration – 7.2 mg/m^3;
- treatment time – 6 minutes;
- number of treatments – 2 times in four days.

References

[1] Normov D A 2003 Sushil'nyye i bakteritsidnyye svoystva ozona [Drying and bactericidal properties of ozone] *Fiziko-tekhnicheskiye problemy sozdaniya novykh tekhnologiy v APK: sbornik nauchnykh trudov* [Physical and Technical Problems of Creating New Technologies in the Agro-Industrial Complex: Collection of scientific papers] (Stavropol: SSAU) 219-20 [In Russian]

[2] Mendes F, Maier D E, Mason L J and Woloshuk C P 2003 Penetration of ozone into columns of stored grains and effects on chemical composition and processing performance *J. Stored Prod. Res.* **39(1)** 33-44

[3] Tzortzakis N G, Taybi T, Antony E, Singleton I, Borland A and Barnes J 2013 Profiling shifts in protein complement in tomato fruit induced by atmospheric ozone-enrichment and/or wound-inoculation with Botrytis cinerea *Postharvest Biology and Technology* **78** 67-75

[4] Yeoh W K, Ali A and Forney C F 2014 Effects of ozone on major antioxidants and microbial populations of fresh-cut papaya *Postharvest Biology and Technology* **89** 56-8

[5] Sun J, Feng Z and Ort D R 2014 Impacts of rising tropospheric ozone on photosynthesis and metabo-lite levels on field grown soybean *Plant Sci* **226** 147-61

[6] Alexandrov Y A and Tarunin B I 1982 Kinetics and mechanism of oxidation of organoelement compounds by ozone *J. of Organometallic Chemistry* **238(2)** 125-57

[7] Da Silva L.M, De Faria L A and Boodts J F C 2001 Green processes for environmental application *Electrochemical ozone production, Pure and Applied Chemistry* 73(12) 1871-84

[8] Gottschalk C, Libra J A and Saupe A 2010 *Ozonation of Water and Waste Water: A Practical Guide to Understanding Ozone and its Applications, 2nd Edition* (Wiley-VCH)

[9] Tippe H and Mauch W 1998 Maximizing the solids reduction by ozonization treatment before anaerobic thermophilic hydrolysis of ligno-cellulosic wastes *Chemical Engineering and Technology* **21(7)** 584-8

[10] Rodrigues F H A, Santos E F, Feitosa J P A, Ricardo N M P S and Heatley F 2004 Ozonation of unstretched natural rubber film from hevea brasiliensis studies by ozone consumption *Polymer International* **53(6)** 733-9

[11] Turkay O, Inan H and Dimoglo A 2015 Experimental study of humic acid degradation and theoretical modelling of catalytic ozonation *Environmental Science and Pollution Research* **22(1)** 202-10

[12] Yulianto E, Restiwijaya M, Sasmita E, Kinandana A W and Nur M 2019 Power analysis of ozone generator for high capacity production *J. of Physics Conf. Series* **1170(1)** 012013

[13] Hadiyanto H, Digda A, Adetya N P and Evanty M 2019 Pretreatment of Herbal, Tofu and Fertilizer Waste Using UV/Ozon Technique and Its Utilization for Spirulina sp Cultivation *IOP Conf. Series: Earth and Environmental Science* **248(1)** 01207

[14] Iztayev A, Yakiyayeva M, Kulazhanov T, Toxanbayeva B and Chakanova Z 2018 Efficient mathematical models of ion-ozon cavitation treatment for long-term storage of grain legume crops *Acta Technica* **63(1B)** 23

[15] Iztayev A, Urazaliev R, Yakiyayeva M, Toxanbayeva B and Dauletkeldi Y 2018 Regress models of ion-ozon treatment without and with cavitation, describing changes of indicators for grain crops quality *Acta Technica* **63(1B)** 22

IOP Conf. Series: Materials Science and Engineering **791** (2020) 012069 doi:10.1088/1757-899X/791/1/012069

Secondary energy component in energy-saving processes of cement production

O A Panova[1], A A Goncharov[2], V M Konovalov[2], P V Besedin[1] and R A Cherkasov[3]

[1]Department of Glass and Ceramics Technology, Belgorod State Technological University named after V.G. Shukhov, Kostyukov St., 46, Belgorod, 308012, Russia
[2]Department of Cement and Composite Materials Technology, Belgorod State Technological University named after V.G. Shukhov, Kostyukov St., 46, Belgorod, 308012, Russia
[3]Department of Life Safety and Department of Building Materials Science, Products and Structures, Belgorod State Technological University named after V.G. Shukhov, Kostyukov St., 46, Belgorod, 308012, Russia

E-mail: pipp@mail.ru; stapostaaa@yandex.ru

Abstract. Problems of energy saving in the firing of raw sludge is an important factor in the long-term development of cement production. An integrated approach to the study of the properties of raw materials and additives makes it possible to find ways to save fuel and energy resources. Reducing the moisture content in the sludge by the use of plasticizing reagents is one way to solve the problem of reducing energy consumption during firing of sludge. Reducing the moisture content of raw sludge from 41% to 35.5% makes it possible to reduce energy consumption by 23.3 kilograms of equivalent fuel per ton of clinker. The most effective type and concentration of plasticizing reagent was selected. A decrease in the moisture content and an increase in flowability were studied on a model raw sludge based on kaolinite-montmorillonite clay, chalk and pyrite cinder – a coal -alkaline reagent (0.1 wt.%). The studied reagents were arranged in the following sequence according to the effectiveness of increasing the flowability of sludge: coal-alkaline reagent, Polyplast SP-1, calcium lignosulfonate (LST), Viscocrete20HE, Linamix PK.

1. Introduction

In the technology of cement production there are methods that make it possible to significantly reduce energy consumption: reducing costs by changes in the composition of the raw mix; fuel and electricity savings by changes in the composition of cement [1]. Reducing the moisture content in the sludge by the use of plasticizing reagents is one way to solve the problem of reducing energy consumption during firing of sludge. It has been repeatedly proved by scientific research that by introducing special additives into the mixture, the moisture content of the sludge is reduced. Currently, preference is given to super - and hyperplasticizers [2-7].

The introduction of plasticizing reagents into the raw material mixture makes it possible to reduce the viscosity of the sludge, increase its fluidity, and also lower the calcite decarbonization temperature ($CaCO_3$), which subsequently will shift the processes of interaction of substances in the solid phase to lower temperatures 1250–1270°C [8].

IOP Conf. Series: Materials Science and Engineering **791** (2020) 012069 doi:10.1088/1757-899X/791/1/012069

Cement industry seek to improve the thermotechnical characteristics during firing of raw cement slurry [9, 10].

Rheology methods are used to study the structure and description of the viscous flow properties of disperse systems [11, 12]. Understanding the ongoing changes in these properties makes it possible to optimize the technological process of obtaining materials with the necessary properties, at the minimum cost of energy resources.

The aim of the study was a comparative assessment of the effect of plasticizing reagents on the possibility of energy saving during firing of cement clinker.

2. Materials and Methods

Superplasticizers Polyplast SP-1 (mixture of sodium salts polimetilennaftalinsulfo acids) Linamix PK (based of polyoxyethylene derivatives of polymethacrylic acid) the hyper plasticizer Viscocrete 20HE (based on polycarboxylate ethers), coal-alkaline reagent (CAR), calcium lignosulfonate (LST) were used in the work.

The minimum possible dosages of plasticizing reagent were determined while maintaining the flowability of the sludge samples, according to the technological regulations, at a level of 54 ± 2 mm on the MKHTI TN-2 device.

The study of the structural and mechanical properties of cement raw sludge was performed on a rotational viscometer Brookfield DV-II+.

The mineralogical composition of the initial raw materials according to x-ray phase analysis performed on an ARL X'TRA diffractometer is shown in the Table 1.

Table 1. The mineralogical composition of the initial raw material components, wt. %.

Component names	Calcite	Montmorillonite	Kaolinite	The feldspars	Quartz
Chalk	96.1	–	–	–	3.4
Clay 1	–	14.1	82.2	3.7	–
Clay 2	5.4	15.9	13.7	17.0	45.6

The processes occurring during heating and cooling of the samples were studied on a synchronous thermal analysis device STA 449 F3 Jupiter – combined TG-DSC that allows measurements of mass changes and thermal effects. The DSC-TG measurement process was carried out in an atmosphere simulating an air environment with a heating rate of 10 °C/min, while the sample was heated to 1450 °C.

3. Results

Raw mix composition for preparing a raw cement slurry is, wt. %: clay 1-2.00; clay 2-14.98; iron supplementation – 2.54; chalk – 80.67. Raw cement slurry was obtained by co-grinding with water in a ball mill of initial raw materials to a residue on sieve No. 008 of 6-8 wt. %. The content of plasticizer in the raw sludge was varied from 0.1 to 1.0 wt. % on a dry matter basis.

The dependence shown in Fig. 1 shows that the required flowability of 54 ± 2 mm is achieved at a moisture content of about 41%. Further research will present the results of reducing moisture to 35.5% with the same flowability. Reducing moisture will reduce the cost of heat for evaporation of moisture from raw sludge for cement production.

The fundamental component that affects the rheo-technological properties of raw sludge is the clay component. Mineral kaolinite practically does not absorb water and does not swell in it, as a result of this raw material sludge has lower water demand. Due to the content of bentonite mineral in natural clay, the moisture demand of raw sludge increases. The adsorption capacity of a clay mineral largely depends on the nature of hydrophobic compounds; therefore, the concentration of surfactants introduced affects the sorption capacity of a clay mineral [13].

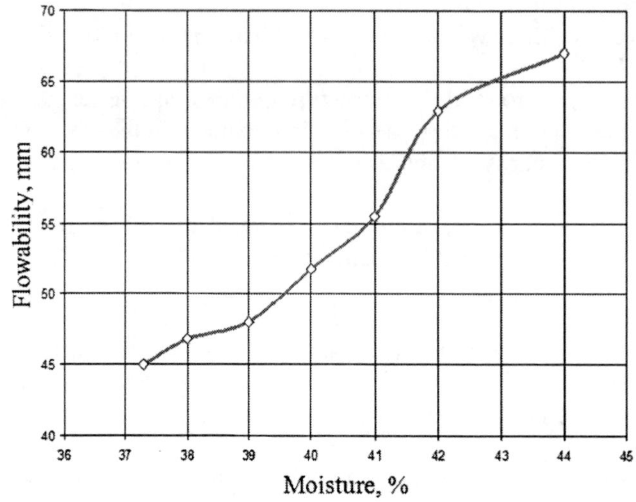

Figure 1. The dependence of moisture to raw cement slurry on flowability.

The results of the studies, which are shown in Figures 2-4, show that there is no clear advantage between superplasticizers and plasticizers. The introduction of 0.1 wt. % LST and Polyplast SP-1 in the composition of the raw material mixture equally increase the flowability to 57.0 and 64.8 mm, respectively. The introduction of Viscocrete hyperplasticizer into the raw mix (Figure 2, curve 3) showed that this reagent exerts its plasticizing effects at a concentration of more than 0.3 wt. %. At low concentrations, clay is adsorbed with respect to the polycarboxylate superplasticizer and its own swelling with water, which leads to a decrease in sludge flow [14]. With the introduction of this reagent, foaming was observed, which occurs due to the air-entraining action on the sludge suspension. This factor is a disadvantage of this plasticizer.

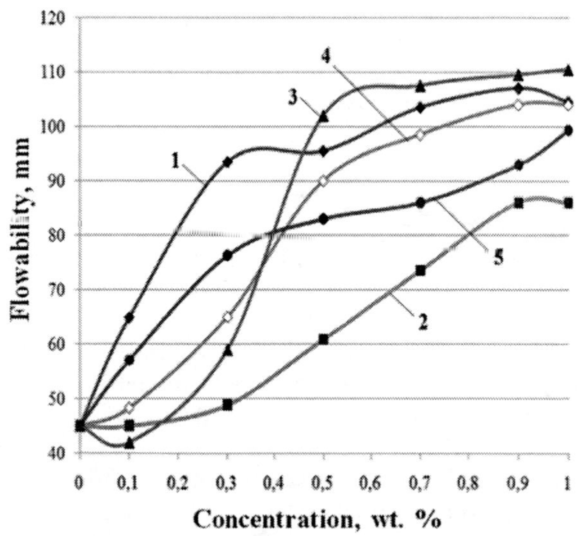

Figure 2. Influence of reagents on flowability of the model slurry: 1 – Polyplast SP-1; 2 – Linamix PK; 3 – Viscocrete; 4 – CAR; 5 – LST.

A decrease in the viscosity of the suspension occurs due to the fact [15] that the surface of the solid phase of the suspension becomes rough, and the adsorbed layer of reagent molecules smoothes it, reducing the coefficient of friction between the particles, and an electric charge of the same name is cre-

IOP Conf. Series: Materials Science and Engineering **791** (2020) 012069 doi:10.1088/1757-899X/791/1/012069

ated on the surface of the particles of the solid phase, which subsequently due to electrostatic forces excludes the possibility of their adhesion.

Coal-alkaline reagent, having an alkali in its composition, peptizes clay particles, thereby contributing to an increase in flowability and a decrease in the plastic viscosity of the slurry suspension (Figures 2, 3, curve 4).

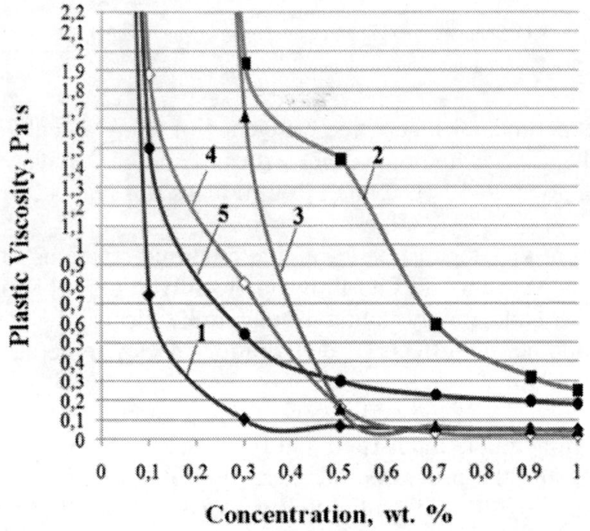

Figure 3. Influence of reagents on plastic viscosity of the model slurry: 1 – Polyplast SP-1; 2 – Linamix PK; 3 – Viscocrete; 4 – CAR; 5 – LST.

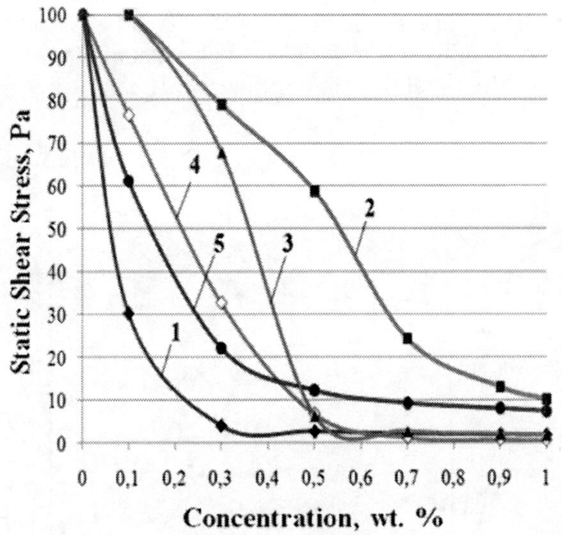

Figure 4. Influence of reagents on static shear stress of the model slurry: 1 – Polyplast SP-1; 2 – Linamix PK; 3 – Viscocrete; 4 – CAR; 5 – LST.

At a minimum concentration (0.1 wt.%), the LST reagent changes the potential of the particles to the negative region and dispersion of the suspension to the smallest particles begins, which prevents the adhesion of individual particles, slows down the coagulation of neoplasms, thereby leading to the release of water. So the mechanism of dilution of the raw mix is manifested. In this regard, the required flowability (Figures 2-4, curve 5) is achieved with less water consumption.

Based on the experimental data, the optimal concentrations of the studied plasticizing reagents were determined (Table 2).

IOP Conf. Series: Materials Science and Engineering **791** (2020) 012069 doi:10.1088/1757-899X/791/1/012069

Table 2. Optimal concentrations of plasticizing reagents for the model raw slurry.

Name of the additive	The optimum concentration of additive, wt. %	Flowability, mm
Without addition	0	45.0
Polyplast SP-1	0.1	64.8
LST	0.1	57.0
CAR	0.3	65.0
Viscocrete20HE	0.3	59.0
Linamix-PK	0.5	60.8

4. Discussion

Works [9, 16] confirmed the theoretical and experimental fact of the efficiency of using a coal-alkali additive, not only as a reagent that improves the rheo-technological properties of sludge, but also as a source of additional heat in the temperature range 200-450 °C, which in turn leads to savings thermal energy.

Analysis of changes in energy consumption at the stage of endo- and exothermic reactions of the raw mixture with the introduction of plasticizing reagents into it, allows us to solve the problems of reducing the moisture content of raw cement slurry and energy saving during clinker firing.

On the thermogram (Figure 5) there are no endothermic effects of dehydration of clay minerals, the endothermic effect of decarbonization occurs at 825.8°C.

For example, when added to raw sludge 0.9 wt. % of the additive containing carbon and sodium hydroxide, the process of dehydration of clay minerals is intensified and the process of decarbonization (803.8°C) is shifted to lower temperatures and the processes of interaction of substances in the solid phase are accelerated. This fact is confirmed by the differential thermal analysis data (Figure 6). When heating samples containing a plasticizer, there is a tendency for the peak of the decarbonization temperature to shift by 22°C.

A decrease in sludge moisture by 1% increases the kiln productivity by 1.5% and gives a saving of 2-2.5% fuel or 5.3 kg of equivalent fuel per ton of clinker, the specific fuel consumption during the wet production method is reached 170 kg of equivalent fuel per ton of clinker [17].

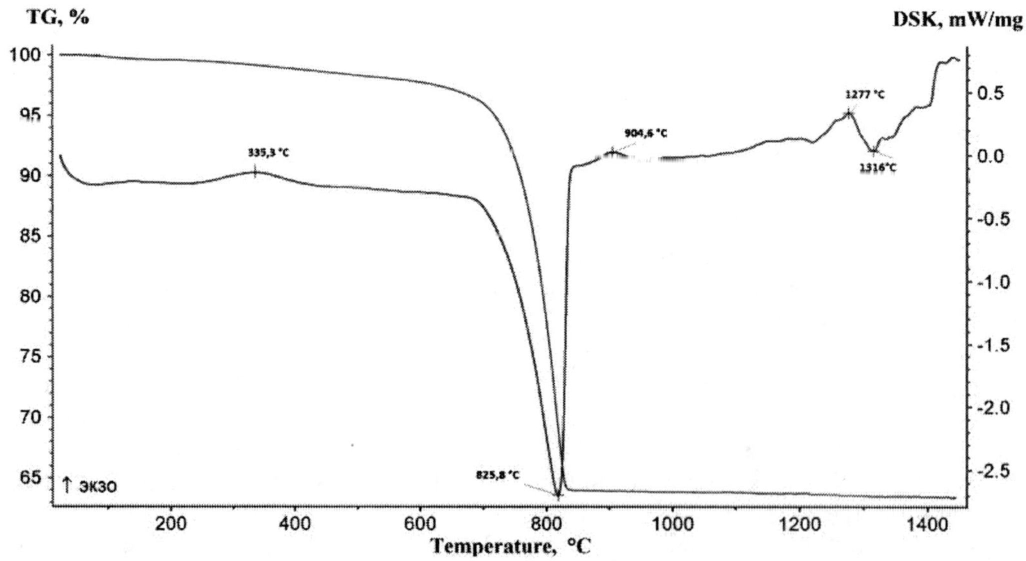

Figure 5. Thermogram of the raw material mixture without the introduction of additives.

IOP Conf. Series: Materials Science and Engineering **791** (2020) 012069 doi:10.1088/1757-899X/791/1/012069

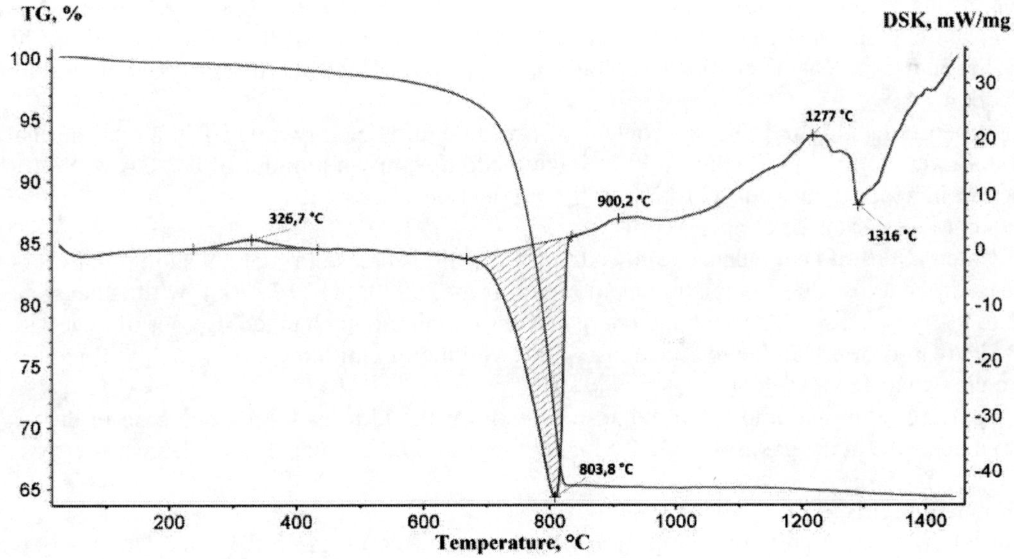

Figure 6. Thermogram of the raw mix with the addition of 0.9 wt. % coal-alkaline reagent.

According to the basic rules for compiling the heat balance [18], we calculated the characteristics of the rotary kiln by the average parameters of the kiln, as well as by the parameters obtained after reducing the moisture content of the raw cement slurry, while maintaining the temperature of the clinker at the outlet of the kiln (Table 3).

Table 3. The heat balance of cement rotary kiln.

Coming of heat	Amount kJ/kg clinker		Consumption of heat	Amount kJ/kg clinker	
Moisture, %	41	35.5	Moisture, %	41	35.5
1. From fuel combustion	6500	5818	1.Thermal Effect of clinker	1841	
2. With raw mix	62	53	2. On evaporation of water	2687	2129
3. Secondary air	1070	997	3. Losses of clinker	1296	
4. With dust return	37	34	4. Losses with the off-gas	1151	998
			5. Losses with the dust off	79	73
			6.Losses to the environment through the enclosure	623	575
Sum total	7668	6902	Sum total	7678	6912

The following technological characteristics were used in the calculations: coefficient of excess air – 1.1; the temperature of the exhaust gases is 210 °C with a change in the kiln productivity from 70 t/h when firing sludge with a humidity of 41% and 75.8 t/h with a moisture content of sludge of 35.5%. According to the methodology [18], the calculation of losses to the environment was performed. From the heat balance equations, the specific fuel consumption for firing 1 kg of clinker (Xt) was determined. With a raw material cement slurry moisture content of 41%, Xt amounted to 221.8 kg of equivalent fuel per ton of clinker; for a sludge with a moisture content of 35.5% Xt, it was 198.5 kg of equivalent fuel per ton of clinker. The obtained values were substituted into the input and account items of the thermal balances of a rotary kiln. The results are presented in table 3.

5. Conclusion

1. The results presented in this study are valid for the proposed model raw sludge and specific additives used in the experimental methodology. Other combinations of initial raw materials and additives will have different final characteristics (rheological, technological and thermal) of raw sludge for cement production.

2. When reducing the moisture content of raw sludge by 5.5% while maintaining normal fluidity 54±2 mm, due to the introduction of plasticizing additives in an amount of 0.1-0.4 wt. %, provides a reduction in specific fuel consumption in the production of cement by the wet method by 23.3 kg of equivalent fuel per ton of clinker.

3. Calculations of heat balances showed that the main decrease in fuel consumption occurs due to a decrease in heat loss due to moisture evaporation from 2687 to 2129 kJ / kg, with exhaust gases from 1151 to 998 kJ / kg, with the fuel and energy sector remaining unchanged. Losses of heat with exhaust gases decreased by 13%, due to a decrease in the volume of emitted gases, which will reduce the load on smoke exhausters.

4. A decrease in the amount of exhaust gases from the kiln leads to a decrease in dust extraction due to a decrease in the gas flow rate. Reducing the load on the exhaust fan will save energy.

Acknowledgments

The article was prepared within development program of the Flagship Regional University on the basis of Belgorod State Technological University named after V.G. Shukhov.

References

[1] Zdorov A I and Zlatkovsky A B 2007 Sovremenny`e napravleniya e`konomii e`nergoresursov v cementnoj promy`shlennosti Ukrainy [Modern directions of energy saving in the cement industry of Ukraine] *Sbornik «Stroitel`ny`e materialy` i izdeliya* URL http://sbcmi.ru/sovremennye-napravleniya-ekonomii-energoresursov-v-tsementnoj-promyshlennosti-ukrainy/ [In Russian]

[2] Palacios M, Puertas F, Bowen P and Houst Y F 2009 Effect of PCs superplasticizers on the rheological properties and hydration process of slag-blended cement pastes *J. of Materials Science* **44(10)** 2714-23

[3] Chen B 2015 Effects of an AMPS-Modifi ed Polyacrylic Acid Superplasticizer on the Performance of Cement-based Materials *J. of Wuhan University of Technology-Mater Sci. Ed* **3** 109-16

[4] Besedin P V, Trubaev P A, Panova O A and Grishko B M 2011 Nekotoryye napravleniya energosberezheniya v proizvodstve tsementa [Some directions of energy saving in cement production] *Tsement i yego primeneniye* **2** 130-4 [In Russian]

[5] Trubaev P A 2006 Exergy analysis of thermal processes in the building materials industry *Theoretical Foundations of Chemical Engineering* **40(2)** 175-82

[6] Trubaev P A and Besedin P V 2005 Criteria for the thermodynamic efficiency of cement clinker production from natural raw material *Theoretical Foundations of Chemical Engineering* **39(6)** 628-34

[7] Trubaev P A 2006 Exergy analysis of thermal processes in the building materials industry *Theoretical Foundations of Chemical Engineering* **40(2)** 175-82

[8] Bushueva N P, Panova O A and Besedin P V 2017 Intensification of synthesis processes of klinker minerals in presence of coal-alkaline additive *Advances in Eng. Research Proc. of the Int. Conf. «Actual Issues of Mechanical Engineering»* vol 133 pp 579-84

[9] Il'ina T N 2008 Snizhenie py`leunosa iz vrashhayushhejsya cementnoj pechi [Reduction in dust carry-over from a rotary cement kiln] *Ximicheskoe i Neftegazovoe Mashinostroenie* **10** 36-40 [In Russian]

[10] Dunuweera S P and Rajapakse R M G 2018 Cement Types, Composition, Uses and Advantages of Nanocement, Environmental Impact on Cement Production, and Possible Solutions

IOP Conf. Series: Materials Science and Engineering **791** (2020) 012069 doi:10.1088/1757-899X/791/1/012069

Advances in Materials Science and Engineering **2018** 4158682

[11] Monicard R 1982 *Drilling Mud and Cement Slurry Rheology Manual* (Paris: Springer-Science+Business Media, BV)

[12] Shahriar A and Nehdi M L 2012 Optimization of rheological properties of oil well cement slurries using experimental design *Materials and Structures* **45(9)** 1403-23

[13] Nafees M, Waseem A and Khan A R 2013 Comparative Study of Laterite and Bentonite Based Organoclays: Implications of Hydrophobic Compounds Remediation from Aqueous Solutions *The Scientific World Journal* **2013** 681769

[14] Wang Y, Zhang H, Zhang Z and Wang J 2015 Development of an Evaluating Method for Carbon Emissions of Manufacturing Process Plans *Discrete Dynamics in Nature and Society* **2015** 784751

[15] Winhab E J 2000 Fluid immobilization – a structure-related key mechanism for the viscous flow behaviour of concentrated suspension system *Applied Rheology* **1002(10)** 134-44

[16] Bushueva N P and Panova O A 2016 Energeticheskoye vozdeystviye ugol'no-shchelochnykh dobavok na obezvozhivaniye glinistykh mineralov [Energy impacts of coal-alkaline additives on the dehydration of clay minerals] *Vestnik BGTU im. V G Shukhov* [*Bulletin of BSTU named after V G Shukhov*] **8** 180-4 [In Russian]

[17] Schorcht F, Kourti I, Scalet B M, Roudier S and Delgado S L 2010 Best Available Techniques (BAT) Reference Document for the Production of Cement, Lime and Magnesium Oxide (Luxembourg: Publications Office of the European Union)

[18] Farag L M 2017 Assessment of the Application of a Ca-Looping Cycle for CO_2 Capture in an Egyptian Cement Kiln Plant *Building materials* **66(5)** 157-65

IOP Conf. Series: Materials Science and Engineering **791** (2020) 012070 doi:10.1088/1757-899X/791/1/012070

On the potential use of basalt waste as mineral fillers

V B Petropavlovskaya[1], T B Novichenkova[1], M Y Zavadko[1] and K S Petropavlovskii[1, 2]

[1]Department of building products and constructions, Tver State Technical University, 22 Af. Nikitin, Tver, 170026, Russia
[2]Research Center, Moscow State University of Civil Engineering (MGSU), 26 Yaroslavskoe chaussee, Moscow, Russia

E-mail: victoriapetrop@gmail.com

Abstract. Producing present-day constructional materials on mineral binders is directly connected to the use of micro- and ultra-disperse technogenic modifying agents. They allow to obtain the maximum particle-particle packing in composition, provide rational raw mix design, contribute to structure formation process. The present article is devoted to the investigation of searching the most rational solution for the problem of basalt waste utilization, with the basalt waste being the valuable secondary raw material resource. The studies have shown that it is possible to use basalt dust as admix to gypsum composites. The basalt waste, by changing the composition grain-size distribution, provides the better particle-particle packing both for fine-grained and coarse waste fractions, thus having positive effect on modified gypsum rock structure. Optimization of internal structure also contributes to the increase of gypsum rock strength properties. The obtained gypsum composite can be profitably employed for manufacturing small-piece walling gypsum materials by casting process.

1. Introduction

Producing present-day constructional materials and concretes based on mineral binders is directly connected to the use of micro- and ultra-disperse modifying agents. They allow to obtain the maximum particle-particle packing in composition [1, 2, 3], provide rational raw mix design, contribute to structure formation process, and they are also able to perform other functions [4-7]. Such modifying agents can be obtained from natural [7] and technogenic sources [8-10], with the high priority of using technogenic raw materials as such admixes. This is due to the availability of such resources, their low cost, as well as to the ecological aspects of their use. Development of resource-saving technologies allows to preserve the natural environment for the future generations.

Today the resource-saving technologies mostly mean the technologies providing the minimal use of primary natural resources during all the production stages. But they should also provide the processing of waste into secondary raw material resources, first of all the large amounts of waste generated at production units. It is the waste processing that is not always properly implemented at production units. That is why it cannot be rated as resource-saving, even with the availability of studied and cost-effective production technology.

The process of basalt fiber production was conceived and improved in 1961, in the Soviet Union. At present it might be classified as insufficiently power-intensive, as during the last 20 years the special low-power consumption units for melting and stretch-forming of basalt fiber have been developed and implemented [1]. However, even this process of basalt fiber production still cannot be called a resource-saving one. Presently up to 15% of waste from the total amount of basalt wool and fiver production is land-buried.

At the same time, it is exactly the basal fiber consumption in our country that has been increasing with every year. The reason is its obvious advantage over other fiber variants in physical-mechanical, performance, economical and other factors.

The areas of waste burial sites have been accordingly increasing with every year as well. The large allocation of land for waste burial sites cannot help affecting the whole ecology situation [2, 3]. Such waste materials, as well as fiber itself, are fire-resistant, chemically stable and safe for humans. They are suitable for realization as secondary raw material resources.

It is also necessary to take into account the world tendency in waste management, that consists of partially replacing construction material components by waste materials [4].

If we follow this tendency, it will be possible to make the process of basalt fiber production a resource-saving one to the full extent.

Basalt fiber itself as reinforcement material has proved itself to be good and it is widely used in various industries: machine-building, car manufacturing, building and construction, an even in sport equipment manufacturing.

Basalt fiber production is accomplished in the following stages: fragmentation of basalt, melting of basalt chips in melting furnaces, and continuous fiber drawing. Later on, fiber can be used as finished product or can be woven into fabric, mats and nets. Fiber and fiber-based materials are lightweight, durable, chemically resistant to aggressive environments. The production costs are less for basalt fiber than for alum and carbon-fiber. The use of basalt materials provides environmental and fire safety for materials and environment. Thus, they can be used, without restrictions, in all the production and consumption spheres [1].

Many authors have studied the potential of using various rocks, their fragmentation waste products as fillers in construction mixes, and also as raw materials for mineral binders, including gypsum binders [13, 14].

Gypsum is widely used as a binder. He has several superior properties, including; low density, good heat and sound insulation, volumetric stability, thermal and fire resistance, easy fabrication, manufacturability, low price and attractive appearance [15].

However, despite these superior properties of gypsum, so is concern regarding its strength, setting time, water absorption and strength. To tackle such inferior properties, attempts have been tried by numerous research work, in which the focus has primarily been on modifications of gypsum matrix using different mineral admixtures and reinforcing materials – to the creation of composite binders [15].

The addition of different dispersed minerals admixtures (e.g. clay minerals, lime, silica fume, microcalcite, pounded glass, Portland cement, cork) can assist in modifying the chemical and physical properties of gypsum, resulting in positive influences on its fresh and hardened properties [15].

The authors of article [8] view the particulars of using technogenic raw materials in composite binders on the gypsum. Article [9] deals with researching heat insulating materials with the similar waste used as components. It has been established that mineral modifiers make it possible, above all, to increase such quality characteristics of the resulting mixes and concretes as strength and density [10-14], and the use of waste in them allows to achieve economic effect as well [16-19]. Thus, at present there are two guidelines for utilization of dust-like mineral waste: to be used as a filler or as a binder (its component), subject to its activity excitation [17-18, 20-24].

The present article is devoted to the investigation of searching the most rational solution for the problem of basalt waste utilization, with the basalt waste being the valuable secondary raw material resource.

2. Materials and Methods

The article deals with the α-modification gypsum binder of G-16 grade by Samara gypsum plant of Samara region.

The α-modification gypsum binder of G-16 grade by Samara gypsum plant is characterized by low water demand – 35-40 %, initial setting time of 4-5 minutes and final setting time of 20 minutes.

Grain-size composition is specified by GOST 125 by residue on the sieve of 0,2 mm – 0,5 %. Absorption of water by GOST comprises 18 %. Volumetric expansion does not exceed 0,25 %.

Compressive strength is 16 MPa, bending tensile strength is 6-7 MPa.

Metal impurity content is maximum 10 mg per 1 kg of gypsum.

Size distribution index of dust removal waste of LLC "Paroc" basalt:

- Mean particle size of dust-like waste d50 – 24.866 micron;
- Maximum particle size of waste particles d99 – 111.743 micron;
- Particles less than 2 micron – 9.66 % by mass.

Basalt dust is represented by minerals of plagioclase group, in particular by quartz, calcite, anorthite. Dust-like waste mass is glassy and laced with many pyroxene and plagioclase crystals (Figure 2).

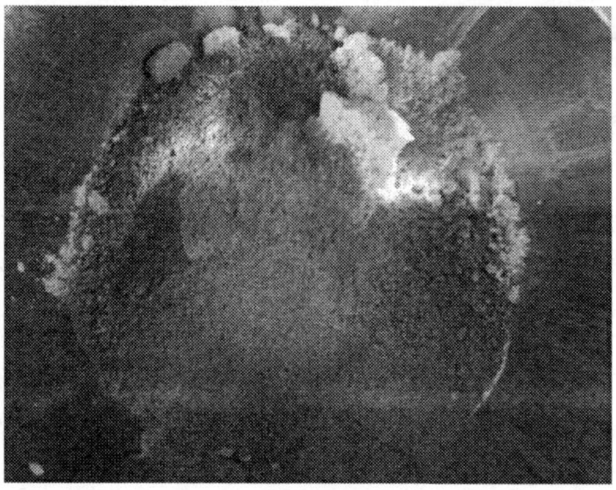

Figure 1. Basalt dust –basalt fiber waste of LLC "Paroc" production unit.

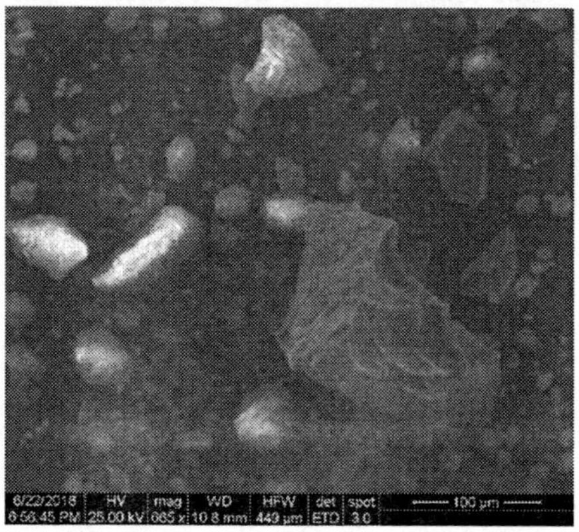

Figure 2. Microstructural photographs of basalt dust of LLC "Paroc".

The studies of basic physical mechanical properties of gypsum composites were conducted in accordance with the requirements of standards GOST 23789, GOST 8462, GOST 7025.

Gypsum paste and gypsum paste-based mixes were prepared by mixing gypsum binder and basalt fiber waste (Figure 3) in the dry state by hand mixer, and then by tempering dry uniform mix with water.

Figure 3. Basalt dust. Appearance of fracture.

For determining the normal consistency of gypsum paste, a clean container was filled with the required amount of tempering water, then the weight of gypsum binder dry powder (or binder with additives) in the amount of 300g was added to this water. The paste was mixed by hand mixer until becoming uniform and then was poured into Suttard form. After 15 seconds from the tempering moment the viscosimeter rose upward vertically. If the flow diameter of the formed gypsum cake comprised (180 ± 5) mm, then the value of water-solid ratio was taken for normal consistency, otherwise the testing continued by changing water content until obtaining the required flow diameter of gypsum paste.

For determining compressive strength limit and bending tensile strength limit for modified gypsum rock, standard test beams were used of 40x40x160 mm dimensions (Figure 4 a). Standard beam testing was conducted by standard methods in accordance with GOST requirements (Figure 4 b, c).

Basalt dust content varied from 0 to 12 % at 2 % interval. Water-gypsum ratio was taken as corresponding to the normal consistency of gypsum paste.

a b c

Figure 4. Materials and methods of testing gypsum binder modified by basalt waste.

3. Results and Discussion

The study results for modifying structure and properties of gypsum rock with basalt waste admix are presented in Figure 5-7.

The study of the basalt waste effect on bending tensile strength limit of gypsum rock is presented in Figure 5.

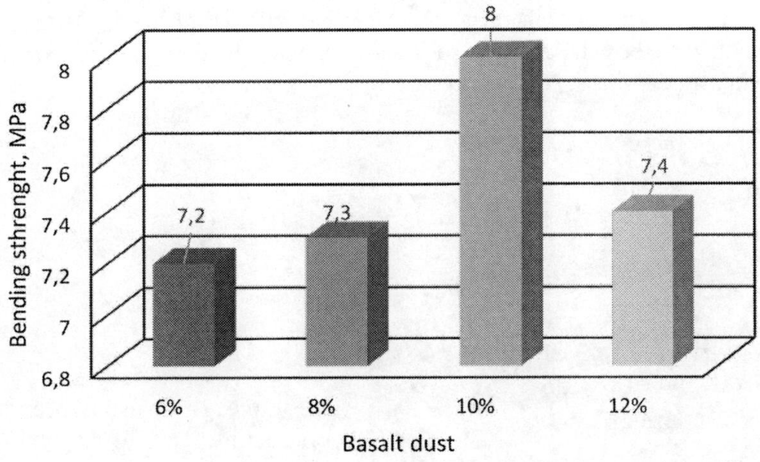

Figure 5. Effect of basalt waste content on bending tensile strength limit.

IOP Conf. Series: Materials Science and Engineering **791** (2020) 012070 doi:10.1088/1757-899X/791/1/012070

The resulting data on the evaluation of waste effect on compressive strength limit of modified gypsum rock is presented in Figure 6.

The microstructure of gypsum rock modified by waste is presented in Figure 7. The sample with basalt waste content of 10% by gypsum binder mass had been selected for studying the modified gypsum microstructure.

Inclusion of basalt dust into gypsum binder composition did not considerably affect the bending tensile strength of the obtained rock. The maximum value of bending tensile strength limit is achieved with the admix content of 10% and comprises 7,9 MPa. It is only by 7 % on the average higher than the tensile strength of the control samples (Figure 5), which, on the average, was in the order of 7 MPa.

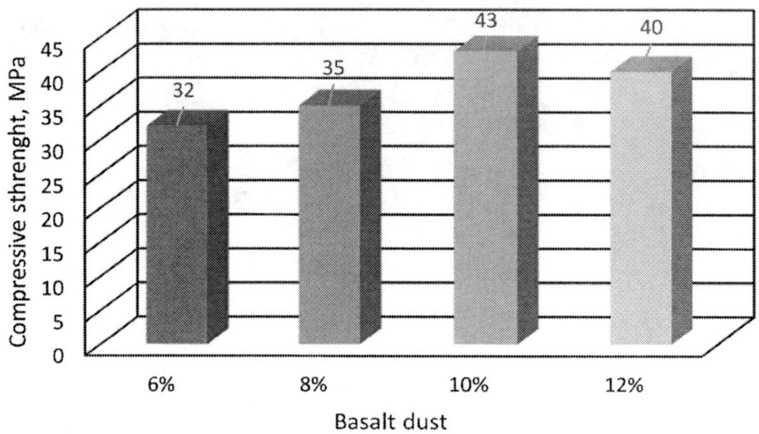

Figure 6. Effect of basalt waste on compressive strength.

To a greater extent the basalt waste affects compressive strength limit. The compressive strength of the samples without admixing basalt dust comprised 30 MPa (Figure 6). The maximum compressive strength was demonstrated by the gypsum rock samples with the composition containing 10% of basalt duct by the gypsum binder mass.

Figure 7. Gypsum rock structure modified by basalt waste.

4. Conclusion

In summary, the studies have shown that it is possible to use basalt dust as admix to gypsum composites. In the context of using this dust for gypsum rock modification, the most efficient use will be with the basalt dust content of 10% by gypsum binder mass. This contributes, first of all, to the increase of

IOP Conf. Series: Materials Science and Engineering **791** (2020) 012070 doi:10.1088/1757-899X/791/1/012070

compressive strength properties of gypsum rock. Evidently it could be explained by the fact that basalt waste, by changing the composition grain-size distribution, provides the better particle-particle packing both for fine-grained and coarse waste fractions, thus having positive effect on modified gypsum rock structure. Optimization of internal structure also contributes to the increase of gypsum rock strength properties. The obtained gypsum composite can be profitably employed for manufacturing small-piece walling gypsum materials by casting process.

References

[1] Buryanov A F, Petropavlovskaya V B, Novichenkova T B and Petropavlovskii K S 2017 Simulating the structure of gypsum composites using pulverized basalt waste *MATEC Web Conf.* Vol 117 00026

[2] Petropavlovskaya V B, Novichenkova T B, Buryanov A F, Solov'ev V N and Petropavlovskii K S 2017 Utilizatsiya otkhodov mineral'nogo volokna v proizvodstve gipsovykh izdeliy [Mineral fiber recycling in the production of gypsum products] *Vestnik MGSU* **12(111)** 1392-8

[3] Sevostyanov V S, Shatalov A V, Sevostyanov M V, Krutov V S and Shatalov V A 2018 Technics and Technology of Processing Basalt Waste *J. of Physics: Conf. Series* **1066(1)** 012027

[4] Thiago Melanda Mendes, Leonardo Guerra and Gilson Morales 2016 Basalt waste added to Portland cement *Acta Scientiarum Technology* **38(4)** 431

[5] Strokova V, Zhernovsky I, Ogurtsova Y, Maksakov A, Kozhukhova M and Sobolev K 2014 Artificial aggregates based on granulated reactive silica powders *Advanced Powder Technology* **25(3)** 1076-81

[6] Sokolova Y, Ayzenshtadt A, Frolova M, Strokova V and Kobzev V 2018 Energy characteristics of finely dispersed rock systems *IOP Conf. Ser.: Mater. Sci. Eng.* **365(3)** 032036

[7] Pavlenko N V, Strokova V V, Kapusta M N and Netsvet D D 2014 About application prospectivity of rocks with different geological and morphological features as basic raw component for free-cement binder production *Applied Mechanics and Materials* **670-671** 462-5

[8] Alfimova N I, Sheychenko M S, Karatsupa S V, Yakovlev E A, Kolomatskiy A S and Shapovalov N N 2014 Features of application of high-mg technogenic raw materials as a component of composite binders *Advances in Environmental Biology* **8(13)** 134-8

[9] Zagorodnyuk L H, Koryakina A A, Sevostyanova K I and Khaheleva A A 2018 Heat Insulating Composite Mixtures with Technogenic Materials *J. of Phys. Conference Series* **1066(1)** 12011

[10] Nelyubova V V, Sivalneva M N, Bondarenko D O and Baskakov P S 2018 Study of activity of polydisperse mineral modifiers via unstandardized techniques *J. of Phys. Conference Series* **1118(1)** 012029

[11] Nelubova V V, Strokova V V, Popov A L and Bezdrodnykh A A 2018 Investigation of the structural topological characteristics of mechanically activated Sialic raw materials *Materials Science Forum* **945** 182-7

[12] Nelyubova V V, Strokova V V, Sumin A V and Jernovskiy I V 2017 The structure formation of the cellular concrete with nanostructured modifier *Key Engineering Materials* **729** 99-103

[13] Petropavlovskaya V B, Buryanov A F and Novichenkova T B 2014 Structuring in Systems on the Basis of Calcium Sulfate Dihydrate Applied Mechanics and Materials *Materials Science and Mechanical Engineering* **467** 91-6

[14] Petropavlovskaya V B, Novichenkova T B, Buryanov A F and PoleonovaY Y 2013 Application of mathematical and computer modeling methods to manufacture high-strength unfired gypsum materials *Procedia Engineering* **57** 906-13

[15] Dhaheer A, Hameed K K, Ammash M S and Ammash H K 2018 Improvement of ordinary and pure gypsum properties by using polyvinyl alcohol (PVA) *Int. J. of Civil Engineering and*

Technology (IJCIET) **9(9)** 323-34

[16] Fediuk R S, Lesovik V S, Liseitsev Y L, Timokhin R A, Bituyev A V, Zaiakhanov M Y and Mochalov A V 2019 *Magazine of Civil Engineering* **85(1)** 28-38

[17] Chernysheva N V, Lesovik V S, Drebezgova M Yu, Shatalova S V and Alaskhanov A H 2018 Composite Gypsum Binders with Silica-containing Additives *IOP Conf. Ser.: Mater. Sci. Eng.* **327(3)** 032015

[18] 17.Zagorodnyuk L K, Lesovik V S, Sumskoy D A, Elistratkin M Y and Makhortov D S 2018 Peculiarities of binding composition production in vortex jet mill *IOP Conf. Ser.: Mater. Sci. Eng.* **327(4)** 042128

[19] Kozhukhova N I, Teslya A Y, Kozhukhova M I, Zhernovsky I V, Yermak S N and Ogurtsova Y N 2019 In-service performance of hybrid geopolymer binders based class F fly ash *IOP Conf. Ser.: Mater. Sci. Eng.* **552(1)** 012035

[20] Kozhukhova N I, Fomina E V, Zhernovsky I V, Strokova V V and Chizhov R V 2014 The utilization efficiency of natural alumosilicates in composite binders *Applied Mechanics and Materials* **670-671** 182-6

[21] Petropavlovskaya V, Zavadko M, Petropavlovskii K, Buryanov A and Novichenkova T 2019 Improving the properties of filled gypsum composites by carbonization *Alternative and smart energy TPCASE 2018* (Voronezh: Voronezh state technical university) pp 218-23

[22] Buryanov A F, Petropavlovskaya V B, Novichenkova T B and Petropavlovskii K S 2018 Increase in efficiency of formation structure process of the modified gypsum by carbonation *IOP Conf. Ser.: Mater. Sci. Eng.* **365** 032061

[23] Petropavlovskaya V B, Buryanov A F, Novichenkova T B and Petropavlovskii K S 2018 Gypsum composites reinforcement *IOP Conf. Ser.: Mater. Sci. Eng.* **65** 032060

[24] Petropavlovskaya V B, Novichenkova T B, Belov D V and Barkaya A T 2019 Deformacii polzuchesti gipsovy`x kompozitov [Deformations of the creep of plaster composites] *High-tech technologies and innovations (XXIII scientific readings)* (Belgorod: BSTU im V G Shukhov) vol 4 pp 330-3

IOP Conf. Series: Materials Science and Engineering **791** (2020) 012071 doi:10.1088/1757-899X/791/1/012071

Heat transfer efficiency in PEBC synthesis

V V Pliushchev[1], O S Kharitonova[2], M A Nazarova[1], V V Bronskaya[1], G V Manuyko[1], R S Shaikhetdinova[1], G A Aminova[1] and D S Balzamov[3]

[1]Department of Technology of Engineering Materials, Kazan National Research Technological University, Karl Marx St., 68, Kazan, 420015, Russia
[2]Department of Chemical Technology of Petroleum and Gas Processing, Kazan National Research Technological University, Karl Marx St., 68, Kazan, 420015, Russia
[3]Department of Power Supply of Enterprises and Energy Resource Saving Technologies, Kazan State Power Engineering University, Krasnoselskaya St., 51, Kazan, 420066, Russia

E-mail: olga.220499@mail.ru

Abstract. The kinetics of propylene and ethylene copolymerization in a fluidized bed reactor is considered. Dynamic reactor model based on material balance of monomers, copolymer and copolymerization kinetics has been developed in order to obtain predicted data on monomer flow, copolymer concentration under different process conditions. The complex two-phase model is presented based on two models: well-mixed and two-phase model. The profiles of concentrations and temperatures in microparticles are obtained, and the average concentrations and temperatures at the active sites of the catalyst are determined. The dependences of temperature on the height of the fluidized bed of the reactor at different gas incoming velocity and at different temperature of the monomer input are obtained. Mathematical model that is basis of creating expert control complex of PEBC synthesis process and allow to make quantitative connection between technological parameters of reactors which allows to control technological mode with the acceptable accuracy is presented. Experiment catalytic system is microspheric titanium trichloride (MS-$TiCl_3$) and co-catalyst-diethylaluminum chloride.

1. Introduction

Propylene-ethylene block copolymer (PEBC) is an inexpensive, environmentally friendly product. Since the opening of new ion-coordinating catalysts, production has been rapidly developing. The demand for the production of Propylene-Ethylene Block Copolymer (PEBC) is related to the value of the properties of this material: high strength, stress cracking resistance in aggressive environments, heat resistance, frost resistance, low specific weight, ability to pass ultraviolet rays and absorb radioactive radiation , good dielectric properties. Carrying out copolymerization of polypropylene and ethylene in the gas phase on complex organometallic catalysts allows to significantly simplify and improve the technology of production of PEBC in the result of the elimination or significant reduction of solvent consumption, the elimination of polymer washing and drying operations, and solvent regeneration. The elimination of polymer washing stage in the gas-phase process is achieved by maximizing the use of catalyst.

A fluidized bed reactor is one of the most common industrial reactors for producing polyolefins and copolymers of olefins due to good mixing of particles, excellent possibilities of heat and mass transfer [1]. The analysis of literature showed the prospects of synthesis of propylene copolymers us-

IOP Conf. Series: Materials Science and Engineering **791** (2020) 012071 doi:10.1088/1757-899X/791/1/012071

ing a number of modern catalysts for obtaining both traditional and new polymer materials with a set of thermophysical and physico-chemical properties [2-8]. Consistent processes of homo- and copolymerization of olefins allow to regulate the composition, molecular mass characteristics, properties and supermolecular structure of polyolefins, to obtain modified polyolefins with a set of properties. There is a perforated grille for uniform distribution of the fed ethylene in the lower part of reactor and there is an extended zone designed to reduce gas speed and absorb the bulk of polyethylene in the upper part. The heat removal of the reaction is carried out by ethylene cooled in the circulating circuit heat exchanger. The technological scheme provides for automatic feeding of catalyst, monomer, somomer, hydrogen and automatic discharge of polymer with the use of logical control systems.

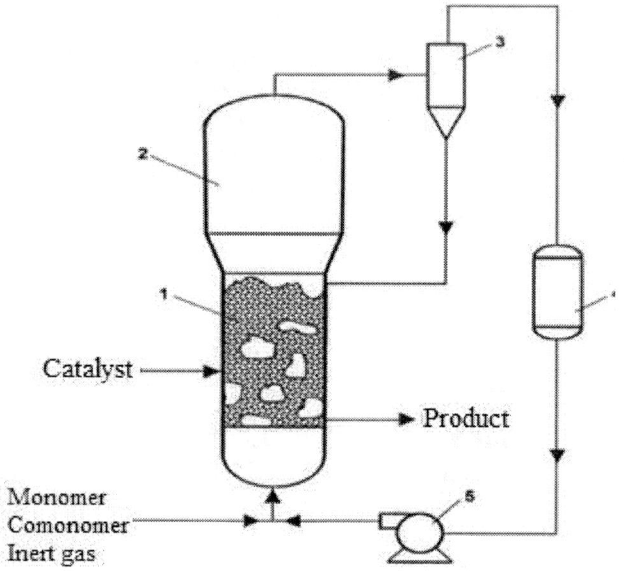

Figure 1. Diagram of ethylene gas phase polymerization in the catalyst fluidized bed: 1 - reaction zone; 2 - separation zone; 3 - cyclone; 4 - heat exchanger; 5 - compressor.

However, today there are a number of technological problems, the main of which is the heat sink. Since the polymerization reaction is highly exothermic, the localized overheating points and, as a result, an abnormality of the fluidization regime may be formed with insufficient heat dissipation. The complexity of modeling such a process consists in the need to take into account many interactions between phases. A distinctive feature of considered gas-phase polymerization is that this system does not involve the gas phase in the polymerization zone. Polymerization occurs in the boundary layer between the solid catalyst and the polymer matrix, which swells due to monomers during polymerization. The role of the gas phase is to transport monomers, mix polymer particles and remove the reaction heat. These problems cannot be solved without the apparatus of mathematical modeling [9-16].

2. Materials and Methods

The mathematical model proposed by K B McAulie, J P Taibot and T J Harris was taken for solving problems [1]. The complex two-phase model is based on two models: well-mixed and two-phase model, according to which some assumptions were introduced:

1. A fluidized bed consists of two phases (bubble phase and continuous phase);
2. The reaction occurs only in the continuous phase in the minimal fluidization regime;
3. The excess gas passes through the bubble phase layer.
4. Bubbles grow only to the maximum stable size. Because bubbles reach their maximum size at the base of the layer, they are assumed to have the same size. They pass through the layer with a constant speed, heat and mass transfer coefficients are average in the height of the layer.
5. The continuous phase acts in perfect mixing mode.

IOP Conf. Series: Materials Science and Engineering **791** (2020) 012071 doi:10.1088/1757-899X/791/1/012071

6. Radial gradients of concentration and temperature can be neglected, as the bubbles are distributed evenly over the bottom.
7. There is a slight resistance to heat and mass transfer between the gas and the solids in the continuous phase. This assumption is true when the catalyst particles are small enough and the catalyst activity is not very high.
8. Average particle size is used and particle agglomeration is not considered in the model.

It is assumed that the activation of catalyst instantaneously occurs, the speed constant value for each stage does not depend on the chain length, the polymer chain activity depends on a last monomer added to the active centre of the polymeric chain. The chain transfer reaction is assumed to occur through the active chain reaction with hydrogen. The copolymerization reaction mechanism can be presented as the following kinetic scheme. $TiCl_4/MgCl_2$ + PEEB + AlR_3 (PEEB -para-ethoxy-ethylbenzoate) is used as the catalyst.

According to the kinetic scheme the kinetic equations were offered:

$$r_{C^*} = -k_i^1 M_1 \cdot C^* - k_i^2 M_2 \cdot C^* - k_d \cdot C^* + k_{tr}^1 P^*[H_2] + k_{tr}^2 Q^*[H_2] \quad (1)$$

$$r_{M_1} = -k_i^1 C^* M_1 - k_p^{11} P^* M_1 - k_p^{21} \cdot Q^* M_1 \quad (2)$$

$$r_{M_2} = -k_i^2 C^* M_2 - k_p^{12} P^* M_2 - k_p^{22} \cdot Q^* M_2 \quad (3)$$

$$r_{P^*} = -k_i^1 C^* M_1 + k_p^{21} Q^* M_1 - k_p^{12} \cdot P^* M_2 - k_{tr}^1 P^*[H_2] - k_d^1 P^* \quad (4)$$

$$r_{Q^*} = -k_i^2 C^* M_2 + k_p^{12} Q^* M_2 - k_p^{21} \cdot Q^* M_1 - k_{tr}^2 Q^*[H_2] - k_d^1 Q^* \quad (5)$$

$$r_{H_2} = -k_{tr}^1 P^*[H_2] - k_{tr}^2 Q^*[H_2] \quad (6)$$

where $P^* = \sum_{S=1}^{\infty}\sum_{n=1}^{\infty} P_{S,n}^*,\ Q^* = \sum_{S=1}^{\infty}\sum_{n=1}^{\infty} Q_{S,n}^*.$

Dependencies of speed constants on temperature are represented by Arrhenius equation [6]:

$$k = k_0 e^{-E/RT} \quad (7)$$

Propylene, ethylene and hydrogen are fed into the fluidization reactor and the polymerization reaction is initiated:

1) The monomer concentration in the reactor is homogeneous and changes only over the copolymerization time.
2) The pressure influence in the reactor on the copolymer density is not taken into account.
3) The diffusion effect of reagents between phases is ignored. It means that the reaction system in the fluidization reactor is considered as a pseudohomogeneous phase.

The material balance equation for catalyst particles in the fluidization reactor is as follows:

$$V_0 \frac{dC^*}{dt} = G_{solid} / M_{solid} - QC^* + r_{C^*} V_0 \quad (8)$$

Similar equations can be obtained for propylene, ethylene, active polymeric chains, hydrogen and the solids copolymer phase:

$$V_0\frac{dM_1}{dt} = \frac{G_{M_1}^{input}}{M_{M_1}} - QM_1 + r_{M_1}V_0 \tag{9}$$

$$V_0\frac{dM_2}{dt} = \frac{G_{M_2}^{input}}{M_{M_2}} - QM_2 + r_{M_2}V_0 \tag{10}$$

$$V_0\frac{dP^*}{dt} = -QP^* + r_{P^*}V_0 \tag{11}$$

$$V_0\frac{dQ^*}{dt} = -QQ^* + r_{Q^*}V_0 \tag{12}$$

$$V_0\frac{d[H_2]}{dt} = \frac{G_{H_2}^{input}}{M_{H_2}} - Q[H_2] + r_{H_2}V_0 \tag{13}$$

$$m\frac{dW}{dt} = G^{input}W^{input} - \frac{m}{V_0}QW + V(r_{M_1}M_{M_1}M_1 + r_{M2}M_{M_1}M_2)V_0 \tag{14}$$

The molar volume of gas mixture in the reactor is determined by the formula (15):

$$V = \frac{1}{M_1 + M_2} \tag{15}$$

The material balance equations for a fluidized bed reactor are given by:

$$\frac{dm}{dt} = G^{input} - \frac{m}{V_0}Q \tag{16}$$

The investigated two-phase model is represented by the system of material and thermal balances (17)-(21):

$$C_{bc} = C_c + \exp^{Kbc\frac{H}{Ub}\frac{Ub}{Ub\,Kbc * H}} \tag{17}$$

$$U_c\frac{C_0 \quad C_c}{H} \quad Q\frac{C_c}{V_c\rho_p} \quad k_{p0}e^{\frac{E_a}{RT_0}} \tag{18}$$

$$C_c X_{kt}\rho_p\frac{1-\varepsilon_{mf}}{\varepsilon_{mf}} + \Delta K_{bc}\frac{C_{bc}-C_c}{(1-\Delta)}\varepsilon_{mf} = 0 \tag{19}$$

$$T_{bc} = T_c + (T_0 - T_c)\left(U_b\frac{c_{pg}}{H_{bc}H}\right) \tag{20}$$

$$AU_c(1-\Delta)C_0\varepsilon_{mf}c_{pg}(T_0-T_c)+c_{pg}k_{bc}(c_b-c_c)V_b(T_{bc}-T_c)+$$
$$+H_{bc}V_b(T_{bc}-T_c)+\Delta Hk_{p0}\exp_c^{\frac{-E}{RT_0}}C_c*X_{kt}\rho_p\frac{1-\varepsilon_{mf}}{\varepsilon_{mf}}=0 \tag{21}$$

where T_0 temperature of the incoming gas, c_{pg} heat capacity of the gas phase, ρ_p density PEBC, U_b bubble phase velocity, U_c reaction rate in a solid medium, H reactor height, K_{fc} mass transfer coefficient of boundary layer in solid phase, K_{bc} mass transfer coefficient from bubble phase to continuous phase, H_{bc} heat transfer coefficient from the bubble phase to the continuous phase, ε_{mf} void content in layer at minimum fluidization rates, T_c temperature in the continuous phase, V_c volume of the continuous phase, C_b concentration of the gaseous monomer in the bubble phase, C_c concentration of the gaseous monomer in the continuous phase.

The system is supplemented with equations for finding the average values of concentration and temperature of the monomer, as well as equations of material and heat balances for a single bubble (22)-(25):

$$\bar{C}_b=C_c+\exp^{\frac{Kbc*H}{Ub}}\frac{U_b}{K_{bc}H} \tag{22}$$

$$\bar{T}_b=T_c+(T_0-T_c)\left(\frac{U_bC_{pg}}{H_{bc}H}\right)\left(1-\exp^{\frac{-Hbc*H}{Ub*H}}\right) \tag{23}$$

$$T_b=T_c+\exp^{\frac{Hbc*H}{Ub*Cb*cpg}} \tag{24}$$

$$C_b=C_c+\exp^{\frac{Kbc*H}{Ub}} \tag{25}$$

where *Cb* concentration of gaseous monomer in the bubble phase, *Cc* concentration of gaseous monomer in the continuous phase, *Hbc* heat transfer coefficient from the bubble phase to the continuous phase, *H* reactor height, *Ub* velocity of the bubble phase, c_{pg} heat capacity of the gas phase, T_0 temperature of the incoming gas, *Tc* temperature in the continuous phase.

3. Results

Equations (1)-(15) make up a dynamic model of propylene and ethylene copolymerization reaction, by which we determine the molar volume of gas mixture, concentration of monomer and solid phase in the fluidized bed reactor. The values of kinetic constants depend not only on the catalytic system composition, technological conditions of the copolymerization process, but also on used mathematical model so it is necessary to be identified them. Calculations using identified parameters $k^1_i=5*10^{-5}$ $m^3/(mol*s)$, $k^2_i=1*10^{-4}$ $m^3/(mol*s)$, $k^{11}_p=k^{21}_p=5*10^{-5}$ $m^3/(mol*s)$, $k^{12}_p=k^{22}_p=1*10^{-4}$ $m^3/(mol*s)$, $k^1_{tr}=k^2_{tr}=1.6*10^{-4}$ $m^3/(mol*s)$, $G_{M1}^{enter}=252$ kg/h. $G_{M2}^{enter}=176$ kg/h, $G_{H2}^{enter}=45.7$ kg/h (figure 2,3) are carried out according to the presented model.

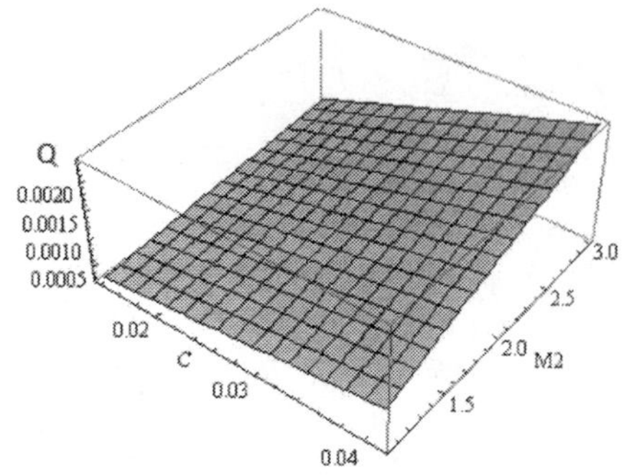

Figure 2. Dependence of active copolymer chains concentration (with the ethylene end link) on concentration of catalyst and ethylene at identified parameters.

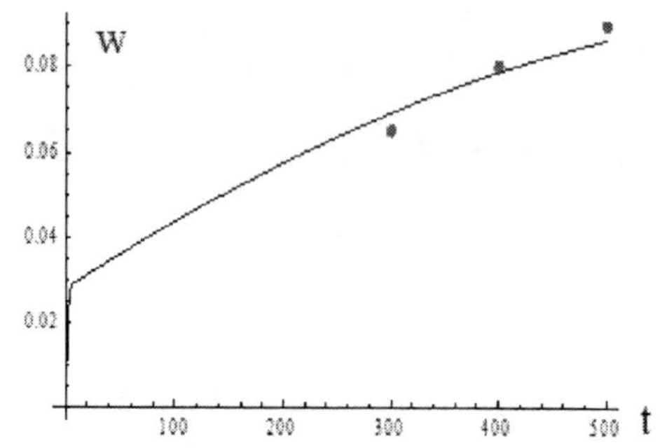

Figure 3. Dependence of a part of the solid copolymer phase on the residence time in the reactor (points — experiment, line — model calculation).

The dependences of temperature on the height of the fluidized bed of the reactor at different gas incoming velocity and at different temperature of the monomer input are obtained.

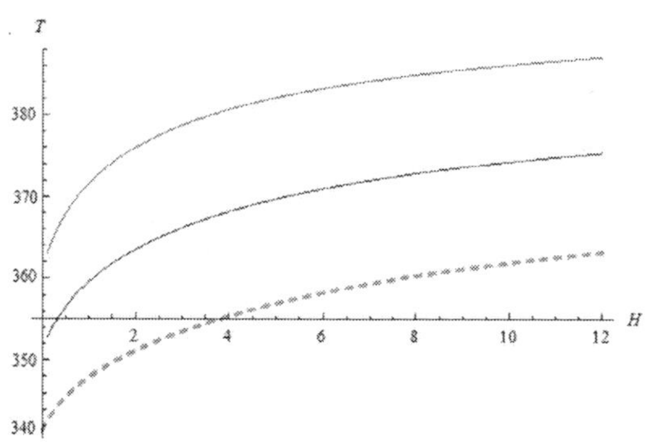

Figure 4. Dependence of the monomer temperature on the height of the continuous phase in the device at $T_0 = 340$, $T_{01} = 350$, $T_{02} = 360$, $C_0 = 20$ kg/m^3, $U_0 = 0.34$.

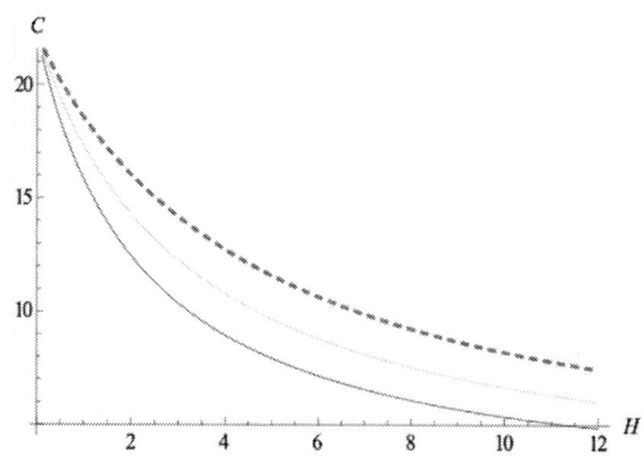

Figure 5. Dependence of monomer concentration on the height of the bubble phase in the apparatus at T_0=340, T_{01}=350, T_{02}=360, C_0=20 kg/m^3, U_0=0.34.

The system of equations of the two-phase model was solved by Runge-Kutta method. Profiles of concentrations and temperatures in microparticles were obtained, as well as the average concentrations and temperatures at the active catalyst centers were determined.

4. Discussion

There is a tendency of increase of active copolymer chains concentration with increase of concentration of monomers and catalyst (figure 2). Kinetic constants, the comparison of the experiment and calculated values of a part of the solid copolymer phase in the reactor are obtained as a result of parametric identification and presented in figure 3. It was also revealed as a result of modeling that the increase of hydrogen concentration leads to an increase of the ethylene links addition speed. A dynamic reactor model has been developed to predict the speed of consumption of monomer and hydrogen over time during propylene and ethylene copolymerization in a fluidized bed reactor. According to the results of simulation it was established that the predicted data on monomer flow, copolymer concentration and slurry density under different technological conditions are consistent with the experimental data.

5. Conclusion

The presented profiles show the nature of the change in concentration and temperature in microparticles due to monomer diffusion and chemical reaction. Monomer concentration decreases because of the monomer consumption and temperature increases due to exothermic nature of ethylene polymerization. That is because of the need to take into account many parameters when describing heterogeneous systems, the mathematical model of ethylene gas phase polymerization process will be supplemented by equations taking into account the influence of growth polymeric particles on the nature of the flow structure and changes in catalyst activity.

To obtain high-quality PEBC in the framework of the existing technological process, it is possible to ensure conditions of synthesis, containing up to 30% of the mass. elastomeric component, with a yield of at least 20% of PEBS, increasing the efficiency of the reactor heat removal system in order to reduce the maximum copolymerization temperature to an acceptable level.

Designation

C — concentration of deactivated catalyst [mol/m^3], C^* = concentration of active catalyst [mol/m^3], G_{M1}^{enter}-mass propylene flow [kg/s], G_{M2}^{enter}-mass ethylene flow [kg/s],G_{H2}^{enter} -mass hydrogen flow [kg/s], G^{enter}-mass flow of materials [kg/s], G_s-mass flow of solid particles [kg/s], $[H_2]$ -hydrogen concentration [mol/m^3], $k^1{}_i$- propylene chain initiation speed constant [m^3/ (mol^{0}s)], $k^2{}_i$ - ethylene chain initiation speed constant [m^3/ (mol^{0}s)], $k^{11}{}_P$-constant of addition speed of propylene to the chain with the propylene end link [m^3/ (mol^{0}s)], $k^{21}{}_P$-constant of addition speed of propylene to the chain with the ethylene end link [m^3/ (mol^{0}s)], $k^{12}{}_P$ - constant of addition speed of ethylene to the chain with the propylene end link [m^3/ (mol^{0}s)], $k^{22}{}_P$-constant of addition speed of ethylene to the chain with the ethylene end link [m^3/ (mol^{0}s)], $k^1{}_{tr}$ - constant of chain transfer (with the propylene end link) to hydrogen [m^3/ (mol^{0}s)], $k^2{}_{tr}$- constant of chain transfer (with the the ethylene end link) to hydrogen [m^3/ (mol^{0}s)], $k^1{}_d$-chain break constant (with the propylene end link) [m^3/ (mol^{0}s)], $k^2{}_d$-chain break constant (with the ethylene end link) [m^3/ (mol^{0}s)], k_d-constant of deactivation speed of active catalyst [m^3/ (mol^{0}s)], m-total mass of a mixed system in a fluidized bed reactor [kg], M_1 = propylene concentration [mol/m^3], M_2 = ethylene concentration, $P^*_{s,\,n}$-active copolymer chain (with the propylene end link) containing s propylene links and n ethylene links [mol/m^3], $P_{s,\,n}$-dead copolymer chain [mol/m^3], $Q^*_{s,\,n}$ -active copolymer chain (with the ethylene end link) containing s propylene links and n ethylene links [mol/m^3], $Q_{s,\,n}$-dead copolymer chain [mol/m^3], R-gas constant, r_{C*}-active catalyst flow speed [mol/m^3 s], r_{M1}- propylene flow speed [mol/m^3 s], r_{M2}-ethylene flow speed [mol/m^{3}s], r_{P*}-growth speed of active chains (with the propylene end link) [mol/m^3 s], r_{Q*}-growth speed of active chains (with the ethylene end link) [mol/m^3 s], r_{H2}- hydrogen flow speed [mol/m^3], T-temperature [K], V_0-effective reactor volume [m^3], W-part of the solid phase in the reactor,M_{M1} -molar mass of propylene [kg/kmol], M_{M2} -molar mass of ethylene [kg/kmol], M_{solid} -molar mass of catalyst [kg/kmol], M_{gas} - the molar mass of gas mixture [kg/kmol].

References

[1] McAulie K B, Talbot J P and Harris T J 1994 A Comparison of two-phase and well-mixed models for fluidized bed polyethylene reactors *Chem.Eng. Sci.* **13** 2035-45

[2] Aminova G A, Manuiko G V, Ignashina T V, Bronskaya V V, Zakharova O V, D'Yakonov G S and Arkhireev V P 2005 Investigation of Simultaneous Chemical-Transformation and Heat-Exchange Process Occurring in the Process of Synthesis of Divinyl Rubber on the Basis of a Cobaltous Catalyst in a Cascade of Two Continuous-Operation Reactors *J. of Eng. Physics*

and Thermophysics **78(3)** 532-40

[3] Aminova G A, Manuiko G V, Bronskaya V V, Ignashina T V, Ismagilova A I and D'Yakonov G S 2006 Mathematical modeling of butadiene rubber synthesis process on neodymium-containing catalyst system *Polymer Science Series A* **48(8)** 881-6

[4] Zacca J and Ray W 1993 Modelling of the liquid phase polymerizatiom of olefins in loop reactors *Chem. Eng. Science* **48(22)** 3743-65

[5] Kharitonova O S, Bronskaya V V, Ignashina T V, Al-Muntaser A A and Khairullina L E 2019 Modeling of absorption process using neural networks *IOP Conf. Series: Earth and Environmental Science* **315** 032025

[6] Kissin Yu V and Rishina L A 2008 Kinetika polimerizacii propilena i e`tilena na geterogenny`x katalizatorax Ciglera-Natta: novy`e rezul`taty` [Kinetics of the polymerization of propylene and ethylene on heterogeneous Ziegler-Natta catalysts: new results] *Vy`sokomolekulyarny`e soedineniya Seriya A* **50(11)** 1911-34 [In Russia]

[7] Rishina L A, Galashina N M, Nedorezova P M, Klyamkina A N, Aladyshev A M, Tsvetkova V I and Kleiner V I 2006 Gomo- i sopolimerizaciya vinilciklogeksana s olefinami v prisutstvii geterogenny`x i gomogenny`x kataliticheskix sistem [Homo and copolymerization of vinylcyclohexane with olefins in the presence of heterogeneous and homogeneous catalytic systems] *Vy`sokomolekulyarny`e soedineniya Seriya A* **48(1)** 25-33 [In Russia]

[8] Aminova G A, Manuyko G V, Bronskaya V V, Ignashina T V, Zakharova O V, and Dyakonov G S 2008 Matematicheskoe modelirovanie processa sinteza butadienovogo kauchuka na kobal`t- i neodimsoderzhashhix kataliticheskix sistemax v kaskade reaktorov s teplootvodom [Mathematical modeling of synthesis of butadiene rubber on cobalt and neodymium-containing catalytic systems in the cascade of heatsink reactors] *Vy`sokomolekulyarny`e soedineniya Seriya B* **50(11)** 2029-34 [In Russia]

[9] Chen Y and Liu X 2005 Modeling mass transport of propylene polymerization on Ziegler Natta catalyst *Polymer* **46(22)** 9434-42

[10] Chai CJ and Zhang GL 2004 *Chemical engineering fluid flows and heat transport* (Beijing: Chemical Engineering Press)

[11] Yiannoulakis H, Yiagopoulos A and Kiparissides C 2001 Recent developments in the particle size distribution modeling of fluidized-bed olefin polymerization reactors *Chem Eng Sci* **56(3)** 917-25

[12] Hong DY 2002 Polypropylene-principle, process and technology *Beijing: China Petrochemical Press* (Chinese)

[13] Zheng Z W, Shi D P, Su P L, Luo Z H and Li X J 2011 Steady-state and dynamic modeling of Basell multi-reactor olefin polymerization process *Ind Eng Chem Res* **50(1)** 322-31

[14] Urdampilleta I, Gonzalez A, Iruin JJ, de la Cal J C and Asua J M 2006 Origins of product heterogeneity in the Spheripol high impact polypropylene process *Ind. Eng. Chem. Res.* **45(12)** 4178-87

[15] Luo Z H, Wen S H and Zheng Z W 2009 Modeling the effect of polymerization rate on the intraparticle mass and heat transfer during propylene polymerization in a loop reactor *J. Chem. Eng. Jpn.* **42(8)** 576-80

[16] Hatzantonis H, Goulas A and Kiparissides C 1998 A comprehensive model for the prediction of particle-size distribution in catalyzed olefin polymerization fluidized-bed reactors *Chem. Eng. Sci.* **53(18)** 3251-67

[17] Young M, Chang W and Ma C 2003 Polymerization kinetics and modeling of a metallocene cyclic olefin copolymer system *Eur. Ploymer J.* **39(1)** 165-71

IOP Conf. Series: Materials Science and Engineering **791** (2020) 012072 doi:10.1088/1757-899X/791/1/012072

Energy efficiency increase of the initial milk processing

G N Samarin[1], A N Vasilyev[1], D A Normov[2], E A Eventeva[3] and A K Mamahay[1]

[1]Department of Automated Electrical and Thermal Technology, Federal Scientific Agro-Engineering Center VIM, 1st Institutsky passage, House 5, 109428, Moscow, Russia

[2]Physics department, FSBEI HE Kuban State Agrarian University, 350012, Krasnodar, 13, Azovskaya St., Flat 40, Russia

[3]Department Automobiles, tractors and agricultural machines, Velikie Luki State Agricultural Academy, 182112, Pskov region, Velikie Luki, Lenina prospect, 2, Russia

E-mail: samaringn@yandex.ru

Abstract. Condition of the dairy industry in the Russian Federation over the last years is characterized by a long-standing and continuously deepening downtrend in dairy production and cow population decrease. From the present day perspective, main objective for Russian livestock farmers, in the context of an import phaseout, is a dairy production increase. Presently at dairy farms in the Russian Federation 99% of freshly drawn milk is being cooled to 4.5°C by a refrigeration method. Refrigerators are complex machines that require skilled workforce for maintenance and operation, consume large amount of electrical energy (3.4 kWh per 100 kg of milk to cool down from 40 to 5°C). Price increase for energy resources, including electricity, is happening in all countries. Price increase is an objective process since commercial mineral deposits run short, their extraction decline, and price tag for extraction and delivery to the place where it can be processed and consumed is increasing. Within this context, any research study that focuses on specific energy resources decrease, including energy consumption for milk production and processing, is a critical task for the present day and for the foreseeable future.

1. Introduction

Currently, the Russian Federation is a world`s largest milk and dairy products manufacturer, but market value milk is a comparably small part in overall production – 57%, and livestock productiveness is twice as small as in developed countries [1]. Increase of market value milk production is a major task for any country, that has influence on the agricultural sector economy and on food supply level for the population.

At the present day, dairy industry of the Russian Federation has two major goals. First – de-crease imported dairy products dependence. Second - ensure increase of dairy products consumption by the population. First problem can be solved by substitution of imported products for domestically made products (moreover, in the Food Security Doctrine minimum threshold of domestic products share must be 90% in the overall commodity resources volume), and to achieve this goal it is necessary to increase market value milk production [1].

The second problem is clearly an extension of the first, therefore, it is necessary to increase production level of raw, primarily market value, milk, preserve demand for it at the same time [1, 2].

It is worth noting that cooled milk is still contaminated with bacteria, that fall into anabiosis state waiting for suitable temperature conditions, and some mesophilic bacteria types can repro-duce at

IOP Conf. Series: Materials Science and Engineering **791** (2020) 012072 doi:10.1088/1757-899X/791/1/012072

10°C. Bacteria begin rapid reproduction during temperature rise. Therefore, product quality suffers and its price drops [3-6].

In order to solve all formulated earlier problems, it is necessary to use special methods and technical means to eliminate bacteria.

Different economic zones have the following basic regulatory documents in use:

For the Russian Federation: FZ-88 and FZ-163 «Technical regulations on milk and milk products» - general and specific safety requirements; GOST R 52054-2003 «Cow`s milk raw» - general requirements for production of all milk product types [1].

For European Economic Union: «Council Directive 92/46 EEC laying down the health rules for the production and placing on the market of raw milk, heat-treated milk and milk-based products; Council Directive 2002/99 EC of 16 December 2002, Regulation (EC) 852/2004 of the European Parliament; Regulation (EC) 853/2004 of the European Parliament; Regulation (EC) 854/2004 of the European Parliament; Regulation (EC) 882/2004 of the European Parliament» [7,8].

For Custom Union: «TR CU 033/2013 Custom Union Technical Regulation. On Safety of Milk and Dairy Products» [1].

Table 1 presents a general safety criterion of raw milk in different economic zones.

Quantity determination of mesophilic aerobic and facultative anaerobic microorganisms (QMAFAnM) is applicable to quantity evaluation of sanitary indicator microorganisms' group. QMAFAnM is consist of different taxonomical groups of microorganisms – bacteria, yeast, mold fungi. Their total count is the evidence of hygienic and sanitary food condition, microbial content level. Optimum temperature for QMAFAnM growth is 35…37°C (in aerobic conditions); temperature limit of their growth is within 20…45°C [1, 7–11].

Abbreviation CFU stands for «colony forming unit» and represents quantity of bacteria that can develop full strength microbial colony [1, 7–11].

In order to reduce the number of bacteria in milk it is necessary to comply with sanitary rules of machine milking. However, as we know from papers [12-16], even small number of bacteria in milk can change its storage life, therefore, there is a need to take measures in order to eliminate or at least suspend bacteria growth in it.

The most effective and accessible for livestock farms is considered method of milk cooling and heating.

Table1. General safety criteria of raw milk in different economic zones [1, 3–5, 7–9, 19–23].

Country, economic zone	Class of reductase test	QMAFAnM, (CFU/gr)
European Union		$< 100\times10^3$
USA	A/	$< 100\times10^3$/
	B	$100\times10^3 \ldots 300\times10^3$
Russian Federation	High grade/	$< 300\times10^3$
	I/	$300\times10^3 \ldots 500\times10^3$
	II	$500\times10^3 \ldots 4000\times10^3$
Customs Union: (Republic of Armenia; Republic of Belarus; Republic of Kazakhstan; Kyrgyz Republic; Russian Federation)		$100\times10^3 \ldots 4000\times10^3$
Ukraine	High grade /	$< 300\times10^3$
	I/	$300\times10^3 \ldots 500\times10^3$
	II	$500\times10^3 \ldots 4000\times10^3$
China		$< 2000\times10^3$
Australia, New Zealand		$< 100\times10^3$

IOP Conf. Series: Materials Science and Engineering **791** (2020) 012072 doi:10.1088/1757-899X/791/1/012072

If milk is cooled down to 8…10°C, most of, found in milk, microorganisms' development sharply slows down, and if cooled down to 2…3°C – growth almost completely stops. When temperature goes down to minus 20°C and abrupt temperature change from plus 20°C to minus 20°C majority of micro-organism dies [17,18].

When milk is heated up to 50°C - 68% of bacteria dies, if milk stays for 20 minutes at this temperature more than 99% of all contained in it microorganisms die. For complete microorganisms and its spores elimination it is necessary to bring milk temperature to 120°C and keep at this level for 15…20 minutes [17, 18].

Basic technological technics for initial milk processing at farms consist of a set of actions, including purification, cooling, and, if necessary, pasteurization. All of this results in an original milk quality preservation and contained microorganisms suppression.

The most significant factor that influences bacteria growth and development – positive temperature. On the other hand, freezing results in slow product destruction, because ice crystals damage cell walls [17, 18, 24].

Light is essential for bacteria development, however, in the presence of ultraviolet light (sun light) bacteria die [25-27].

Level of acidity is very important for bacteria functioning – level of pH in milk has a key role and titratable acidity is not [25, 26].

Depending on oxygen demand microorganisms can be classified in the following way: aerobic, anaerobic, facultative aerobic/anaerobic – typical example is an ordinary lactic acid bacterium, microaerophilic.

Water is a major component in bacterial cell, besides water content in product, osmotic pressure of water has very important role as well. Nutrients are needed for the microorganisms development [28-30].

Milk cooling process consumes a large amount of energy, requires special equipment and its centralized maintenance.

There are several alternative methods of milk processing exists: treatment by ultraviolet light, ultrasound, electric infrared heating, electro treatment (electrochemical treatment), ultra-high pressure, bactofugation, sterilization and others [1, 18].

Therefore, goal of this paper is development and justification of design-and-technology parameters and operating mode of the batch type milk decontamination plant in the context of rational use of energy.

On the basis of the desired goal, we outlined following tasks:

- develop a technological scheme of milk decontamination by repeated ultrasound treatment;
- find rational operating modes and complex of design-and-technology parameters for the plant, that provide milk microbiologic characteristics improvement with as minimum as possible en-er-gy consumption.

2. Materials and Methods

Research study was conducted in accordance with the developed specific methods using modern equipment and measuring instruments. Main calculations and analysis of the research study results were run using mathematical statistic methods and regression analysis using active experimental design theory. All studies on physicochemical characteristics of milk were conducted in accordance with standard methods and in compliance with GOSTs:

- GOST R 52054-2003 «Cow`s raw milk. Specifications»;
- GOST 3624-92 «Milk and milk products. Titrimetric methods of acidity determination»;
- GOST 32901-2014 «Milk and milk products. Methods of microbiological analysis».

Resazurin reduction test was used to determine bacteria quantity. According to the method described in GOST 32901-2014 «Milk and milk products. Methods of microbiological analysis» bacterial count level was determined. Method is based on resazurin reduction by oxidation-reduction enzymes, discharged by microorganisms into milk.

IOP Conf. Series: Materials Science and Engineering **791** (2020) 012072 doi:10.1088/1757-899X/791/1/012072

Reductase test (also methylene blue reaction, resazurin test) - test method for bacterial contamination of non-pasteurized milk, based on dye discoloration in presence of bacteria metabolism byproducts. GOST 32901-2014 «Milk and milk products. Methods of microbiological analysis» allowed to use reactions with resazurin and methylene blue. Methylene blue reduction test is less accurate and lost popularity. Since 1929 resazurin is being used as an indicator of bacterial contamination in Persh and Simmert method (1929) [31].

3. Results

Study of layer-by-layer milk spoilage in the unsealed tank without an agitator system was carried out during summer period at a natural ambient air temperature of 16...20°C. Three tanks of 7 liters each were used. Objective of this study is milk spoilage character determination in differ-ent layers of liquid. Two main milk parameters were measured during the experiment: tempera-ture and quantity of CFU per unit volume of liquid. Table 2 presents experiment results.

Mathematical treatment of the experimental findings [32-34] allowed us to obtain adequate regression equation, which describes dependence of quantitative content of CFU per unit vol-ume of milk`s bottom layer liquid (Y, mln bacteria/ml) on storage life (X, h) with certainty R2 = 0.982.

$$Y = 0{,}0261X^5 - 0{,}5755X^4 + 4.641X^3 - 16.688X^2 + 26.077X - 13.292 \quad (1)$$

Table 2. Experimental study results.

Sampling time (h)	Milk temperature in the top layer (TL) (°C)	Milk temperature in the bottom layer (BL) (°C)	Bacteria quantity in milk TL (mln bacteria/ml)	Bacteria quantity in milk BL (mln bacteria/ml)
10	16.05	16.32	0.3	0.3
14	18.30	17.89	1.75	0.4
18	18.74	19.38	8.0	0.4
22	19.25	19.69	20.0	0.4
6	18.27	19.09	20.0	1.75
10	18.74	19.09	20.0	1.75
14	18.39	19.44	20.0	1.75
18	18.61	19.05	20.0	1.75
22			20.0	20.0

Figure 1 presents results of this experiment in graphical form.

4. Discussion

From the equation (1) and graphical interpretation (Figure 1) follows that top layer of milk gets spoiled faster. Therefore, during a study of milk treatment by ultrasound, ultraviolet light it is necessary to start milk radiation from the top and then go down to the bottom.

Down the line, this information can be used in the following developments: new technological scheme of milk decontamination by ultrasound and ultraviolet light repeated exposure; plant, with necessary number of radiators, which allows to decontaminate milk in the process of multiple decontaminations, and also to find rational operating modes and complex of design-and-technology parameters of the plant, that provide milk microbiologic characteristics improvement with as minimum as possible energy consumption.

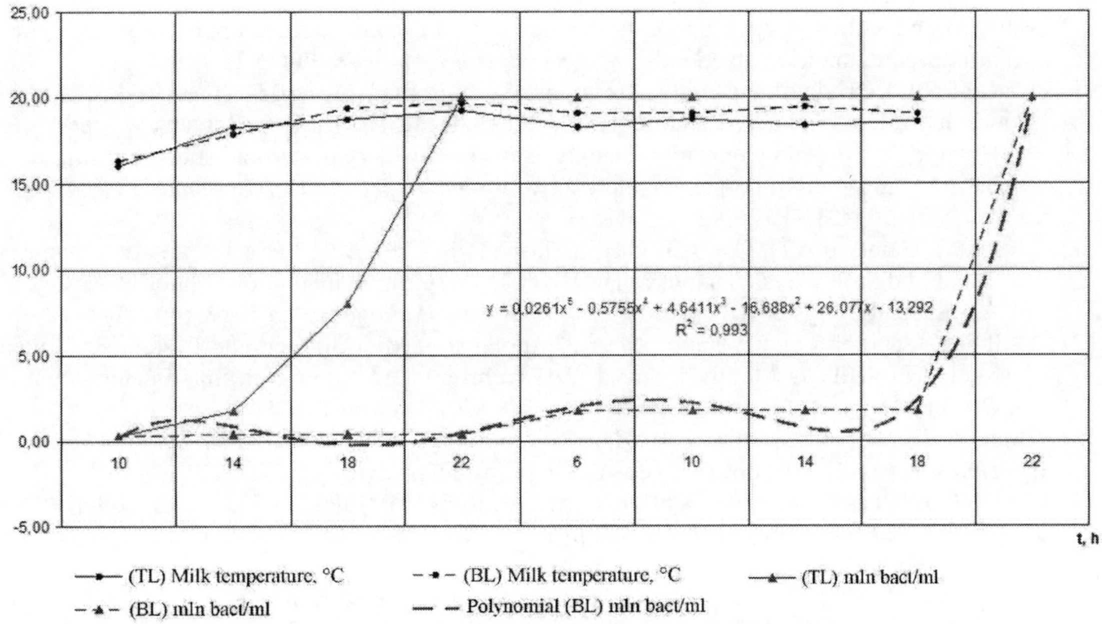

Figure 1. Dependence of bacteria quantity change in bottom a top milk layers in the tank at a specific temperature in the course of time.

5. Conclusion

On the daily basis milk cooling in the farms requires on average 1.37 kW/h (t/day) or for the farm OJSC «Krasnoe Znamya» Novosokolnichesky district of Pskov region of the Russian Federation – over the year period 3903 t of milk requires 132509 kW/h for preservation.

Milk in the unsealed tank begins to spoil from the top, therefore, experiments has confirmed where devices for alternative milk preservation methods (ultrasound, ultraviolet light and others) should be placed. In doing so, we can optimize their number and then fulfill power saving objective.

Milk spoilage process is described by the regression equation (1).

References

[1] Samarin G N, Ruzhiev V A and Shilin E V 2017 *Obrabotka moloka na malyh predpriyatiyah al'ternativnymi metodami* [Milk processing at small enterprises by alternative methods] *Vestnik APK Stavropol* **1 (25)** 49-54 [In Russian]

[2] Samarin G N 2008 *Upravlenie sredoj obitaniya sel'skohozyajstvennyh zhivotnyh i pticy: monografiya* [Habitat management of farm animals and birds: monograph] (Velikiye Luki: FGOU VPO Velikolukskaya gosudarstvennaya sel'skokhozyaystvennaya akademiya) [In Russian]

[3] Boor K J 1997 Pathogenic Microorganisms of Concern to the Dairy Industry *Dairy, Food and Environ. Sanitation* **17** 714-7

[4] Chambers J V 2002 The Microbiology of Raw Milk *Dairy Microbiology Handbook* ed R K Robinson (New York: Wiley-Interscience) pp 39-90

[5] Ye A, Cui J, Dalgleish D and Singh H 2017 Effect of homogenization and heat treatment on the behavior of protein and fat globules during gastric digestion of milk *J. Of Dairy Science* **100(1)** 36-47

[6] Liang L, Qi C, Wang X, Jin Q and McClements D J 2017 Influence of Homogenization and Thermal Processing on the Gastrointestinal Fate of Bovine Milk Fat: In Vitro Digestion Study *J. Of Agricultural And Food Chemistry* **65(50)** 11109-17

[7] Bisig W et al 2019 *The technology of pasteurisation and its effects on the microbiological and nutritional aspects of milk* (Brussels: IDF)

[8] Mack G and Kohler A 2018 Short and long-rung policy evaluation: support for grassland-based milk production in Switzerland *J. of Agricultural Economics* **70(1)** 215-40

[9] Carlin F 2011 Origin of bacterial spores contaminating foods *Food Microbiology* **28(2)** 177-82

[10] Coorevits A, De Jonghe V, Vandroemme J, Reekmans R, Heyrman J, Messens W, de Vos P and Heyndrickx M 2008 Comparative analysis of the diversity of aerobic spore-forming bacteria in raw milk from organic and conventional dairy farms *Systematic And Applied Microbiology* **31(2)** 126-40

[11] Cusato S, Gameiro A H, Corassin C H, Sant'ana A S, Cruz A G, Faria J A and de Oliveira C A 2013Food Safety Systems in a Small Dairy Factory: Implementation, Major Challenges, and Assessment of Systems' Performances *Foodborne Pathogens And Disease* **10(1)** 6-12

[12] Doll E V, Scherer S and Wenning M 2017 Spoilage of Microfiltered and Pasteurized Extended Shelf Life Milk Is Mainly Induced by Psychrotolerant Spore-Forming Bacteria that often Originate from Recontamination *Frontiers In Microbiology* **8** 135

[13] Boor K J and S C Murphy 2002 The Microbiology of Market Milks *Dairy Microbiology Handbook* ed R K Robinson (New York: Wiley-Interscience) pp 91-122

[14] Doyle C J, Gleeson D, Jordan K, Beresford T P, Ross R P, Fitzgerald G F and Cotter P D 2015 Anaerobic sporeformers and their significance with respect to milk and dairy products *Int. J. of Food Microbiology* **197** 77-87

[15] Driehuis F, Hoolwerf J and Rademaker J L W 2016 Concurrence of spores of Clostridium tyrobutyricum, Clostridium beijerinckii and Paenibacillus polymyxa in silage, dairy cow faeces and raw milk *Int. Dairy J.* **63** 70-7

[16] O'Riordan N, Kane M, Joshi L and Hickey R M 2014 Structural and functional characteristics of bovine milk protein glycosylation *Glycobiology* **24(3)** 220-36

[17] Huck J R, Sonnen M, Boor K J 2008 Tracking heat-resistant, cold-thriving fluid milk spoilage bacteria from farm to packaged product *Journal Of Dairy Science* **91(3)** 1218-28

[18] Chavan R S, Chavan S R, Khedkar C D and Jana A H 2011 UHT Milk Processing and Effect of Plasmin Activity on Shelf Life: A Review *Comprehensive Reviews In Food Science And Food Safety* **10(5)** 251-68

[19] Aiassa E et al 2015 Applicability and Feasibility of Systematic Review for Performing Evidence-Based Risk Assessment in Food and Feed Safety *Critical Reviews In Food Science And Nutrition* **55(7)** 1026-34

[20] Finnegan W, Yan M, Holden N M and Goggins J 2018 A review of environmental life cycle assessment studies examining cheese production *Int. J. Of Life Cycle Assessment* **23(9)** 1773-87

[21] Djekic I, Miocinovic J, Tomasevic I, Smigica N and Tomica N 2014 Environmental life-cycle assessment of various dairy products *J. Of Cleaner Production* **68** 64-72

[22] Depping V, Grunow M, van Middelaar C and Dumplerc J 2017 Integrating environmental impact assessment into new product development and processing-technology selection: Milk concentrates as substitutes for milk powders *J. Of Cleaner Production* **149** 1-10

[23] Bacenetti J, Bava L, Schievano A and Zucali M 2018 Whey protein concentrate (WPC) production: Environmental impact assessment *J. Of Food Eng.* **224** 139-47

[24] Mafart P, Leguerinel I, Couvert O and Coroller L 2010 Quantification of spore resistance for assessment and optimization of heating processes: A never-ending story *Food Microbiology* **27(5)** 568-72

[25] Coorevits A, De Jonghe V, Vandroemme, Reekmans R, Heyrman J, Messens W, de Vos P and Heyndrickx M 2008 Comparative analysis of the diversity of aerobic spore-forming bacteria in raw milk from organic and conventional dairy farms *Systematic And Applied Microbiology* **31(2)** 126-40

[26] Giffel M C T, Wagendorp A, Herrewegh A and Driehuis F 2002 Bacterial spores in silage and

raw milk *Antonie Van Leeuwenhoek* **81(1-4)** 625-30

[27] Luecking G, Stoeckel M, Atamer Z, Hinrichs J and Ehling-Schulz M 2013 Characterization of aerobic spore-forming bacteria associated with industrial dairy processing environments and product spoilage *Int. J. Of Food Microbiology* **166(2)** 270-9

[28] Ben Jemaa M, Falleh H, Neves M A, Isoda H, Nakajima M and Ksouri R 2017 Quality preservation of deliberately contaminated milk using thyme free and nanoemulsified essential oils *Food Chemistry* **217** 726-34

[29] Calo J R, Crandall P G, O'Bryan C A and Ricke S C 2015 Essential oils as antimicrobials in food systems - A review *Food Control* **54** 111-9

[30] Gouvea F dos S, Rosenthal A and Ferreira E H da R 2017 Plant extract and essential oils added as antimicrobials to cheeses: a review *Ciencia Rural* **47(8)** e20160908

[31] Pesch K L and Simmert U 1929 *Milchw. Forsch* **8** 551

[32] Samarin G N, Vasilyev A N, Zhukov A A, Soloviev S V 2018 Optimization of Microclimate Parameters Inside Livestock Buildings *Intelligent Computing & Optimization. Advances in Intelligent Systems and Computing* ed Vasant P, Zelinka I, Weber G W vol 866 (Springer)

[33] Ben-Tal A and Nemirovski A 2001 *Lectures on modern convex optimization: analysis. algorithms. and engineering applications MPR-SIAM Series on optimization* (Philadelphia: SIAM)

[34] Ben-Tal A and Nemirovski A 2002 Robust optimization – methodology and applications *Mathematical Programming* **92(3)** 453–80

Analysis of the effect of detergent additives on fuel on the performance of a diesel engine

A Yu Shabanov, Yu V Galyshev, A B Zaitsev and A A Sidorov

Higher School of Power Engineering, Peter the Great Saint-Petersburg Polytechnic University, Polytechnicheskaya St., 29, St.Petersburg, 195251, Russia

E-mail: turbo_motor@mail.ru

Abstract. In work questions of influence of composition of basic diesel fuel on overall performance of the washing additives on the basis of results of motor tests are considered. The received results show need of individual selection of concentration of additive depending on composition of basic fuel. The analysis of results allowed to draw a conclusion on existence of an optimum ratio of concentration of additive and composition of basic fuel at which receiving the maximum effect of input of additive at preservation of the acceptable prime cost of the final product of fuel is possible. This conclusion is supported by the results of comparative stand tests of diesel engine firing fuels that significantly differ in basic physicochemical parameters. The tests have been conducted on laboratory test stand equipped with hydraulic dynamometer coupled with full-sized diesel engine type KAMAZ-740.10. Such arrangement enabled operation at any point within allowed zone of engine load/speed map.
Keywords: The diesel engine, basic diesel fuel, the washing additive, kinematic viscosity, toxicity, smoke of the fulfilled gases.

1. Introduction

State-of-the-art fuels for modern high-powered high-speed diesel engines, particularly those featuring direct fuel injection and turbocharging, must, inter alia, have anti-sludge properties and high cleaning capacity [1-4].This goal may be achieved by improvement in base fuel quality and use of dedicated detergent additives [5-7].

This paper discusses the results of investigation into detergent additive performance as a function of its concentration and base fuel composition.

This investigation was targeted at:

- determination of correlation between fuel composition and engine combustion efficiency, toxic emission and exhaust opacity;
- determination of optimum detergent additive concentration in terms of engine fuel efficiency and environment performance, depending on base fuel type

2. Materials and Methods

A series of experiments were conducted to study correlation between fuel composition and detergent additive concentration, on the one hand, and engine performance, on the other hand.

The tests have been conducted on laboratory test stand equipped with hydraulic dynamometer coupled with full-sized diesel engine type KAMAZ-740.10. Such arrangement enabled operation at any point within allowed zone of engine load/speed map.

IOP Conf. Series: Materials Science and Engineering **791** (2020) 012073 doi:10.1088/1757-899X/791/1/012073

Test procedure included preliminary partial engine disassembly, photofixation of initial contamination level, weighing of inlet & outlet valves and injector atomizers with analytical balance so as to register initial level of deposit. Weighting accuracy was within 0,001 g. After reassembly and short-term running-in, a number of engine initial parameters were measured, including power, fuel consumption, exhaust opacity, noxious emission content and exhaust gas temperature. All the measurements were carried out on engine loaded according to 2 different load curves. Where upon the engine ran for 20 hours at specified loads firing fuel under test.

On completion of the test cycle engine parameters at the same load points have been measured once more. Then partial engine disassembly followed by visual examination and weighing of selected parts was repeated, which enabled evaluation of changes in sedimentation level of their critical surfaces.

Three types of winter diesel fuel from different producers have been selected for testing. The results of testing of each base fuel type are presented in table 1.

Table 1. Principal physicochemical properties of selected base fuel types.

№	Parameter/Characteristic	Diesel fuel sample		
		№1	№2	№3
1	Cetane number, units	57.1	54.5	51.5
2	Fractional composition:			
	is rectified at temperature 250°C, vol.%	35	28	45
	is rectified at temperature 350°C, vol.%	95	91	99
	95% vol. is rectified at temperature, °C	350	358	298
3	Flash point – Pensky-Martens closed cup method, °C	60	55	52
4	Kinematic viscosity, mm^2/s	2.94	2.18	1.45
5	Mass fraction of polycyclic aromatic hydrocarbons, %	1,0	1,6	0,6
6	Cold filter plugging point, °C	-31	-41	-41
7	Mass fraction of sulfur, mg/kg	3.1	4.9	2.9
8	Density at 15 °C	818	812	804

3. Results

The analysis of the test results witnesses to compliance of all the three types of base fuels have obvious spread in some principal fuel characteristics. Thus, spread in cetane number, a main parameter characterizing fuel combustibility, was as high as 5.6. Furthermore, there is large spread in fractional composition that characterizes fuel volatility. Thus, temperature at which 95% fuel evaporates, varies from 298° to 358°, i.e. its variation practically amounts to 17%. Vaporization temperature largely determines fuel volatility and its scaling tendency. Flash point of selected fuels varies from 52° to 60°C, i.e. by 13%. Most prominent is spread in kinematic viscosity, being practically twofold. All the above parameters affect mixture formation and fuel combustion. This was demonstrated by the results of the first test cycle, which included comparative engine stand tests with each of three selected base fuel types.

The diagrams figure 1 and figure 2 present fuel consumption and exhaust gas opacity curves vs. engine torque at constant engine speed 1500 RPM for all the three tested fuel types.

The difference in fuel consumption on the average was about 2.2%, while in low load area it amounted to 8...9% (figure 1).

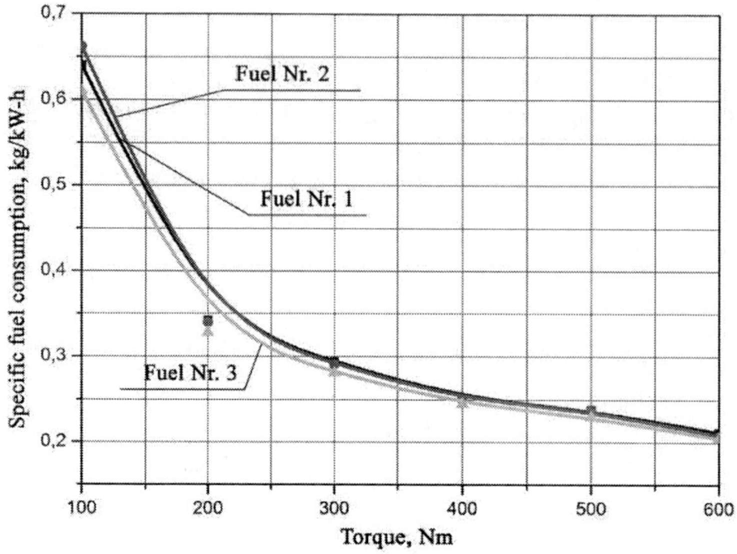

Figure 1. Specific fuel consumption vs. engine load depending on fuel type.

Greater difference may be seen in exhaust gas opacity (up to 20…25% in maximum load zone) and up to 12…15% on the average (figure 2). Supposedly, the main reason for such variation is difference in fuel hydrocarbon-type composition, which entails as big difference in fuel combustibility and volatility.

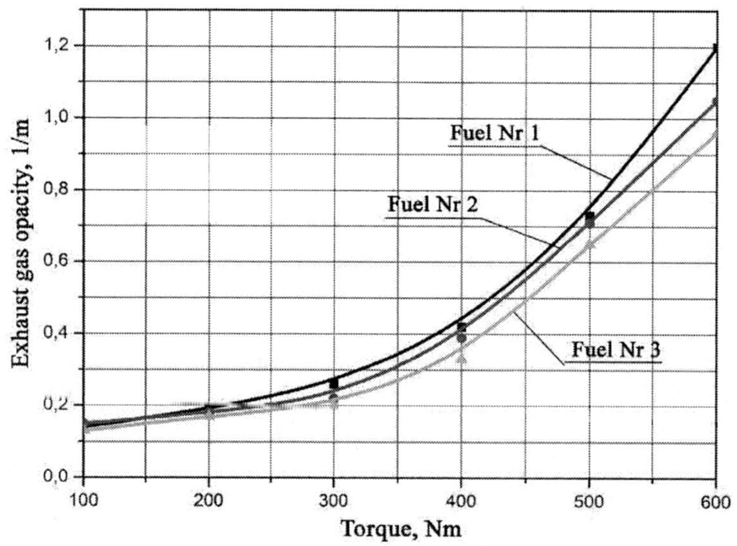

Figure 2. Exhaust gas opacity vs. engine load depending on fuel type.

It should be noted that the engine under test features mechanical fuel injection system, and is optimized for low cetane number fuels. This accounts for the best performance of base fuel Nr. 3 (in terms of fuel consumption and exhaust gas opacity), whose cetane number is the least among tested fuels. Moreover, this fuel is the best in terms of volatility as well, as follows from its fractional composition.

Also, the results of representative part weighing show definite correlation between fuel composition and its scaling tendency (table 2).Respective difference in averaged masses of deposit on valves and injectors is quite perceptible, though relatively small.

IOP Conf. Series: Materials Science and Engineering **791** (2020) 012073 doi:10.1088/1757-899X/791/1/012073

Table 2. Increment in representative part mass for the test cycle period.

Nr.	Parts	Increment in averaged deposit mass for the test cycle period (mg / percent in relation to initial values)		
		Diesel fuel sample		
		Nr.1	Nr.2	Nr.3
1	Injector atomizers	+1 mg/ +2.6%	+2 mg/ +5.8%	0 mg/0.0%
2	Inlet valves	+10 mg/+ 3.6%	+12 mg/ +4.3%	+8 mg/ +2.9%
3	Outlet valves	+7 mg/ +4.3%	+9 mg/ +5.6%	+6 mg / +3.8%

Particularly noteworthy is distinct similarity of dependences of fuel economy and fuel scaling tendency on its composition. Fuel sample Nr. 3 that showed maximum combustion efficiency, demonstrated less scaling tendency as well.

The second part of this work included experimental investigation of correlation between detergent additive concentration, on the one hand, and engine fuel efficiency, emission performance, and level of deposits on combustion chamber surfaces, on the other hand.

In these tests a multi-component additive was used, originally developed as a part of premium commercial grade fuels, with a view to improve fuel detergency and combustibility. Additive mass concentration recommended by manufacturer equals 100 ppm, regardless of base fuel sort and composition. Confirmation of the above recommendation was one of the goals of this investigation.

This test cycle, like the previous one, was carried out according to the test procedure described earlier. Base fuel types used in these tests were modified with the above-mentioned multi-component additive of varying concentration (50, 100, 150, 200 ppm).

The results of these tests demonstrated undoubted effectiveness of the multi-component additive. It should be noted, however, that its performance depends considerably on base fuel composition and additive concentration. Shown in figure 3 and figure 4 are functional dependences of fuel saving and decrease in exhaust gas opacity (averaged across all the measurement points) on multi-component additive concentration. In each case a respective value measured at the final test stage for each fuel type with no additive was taken as a comparison base.

It may be noted that for each fuel type there is optimum concentration of the additive, at which its effectiveness reaches maximum.

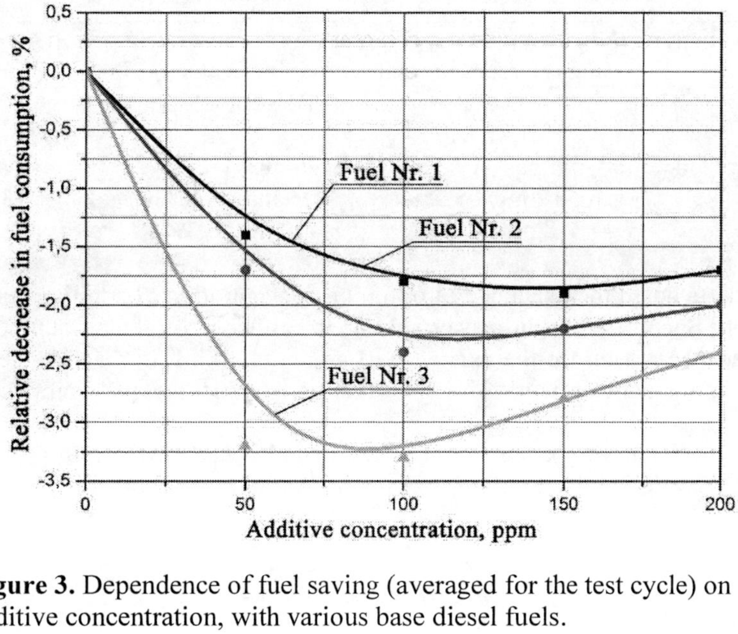

Figure 3. Dependence of fuel saving (averaged for the test cycle) on additive concentration, with various base diesel fuels.

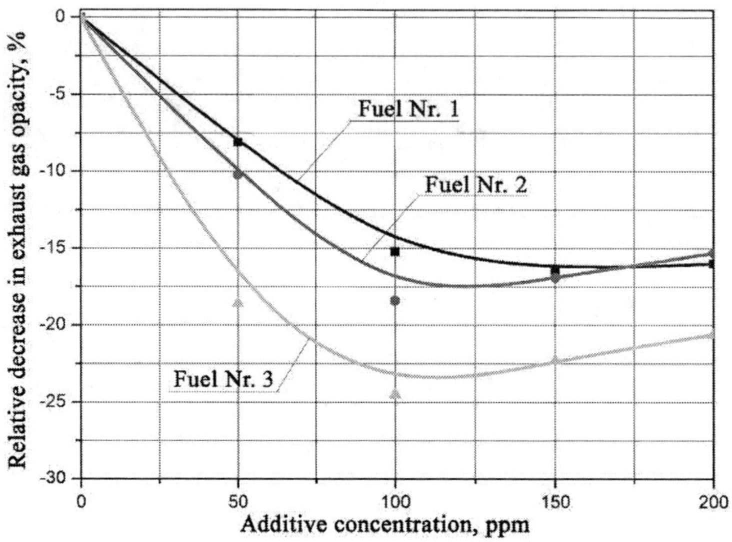

Figure 4. Dependence of decrease in exhaust gas opacity (averaged for the test cycle) on additive concentration, with various base diesel fuels.

The latter reveals itself in the biggest decrease in fuel consumption and exhaust gas opacity. Thus, for base fuel Nr. 1 such optimum point lies in the range 140...150 ppm, for base fuel Nr. 2 – in the range 110...120 ppm, for base fuel Nr. 3 it drops down to 75 ppm. The same is largely true for exhaust gas opacity trend. Use of the additive in optimum concentration results in decrease in fuel consumption by 1.8...3.5%, in exhaust gas opacity – by 17...22% as compared with respective base fuel. The best results were shown with base fuel Nr. 3.

Both weighing and visual inspection of selected parts confirmed effectiveness of the additive in terms of detergency (table 3) and existence of optimum concentration, at which the cleaning capacity reaches its maximum (figure 5).

Table 3. Decrement in representative part mass for the period of test cycle for fuel modified with optimum amount of the multi-component additive.

		Increment in averaged deposit mass for the test cycle period (mg / percent in relation to initial values)		
Nr.	Parts	Diesel fuel sample		
		Nr.1	Nr.2	Nr.3
1	Injector atomizers	-6 mg/ -17.5%	-6 mg/-18.4%	-9 mg/ -26.2%
2	Inlet valves	-12 mg /-5.2%	-15 mg /-5.8 %	-18 mg/-6.9%
3	Outlet valves	-8 mg /-5.0%	-10 mg/-6.3%	-12 mg /-7.4%

A number of authors noted the existence of optimum concentration of additive containing combustion intensifier [8-10].Such effect is in a way similar to influence of cetane number on combustion process. In both cases the reason for the existence of an optimum lies in reaching certain balance between compression and expansion work of engine thermodynamic cycle as combustion rate changes [11-12].

Obviously, fuel activation with combustion intensifier as a part of the multi-component additive modifies ignition lag period, and, therefore, balance of kinetic and diffusion phases of combustion process [13-16].

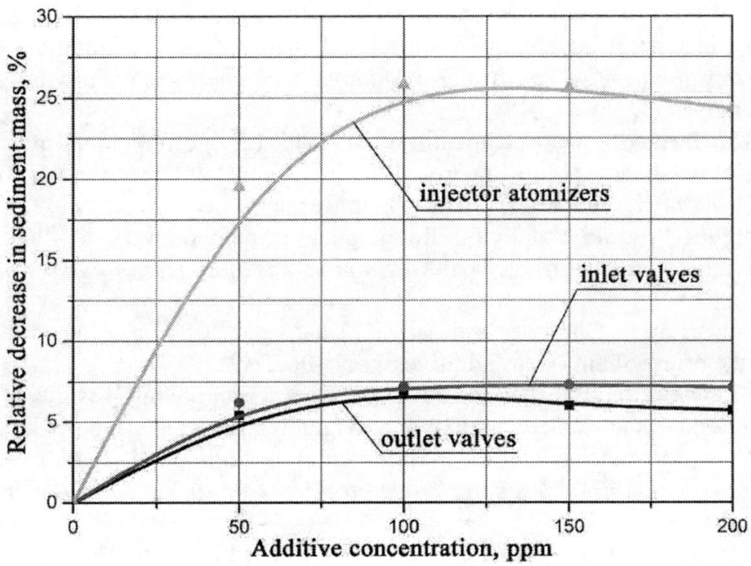

Figure 5. Correlation between cleaning effect (relative decrease in mass of sediments on representative part surfaces) and concentration of the additive for base fuel Nr.3.

Fortunately, both optimum additive concentrations – based on combustion efficiency and minimum deposit – closely agree, which makes it possible to considerably simplify time-consuming and costly multifactorial experiments.

4. Conclusion

The following conclusions may be derived from this investigation:

Nomenclature of base diesel fuels commercially available in Russian Federation is characterized by wide range of principal physicochemical parameters, such as cetane number, fractional composition, kinematic viscosity, flash point and density, considered acceptable. These are parameters that strongly influence mixture formation and fuel combustion processes, and therefore, engine power, fuel efficiency, exhaust gas toxicity and opacity.This conclusion is supported by the results of comparative stand tests of diesel engine firing fuels that significantly differ in basic physicochemical parameters.Mean spread in fuel consumption averaged for the test cycle amounts to 2.2%. Similar spread in exhaust gas opacity is still more pronounced – up to 12…15%.

Testing of diesel fuels modified with multi-component additive revealed that effectiveness of the additive generally depends on is concentration. Optimum additive concentration widely varies (75…160 ppm), depending on base fuel composition. Total effect of the additive usage is noticeable, amounting to 3.5% in terms of fuel economy and to 22% in terms of exhaust opacity reduction. Besides, the additive considerably lowers sedimentation on exposed surfaces of combustion chamber, inlet and outlet valves. Thus modification of base fuel with the multi-component additive appreciably improves its performance.

References

[1] Danilov A M 2014 Additives as a Solution to Chemmotological Problems *Chemistry and Technology of Fuels and Oils* **50(5)** 406–10

[2] Grażyna Żak, Leszek Ziemiański, Zbigniew Stępień and Michał Wojtasik 2014 Engine testing of novel diesel fuel detergent–dispersant additives *Fuel* **122(15)** 12-20

[3] Alexanyan K, Stokolos O, Solodova E, Zaytceva Y, Salmanov S Y, Yarullin N, Naletova A and Mikhailov E 2018 Istoriya razvitiya i primeneniya antiokislitel`ny`x prisadok dlya topliv i masel [The history of development and applying of antioxidant additives for fuel and oil] *Izvestiya vysshikh uchebnykh zavedenii Khimiya khimicheskaya tekhnologiya* **61(9-10)** 120-5 [In Russian]

[4] Borzaev B Kh, Karpov S A and Kapustin V M. 2007 Mnogofunkcional`ny`e dobavki k avtomobil`ny`m benzinam [Multifunctional additives for automotive gasolines] *Khimiya i tekhnologiya topliva i masel* **43(2)** 108-13 [In Russian]

[5] Danilov A M 2001 Prisadki k toplivam. Razrabotka i primenenie v 1996—2000 gg. [Fuel Additives: Evolution and Use in 1996-2000] *Khimiya i tekhnologiya topliva i masel* **6** 43-50 [In Russian]

[6] Shilpi Agarwal, Vijay K. Chhibber and Ajay K. Bhatnagar 2013 Tribological behavior of diesel fuels and the effect of anti-wear additives *Fuel* **106** 21-9

[7] Shabanov A, Zaitsev A and Mokhnatkin E 2008 Svyaz' gruppovogo sostava topliva i motornyx i e'kologicheskix pokazatelej benzinovogo dvigatelya [Correlation between fuel composition and petrol engine mechanical & environmental performance] *Trudy V Mezhdunarodnoj nauchno-prakticheskoj konferencii "Novye topliva s prisadkami"* (St. Petersburg) pp 60-3 [in Russian]

[8] Mokhnatkin E, Smerechuk V and Shabanov A 2008 Stendovye motornye ispytaniya avtomobil'nyx topliv kak vazhnejshaya chast' sistemy dobrovol'noj sertifikacii goryuche-smazochnyx materialov [Test-bed measurements of automotive fuel properties as a core part of fuels and lubricants voluntary certification system] *Trudy V Mezhdunarodnoj Nauchno-Prakticheskoj Konferencii "Novye topliva s prisadkami"* (St. Petersburg) pp 31-6 [in Russian]

[9] Poletaeva O Y, Movsumzade E M, Kolchina G Y and Bakhtina A Y 2018 Vliyanie stroeniya molekul vy`sokooktanovy`x komponentov benzinov i antiokislitel`ny`x prisadok k toplivam na e`ffektivnost` ix dejstviya [Influence of molecular structure of high octane gasoline components and fuel antioxidant additives on efficiency of their action] *Izvestiya Vysshikh Uchebnykh Zavedenii Khimiya Khimicheskaya Tekhnologiya* **59(12)** 49-56 [In Russian]

[10] Jehlik F, Lohse-Busch H, Iliev S and Hall C 2019 Capturing the Impact of Fuel Octane Number on Modern Gasoline Vehicles with Octane Indices *SAE Int. J. Fuels Lubrricants* **12(2)** 63-86

[11] Kontses A, Dimaratos A, Keramidas C, Williams R, Hamje H, Ntziachristos L and Samaras Z. 2019 Effects of fuel properties on particulate emissions of diesel cars equipped with diesel particulate filters *Fuel* **255(1)** 115879

[12] Spirkin V G, Lykov O P and Bel'dii O M 2001 E`kologicheski bezopasny`e dobavki dlya dizel`ny`x topliv [Environmentally Safe Additives for Diesel Fuels] *Khimiya i tekhnologiya topliva i masel* **6** 29-31 [In Russian]

[13] Fayyazbakhsh A, Pirouzfar V 2016 Investigating the influence of additives-fuel on diesel engine performance and emissions: Analytical modeling and experimental validation *Fuel* **171(1)** 167-77

[14] Zhang W, Ding H, Shuai S, Zheng B, Cantlay A, Natarajan V, Zhan Z S and Pu Y 2017 Effect of Fuel Detergent on Injector Deposit Formation and Engine Emissions in a Gasoline Direct Injection (GDI) Engine *SAE Technical Paper 2017-01-2247*

[15] Kak A, Kumar N, Gurjar B, Singh S and Gupta D 2015 Comparative Study of Emissions and Performance of Hydrogen Boosted SI Engine Powered by Gasoline Methanol Blend and Gasoline Ethanol Blend *SAE Technical Paper 2015-01-1677*

[16] Rockstroh T, Floweday G and Wilken C 2016 Options for Use of GTL Naphtha as a Blending Component in Oxygenated Gasoline *SAE Int. J. Fuels Lubrricants* **9(1)** 191-202

IOP Conf. Series: Materials Science and Engineering **791** (2020) 012074 doi:10.1088/1757-899X/791/1/012074

The control of lighting up regime of greenhouse plants with LED irradiators

S K Sheriazov[1], S A Popova[1] and I I Karimov[2]

[1]Department of Energy Engineering, South Ural State Agrarian University Lenin Avenue, 75, Chelyabinsk, 454080, Russia
[2]Department of Energy Engineering, Bashkir state agrarian University34, 50-letia October St., Ufa, 450001, Russia

E-mail: sakenu@yandex.ru

Abstract. A system for controlling the lighting up regime is proposed in the article, which gives the opportunity to reduce the cost of consumed electricity. The main element of the proposed control system is a processor, which calculates the necessary power of LED irradiators and the lighting up regimes according to a specially programmed program. For selecting the power and the regime of radiation of a given spectrum, a mathematical model of the geometry of the plant stalk in the process of its growth, obtained from the experimental data, is given. A two-factor experiment was carried out to determine the quantitative effect of the red and blue spectra emitted by LED sources on the geometry of the formed stalk of seed potatoes. The paper presents the results of the experimental studies on a genetically homogeneous material of potatoes grown from meristematic cells, making it possible to obtain reliable responses to different spectral composition of blue and red radiation of LED irradiators. The given system of controlling the lighting up regime allows growing the healthy plants in optimal photosynthesis with the minimal costs of electricity.

1. Introduction

The main source of light of modern greenhouses is sodium lamps, which limit the control of the process of lighting up though it is possible. [1, 2]. The lighting up process lasts for 20 hours which is not economically correct.

In one's time, the experiments were carried out to study the intensity of photosynthesis of cucumber plants at the development stage of 2–4 leaves and the diagrams were obtained that clearly showed a sharp decrease in photosynthesis after several hours of intensive lighting up [3, 4]. At the same time, the plant breathes intensively and consumes accumulated nutrients. Consequently, it is necessary to find the correct lighting up regime and to optimize it using more modern irradiators which permit to control the process.

Nowadays, there has been a tendency to use more modern and economical light sources, the production technology of which is being improved, and the production becomes cheaper every year. These are LED-irradiators, consisting of a certain number of LED elements. They have already come to industrial greenhouses and they convert electricity not into heat, but into a certain emission spectrum, the one that plants need most of all [5, 6]. However, the experiments with LED irradiators are still at the initial stage, and so far a small number of plants have passed through the stage of the primary researches. However traditional long-lived greenhouse vegetable plants with a growing season of up

IOP Conf. Series: Materials Science and Engineering **791** (2020) 012074 doi:10.1088/1757-899X/791/1/012074

to 10-12 months are still little studied and the information on this subject appears in literary sources rarely [7-9].

The purpose of lighting up of the plants is to attain the correct geometry of the stalk in the process of its growing. Only then we can be sure that the quality of the seeds from the healthy plant will be the best.

2. Materials and Methods

In the process of growing greenhouse plants it is important to control the process of lighting up. The process of irradiation is the lighting up the plants with a given spectrum and the power of emission.

The research of the influence of the red-blue emission spectrum of LED in the process of lighting up the plants has shown the regularities. Maximum photosynthesis is possible only with the optimal combination of the percentage of blue and red spectra with the obligatory radiation from the lamps of a wide solar spectrum [10-12].

The emission spectra in the process of irradiation have a different effect on the geometry of the greenhouse plants. So an unreasonable lighting up with given emission spectra can make the stalk of a plant oblong. Then the number of leaves will be decreased, there will be a large distance between the leaves, which reduces the intensity of photosynthesis. It is also possible that the plant can slow down its growth, as the number of fruit ovaries will decrease.

Then it is necessary to control the emission parameters according to a given algorithm. It requires the development of the mathematical model that relates emission parameters to plant growth rates.

The mathematical models can and should be used to control the technological processes. This feature of automatic control systems permits to relate the consumed resources properly to gain high-quality products.

Modern capabilities of science and technology allow gaining the mathematical models of various processes occurring in biological objects. Some models are incredibly complex, but rather accurately allow us to predict the rate of plant development, a set of their biological mass, and the geometry of the stalk and fruit changes [13]. The others include regression-type models [14], and they are able to quantify the effect of various combinations of environmental factors on the vital activity of plants, without in-depth analysis of the biological processes occurring in them. At the same time, not all factors can be used, but only a part of them, leaving others outside the plan of experimental studies. These models include the mathematical models of the intensity of photosynthesis or dark breathing [15].

The factors affecting the vital activity of plants, but not included in the experiment, are nominally present in it as averaged parameters of the habitat without the possibility of their variation. In this case, the experiment will be recognized as correct, and the results of the study calculated from the model can be used to regulate the conditions of plant growing in the greenhouses [16, 17].

At present, there is no recommendation on the lighting up of greenhouse plants, the implementation of which would allow growing the healthy plants with optimal photosynthesis with the minimal costs on electricity [18]. Then it is relevant to study the influence of emission spectra on the vital activity of plants, and to learn how to control them to minimize the necessary costs for the production of greenhouse products

2.1. Formulation of the problem

The experimental researches are needed to study the influence of the spectral composition of LED irradiators on the vital activity of greenhouse plants. The experimental technique should include special climatic chambers or phytotrons.

According to the experimental data, it is necessary to develop the mathematical model of plant growth, depending on the emission spectrum and the irradiation regime. They will further optimize the regimes and control the process of irradiation.

IOP Conf. Series: Materials Science and Engineering **791** (2020) 012074 doi:10.1088/1757-899X/791/1/012074

Today, greenhouses are not massively equipped with LED-sources. Therefore, to determine the optimal spectral composition of emission from LED light sources, the expensive experimental studies have been required.

To study the behavior of a potato plant at different levels of the blue-red emission spectrum of irradiators based on LED-elements, a photo set was created. The final aim of the experiment is to attain the mathematical model of the growth and the diameter of the stalk. On the basis of the obtained model, it is necessary to create an algorithm for controlling the growth characteristics of a plant in the course of its vital activity with the minimal lighting up costs [19-21].

2.2. Research methods.
During the research, it is important to identify the emission spectra necessary for a plant life. At the same time, plants with a short lifespan and a rapid onset of the generative phase of development have been studied. Most of the experiments were carried out on leafy green plants or potatoes, to obtain elite seeds [22-25].

The setup shown in Figure 1 was used for the research; it consists of LED lights, a control unit, as well as a measuring complex based on the ACP ZET–210. Plant growth (height) was determined using the ODROID-C1 + minicomputer, which is capable of processing frames captured by a web-camera. The level of irradiance of the space above the plants was regulated by means of a controller and a specially created computer program in accordance with the developed algorithm. The processing of the information obtained during the experiment was carried out by certified programs [26].

Figure 1. The general view of the experimental setup.

To take the main characteristics of the lighting fitting, a certified instrument TKA-Spectrum Spectrophotometer was used. The level of irradiation was measured with an APDS – 9250 light sensor which was calibrated with a spectrophotometer.

The emission level of the LED-elements was regulated by the number of switched on LEDs in the work, while it was possible to provide the required spectral composition. The microclimate inside the phytobox was maintained by a special control unit.

IOP Conf. Series: Materials Science and Engineering **791** (2020) 012074 doi:10.1088/1757-899X/791/1/012074

The experiment was realized on a genetically homogeneous material of potatoes grown from meristematic cells and the research materials are given in [27]. This approach permits to gain reliable responses to the influence of different spectral composition of the blue-red emission of LED - irradiators.

A two-factor experiment was carried out to determine the quantitative effect of the red and blue spectra emitted by LED sources on the geometry of the formed stalk of seed potatoes. Table 1 shows the planning matrix of the experiment in coded units, and Table 2 shows the values of the levels and the intervals of variation of the specified factors.

Table 1. The levels of factors variation.

№ experiment	Levels of variation red X1	blue X2
1	–1	–1
2	+1	–1
3	–1	+1
4	+1	+1
5	–1	0
6	+1	0
7	0	–1
8	0	+1
9	0	0

Table 2. Emission intensity for a given level of variation.

Name and designation of factors		Emission intensity, % –1	0	1	Variation intervals
The intensity of red spectrum, %	X1	40	60	80	20
The intensity of blue spectrum, %	X2	40	60	80	20

The experiments carried out with the purpose of gaining the mathematical models are justified if they allow the use of modeling results for the purpose of controlling the most important plant growth factors.

3. Results

In the process of experimental researches, the influence of LED - irradiation on the potatoes, the data array was obtained. After processing the results, the dependences shown in Figure 2 were obtained.

As can be seen from the given graphs the dependences of the growth of the length of the plant stalk has the form of an inverted parabola and the graph of the diameter of the stalk has the appearance of a conventional parabola. A combined solution to optimize the two radiation spectra is not possible.

Then, using the experimental data, it is necessary to obtain the mathematical models for each considered geometric size of the stalk, depending on the emission spectrum. The mathematical models permit to investigate how the geometry of a potato stalk will vary with different combinations of the examined spectra of LED emission sources [27].

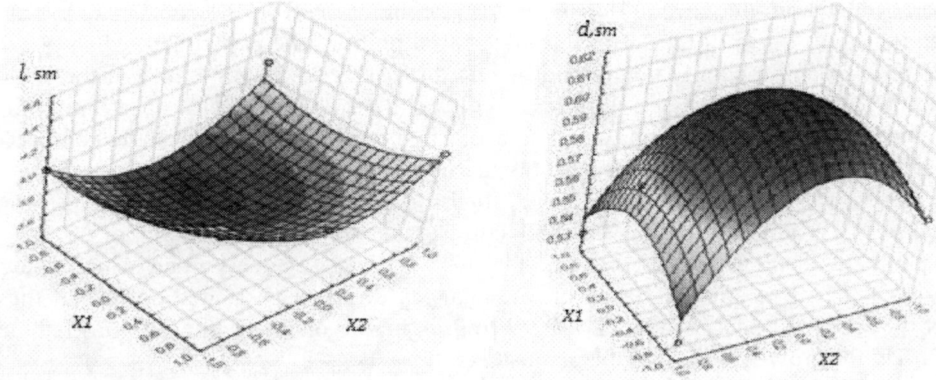

Figure 2. The dependences of the growth of the length (a) and the diameter (b) of the plant stalk with lighting up of LED sources.

As a result of processing the experimental data, a mathematical model of the stalk length increase of the following type was developed:

$$l_{st}(X1,X2) = 3.6222 + 0.1667\,X1 - 0.1\,X2 + 0.1667\,X1^2 - 0.075\,X1\,X2 + 0.366\,X2^2 \quad (1)$$

and for defining the diameter of the incremental segment of the stalk:

$$d_{st}(X1,X2) = 0.6089 - 0.0067\,X1 - 0.0033\,X2 - 0.0233\,X1^2 + 0.005\,X1\,X2 - 0.0433\,X2^2 \quad (2)$$

The given mathematical models allow us to construct the graphical dependencies of the frontal projections and to analyse them. Figure 3 shows the frontal projection of graphs of the increment of the length of the new plant stalk segments (the curves 1, 2, 3) and the diameters of these segments depending on the change in the amount of the red component of the LED spectrum — X1 emission. In this case, the values of emission in the blue spectrum are strictly fixed at the conditional points of the experiment, i.e. when X2 = –1; 0 and +1 in accordance with the adopted plan, shown in the table 1.

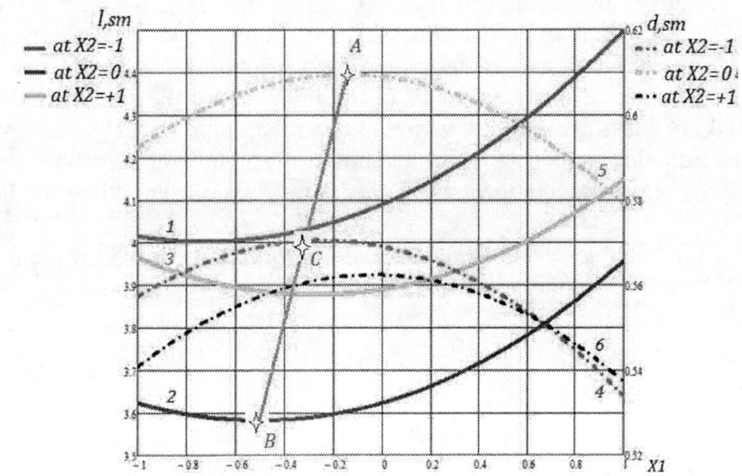

Figure 3. The dependences of the increment l_{st} and d_{st} from the coded factors X1 and X2.

IOP Conf. Series: Materials Science and Engineering **791** (2020) 012074 doi:10.1088/1757-899X/791/1/012074

4. Discussion

The experimental curves 1, 2, 3 increments in the stalk length of potatoes have a clear shown minimum. Moreover, the minimum point of the graphs begins to shift in the coordinate system from the conditionally negative area of the X1 axis to the central point of the experiment when the X2 parameter for the blue spectrum changes. It should be noted that the maximum length of the stalk growth is expected with the parameters X2 = –1 and X1 = +1, i.e. the smaller the blue spectrum in the emission of the LED source; the less the plants are stretched. Therefore: the red spectrum helps to accelerate the growth of the stalk, and the blue spectrum inhibits the ability of plants to stretch.

The arrangement of the group of the curves for lst (1,2,3) indicates that the smallest growth of the stalk (the curve 2) is achieved if X2 = 0, which corresponds to the central point of

the experiment. The curve (3), corresponding to X2 = +1, is much higher, which does not correspond to the logic of reasoning. Based on the analysis of the curves, it follows that the minimum growth of the stalk will occur when X2 = 22 W/m2. A higher or lower intensity of the blue spectrum does not affect the growth indices of the plant stalk.

The graphical interpretation of the equation (2) is presented in Fig. 3. The curves 4, 5, 6, characterize changes in the diameter of the stalk increment depending on the intensity of the LED emission in the red and blue spectra.

The analysis of the data shows that the group of graphs looks like an inverted parabola. Optimum graphs are the maximum values of the diameters of the incremental portion of the stalk. The curve 4, at X2 = 0 is located above the rest and the optimum of the curves 5 and 6 are shifted along the X1 axis relative to the optimum of the curve 4 to the left and to the right, but the stalk diameter is almost at the same level.

It can be preliminarily concluded that the intensity of LED emission in the blue spectrum, as in red, does not contribute to an excessive increase in the diameter of the stalk. The optimum at X1 = +1, less than optimum at X1 = –1, although not significantly, but less than optimum at the center of the experiment (see tab.) At X1 = 0. It means that in reasonable doses (X1 = 0) the blue spectrum also affects the geometry of the stalk, thickening its diameter.

For practical application of the results of the experiment, a compromise should be found between the amount of emission of LED-irradiators in the red and blue spectrum, in particular, to obtain a potato plant with the suitable stalk geometry. At the same time, the mathematical models of the length and the diameter of the stalk do not have the point of intersection that satisfies the solution for optimizing the lighting up regimes.

To solve this problem, we use the method of the direct compromise, examined in several articles [28]. Graphically, it looks like this. From the optimum of the stalk diameter graph dst at X2 = 0 (the curve 5) to the minimum of the stalk growth graph lst (the curve 2), we draw a straight line segment AB (see figure 3). Then determine the coordinates of the point C, which is the geometric midpoint of the segment AB On the abscissa axis, determine the corresponding value X1 = 0.3, and on the ordinate axis = 4 cm. The values of the coordinates of the midpoint of the segment AB determined by the compromise method are substituted into equation (1) and the value of X2 is determined from the solution of the quadratic equation, which satisfies the compromise and which is the third coordinate.

In fact, the middle point of the compromise C lies in the area of the optima of all the other curves 1, 3 and 4, 6.

Express the coordinate values calculated by the compromise method into the model (1) and determine the value of X2 from the model equation. Provided that X1 = - 0.3, and lst = 4cm, we get:

$$4 = 3.62 + 0.17\,(-0.3) - 0.1\,X2 + 0.17\,(-0.3)2 - 0.06\,(-0.3)\,X2 + 0.37\,X2^2$$

so

$$0.37\,X2_2 - 0.08\,X2 - 0{,}45 = 0$$

$X2_1 = 1.2$ (relative unit) or $X2_2 = -1.05$(relative unit)

With the help of the proportions and the data in the Table 1, you can recalculate and get that:

$$X2_1 = 96\% \text{ or } X2_2 = 42\%.$$

The results of the research show that for the growth of a plant it is offered the lighting up with a blue spectrum using 96% of all the LED elements of the corresponding spectrum, or using only 42%. Naturally, to save the energy, it will be enough to turn on a smaller number of LED elements, and the plant will not be harmed. This is the principle of control according to the mathematical model.

Thus, the percentages of emission are transferred into the energy of one or another emission or to the number of LED-elements included in the lighting up device. The plant will receive a sufficient amount of special emission for its safe growth with the minimal costs on the lighting up.

For the realization of the research results of the influence of the blue and red spectra on the growth and the development of any greenhouse culture, a system of automatic control of the lighting up regime has been developed (Figure 4).

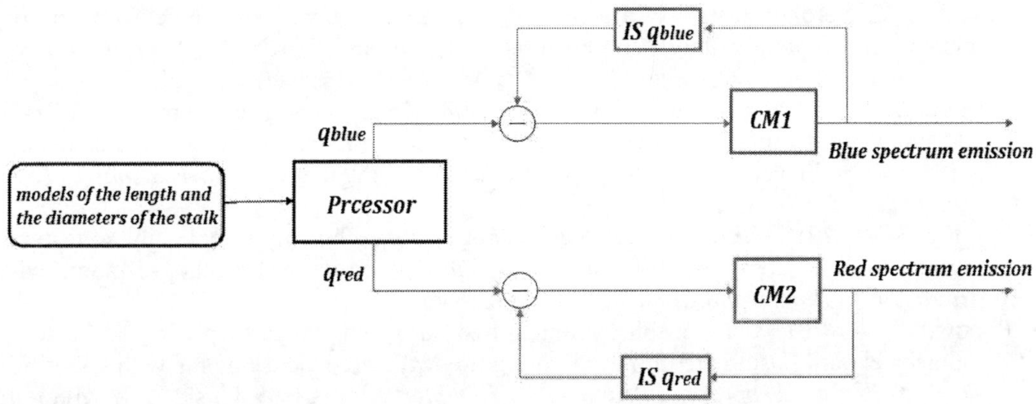

Figure 4. The system of control the optimal LED–lighting up. Initial sensor (IS) qblue and qred – blue and red emission sensor, respectively; CM – control module LED–emission sources; qtbl and qtred – CPU-calculated tasks for optimal emissions in the blue and red spectra.

The important element in the control system is the processor, which, on the basis of the offered mathematical model (1 and 2) and according to a specially programmed program, calculates the power of blue and red emission LED necessary for lighting up. The mathematical models for the growth of potato stalks can be obtained for other crops, like tomatoes, cucumbers and flowers.

The offered system contains sensors for each spectrum of emission from the LED-irradiator. In principle, these sensors are able to determine the total radiation coming from the sun and LED-irradiator. This approach will realize the rational irradiation of plants.

The processor generates a signal to turn on a certain number of blue and red LEDs, the emission of which will contribute to the correct geometry of the formation of new segments of plant stalks. At the same time, the general light can be turned off after 12 hours of operation, which makes it possible to save energy. It is necessary to turn on the LED lighting up on the task calculated by the processor in order to shorten the dark period when breathing and burning of accumulated carbohydrates is carried out.

5. Conclusion

For greenhouse plants it is necessary to light up different emission spectrum, contributing to the proper growth of the plant stalk in the process of its cultivation. The research has revealed that the most important components of the spectrum necessary for a plant life are the red and blue emission of LED

IOP Conf. Series: Materials Science and Engineering **791** (2020) 012074 doi:10.1088/1757-899X/791/1/012074

irradiators. At the same time, plants with a short lifespan and a rapid onset of the generative phase of the development were studied and the experiments were carried out on deciduous green crops or potatoes to receive elite seeds.

The researches of the influence of the red-blue emission spectrum of LED in the process of lighting up the plants have revealed regularity. Thus, maximum photosynthesis is possible only with the optimal combination, in percentage, of the blue and red spectra with the required presence of radiation from the lamps of a wide solar spectrum. The researches of their influence are important, since the blue and red spectra have different effects on the formation of the geometry of the plant stalk.

Processing the experimental data, we have not obtained a single-valued model that allows optimizing the growth processes of the plant. In this case, a compromise decision was made to adapt the obtained models to the intensity of photosynthesis. The results can serve to control the environmental factors depending on the level of emission in the red and blue spectrum for the lighting up and the natural source of light, solar radiation.

References

[1] Kononenko A N 2016 Vliyanie razlichnych istochnikov sveta na razvitie mini-rastenij v usloviyah sverokul'tury [The Influence of various light sources on the development of mini-plants of potatoes under photoculture conditions] *Izvestiya Sankt-Peterburgskogo Gosudarstvennogo Agrarnogo Universiteta* **45** 50-6 [In Russian]

[2] Guzhov S, Polishchuk A and Turkin A 2008 Koncepciya primeneniya svetil`nikov so svetodiodami sovmestno s tradicionny`mi istochnikami sveta [The concept of using lamps with LEDs in conjunction with the traditional light sources] *Sovremenny`e Texnologii Avtomatizacii* **1** 14-8 [In Russian]

[3] Popova S A 2017 Obosnovanie prery`vistogo dosvechivaniya teplichny`x rastenij SD-obluchatelyami [Reasoning of interrupted lighting up of the greenhouse plants with CD-irradiators] *Vestnik IrGSXA* **80** 118-25 [In Russian]

[4] Popova S A 2017 Primenenie impul`snogo dosvechivaniya teplichny`x rastenij SD-obluchatelyami [Using the impulse lighting up of the greenhouse plants with CD-irradiators] *E`nergetika Agropromy`shlennomu Kompleksu* (Chelyabinsk: Yuzhno-ural`skij gosudarstvenny`j agrarny`j universitet) pp 147-56 [In Russian]

[5] Hogewoning S W, Trouwborst G, Engbers G J, Harbinson J, vanIeperen W, Ruijsch J and van Kooten O 2007 Plant Physiological Acclimation to Irradiation by Light- Emittitn Diodes (LEDs) *Proc. XXVII IHC-S6 High-Qual. Crop Prod. under Protect. Cultiv* Ed D J Cantliffe (ISHS Acta Horticulturae vol 761) pp 183-92

[6] Dueck T A, Janse J, Eveleens B A, Kempkes F L K and Marcelis L F M 2012 Growth of tomatoes under hybrid LED and HPS lighting systems *Proc. IS on GreenSys2011* Ed. C. Kittas et al. (ISHS Acta Horticulturae vol 952) pp 335-42

[7] Hogewoning S W, Trouwborst G, Maljaars H et al. 2010 Blue light dose-responses of leaf photosynthesis, morphology, and chemical composition of Cucumis sativus grown under different combinations of red and blue light *J. of Experimental Botany* **61(11)** 3107-17

[8] Poudel P R, Kataoka I and Mochioka R 2008 Effect of red- and blue-light-emitting diodes on growth and morphogenesis of grapes *Plant Cell, Tissue and Organ Culture* **92** 147-53

[9] Galiullin R R and Karimov I I 2016 E`ffektivnost` ispol`zovaniya svetodiodny`x svetil`nikov v teplichny`x xozyajstvax [Efficiency of using LED lamps in the greenhouses] *E`lektrotexnicheskie i Informacionny`e Kompleksy` i Sistemy`* **1** 34-9 [In Russian]

[10] Cummings I J , Reid J B and Koutoulis A 2007 Red to far-red ratio correction in plant growth chambers - growth responses and influence of thermal load on garden pea *Physiologia Plantarum* **131(2)** 171-9

[11] Hogewoning S W, Trouwborst G, Maljaars H, Poorter H, van Ieperen W and Harbinson J 2010 Blue light dose-responses of leaf photosynthesis, morphology, and chemical composition of Cucumis sativus grown under different combinations of red and blue light *J. of*

Experimental Botany **61(11)** 3107-17

[12] Golovatskaya I F, Minich A S, Minich I B and Bolshakova M A 2012 Regulyaciya rosta i razvitiya rastenij Brassica oleracea L. s pomoshh`yu korrekcii solnechnogo izlucheniya [Regulation of the plant growth and the development of Brassica oleracea L. by means of solar radiation correction] *Vestnik Tomskogo Gosudarstvennogo Universiteta. Biologiya* **2(18)** 151-65 [In Russian]

[13] Belonogov A N 1982 *Matematicheskoe modelirovanie produkcionnogo processa ogurcza v tepliczax* [Mathematical modeling of the production process of cucumber in the greenhouses] (Chelyabinsk: UralNIISX) [In Russian]

[14] Popova S A 2010 Matematicheskoe modelirovanie produktivnosti rastenij kak sredstvo povy`sheniya e`ffektivnosti e`nergosberezheniya [The mathematical modeling of the productivity of plants as the means of the increasing the effectiveness of energy-saving] *Vestnik KrasGAU* **7** 141-5 [In Russian]

[15] Sysoeva M I and Markovskaya E F 2001 Temperaturnaya regulyaciya skorosti razvitiya rastenij ogurcza v ontogeneze [Temperature regulation of the development rate of cucumber plants in ontogenesis] *Vestnik Bashkirskogo Universiteta* **2(1)** 164-5 [In Russian]

[16] Popova S A, Rychkova N M and Suprun M A 2016 Sposob avtomaticheskogo upravleniya sveto-temperaturny`m rezhimom v teplice [The method of automatic control of the light-temperature regime in the greenhouse] *Patent №2586923* [In Russian]

[17] Sheryazov S K and Popova S A 2017 Mnogomernaya sistema optimizacii mikroklimata teplicy [Multidimensional system for optimization of greenhouse microclimate] *Nauchno-texnicheskoe obespechenie APK Sibiri* (Novosibirsk: SFNCzA RAN) vol 2 pp 148-53 [In Russian]

[18] Hirai T, Amaki W and Watanabe H 2006 Action of blue or red monochromatic light on stem internodal growth depends on plant species *V International Symposium on Artificial Lighting in Horticulture* (ISHS Acta Horticulturae vol 711) pp 345-50

[19] KilianaD G, Bartkowiak, Kazda M and Kaufmann D 2014 Modelling the growth of plants with a uniform growth logistics *J. of Theoretical Biology* **349** 57-65

[20] Tamulaitis G, Duchovskis P, Bliznikas Z, Brazaityte A, Novickovas A and Zukauskas A 2005 High-power light-emitting diode based facility for plant cultivation *J. of Phys. D: Appl. Phys.* **38** 3182-7

[21] Gomez C, Morrow R C, Bourget C M, Massa G D and Mitchell C A 2013 Comparison of intracanopy light-emitting diode towers and overhead high-pressure sodium lamps for supplemental lighting of greenhouse-grown tomatoes *HortTechnology* **23(1)** 93-8

[22] Urbonaviciute A, Pinho P, Samuoliene G, Duchovskis P, Vitta P, Stonkus A, Tamulaitis G, Zukauskas A and Halonen L 2007 Effect of Short-Wavelength Light on Lettuce Growtli and Nutritional Quality *In Scientific Works of the Lithuanian Institute of Horticulture and Lithuanian University of Agriculture - Sodininkyste ir Daržininkyste* vol 26 pp 157-65

[23] Martirosyan Yu Ts, Polyakova M N, Dilovarova T A and Kozobryukhov A A 2013 Fotosintez i produktivnost` rastenij kartofelya v usloviyax razlichnogo spektral`nogo oblucheniya [Photosynthesis and productivity of potato plants under various spectral irradiation] *Sel`skoxozyajstvennaya biologiya* **1** 811-22 [In Russian]

[24] Yingchao Xu, Yongxiao Ch, Guanyu Ch and Hongyi L 2016 The research on LED supplementary lighting system for plants *Optik – Int. J. for Light and Electron Optics* **127(18)** 7193-201

[25] Fokin A A and Gordeev A S 2013 E`ksperimental`ny`e issledovaniya vliyaniya parametrov svetodiodny`x svetil`nikov na urozhajnost` zelennogo luka pri e`lektrodosvechivanii [Experimental studies of the effect of the parameters of LED lamps on the yield of green onions during electric lighting] *Vestnik Michurinskogo Gosudarstvennogo Agrarnogo Universiteta* **5** 59-62 [In Russian]

[26] Popova S A and Karimov I I 2017 E`ksperimental`naya fitokamera dlya issledovaniya faktorov

rosta rastenij v usloviyax zakry`togo grunta [Experimental phytocamera for the study of the plant growth factors in the greenhouse conditions*]* *Vestnik IrGSXA* **81-1** 153-60 [In Russian]

[27] Karimov I I 2017 *Razrabotka i issledovanie svetodiodny`x svetil`nikov s izmenyaemy`m spektrom izlucheniya dlya teplichny`x xozyajstv* [The development and the research of LED lamps with variable radiation spectrum for the greenhouses] (Troiczk: Yuzhno-Ural`skij gosudarstvenny`j agrarny`j universitet) [In Russian]

[28] Izakov F Ya 2003 *Planirovanie e`ksperimenta i obrabotka opy`tny`x danny`x* [Experiment planning and experimental data processing] (Troiczk:: Yuzhno-ural`skij gosudarstvenny`j agrarny`j universitet) [In Russian]

IOP Conf. Series: Materials Science and Engineering **791** (2020) 012075 doi:10.1088/1757-899X/791/1/012075

The processing of kieselguhr sludge with obtaining a new end product as reserve for reducing resources consumption of vegetable oil manufacture

I V Starostina, E V Porozhnyuk and D V Stolyarov

Department of Industrial Ecology, Belgorod State Technological University named after V.G. Shukhov, Kostyukov St., 46, Belgorod, 308012, Russia

E-mail: starostinairinav@yandex.ru

Abstract. The article presents the research findings of processing kieselguhr sludge – a waste product of refined sunflower oil production, generated at winterization stage, when vegetable waxes, waxlike substances and other components, which crystallize at low temperatures and cause the oil turbidity, are removed. The resulting waste product is a slow stock of organo-mineral composition with organic matter content 65-70%. This sludge is virtually not recycled, but dumped at municipal solid waste disposal sites. This results in the environment pollution and non-recoverable losses of high-quality mineral stock and valuable organic components. It has been determined that after thermal modification of sludge waste the organic substances in it are partially oxidized with the formation of carbon layer on the surface of diatomite particles. The resulting product is a thermally modified kieselguhr sludge, which is characterized with mesoporous structure, high water repellency, and oil-receptivity and can be used as an adsorption material for purifying wastewaters from emulsified petroleum products in dynamic conditions. The secondary use of kieselguhr sludge would allow reducing the amounts of waste accumulation and improving the efficiency of raw material resources utilization.

1. Introduction

Vegetable oil extraction is one of the most intensively developing branches of food industry. There are four major producers of vegetable oil in the world market - Ukraine, Russia, European Union and Argentina, which account for about 70% of the global oil production. In the structure of global vegetable oil production the palm oil prevails - 38%, while the soya oil ranks second in popularity, its share amounting to 33% in 2016. The sunflower and coleseed oils, produced mostly in the countries of the Eurasian Economic Union (EAEU), make up 10 and 16% of the global production, respectively (figure 1).

The vegetable oil market in the Russian Federation has its own pronounced specifics: the largest share of production and consumption accounts for sunflower oil. In 2016 the amount of its production was 3.8 mln tons. To obtain the high-quality cooking oils they are refined – i.e. purified from related substances, mechanical impurities, phosphatides, vegetable waxes, waxlike substances and other components [1-3], which crystallize at low temperatures and cause the oil turbidity.

Refining means a complicated multi-stage process, which consists of such individual operations as hydration, neutralization, bleaching, winterization and deodorization. Waxes and waxlike substances are removed by means of low-temperature fractionation method, which is called winterization or freezing-out.

IOP Conf. Series: Materials Science and Engineering **791** (2020) 012075 doi:10.1088/1757-899X/791/1/012075

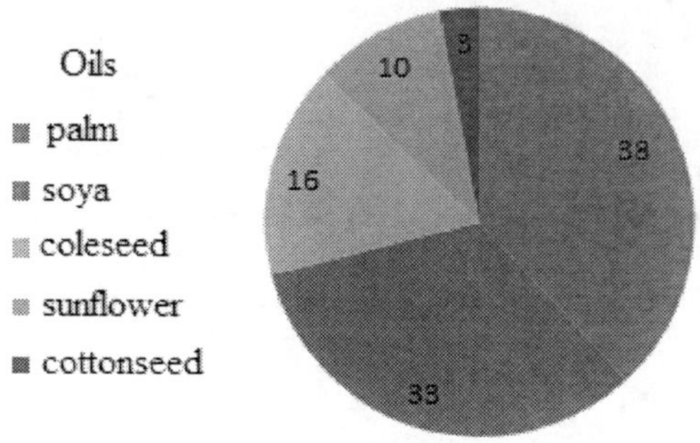

Figure 1. Global vegetable oil production structure in 2016, %

This method consists in the slow cooling of oil to the waxes' crystallization temperature (8-12°C) and exposure to this temperature with the subsequent filtration [3].

Then the organic impurities are removed in pre-coat filtration systems, in which as a filtering element the natural sorption materials are used – kieselguhr, perlite, bleaching clay and others. The sorption materials consumption makes up from 0.1 to 1% of the weight of the oil. As a result, the refined oil is obtained and the waste products are formed, which are called paste-like filter masses or waste sludges.

Together with the extracted impurities, a certain amount of oil is also removed at filtering. The total content of organic impurities in waste sludge can amount to 70%, and the waxlike impurities can make up from 1.5 to 12% depending on the properties of the vegetable oil to be refined. The wax content in the raw sunflower oil depends on the species and varieties of the oil-bearing crop, as well as on the location of the growing area, temperature and oil treatment technology, and can vary from 0.02 to 0.35% [4, 5]. After refining, the content of waxes and waxlike substances is reduced to 0–0.06% [6-8].

Unfortunately, the amounts of generated filter masses at oil-extraction industry enterprises are virtually not accounted by statistics. But, bearing in mind that the average yield of waste sludge can reach 150 kg per 1 ton of refined oil [9], we can calculate the volume of such waste generation at this or that enterprise (table 1).

Table 1. Approximate volume of filter masses generation at some fat-and-oil enterprises of the Russian Federation.

Enterprise	Oil production, mln tons per year	Filter masses formation volume, thousand tons per year
J-SC «Aston»	3.3	490
GC «Sodruzhestvo»	2.9	420
GC «Yug Rusi»	2.8	400
LLC «Solnechnye producty»	1.6	240

The «EFKO» company alone, which ranks second in the Russian Federation in terms of sunflower seeds processing, produces 500 tons of sunflower oil a day, and, consequently, generates up to 70 tons of waste kieselguhr sludge a day.

Filter masses, according to waste classification catalogue, belong to low-hazardous wastes (hazard category 4), which allows the enterprises dumping them at municipal solid waste disposal sites. This results in considerable non-recoverable losses of high-quality mineral stock and valuable organic components.

The problem of rational utilization of oil-extraction industry waste sludges has been paid great attention. They are known to be used as fat components, minor-nutrient elements and biologically active substances in mixed feed formulas for laying hens, as added fat in raising broiler chickens and cattle milking herds [10]. This helps increasing the productivity of animals by 10-20%.

IOP Conf. Series: Materials Science and Engineering **791** (2020) 012075 doi:10.1088/1757-899X/791/1/012075

There are technologies of extracting from the waste filter masses vegetable waxes and oils, which can be used for drying oil production. By using saponification method of waste sludge, the bar soap and soapy pastes were obtained [11]. Filter masses are known to be used as vulcanization activators in industrial rubber goods production technologies [12]. In Europe the oil-extraction industry waste sludges, due to their high energy content, are used as alternative sources of energy instead of the costly hydrocarbon fuel at brick and cement factories. A water-repellent emulsion for protecting construction products and structures from adverse external factors etc. can be also obtained from this waste [13].

There is a method of extracting vegetable wax from the waste sludge [14]. The greater part of oil (up to 80%), contained in the sludge, is removed by water-and-soap solution treatment with the subsequent centrifuging, and the wax particles are extracted with a solvent. The pureness of the obtained wax amounts to 96%.

The oil-contaminated diatomite is known to be used together with galvanic sludge and glass scrap in ceramic materials production [15]. It has been demonstrated that the obtained samples of ceramic bricks have the adequate compressive strength, low water adsorption and no heavy metal leaching, which determines the economic attractiveness of such materials.

The usage of oil-extraction industry waste kieselguhr sludge as a sorption material for removing various pollutants from wastewater is not given much coverage in literature. Chinese scientists have carried out research concerning the usage of waste diatomite as a sorption material for removing herbicides and dyestuffs from water media [16, 17], but they considered the diatomite waste of winemaking industry.

But the above-mentioned variants of using waste sludge don't solve the problem of recycling the whole amount of the generated waste. Besides, it should be taken into account that the permanent storage of filter masses, having high specific surface and organic matter content, at the open area can result in the intensive biodeterioration of impurities, oxidation of hydrocarbonic components with the formation of highly-toxic compounds, or spontaneous combustion of waste and the occurrence of fire-hazardous and explosion-hazardous situations at disposal sites.

Taking into account the growing rates of oil-bearing crops processing and the increasing amounts of vegetable oil, especially sunflower oil, production, the generation and accumulation of sludge wastes is going to increase. This circumstance makes it necessary to rapidly recycle such wastes. So, searching for the new areas of refined vegetable oil production waste filter masses utilization is a crucial task.

The purpose of this work is obtaining on the basis of refined oil production waste sludge a new end product – a carbonaceous sorption material to use it in industrial wastewater purification from petroleum products.

2. Materials and Methods

As an object of research the oil extraction industry waste kieselguhr sludge (WKS) of the «EFKO» plant, Alekseevka city, Belgorod region, Russia, was used. Waste kieselguhr sludge is a dingy-white slow stock with organic impurities content up to 70%. The mineral part of the sludge is represented with diatomite – sedimentary rock, formed by siliciferous frustules of microphytic diatomic algae – diatoms and radiolarians. The main component of the siliciferous frustule (skeleton) is presented with amorphous siliceous hydrates of various hydration degree – opal varieties of the type $mSiO_2 \cdot nH_2O$; the crystalline component is presented with quartz impurities.

The chemical composition is characterized with the following oxides content, wt. %: CaO – 5.85, Na_2O_3 – 2.9, Al_2O_3 – 2.4, Fe_2O_3 – 0.89, MgO – 0.54, K_2O – 0.36, TiO_2–0.14, P_2O_5 – 0.13, SO_3 – 0.06.

The thermal modification of WKS was carried out in an electrical furnace SNO 25/12 within 1 hour, in conditions of low oxygen.

3. Results and Discussion

The research findings, presented in the earlier work, have demonstrated [18], that at the heat treatment of WKS at temperatures from 430 to 600°C within 1 hour in conditions of low oxygen the organic impurities are partially oxidized, with the formation of soot-graphite compositions on the surface of the principal mineral – diatomite – and the generation of a new product – the carbonaceous thermally-modified kieselguhr sludge (TKS). The obtained material has shown high efficiency of extracting heavy metal ions from wastewaters.

This paper considers the possibility of using TKS for removing emulsified petroleum products from wastewaters. It should be mentioned that the heat treatment changes the color of the end product – from brownish gray (at 430°C) through black (at 450-550°C) to dingy white (at 600°C and more). The color of the obtained TKS correlates with the carbon content, forming on the surface of kieselguhr (diatomite) particles (figure 2), and water adsorption of the formed surface (table 2).

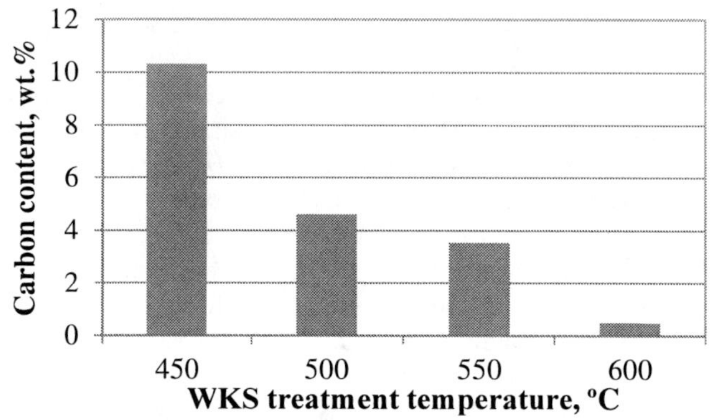

Figure 2. The influence of WKS treatment temperature on the carbon content in TKS.

Table 2. Water adsorption of TKS, obtained at various temperatures.

Parameter	Measuring unit	WKS modification temperature, °C				
		Initial	450	500	550	600
Water adsorption	g/g	0.7	2.0	12.2	21.6	24.2

The highest carbon content – 10.31% is formed at temperature 450°C, which provides the lowest water adsorption. Increasing the temperature up to 600°C results in the disrupting of carbon – it is oxidized to CO_2 and its content is reduced to the lowest value – 0.49% (figure 2). As a result, the hydrophilic properties of the obtained material are increased – water adsorption reaches 24.2 %.

Further on, analyzing the influence of WKS heat treatment on the chemical nature of the resulting products' surface was of interest. The qualitative analysis of functional groups was carried out by means of infrared spectroscopy, and the quantitative analysis – by means of selective neutralization according to [19-21]. The infrared spectra of WKS, treated at temperatures 450 and 500°C and designated TKS_{450} and TKS_{500}, respectively, are presented in figure 3. At the infrared spectra the bands with high intensity at 471, 791 and 1088-1092 cm-1 are registered, which can refer to siloxane groups of Si-O-Si type, which is in agreement with natural diatomite research findings [3].

The absorption bands are found, which can refer to phosphates (1000-1100 cm-1), and to C-OH groups [22]. The band at 791 cm-1 can be associated with valence vibrations νs(Al-O) of clay impurity minerals, contained in diatomite. The high-intensity bands at 2926, 2853 cm-1, which corresponds to the range (2850 – 3000 cm-1), can refer to valence vibrations, and the medium-intensity band at 1470 cm-1 – to bending vibrations of methylene groups, characterizing the presence of Csp3-H bond on the carbonaceous surface of TKS_{450}. With the increase of the treatment temperature up to 500°C their intensity goes down to zero, which indicates the burning-out of the surface carbon forms, containing fragments of C-H.

IOP Conf. Series: Materials Science and Engineering **791** (2020) 012075 doi:10.1088/1757-899X/791/1/012075

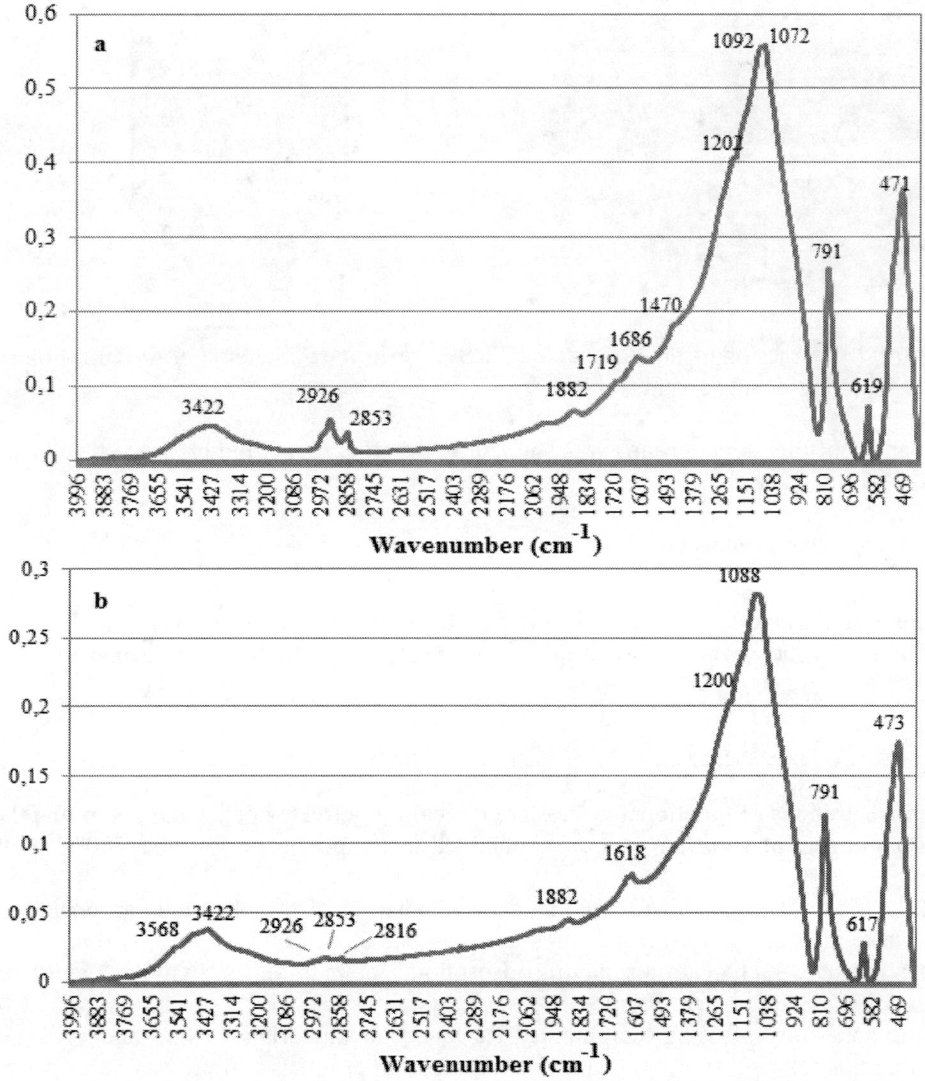

Figure 3. Infrared spectra of carbonaceous materials - TKS_{450} (a) and TKS_{500} (b).

The valence vibrations at 1719 cm-1 characterize the C=O bonds in compound ethers, contained in vegetable waxes and waxlike substances [23, 24], which provides TKS_{450} with high surface hydrophobicity – water adsorption is no more than 2% (table 2). In figure 4 we can see that at applying water on the surface of the compacted material a sitting drop is formed, while petroleum soaks in. So, the material has oleophilic properties.

Increasing the kieselguhr sludge modification temperature up to 500°C results in decomposition of waxlike substances, which influences the infrared spectrum – the valence vibrations in the range 1700-1750 cm-1 are not registered, and the hydrophobicity degree of TKS_{500} particles surface is reduced – water adsorption amounts to 12.2% (table 2).

The infrared spectrum range 3100-3700 cm-1 characterizes the valence vibrations of various types of hydroxyl groups, including those in the water molecules, adsorbed on hydroxyl groups. The band 3422 cm-1 can probably refer to both C-O-H groups, and to silanol groups Si-O-H [22], which can effectively bind heavy metals by the ion exchange mechanism [21].

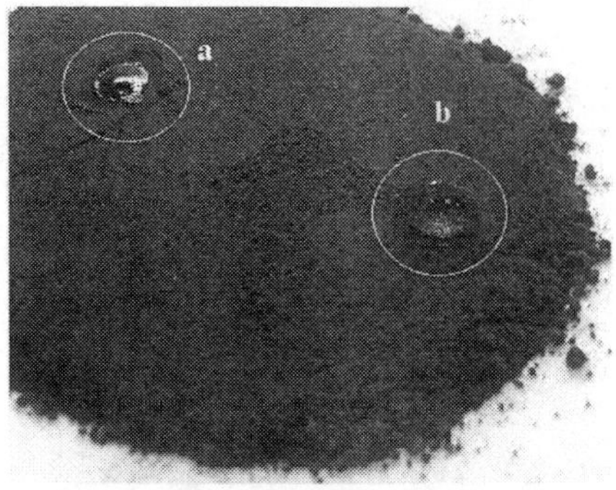

Figure 4. Interaction of the compacted TKS_{450} layer with petroleum (a) and water (b).

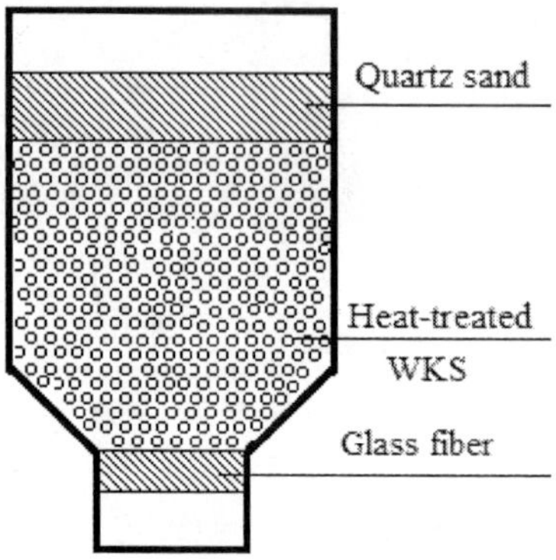

Figure 5. Scheme of the filtration plant.

The quantity of functional groups was studied by the Boehm method [19]; the results are presented in table 3.

Table 3. Functional groups content on the surface of TKS, obtained at various temperatures (10^{-3} mg eq/g).

Temperature of WKS treatment (°C)	Carboxyl groups	Total carboxyl and lactonic groups	Lactonic groups	Total carboxyl, lactonic and hydroxyl groups	Hydroxyl groups
450	4.967	7.336	2.369	8.775	1.439
500	0.503	1.147	0.644	3.477	2.330

Low concentrations of functional groups are typical for activated coals and determine the similarity of the carbonic part of sorption materials, obtained at temperatures 450 and 500°C, with activated coals.

So, the carried-out research of the chemical composition of the obtained carbonaceous materials, based on waste kieselguhr sludge, indicates the presence of carbon and various functional groups on the particles surface. Such materials, having adsorption centers of various natures, would exhibit activity to both polar and nonpolar substances.

Then the possibility of using the obtained carbonaceous materials TKS_{450} and TKS_{500} as sorbents for extracting nonpolar substances – emulsified petroleum products – from wastewaters was analyzed. The sorption properties of TKS_{450} and TKS_{500} to petroleum products were studied in dynamic conditions by filtering model emulsions through a stationary bed of a sorbing agent. For this purpose a filtration plant was prepared, the scheme of which is presented in figure 5.

Sand and adsorbents were placed in a column 30 mm in diameter; the lower layer of the absorption material was 40 mm thick, and the upper sand layer - 2 mm thick. The necessity of using sand as a weight is explained by the high buoyancy of the sorbents under study. The sand was preliminarily washed in distilled water and tempered in muffle furnace at temperature 500°C within 3 hours. The absorption materials - TKS_{450} and TKS_{500} were used in fractions 0.63-1 mm.

As a model wastewater, the petroleum products emulsion with concentration 15 mg/dm3 was prepared, by mixing a precisely weighed quantity of industrial oil I-20A with density 0.818 g/dm3 and the distilled water within 48 hours.

The flow of model wastewater in the filter was performed from top downward with throughput rate 2.5 cm3 per min.

The efficiency of model wastewater purification from emulsified petroleum products was evaluated by the alteration of emulsions turbidity.

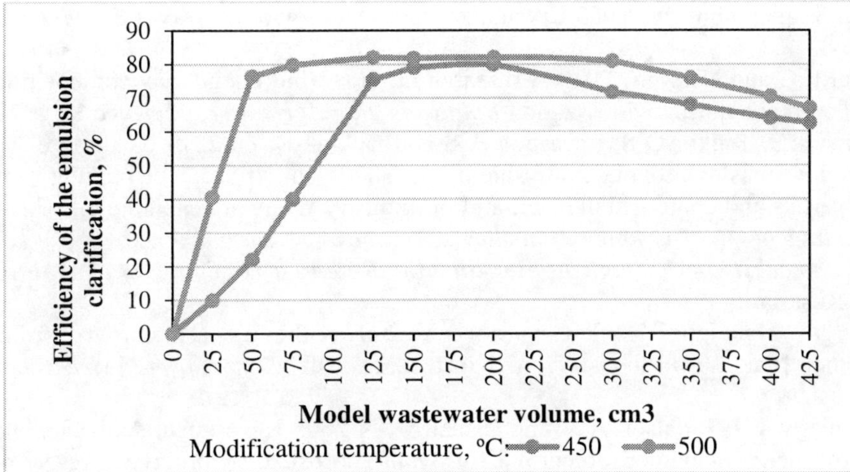

Figure 6. The influence of WKS modification temperature on the efficiency of extracting emulsified petroleum products in dynamic conditions.

The research findings, presented in figure 6, have shown that the temperature of WKS modification has no substantial impact on the maximum efficiency of extracting emulsified petroleum products. This characteristic amounted to 80% and 82% for TKS_{450} and TKS_{500}, respectively. But when using TKS_{500} the adsorption filter reaches its efficient mode after the throughput of 75 ml of model wastewater, and when using TKS_{450} the high purification efficiency (79%) is reached after wastewater throughput in amount of 150 ml.

So, the thermally-modified oil-extraction industry waste kieselguhr sludge can be used as a carbon-containing sorbent for wastewater purification from petroleum and petroleum products in dynamic conditions. The highest sorption properties to emulsified petroleum products are demonstrated by the materials, obtained at temperatures 450 and 500°C, designated TKS_{450} and TKS_{500}, respectively.

The secondary use of kieselguhr sludge would allow reducing the amounts of waste accumulation, decreasing the anthropogenic load on the environment and improving the efficiency of raw material resources utilization.

Acknowledgments
The work is realized in the framework of the Program of flagship university development on the base of the Belgorod State Technological University named after V.G. Shoukhov, using equipment of High Technology Center at BSTU named after V.G. Shoukhov.

References

[1] Turkulov J, Dimic E, Karlovic Dj and Vuksa V 1986 The effect of temperature and wax content on the appearance of turbidity in sunflowerseed Oil *J. Am. Oil Chem. Soc.* **63** 1360-3

[2] Coutinho C et all 2009 State of art of the application of membrane technology to vegetable oils: A review *Food Res. Int.* **42** 536-50

[3] Butina E A 2012 Primenenie otbelivayushchih zemel na osnove diatomita dlya otbelki rastitelnyh masel [The usage of diatomite-based bleaching earth for vegetable oils bleaching] *Masla i zhiry* [*Oils and Fats*] **2** 17-9 [In Russsian]

[4] Morrison W H 1983 Variation in the Wax Content of Sunflower Seeds with Location and Hybrids *J. Am. Oil Chem. Soc.* **60** 1013–5

[5] Carelli A A, Frizzera L M, Forbito P R and Crapiste G H 2002 Wax composition of sunflower seed oils *J. Am. Oil Chem. Soc.* **79** 763-8

[6] Martini S and Anon M C 2000 Determination of Wax Concentration in Sunflower Seed Oil

J. Am. Oil Chem. Soc. **77** 1087-93

[7] Martini S and Anon M C 2003 Crystallization of sunflower oil waxes *J. Am. Oil Chem. Soc.* **80** 525-32

[8] Ergonul P G and Nergiz C 2013 A research on the turbidity and wax composition of sunflower and corn oils during winterization *Bulgarian J. of Agricultural Science* **19** 497-502

[9] Kosulina T P, Tsokur O S, Levashov A S and Lukina D Yu 2013 Nekotorye svojstva i sostav othodov maslozhirovoj promyshlennosti stadii vinterizacii rastitelnogo masla [Certain properties and composition of fat-and-oil industry waste at vegetable oil winterization stage] *E`kologicheskij vestnik nauchny`x centrov chernomorskogo e`konomicheskogo sotrudnichestva* [*Ecological Bulletin of Research Centers of the BSEC*] **4** 69-75 [In Russsian]

[10] Lisitsin A, Machigin V and Grigorieva V 2007 Othody maslozhirovoj promyshlennosti v kormah [Fat-and-oil industry waste in feeding stuff] *Kombikorma* [*Mixed fodders*] **1** 74 [In Russsian]

[11] Pochernikov V I, Rafalson A V and Lisitsin A N 2005 K voprosu ispolzovaniya otrabotannyh zhirosoderzhashchih adsorbentov i filtrovalny poroshkov v proizvodstve tverdogo myla [On the question of using waste fat-containing adsorbents and filter powders in hard soap production] *Vestnik Vserossijskogo nauchno-issledovatel`skogo instituta zhirov* [*Bulletin of RNSRIF*] **2** 9–13 [In Russsian]

[12] Ososhkin I A, Shutilin Yu F and Karmanova O V 2007 *Proizvodstvo rezinovyh tekhnicheskih izdelij* [Industrial rubber products manufacturing: study guide] (Voronezh: VSTA publishing office) [In Russsian]

[13] Chukhlanov V Yu, Nikonova N Yu and Alekseenko A H 2003 Gidrofobiziruyushchaya zhidkost dlya betonnyh i zhelezobetonnyh konstrukcij [Water-repelling liquid for concrete and reinforced concrete structures] *Stroitel`ny`e materialy`* [*Building materials*] **12** 38-9 [In Russsian]

[14] Ramos A F and Berbe F R 1985 A Study on the Recovery of Byproducts from the Dewaxing of Sunflower Oils of National Production *Grasas Aceites* **36** 1–4

[15] Mymrine V et all 2013 Oily diatomite and galvanic wastes as raw materials for red ceramics fabrication *Construction and Building Materials* **41** 360-4

[16] Tsai W T, Hsien K J, Chang Y M and Lo C C 2005 Removal of herbicide paraquat from an aqueous solution by adsorption on to spent and treated diatomaceous earth *Bioresource Technology* **96** 657-63

[17] Tsai WT, Chang Y M, Lai C W and Lo C C 2005 Adsorption of ethyl violet dye in aqueous solution by regenerated spent bleaching earth *J Colloid Interface Sci.* **289** 333-8

[18] Starostina I V, Stolyarov D V and Kosukhin M M 2017 Carbonic and Mineral Adsorbent, Received out of a Slurry Withdrawal of Oil-Extracting Production *Solid State Phenomena* **265** 319-24

[19] Boehm H P 1966 Chemical identification of surface groups *Advances in catalysis and related subjects* **16** 197-274

[20] Boehm H P 2002 Surface oxides on carbon and their analysis: a critical assessment *Carbon* **40** 145-9

[21] Oickle A M, Goertzen S L and Hopper K R 2010 Standardization of the Boehm titration: Part II. Metod of agitation, effect of filtering and dilute titrant *Carbon* **48** 3313-22

[22] Nakamoto K 1966 *Infrakrasnye spektry neorganicheskih i koordinacionnyh soedinenij* [Infrared spectra of inorganic and coordination compounds] (Moscow: Mir) [In Russsian]

[23] Brevedan M I V, Carelli A A and Crapiste G H 2000 Changes in composition and quality of sunflower oils during extraction and degumming *Grasas Aceites* **51** 417-23

[24] Kanya T C S, Rao L J and Sastry M C S 2007 Characterization of wax esters,free fatty alcohols and free Fatty acids of crude wax from sunflower seed oil refineries *Food Chem.* **101** 1552-7

Technological recommendations for the use of leaf litter based adsorption material to remove an oil slick from water bodies surface

S V Stepanova[1], A A Alekseeva[1] and L Y Khafizova[2]

[1]Department of Engineering Ecology, Kazan National Research Technological University, K. Marx St., 68, Kazan, 420015, Russia
[2]Department of Foreign Languages for Professional Communication, Kazan National Research Technological University, K. Marx St., 68, Kazan, 420015, Russia

E-mail: annank90@mail.ru

Abstract. The paper presents the study results on properties of leaf litter based adsorption material. The indicators of mixed leaf litter, characterizing it as an adsorption material, are established. Measurement of samples wetting angle with oil and water showed that the original material has hydrophobic properties. The oil intensity was about 10 g/g, and water absorption was 5-6 g/g, in addition, moisture, ash, total pore volume, bulk density were determined. Degree of water purification from oil was more than 98%, and biotest of purified water showed a decrease in toxicity 2.5 times. Based on this, technical options, are proposed for using mixed leaf litter for accidental oil spills response on the water surface with the help of mats, booms and universal design booms. Thermal combustion is proposed as a method of spent adsorption material disposal.

1. Introduction

The increase in production and transportation, refining and consumption of oil and its derivatives lead to a global environmental degradation. Oil and its derivative products adversely affect all links of the biological chain.

One ton of spilled oil pollutes almost 12 km^2 of water surface. Oil prevents oxygen from entering the water, which leads to the loss of a water body properties as a full-fledged favorable environment component. Oil slicks can significantly disrupt exchange of energy, heat, moisture, gases between the ocean and the atmosphere. The cessation of aeration can be detrimental to all living organisms in the pond.

Oil-containing waters are divided into two different groups: the first is the waters of natural reservoirs contaminated as a result of accidental and unauthorized discharges of oil products, and surface runoff from urban and industrial sites, seaports, etc.; the second is wastewater generated as a result of technological processes at oil production, storage, processing and transportation facilities, washing of any type of transport, etc. [1].

Cleaning the water bodies surface from contamination involves removing an oil slick by mechanical and (or) physico-chemical methods. The most promising and environmentally sound way is considered to be a method of removing oil slick products using oil sorbents [2].

The literature describes numerous methods for producing sorbents and filtering materials for water purification from oil products and technological schemes for their application. Presently, the sorbents are imposed with a number of maximum versatility requirements. First of all, this is high oil absorption, then environmental safety, buoyancy and minimal water absorption [3]. The development and study of sorption materials that would meet these requirements and have a low price is an urgent task.

A variety of raw materials are used for production of sorbents [4-10]. The most promising sorbents are production and consumption waste, as it allows to simultaneously solve a disposal problem. In particular, trees leaf litter collected in recreational city areas can be used as adsorption material. This waste cannot be used as basis for mineral fertilizers and compost, since pollutants (car wastes, ben-

IOP Conf. Series: Materials Science and Engineering **791** (2020) 012076 doi:10.1088/1757-899X/791/1/012076

zopyrene, anti-icing agents, etc.) penetrate the stomata, are fixed in the leaves, and can enter the plant root system, causing its death, in addition, they contribute to biological pollution of soils. Therefore, it is to be disposed in urban landfills. Use of plant polymer waste in order to obtain new materials is an important task [11]. Presently, there are many technological methods for processing plant polymer waste into composite materials [12]. Using the material to clean the water surface from oil spills would solve two environmental problems at once: production of cheap adsorbent and reducing the negative impact on ecological systems.

The aim of the work is to study the processes of oil removal from water surface by the leaf litter based adsorption material and development of technological recommendations for its use in the accidental oil spills response.

2. Materials and Methods

2.1. A subsection

Model water contaminated with oil slick was the objects of the study for the work.

The high-sulfur Devonian oil of the Tumutukskiy field was considered as a pollutant.

The mixed leaf litter of trees collected from recreational zones in city of Kazan (Russia) was used as sorption material. The collected material was dried at a temperature of 20 ± 5°C and ground to particles with a size not exceeding 10 mm2. The average composition of mixed leaf litter: birch - 34-29%, poplar - 31-28%, linden - 10-9%, maple - 10-8%, others - 15-14%.

2.2. Methods for Determining Main Sorption Characteristics

Defining main characteristics, which are total pore volume, poured density, ash content, moisture, buoyancy of the samples, was carried out according to standard methods [13, 14].

Sorption capacity for iodine: a sample of sorption material weighing 10g was previously boiled for 10 minutes in 200 cm^3 0.2 N. HCl solution, then it was washed with distilled water and dried for 1 hour at a temperature of 100-110°C. To determine 1g of the material, it was shaken for 15-30 minutes with 100 cm^3 of 0.1 N solution of iodine in KI (25 g/dm^3), then a 10 cm^3 sample was titrated with 0.1 N solution of sodium thiosulfate (indicator is starch). Iodine value was determined by the formula:

$$E_1 = 12{,}7 \cdot (V_1 - V_2) / m_y, \tag{1}$$

where, V_1 and V_2 are volume of 0.1 N sodium thiosulfate solution before and after addition of sorption material;

m_y is the mass of the sample (g).

Sorption capacity for oil under static conditions: 50 dm^3 of Devonian oil was poured, and a 40 mm × 40 mm × 10 mm brass mesh box was placed into petrie dishes, in which there was a test sample weighing 1g. After 1, 5, 15, 30, 45, and 60 minutes, the brass box with the test material was removed, some of the oil was drained, and the mass of the sample was determined on an HR-200AZ laboratory balance. The experiment was carried out at an ambient temperature of 20 ± 5°C.

Static oil intensity was determined by the following relation:

$$A = m_{absorb} / m_{sorption}, \tag{2}$$

where, m_{absorb} is the mass of absorbed oil (g);

$m_{sorption}$ is the mass of sorption material (g).

For the experimental determination of water absorption by mixed leaf litter, a sample weighing 1g was put in 40 mm × 40 mm × 10 mm brass mesh box, which was placed on the water surface. After 1, 5, 15, 30, 45, 60 minutes, the box with the test sample was removed, excess water drained, and it was weighed on HR-200AZ laboratory balance [13].

Water absorption was determined by the following ratio:

$$B = m_{absorb} / m_{sorption}, \tag{3}$$

IOP Conf. Series: Materials Science and Engineering **791** (2020) 012076 doi:10.1088/1757-899X/791/1/012076

where, m_{absorb} is the mass of absorbed oil (g);
$m_{sorption}$ is the mass of sorption material (g);
B is water absorption (%).

The residual oil content in the water was determined by the extraction method. The contents of the petrie dishes (oil-contaminated water) were poured into the separatory funnel, and 6 cm^3 of extractant, which was CCl_4, and 1 ml of 0.1 N HCl solution were added. The contents were shaken vigorously three times for 10 minutes, and settled until its complete stratification into extract (CCl_4 + oil) and refined water. The separated extract layer, previously passed through an ashless filter, was placed in a crucible. When the contents of the crucible were heated, CCl_4 was evaporated. The residual oil content was determined by the difference in the mass of the crucible before and after extraction.

2.3. Instrumental Methods

For visual observation of the oil absorption process (spreading on the surface, absorption), and samples of the materials before and after sorption, Micromed R-1 microscope with a dcm 130 video eyepiece was used. 100x magnification photographs were taken with use of the proposed software.

Biotesting is a procedure for determining environmental toxicity with the use of test objects that signal danger, regardless of what kind and combination substances cause changes in vital functions of the test objects. The biotesting was carried out in a accredited analytical test center laboratory of Kazan (Volga region) Federal University. The test procedure was carried out in accordance with Federal Registry.1.39.2007.03222 and Federal Environmental Regulation 14.1: 2: 4.12-06. Daphnia magna Straus was used as a test object; observation duration was 48 hours.

The determination of chemical elements (in mass fractions) in the ash produced after burning the spent oil-saturated sorption material was carried out in a SUR-02 "Renom FV" spectrometer (a universal X-ray spectrometer СУР-02 «Реном ФВ»), using an X-ray fluorescence method. The working wavelength is 0.001-2 nm.

3. Results

3.1 Main Characteristics of Leaf Litter as a Sorption Material.

Table 1 demonstrates values of the total pore volume, poured density, ash content, moisture, buoyancy and sorption capacity of the materials under study towards iodine.

Sorption activity largely depends on the moisture content of a sorbent. The moisture content of mixed leaf litter was 8-10%, that indicates no need in additional heat treatment. This moisture value will not affect the oil absorption from water surface.

The ash content of the sorption material shows how much sorption material remains after complete thermal burning. Low ash content can significantly simplify the process of spent sorption material disposal.

The value of the total pore volume of an adsorbent for water gives an idea of the maximum volume of a substance which this adsorbent can absorb. This indicator is quite high (table 1) for mixed leaf litter, while for mineral sorbents it is at the level of 0.5-2 cm^3/g.

Table 1. Main Characteristics of the Mixed Leaf Litter.

Parameter	Unit of Measurement	Mixed Leaf Litter
Moisture	%	10.01
Ash Content	%	0.21
Total Pore Volume for Water	cm^3/g	7.58
Poured Density	g/cm^3	0.08
Sorption Capacity for Iodine	%	27.80
Buoyancy for 24 h	%	92.00
Buoyancy for 72 h	%	69.70

IOP Conf. Series: Materials Science and Engineering **791** (2020) 012076 doi:10.1088/1757-899X/791/1/012076

To determine the presence and quantity of micropores in the sorbent, the iodine sorption capacity was determined. The experimental data shows that mixed leaf litter has a highly developed surface that combines various types of pores.

The buoyancy values of the samples indicate that the studied adsorption material is technologically applicable and can stay on the water surface during sorption process without any additional holding devices.

The main indicator of any adsorption material is its ability to absorb a specific pollutant. In the case of an oil spill response, the leaf litter has not only to absorb the oil, but also has to be hydrophobic, because under favorable weather conditions (air and water temperatures are above 20°C), oil can spread over the water surface as a thin slick [14]. If the oil slick thickness is less than 3 mm, the leaf litter is in direct contact with water. If the material is hydrophilic, the pores will be filled with water, which will cause decrease in buoyancy and the adsorption capacity of the samples towards oil. Figure 1 shows that the sorption capacity for oil with a contact time of 5 to 60 minutes for mixed leaf litter is more than 10 g/g, which is averagely 60% higher than the sorption capacity for water. Thus, the material is sufficiently hydrophobic. To conduct an experiment on measuring the leaf litter sorption capacity, imitation of an oil spot was created: oil was put on water (V = 7 ml) so that the thickness of the spot was 5-7 mm, while the sorption capacity of the material for oil decreased slightly. Obviously, this is due to the incomplete filling of pores with oil.

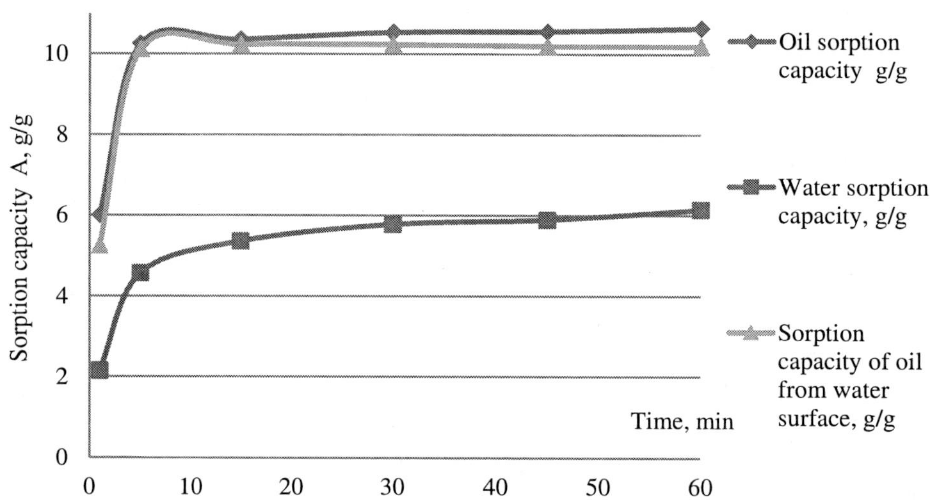

Figure 1. Sorption capacity of mixed leaf litter for oil, water, and oil from water surface.

Analyzing of these dependences let us conclude that oil absorption from the water surface is uneven. With increasing holding time, the samples capacity increases and their full saturation occurs after 30 minutes. It is explained by filling of the samples pores with water and oil, and the equality of rate values of the sorption and desorption processes. When sorption material is put on a contaminated surface, contact with oil occurs first. The components included in oil wet the leaf litter surface more actively, than water and the pores are filled with contaminants.

Figure 2 demonstrates photographs of leaf litter samples before and after oil sorption taken with Micromed R-1 (Микромед Р-1) microscope. Since the main part of the mixed leaf litter contains birch and poplar leaves, these samples were examined under a microscope in particular.

IOP Conf. Series: Materials Science and Engineering **791** (2020) 012076 doi:10.1088/1757-899X/791/1/012076

Figure 2. Microphotographs of leaves samples 1 – birch leaf litter, 2 – poplar leaf litter (a – original, b – covered with oil slick).

The micrographs demonstrate that the oil not only adheres to the surface of the leaf litter, but also goes deep into some areas, as evidenced by darker spots in the tubules of the leaf. Therefore, the leaf litter can be considered as a plant adsorbent.

For a more detailed study of the interaction of leaf litter with water, the hydrophobicity of the original and oil-wetted samples was determined by measuring the wetting angle. Figure 3 shows time dependences of changes in wetting angle of the original and oil-wet samples.

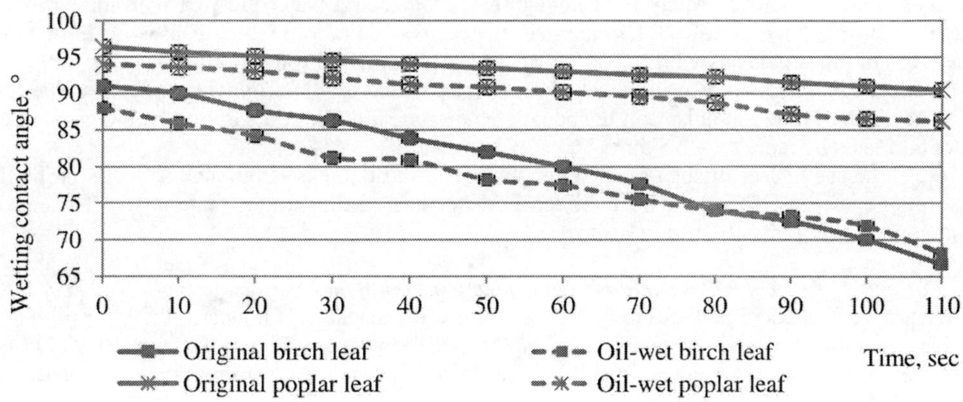

Figure 3. Dynamics of the original and oil-wet samples' surfaces wettability by a distilled water drop.

IOP Conf. Series: Materials Science and Engineering **791** (2020) 012076 doi:10.1088/1757-899X/791/1/012076

The difference between the original and oil-wet samples is insignificant. Lower contact angles (4-5°) of the wetted samples indicate that, when the leaves contact with oil and water, oil is absorbed first, forming a slick on the surface of the material, which prevents filling pores with water, and oil fills all the pores. This can be explained by the lower oil density, due to which the adsorbate rate of diffusion in the interport space is higher than that of denser water. The presented results can be confirmed by the fact that the leaf surface has a more pronounced affinity for oil than for water.

After removing the oil slick from the water surface, the residual oil content in the water was determined by the extraction method. It is possible to calculate the purification degree of polluted water from oil using the known residual concentration. The purification degree (CO, %) shows the fraction of the substance absolute amount that is captured by the sorbent and gives a fairly complete picture of the process nature. This indicator is an important criterion in determining optimal conditions of the adsorption process and is calculated by the following formula:

$$CO = \frac{C_o - C_{res}}{C_o} \cdot 100\%, \tag{4}$$

where, C_0 is initial oil concentration in water, g / dm^3;
C_{res} is resudial oil concentration in water, g / dm^3.

As can be seen from table 2 the purification degree of the water surface from oil at a contact time of more than 5 minutes is 98 – 99%. Obviously, the main part of oil is absorbed during the first five minutes, and desorption of oil from the material pores is not observed during 60 minutes.

Table 2. Time Dependance of Purification Degree of Water from Oil.

Time, min	Degree of Purification, %
1	93.5
5	98.7
15	98.9
30	98.9
45	98.8
60	98.9

The data in table 2 shows that the water purification from oil is not complete, therefore, the toxicity analysis of purified water by the biotest method was carried out. Crustaceans *Daphnia magna straus* were used as test objects. Analysis of contaminated water and water purified with mixed leaf litter samples showed decrease of the harmless substances concentration, which causes death of no more than 10% of test organisms with $HCWD_{10\text{-}48}$ = 717 in not purified water to $HCWD_{10\text{-}48}$ = 260 (HCWD is harmless concentration of water dilution) in water purified by the original sample of mixed leaf litter. The death of microorganisms in the aquatic environment during oil sorption by the original mixed litter decreases by more than 2.5 times.

Thus, the use of leaf litter to clean water surface from oil slicks is rational, as it has properties of adsorption materials. Therefore, it can be recommended as a bulk sorption material, a filler for booms, mats for collecting oil from water surface.

3.2 Variants of Technological Design of Oil Collection from Water Surface Using Mixed Leaf Litter.
There are many modern methods to collect oil from water surface [16]. One of the most used technological solutions is use of the examined samples as filler in mats (Figure 4) sized 10,250.05 m and weighing of 1 kg. Polymer net resistant to oil and with a mesh sized 66 mm in the transverse and longitudinal directions is selected as a shell. The dimensions and loading of the mat may vary depending on the area of contamination.

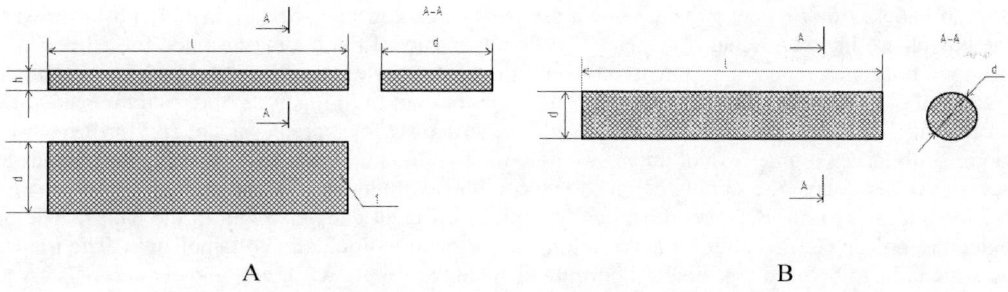

Figure 4. A – Absorbent mat (l=1000-3000 mm, d= 250-2000 mm, h=50-400 mm): 1 – polymer net; **B** – Sorbing bolster (l=1000-3000 mm, d= 120-500 mm).

The use of mixed leaf litter samples as stuffing in mats was tested on a water body (untitled stream) in the area of company Kara Altyn Enterprise activity. The stream was blocked by two sorption mats set across the entire width of the water body. Sampling during the testing was carried out 1 m higher and lower the mats in the water. The oil content before passing through the sorbent boom was 99.68 g / dm3, and after sorption - 0.09 g / dm3. Hence, the purification degree of the water surface from oil slick during industrial tests was 99%.

As the next technological solution for localization of oil slicks on the water surface, as well as preventing oil from entering the coastline, defence booms and shoreline shielding booms are used. Adsorption material based on the leaf litter is recommended for use in a universal design boom. The boom structure contains a bolster made of sorption material, shown in Figure 4 B, as well as a mat shown in Figure 4 A.

The sorbing bolster is made similarly to the mat and can be used both as an independent sorbing agent and as part of a universal boom, the design of which is shown in Figure 5.

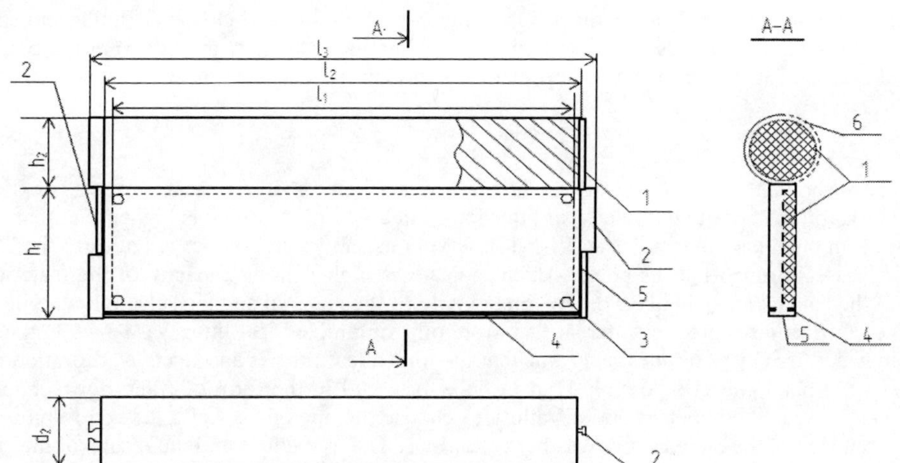

Figure 5. Universal sorbing boom 1 – sorbing element (leaf litter), 2 – bracing, 3 – bracing for a sorbing mat, 4 – ballast, 5 – shell, 6 – polymer net.

IOP Conf. Series: Materials Science and Engineering **791** (2020) 012076 doi:10.1088/1757-899X/791/1/012076

The sorbing boom is made of polyvinyl chloride (PVC) shell 5, which is resistant to oil, radiation and microorganisms. The sorbing mat size can be selected depending on the water body and the nature of the oil slick. Iron or steel galvanized chains, cables, etc. can be used as a load 4 and for trawling. The booms are interconnected by bracing 2. The upper part of the boom, designed for a filler in the form of a bolster, is made of two materials: one side of the circle is made of PVC, which prevents the passage of oil, the second half is made of a large polymer net 6. In this type of design the ready made bolster with sorption material 1 is placed inside the structure, for its replacement and further reuse of boom. With large volumes of pollution and high oil slick thickness, as well as with strong winds and waves over 2-4 points, the sorbing boom can be equipped with a mat as an additional sorbing element. Bracing 3 allows to place the boom vertically, with a sufficient counterweight of the ballast. The presented design can be used both for localization and for parallel liquidation of an oil spill. The frame of the structure can be used repeatedly if sorbing elements are replaced. Used oil-saturated bolsters and mats must be regenerated or disposed of.

One of the main aspects of industrial ecology is the absence of secondary pollution during industrial technological processes, therefore it is necessary to develop a technology for the safe disposal or regeneration of used adsorption materials.

For this purposes, a possibility of extracting sorbed oil from the sorbent pores was examined by determining the retention capacity of mixed leaf litter, which was 2.713 g / g. This indicates a slight desorption of the pollutant and futility of its extraction and regeneration. High consumption and cost of regenerating materials make this procedure economically inefficient, in addition, there is an issue with disposal or regeneration of spent solvents, which leads to repeated environmental pollution. Thus, it is proposed to utilize the spent mixed leaf litter in oil sludge thermal processing plants for complete thermal decomposition of oil and leaf litter. After burning 1 kg of spent adsorption material 95 mg of ash is produced. In order to determine its hazard class, an elemental analysis was performed (table 3).

Table 3. Concentration of elements in the ash, in mass fractions over 100%.

Zn	Cu	Fe	Mn	Ni	Ti	V	Ca	K	Cl	S	P	Si
0.2	0.2	4.5	1.6	0.1	0.2	0.5	51	12	3.7	23	0.1	2

Thus, heavy metals are found in the ash in a low concentration, and elemental sulfur and calcium, which is the basis of the waste, are non-toxic and of low hazard. Thus, the ash from the burning of spent oil saturated leaf litter has a hazard class IV and can be disposed of at solid municipal waste landfills.

Conclusion

Studies have shown that mixed leaf litter, which is a waste from urban areas, can act as adsorption material. The sorption capacity of mixed leaf litter for oil is 8 - 10 g / g, which is higher than that of most adsorbents on plant and mineral basis. Visual observations show that oil does not stick to the leaf surface, but passes through into its pores, which indicates a high holding capacity of the material. The values of the contact angle indicate the hydrophobicity of the material and the selectivity to oil absorption. The purification degree of water surface from oil using mixed leaf litter was 98 - 99% and a decrease in water toxicity (2.5 times) confirms that the mixed leaf litter is an effective adsorption material. Use of the plant waste is possible both in bulk form and in the form of fillers for technological shells. The presented equipment options allow to choose methods for specific cases of spills, taking into account size of the oil slick and weather conditions. The ash values and the results of the analysis of ash after thermal combustion of oil-saturated waste show that the resulting waste is non-hazardous and is formed in small quantities. Thus, mixed leaf litter can be considered as an effective and technologically applicable adsorption material for the accidental oil spills response on water surface.

References

[1] Liu Y, Li N, Lou Y and Zhao X 2019 Effect of oil spill stress on fatty acid stable carbon isotope composition of ulva pertusa *The Science of the Total Environment* **649** 1443-51

[2] Litvinova T E, Sulimova M A and Cheremisina O V 2017 The usege of a multifunctional sorbent based on ferromanganese nodules for neutralizing wastewater from oil refineries *Int. Multidisciplinary Scientific Geoconf. Surveying Geology and Mining Ecology Management, Sgem* (Albena) pp 1067-74

[3] Fazullin D D, Kharlyamov D A, Mavrin G V , Stepanova S V, Shaikhiev, I.G., Shaimardanova A S and Alekseeva A A 2016 The use of leaves of different tree species as a sorption material for extraction of heavy metal ions from aqueous media *Int. J. of Pharmacy and Technology* **8(2)** 14375-91

[4] Kumar A, Lingadurai S, Jain A and Barman N R 2010 Erythrina variegata linn: a review on morphology, phytochemistry, and pharmacological aspects *Pharmacogn Rev* **4(8)** 147-52

[5] Rao K, Anan, S and Venkateswarlu P 2010 Equilibrium and kinetic studies for Cd(II) adsorption from aqueous solution on Terminalia catappa Linn leaf powder biosorbent *Indian J. of Chemical Eng.* **13** 329336

[6] Harikishore K R, Harinatha Y and Seshaiaha K 2010 Biosorption of Pb(II) from aqueous solutions using chemically modified Moringa oleifera tree leaves *Chemical Eng. J.* **162(2)** 626-34

[7] Harikishore K, Harinatha Y, Seshaiaha K and Leec S 2012 Optimization of Cd(II), Cu(II) and Ni(II) biosorption by chemically modified Moringa oleifera leaves powder *Carbohydrate Polymers* **88(3)** 1077-86

[8] Sharma A and Bhattacharyya K 2005 Azadirachta indica (Neem) leaf powder as a biosorbent for removal of Cd(II) from aqueous medium *J. of Hazardous Materials* **125** 102-12

[9] Pandhare G, Trivedi N, Pathrabe R and Dawande S 2013 Adsorption of cadmium (II) and lead (II) from a stock solution using Neem leaves powder as a low-cost adsorbent *Int. J. of Innovative Research in Science, Eng. and Technology* **2(10)** 5752-61

[10] Awwad A and Farhan A 2012 Equilibrium, kinetic and thermodynamics of biosorption of lead (II) copper (II) and cadmium (II) Ions from aqueous solutions onto olive leaves powder *American J. of Chemistry* **2(4)** 238-44

[11] Wan Ngah W and Teong L 2011 Adsorption of dyes and heavy metal ions by chitosan composites *A review Carbohydrate Polymers* **83** 1447-56

[12] Sevostyanov V S, Goryagin P Yu, Obolonsky V V and Ermilov R A 2018 Resource-Saving Technological Complex for Polymer Waste Processing *J. Phys.: Conf. Ser* **1066** 012025

[13] Alekseeva A A and Stepanova S V 2015 The use of sorbent based on the leaves of trees to remove the oil film on the surface of water at liquidation of emergency floods *5th Int. Conf. on Science and Technology. Ecology* (London: Scieuro) pp 220-9

[14] Smirnov A D 1982 *Sorbtsionnaya ochistka vody* [Sorption water treatment] (Leningrad: Himiya) [In Russian]

[15] Ismail A 2015 Preparation and Evaluation of Fatty-Sawdust as a Natural Biopolymer for Oil Spill Sorption *Chemistry J.* **05(5)** 80-5

[16] AL Zubaidi I A and AL Tamimi A K 2016 Remediation of water from crude oil spill using a fibrous sorbent *Environmental Technology and Innovation* **6** 105-14

IOP Conf. Series: Materials Science and Engineering **791** (2020) 012077 doi:10.1088/1757-899X/791/1/012077

A pigment–filler in rubber compounds based on calcium carbonate containing waste

G I Tarasova

Department of Industrial Ecology, Belgorod State Technological University named after V.G. Shukhov, Kostyukov St., 46, Belgorod, 308012, Russia

E-mail: taga307@yandex.ru

Abstract. The article presents theoretical background and experimental data on the use of thermally modified waste of the sugar industry – sugar defecate as a black pigment filler in rubber compounds instead of carbon black and chalk. In previous studies, a method for processing calcium carbonate-containing raw material by thermal modification of its surface was suggested. As a result of heat treatment at a temperature of 600°C, a thermolysis sugar defecate TD_{600} - a fine black powder - was obtained. It was established that on the surface a layer of carbon (soot) coating $CaCO_3$ particles is formed. The purpose of this article is to study the possibility of using the thermolysis defecate TD_{600} as a pigment-filler in rubber compounds. To establish the possibil-ity of using the TD_{600} thermolysis defecate as a pigment-filler and an amplifier of polymer compositions, its physicochemical parameters were studied and compared with traditional pig-ments. A comparative analysis of physicochemical properties of fillers used in rubber com-pounds and thermolysis defecate TD_{600} was carried out. Properties of the pigment filler corre-spond to the parameters of materials used in rubber compounds; its oil absorption (80 ml/100g) is similar to that of classic carbon P514. The use of TD_{600} a black pigment-filler will solve two problems: saving raw materials and energy and improving the ecological situation.
Key words: pigments-fillers, thermality defecate TD_{600}, carbon black, carbon black, rubber compound, the Mooney viscosity, the butyl rubbe.

1. Introduction

Analysis of the market for the production and consumption of carbon black and carbon black in the world and Russia showed that the quantity of the product produced in the Russian Federation does not cover demand, and foreign analogues are too expensive. Figure 1 shows the main producers of carbon black in Russia.

A tire tread rubber contains about 45 parts of carbon black per 100 parts of rubber. Today, carbon black consumption is more than 6 800 000 tons. Most of this volume is used in rubber. Only 550 000 tons are used as pigment soot.

About 70% of all manufactured carbon black is used in the production of tires, ~ 20% in the production of rubber products. The remaining amount is used as a black pigment; a plastics antidegradant; a component that gives plastics special properties (electricity conductivity, absorption of ultraviolet radiation, radiation of radars).

IOP Conf. Series: Materials Science and Engineering **791** (2020) 012077 doi:10.1088/1757-899X/791/1/012077

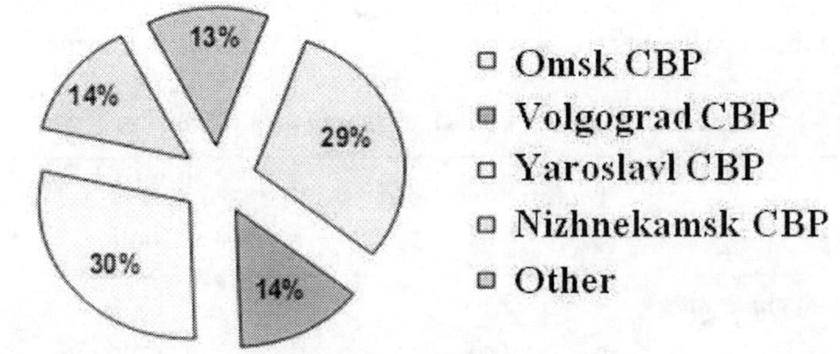

Figure1. Distribution of carbon black production in Russia.

From [1-10], it is known that carbon black is the main dispersed fillers used to strengthen elastomers.

In previous studies [11–17], it was shown that TD_{600} thermolysis defecate is used as a sorbent for purifying wastewater from various organic and inorganic pollutants, a filtration material for cleaning from soluble organic impurities, and a pigment-filler in silicate paints. The thermolysis defecate TD_{600} is used as a black pigment - filler in rubber compounds instead of carbon black and chalk. This paper presents studies of physicochemical properties of the pigment-filler based on TD_{600} and comparative characteristics of the carbon black-based pigments.

2. Materials and Methods

To establish the possibility of using the TD_{600} thermolysis defecate as a pigment-filler and an amplifier of polymer compositions, its physicochemical parameters were investigated and compared with traditional pigments.

The physical and chemical properties of the pigment-filler were determined depending on the temperature, pH of the aqueous suspension according to GOST 21119.3-91; density according to GOST 21119.5-75.

Determination of oil number or oil absorption. Five g of the pre-dried pigment were put into a porcelain cup with a capacity of 50 ml. The content was mixed with a glass rod with a melted (rounded) end; using a microburette with a capacity of 5 ml, 4-5 drops of oil were addedcarefully mixing the mass after each addition; after the formation of non-adherent lumps, the oil was added. To determine the absorption of oil, "Oksol" was used.

3. Results and Discussion

Oil absorption (in g/100 g) was calculated by formula: $M^1 = Vp_m/m_n$, where V is the volume of oil consumed, ml; m_n – pigment weighed, g; p_m – oil density; p – 0.778 g/cm^3. The results of experimental calculations are presented in Table. 1. The light resistance of the pigment was 6 months. Preliminary experiments showed the possibility of good dispersibility of TD_{600} in butyl rubber.

The results are presented in Tables 1 and 2.

Table 1. Physical and chemical properties of the defecate depending on the processing temperature.

Temperature, °C	Color of the thermolysis product	Value of the pH water suspension	Density, kg/m^3	Bulk volume, kg/m^3	Average particleradius um	Specific surface area, m^2/g	Oilnumber, ml/100g
105	TDbeige	9.8	2590	1540	2.0	14.08	-
500	TD grey	8.5	2545	1060	0.091	18.6	60
600	TD black	7.1	2530	1040	0.08	22.22	80
650	TD grey	8.4	2500	1050	0.098	23.5	70

Many authors have studied the effects of "binary" rubber-based fillers [8, 18]. We suggest using a ready-made "two in one" pigment - a filler consisting of calcium carbonate with a surface chemically bound layer of carbon black which is formed during heat treatment of the defecate at t = 600 °C.

Table 2. Physical and chemical properties of carbon fillers.

Indicators	Technicalcarbon P514	Naturalfillers			TD$_{600}$
		Graphite GLS-3	chalk	talc	
Specific adsorption surface m^2/g	52	23.0	18.1	12.0	22.22
Density, kg/m^3	1860	2130	2650	2750	2530
Oil number mg/100g	89	82.5	48	45	80
pH of water suspension	8.2	9.3	8.7	9.0	7-7.5

Analysis of the physicochemical properties of the pigment TD$_{600}$ showed that it has similar properties as widely used fillers.Its oil absorption property is similar to that of P514 technical carbon.

Thus, the thermolysis defecate t = 600°C ($CaCO_3$ + C_{soot}) can be recommended as a "binary" filler in rubber compounds.

3.1. Analysis of dispersibility of TD600 filler in butyl rubber

Since the main indicator for the composition of rubber mixtures is dispersibility of the filler in rubber, we investigated the dispersion of the pigment TD$_{600}$ in butyl rubber. The method for assessing the degree of dispersion of fillers [18-20] is used to evaluate the quality of rubber mixing,structure parameters and properties of rubber compositions.

It is known that the rate and degree of dispersion of fillers in rubber mixtures depend on the nature of the polymer used, mixing conditions and properties of the filler. Rubber compounds based on industrial butyl rubber containing 50 weight parts of carbon black filler (TU) and TD$_{600}$ thermolysis defecate were used as research objects. The mixture was prepared using a laboratory mixer 320×160 during 8; 10; 12; 16, 20 and 30 minutes of mixing carried out sampling for analysis (Mooney viscosity). The results are presented in Figure 2.

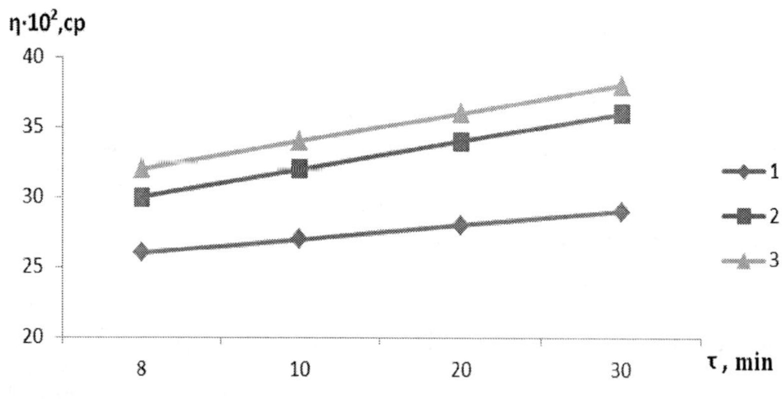

Figure2. The dependence of the Mooney viscosity in TU P514 (1) - based mixtures n; TD600 (2); CAD ground (3) on the duration of mixing.

The mixing time for the formation of a homogeneous compositionat which filler particles in rubber are distributed evenly turned out to be the same for all fillers (20-30 min). Mooney viscosity is intermediate between technical carbon and powdered CAD ground; however, a high degree of dispersion was observed for TD_{600} (95%). In fact, a chemical bond is formed between the polymer and the filler, and homogeneous floccules are formed during solidification which contributes to the thermo-mechanical high stability of the rubber mixture.

The advantage of the material is that if ordinary soot and black cannot be used to form suspensions due to low bulk density (100 ÷ 890 kg/m^3), they float on the surface of the dispersion medium (water or oil) when preparing suspensions. Pigment filler TD_{600} lacks this disadvantage. The indicators obtained on the basis of the defecate pigment compared with soot are presented in Table 3.

Table 3. Comparative indicators of pigment TD_{600} with carbon black.

T of firing, °C	Pigmentcolor	Typeofdestruction Colorchange	Bulkdensity, kg/m ^{3}
550	Dark grey	Barely noticeable	1060
600	saturated black	missing	1040
650	Gray	hardly visible	1030
Soot	black	invisible	100-500
black	black	inconspicuous	850-890

Analysis of the data shows that the black pigment corresponds to the typical sample of carbon black oby color, but it has 1.5-2 times higher bulk density which makes it possible to produce both aqueous and oily suspensions and use the pigment-filler for producing building materials and paints.

The study of the chemical composition of soot on the surface of $CaCO_3$ showed that its main component is carbon – 99.65%; hydrogen – 0.28%; oxygen – 0.08%; inorganic substances (ash) – 0.01%. Oxygen and hydrogen are present in soot due to adsorption of combustion products of carbohydrates on soot particles. Therefore, carbon can be presented as crystallites (C_xO_y).

The soot on the surface of the defecate is highly resistant to acids, alkalis, and light. The true density of the black pigment ranges from 2545 to 2590 kg/m^3. The bulk density of the pigment is: 1040-1060 kg/m^3. According to microscopic studies, each crystal consists of 3-7 parallel flat carbon lattices (Figure 3).

The presence of C_xO_y systems has a great influence on the surface properties. In cases where the amount of soot on the surface is not large, the solution becomes neutral (pH = 7.1); loss due to ignition at t = 600 °C is equal to 18.25%.

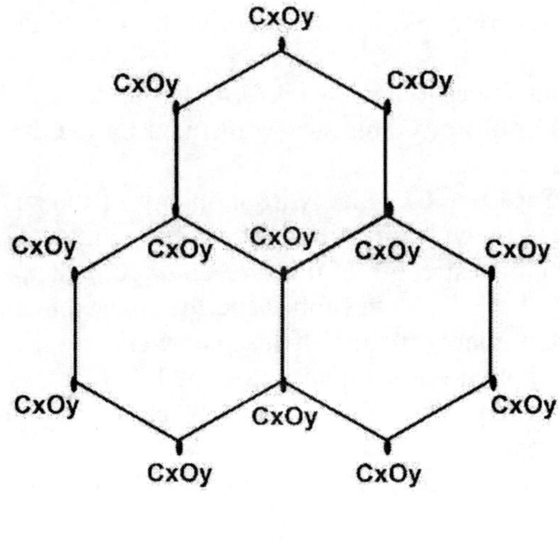

Figure3. The structure of inertness, TD_{600} can be recommended as a pigment-filler in rubber mixtures.

IOP Conf. Series: Materials Science and Engineering **791** (2020) 012077 doi:10.1088/1757-899X/791/1/012077

Rubber is a multi-component system consisting of a polymer base (rubbers) and various chemical additives. To produce rubber, high molecular weight polymers with a low boiling point (room) and high elasticity are used. We used butyl rubber (polyisobutylene) with the formula:[-C(CH_3)$_2$- CH_2-]n-[-CH_2C(CH_3)=CH-CH,-]m, with the following parameters: ρ=910-930 kg/m^3; $t_{fluxing}$=100 °Catμ=70000–225000 (molecular mass).

When mixing, the solid phase of the pigment is evenly distributed in the dispersion medium (rubber). TD_{600} particles form hydrogen bonds with the hydrogen of the isoprene group of butyl rubber. The binding energy determined by the calorimetric method for the heat of wetting in benzene was 35–40 kJ/mol, and the length of the hydrogen bond was 0.17–0.14 nm [21–23].

As a result, the mixing process is uniform and stable, the sludge suspension is zero. Thanks to this “crosslinking”, the bond of the polymer and pigment particles TD_{600} is stable. Soot particles on the surface of $CaCO_3$ have a spherical shape which is in compliance with literature data [7].

The rubber contains various ingredients: vulcanizing agents, accelerators and vulcanization activators, plasticizers, fillers, antioxidants and other components that give it specific properties. The main components of rubber products (RTI) are pigments - carbon black fillers and inorganic fillers (chalk, clay, etc.).

We have suggested using the TD_{600} thermolysis defecate instead of chalk and soot which will make it possible to reduce the price of rubber products and recycle 80% of waste of the sugar industry improving improve the environmental situation.

Preliminary experiments showed good dispersibility of TD_{600} in butyl rubber. Comparative physicochemical properties of pigment-fillers (Table 2) showed that TD_{600} have properties similar to those of traditional. Itsoil absorption (80 ml/100g) is similar to that of classic carbon P 514.

Physical and mechanical properties of the rubber compound (butyl rubber –100 w.f.; pigment filler ($CaCO_3$ +C_{soot}) – 55-56 w.f.; sulfur - powder –2 w.f.) are presented in table 4.

Table 4. Physical and mechanical properties of the rubber compound

Brand TU	Conditional stress during elongation of 300%, kgf/cm^2		Conditionaltensile strength, $^{kg\cdot}$s/cm^2		Relativeextension, %	
	min	max	min	max	min	max
TD_{600}	92	97	170	194	510	586
P 514	84	96	174	190	530	559

P 514 was used as a reference filler. Table 4 shows that physicomechanical indicators of our products are the same level as those of rubber containing P 514. At the same time, carbon chemically bound to particles of calcium carbonate improves compatibility of fillers with other hydrocarbon ingredients of the rubber mixture.

Thus, the combined effect of these factors contributes to a more uniform distribution of the pigment in the rubber mixture increasing its density and uniformity.This was confirmed by the data of microscopic analysis of rubber mixtures.

The great advantage of using the pigment ($CaCO_3$ + C_{soot}) is hydrophobicity of the surface which helps reduce the cracking of rubber compounds in the vulcanization process and reduce the amount of scrap due to a decrease in the physical and mechanical strength of mixtures and products.

In order to establish the uniform distribution of the TD_{600} in rubber and the shape and size of particles of mixture components, optical studies of the microstructure of samples were carried out using a BIOLAM optical microscope. The scale of the photograph is equal to the total increase in the system of applied objects and eyepieces. An eyepiece 10×, a lens 9×0.2, and an objective micrometer with a division value equal to 10 μm = 0.01 mm were used.

An increase in micrographs was determined by dividing the size of a crystal by the true size of the same crystal previously measured under the microscope. The magnification was 200. Figure 4 shows

IOP Conf. Series: Materials Science and Engineering **791** (2020) 012077 doi:10.1088/1757-899X/791/1/012077

micrographs of the initial defecate and the TD_{600} pigment which is black with structural networks of soot particles.

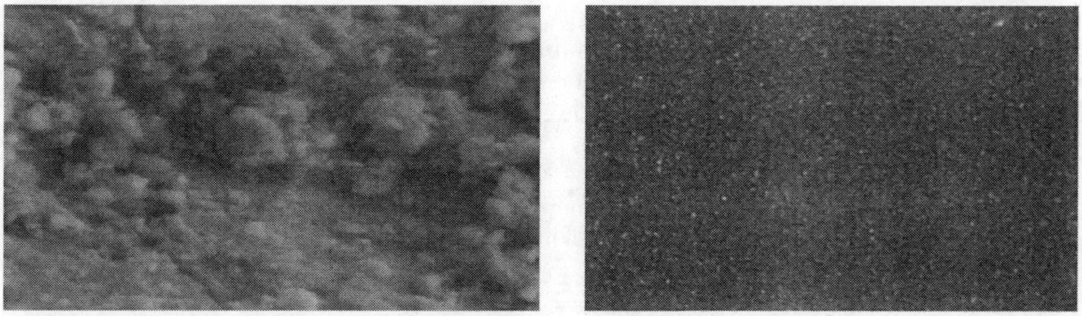

a b

Figure 4. Photomicrographs of samples (×200): initial defecate (a); dry pigment ($CaCO_3$ + C_{soot}) (b).

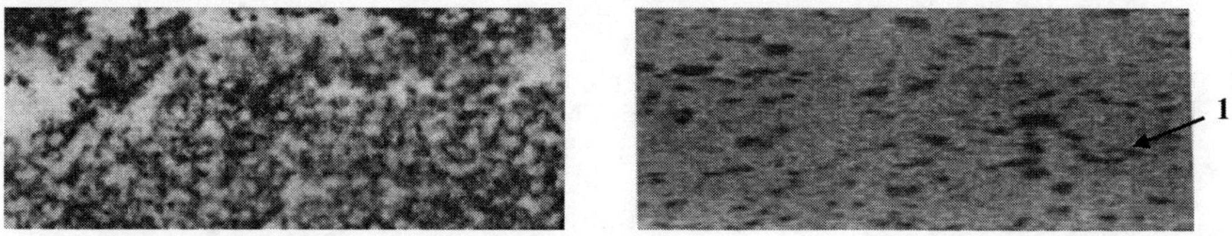

a b

Figure 5. Pigment ($CaCO_3$ +C_{soot}) inthewater drop, ×200 (a); the mixture product in butyl rubber1 - ($CaCO_3$+ C_{soot}) particle, ×200 (b).

The micrograph of the pigment TD**600** (Figure 5b) shows that it is evenly distributed in the polymer base throughout the entire volume of the dispersion medium. Soot particles are visible on the $CaCO_3$surface. Due to its high lightfastness, structure, high specific surface area and inertness, the pigment was recommended and tested as a pigment-filler produced byKurskrezinotehnik. According to the laboratory data, the physico-technical indicators of the rubber product meet the federal standards. Mooney viscosity at 100 °C is 25-35 s.u.; conditional elongation at the 300% elongation is 45 kg.s./cm^2; conditional tensile strength is 45 cm; Shore hardness is 63, humidity is 0.2%.

Rubber mixtures were manufactured on laboratory rollers according to the regimes adopted for serial mixtures; there were no complications in technological processes and mixing parameters were not adjusted. TD_{600} was well distributed in the rubber matrix, without forming a mixture of foreign inclusions.

4. Conclusion

Comparative analysis of physicochemical properties of fillers used in rubber mixtures and a TD_{600} thermolysis defecatewas carried out. Properties of the pigment-filler correspond to the parameters of materials used in rubber compounds;its oil absorption (80 ml/100g) is similar to that of classic carbon P514.

Analysis of dispersion and mixing in butyl rubber was carried out. Mooney viscosity was determined (25-30 units). It meets the requirements of GOST for rubber compounds pigments. The TD_{600} can be recommended as a "dual-binary" filler in rubber compounds.

Comparative analysis of properties of the pigment-filler and traditional carbon black pigments made it possible to establish that the TD_{600} pigment corresponds to a typical soot sample, but has 1.5-2 times higher bulk density and will not float in the volume polymer base when mixed. It can improve homogenization of the mixture.

The assessment of stability of quality indicators and properties of TD_{600} rubber compounds in the Central laboratory of Kursk plant "Rezinotekhnik" showed that their indicators correspond to the P514 pigment.

The main consumer of carbon black and chalk is the rubber industry which uses more than 80% of world soot products. The use of the developed pigmentwill allow for 80% utilization of calcium carbonate-containing waste and improvement of the ecological situation.

Acknowledgments
The work was performed as part of the Program of flagship university development in Belgorod State Technological University named after V G Shukhov, using equipment of the High Technology Center of BSTU named after V G Shukhov.

References

[1] Treboganova V M 1979 *Primenenie texugleroda v promy`shlennosti RTI* [The use of carbon black in the industry of rubber products] (Moscow: TsNIITEneftekhim) [In Russian]

[2] Pechkovskaya K A 1968 Sazha kak usilitel` kauchuka [Soot as an amplifier rubber] (Moscow: Himiya) [In Russian]

[3] Kraus G 1966 *Reinforcement of elastomers* (New York: John Wiley & Sons Inc)

[4] Orekhov S V 1973 Perexod shinnoj promy`shlennosti na ispol`zovanie texugleroda v sootvetstvii s nomenklaturoj i trebovaniyami mezhdunarodny`x standartov [The transition of the tire industry to the use of carbon black in accordance with the nomenclature and international standards] *Sy`r`e i materialy` dlya rezinovoj promy`shlennosti* [*Raw materials and materials for the rubber industry*] (Moscow: GUPNIISHIP) pp 89-91 [In Russian]

[5] Verbas V V 2002 Osobennosti primeneniya novy`x marok texnicheskogo ugleroda v shinny`x rezinax [Features of the use of new grades of carbon black in tire rubber] *Voprosy` ximii i ximicheskoj texnologii* [Chemistry and chemical technology issues] **5** 50-6 [In Russian]

[6] Belozerov I V 1964 *Texnologiya reziny`* [Rubber technology] (Moscow: Himiya) [In Russian]

[7] Koshelev F F, Kornev A E and Bukanov A M 1978 *Obshhaya texnologiya reziny`* [General technology of rubber] (Moscow: Himiya) [In Russian]

[8] Goncharov V M and Khodov V V 1988 Vliyanie sostava i sposoby` polucheniya binarny`x napolnitelej na osnove texnicheskogo ugleroda i svojstva rezin [The influence of the composition and methods of production of binary fillers based on carbon black and properties of rubbers] *Kauchuk i rezina* [*Rubber*] **7** 30-4 [In Russian]

[9] Yashunskaya F I et al 1974 *Spravochnik rezinshhika* [Handbook of the rubber producer] ed P A Zaxarchenko et al (Moscow: Himiya) [In Russian]

[10] Estrin R I and Pesin O Yu 1994 Vzaimosvyaz` mezhdu strukturny`mi pokazatelyami texnicheskogo ugleroda i svojstvami rezinovy`x smesej i vulkanizatorov [The relationship between structural indicators of carbon black and properties of rubber compounds and vulcanizers] *Kauchuk i rezina* [*Rubber*] **4** 12-6 [In Russian]

[11] Lupandina N S, Sapronova Z A and Lesovik V S 2014 Copper and nickel substances extraction from water mediums by waste of disaccharide production *J. of Eng. and Appl. Sci.* **9(8)** 310-5

[12] Starostina I V, Sverguzova S V, Ovcharova I V and Pendurin E A 2015 Recycling of microbiological industry waste with the obtaining of foaming agents for building industry *Int. J. of Applied Eng. Res.* **10(21)** 42701-6

[13] Sverguzova S V, Starostina I V, Tarasova G I and Fomina E V 2015 Using Ferruginous Quartzite Tailings In Dry Building Mixes *Int. J. of Applied Eng. Res.* **10(24)** 45082-90

[14] Sverguzova S V, Tarasova G I, Starostina I V and Fedorina M Y 2015 Composite pigment-filler on the basis of sludge of heat power plant and tails of engichment of ferruterousquartzites *Res. J. of Appl. Sci.* **10(12)** 827-31

IOP Conf. Series: Materials Science and Engineering **791** (2020) 012077 doi:10.1088/1757-899X/791/1/012077

[15] Sapronova Z A and Starostina I V 2016 Disposal of Synthetic Surfactants-Containing Wastewater Treatment Sludge in the Ceramik Brick Production *Procedia Eng.* **150** 1610–6

[16] Tarasova G I and Starostina I V 2016 Special features of processing siliceous rocks *Int. J. of Pharmacy and Technology* **8(4)** 24770-80

[17] Sukhanov E V, Sapronova Z A and Fomina E V 2017 Nekotory`e osobennosti koagulyacionnoj ochistki vody` s pomoshh`yu py`li e`lektrostaleplavil`nogo proizvodstva [Some features of coagulative water purification with the help of electric steelmaking dust] *E`kologiya i Promy`shlennost` Rossii* [*Ecology and industry of Russia*] **21(1)** 24-9 [In Russian]

[18] Harris E V, Jones F E and Church F C 1976 *Carbon blacks in a variety of elastomers. Cabot Corp. Technical Report RG-130* (Alpharetta: Cabot Corp.)

[19] Dmitrenkov AI, Shumilin YF 2002 Analiz dispergirovaniya napolnitelej v kauchukax [Analysis of dispersion of fillers in rubbers] *Voprosy` ximii i ximicheskoj texnologii* [*Chemistry and chemical technology issues*] **5** 111-6 [In Russian]

[20] Grishin O I 1980 Ocenka stepeni dispergirovaniya [Estimation of the degree of dispersion] *Kauchuk i rezina* [Rubber] **10** 24-7 [In Russian]

[21] Epstein L M and Shubina E S 2003 Mnogolikaya vodorodnaya svyaz` [Multi-sided hydrogen bond] *Priroda* **6** 40-5 [In Russian]

[22] Maes G and Smets J 1993 Cooperative hydrogen bond. Quantitative study using IR Fourier spectroscopy of matrix isolation *J. Phys. Chem.* **97(9)** 1818-25

[23] Tarasova G I 2018 Razrabotka sostavov silikatny`x krasok na osnove termoliznogo transporterno-moechnogo osadka - otxoda saxarnoj promy`shlennosti [The development of compositions of silicate-based paints thermalizing conveyor-washing sediment – waste of the sugar industry] *Stroitel`ny`e materialy` i izdeliya* [*Construction Materials and Products*] **1(1)** 21–31 [In Russian]

[24] Tarasova G I 2012 Racional`ny`j sposob polucheniya pigmentov-napolnitelej iz metallosoderzhashhix promy`shlenny`x otxodov [Rational method of producing pigments – fillers from metal-containing industrial waste] *Vestnik BGTU im V G Shukhova* [*Bulletin of BSTU named after V G Shukhov*] **2** 128-32 [In Russian]

Theoretical and experimental research on pre-sowing seed treatment

A A Vasilyev[1], A N Vasilyev[1], A K Dzhanibekov[1], G N Samarin[1] and D A Normov[2]

[1]Office of Automated Electric and Heat Technology, Federal Scientific Agro-Engineering Center VIM, 1st Institutsky passage, House 5, 109428, Moscow, Russia
[2]Physics department, FSBEI HE Kuban State Agrarian University, 350012, Krasnodar, 13, Azovskaya street, Flat 40, Russia

E-mail: vasilev-viesh@inbox.ru

Abstract. Grain is a biological object, with all specific responses to external actions, that reside in this category. To define principles that are needed for pre-sowing seed treatment, possible seed responses to external actions were examined. For this, the analysis of the seed's adaptive response was used as an approach. Main rules to organize seeds stimulus have been established. Use of an informational approach for analysis of bio-logical seed response on external action allowed to find utilizing rules for electrophysical stimulation during a pre-sowing treatment. The idea behind them is to use the following sequence of operations at the pre-sawing seeds treatment: determine the lowest external action energetic at which grains response; organize cyclical action of external factor on seeds; change of action amplitude in each cycle; change of external action type. Field experiments have been conducted for several years, during which were determined operational modes of aerated grain bin electrical equipment, when used for pre-sowing seed treatment.
Keywords: pre-sowing seed treatment; air-heat treatment; electroactivated air; ions; electrphysical action; germination; germinative energy; yield; adaptive reaction.

1. Introduction

Pre-sowing seed treatment is a very important technological operation that improves seed material quality, which provides protection from pests and boosts sowing seed quality. The present level of these technologies offers widespread use of electrophysical methods of seed treatment, which allow to increase crop yield at relatively low energy consumption during treatment. This promotes research on the subject, but the implementation of such equipment is not widespread yet. There are several reasons, and the main is that grain is a biological object with all specific responses to ex-ternal actions, that reside in this category. Manufacturer, who decides to use electrophysical stimulation for pre-sowing treatment, wants to be certain in several aspects: receive guaranteed declared result; absence of side effects; stability of results, that are not affected by climatic factors. Let`s consider some elements of the electrophysical treatment influence on seed characteristics that can help manufacturers to get answers.

2. Biological aspects of sowing seed quality change under the influence of different factors

Observation of pre-sowing seed treatment methods demonstrates that there is a general scheme of living organism response, and grains, in particular, on the external action. Response can be directed at

IOP Conf. Series: Materials Science and Engineering **791** (2020) 012078 doi:10.1088/1757-899X/791/1/012078

both the preservation of initial grains condition (resistance, resiliency) and adaptation to a new condition.

Plant cells response to different external actions, for example, to temperature, were similar to other organisms' responses (animals, humans) [1-4], therefore, further consideration of this issue will be based on data not only from plants study but also from the other subjects.

In some papers [5, 6], any action on plants considered as stress, that leads to adaptation and development of stability. Specific responses evolve gradually during plant stability development. Fist to arise are nonspecific stress responses, such as a change of membrane permeability, increase of cell cyttropism to dye, pH shift, reduction of membrane potential, which change cytoplasm structure. Under the influence of these quick responses arise intermediate responses (acute-phase proteins, photosynthesis and growth arrest).

Living organism's evolution study tells us that there are a limited number of standard single-type responses to a large variety of actions. Grain response is almost the same to all types of seed treatments (UHF, laser radiation, heat and so on). The difference in responses can depend on the penetration depth of action (acts differently on different parts of grains), hardness and exposure.

Stress is a response to a strong stimulus, however, clear separation of stimulus based on its strength hasn`t been done by anybody. Fundamental research on stress influence on the living organism had been conducted by Selye [7, 8].

Garkavi L.H. and coauthors have shown that a strength of action defines the type of an adaptation response. There are three types of such responses:

- response to a weak action is called training.
- response to a moderate action is called activation.
- strong or extreme action leads to Selye stress.

The first stage, after action onset, is a «going through orientation» of the biological object when it decides further behavior scheme. This stage is called «orientation». If acting factor strength continues to be weak, then the biological object stops taking it as a dangerous and doesn`t accept it. If this situation is transferred on a pre-sowing seed treatment process, then it explains why a weak level of different factors action doesn`t change sowing characteristics. Therefore, it is necessary to know «the lowest» limit of every action, that seeds will respond to. To ensure biological objects staying in the «training» mode, it is important to change the intensity of external action in a cyclical way. In this case, the next stage will start developing, so-called «recombination».

The systematic repetition of weak but gradually increasing stimulus, after some time leads to a defending system enhancement, for this reason, the organism becomes resistant to damaging actions not only because of an inhibition development but also due to defending system activation increase. It is notable, that whatever action was used for «training», resistance increase happens not only to this action but also to the others [10]. Referring to the pre-sowing seed treatment by air-heat processing [11-13], it is reasonable to make an assumption, that positive effect at such pro-longed way of treatment shows up precisely because temperature and relative air humidity are changing during the day. Moreover, changes are happening, for the most part, gradually. There-fore, weak stimulus repetition becomes systematic. The systematic repetition of the moderate strength stimulus, over a period of time, leads to the development of «long-lasting activation».

The facts mentioned above allow assuming that if air-heat treatment of seeds will be carried out in a similar mode, then there is a possibility to get the stable maximum effect. The question is that how the level of seed resistance is connected to its sowing characteristics, what type and what lev-el of external action should be, to ensure seeds «move up» to the «long-lasting activation» level of response. Garkavi L.H. [10] on this problem explains that the level of external action is a relative value. The reference point should start from the action strength when the response of the biological object on the given type of action is evident [10]. Therefore, during the setting of the pre-sowing seed treatment measure of acting factor is not so much important as the increment of its measure in relation to the initial condition. However, there are natural boundary conditions exist, that cannot be omitted, otherwise, the effect might be negative. For example, it is known, that grains shouldn`t be heated above the temperature of 55-60°C. When this is not the case, their sowing quality degrade.

Referring to the factor`s level, which ensures grains transition to the required response, paper [10] says that «... during the experiment, to move from one response to the other (adjacent), it is enough to change acting factor value on 20%». In this regard, main adaptation responses – «training», «mild activation», «extreme activation», and stress – can be triggered in a narrow dose range (strength) of acting factor.

Based on the example of grains temperature change during its aeration treatment, let`s examine the condition of the grain layer.

Let`s define the initial grain temperature at 20°C. Heating of the grain to the temperature of 22-24°C transfers it to the training mode. To the mild activation – 26.4-28.8°C. Extreme activation mode requires a grain temperature at 31.7-34.6°C. Initial temperature decrease will change the temperature of the transition thresholds.

Based on the above, it can be concluded, that seeds resistance increase (and as a consequence, sowing quality) requires the following conditions [14, 15]:

- for adaptation responses occurrence on a level of functional changes but not on a structural lev-el, acting factor should be gradually increasing, longstanding, alternating (cyclical);
- to receive an activation response, it is necessary to start with a physical stimulus of a certain value, then it should be decreased or stopped. Next step is an increase of stimulus strength on 10-20% in relation to the initial value;
- the initial value of a stimulus should be minimal to ensure adaptation responses occurrence at the lower floors of responsiveness. This ensures the highest responsiveness level;
- maximum effect from the physical stimulus can be achieved by a combination of nonspecific responses with specific.

Authors have attempted to describe grain's behavior, as a biological object, under the external influence on it by using the informational approach [16]. Usefulness of grains response was taken as a criterion of stimulus effectiveness. Used by the authors principle of maximum information have shown that usefulness of seeds response to the external action will be maximum at a changeover of different acting factors or at cycles «action - rest». Usefulness of response gets higher when the energetic level of external action goes down. Longstanding low energetic action and short term high energetic action gives different final usefulness of grains response and different pre-sowing treatment results. This will be reflected in specific grain's responses to the external action, which can be observed at different stages of plant development, spike formation, grain filling. In order not to create an extreme inhibition response in grain, it is preferable that the energy level of external action was as low as possible, but still, be accepted by the grains.

The use of the maximum information principle helped to formulate electro-technologies implementation approaches for technological processes development, which are directed at sowing seeds quality improvement and yield structure adjustment. [17]:

- determination of the lowest level of electrophysical value, that ensures seeds response;
- cyclical change of the action amplitude;
- changeover of the external action.

3. Experimental determination of the air-heat processing modes of pre-sowing seed treatment

Experiment on pre-sowing seed treatment by the air-heat processing has been conducted, to confirm presented theoretical statements. Determination of an optimal treatment mode requires a clear understanding of being examined factors range change on the existing facilities during the standard technological process. Aeration has been taken as an example of such a process.

The preliminary experiment has been conducted at the first stage, the objective of which was the determination of what seed treatment mode gives the best effect. Preliminary experiment results have shown that the best seed treatment mode is the cyclical mode: «heating» – «cooling» – «heating». The best results have been obtained at airspeed which is going through grain layer Va = 0.3 m/s, air temperature Ta = 29 - 30°C, time of half-cycle treatment tp = 0.4 hr. Holding time – 2 days. Half-cycle treatment time is the time during which the heating or cooling phase happens once. Obtained results have confirmed theoretical statements for the selection of the best modes for pre-sowing treatment.

However, these results were obtained during seeds treatment on the laboratory experimental equipment, where the layer of treated seeds was 0.15 m and 0.45 m. When the treated grain layer is thin, then it can be considered that the temperature of air and grain is uniform throughout the entire layer. In the existing aerated grain bins, air should penetrate the grain layer of 1.2-1.5 m.

During treatment by the aeration of grain layer with thickness of 1,2 m, airspeed slows from 0.7 m/s at the entrance to 0.1-0.2 m/s at the exit from the grain layer. Air temperature and temperature of grain heating are also changing throughout the layer. Therefore, to speak about adequate accuracy of the effectiveness of one or another treatment mode, it is necessary to obtain the dependence of sowing seed qualities on operational modes of existing aerated grain bins electrical equipment, expressed in air temperature and speed. Multifactorial experiment allowed to obtain such dependence. The experiment has been conducted for three factors: grain temperature increment DQ_c in %, airspeed V_a in m/s, time of half-cycle treatment t_p in hr. Corresponding regression equations have been obtained.

For germinative energy:

$$EG = -27.5736 + 0.693\Delta\Theta_c + 62.256V_a + 20.736t_p + 0.0864\Delta\Theta_c V_a + 0.0392\Delta\Theta_c t_p - 4.08V_a t_p - 0.007\Delta\Theta_c^{\,2} - 69.44V_a^{\,2} - 9.8t_p^{\,2}. \tag{1}$$

For germination:

$$G = -23.905 + 0.598\Delta\Theta_c + 50.488V_a + 19.405t_p + 0.082\Delta\Theta_c V_a + 0.035\Delta\Theta_c t_p - 3.9V_a t_p - 0.0067\Delta\Theta_c^{\,2} - 56.32V_a^{\,2} - 9.8t_p^{\,2} \tag{2}$$

Based on the multifactorial experiment results and their processing, family of curves of two-dimensional section of response surface were obtained. Contour plots were constructed for the equation (2) $G = f(\mathrm{DQ_c}, t_p)$, Figure 1 and $G = f(\mathrm{DQ_c}, V_a)$, Figure 2.

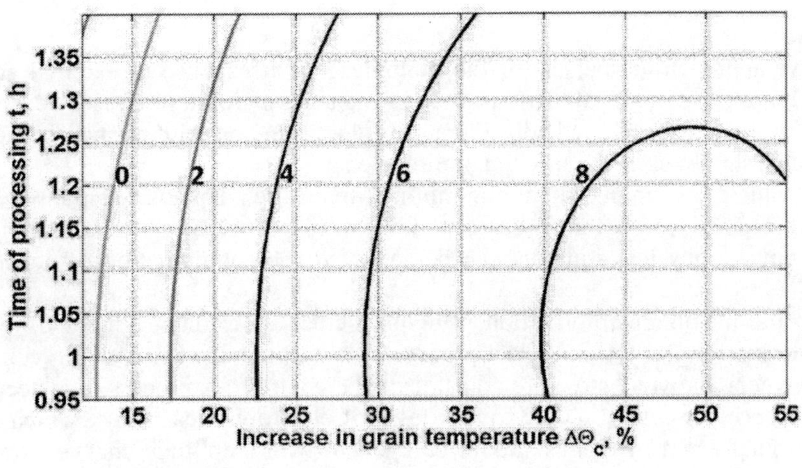

Figure 1. Contour plot of the dependence $G = f(\mathrm{DQ_c}, t_p)$.

Analysis of the contour plots shows that the smallest half-cycle of pre-sowing treatment, that was used in the experiment (0.95 hr.), allows to ensure the biggest seed germination increment (8%). However, this requires the temperature of the grain heating increment up to 40%. This mode can be realized at initial grain temperature not higher than 21°C and atmospheric air not lower than 25°C. The reason for this is that electric air heater capacity in the existing aerated grain bins pro-vides air heating on 7-8°C only. Lower initial grain temperatures (12-18°C) allow much easier im-plementation of such modes in the existing equipment. The best half-cycle time is 1,05 hr., at which grain temperature in-crement can be lowered to 39%.

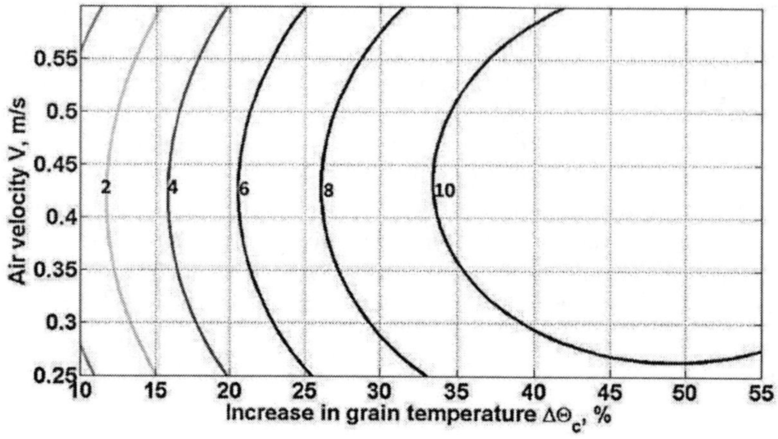

Figure 2. Contour plot of the dependence $G = f(\mathrm{DQ}_c, V_a)$.

Analysis of the contour plot in Figure 2 allows to suggest that if the airspeed in the intergrain space stays at 0,42 m/s than to achieve maximum value of the germination, it is enough to limit grain heating temperature increment to 33%.

The use of grain bins with aeration for pre-sowing seed treatment requires consideration of their design specifics (airspeed change throughout the layer and air heater capacity), which impose con-straints on the pre-sowing treatment modes. However, obtained dependences (1) and (2) reflect pa-rameter relationship for the other facilities also, which can use discussed treatment modes. Grain dry-ers, for example. The maximum grain heating temperature, in this case, is the main restriction. It shouldn`t exceed 45°C.

4. Conclusion

Use of adaptive approach to analysis of the biological seed response to external action during pre-sowing treatment allowed to establish that seeds resistance increase (and as a consequence, sow-ing qualities) requires action to be gradually increasing, longstanding, alternating (cyclical). At the same time, the stimulus should be cyclical with amplitude change.

The initial value of a stimulus should be minimal to ensure adaptation responses occurrence at the lower floors of responsiveness. In this case, the level of responsiveness will be the biggest. The maxi-mum effect from the physical stimulus can be achieved by a combination of nonspecific responses with specific.

The use of the maximum information principle helped to formulate methods of electrophysical stimulus implementation for technological processes development, which are directed at sowing seeds quality improvement and yield structure adjustment. To control this process, it is necessary to observe the following algorithm: determine the lowest level of electrophysical value, which leads to the seed response to external action; stimulus should be cyclical with amplitude change; external action type changeover should be implemented during seed treatment.

The result of the experiments confirmed theoretical researches and determined modes of the air-heat pre-sowing seed treatment on aerated grain bins. It was established that to obtain maxi-mum seed germination increment, it is necessary to aerate grains with air at speed of 0,45 m/s, the temperature which ensures grain temperature increment 33-40%, with half-cycle treatment time of 0,95-1,1 hr.

IOP Conf. Series: Materials Science and Engineering **791** (2020) 012078 doi:10.1088/1757-899X/791/1/012078

References

[1] Rubin A and Riznichenko G 2014 *Mathematical Biophysics* (New York: Springer)

[2] Mirzaei Ali et al. 2019 Synergistic effects of polyaniline and pulsed electromagnetic field to stem cells osteogenic differentiation on polyvinylidene fluoride scaffold *Artificial Cells Nanomedicine and Biotechnology* **47(1)** 3058-66

[3] Arjmand Monireh et al. 2018 Osteogenic differentiation potential of mesenchymal stem cells cultured on nanofibrous scaffold improved in the presence of pulsed electromagnetic field *J. of Cellular Physiology* **233(2)** 1061-70

[4] Shapovalenko N S, Dorovskih V A, Korshunova N V, Shtarberg M A, Slastin S S and Nevmyvako E E 2011 Vliyanie holodovogo stressa na intensivnost' perekisnogo okisleniya lipidov i antioksidantnuyu sistemu tkanej ehksperimental'nyh zhivotnyh [Vliyanie holodovogo stressa na intensivnost' perekisnogo okisleniya lipidov i antioksidantnuyu sistemu tkanej ehksperimental'nyh zhivotnyh] *Byulleten' Fiziologii i Patologii Dyhaniya* **39** 22-5 [In Russian]

[5] Garkavi L Kh 2015 Periodicheskaya sistema nespetsificheskikh adaptatsionnykh reaktsiy i yeye znacheniye dlya teorii i praktiki [Periodic system of unspecific adaptational reactions and its significance for theory and practice] *Cardiometry* **7** 51-6 [In Russian]

[6] Kim Ju Hee, Lim Sung Don and Jang Cheol Seong 2019 Oryza sativa heat-induced RING finger protein 1 (OsHIRP1) positively regulates plant response to heat stress *Plant molecular biology* **99(6)** 545-59

[7] Selye H 1954 On the nature of disease *Texas Reports on Biology and Medicine* **2** 390-422

[8] Selye H 1976 Forty Years of Stress Research: Principal remaining Problems and Misconceptions *Canadian Medical Association J.* **15(1)** 53-6

[9] Garkavi L H, Kvakina E B and Ukolova M A 1990 *Adaptacionnye reakcii i rezistentnost' organizma* [Adaptive reactions and resistance of the organism] (Rostov na Donu: Izdatel`stvo Rostovskogo universiteta) [In Russian]

[10] Garkavi L H, Kvakina E B, Kuz'menko T S and Shihlyarova A I 2002 *Antistressornye reakcii i aktivacionnaya terapiya. Reakciya aktivacii kak put' k zdorov'yu cherez processy samoorganizacii* [Antistress reactions and activation therapy. Reactionary activation as a way to health through self-organization processes] (Ekaterinburg: Filantrop) [In Russian]

[11] Luna Belen, Chamorro Daniel and Perez Beatriz 2019 Effect of heat on seed germination and viability in species of Cistaceae *Plant ecology & diversity* **12(2)** 151-8

[12] Dias L S, Pereira I P and Dias A S 2019 Seed Germination in Cistus ladanifer: Heat Shock, Physical Dormancy, Soil Temperatures and Significance to Natural Regeneration *Plants* **8(3)** 63

[13] Vasil'ev A N, Udincova N M and Yaltanceva V V 2008 Predposevnaya obrabotka semyan na ustanovkah aktivnogo ventilirovaniya [Pre-sowing seed treatment at active ventilation plants] *Pererabotka i Hranenie s.h. Produkcii* **1** 19-21 [In Russian]

[14] Vasilyev A N 2013 Elektrotechnologies as controlling influence on biological objects in agricultural production *Research in Agricultural Electric Engineering* **3** 96-102

[15] Vasil'ev A N and Kononenko A F 2008 *Ehlektrotekhnologiya i upravlenie v realizacii adaptivnyh rezhimov predposevnoj obrabotki zerna aktivnym ventilirovaniem* [Electrotechnology and control in the implementation of adaptive modes of pre-sowing grain processing by active ventilation] (Rostov-na-Donu: Terra Print) [In Russian]

[16] Vasiliev A N, Ospanov A B, Budnikov D K., Karmanov D K, Salginbayev D B and Vasilyev A A 2016 Controlling reactions of biological objects of agricultural production with the use of electrotechnology. *Int. J. of Pharmacy & Technology* **8(4)** 26855-69

[17] Budnikov D, Vasiliev A N, Vasilyev A A, Morenko K S, Mohamed I S and Belov A 2019 The Application of Electrophysical Effects in the Processing of Agricultural Materials *Advanced Agro-Engineering Technologies for Rural Business Development* (Hershey PA: IGI Global) pp 1-27

AUTHOR INDEX

Institute of Physics
Dirac House, Temple Back
Bristol BS1 6BE UK

CURRAN ASSOCIATES INC.
proceedings.com

ISSN: 1757-8981
ISBN 978-1-7138-4720-5